Handbook of Experimental Pharmacology

Continuation of Handbuch der experimentellen Pharmakologie

Vol. 59/I

Mediators and Drugs in Gastrointestinal Motility I

Morphological Basis and Neurophysiological Control

Contributors

H. G. Baumgarten · A. Bennett · G. Bertaccini · R. Buffa
C. Capella · R. Caprilli · C. F. Code · E. Corazziari
M. Costa · G. Frieri · J. B. Furness · R. A. North
C. Roman · G. J. Sanger · E. Solcia · P. Tenti · T. Tomita
L. Usellini · P. Vernia

Editor

G. Bertaccini

Springer-Verlag Berlin Heidelberg New York 1982

Professor GIULIO BERTACCINI, M.D.
Head of the Department of Pharmacology,
School of Medicine, University of Parma,
I-43100 Parma

With 80 Figures

ISBN-13: 978-3-642-68439-5 e-ISBN-13: 978-3-642-68437-1
DOI: 10.1007/978-3-642-68437-1

Library of Congress Cataloging in Publication Data. Main entry under title:
Mediators and drugs in gastrointestinal motility. (Handbook of experimental pharmacology; v. 59)
Contents: 1. Morphological basis and neurophysiological control – 2. Endogenous and exogenous agents.
Includes bibliographies and index. 1. Gastrointestinal system – Motility. 2. Gastrointestinal agents. 3. Neurotransmitters. 4. Gastrointestinal hormones. I. Baumgarten, H.G. II. Bertaccini, G. (Giulio), 1932–. III. Series.
[DNLM: 1. Gastrointestinal motility – Drug effects.
W1 HA51L v. 59 pt. 1/WI 102 M489] QP905.H3 vol. 59 [QP145] 615'.1s [612'.32] AACR2 81-21349
ISBN-13: 978-3-642-68439-5

Softcover reprint of the hardcover 1st edition 1982

Typesetting, printing, and bookbinding: Brühlsche Universitätsdruckerei Giessen
2122/3130-543210

List of Contributors

Professor H. G. BAUMGARTEN, Freie Universität Berlin, Institut für Anatomie und Elektronenmikroskopie, Königin-Luise-Str. 15, D-1000 Berlin 33

Professor A. BENNETT, Department of Surgery, King's College Hospital, Medical School, University of London, Denmark Hill, GB-London SE5 8RX

Professor G. BERTACCINI, M.D., Head of the Department of Pharmacology, School of Medicine, University of Parma, I-43100 Parma

Dr. R. BUFFA, Istituto Anatomia Patologica and Centro Diagnostica Isto-patologica, University of Pavia, I-27100 Pavia

Dr. C. CAPELLA, Istituto Anatomia Patologica and Centro Diagnostica Isto-patologica, University of Pavia, I-27100 Pavia

Professor R. CAPRILLI, Cattedra di Gastroenterologia, II Clinica Medica, Policlinico Umberto I, University of Rome, I-00100 Roma

C. F. CODE, M.D., Section Gastroenterology, San Diego VA Medical Center, 3350 La Jolla Village Drive, San Diego, CA 92161/USA

Dr. E. CORAZZIARI, Cattedra di Gastroenterologia, II Clinica Medica, Policlinico Umberto I, University of Rome, I-00100 Roma

Dr. M. COSTA, School of Medicine, Department of Human Physiology, The Flinders University of South Australia, Bedford Park, South Australia, 5042, Australia

Dr. G. FRIERI, Cattedra di Gastroenterologia, II Clinica Medica, Policlinico Umberto I, University of Rome, I-00100 Roma

Dr. J. B. FURNESS, Department of Human Morphology, The Medical School, Flinders University, Bedford Park, South Australia, 5042, Australia

Dr. R. A. NORTH, Department of Nutrition and Food Science, Massachusetts Institute of Technology, Room 16-321, Cambridge, MA 02139/USA

Professor C. Roman, Faculté des Sciences et Techniques Saint-Jérôme, Département de Physiologie et Neurophysiologie, Rue Henri-Poincaré, F-13397 Marseille Cedex 4

Dr. G. J. Sanger, Beecham Pharmaceuticals, Medical Research Centre, GB-Harlow, Essex CM 19 5AD

Professor E. Solcia, Istituto Anatomia Patologica and Centro Diagnostica Istopatologica, University of Pavia, I-27100 Pavia

Dr. P. Tenti, Istituto Anatomia Patologica and Centro Diagnostica Istopatologica, University of Pavia, I-27100 Pavia

Dr. T. Tomita, Professor of Physiology, Nagoya University School of Medicine, 65 Tsuruma-scho, Showa-ku, Nagoya 466, Japan

Dr. L. Usellini, Istituto Anatomia Patologica and Centro Diagnostica Istopatologica, University of Pavia, I-27100 Pavia

Dr. P. Vernia, Cattedra di Gastroenterologia, II Clinica Medica, Policlinico Umberto I, University of Rome, I-00100 Roma

Preface

Since the exhaustive Handbook of Physiology (Alimentary Canal, Section 6, Motility) edited by CHARLES F. CODE in 1968, no complete survey of the morphological basis and the physiological control of intestinal motility has been published, in spite of the enormous amount of new data in the literature on this topic. The new techniques and methodologies, the use of electron microscopy, radioimmunoassay and binding techniques, as well as ever more sophisticated electrophysiological procedures have made possible a real flood of discoveries in this field. Moreover, the possibility of new studies of the endocrine cells in biopsies of human intestinal mucosa even during routine endoscopies, has opened new horizons for gastroenterologists and generated a number of important contributions to our knowledge of the morphology and physiopathology of the gut. As usual, new discoveries have also revealed both ignorance and many new problems. For this reason, although many of the data reported in this volume can be considered as firmly established, others still require confirmation, and the results of new research in this field are awaited with extreme interest.

Since advances are occurring so rapidly, even experts in the specific topics need frequent comprehensive reviews. To avoid an excessively large volume, considerations of the pancreas, liver, and biliary system were not included in this Handbook, which, nevertheless, has attempted to offer the reader the essence of more than 1,500 papers. In a volume with so many contributors treating different aspects of the subject, some overlapping was unavoidable; however, as Editor I tried to reduce this to a minimum. The diversity of views expressed by different authors actually may represent an advantage for workers in the field and open new pathways for exploring the same subject. The authors' major goal has been to interpret and clarify concepts derived from different disciplines and to provide not only an exhaustive compilation of data but also a synthesis (at times critical) of information that should enable the reader to appreciate the significance of the advances in this field.

This is a Handbook of Experimental *Pharmacology*, but we deemed it necessary to deal extensively with the anatomical, histological, biochemical, and physiological bases for the understanding of the pharmacology of the gastrointestinal tract (this is included in Part II, Endogenous and Exogenous Agents). Separate chapters deal with various techniques that are usually employed either in *vitro* or *in vivo* in the study of gut motility. Tables and selected figures help the reader to focus upon the most impressive data.

I should like to express my profound gratitude especially to Professor H. Herken, member of the Handbook's Editorial Board, for the opportunity to serve as Editor of this volume.

Despite some lack of punctuality in delivery of manuscripts, which delayed publication, all authors deserve my thanks for their participation in this work and for their exceptionally competent contributions.

GIULIO BERTACCINI

Contents

CHAPTER 4

Ionic Basis of Smooth Muscle Action Potentials. T. TOMITA. With 1 Figure

CHAPTER 5

Electrophysiology of Intestinal Smooth Muscle. R. CAPRILLI, G. FRIERI, and P. VERNIA. With 18 Figures

CHAPTER 6

Electrophysiology of the Enteric Neurons. R. A. NORTH. With 10 Figures

CHAPTER 7

In Vivo Techniques for the Study of Gastrointestinal Motility. E. CORAZZIARI
With 7 Figures

CHAPTER 8

In Vitro Techniques for the Study of Gastrointestinal Motility
G. J. SANGER and A. BENNETT. With 6 Figures

CHAPTER 9

Nervous Control of Esophageal and Gastric Motility
C. ROMAN. With 18 Figures

CHAPTER 10

Nervous Control of Intestinal Motility
M. Costa and J. B. Furness

CHAPTER 11

Identification of Gastrointestinal Neurotransmitters

J. B. FURNESS and M. COSTA

Contents

Part II: Endogenous and Exogenous Agents

CHAPTER 1

Historical Perspective

C. F. CODE

We are just beginning to find out how the motor functions of the stomach and bowel are accomplished. The system is complicated, much more so than I thought forty years ago as I sat with DONALD/M. DOUGLAS in the attic of the Mayo Institute of Experimental Medicine, watching the beautifully rhythmic bursts of small bowel contractions of conscious, trained, fasted dogs (DOUGLAS and MANN 1939). ALVAREZ (1914) had seen the same rhythm earlier in the bowel of anesthetized rabbits when he worked in CANNON's laboratory at Harvard. He based his concept of the intestinal gradient upon the frequency of these contractions (1928). Much earlier, near the turn of the century, BOLDIREFF (1905) had seen rhythmic bursts of motor activity in the stomachs of his fasted, conscious dogs – he called them "work periods" and he knew their action was propulsive. Much later, in the 1950s, young scholars working with me recorded these rhythmic bursts in human beings and dogs and we called them "basic rhythm" (FOULK et al. 1954). Colleagues of BASS (REINKE et al. 1967) recorded them with strain gauges and called them "burst activity". But no matter by what name, they were all the same! It was not until the 1970s that a group of us, led by JOSEPH SZURSZEWSKI (SZURSZEWSKI and CODE 1968; SZURSZEWSKI 1969) recognized that these bursts of rhythmic motor and electrical activity migrated distally down the small bowel from stomach to the ileocecal valve, as a part of the interdigestive complex (SZURSZEWSKI 1969; CODE and MARLETT 1975). We later discovered that the rhythmic bursts were composed of peristaltic contractions which provided the "housecleaner" component of the complex (CODE and SCHLEGEL 1974; SCHLEGEL and CODE 1976). Just think for a moment – over 70 years had passed from the time of the first recording to its recognition as a part of a pattern – the pattern of interdigestive motility of mammals.

The longer I work in the canal, and there is not too much time left, the more I perceive the stomach and bowel as responding to each set of different conditions with the development of standard, stereotyped patterns. The importance of the recognition of these patterns in the advancement of knowledge cannot be overemphasized. Once the pattern is identified it can be used and reused and counted upon as a tool for those exploring the details of control mechanisms. Recall for a moment how often the peristaltic reflex has been employed in such explorations.

With the pattern in place, in gear so to speak, the experimenter can identify the associated electrical activity in the effector organs of the system – the smooth muscle cells, in the neural cells of the myenteric and prevertebral plexuses, and in the central nervous system (CNS). A limb of the control system, neural or hormonal, can be removed or stimulated and the effects measured. The actions of paracrine agents, hormones, other chemical regulators and drugs of all kinds, can be tested against a reliable, reproducible background. There is no use "bird-watching" in the gut – nature gives up her secrets too grudgingly. "Attack, perturb," people like to say these days – whatever, the system has to be challenged!

Up until the 1950s I thought the control system of gastrointestinal motility was simple. That was before I knew anything about it! The myenteric and prevertebral ganglia (the celiac and superior and inferior mesenteric ganglia) were relay and distribution centers, the commands were generated elsewhere presumably mostly in the CNS. The onyl chemical messengers were adrenergic, adrenaline and noradrenaline, although the importance of the latter was not understood. Not until the late 1930s and early 1940s was the Nobel prize-winning work of OTTO LOEWI and Sir HENRY DALE fully appreciated. They, you will recall, established the chemotransmitter actions of cholinergic and anticholinergic substances in sympathetic ganglia and at myoneural junctions. I was working in DALE's laboratory the morning the award was announced. In celebration, he took the day off from his numerous administrative duties to spend it in the laboratory doing an experiment with LINDOR BROWN.

Just look what has happened in the interval – what an explosion of knowledge! You will see it in E. DANIEL's review (this Handbook, Vol. 59/II, Chap. 5). Consider the breadth of the receptor science which has developed, the number of different receptors in the gut system alone, each with its specific agonist and antagonist. And the number of peptides, reviewed by G. BERTACCINI, M. COSTA and G. B. FURNESS (Chaps. 10 and 11), present within the system and outside it (brain), with hormonal and paracrine capabilities, all of which may influence the system's performance. Notice too, please, that my old friend and fickle mistress, histamine is there too, with 5-hydroxytryptamine and with that new and exciting family of "good guys and bad guys," the prostaglandins, which are presented by A. BENNETT (Chap. 8).

In the face of such complexity should we throw up our hands in horror and "await developments?" Not at all, the conglomerate has now become really interesting. Reflect on the system; what exciting possibilities, multiple receptor sites on the end organs, the effectors of the system, the smooth muscle cells, the generation and degeneration of these receptor sites and the multiple sites on at least some of the ganglion cells of the neural control system. What opportunities for variety in the product (motility), for cooperation and for conflict within the system! The approach clearly has to be, "anything is possible!"

But there are some rules to the game and flights of the imagination must be within them. I am reminded of an occasion when, I was attending a scientific meeting with Sir HENRY DALE. He turned to me after listening to some extrapolations by one of our colleagues in directions quite opposite to those prescribed by available facts and said, "what can be done with —, he is so undisciplined!" Sir HENRY's steps forward, and some were giant, were based on what he perceived as the best available knowledge. His emotinal ties were constrained, his factual associations were powerful and overwhelming.

In the stomach and small bowel, the pacesetter potential, the control potential of the smooth muscle cells themselves, prescribes constraints to the patterns of contraction which may be composed. The pacesetter potential sets the maximum frequency of contraction, it dictates the direction and velocity of propagation of contraction, and it forces simultaneous, coordinate contraction of the circular smooth muscle cells lying in the same plane, around the hollow viscus (CODE et al.

1968; DANIEL and CHAPMAN 1963; SARNA et al. 1971, 1972). But then, to keep all options open, as nature so often does, these regulations may be avoided by dissolving the pacesetter potential, in respone to commands which are unknown today and a contraction is produced which, I think, must be phylogenetically very primitive (WEISBRODT 1974; CODE 1979a). Its governance is not known; it looks like peristalsis of the earthworm to me!

The rules and regulations of the neural control system are just emerging (WOOD 1970; SZURSZEWSKI and WEEMS 1976), and I am not an expert in this area. I do recall in the 1950s pleading with colleagues in CNS neurophysiology that they abandon the brain and its waves to study instead the neural pathways and the electrical activity related to control of the gut, using recordings from muscle, to mural ganglia, to prevertebral ganglia, and back again to muscle, for here, I said, was a neural control system, a little brain if you will, capable of functioning independently of the CNS. Simple, I thought – well, others have shown it isn't but that it is tremendously intriguing (WOOD 1975; KREULEN and SZURSZEWSKI 1979).

Even the layout of the neural plexus between the muscle layers of the gut is provocative. It reminds me of a road map of a North American prairie province or state, its straight lines of intercommunication between collections of neurons placed like villages at points of intersection. Both have direction and orientation – according to the compass on the map and according to the direction of transport of contents in the gut (CODE 1979b).

The important capabilities of some of the neurons of the ganglia at intersections in the mural plexus and in the prevertebral plexus, are just emerging. The prospects are exciting. My imagination, I fear, has the better of me.

Some of these cells, no doubt perform the simple relay function (LANGLEY 1921) which many of us thought was their only function. Such cells may be the most highly developed. But it is the more primitive cells lying in this system which hold such fascination. These cells have more than one type of receptor. They respond to a number of neural, paracrine, or humoral transmitters. How many different receptors? The limit is not known. These cells have been most clearly recognized in the prevertebral ganglia (KREULEN and SZURSZEWSKI 1979; SZURSZEWSKI 1977). The efferent messages they generate are not the consequence of a single input – they are not simple relay stations. The commands that leave these cells may be different from those which entered, depending upon other messages the cell is receiving. The different inputs appear to be recognized, weighed, a consensus reached, and the resultant then exits via the efferent pathway. Surely this is a primitive form of decision making. The cells of the little brain do not "think" it through. But their output is affected by their receptive state, something like mood in the big brain, which also seems to be under chemotransmitter regulation. There may be some relationship between the two! After all, many of the same chemoregulators are present in both systems.

No matter how they function, these cells are the neural units comprising the "little brain", controlling, in part, the development and maintainence and the span in time and distance, of patterns of motor activity along the tract. It can function without the big brain but loss of intercommunications with the CNS does produce control limitations which are concentrated at the ends of the tract and may be serious, although useful in some conditions (e.g., highly selective vagotomy for duodenal ulcer).

Another property of the little brain is that it sends out most of its messages over an expanding network of neural fibrils so that a wide area of the gut is affected by activity initiated in relatively few neural elements. This is the multiplier in the system. Neural control in the smooth muscle system is not pinpoint specific as it can be in the skeletal muscle system.

The infusion of new techniques during the last few decades has greatly expanded the real and potential usefulness of biopsies of tissue in the diagnosis of various diseases and in exploring their mechanisms. The application, for example, of immunoflourescent staining of tissues opens a new opportunity for the pathologist to assess normal and abnormal losses or accumulations of highly specific and physiologically potent chemical compounds. But even more exciting are the prospects of the physiologic and pharmacologic biopsy. The modern array of specific agonists and antagonists (see this Handbook, Vol. 59/II, Chaps. 3 and 5) gives the investigator an arsenal of compounds with which to explore normal and abnormal performance of individual or clusters of effector or neural cells. And nowhere is the search for mechanisms, both normal and awry, likely to be more fruitful than with the smooth muscle and ganglion cells of the alimentary canal. Investigations are at work in these areas (SANDERS et al. 1979).

Finally, the perspective of years should provide some warnings – of past mistakes to be avoided, things along the road of progress to be wary of. I see two. My mentor and superb scientist, FRANK C. MANN, maintained that in testing a hypothesis one of three answers should be obtained. "Yes, it works that way; no, it doesn't; or you cannot find it out that way." I have wasted time persisting with the last – studying the phenomenon which was not there at all or fiddling on the fringes of accuracy of a method which could not tell me differences I was attempting to define. Do not do as I have. When you have exhausted your talents with the method and are getting nowhere, set the project aside and wait for new approaches, new methods, or develop them yourself.

Secondly, how important is the question you are seeking to answer? Your own investigations may have beguiled you into the study! The answer to this is up to you, but research consultation may be required to provide the necessary perspective. Sometimes though, warnings may be seen ahead and in the area of chemotransmitter substances, I see a prospective complication – a cloud on the distant horizon. Experience has led me to the conclusion that in the development of a species, the forces at work tend to keep all options open. A mechanism, a chemical compound, useful at one stage in the evolutionary process, but superseded by another may persist in vestigial form. It becomes a reminder of the past, a footprint on the way, an option available but little used. The human appendix is an anatomic example. Do some of the chemotransmitters being identified these days in such numbers, particularly in the gut and now in the older parts of the brain, represent such vestiges of the past? It is a disturbing thought. It could be true.

There is much to be done. The specifics presented in this handbook should help you to start or to extend your researches into new areas.

References

Alvarez WC (1914) Functional variations in contractions of different parts of the small intestine. Am J Physiol 35:177–193

Alvarez WC (1928) The mechanism of the digestive tract. Hoeber, New York

Boldireff WN (1905) Le travail périodique de l'appareil digestif en dehors de la digestion. Arch Sci Biol 11:1–157
Code CF (1979a) Diarrheogenic motor and electrical patterns of the bowel. In: Janowitz HD, Sachar DB (eds) Frontiers of knowledge in the diarrheal diseases. Projects in Health, Upper Monclair, NY, pp 227–241
Code CF (1979b) Conclusions reached by CF Code after hearing a group of reports on Distribution of peptides in the nervous system, Chairman Dr. Jeffrey Barker, presentations by Dr. Robert Elde, Dr. John Furness and Dr. Michael D. Gershon at a National Institutes of Health Workshop on Functional Disorders of the Gastrointestinal Tract. NIH, Bethesda Maryland (1979b)
Code CF, Marlett JA (1975) The interdigestive myoelectric complex of stomach and small bowel of dogs. J Physiol (Lond) 246:289–309
Code CF, Schlegel JF (1974) The gastrointestinal interdigestive housekeeper. Proceedings of the 4th International Symposium on Gastrointestinal Motility. Mitchell, Vancouver, pp 631–633
Code CF, Szurszewski JH, Kelly KA, Smith IB (1968) A concept of control of gastrointestinal motility. In: Code CF, Heidel W (eds) Alimentary canal. American Physiological Society, Washington DC (Handbook of physiology, vol V, sect 6, pp 2881–2896)
Daniel EE, Chapman KM (1963) Electrical activity of the gastrointestinal tract as an indication of mechanical activity. Am J Dig Dis (1963) 8:54–102
Douglas DM, Mann FC (1939) An experimental study of the rhythmic contractions in the small intestine of the dog. Am J Dig Dis 6:318–322
Foulk WT, Code CF, Morlock CF, Bargen JA (1954) A study of the motility patterns and the basic rhythm in the duodenum and upper part of the jejunum of human beings. Gastroenterology 26:601–611
Kreulen DL, Szurszewski JH (1979) Electrophysiologic and morphologic basis for organization of neurons in prevertebral ganglia. In: Janowitz D, Sachar DB (eds) Frontiers of knowledge in the diarrheal diseases. Projects in Health, New York, pp 211–225
Langley JN (1921) The autonomic nervous system, part I. Heffer
Reinke DA, Rosenbaum AH, Bennett DR (1967) Patterns of dog gastrointestinal contractile activity monitored in vivo with extraluminal force transducers. Am J Dig Dis 12:113–141
Sanders K, Menguy R, Chey W et al. (1979) One explanation for human antral tachygastria. Gastroenterology 76:1234
Sarna SD, Daniel EE, Kingma YJ (1971) Simulation of slow-wave electrical activity of small intestine. Am J Physiol 221:166–175
Sarna SK, Daniel EE, Kingma YJ (1972) Simulation of the electric-control activity of the stomach by an array of relaxation oscillators. Am J Dig Dis 17:299–310
Schlegel JF, Code CF (1976) The gastric peristalsis of the interdigestive housekeeper. In: Vantrappen G (ed) Proceedings of the Fifth International Symposium on Gastrointestinal Motility. Typoff, Herentals, p 321
Szurszewski JH (1969) A migrating electric complex of the canine small intestine. Am J Physiol 217:1757–1763
Szurszewski JH (1977) Toward a new view of prevertebral ganglion. Brooks, FP Evers, PW (eds) Nerves and the gut. Slack
Szurszewski JH, Code CF (1968) Activity fronts of the canine small intestine. Gastroenterology 54:1304
Szurszewski JH, Weems WA (1976) Control of gastrointestinal motility by prevertebral ganglia. In: Bulbring E, Shuba MF (eds) Physiology of smooth muscle. Raven, New York; pp 379–383
Weisbrodt NW (1974) Electrical and contractile activities of the small intestine of the cat. Am J Dig Dis 19:93–99
Wood JD (1970) Electrical activity from single neurons in Auerbach's plexus. Am J Physiol 219:159–169
Wood JD (1975) Neurophysiology of Auerbach's plexus and control of intestinal motility. Physiol Rev 55:307–324

CHAPTER 2

Morphological Basis of Gastrointestinal Motility: Structure and Innervation of Gastrointestinal Tract

H. G. BAUMGARTEN

A. Introduction

The general architecture, regional differences in gross morphology, histology, and ultrastructure of gastrointestinal smooth muscle and its innervation have been reviewed by SCHOFIELD (1968), CAMPBELL and BURNSTOCK (1968), and KOSTERLITZ (1968). The structural and functional properties of smooth muscle in general and of gastrointestinal smooth muscle in particular have been described in detail in several articles in the books edited by BÜLBRING et al. (1970) and BÜLBRING and SHUBA (1976). The literature published on these subjects before 1968 is comprehensively treated in these reviews and will not be repeated in this chapter which presents more recent findings on the structure and innervation of gastrointestinal smooth muscle.

B. Structure of Smooth Muscle

I. Arrangement and Ultrastructure of Smooth Muscle Cells

Present-day knowledge on the structure of intestinal smooth muscle derives from a series of papers by GABELLA (1974, 1976, 1979 a, b, c, d, e) which cover qualitative and quantitative aspects of smooth muscle cells, their ultrastructure, and interrelationships. In these studies and in earlier papers (PROSSER et al. 1960; YAMAUCHI 1964; NAGASAWA and SUZUKI 1967; BENNETT and ROGERS 1967; DEWEY and BARR 1968; NISHIHARA 1970), the taenia coli (mostly of the guinea pig) has been the preferred object of investigation and has served as a model for the architecture and structure of gastrointestinal smooth muscle in general, despite known differences in regional variations of the organization of smooth muscle bundles and layers in the different portions of the alimentary tract. The overall picture of the structure of intestinal smooth muscle that emerges from these studies is that the functional unit consists of bundles of cells (in the taenia and more evident in the adjacent circular smooth muscle layer), separated by connective tissue septa carrying blood vessels and nerves; however, serial sections demonstrate extensive exchange of muscle cells between the neighboring bundles in the taenia, indicating rapid reshaping of groups of cells along the longitudinal axis of the taenia. Bundles of smooth muscle cells are more clear-cut in the circular layer and septa may thus extend across the entire thickness of the layer. A special layer of thin, dark, smooth muscle cells separates the ordinary smooth muscle cells of the circular layer from the submucosa (GABELLA 1974); this layer is present in the cecum and ileum of the guinea pig and the ileum of several mammalian species (rat, rabbit, mouse, sheep, dog, and cat) (GABELLA 1974). The longitudinal (including the taenia) and circular muscle layers are connected by small bundles of smooth muscle originating in either layer and running obliquely across the interlayer connective tissue septum, which is largely occupied by the ganglia and connecting strands of the myenteric plexus.

According to the findings of GABELLA (1976), smooth muscle cells of the guinea pig taenia coli (fixed under 1 g load) are 515 μm long and their volume is calculated to be 3,500 μm^3; the mean surface area of muscle cells is 5,300 μm^2. Each muscle cell has about 168,000 caveolae at its surface which thus account for about 30% of the total surface area of individual smooth muscle cells. Characteristically, these invaginations are not distributed at random but form rows at the cell surface (see also WATANABE and YAMAMOTO 1974) interposed between dense bands that may be attached to the inner leaflet of the cell membrane and serve as attachement structures or thin actin filaments (7 nm diameter) and intermediate filaments (10 nm diameter; see also BAUMGARTEN et al. 1971). The caveolae are often associated with tubules and sacs of smooth endoplasmic reticulum. Though their exact functional role is unknown they may be involved in Ca^{2+} transport and thus in electromechanical coupling. Dense bands or patches also occur among the filament-rich portions of the cytoplasm, i.e., without detectable connection to the plasmalemma. In addition to the thin and intermediate filaments, smooth muscle cells from the taenia, when fixed under optimum conditions, contain thick (myosin) filaments measuring 12–18 nm in diameter; in transverse sections they have an ill-defined outline. They are irregularly distributed among the actin und intermediate filament bundles. The intermediate filaments often form rosettes around the dense patches or bands already mentioned. The dense bands may occupy up to 50% of the cell at the level of the nucleus and constitute about 4% of the total cell volume. The muscle cells of the taenia coli contain abundant sarcoplasmic reticulum.

II. Types of Contact Between Muscle Cells

Different types of junctional structures are present in the taenia coli of the guinea pig and the circular smooth muscle layer of the guinea pig ileum (GABELLA 1976, 1979a; GABELLA and BLUNDELL 1979) and of dog stomach and intestine (DANIEL et al. 1976); these include: intermediate junctions, nexuses, and interdigitations. The intermediate junctions are characterized by: (a) an intercellular cleft of 30–40 nm (or more) with a band of dense material that merges with the basal lamina; and (b) electron-dense material attached to the inner aspect of the two opposing membranes. The junctions may extend for 1–2 μm. Thin and intermediate filaments are seen to anchor in the electron-dense intracellular bands of these junctional complexes. In addition, small rounded intermediate junctions are present in the taenia coli which have a narrow intercellular cleft (15–20 nm); the electron-dense patches at the inner aspect of the plasmalemmata are devoid of filament insertions. As already mentioned by BURNSTOCK (1970), similar junctional structures are rarely observed between a nerve ending and a smooth muscle cell (GABELLA 1979a). Interdigitations occur between the laminar or cylindrical processes of muscle cells at their tapering ends. Protrusion of an isolated cell process into an invagination of the facing smooth muscle cell is another form of intercellular connection. Gap junctions or nexuses are common between smooth muscle cells of the circular muscle layer of the guinea pig ileum and occupy about 0.2% of the cell surface (GABELLA and BLUNDELL 1979), but they are rare in the taenia coli (FRY et al. 1977; GABELLA 1976, 1979a). They have not been found in the longitudinal muscle of the dog duodenum (HENDERSON et al. 1971), stomach (DANIEL et al. 1976), or the longitudinal muscle of the guinea pig ileum (GABELLA and BLUNDELL 1979). Since electrical coupling occurs in gastrointestinal muscle layers which lack gap junctions, nexuses are not necessary for electrical communication as previously suggested (DEWEY and BARR 1968; BARR et al. 1968; BURNSTOCK 1970).

III. Muscle Connective Tissue Links

Tangential sections of smooth muscle cells in different organs reveal that microfibrils of the basal lamina which borders on the plasmalemma establish links to collagen fibrils or elastic fibers that extend into the neighboring connective tissue (see BAUMGARTEN et al. 1971: ductus epididymidis; HOLSTEIN et al. 1974: tunica dartos; GABELLA 1979a: taenia coli). At sites of anchoring of collagen fibrils to the microfibrillar web of the basal lamina, the plasmalemma reveals dense plaques inside the muscle cell with thin actin filaments and intermediate filaments attached to the dense bands, suggesting that force generated by myofilament shortening is transmitted to the stroma "microtendons" at these points in visceral smooth muscle. These microtendons are in turn linked to the collagen fibers in the connective tissue septa of the muscle which thus may be considered as "floating laminar intramuscular tendons" (GABELLA 1979a).

C. Innervation of Gastrointestinal Smooth Muscle

In addition to the review article on the structure of smooth muscle and its innervation by BURNSTOCK (1970), a number of critical reviews on selected issues of neurotransmitters and of the innervation of the gastrointestinal tract have appeared (extrinsic and intrinsic innervation of the vertebrate gut, see GABELLA 1979f; nonadrenergic, noncholinergic neurotransmission, see BURNSTOCK 1979a, b, BURNSTOCK and SZURSZEWSKI 1979, HÖKFELT 1979; BURNSTOCK and HÖKFELT 1979, pharmacologic, electrophysiologic, histochemical, and ultrastructural differentiation of types of nerve in the enteric nervous system, see FURNESS and COSTA 1980; distribution of peptide-containing and catecholamine-containing neurons in the gastrointestinal tract, see SCHULTZBERG et al. 1980; distribution and function of neuronal peptides in the intestine, see FURNESS et al. 1980; cellular aspects of catecholaminergic neurons, see GEFFEN and JARROTT 1977; biochemistry and physiology of cholinergic transmission, see COLLIER 1977; biochemistry and physiology of serotonergic transmission, see GERSHON 1977; peripheral autonomic transmission, see BURNSTOCK and BELL 1974; adrenergic innervation of the gut, see FURNESS and COSTA 1974) which contain up-to-date and extensive bibliograpies; the reader of this chapter is referred to these articles for more detailed information on the different aspects of autonomic innervation.

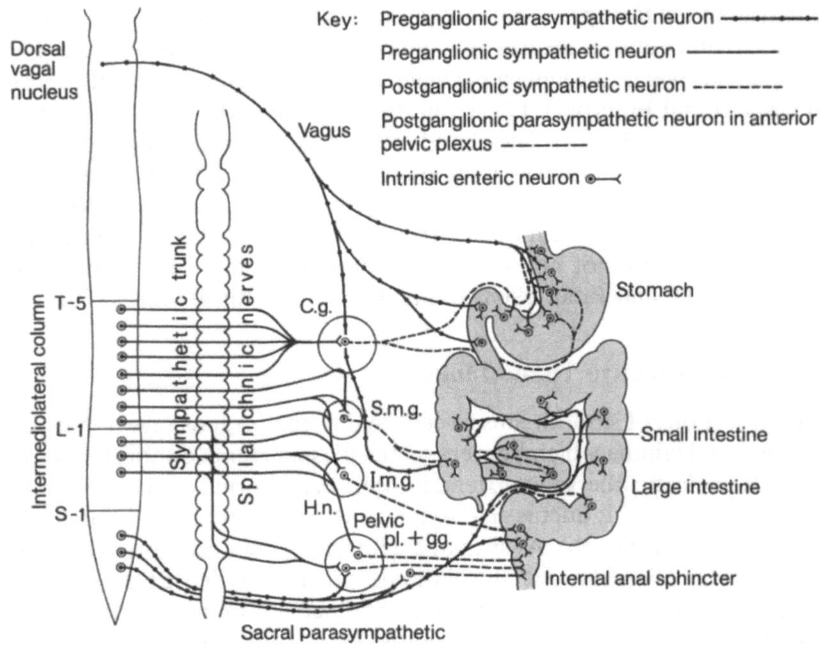

Fig. 1. Autonomic innervation in the mammalian gut. A detailed discussion of the figure is given in the text

I. Extrinsic Innervation

1. Parasympathetic and Sympathetic Innervation

There is general agreement that, in most mammals including humans, the vagus carries preganglionic parasympathetic fibers to the stomach, small intestine, and proximal parts of the large intestine; in some species including cats and humans, the vagus may extend as far distally as the middle of the transverse colon, but the number of functionally important contacts to postganglionic neurons in the colon is small compared with the influence of the preganglionic fibers arising in the sacral pelvic nerves. The sacral parasympathetic nerves provide an input to the entire large intestine including the internal anal sphincter (HULTÉN 1969).

Figure 1 is a schematic representation of the extrinsic autonomic (efferent) input to the enteric neurons of the gut and to the internal anal spincter (based on data in NOBACK and DEMAREST 1975; FURNESS and COSTA 1974; GERSHON 1979). Preganglionic parasympathetic fibers originate in the lower medulla oblongata (dorsal[motor]nucleus of the vagus) and are carried by the vagus to the stomach, small intestine, and proximal portion of the large intestine; the entire colon receives an important parasympathetic input from the sacral spinal cord (S-1–S-3) via the pelvic nerves (sacral parasympathetic). Except for the preganglionic axons of the sacral parasympathetic innervating the sphincter internus which synapse with neurons of the pelvic ganglia, almost all preganglionic parasympathetic fibers ter-

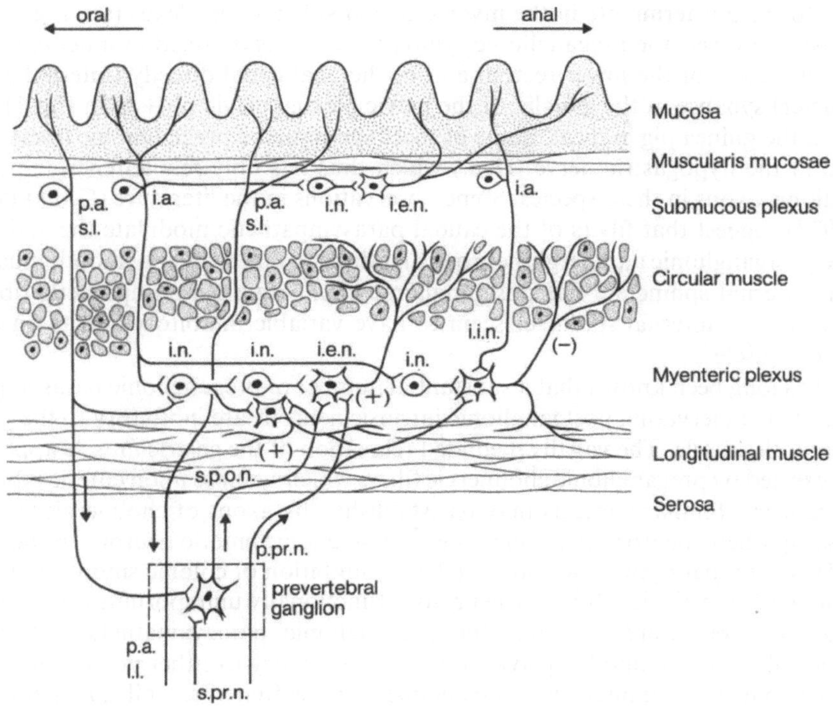

Fig. 2. Schematic portrayal of the various extrinsic and intrinsic (enteric) neurons that are involved in the control of gastrointestinal smooth muscle and their possible interconnections. See text for details.
Symbols indicate: +, excitatory; −, inhibitory; *i.a.*, intrinsic afferent; *p.a.s.l.*, primary afferent short loop; *p.a.l.l.*, primary afferent long loop; *i.n.*, interneuron; *i.e.n.*, intrinsic excitatory neuron; *i.i.n.*, intrinsic inhibitory neuron; *s.pr.n.*, sympathetic preganglionic neuron (or nerve); *s.p.o.n.*, sympathetic postganglionic neuron (or nerve); *p.pr.n.*, parasympathetic preganglionic neuron (or nerve)

minate on intrinsic enteric neurons of the gut (see Fig. 2 for details). The preganglionic sympathetic fibers arise in neurons of the intermediolateral column (extending from T-5 to L 2-3 and pass through the sympathetic chain without synaptic relay; as shown in the diagram, axons derived from preganglionic neurons in segments T-5–T-9 reach the celiac ganglion (C. g.) via the greater splanchnic nerve and synapse with postganglionic adrenergic neurons that are distributed to the stomach and proximal small intestine in company with arteries. Some preganglionic axons (from T-10 and T-11) are conveyed to the celiac ganglion by the lesser and least splanchnic nerve; however some axons pass the celiac ganglion which receives additional preganglionic input from neurons of the visceromotor column at T-12–L-1. Postganglionic noradrenergic neurons in the superior mesenteric ganglion (S. m. g.) provide the postganglionic input to the distal small intestine and proximal colon (via the corresponding artery). The remaining portions of the colon receive postganglionic noradrenergic fibers from the inferior mesenteric ganglion (I. m. g.); preganglionic input from L-1–L-3. Most of the postganglionic adrenergic fibers to the gut terminate in the myenteric and submucous plexus (see Fig. 2 for details). However, the preganglionic sympathetic axons destined to innervate the smooth muscle of the lower rectum and of the anal canal directly (internal anal sphincter) synapse in the ganglia of the pelvic plexus (pelvic pl. + g. in Fig. 1), at least in the guinea pig and cat. Some of these sympathetic preganglionic fibers descend in the hypogastric nerve (H. n.) which thus has only few adrenergic postganglionic axons in these species. Some observations in the literature (GARRETT et al. 1974) suggest that fibers of the caudal parasympathetic modulate the activity of the postganglionic adrenergic neurons that directly innervate the smooth muscle of the internal sphincter. The postganglionic cholinergic parasympathetic fibers that reach the internal sphincter seem to have variable inhibitory effects on the smooth muscle.

It has long been known that a substantial number of preganglionic parasympathetic fibers innervating postganglionic intrinsic neurons are inhibitory to the gastric smooth muscle. The vagally mediated relaxation of the gastric muscle appears to be exerted by preganglionic cholinergic fibers which synapse upon enteric inhibitory neurons. Similar contacts may be established by axons of cholinergic sacral parasympathetic neurons that innervate postganglionic enteric neurons in the colon. However, parasympathetically mediated inhibition of colonic smooth muscle by contacts to intrinsic inhibitory neurons is functionally unimportant when compared with the stomach. Thus, the preganglionic parasympathetic extrinsic neurons of the vagus and the pelvic nerves are excitatory to either postganglionic, enteric cholinergic stimulatory or noncholineric, nonadrenergic inhibitory neurons (ABRAHAMSSON 1973; GONELLA 1978).

Figure 2 is a highly schematic portrayal of the various extrinsic and intrinsic (enteric) neurons that participate in the control of gastrointestinal smooth muscle (with the exception of the internal sphincter) and their possible interrelationship. The gut contains intrinsic afferent (mechanoceptor) neurons which project to interneurons in the myenteric plexus; the interneurons, some of which may be polarized neurons (such as the somatostatin-positive neurons), form descending rows of interconnected neurons that eventually impinge upon intrinsic excitatory neurons (+); the latter mediate the preparatory and the propulsive (or emptying) phase of the peristaltic reflex, i.e., sequential contraction of the longitudinal and circular smooth muscle layer (see KOSTERLITZ 1968 for details). Both types of excitatory neurons, the one contracting the longitudinal muscle and the other contracting the circular muscle, involve cholinergic mechanisms; the excitatory neurons to the longitudinal muscle are in part atropine sensitive. Some of the neurons that contract the circular muscle appear to be atropine resistant but sensitive to nicotinic blockers. Atropine-resistant contractions of the longitudinal muscle can be obtained by substance P (FRANCO et al. 1979 a, b) or serotonin (5-hydroxytryptamine, 5-HT; COSTA and FURNESS 1979; FURNESS and COSTA 1973); the circular muscle of the proximal colon is also responsive to serotonin and 5-HT-mediated contractions are blocked by methysergide (COSTA and FURNESS 1979). The existence of noncholinergic, serotonin-releasing neurons in the myenteric plexus of the ileum and proximal colon has been postulated (GERSHON et al. 1965) which synapse with enteric neurons (see WOOD et al. 1980) and elicit slow e. p. s. p. During peristalsis, a wave of descending inhibition moves aborally in front of the contraction of the circular smooth muscle (JANSSON 1969; FURNESS and COSTA 1973); the neurons responsible for the descending inhibition (–) are intrinsic to the intestine and have been said to use ATP or a related nucleotide (BURNSTOCK 1972, 1977) or vasoactive intestinal peptide (VIP; FAHRENKRUG et al. 1978; FURNESS et al. 1980). These nerves also participate in vagovagal receptive relaxation of the stomach (ABRAHAMSSON 1973); the diagram thus shows connections of the extrinsic preganglionic parasympathetic fibers to enteric inhibitory neurons. The postganglionic sympathetic adrenergic neurons are involved in short-loop intestinointestinal inhibitory reflexes that operate via the prevertebral ganglia (SZURSZEWSKI 1977, 1979; WEEMS and SZURSZEWSKI 1978) or in long-loop inhibitory reflexes that operate via the spinal cord (ABRAHAMSSON 1973; SZURSZEWSKI 1979; for review see FURNESS and COSTA 1974). The role of the remaining peptide-containing enteric neurons in gut motility remains to be elucidated.

It has been suggested that 5-HT may be involved in the vagally mediated inhibition of gastric smooth muscle (BÜLBRING and GERSHON 1967). Very recent evidence, using 5-HT ^{3}H autoradiography, indicates that the nodose ganglion may contain serotonergic neurons (SEGU et al. 1981), the axons of which descend in the vagus. Whether these neurons are involved in the aforementioned relaxation of the stomach is unknown. These neurons thus differ from the intrinsic serotonergic neurons of the gut characterized by GERSHON (1977); the location of the latter neurons appears to be restricted to the myenteric plexus of (mainly) the ileum and proximal colon and these neurons mediate excitation rather than inhibition of smooth muscle, primarily by acting on intrinsic neurons (WOOD and MAYER 1979a, b).

It has long been known that the vagus nerve contains many afferent fibers, some of which may originate from neurons of the gut while others represent the peripheral processes of neurons located in the sensory ganglia (GABELLA 1979f). Both types of afferent neurons, those having their perikarya in the gut itself and those having their cell bodies in the two sensory ganglia of the vagus, may in part use peptides (VIP, gastrin/cholecystokinin, substance P, and somatostatin) as transmitters (HÖKFELT 1979; SCHULTZBERG et al. 1980; see also Table 1). These neurons and their axons may contribute the afferent link in short- or long-loop reflexes that control gastric and/or gastrointestinal smooth muscle activity by modulating the firing rate of efferent parasympathetic neurons (at the level of the gut itself or in the central nervous system). By histochemistry, it has repeatedly been shown that the vagus nerve carries some noradrenergic postganglionic fibers (BAUMGARTEN and LANGE 1969; LUNDBERG et al. 1976; GONELLA et al. 1977) the exact termination of which has not been assessed. The same applies to parts of the pelvic plexus which contains both noradrenergic perikarya and axons. The source of some of these fibers appears to be the sacral sympathetic chain whereas the perikarya belong to peripheral, i.e., organ-near postganglionic adrenergic neurons (see Fig. 1) which supply visceral organs in the pelvic cavity, including the rectum and sphincter internus (COSTA and FURNESS 1973; FURNESS and COSTA 1974). These finding have challenged the classical concept and textbook schemes of the location of the postganglionic noradrenergic neurons innervating the pelvic viscera in the male and female and demonstrate that the hypogastric nerve (or nerves) are largely composed of preganglionic fibers (SJÖSTRAND 1965) and that the synapse between

Table 1. Localization and number of peptide-containing cell bodies in some parts of the peripheral nervous system. (Adapted from HÖKFELT 1979)

	SP	SOM	ENK	VIP	GAS/CCK
Primary sensory neurons (rat) including viscerosensory neurons	+++	+	–	+[c]	++
Sympathetic ganglia (guinea pig)					
Superior cervical ganglion	–	(+)	+	(+)	–
Prevertebral ganglia	–	++++[a]	–	+[b]	–
Parasympathetic ganglia (guinea pig)					
Pelvic plexus	–	++	–	+++[b]	–

SP = substance P-like immunoreactivity; SOM = somatostatin-like immunoreactivity; ENK = enkephalin-like immunoreactivity; VIP = vasoactive intestinal p peptide-like immunoreactivity; GAS/CCK = Gastrin/Cholecystokinin-like immunoreactivity. The table gives a very rough estimate of the approximate numbers of immunoreactive cells in various ganglia: ++++ very high number; +++ high number; ++ medium number; + low number; (+) single cells; – no cells observed

[a] Many SOM cells contain NA

[b] Some VIP cells are cholinergic vasodilator neurons (LUNDBERG et al. 1979)

[c] Some VIP cells are located in the gut and project to the paravertebral ganglia

pre- and postganglionic adrenergic neurons of lumbosacral origin is not exclusively localized at the level of the inferior mesenteric ganglion and plexus, but rather at the level of the sympathetic chain ganglia (mainly for blood vessel innervation), within parts of the pelvic plexus, and close to or even within the adventitia of the target organs (for innervation of the intrinsic enteric neurons of the lower rectum and the smooth muscle of the lower rectum and internal anal sphincter in the guinea pig and the cat). As pointed out by FURNESS and COSTA (1974), there may be considerable species variations in the location of the synapse between the pre- and postganglionic sympathetic neurons innervating the internal anal sphincter.

The sympathetic innervation of the more proximally located partions of the alimentary tract corresponds to the classical scheme as elaborated by GASKELL (1916) and LANGLEY (1921). However, since most postganglionic adrenergic axons that innervate the gastrointestinal tract synapse with and end at the dendrites and perikarya of intrinsic enteric neurons rather than at the smooth muscle, the two-chain efferent sympathetic pathway has only indirect effects on the gastrointestinal smooth muscle by modulating the activity of intrinsic neurons (JACOBOWITZ 1974).

In the light of recent physiologic data on colocolonic and rectosphincteric reflex pathway organization (KREULEN and SZURSZEWSKI 1979a, b; DE GROAT and KRIER 1978; GARRETT et al. 1974), one has to postulate that, in analogy to what has been stated above for the vagus, most sympathetic (efferent) fibers that innervate the gastrointestinal tract (i.e., the thoracic splanchnic nerves, the lumbar splanchnic nerves, the hypogastric nerves, and the pelvic nerves and plexuses) contain axons of afferent neurons in addition to nonadrenergic, noncholinergic efferent neurons (such as peptidergic neurons, see Table 1) which modulate the activity of the pre- or postganglionic sympathetic neurons at the level of the central nervous system (CNS) or in the prevertebral ganglia and may constitute autonomic loops involved in gastrointestinal reflex pathways.

II. Intrinsic Innervation

It is now recognized that gastrointestinal smooth muscle receives – with the exception of the sphincteric region – only a minor direct supply from postganglionic sympathetic adrenergic axons, consequently, most functionally important contacts to smooth muscle cells are derived from axons of intrinsic enteric neurons. These are organized in the myenteric (Auerbach's) plexus and submucous (Meissner's) plexus (see Fig. 2) and neuronal perikarya occur only infrequently in the remaining plexuses of the wall of the digestive tract. The two ganglionated plexuses (myenteric and submucous) which are located in the connective tissue between the longitudinal and circular smooth muscle layers and the mucosa, respectively, extend from the stomach to the upper limits of the internal anal sphincter with some region-specific and species-specific differences in their prominence; these are reflected in the results of nerve cell counts (see GABELLA 1979f). Within the myenteric and submucous plexus, the nerve cell bodies are concentrated in ganglia which are interconnected by nerve strands. When projected into a plane, the ganglia form the nodes in a network of connecting (internodal) strands (Fig. 3); as pointed out by GABELLA (1979f), the shape and size of the meshes are different in thy myenteric and submucous plexus and in various parts of the gastrointestinal tract; in addition their architectural features are species specific. Finally, the number and neuron density of nodes per unit area of the myenteric plexus varies along the circumference of the digestive tube: it is particularly high at the lesser curvature of the stomach (rat and guinea pig), beneath and adjacent to the mesenteric attachment (most species investigated), and underneath the taeniae coli (e.g., in the guinea pig). Nerve fiber bundles are seen to separate from the nodes and internodal strands (often called the primary meshes of the plexus) and to penetrate into the adjacent muscle layers where they may establish secondary, and by further ramification and anastomosing of axons into different Schwann units, so-called tertiary meshes in the muscular plexus of the circular muscle layer; nerve fiber bundles in the longitudinal smooth muscle layer, and particularly in the taeniae, run parallel to the smooth muscle cells and do not constitute a true plexus (BENNETT and ROGERS 1967). The outermost, surface-near fiber bundles have been considered part of a so-called subserous plexus (in the stomach and intestine close to the mesenteric attachment; SCHABADASCH 1930; FERRI and OTTAVIANI 1966) which is, however, difficult to separate from the peripheral ramifications of the myenteric plexus. Where present, the subserous plexus contains mainly branches of extrinsic nerves that connect with the intrinsic plexuses. As pointed out by BAUMGARTEN et al. (1973a), bundles comprising nerve fibers of extrinsic origin (including those leaving the gut to run toward the prevertebral ganglia) can easily be distinguished in both the light microscope and electron microscope since they have endoneural and perineural collagen and a perineural epithelial sheath in contrast to interconnecting strands and branches of the myenteric and submucous plexus which lack these supporting structures and consist of tightly packed Schwann units.

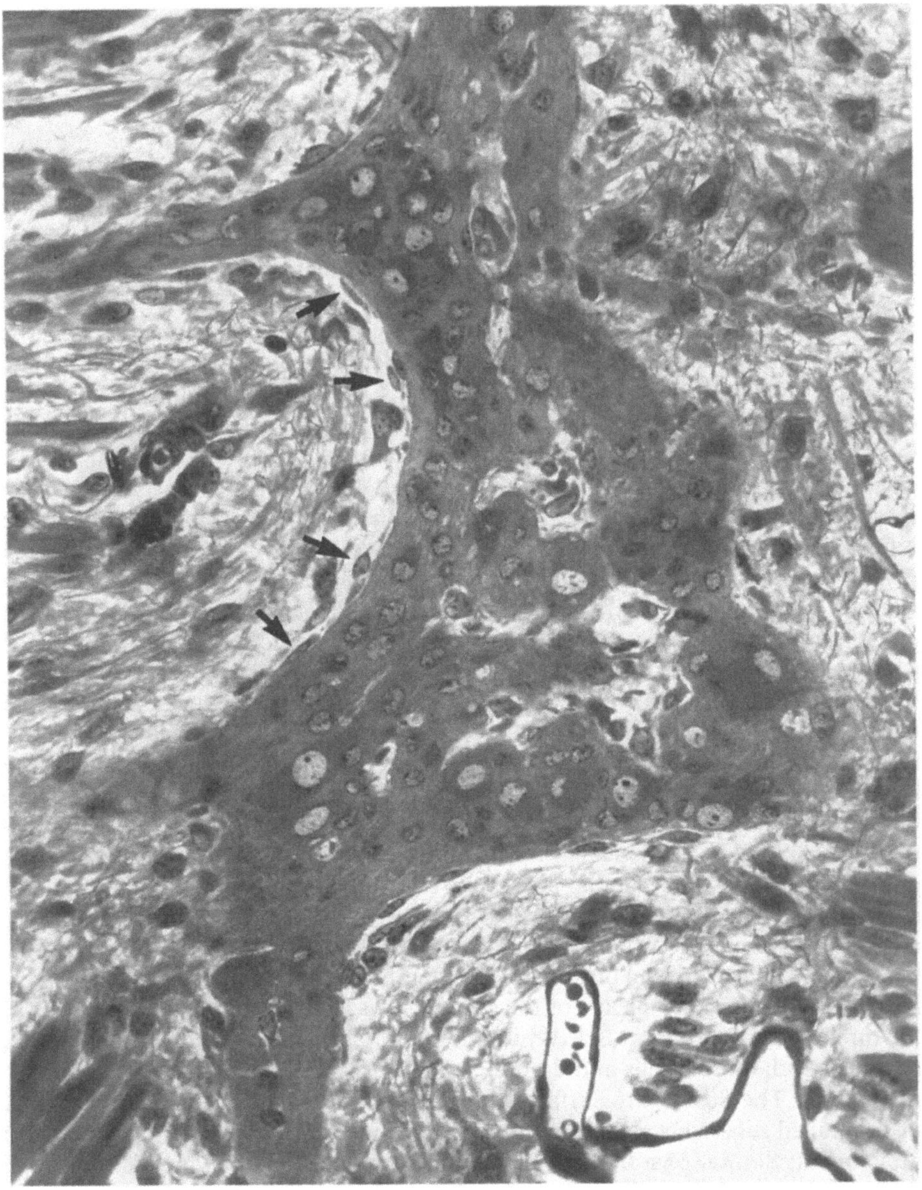

Fig. 3. Semithin section (1 μm) from guinea pig cecum, illustrating the architecture of a myenteric ganglion (node of plexus) with its internodal conncetives (comprising mainly neuropil structures). The large ganglion (underneath the taenia) is embedded in connective tissue; the connective tissue space has access to the extracellular space of the ganglion since the capsule cells (modified fibroblasts indicated by *arrows*) form only an incomplete sheath at the surface. Inside the ganglion, neuronal perikarya showing large, round, transparent nuclei and dark, basophilic cytoplasm are easily distinguished from the ovoid, small nuclei of glial cells which have granular heterochromatin. (BAUMGARTEN et al. 1970) × 1,200

The fiber plexus in the circular muscle comprises:

1) Multiaxonal Schwann units forming nerve bundles that mainly course parallel to the smooth muscle bundles.

2) Small nerve bundles with oligoaxonal Schwann units that interconnect the multiaxonal bundles (to constitute meshes in the muscular plexus) or give rise to axons that terminate in between the muscle cell.

3) Small bundles that penetrate most of the depth of the circular muscle to merge into an elaborate fiber plexus close to the inner (mucosal) aspect of the circular muscle (plexus muscularis profundus) which has specialized dark smooth muscle cells (GABELLA 1974).

4) Larger bundles of nerve fibers that run obliquely across the circular muscle layer, sometimes in company with blood vessels to interconnect the myenteric and the submucous plexus.

The architecture of the submucous plexus (formerly also called Meissner's plexus and subdivided into a plexus submucosus externus and internus by HENLE and SCHABADASCH: see SCHABADASCH 1930; GABELLA 1979f) is similar to that of the myenteric plexus; the submucous plexus sends branches into the muscularis mucosae and into the mucosa itself (mucosal plexus), the mucosal plexus may be divided into a subglandular, periglandular, and villous component (for details see the review by FURNESS and COSTA 1980).

III. Neurons and Nerve Fibers Involved in Gastrointestinal Motility

With the exception of nerve fibers that innervate vascular smooth muscle and gland cells, all the remaining intrinsic enteric neurons and fibers and axons of extrinsic origin that terminate in the gut can be assumed to participate in the control of gastrointestinal smooth muscle activity either by direct actions, via interneurons, or by modulating the activity of inhibitory or excitatory gut neurons (i.e., the final executive neurons which contact smooth muscle; see Fig. 2 and Table 4).

Several criteria and methods have been used to define and classify the extrinsic and intrinsic fibers in the gut and their synaptic or synaptoid connections, such as morphological and staining characteristics, pharmacologic properties, and histochemical and immunohistochemical reactivity; but it has been more difficult to classify the neuronal perikarya that give rise to the different types of fibers. The reasons are: (a) their morphological similarity, i.e., the perikarya lack transmitter-specific characteristics for light microscopy or electron microscopy; and (b) their poor immunoreactivity which may be due to storage of large precursor proteins rather than the final oligopeptide which is cleaved from the precursor only in the terminal portion of the telodendron. Thus, the various types of neural perikaryon in the gut known to contain a variety of transmitters and peptides are not unequivocally identified in the electron microscope as yet. The light microscôpic classification of nerve cells in the intestine (mainly by silver impregnation methods and by acetylcholinesterase histochemistry) have no clear correlation to the histochemically and immunohistochemically identified types of cells in the intestine.

1. Neuronal Perikarya

As pointed out in the electron microscopic studies published on this issue, the nerve cell bodies have different sizes (10–12 μm up to 30–35 μm) and very variable amounts of organelles (see BAUMGARTEN et al. 1970; GABELLA 1972a; COOK and BURNSTOCK 1976a) but it is uncertain whether this merely reflects differences in the functional activity of the neurons at the time of fixation or whether it reflects the existence (as proposed by COOK and BURNSTOCK 1976) of nine types of neurons. While the neurons in both the myenteric and submucous plexus have all the features typical of neurons in general (Fig. 4), they differ from those in autonomic and peripheral sensory ganglia by the following criterion: part of their surface is not covered with satellite glia and may abut directly on the basal lamina ensheathing the cells at the surface of the intramural ganglia. Characteristically, the ganglia also lack a continous layer of cells and processes that would constitute a barrier against the surrounding collagen-filled extracellular space that borders on the adjacent smooth muscle cells. Where present, these discontinuous cells and their processes (Fig. 3) resemble modified fibroblasts and may be identical with interstitial cells and so-called capsule cells (RICHARDSON 1960; HAGER and TAFURI 1959; TAXI 1959, 1965; YAMAUCHI 1964; ROGERS and BURNSTOCK 1966; BAUMGARTEN et al. 1970; GABELLA 1972b; COOK and BURNSTOCK 1976b). The concept of the absence of a barrier being equivalent to the perineurium is in harmony with the results of tracer studies (JACOBS 1980).

As noted by several investigators (TAXI 1958, 1959, 1961, 1965; BAUMGARTEN et al. 1970; GABELLA 1972a; COOK and BURNSTOCK 1976a, b; YAMAMOTO 1977; HOYES and BARBER 1980), part of the surface of the neuronal somata and their dendritic intraganglionic extensions are covered by glia cell processes, synaptic, and nonsynaptic (synaptoid) contacts of axon varicosities. The plexus neuropil is tightly packed with only few interaxonal glial cell processes interspersed. Taken together, the structure of enteric neurons and their intraganglionic surroundings resembles a CNS neuropil (Fig. 4). By histochemistry, immunohistochemistry, and autoradiography, certain types of neuronal perikarya have been identified as intrinsic aminergic and peptidergic; these include; catecholaminergic; amine-precursor-uptake-decarboxylating (APUD); amine-handling, acetylcholinesterase-containing; VIP-, substance P-, somatostatin-, enkephalin-, gastrin/cholecystokinin-antibody-reactive; quinacrine- and γ-aminobutyric acid ^{3}H-accumulating neurons (Table 4).

a) Intrinsic Noradrenergic Perikarya

The existence of intrinsic noradrenergic perikarya in the gut has been demonstrated by histofluorescence (COSTA et al. 1971; FURNESS and COSTA 1971), by dopamine-β-hydroxylayse (DBH) antibody staining (SCHULTZBERG et al. 1980), and by indirect biochemical techniques (GABELLA and JUORIO 1975), i.e., persistence of noradrenaline (NA) in the myenteric plexus following denervation of the extrinsic noradrenergic fibers. However, such intrinsic NA perikarya seem to occur only exceptionally in mammals, such as the guinea pig, where they are mainly found in the proximal colon and the rectum (FURNESS and COSTA 1973); they account for 1% or less of the total number of neurons in the myenteric plexus. Such neurons

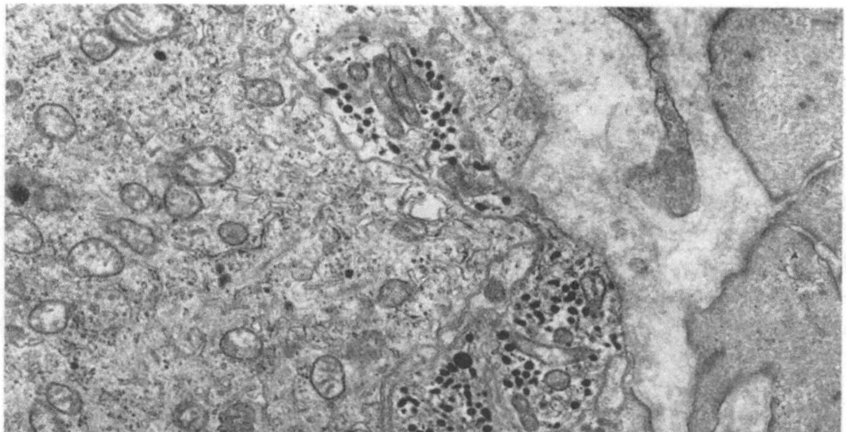

Fig. 4. Low power electron micrograph illustrating the periphery of a myenteric ganglion from the rectum of a rhesus monkey pretreated with i. p. 5-OH-DA. On the left: perikaryon of a nerve cell containing mitochondria, free and membrane-bound ribosomes, and a few dense bodies. The nerve cell is embraced by a complex neuropil consisting of nonterminal intervaricose thin axonal profiles and of varicose, vesicle-filled axon terminals. Axonal beads of adrenergic fiber(s) are characterized by medium-sized and large granular vesicles showing highly electron-dense cores. The adrenergic boutons form passing-by close contacts on the somata of enteric nerve cells and lack synaptic membrane specializations; one of these varicosities borders on the periganglionic space which merges with the connective tissue space surrounding the smooth muscle cells. At this point, there is a gap between the processes of capsular fibroblasts, indicating absence of a barrier between the neuronal and connective tissue extracellular space. Part of the plexus neuropil is ensheathed by glial cell processes (BAUMGARTEN et al. 1970) ×14,000

are more frequent in the gizzard of the chicken (BENNET et al. 1973) and have also been seen in the colon of the lizard (READ and BURNSTOCK 1968). Noradrenergic neurons are sparse in the wall of the lamprey gut and most intrinsic perikarya that react with formaldehyde to yield an intense green fluorophore are dopamine-containing neurons (BAUMGARTEN et al. 1973c) as evidenced by microspectrofluorimetry, by chemical determination of catecholamines in the proximal and distal parts of the gut, and by thin layer chromatography of perchloric acid extracts of the whole lamprey intestine. The preferential occurrence of dopamine as a possible transmitter substance in intrinsic neurons of the gut in the lamprey, which differs from the situation in mammals, resembles the conditions found in invertebrates (BURNSTOCK and ROBINSON 1967; CAMPBELL and BURNSTOCK 1968; MYHRBERG 1972; JUORIO and KILLIK 1972; WELSH 1972).

b) Decarboxylating and Amine-Handling Neurons

Amine-precursor-uptake-decarboxylating and so-called amine-handling neurons in both the submucous and myenteric plexus have been identified by COSTA et al. (1976), FURNESS and COSTA (1978), GERSHON et al. (1965), and DREYFUS et al. (1977a) in the guinea pig intestine by precursor loading (L-dopa; L-tryptophan), histochemistry (uptake of dopamine, 6-hydroxytryptamine) or autoradiography (uptake of 5-hydroxytryptophan ^{3}H TAXI and DROZ 1966; uptake of 5-HT ^{3}H; ROBINSON and GERSHON 1971; ROTHMAN et al. 1976; GERSHON et al. 1976) following prior removal of noradrenergic fibers by sectioning of the mesenteric nerves or by 6-hydroxydopamine, or blockade of the NA-uptake mechanism (e.g., by desipramine). Finally, 5,7-dihydroxytryptamine (5,7-DHT), which yields a bright yellow fluorescent β-carboline upon treatment with formaldehyde in the presence of protein, has been used by GERSHON et al. (1980b) to label "enteric serotonergic neurons" in the gut of both guinea pigs and mice.

Collectively, these studies reveal that there exists a population of intrinsic neurons in the gut which take up exogenously applied amine precursors and have the ability to decarboxylate these precursors (i.e., these neurons possess aromatic L-amino acid decarboxylase). The corresponding amines are normally rapidly degraded by monoamine oxidase (MAO) and can be visualized satisfactorily only when MAO is blocked. Under physiologic conditions, these neurons do not store sufficient amounts of aromatic amines and/or precursor acids to react with gaseous formaldehyde or glyoxylic acid and yield a detectable fluorophore. From their studies using L-dopa, dopamine, and 6-HT, respectively, as markers for this peculiar type of intrinsic neuron, FURNESS and COSTA (1978) conclude that the same type of neuron (and its processes) is responsible for precursor amino acid and for amine uptake. Furthermore, they discuss the possibility that these precursor-uptake-decarboxylating and amine-handling neurons might be identical with intrinsic neurons, the axons of which concentrate exogenously applied 5-HT ^{3}H (ROBINSON and GERSHON 1971; GERSHON et al. 1976; DREYFUS et al. 1977b).

c) Enteric Serotonergic Neurons

GERSHON and his associates (GERSHON et al. 1965, 1965, 1980a, b; GERSHON and ROSS 1966; DREYFUS et al. 1977a, b; JONAKAIT et al. 1977, 1979) proposed the idea that the mammalian gut might possess a system of intrinsic "enteric serotonergic neurons" that are preferentially localized in the myenteric and submucous plexus of the ileum and proximal colon. Similar neurons are said to occur throughout the entire alimentary canal of the chicken (GERSHON et al. 1980a) with most of the cells concentrated in the midgut; provided there is a correlation between the concentration of serotonin in the mucosa-free portion of the gut wall and the ability of the myenteric and submucous plexus neuropil to accumulate 5-HT ^{3}H via a high affinity uptake mechanism, then the rectum of the chick is also equipped with a rather high number of such neurons. In the adult mammalian gut, the perikarya (and processes) of this neuron system can only be identified after loading the tissue with L-tryptophan and prior inhibition of MAO to prevent metabolism of the newly formed serotonin. This indicates that the so-called enteric serotonergic neurons do not store amounts of serotonin sufficient to allow detection by the Falck–Hillarp procedure. The question thus arises whether these neurons utilize an indole as transmitter (such as melatonin or *N,N*-dimethyltryptamine) which cannot be rendered fluorescent with the formaldehyde condensation method. This question remains to be explored in future studies.

As pointed out by GERSHON et al. (1965, 1976, 1980a, b), the cell bodies do not take up sufficient amounts of 5-HT ^{3}H to permit identification of the perikarya in the adult animal. However, at early stages of migration and differentiation, the cell bodies accumulate serotonin ^{3}H, indicating that the membrane-bound, high affinity uptake mechanism is shifted from the perikaryon to the cell processes during development (GERSHON et al. 1980a). This developmental loss of uptake capacity at the level of the perikaryon membrane in the enteric neurons is surprising since brain serotonergic cells display uptake properties for 5-HT ^{3}H (CHAN-PALEY 1975), 5,6-DHT ^{14}C, and 5,7-DHT ^{14}C (H.G. BAUMGARTEN unpublished work 1980, 1981), even at adulthood. The developmental decrease in retention of radiolabel after exposure to 5-HT ^{3}H or 5,6- and 5,7-DHT ^{14}C registered in perikarya of the central serotonergic system is easily explained by the increasingly unfavorable surface/volume relationships in the growing neurons and maturation of enzymatic mechanisms for metabolism of 5-HT and 5,6- and 5,7-DHT. There is, however, more specific evidence for the seronotonergic nature of the enteric 5-HT-accumulating neurons, such as immunoreactivity to tryptophan hydroxylase (GERSHON 1977) and to serotonin (M.D. GERSHON and W.H.M. STEINBUSCH unpublished work 1980). The presence of tryptophan hydroxylase in the myenteric plexus is also suggested by the ability of longitudinal smooth muscle–myenteric plexus preparations to convert tryptophan ^{3}H into 5-HT ^{3}H (DREYFUS et al. 1977a); however, it should be mentioned that we have failed to demonstrate tryptophan hydroxylating capacity in homogenates of this preparation in guinea pigs (H.G. BAUMGARTEN, J.S. VICTOR and W. LOVENBERG unpublished work 1973) as have G.J. LEES and co-workers (quoted in FURNESS and COSTA 1980 p.9).

d) Acetylcholinesterase-Positive Neurons

Histochemically, acetylcholinersterase activity has been demonstrated in neurons of the submucous and myenteric plexus of the cat (TAXI 1965), mouse (TAXI 1965), rat (KOELLE 1951; LEAMING and CAUNA 1961; JACOBOWITZ 1965; TAXI 1965; GUNN

1968; VAN DRIEL and DRUKKER 1973), pig (GUNN 1968), and ram. Their number and staining intensity was found to vary considerably in the submucous and myenteric plexus of different regions of the gastrointestinal tract and in different species. The variability in the staining may be due to the methodology and to changes in the production of this degradative enzyme in different neurons – apart from possible differences in the number and topography of the acetyl-cholinesterase-positive neurons in different regions of the gastrointestinal tract of different species. It has been shown that acetylcholinesterase is present in cholinergic and noncholinergic, e.g., adrenergic neurons, and that it is no reliable marker for the presence of cholinergic cells and axons. However, there is considerable pharmacologic evidence for the existence of intrinsic cholinergic neurons in the mammalian intestine (PATON and ZAR 1968) and cholinergic vesicles have been successfully isolated from the myenteric plexus–longitudinal smooth muscle preparation (DOWE et al. 1980). The exact anatomic mapping of these intrinsic neurons requires autoradiography following choline ^3H administration and/or immunohistochemistry of choline acetyltransferase which has not convincingly been achieved so far.

e) Peptide-Containing Neurons

The antigenic similarities of certain gastrointestinal peptides, which are partly due to the presence of related or identical amino acid sequences in the different peptide species, may cause cross-reactivity of the antibodies used in the immunohistochemical procedures to testify the occurrence of peptidergic neurons in the gut; consequently, at the present status of knowledge, the term "peptide-like immunoreactivity" should be used to describe the results of antibody-mapping studies. For optimum visualization of neuronal perikarya, inhibition of axoplasmic flow by ligation or sectioning of axons or pretreatment with tubulin-depolymerizing drugs is required to produce sufficient accumulation of immunoreactive material (SCHULTZBERG et al. 1980). Of all the mammalian species studied so far, the guinea pig has served as the preferred animal, mainly because the physiology and pharmacology of gastrointestinal motility are particularly well understood in this species. In addition, data on peptide-containing fibers and cells of the rat alimentary tract are available which show that there is some species variability in the number and staining intensity of immunoreactive structure along the digestive tube (SCHULTZBERG et al. 1980). To facilitate the appreciation of the relative proportion of immunoreactive neuronal perikarya in the different parts of the myenteric and submucous plexus of the rat und guinea pig alimentary canal, data from the paper by SCHULTZBERG et al. (1980) are reproduced in Table 2. It is evident that the density of the different types of immunoreactive perikarya varies considerably in: (a) the different portions of the alimentary tract in both species; and (b) the submucous and myenteric plexus of any given region of the gut wall.

A craniocaudal increase in the proportion of cells along the gut is noted in case of VIP-like immunoreactivity in the submucous plexus of the rat but not of the guinea pig. Moderate numbers of enkephalin-like immunoreactive cells are present in all parts of the myenteric plexus of the gastrointestinal tract in both species; similar cells seem to be lacking in the submucous plexus of either species. In the rat, the regional density of somatostatin-like immunoreactive cells in the myenteric and

Table 2. Proportion of peptide-containing cell bodies in the sumbmucous and myenteric plexus in the corpus of the stomach, the distal ileum, and the proximal colon of rat and guinea pig (SCHULTZBERG et al. 1980)

	SP		VIP		ENK		SOM	
Plexus	s	m	s	m	s	m	s	m
Rat								
Corpus	0 (6)	16.7 (168)	33.0 (3)	6.3 (255)	0 (3)	7.5 (281)	0 (3)	0 (146)
Ileum	20.4 (387)	16.4 (1,103)	52.5 (533)	2.8 (1,338)	0 (488)	3.9 (1,077)	19.2 (626)	2.0 (1,072)
Colon	1.7 (118)	13.0 (300)	65.9 (992)	5.6 (2,670)	0 (453)	4.9 (2,667)	3.2 (1,160)	14.4 (3,332)
Guinea pig								
Corpus	0 (5)	11.7 (197)	20.0 (20)	6.6 (517)	0 (6)	12.0 (683)	0 (2)	1.4 (364)
Ileum	5.1 (195)	2.8 (1,053)	27.3 (711)	7.8 (1,434)	0 (812)	15.2 (1,762)	19.7 (751)	3.8 (1,500)
Colon	4.3 (1,088)	12.9 (1,241)	27.0 (1,525)	2.3 (1,631)	0 (1,255)	12.9 (2,442)	20.4 (1,621)	2.6 (1,793)

The proportion of peptide-containing cell bodies is given as the percentage of total number of cell profiles counted; total numbers are given in parentheses. Abbreviations: SP = substance P-like immunoreactivity; VIP = vasoactive intestinal peptide-like immunoreactivity; ENK = enkephalin-like immunoreactivity; SOM = somatostatin-like immunoreactivity; s = submucous; m = myenteric.

0 indicates that no immunoreactive cells have been observed in the sections analyzed quantitatively, but does not exclude the possibility that single cells have been observed in other sections (see text). Although only certain parts of the gastrointestinal tract have been selected for quantitative analysis, the immunoreactive cells in the remaining parts have been studied semiquantitatively and were mostly found to occur in similar numbers as in the selected regions. Exceptions to this are mentioned in the text.

The following points should be considered when reading this table. (1) the treatment with mitotic inhibitor may not reveal all peptide-containing cell bodies; (2) dense networks of peptide-containing nerve terminals may obscure the cell bodies – this is a particular problem for substance P in the myenteric plexus of the guinea pig intestine; (3) it cannot be excluded that in the light microscopic analysis some glial cell profiles have been counted leading to an underestimation of the proportion of peptide-containing neurons; (4) the number of submucous ganglia in the stomach is very low and the figures for these ganglia should therefore be used with particular caution.

The quantitative results presented were added after submission of the original manuscript. In the latter, the number of peptide-containing cell bodies was presented as number of cells in each section. In the present version, we have adopted the principles used by FURNESS, COSTA and collaborators (FURNESS and COSTA 1980), giving the proportion of immunoreactive cells in total cell numbers. This seems to give a more accurate picture of the gastrointestinal peptide-containing neurons. Furthermore, this allows comparison between the two sets of results, although FURNESS and COSTA (1980) analyzed whole mount preparations

submucous plexus varies considerably; conversely, their relative proportions show little regional variation in the myenteric plexus of the guinea pig. Somatostatin-like positive cells are absent in the submucous plexus of the stomach of either species, numerous in the ileum of both species, and frequent in the colon of the guinea pig, but rare in the colon of the rat. Substance P-like immunoreactive cells constitute an important population of the total number of cells in the myenteric plexus whereas VIP-positive cells predominate in the submucous plexus of either species.

By incubating consecutive sections of the gut with antisera raised against Met-enkephalin and Leu-enkephalin and β-endorphin, SCHULTZBERG et al. (1980) have been able to show that cells with similar distribution characteristics react positively towards the three antisera, suggesting that one and the same type of neuron contains all three related peptides. Using the same technique, SCHULTZBERG et al. (1980) presented evidence for a separate localization of substance P-like, VIP-like and enkephalin-like immunoreactivity in distinct nerve cell populations and for potential coexistence of somatostatin-like and gastrin/cholecystokinin-like immunoreactivity in one and the same type of intrinsic enteric neuron. The latter cells occurred only randomly in the rat colon, indicating that but a fraction of the somatostatin-like positive cells contained a gastrin/cholecystokinin-like peptide. In the guinea pig, gastrin/cholecystokinin-positive cells were more numerous in the colon and mainly confined to the submucous plexus.

f) Dopamine-β-hydroxylase-Immunoreactive Cells

Only few DBH-immunoreactive cells are present in the myenteric plexus of the guinea pig and rat stomach and the rat colon, in harmony with the fluorescence microscopic observations already reported (SCHULTZBERG et al. 1980); see Sect. C. III. 1.a.

g) Quinacrine-Accumulating Cells

OLSON et al. (1976) reported that the fluorescent antimalarial acridine derivative, quinacrine, binds selectively to a certain population of nerve cells and fibers in the gut wall; they suggested that the peptidergic neurons might be the ones responsible for quinacrine uptake. Recently, EKELUND et al. (1980) produced further evidence for the latter concept by showing that quinacrine is accumulated and stored in the secretory granules of certain peptide hormone-synthesizing cells. It is not settled whether ATP represents one of the important potential binding sites for quinacrine in vivo as proposed by IRWIN and IRWIN (1954) based on in vitro studies, particularly since it has previously been demonstrated that this drug has a specific affinity for lysosomes (ALLISON and YOUNG 1964). Although ATP is present e.g., in the secretory granules of insulin-producing cells (LEITNER et al. 1975) and chromaffin granules of adrenomedullary cells (LOREZ et al. 1975; PLETSCHER 1975), and both types of granules appear to retain quinacrine; this does not necessarily imply that most of the accumulated quinacrine is bound to this nucleotide.

As pointed out by EKELUND et al. (1980), many peptide-containing endocrine cells in the intestinal mucosa, including the enterochromaffin cells (the 5-HT storage granules of which may also contain ATP) failed to show any affinity for quinacrine. This finding suggests that only certain types of peptide-producing cells and neurons have transport and binding mechanisms for quinacrine and that the quinacrine-positive neurons in the gut represent only a fraction of the total number of peptidergic neurons in the gut. It is thus unclear at present whether quinacrine stains ATP-containing secretory granules and lysosomes, or both, or whether it is also bound to secretory granules devoid of ATP. Further research on this problem is necessary to answer the questions: which of the various types of peptide-containing intestinal neurons are labeled by quinacrine and can quinacrine be used as a tool to prove or disprove the existence of ATP-releasing, purinergic neurons, as postulated by BURNSTOCK (1972, 1974, 1975, 1977, 1979a, b)?

h) γ-Aminobutyric-Acid-Accumulating Cells

Neurons derived from the myenteric plexus of the newborn guinea pig and cultured in vitro possess high affinity uptake sites for tritium-labeled γ-aminobutyric acid (GABA ^{3}H; JESSEN et al. 1979) and synthesize GABA ^{3}H from glutamic acid ^{3}H, suggesting that a small population of GABA-synthesizing intrinsic interneurons exist in the vertebrate gut. However, the unequivocal demonstration of GABA-ergic neurons requires immunohistochemical studies with glutamic acid decarboxylase (GAD) antibodies since it is known that some neurons in the CNS accumulate GABA that are not GABA-ergic (for review see ROBERTS et al. 1976).

2. Nerve Fibers Originating from Extrinsic and Intrinsic Neurons

By means of histochemistry, immunohistochemistry, autoradiography, and culture techniques, combined with surgical or chemical denervation methods and pharmacologic inhibition of axoplasmic transport, certain types of nerve fibers have been identified in the gut at the light and/or electron microscopic level. The correlation between the ultrastructural features of nerve fibers in the intestine and the different neutransmitters or neuromodulators used by these neurons is still controversial and, in many cases, circumstantial and indirect. This has been clearly pointed out in a recent review by FURNESS and COSTA (1980). Table 3 gives an account of the different types of morphologically distinguishable axons in the intestine.

a) Cholinergic Fibers

Most of the evidence for the existence of intrinsic cholinergic axons derives from pharmacologic studies which indicate that the major proportion of acetylcholine released by electrical stimulation from the myenteric plexus is of intrinsic neuronal origin, since extrinsic denervation of the preganglionic cholinergic input to the gut does not significantly alter the releasable quantities of acetylcholine (FELDBERG and LIN 1950; PATON and ZAR 1968; KOSTERLITZ et al. 1970). In harmony with this concept are findings by FILOGAMO and MARCHISO (1970), indicating that the activity of choline acetyltransferase – the key enzyme in acetylcholine biosynthesis – is unaltered after extrinsic denervation of the rabbit ileum. Furthermore, the majority of nerve endings in the myenteric plexus of the guinea pig ileum remain morphologically intact following extrinsic denervation (GABELLA and JUORIO 1973). As pointed out earlier, acetylcholinesterase positivity is no reliable indicator of the presence of cholinergic nerves since noncholinergic neurons also synthesize this enzyme. At present, the most reliable way of identifying presumed cholinergic axons is careful electron microscopy of serially sectioned gut, since it is known from ultrastructural analysis of targets receiving a purely cholinergic input (such as striated muscle, postganglionic sympathetic neurons, and the adrenal medulla) that cholinergic varicosities have a majority of electron-lucent, small synaptic vesicles

Table 3. Morphologically distinct types of axons in the intestine[a]

Type	Comments
1. Axons with round s.c.v. (30–60 nm)	Cholinergic axons (type 1a of HOYES and BARBER 1980 in 5-OH-DA-treated, immersion-fixed material) are probably in this class but other types might be (in perfusion-fixed material, HOYER and BARBER describe 3 subclasses of axons with small empty vesicles, types 1a–c). GABELLA distinguishes 3 subgroups (1972a); most intrinsic; also contain small percentage of l.g.v.
2. Axons with round s.g.v., (30–60 nm), the cores frequently being eccentric or ring-like	Intrinsic to intestine; although formerly assumed to be, these are not noradrenergic; less common than type 1, but encountered in nearly all ganglia; form axodendritic and axosomatic synapses; also contain a few l.g.v.; correspond to type 1b of HOYES and BARBER (1980)
3. Axons with round (30–60 nm) and flattened (50–70 × 15–40 nm) vesicles in variable proportions	Almost certainly noradrenergic; vesicles usually do not have granular cores in control tissues; extrinsic in origin, load with 5- or 6-OH-DA, give positive chromaffin reaction, contain dopamine-β-hydroxylase; slightly more common than type 2; form few synapses; contain few l.g.v.; correspond to type 1b of HOYES and BARBER (1980)
4. So-called p-type axons, nomenclature adopted from BAUMGARTEN et al. (1970)	Likely to encompass several nerve types, including peptide nerves and enteric inhibitory nerves; form axosomatic and axodendritic synapses; mostly intrinsic; commonly also contain many round s.c.v. (30–60 nm); correspond to type 3 of HOYES and BARBER (1980)
5. Uranaffin-positive axons; round vesicles (30–60 nm)	Could be a subclass of type 2 or equivalent to type 2; comparable number to type 2; intrinsic to intestine; contain some unreactive l.g.v.
6. Axons with many small mitochondria	Rare; possibly sensory
7. Axons crowded with glycogen	Rare; nature unknown
8. Axons with dense cytoplasm	Rare; possibly degenerating axons (spontaneous degeneration of nerve endings occurs in the intestine)
9. Axons with many lysosomes and dense bodies	Rare; possibly degenerating axons
10. Axons with many long vesicles	Only reported in cecum (COOK and BURNSTOCK 1976a); no synapses found
11. Intermediate or unclassifiable axon types	Because there is more than one vesicle morphology in each of the axon types, in particular types 1–4, axons containing only a small number of vesicles often cannot be classified; moreover, in type 4 axons the l.g.v. and s.c.v. are sometimes in separate groups

[a] Abbreviations: s.c.v. = small clear vesicles; s.g.v. = small granular vesicles; l.g.v. = large granular vesicles; OH-DA = hydroxydopamine
Modified from FURNESS and COSTA (1980)

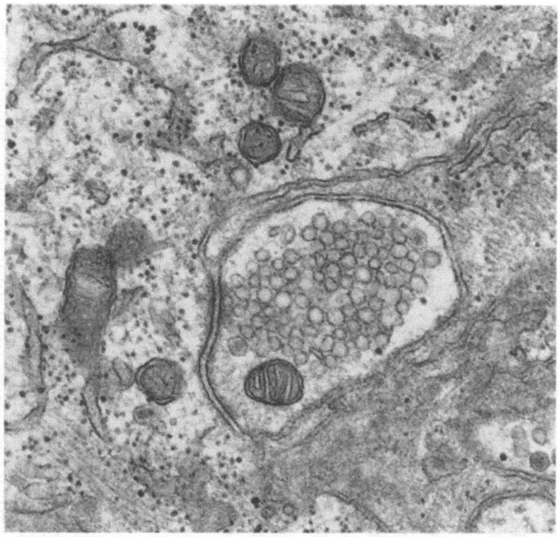

Fig. 5. Guinea pig cecum pretreated with 6-OH-DA. Myenteric plexus. Synaptic contact of a glial cell-ensheathed bouton to a dendritic profile of an enteric ganglion cell. The bouton is mainly filled with small empty vesicles and may be cholinergic. (BAUMGARTEN et al. 1970) ×30,000

(35–60 nm diameter) and a minority of large dense core vesicles (80 nm diameter; see Table 4).

Sites of contact of cholinergic varicosities to somata or dendrites of intrinsic enteric neurons (Fig. 5) show pre- and postsynaptic densities (GABELLA 1972a; BAUMGARTEN et al. 1970; COOK and BURNSTOCK 1976a, b; HOYES and BARBER 1980) whereas contacts to smooth muscle cells (Fig. 6) are devoid of such "synaptic" specializations (BENNETT and ROGERS 1967; GABELLA 1972b). The designation of axons of this type as cholinergic requires serial sectioning because accidental clustering of apparently empty vesicles of similar diameter range may occur in noncholinergic axons, e.g., in peptide-containing and in adrenergic axons (if fixed by perfusion of the animals with warm glutaraldehyde, 3%, in 0.1 *M* cacodylate buffer; HOYES and BARBER 1980).

Appropriate pharmacologic pretreatment of the experimental animals or incubation of fresh tissue in solutions containing false adrenergic transmitters (see Sect. III. 2. b) before fixation or selective chemical degeneration of the adrenergic axons by 6-hydroxydopamine (6-OH-DA) prior to electron microscopic analysis may help to avoid misclassification of adrenergic axons as cholinergic ones. Problems in classification of axons as cholinergic may, therefore, result from the condition of the tissue at the time of fixation and the type and mode of administration of the fixative used. Finally, problems in classification may arise from the presence of axons with unknown neurotransmitters and similar vesicle populations, such as type 2 axons in the myenteric plexus, as characterized by HOYES and BARBER (1980) in the myenteric plexus of the guinea pig stomach which have a network of flattened tubules resembling those in the smooth endoplasmic reticulum, in addition to small clear vesicles. The results published by HOYES and BARBER (1980) suggest that cho-

Table 4. Types of neuron, their transmitters and their possible function in gut motility

Neuron type	Suspected neurotransmitter	Proposed function
Intrinsic interneurons, excitatory (myenteric plexus)	Acetylcholine	Peristaltic reflex elicited by gut distension; antagonized by nicotinic blockers; fast e.p.s.p.
Intrinsic interneurons, excitatory (myenteric plexus of the ileum and proximal colon)	Serotonin or a closely related indolealkylamine; costorage with substance P?	Noncholinergic excitatory input to enteric neurons; slow e.p.s.p; blocked by methysergide; role unknown; results by WOOD et al. (1980) suggest that substance P is not responsible for the slow e.p.s.p.
Intrinsic interneurons, inhibitory to cholinergic neurons or Excitatory to enteric inhibitory neurons (myenteric and submucous plexus)	Somatostatin	Interneuron in descending pathways in the myenteric plexus; inhibitory control of peristalsis by stimulation of enteric inhibitory neurons (VIP neurons?) and inhibition of acetylcholine release from enteric excitatory neurons
Intrinsic excitatory neurons	Acetylcholine?	Atropine-resistant excitation of smooth muscle during the peristaltic reflex; contraction of muscle by a direct action, physiologic role unknown, resistant to hyoscine
Intrinsic excitatory neurons (myenteric plexus; projecting to smooth muscle)	Substance P?	
Intrinsic inhibitory neurons (myenteric and submucous plexus; projecting to other neurons and to the muscle)	ATP or related nucleotides	Evidence mainly based on pharmacologic results (BURNSTOCK 1972, 1975)
	VIP	Evidence based on immunohistochemistry and on pharmacologic findings (LARSSON et al. 1976; SUNDLER et al. 1977; COSTA et al. 1980b; FAHRENKRUG et al. 1978) involved in the descending inhibition that moves distad in front of a peristaltic contraction (COSTA and FURNESS 1974; FURNESS and COSTA 1977)
Intrinsic afferent neurons, primary afferent short-loop neurons (submucous plexus) Primary afferent long-loop neurons (cell bodies in spinal ganglia)	Peptides such as VIP and substance P	Mechanoceptor neurons involved in intrinsic reflexes of the gut, involved in inhibitory control of the gut cranial and caudal to a moving peristaltic wave via synaptic relay in the prevertebral ganglia (short-loop reflex) or the spinal cord (long-loop reflex)

Postganglionic sympathetic noradrenergic neurons (cell bodies in prevertebral ganglia and pelvic plexus projecting towards the myenteric and submucous plexus and circular muscle, particularly the internal anal sphincter)	Noradrenaline (and costorage of somatostatin)	Efferent link in sympathetic reflexes modulating gut motility: inhibition of gastrointestinal smooth muscle by reduction of ACh output and direct inhibition of smooth muscle; contraction of the internal anal sphincter; somatostatin acts by inhibiting ACh output and by stimulating enteric inhibitory neurons
Preganglionic parasympathetic neurons (cell bodies in the dorsal motor nucleus of the vagus and in the visceroefferent column of the spinal cord segments S-1–S-3, projecting to enteric cholinergic excitatory and noncholinergic, nonadrenergic enteric inhibitory neurons)	Acetylcholine	Facilitating propulsive activity in the stomach and the large intestine; involved in the defecation reflex; preganglionic parasympathetic component in vagovagal inhibitory reflexes mediating, e.g., receptive relaxation of the stomach

Compiled from data in FURNESS and COSTA 1980; see text discussion of Fig. 2 for types of neurons and the terminology used here

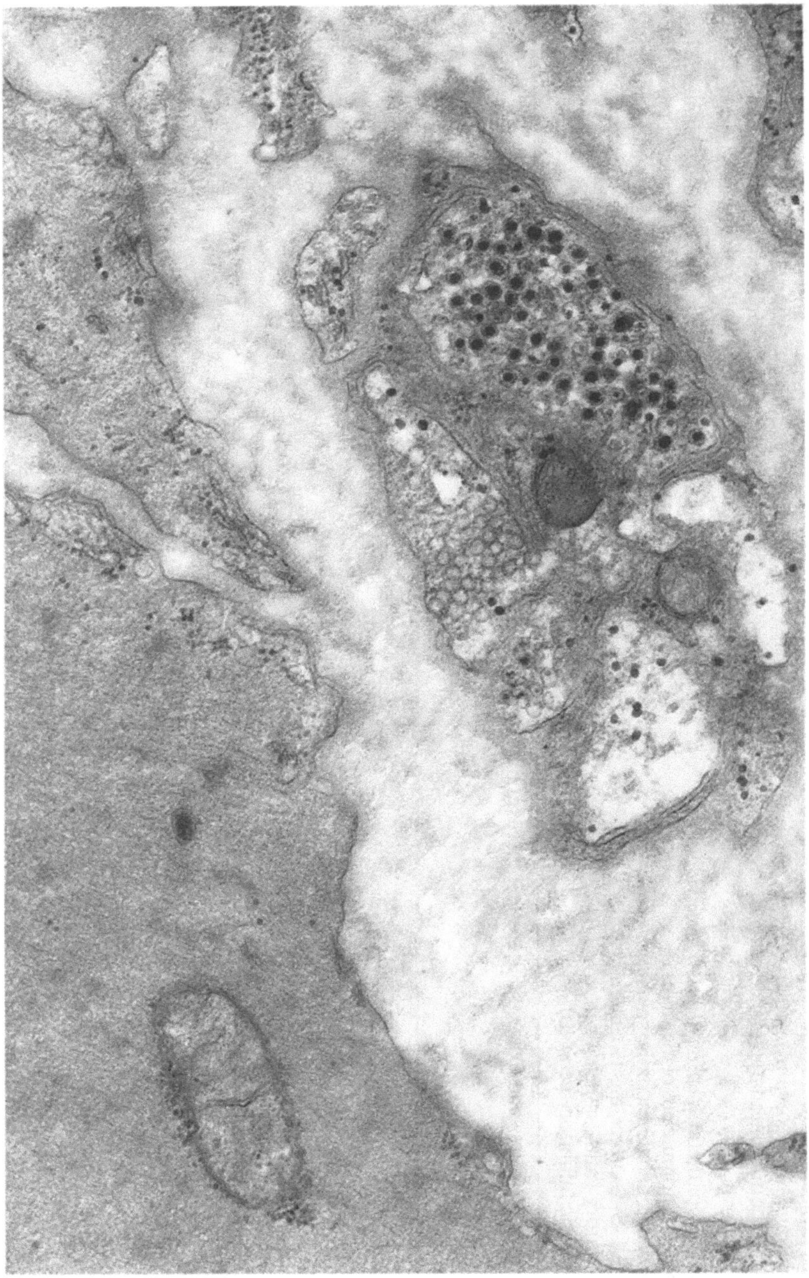

Fig. 6. Circular smooth muscle layer of the guinea pig cecum treated with 5-OH-DA. Oligoaxonal Schwann unit (small nerve fiber bundle) containing varicose swellings of a cholinergic profile (small empty vesicles) and an adrenergic axon (small and medium-sized dense core vesicles). Note the wide cleft between the Schwann cell-denuded axonal varicosity and the facing muscle cells. ×40,000

linergic axons represent an important fraction of the total number of varicose fiber profiles in the myenteric plexus of the guinea pig stomach, most of which are presumably of intrinsic origin.

The literature does not contain semiquantitative data on the relative proportion of cholinergic fibers in the muscle layers. It is worth mentioning that we have seen increased numbers of apparently cholinergic (and adrenergic) varicose profiles in the enlarged prestenotic smooth muscle layers of the colon removed from patients with Hirschsprung's disease (Baumgarten et al. 1973a); this correlates with increased numbers of cholinesterase-positive nerves (Kamijo et al. 1953; Niemi et al. 1961; Meier-Ruge and Morger 1968). This increase in the number of cholinesterase-reactive fibers and the associated absence of intramural ganglion cells serves as a preoperative diagnostic test in biopsy material of the human colon and rectum (Meier-Ruge and Morger 1968).

b) Adrenergic Fibers

Owing to the fact that a number of rather selective methods for the cellular localization of noradrenaline and related biogenic amines are available, data on the distribution and termination of adrenergic axons in the intestine of mammal (and of nonmammalian species) are comprehensive. These methods include: histofluorescence techniques (combined with microspectrofluorometry for differentiation of noradrenaline, dopamine, adrenaline, and 5-hydroxytryptamine), i.e., the Falck–Hillarp formaldehyde condensation technique (see Falck and Owman 1965 for details) and the more recent, sensitive modifications of this method, i.e., the magnesium–formaldehyde and aluminum–formaldehyde procedures (Lorén et al. 1976, 1980); autoradiography following noradrenaline ^{3}H injection or incubation; dopamine-β-hydroxylase immunohistochemistry; loading of noradrenaline storage sites in adrenergic axons by 5- or 6-hydroxydopamine or α-methylnoradrenaline prior to appropriate aldehyde fixation; chemical destruction of noradrenergic axons by 6-OH-DA pretreatment and subsequent electron microscopic analysis; and positive reaction of axons with the modified chromaffin method by Tranzer and Richards (1976). A survey of the published literature on electron microscopic identification of axons in the intestine reveals that there is disagreement concerning the classification of certain axons as adrenergic (see later in this section for details); this is due to variable conditions of the tissue at the time of fixation and differences in the fixation procedures employed, apart from possible species – and region-specific differences in the electron microscopic features and vesicle equipment of intestinal adrenergic fibers and so-called amine-handling axons which may be confused with adrenergic ones.

The histofluorescence findings on the occurrence and distribution of adrenergic axons in the mammalian intestine are rather clear-cut and consistent and have been summarized by Jacobowitz (1965) and Norberg (1964) and reviewed by Gabella (1979f). With the exception of the proximal colon in the guinea pig, where about 50% of the adrenergic fibers (Gabella and Juorio 1975) derive from intrinsic noradrenergic perikarya (which account for only 1% of the neurons in the myenteric plexus; Costa et al. 1971; Furness and Costa 1971), almost all noradrenergic axons found in the various portions of the gut in different mammals are of extrinsic origin and are thus part of postganglionic sympathetic noradrenergic neurons. The myenteric plexus of all portions of the gastrointestinal tract in most mammals investigated with quantitative fluorometric methods (HOLZBAUER and Sharman 1972; Juorio and GABELLA 1974) and with the Falck–Hillarp method (NORBERG 1964; JACOBOWITZ 1965; BAUMGARTEN 1967; GABELLA and COSTA 1967, 1968; READ and BURNSTOCK 1968; COSTA and GABELLA 1971; FURNESS and COSTA 1974) receives a prominent input of noradrenergic varicose fibers which form basket-like structures around nonfluorescent enteric ganglion cells. It is worth mentioning that the richly beaded adrenergic fibers are particularly numerous in the periphery of the ganglia; noradrenaline released from such varicosities may affect adjacent smooth muscle cells since there is no diffusion barrier at the surface of the intramural ganglia (see Sect. C. III. 1). The interconnecting strands of the myenteric plexus also contain varicose fluorescent adrenergic fibers, but the spacing of the axonal beads may be less close and the size of the swellings more variable.

Along the gastrointestinal tract, the distribution and density of varicose adrenergic fibers in the myenteric plexus appears to be rather uniform. According to the findings of BAUMGARTEN et al. (1970), there is a proximodistal increase in the concentration of noradrenaline of the guinea pig large intestine; the data given by JUORIO and GABELLA (1974), suggest that the longitudinal muscle–myenteric plexus of the guinea–pig colon has the highest noradrenaline content. The concentration range estimated for the guinea pig large intestine by BAUMGARTEN et al. (1970) is similar to that found by JUORIO and GABELLA (1974), namely 0.59–1.18 μg/g fresh tissue weight and 0.56–1.3 μg/g, respectively. Apart from the stomach of the rat, cat, and guinea pig (FURNESS and COSTA 1974), adrenergic fibers have been shown to supply the nonfluorescent ganglion cells of the submucous plexus in all sections of the intestine (NORBERG 1964; JACOBOWITZ 1965; READ and BURNSTOCK 1968; COSTA and GABELLA 1971; FURNESS and COSTA 1973, 1974); fluorescent fibers are also abundant in the interconnecting strands of this plexus.

The number of adrenergic fibers which have no apparent relationship to vessels (arteries, arterioles, venules, and veins) and are destined to innervate smooth muscle cells of the longitudinal and circular layer (of the muscularis propria) and of the tunica muscularis mucosae is comparatively small. There is a sparse adrenergic plexus in the circular muscle layer of the gastrointestinal tract in all mammalian species investigated (GABELLA and COSTA 1967; BAUMGARTEN 1967; FURNESS 1970; SILVA et al. 1971; COSTA and GABELLA 1971; GABELLA and COSTA 1971; HOWARD and GARRETT 1973; FURNESS and COSTA 1973). In the distal portion of the rectum of some species (such as the cat, guinea pig, and human) a gradual increase in the density of this plexus is noted which reaches its maximum in the internal anal sphincter; the adrenergic fibers constitute an irregular network of small bundles which have a twisted arrangement and which course through the collagen-filled wide connective tissue spaces (BAUMGARTEN 1967; GABELLA and COSTA 1968; COSTA and GABELLA 1971; HOWARD and GARRETT 1973; FURNESS and COSTA 1973). An increased density of adrenergic fibers has been found in the ileocolic sphincter of the rat and rabbit (FURNESS and COSTA 1974), the pyloric muscle of the rat (GILLESPIE and MAXWELL 1971) and the gastroesophageal junction of the guinea pig (GABELLA and COSTA 1968). In general, adrenergic fiber bundles are very sparse in the longitudinal smooth muscle of the gut in most mammals investigated; a moderate number of such bundles is present in the guinea pig taenia coli

(ÅBERG and ERÄNKÖ 1967; BENNETT and Rogers 1967; COSTA and Gabella 1971) and the longitudinal smooth muscle layer of the distal rectum in the cat and guinea pig (HOWARD and GARRETT 1973; GABELLA and COSTA 1968; COSTA and GABELLA 1971; FURNESS and COSTA 1973).

Little attention has been paid to the innervation of the muscularis mucosae but adrenergic fibers have been seen in association with this muscle layer in the gastroesophageal junction of the cat and monkey (BAUMGARTEN and LANGE 1969) and of the anal canal of the guinea pig (COSTA and GABELLA 1971). It is suprising to realize that, despite the availability of a number of rather specific methods (see earlier in this section) for their ultrastructural characterization, there is still disagreement concerning the unequivocal identification of adrenergic varicosities and their mode of contact to ganglion cells in the enteric plexuses. This may be illustrated by comparing the size range and morphology of vesicles in axonal varicosities of the myenteric plexus in the large intestine of the rhesus monkey and guinea pig (BAUMGARTEN et al. 1970) with those in the stomach of the guinea pig following 5-hydroxydopamine (5-OH-DA) treatment in vivo and in vitro, respectively, and fixation with intracardiac cold glutaraldehyde (6%) or with warm glutaraldehyde (3%) perfused through the aorta, or incubation with 5-OH-DA in vitro and subsequent fixation by immersion into 3% glutaraldehyde (HOYES and BARBER 1980).

Using in vivo pretreatment with 5-OH-DA and intravascular perfusion fixation with 6% cold glutaraldehyde, BAUMGARTEN et al. (1970) described axonal varicosities in the myenteric plexus of the rhesus monkey and guinea pig which were characterized by small, medium-sized, and occasionally large granular vesicles (40–60, 50–90, and 90–130 nm diameter), often located in the periphery of the ganglia (see Figs. 4, 7–9 and Table 4). Sites of contact to somata and dendrites of nonadrenergic enteric neurons were often devoid of pre- and postsynaptic membrane specializations (Fig. 4) and some of the varicosities were found to face the periganglionic extracellular space at sites where so-called capsule cells (i.e., interstitial cells resembling fibroblasts) were absent and smooth muscle cells with their surrounding connective tissue space were seen to border directly on the surface of the ganglionic neuropil. These varicosities also occurred in the neuropil of the interconnecting strands of the myenteric plexus and were seen to degenerate following intraperitoneal 6-OH-DA pretreatment (50 mg/kg, free base, 48 and 24 h before fixation) thus testifying to their adrenergic nature (Fig. 7). An additional, specific feature of such large adrenergic varicosities in the guinea pig was the abundance of dumbbell-shaped small granular vesicles and of tubules resembling those in the endoplasmic reticulum with an elctron-dense internal matrix (Fig. 8) following 5-OH-DA treatment and fixation by systemically infused glutaraldehyde. In most granular vesicles, a moderately electron-dense halo separated the highly opaque core from the intact limiting membrane of the vesicle (Fig. 9). Finally, some of the mitochondria in the adrenergic varicosities exhibited an increased electron density and the number of dense bodies was increased after repeated 5-OH-DA treatment. In their study on the differentiation of fiber types in the myenteric plexus of the guinea pig stomach, HOYES and BARBER (1980) did not detect varicosities with small or medium sized dense cores after perfusion of glutaraldehyde; they also failed to see major differences in the mean size of the large granular vesicles in the three types of axons distinguished by morphological and cytochemical (i.e., incu-

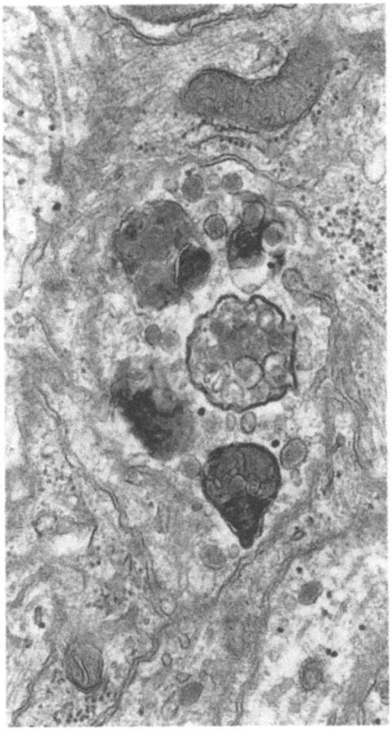

Fig. 7. Guinea pig cecum, treated with 6-OH-DA. Degenerating adrenergic varicosity in the myenteric plexus neuropil. Note autophagocytic disintegration of medium-sized and large granular vesicles. (BAUMGARTEN et al. 1970) × 30,000

bation of gut tissue in Krebs solution containing 5-OH-DA) criteria and classified as type 1 profiles. The mean diameter of the small vesicles was reported to be 47 nm and that of the large granular vesicles 81 nm for axons comprising supposed cholinergic (1 a) adrenergic (1 b) and nonadrenergic, noncholinergic (1 c) profiles (equipped with a peripheral rim or halo of electron-opaque material in the small vesicles). After 5-OH-DA incubation of gut tissue, the following changes were noted in the vesicles: increase in the mean diameter of the smaller vesicles in type 1 a and 1 c axons; together with development of a centrally located dense core in the 1 b profiles; and, occasionally also in those of the 1 c profiles. These changes in the size and electron density of the vesicles in the three types of axons following 5-OH-DA treatment in vitro and subsequent immersion fixation are in part due to nonspecific effects of the incubation and delayed fixation procedure employed and this is also evidenced by the fact that the networks of tubules resembling those in the smooth endoplasmic reticulum in so-called type 2 axons were not detected at all in immersion-fixed specimens. Not only were these fixation-sensitive structures preserved in our perfusion studies but in part found to accumulate electron-dense material after intraperitoneal (i.p.) administration of 5-OH-DA, suggesting that such amine-concentrating tubular structures occur in adrenergic axons and possibly also in so-called amine-handling profiles (WILSON et al. 1979).

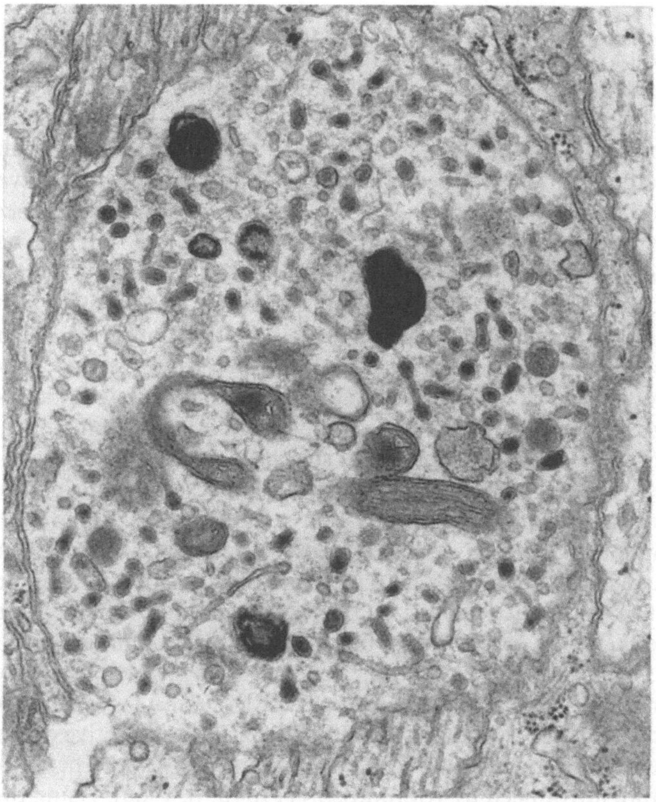

Fig. 8. Guinea pig cecum, fixed by intravascular glutaraldehyde. The large bouton in the myenteric plexus contains small granular vesicles, dumbbell-shaped granular vesicles and profiles of smooth endoplasmic reticulum that are in part filled with a flocculent, moderately electron-dense core. This bouton is probably adrenergic. (BAUMGARTEN et al. 1970) × 30,000

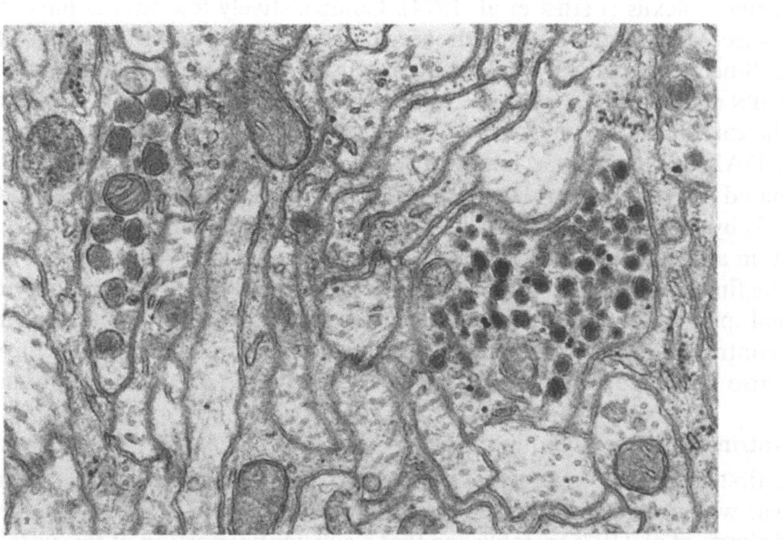

Fig. 9. Myenteric plexus neuropil of rhesus monkey rectum, fixed by intravascular glutaraldehyde. Adrenergic varicosity filled with medium-sized and large granular vesicles (on the right) and p-type varicosity with very large, heterogeneous dense core vesicles. × 30,000

Axons with flattened vesicles have previously been shown to take up catecholamines (e.g., in the cat rectum, HOWARD and GARRETT 1973; axons of type 3 in the rat anococcygeus muscle, GIBBINS and HALLER 1979). Unpublished results (J. B. FURNESS and M. COSTA unpublished work, 1980; J. B. FURNESS unpublished work 1980) indicate that axons in the guinea pig small intestine which have round (30–60 nm) and flattened vesicles (50–70 × 15–40 nm) are adrenergic, since they give a positive chromaffin reaction and contain immunoreactive DBH. In harmony with the findings of HOYES and BARBER (1980) and J. B. FURNESS (unpublished work 1980) are the observations by BAUMGARTEN et al. (1970) that most adrenergic boutons that impinge upon somata and dendrites of nonadrenergic, enteric neurons lack synaptic specializations. In this respect, our results deviate from those of GABELLA (1972a) who consistently found pre- and postsynaptic specializations in adrenergic boutons of the guinea pig myenteric plexus. Some of the discrepancies concerning the vesicle equipment of supposed adrenergic axons in the intestine may be due to species- and region-specific differences in the morphology of adrenergic axons apart from the possibility that so-called amine-handling axons of intrinsic origin may have been classified incorrectly. 6-OH-DA treatment appears to be required to show the true adrenergic nature of axons that take up noradrenaline analog; within certain dose limits, 6-OH-DA precipitates degeneration only in adrenergic axons. However, if the dose of 6-OH-DA exceeds certain limits (more than 100 mg/kg i.p.), nonadrenergic axons, such as serotonergic ones, may be nonspecifically damaged by 6-OH-DA (H. G. BAUMGARTEN and L. LACHENMAYER unpublished work 1979; 1980). Another technique that allows selective demonstration of the noradrenaline storage sites by electron microscopy is the permanganate fixation procedure introduced by RICHARDSON (1964a, b, 1966). Using this technique, which was extensively applied by HÖKFELT to identify central and peripheral monoaminergic boutons and varicosities (HÖKFELT 1968, 1969), profiles which contain small granular vesicles have been identified as adrenergic in the cat myenteric plexus (FEHÉR et al. 1974). Comparatively few studies have dealt with the selective identification of adrenergic axons in the smooth muscle of the intestine. Since these are particularly numerous in the internal anal sphincter, BAUMGARTEN et al. (1972) studied the accumulation of 5-OH-DA in adrenergic profiles in the cat internal sphincter and their degeneration following pretreatment with 6-OH-DA. No synaptic specializations were seen in these contacts which may be designated as synapses par distance. The results obtained by HOWARD and GARRETT (1973) by means of noradrenaline analogs on the cat sphincter are, to a great extent, in agreement with those published by BAUMGARTEN et al. (1972). According to the findings of GARRETT et al. (1974), the noradrenergic fibers supplying the internal sphincter muscle constitute the most important pathway involved in the motor control of the anal sphincter of the cat; cholinergic contractions of the sphincter are also mediated through this adrenergic pathway and are therefore indirect.

c) Intrinsic Amine-Handling Fibers

The first evidence for an uptake of amines into intrinsic fibers of the myenteric plexus was provided by ROBINSON and GERSHON in 1971. ROTHMAN et al. (1976) and COSTA et al. (1976) established that a substantial portion of this uptake is confined to nonadrenergic nerves; that the 5-HT ^{3}H-accumulating fibers are intrinsic to the gut was shown in culture experiments by DREYFUS et al. (1977 b). The finding by DREYFUS et al. (1977 a) that the axons of intrinsic neurons grown in culture are able to synthesize serotonin ^{3}H from tryptophan ^{3}H suggest that these neurons might be serotonergic though other authors have failed to detect tryptophan-hydroxylating capacity in myenteric plexus–longitudinal smooth muscle preparations of the gut (G. J. LEES and co-workers, quoted in FURNESS and COSTA 1980; H. G. BAUMGARTEN, J. S. VICTOR and W. LOVENBERG unpublished work 1973). Using a different approach (loading of the intrinsic neurons and their axons by L-dopa, dopamine, or 6-hydroxytryptamine following chemical denervation of the noradrenergic input to the guinea pig small intestine by 6-OH-DA pretreatment), FURNESS and COSTA (1978) demonstrated neurons in the submucous (11%) and myenteric (0.4%) plexus the axons of which ramified among other intrinsic neurons of either plexus, their principal connections and in the tertiary strands of the myenteric and deep muscular plexus. However, these fibers do not normally store amounts of catechol–or indoleamines sufficient to permit their detection by histofluorescence. Since they do not reveal tyrosine or dopamine-β-hydroxylase activity (FURNESS et al. 1979), these neurons are unlikely to synthesize and store a catecholamine. If they synthesize, store, and release an indoleamine, as proposed by GERSHON et al. (1965), it is unlikely to be serotonin itself, but possibly a closely related substance which remains to be conclusively identified (COSTA and FURNESS 1979).

ROTHMAN et al. (1976) and DREYFUS et al. (1977 b) studied the uptake of serotonin ^{3}H into processes of fetal rabbit ileum and mouse small intestine maintained in culture and found retention of serotonin in large granular vesicles (65–100 nm diameter); the same nerve profiles also contained small clear vesicles. In addition, large granular vesicles ($\leqq$160 nm diameter) in expanded axonal processes were labeled. It is not clear at present whether these labeled profiles belong to the so-called amine-handling neurons discussed earlier in this section or whether they represent peptide-containing axons which are characterized by the coexistence of monoamines and which might thus represent the neuronal counterpart of the peptide-synthesizing diffuse endocrine cell systems (APUD cells of PEARSE 1977). Further work is required to establish whether the uranaffin-positive axons (WILSON et al. 1979) in the intestine belong to this category of intrinsic amine-handling and/or peptide-containing neurons. The uranaffin method has been claimed to stain the amine storage sites (RICHARDS and DA PRADA 1977).

Unequivocal evidence for the existence of intrinsic serotonergic neurons in the gut has so far been obtained only in the gut of the river lamprey (BAUMGARTEN et al. 1973 c) and in the gut of the teleost *Myoxocephalus* (WATSON 1979). In both these instances, the storage sites for serotonin appeared to be mainly large granular vesicles (75–160 nm in the axons of the lamprey gut and 50–150 nm in axons of the teleost gut). Whether or not the 5-HT ^{3}H-concentrating intrinsic axons in the myenteric plexus of mammals have any relationship (possibly phylogenetic) with the serotonergic axons in the gut of lower vertebrates and whether the latter store one or more peptides in addition to 5-HT remains to be elucidated in future studies. From experiments with the indole neurotoxin 5,7-DHT (introduced into neurobiologic research as an experimental tool by BAUMGARTEN and LACHENMAYER in 1972 and BAUMGARTEN et al. in 1973 b), GERSHON et al. (1980 b) have concluded that the enteric serotonergic neurons of the gut resemble central serotonin-containing

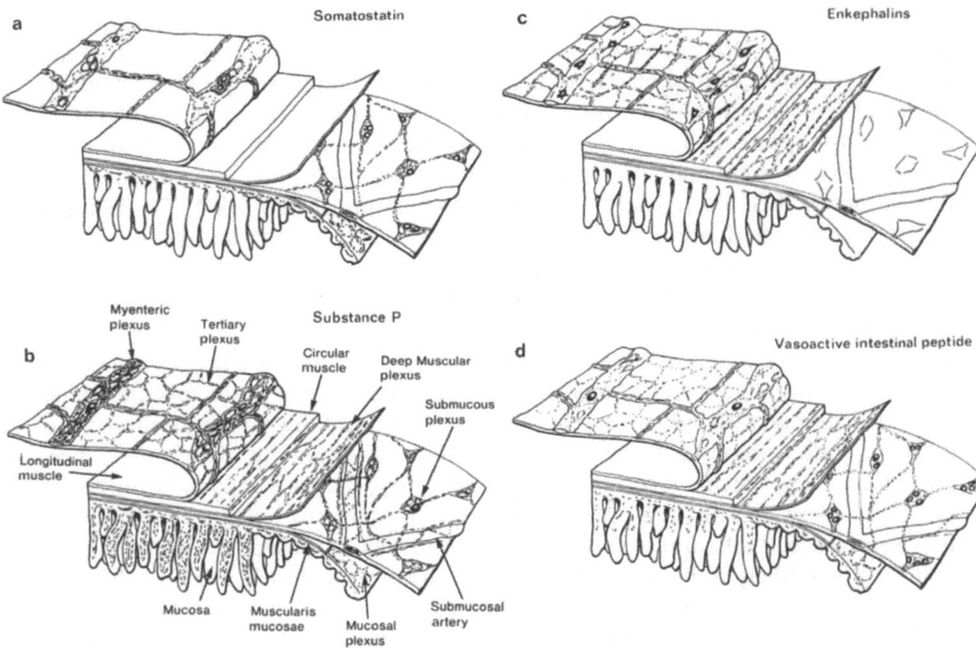

Fig. 10 a–d. Summary of the distributions of peptide-containing nerve cell bodies and axons in the wall of the guinea pig small intestine. The intestinal wall has been drawn with the layers partly separated in the manner used to study the distributions of peptides in whole mounts. Note that each of the peptides has a unique distribution. Nerve cell bodies are drawn as *black rings* in the ganglia in the proportions observed. Axons are drawn as *broken lines* in the appropriate layers. **a** somatostatin; **b** substance P; **c** enkephalins; **d** vasoactive intestinal peptide. (FURNESS et al. 1980)

neurons in mammals, which are well known for their susceptibility to the toxic effects of 5,6- and 5,7-DHT (BAUMGARTEN et al. 1975; BJÖRKLUND et al. 1975).

d) Peptide-Containing Axons

Current knowledge on the distribution of immunoreactive peptide-containing processes of intrinsic neurons in the gut of the rat has been summarized and illustrated by SCHULTZBERG et al. (1980) and their functional significance has been discussed by FURNESS et al. (1980) (see also Fig. 10).

α) Substance P. Substance P-like immunoreactive fibers are prominent in the myenteric plexus and the circular smooth muscle layer along the entire gastrointestinal tract of both rat and guinea pig, although in the guinea pig stomach the density of such fibers is low. However, the smooth muscle layer of the pyloric portion of the stomach is rich in substance P. A moderately dense network of fibers is also observed in the submucous plexus. These findings are, to a great extent, in agreement with those reported by COSTA et al. (1980a).

β) Vasoactive Intestinal Peptide. VIP-like immunoreactive fibers are seen in moderate to large numbers in association with the myenteric and submucous plexus, where they form networks of varicose fibers around unreactive neuronal perikarya. From the myenteric plexus, VIP-positive axons can be followed into the circular muscle layer and the so-called deep muscle fiber plexus. Only few fibers penetrate into the longitudinal muscle (as in the case of substance P-like fibers). The observations by SCHULTZBERG et al. (1980) have been confirmed by COSTA et al. (1980b). Enkephalin-like (both Met- and Leu-enkephalin) and β-endorphin-like immunoreactive material appears to be confined to the same axons in the gut and occurs in moderate to large numbers in fibers of the myenteric plexus and of the circular smooth muscle layer. In both species, enkephalin-positive axons are also observed in the longitudinal smooth layer (particularly large numbers of such fibers are present in the stomach and cecum of the guinea pig). However, the longitudinal muscle of the rat small intestine is devoid of enkephalin-immunoreactive fibers. In the guinea pig, the muscularis mucosae of the stomach and duodenum receive a dense innervation of enkephalin-positive axons. Although the β-endorphin-positive fibers are always found in association with enkephalin-reactive fibers, their number is lower and regions of the gut with few enkephalin-immunoreactive profiles may lack endorphin-reactive axons, such as the myenteric plexus of the guinea pig stomach. In the rat, the number of endorphin-reactive fibers is small in the stomach (longitudinal layer and muscularis mucosae of the fundus and corpus), the large intestine (muscularis mucosae of the colon and rectum, submucous plexus of the colon) and few or single endorphin-positive axons are seen in the guinea pig stomach (muscularis mucosae of the fundus, corpus, and pylorus, longitudinal layer of the fundus and pylorus); small intestine (muscularis mucosae of the duodenum and jejunum and submucous plexus of the duodenum); and large intestine (muscularis mucosae of the colon and rectum, longitudinal muscle of the cecum, distal colon, and rectum, submucous plexus of the cecum).

γ) Somatostatin. Somatostatin-like, immunoreactive fibers are rich in the myenteric and submucous plexus of the small and large intestine of the guinea pig but less abundant in those of the rat. Some somatostatin-positive fibers occur in the longitudinal and circular smooth muscle layer of the gut in both species. The intraganglionic fibers form dense, basket-like aggregations around somatostatin-positive and somatostatin-negative perikarya.

δ) Gastrin/Cholecystokinin. The distribution of gastrin/cholecystokinin-like immunoreactive fibers varies both between species and regions of the gastrointestinal tract. In the rat, the density of fibers is very low in all layers of the stomach and colon, with the exception of the submucous plexus of the ileum and the longitudinal muscle layer of the small and large intestine where such fibers seem to be lacking. In the guinea pig, reactive fibers are confined to the two ganglionic plexuses of the small intestine and stomach and the myenteric plexus of the large intestine.

ε) Neurotensin. Neurotensin immunoreactivity appears to occur only in fibers of the rat gastrointestinal tract (in the myenteric plexus of the stomach and duodenum, in the longitudinal muscle layer of the corpus ventriculi, the circular muscle layer of the cardia and corpus ventriculi, and of the cecum).

e) Coexistence of Peptides and Amines and Possible Function of Peptide Neurons

According to the results of SCHULTZBERG et al. (1980), somatostatin-like and gastrin/cholecystokinin-like immunoreactive substances, possibly a precursor protein, coexist in some intrinsic neurons of the guinea pig proximal colon which was used in the correlation (cross-reaction) studies. In addition, there is evidence for the existence of a somatostatin-like peptide in postganglionic sympathetic neurons (HÖKFELT et al. 1977), the axons of which provide an extrinsic noradrenergic input to enteric neurons of the myenteric and submucous plexus, i.e., the axons which contain immunoreactive DBH. In harmony with the finding of coexistence of a (low molecular weight) classical transmitter with a somatostatin-like peptide in sympathetic postganglionic neurons is the observation by COSTA et al. (1977) that the somatostatin levels in the gut are reduced after extrinsic denervation. It is not established whether VIP occurs together with acetylcholine in fibers of the gut. Coexistence of these two substances has been demonstrated in certain acetylcholinesterase-positive (supposed cholinergic) sympathetic (vasodilator and sudomotor) and parasympathetic neurons of the cat (LUNDBERG et al. 1979), the exact termination of which remains to be demonstrated. Finally, it remains to be elucidated whether the so-called intrinsic serotonergic neurons of the mammalian gut (see Sect. C. III. 1. c) contain peptides in addition to 5-HT or a related indoleethylamine.

Evidence from studies on the descending bulbospinal 5-HT projection neurons in the rat indicate that serotonin coexists with substance P and possibly also enkephalins in some raphe neurons (HÖKFELT et al. 1978; BJÖRKLUND et al. 1979). By analogy it may be postulated that part of the enteric 5-HT neurons store peptides in addition to an indoleamine. The only indirect support for this concept derives from ultrastructural studies on the reaction of the intrinsic enteric "serotonergic" axons to the neurotoxins 5,6- and 5,7-DHT (GERSHON et al. 1980b) which degenerate following subcutaneous administration of high doses of 5,7-DHT and which are characterized by large granular vesicles (mean diameter 120 nm); the latter may represent the storage site for the postulated peptide or peptides. The diameter of these large granular vesicles is within the size range of dense core vesicles said to be typical for p (peptide) fibers (Fig. 11) in the mammalian intestine (BAUMGARTEN et al. 1970; GABELLA 1972a, 1979f; 70–160 nm, mean diameter 120 nm). LARSSON (1977) has shown that VIP-positive axons are equipped with large opaque vesicles of a similar size range. Owing to the variable appearance of their core and membrane, GABELLA (1972a) has termed these vesicles heterogeneous granular vesicles (Table 4).

Axons with these large granular vesicles are not affected by extrinsic denervation of the small intestine (GABELLA 1979f) and are thus intrinsic to the gut. According to the work of BURNSTOCK (1972, 1975, 1977, 1979a; see also ROBINSON and GERSHON 1971; CAMPBELL et al. 1978), these vesicles (termed large opaque vesicles, 80–200 nm diameter) represent the storage site for ATP (or related nucleotides) which is said to be released as a neurotransmitter upon electrical stimulation of enteric nonadrenergic, noncholinergic inhibitory neurons but, unequivocal morphological and histochemical evidence for the concept of purinergic transmission is still lacking.

More recent evidence suggests that VIP could be the transmitter of these enteric inhibitory neurons which are involved in descending receptive relaxation of the muscle of the vertebrate gut during peristalsis (FAHRENKRUG et al. 1978; JANSSON 1969; FURNESS and COSTA 1973, 1974, 1977, 1980). It remains to be shown conclu-

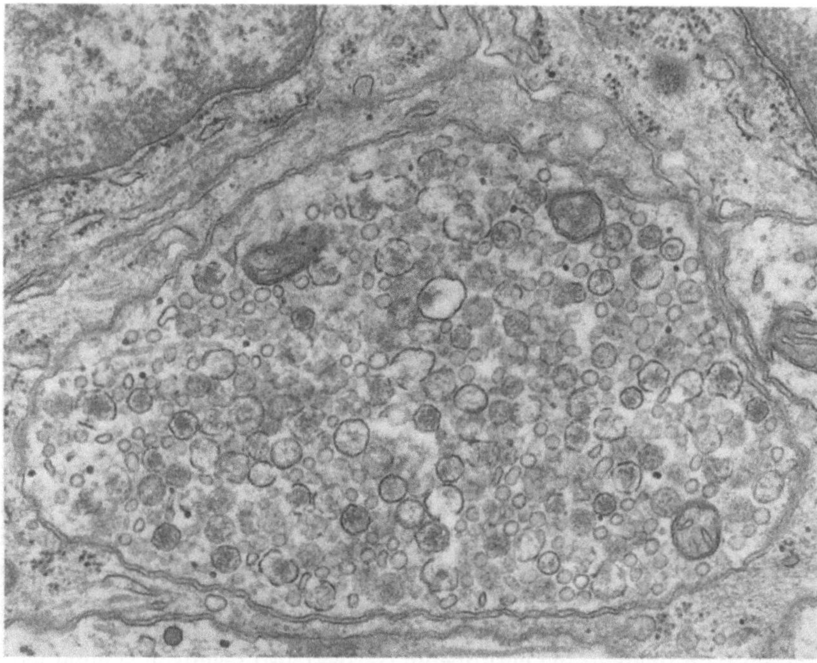

Fig. 11. Guinea pig cecum, myenteric plexus neuropil. Preterminal bouton of p-type fiber ensheathed by glial cell process. Note heterogeneous large dense core vesicle; the membrane of some of the large granular vesicles is fragmented. (BAUMGARTEN et al. 1970) ×30,000

sively whether peptides are localized in the large granular vesicles of the other types of nerve fibers described in Sects. C. III. a–c.

The findings by FURNESS et al. (1980) suggest that the somatostatin-positive neurons are polar interneurons within the myenteric plexus that impinge upon: (a) other somatostatin-reactive neurons, thus forming descending series of neurons; or (b) non-somatostatin neurons, e.g., cholinergic neurons, thus affecting transmission in intrinsic circuits of the enteric nervous system (COHEN et al. 1978; FURNESS and COSTA 1979). VIP neurons in the myenteric plexus project in an anal direction to provide terminals to the circular muscle and nearby distal rows of ganglia. FURNESS et al. (1980) propose that the somatostatin neurons synapse with enteric inhibitory, probably VIP-releasing neurons that control the activity of the circular smooth muscle, and that both are part of the descending inhibitory component of the peristaltic reflex.

Substance P pathways appear to project in both directions (oral and anal) (FURNESS et al. 1980) and this peptide, when released from the intrinsic neurons, seems to be responsible for slow depolarization of myenteric neurons and contraction of the longitudinal muscle (MORITA et al. 1980). Enkephalins and morphine inhibit the peristaltic reflex by hyperpolarization of nerve cells in the myenteric plexus concomitant with a reduction in acetylcholine release (NORTH et al. 1979). FURNESS et al. (1980) speculate that this effect might be due to prejunctional inhibition of cholinergic excitatory axons that supply the circular muscle.

D. Integration of Extrinsic and Intrinsic Neurons into Reflex Circuits

The role of the preganglionic parasympathetic cholinergic fibers in reflex stimulation of gastric motility (which is mainly exerted via enteric cholinergic neurons) is well understood and appears to depend on increases in the amplitude and duration of the plateau of the action potential generated in the muscle of the stomach itself; this myogenic action potential regulates the occurrence of peristaltic contractions and the strength of the moving peristaltic wave (EL-SHARKAWY and SZURSZEWSKI 1978; SZURSZEWSKI 1977, 1979). Sympathetic nerve stimulation reduces the amplitude of the plateau potentials and of phasic contractions of the corpus and antrum; this effect is said to be mediated mainly via α-adrenoceptors located on smooth muscle cells (GERSHON 1967; SZURSZEWSKI 1979); however, it seems that sympathetic postganglionic noradrenergic neurons also inhibit the peristaltic reflex of the stomach and intestine by reducing the acetylcholine output from intrinsic cholinergic neurons (SCHAUMANN 1958, PATON and VIZI 1969; KOSTERLITZ et al. 1970) and, according to the findings of MANBER and GERSHON (1978) and GERSHON (1979), by presynaptic axoaxonal inhibition of acetylcholine release. Adrenergic sympathetic fibers are involved in reflex inhibition of gastric motility elicited from the antrum (ABRAHAMSSON 1973), in the intestinointestinal reflex (PEARCY and LIERE 1926) and in the enterogastric reflex (SEMBA et al. 1964). Stimulation of visceroafferent pain fibers in the peritoneum (serosa) also results in sympathetically mediated inhibition of peristaltic activity in the gastrointestinal tract. Strong, persistent activation of this viscerovisceral reflex may result in adynamic ileus. Stress-provoked increased release of catecholamines from the sympathoadrenal system into the systemic circulation affects gastrointestinal smooth muscle via inhibitory β-adrenoceptors located on smooth muscle in addition to the nerve-mediated effects that operate mainly by α-adrenoceptor mechanisms (including stimulation of release-inhibiting autoreceptors on noradrenaline neurons). The role of adrenergic fibers and mechanisms in the inhibitory control of gastrointestinal motility has been discussed in detail by FURNESS and COSTA (1974).

More recent work (KREULEN and SZURSZEWSKI 1979a, b; SZURSZEWSKI 1979) indicates that extraspinal, i.e., peripheral, reflex arcs participate in the control of colonic smooth muscle activity (intestinointestinal reflex) which involve mechanoceptor afferent fibers that synapse with postganglionic sympathetic noradrenergic neurons located in the prevertebral ganglia. A viscerotopic distribution of the afferent fibers (the cell bodies of which reside in the wall of the intestine) originating in the different portions of the colon has been found by KREULEN and SZURSZEWSKI: those derived from the proximal colon project to the celiac ganglion, those from the midcolon project to the superior mesenteric and from the distal colon to the inferior mesenteric ganglion. Activation of these mechanoreceptor neurons appears to influence the discharge rate of the noradrenergic neurons in a facilitatory way (long-lasting depolarizations, which follow an initial period of after hyperpolarization; WEEMS and SZURSZEWSKI 1978; SZURSZEWSKI 1979).

Blockers of noradrenergic or cholinergic transmission mechanisms do not modify these electrophysiologic effects in the prevertebral ganglia, suggesting that noncholinergic, nonadrenergic transmitters are involved. Since visceroafferent neurons which have their cell bodies in the gut are known to project to prevertebral ganglia and have been shown to store peptides, such as VIP (BRYANT et al. 1976; LARSSON et al. 1976; FUXE et al. 1977; SCHULTZBERG et al. 1978, 1979; HÖKFELT 1979), it may be postulated that peptidergic afferent fibers from mechanoceptors in the gut modulate synaptic transmission in the prevertebral ganglia. It is not yet clear whether interneurons in the prevertebral ganglia (cells of the small, intensely fluorescent type) are integral parts of these peripheral reflex arcs. Transmission in the prevertebral ganglia is further influenced by axon collaterals of primary sensory neurons which have their cell bodies in the spinal ganglia and store substance P and by enkephalin-containing fibers which originate in the spinal cord (HÖKFELT 1979). The functional significance of these noncholinergic, nonadrenergic neurons in the motility of the gut remains to be elucidated. The viscerotopic arrangement of the projections of the enteric mechanoceptor neurons suggests that peripheral reflex arcs may control segmental, peristolic, and peristaltic activity along the colon without involvement of the spinal cord and supraspinal influences.

Nonadrenergic, noncholinergic intrinsic neurons of the gut which receive input from extrinsic vagal and sacral parasympathetic preganglionic neurons are components of inhibitory reflex arcs that facilitate or even enable passage of material in the digestive tract by relaxing sphincters, by increasing the capacity of the stomach (receptive ralaxation) and possibly the rectum, and by causing a descending wave of relaxation in the gastrointestinal tract in front of an advancing bolus (ABRAHAMSSON and JANSSON 1969; OHGA et al. 1970; BEANI et al. 1971; BURNSTOCK and COSTA 1973; BENNETT and STOCKLEY 1973; FURNESS and COSTA 1973; GARRETT et al. 1974; KREULEN and SZURSZEWSKI 1979a, b; FURNESS and COSTA 1980; FURNESS et al. 1980). These neurons are integral parts of the peristaltic reflex which can be elicited, even in the extrinsically denervated gut in vitro (TRENDELENBURG 1917).

The transmitter of this important enteric inhibitory neuron system in the vertebrate gut remains to be identified conclusively and the intrinsic and extrinsic connections of these neurons are not fully understood as yet. BURNSTOCK's proposal that ATP or related nucleotides may be the active compounds released by these neurons is not generally accepted. Immunohistochemical findings by FURNESS et al. (1980) and by SCHULTZBERG et al. (1980) suggest that certain peptides, particularly VIP, are candidates for a transmitter role in these enteric inhibitory neurons. Loss of these neurons may be responsible for failure of the gut smooth muscle to relax in front of a descending boulus, thus causing stenosis and prestenotic hypertrophic dilatation of the colon as in patients with Hirschsprung's disease (megacolon). It has been claimed that noncholinergic, nonadrenergic neurons of the myenteric plexus in the lower rectum and anterior pelvic plexus that project to the internal anal sphincter (GARRETT et al. 1974) are involved in sphincter relaxation during defecation. The defecation reflex is mediated by parasympathethic preganglionic neurons which synapse with enteric cholinergic excitatory neurons of the colon and rectum (HULTÉN 1969; DE GROAT and KRIER 1978) and with noncholinergic, nonadrenergic inhibitory neurons innervating the smooth muscle of the internal anal sphincter (see Sect. C. I. 1). These effects are antagonized by sympathetic adrenergic fibers which relax the colonic smooth muscle and contract the sphincter (HOWARD and GARRETT 1973; GARRETT et al. 1974) mainly through α-adrenergic mechanisms.

E. Conclusions

This chapter has summarized and discussed current knowledge and concepts on the extrinsic and intrinsic innervation of gastrointestinal smooth muscle. With the exception of the internal anal sphincter, autonomic efferent modulation of gut smooth muscle is indirect, i.e., via synaptic relay in enteric ganglia. Adrenergic postganglionic neurons are involved in long-loop, intestinointestinal inhibitory reflexes, involving the spinal cord and in short-loop, inhibitory reflexes cranial and caudal to peristaltic activity, involving the prevertebral ganglia. Maximum activation of the long-loop, inhibitory sympathetic reflexes by intense stimulation of mechanoceptor and pain afferent fibers from the serosa may result in adynamic ileus. Receptive relaxation of the stomach (vagovagal reflex), and the descending inhibition of gut smooth muscle that travels caudad in front of a peristaltic contraction wave, involve intrinsic noncholinergic, nonadrenergic inhibitory neurons the transmitter of which has been claimed to be ATP, related nucleotides, or a peptide such as VIP. These neurons are integrated into an intrinsic peristaltic reflex circuit which also functions in the extrinsically denervated intestine. Loss of these intrinsic inhibitory neurons, may result in failure of the gut to relax, i.e., obstructive stenosis as in Hirschsprung's disease. Parasympathetic preganglionic fibers that synapse with intrinsic excitatory cholinergic neurons enhance intrinsic propulsive activity in the stomach and duodenum (via vagal efferent fibers) and mediate mass propulsive activity in the colon (via the pelvic splanchnic nerves), resulting in defecation. The effect of activation of the preganglionic parasympathetic excitatory neurons is antagonized by sympathetic pathways to the colon and sphincter muscle; the synaptic relay between pre- and postganglionic sympathetic neurons which supply the internal sphincter appears to be located in the pelvic ganglia. Cholinergic parasympathetic inhibition of the sphincter is permissive and indirect, possibly involving cholinergic inhibition of tonic sympathetic adrenergic discharge. Noncholinergic, nonadrenergic relaxation of the sphincter may also be important in the defecation reflex.

Histochemical and immunohistochemical techniques have been used in mapping of noradrenergic fibers and of peptide-containing neurons and nerve fibers throughout the mammalian gut. Conclusive demonstration of intrinsic cholinergic neurons and of indoleaminergic neurons and their projections depends on the development of more specific histochemical techniques; however, a wealth of pharmacologic and electrophysiologic data indicates that these neurons are essential for the transmission of excitatory inputs to the smooth muscle and stimulation of enteric excitatory neurons. Understanding of the role of the various peptidergic enteric neurons in gut motility is only at its beginning and largely hypothetical. Coexistence of peptides and classical small molecular weight chemical transmitters seems to occur (e.g., somatostatin and nordrenaline, VIP and acetylcholine, and substance P and/or enkephalins and indoleamines) but the functional meaning of this coexistence is uncertain and the subcellular sites of storage of these peptides are not clear at present.

The coexistence of classical chemical transmitters and of peptides in enteric neurons and in sympathetic and parasympathetic neurons suggests that these neurons share properties with many diffuse endocrine cells in the mucosa of the gut and respiratory and urinary tracts which are characterized by uptake and decarboxylation of amine precursons (APUD cells), capacity to store amines, and synthesis and release of peptides. In the myenteric and submucous plexus, neurons, and nerve fibers have been identified that take up amines and their precursors (so called amine-handling neurons) which are clearly different from the noradrenergic sympathetic axons; these amine-handling neurons may be identical with those peptide-containing neurons that are characterized by coexistence of peptides.

Unequivocal ultrastructural identification of these different types of nerve fibers remains a problem that could be solved by the development of more specific cytochemical or immunocytochemical methods. At present, such methods are available for the identification of noradrenergic neurons and VIP-containing neurons but these methods have not yet been systematically applied to sections of the various portions of the gastrointestinal tract of different species. As a consequence, detailed qualitative and quantitative data on the synaptic interrelationship of the various types of extrinsic and intrinsic neurons involved in gut motility are lacking. These data are necessary for the reconstruction of intrinsic neuronal circuits and their modulatory input.

Acknowledgement. This work was supported by grants form the Deutsche Forschungsgemeinschaft which are gratefully acknowledged.

References

Åberg G, Eränkö O (1967) Localization of noradrenaline and acetylcholinesterase in the taenia of the guinea-pig caecum. Acta Physiol Scand 69:383–384

Abrahamsson H (1973) Studies on the inhibitory nervous control of gastric motility. Acta Physiol Scand [Suppl] 390:1–38

Abrahamsson H, Jansson G (1969) Elicitation of reflex vagal relaxation of the stomach from pharynx and esophagus in the cat. Acta Physiol Scand 77:172–178

Allison AC, Young MR (1964) Uptake of dyes and drugs by living cells in culture. Life Sci 3:1407–1414

Barr L, Berger W, Dewey MM (1968) Electrical transmission at the nexus between smooth muscle cells. J Gen Physiol 51:347–368

Baumgarten HG (1967) Vorkommen und Verteilung adrenerger Nervenfasern im Darm der Schleie (Tinca vulgaris Cuv.). Z Zellforsch 76:248–259

Baumgarten HG, Lachenmayer L (1972) Chemically induced degeneration of indoleamine-containing nerve terminals in rat brain. Brain Res 38:228–323

Baumgarten HG, Lange W (1969) Adrenergic innervation of the oesophagus in the cat (Felis domestica) and rhesus monkey (Macacus rhesus). Z Zellforsch 95:529–545

Baumgarten HG, Owman C, Holstein AF (1970) Auerbach's plexus of mammals and man: electron microscopic identification of three different types of neuronal processes in myenteric ganglia of the large intestine from rhesus monkeys, guinea-pigs and man. Z Zellforsch 106:376–397

Baumgarten HG, Holstein AF, Rosengren E (1971) Arrangement, ultrastructure and adrenergic innervation of smooth musculature of the ductuli efferentes, ductus epididymidis and ductus deferens of man. Z Zellforsch 120:37–79

Baumgarten HG, Holstein AF, Stelzner F (1972) Unterschiede in der Innervation des Dickdarmes und des Sphincter ani internus bei Säugern und beim Menschen. Ergeb Bd Anat Anz 130:43–47

Baumgarten HG, Holstein AF, Stelzner F (1973a) Nervous elements in the human colon of Hirschsprung's disease. Virchows Arch [Pathol Anat] 358:1113–1136

Baumgarten HG, Björklund A, Lachenmayer L, Nobin A (1973b) Evaluation of the effects of 5,7-dihydroxytryptamine on serotonin and catecholamine neurons in the rat CNS. Acta Physiol Scand [Suppl] 391:1–19

Baumgarten HG, Björklund A, Lachenmayer L, Nobin A, Rosengren E (1973a) Evidence for the existence of serotonin-, dopamine- and noradrenaline-containing neurons in the gut of Lampetra fluviatilis. Z Zellforsch 141:33–54
Baumgarten HG, Björklund A, Nobin A, Rosengren E, Schlossberger HG (1975) Neurotoxicity of hydroxylated tryptamines: structure-activity relationships. 1. Long-term effects on monoamine content and fluorescence morphology of central monoamine neurons. Acta Physiol Scand [Suppl] 429:1–27
Beani L, Bianchi C, Crema A (1971) Vagal non-adrenergic inhibition of guinea-pig stomach. J Physiol (Lond) 217:259–279
Bennet T, Malmfors T, Cobb JLS (1973) Fluorescence histochemical observations on catecholamine-containing cell bodies in Auerbach's plexus. Z Zellforsch 139:69–81
Bennett A, Stockley HG (1973) A study of the intrinsic innervation of human isolated gastro-intestinal muscle using electrical stimulation. J Physiol (Lond) 233:34P–35P
Bennett MR, Rogers DC (1967) A study of the innervation of the taenia coli. J Cell Biol 33:573–596
Björklund A, Baumgarten HG, Horn AS, Nobin A, Schlossberger HG (1975) Neurotoxicity of hydroxylated tryptamines: structure-activity relationships. 2. In vitro studies on monoamine uptake inhibition and uptake impairment. Acta Physiol Scand [Suppl] 429:31–61
Björklund A, Emson PC, Gilbert RFT, Skagerberg G (1979) Further evidence for the possible co-existence of 5-hydroxytryptamine and substance P in the medullary raphe neurones of rat brain. Br J Pharmacol 66:112–113P
Bryant MG, Polak MM, Modlin I, Bloom SR, Albuquerque RH, Pearse AGE (1976) Possible dual role for vasoactive intestinal peptide as gastrointestinal hormone and neurotransmitter substance. Lancet 1:991–992
Bülbring E, Gershon MD (1967) 5-Hydroxytryptamine participation in the vagal inhibitory innervation of the stomach. J Physiol (Lond) 192:823–846
Bülbring E, Shuba MF (1976) Physiology of smooth muscle. Raven, New York
Bülbring E, Brading A, Jones A, Tomita T (eds) (1970) Smooth muscle. Arnold, London
Burnstock G (1970) Structure of smooth muscle and its innervation. In: Bülbring E, Brading A, Jones A, Tomita T (eds) Smooth muscle. Arnold, London, pp 1–69
Burnstock G (1972) Purinergic nerves. Pharmacol Rev 24:509–581
Burnstock G (1974) Innervation of vascular smooth muscle: histochemistry and electron microscopy. Clin Exp Pharmacol Physiol [Suppl] 2:7–20
Burnstock G (1975) Ultrastructure of autonomic nerves and neuroeffector junctions; analysis of drug action. Methods Pharmaco 3:113–137
Burnstock G (1977) Purine nucleotides and nucleosides as neurotransmitters or neuromodulators in the central nervous system. In: Usdin E, Hamburg DA, Barchas ED (eds) Neuroregulators and psychiatric disorder. Oxford University Press, Oxford New York, pp 470–477
Burnstock G (1979a) Non-adrenergic, non-cholinergic autonomic nerves. Neurosci Res Prog Bull 17 3:392–405
Burnstock G (1979b) Putative neurotransmitters: adenosine triphosphate. In: Non-adrenergic, non-cholinergic neurotransmission. Neurosci Res Prog Bull 17 3:406–414
Burnstock G, Bell C (1974) Peripheral autonomic transmission. In: Hubbard JI (ed) The peripheral nervous system. Plenum, New York London, pp 277–327
Burnstock G, Costa M (1973) Inhibitory innervation of the gut. Gastroenterology 64:141–144
Burnstock G, Hökfelt T (1979) Coexistence of transmitters. In: Non-adrenergic, non-cholinergic neurotransmission. Neurosci Res Prog Bull 17/3:460
Burnstock G, Robinson PM (1967) Localization of catecholamines and acetylcholinesterase in autonomic nerves. Circ Res [Suppl 3] 20/21:43–55
Burnstock G, Szurszewski JH (1979) Physiological roles. In: Non-adrenergic, non-cholinergic neurotransmission. Neurosci Res Prog Bull 17/3:396–405
Campbell G, Burnstock G (1968) Comparative physiology of gastrointestinal motility. In: Handbook of physiology, vol 4, sect 6. American Physiological Society, Washington, DC, pp 2213–2266

Campbell G, Haller CJ, Rogers DC (1978) Fine structural and cytochemical study of the innervation of smooth muscle in an amphibian (Bufo Marinus) lung before and after denervation. Cell Tissue Res 194:419–432
Chan-Palay V (1975) Fine structure of labelled axons in the cerebellar cortex and nuclei of rodents and primates after intraventricular infusions with tritiated serotonin. Anat Embryol (Berl) 148:235–265
Cohen ML, Rosing E, Wiley KS, Slater IH (1978) Somatostatin inhibits adrenergic and cholinergic neurotransmission in smooth muscle, Life Sci 23:1659–1664
Collier B (1977) Biochemistry and physiology of cholinergic transmission. In: Handbook of physiology: The nervous system, sect 1, vol 1, Cellular biology of the neuron, part 1. American Physiological Society, Washington, DC, pp 463–492
Cook RD, Burnstock G (1976a) The ultrastructure of Auerbach's plexus in the guinea-pig. I. Neuronal elements. J Neurocytol 5:171–194
Cook RD, Burnstock G (1976b) The ultrastructure of Auerbach's plexus in the guinea-pig. II. Non-neuronal elements. J Neurocytol 5:195–206
Costa M, Furness JB (1973) The origins of the adrenergic fibres which innervate the internal anal sphincter, the rectum and other tissues of the pelvic region in the guinea-pig. Z Anat Entwicklungsgesch 140:129–142
Costa M, Furness JB (1979) On the possibility that an indoleamine is a neurotransmitter in the gastrointestinal tract. Biochem Pharmacol 28:565–571
Costa M, Gabella G (1971) Adrenergic innervation of the alimentary canal. Z Zellforsch Mikrosk Anat 122:357–377
Costa M, Furness JG, Gabella G (1971) Catecholamine containing nerve cells in the mammalian myenteric plexus. Histochemie 25:103–106
Costa M, Furness JB, McLean JR (1976) The presence of aromatic l-amino acid decarboxylase in certain intestinal nerve cells. Histochemistry 48:129–143
Costa M, Patel Y, Furness JB, Arimura A (1977) Evidence that some intrinsic neurons of the intestine contain somatostatin. Neurosci Lett 6:215–222
Costa M, Cuello AC, Furness JB, Franco R (1980a) Distribution of enteric neurons showing immunoreactivity for substance P in the guinea-pig ileum. Neuroscience 5:323–331
Costa M, Furness JB, Buffa R, Said S (1980b) Distributions of enteric nerve cell bodies and axons showing immunoreactivity for vasoactive intestinal polypeptide in the guinea pig intestine. Neuroscience 5:587–596
Daniel EE, Daniel VP, Duchon G, Garfield RE, Nichols M, Malhotra SK, Oki M (1976) Is the nexus necessary for cell-to-cell coupling of smooth muscle?, J Membr Biol 28:207–239
De Groat WC, Krier J (1978) The sacral parasympathetic reflex pathway regulating colonic motility and defeacation in the cat. J Physiol (Lond) 276:481–500
Dewey MM, Barr L (1968) Structure of vertebrate intestinal muscle. In: Handbook of physiology, vol IV, sect 6, American Physiological Society, Washington, DC, pp 1629–1654
Dowe GHC, Kilbinger H, Wihittaker VP (1980) Isolation of cholinergic synaptic vesicles from the myenteric plexus of guinea-pig small intestine. J Neurochem 35:993–1003
Dreyfus CF, Bornstein MB, Gershon MD (1977a) Synthesis of serotonin by neurons of the myenteric plexus in-situ and in organotypic tissue culture. Brain Res 128:125–139
Dreyfus CF, Sherman DL, Gershon MD (1977b) Uptake of serotonin by intrinsic neurons of the myenteric plexus grown in organotypic tissue culture. Brain Res 128:109–123
Ekelund M, Ahrén B, Hakanson R, Lundquist I, Sundler F (1980) Quinacrine accumulates in certain peptide hormone-producing cells. Histochemistry 66:1–9
El-Sharkawy TY, Szurszewski JH (1978) Modulation of canine antral circular muscle by acetylcholine, noradrenaline and pentagastrin. J Physiol (Lond) 279:309–329
Fahrenkrug J, Haglund U, Jodal M, Lundgren O, Olbe L, Schaffalitzky de Muckadell OB (1978) Nervous release of vasoactive intestinal polypeptide in the gastrointestinal tract of cats: possible physiological implications. J Physiol (Lond) 284:291–305
Falck B, Owman C (1965) A detailed methodological description of the fluorescence method for the cellular demonstration of biogenic monoamines. Acta Univ Lund 2:1–23
Fehér E, Csanyi K, Vajda J (1974) Comparative electron microscopic studies on the preterminal and terminal fibres of the nerve plexuses of the small intestine employing different fixation methods. Acta Morphol Acad Sci Hung 22:147–159

Feldberg W, Lin RCY (1950) Synthesis of acetylcholine in the wall of the digestive tract. J Physiol (Lond) 111:96–118
Ferri E, Ottaviani G (1966) Osservationi istoanatomiche sopra il plesso nervoso sottosieroso dello stomaco. Acta Neuroveg 28:339–352
Filogamo G, Marchiso PC (1970) Cholineacetyltransferase activity of rabbit ileum wall. The effect of extrinsic and intrinsic denervation and of combined experimental hypertrophy. Arch Int Physiol Biochim 78:141–152
Franco R, Costa M, Furness JB (1979a) Evidence for the release of endogenous substance P from intestinal nerves. Naunyn Schmiedebergs Arch Pharmacol 306:185–201
Franco R, Costa M, Furness JB (1979b) Evidence that axons containing substance P in the guinea-pig ileum are of intrinsic origin. Naunyn Schmiedebergs Arch Pharmacol 307:57–63
Fry GN, Devine CE, Burnstock G (1977) Freeze-fracture studies of nexuses between smooth muscle cells. J Cell Biol 72:26–34
Furness JB (1970) The origin and distribution of adrenergic nerve fibres in the guinea-pig colon. Histochemie 21:295–306
Furness JB, Costa M (1971) Morphology and distribution of intrinsic adrenergic neurones in the proximal colon of the guinea-pig. Z Zellforsch 120:346–363
Furness JB, Costa M (1973) The ramifications of adrenergic nerve terminals in the rectum, anal sphincter and anal accessory muscles of the guinea-pig. Z. Anat Entwicklungsgesch 140:109–128
Furness JB, Costa M (1974) The adrenergic innervation of the gastrointestinal tract. Ergeb Physiol 69:1–52
Furness JB, Costa M (1977) The participation of enteric inhibitory nerves in accommodation of the intestine to distension. Clin Exp Pharmacol Physiol 4:37–41
Furness JB, Costa M (1978) Distribution of intrinsic nerve cell bodies and axons which take up aromatic amines and their precursors in the small intestine of the guinea-pig. Cell Tissue Res 188:527–543
Furness JB, Costa M (1979) Actions of somatostatin on excitatory and inhibitory nerves in the intestine. Eur J Pharmacol 56:69–74
Furness JB, Costa M (1980) Types of nerve in the enteric nervous system. Neuroscience 5:1–20
Furness JB, Costa M, Freeman CG (1979) Absence of tyrosine hydroxylase activity and dopamine β-hydroxylase immunoreactivity in intrinsic nerves of the guinea-pig ileum. Neuroscience 4:305–310
Furness JB, Costa M, Franco R, Llewellyn-Smith IJ (1980) Neural peptides in the intestine: distribution and possible functions. Biochem Psychopharmacol 22:601–617
Fuxe K, Hökfelt T, Said SI, Mutt V (1977) Vasoactive intestinal polypeptide and the nervous system: immunohistochemical evidence for localization in central and peripheral neurons, particularly intracortical neurons of the cerebral cortex. Neurosci Lett 5:241–246
Gabella G (1972a) Fine structure of the myenteric plexus in the guinea-pig ileum. J Anat 111:69–97
Gabella G (1972b) Innervation of the intestinal muscular coat. J Neurocytol 1:341–362
Gabella G (1974) Special muscle cells and their innervation in the mammalian small intestine. Cell Tissue Res 153:63–77
Gabella G (1976) Quantitative morphological study of smooth muscle cells of the guinea-pig taenia coli. Cell Tissue Res 170:161–186
Gabella G (1979a) Smooth muscle cell junctions and structural aspects of contraction. Br Med Bull 35/3:213–218
Gabella G (1979b) Hypertrophic smooth muscle. I. Size and shape of cells, occurrence of mitoses. Cell Tissue Res 201:63–78
Gabella G (1979c) Hypertrophic smooth muscle. II. Sarcoplasmic reticulum, caveolae and mitochondria. Cell Tissue Res 201:79–92
Gabella G (1979d) Hypertrophic smooth muscle. III. Increase in number and size of gap junctions. Cell Tissue Res 201:263–276
Gabella G (1979c) Hypertrophic smooth muscle. IV. Myofilaments, intermediate filaments and some mechanical properties. Cell Tissue Res 201:277–288

Gabella G (1979f) Innervation of the gastrointestinal tract. Int Rev Cytol 59:129–193
Gabella G, Blundell D (1979) Nexuses between the smooth muscle cells of the guinea-pig ileum. J Cell Biol 82:239–247
Gabella G, Costa M (1967) Le fibre adrenergiche nel canale alimentare. G Accad Med Torino 130:198–217
Gabella G, Costa M (1968) Adrenergic fibres in the mucous membrane of guinea-pig alimentary tract. Experientia 24:706–707
Gabella G, Juorio AV (1973) Changes in the concentration and uptake of noradrenaline in degenerating adrenergic fibres. J Physiol (Lond) 233:3–5 P
Gabella G, Juorio AV (1975) Effect of extrinsic denervation on endogenous noradrenaline and 3H-noradrenaline uptake in the guinea-pig colon. J Neurochem 25:631–634
Garrett JR, Howard ER, Jones W (1974) The internal anal sphincter in the cat: a study of nervous mechanisms affecting tone and reflex activity. J Physiol (Lond) 243:153–166
Gaskell WH (1916) The involuntary nervous system. Longmans Green, London
Geffen LB, Jarrott B (1977) Cellular aspects of catecholaminergic neurons. In: Handbook of physiology: The nervous system, vol 1, sect 1, Biology of the neuron part 1. American Physiological Society, Washington, DC, pp 521–571
Gershon MD (1967) Inhibition of gastrointestinal movement by sympathetic nerve stimulation: the site of action. J Physiol (Lond) 189:317–327
Gershon MD (1977) Biochemistry and physiology of serotonergic transmission. In: Handbook of physiology: The nervous system, vol 1, sect 1, Cellular biology of the neuron, part 1. American Physiological Society, Washington, DC, pp 573–623
Gershon MD (1979) The autonomic nervous system and neuroeffector junctions. Neurosci Res Prog Bull 17/3:384–388
Gershon MD, Ross LL (1966) Radioisotopic studies of the binding, exchange, and distribution of 5-hydroxytryptamine synthesized from its radioactive precursor. J Physiol (Lond) 186:451–476
Gershon MD, Drakontides AB, Ross LL (1965) Serotonin: synthesis and release from the myenteric plexus of the mouse intestine. Science 149:197–199
Gershon MD, Robinson RG, Ross LL (1976) Serotonin accumulation in the guinea pig myenteric plexus: ion dependence, structure-activity relationship and the effect of drugs. J Pharmacol Exp Ther 198:548–561
Gershon MD, Epstein ML, Hegstrand L (1980a) Colonization of the chick gut by progenitors of enteric serotonergic neurons: distribution, differentiation, and maturation within the gut. Dev Biol 77:41–51
Gershon MD, Sherman DL, Dreyfus C (1980b) Effects of indolic neurotoxins on enteric serotonergic neurons. J Comp Neurol 190:581–596
Gibbins IL, Haller CJ (1979) Ultrastructural identification of non-adrenergic, non-cholinergic nerves in the rat anococcygeus muscle. Cell Tissue Res 200:257–272
Gillespie JS, Maxwell JD (1971) Adrenergic innervation of sphincteric and non-sphincteric smooth muscle in the rat intestine. J Histochem Cytochem 19:676–681
Gonella J (1978) La motricité digestive et sa régulation nerveuse. J Physiol (Paris) 74:131–140
Gonella J, Niel JP, Roman C (1977) Vagal control of lower oesophageal sphincter motility in the cat. J Physiol (Lond) 273:647–664
Gunn M (1968) Histological and histochemical observations on the myenteric and submocous plexuses of mammals. J Anat 102:223–239
Hager A, Tafuri WL (1959) Elektronenoptischer Nachweis sogenannter neurosekretorischer Elementargranula in marklosen Nervenfasern des Plexus myentericus (Auerbach) des Meerschweinchens. Naturwissenschaften 46:332–333
Henderson RM, Duchon G, Daniel EE (1971) Cell contacts in duodenal smooth muscle layers. Am J Physiol 221:564–574
Hökfelt T (1968) In vitro studies on central and peripheral monoamine neurons at the ultrastructural level. Z Zellforsch Mikrosk Anat 91:1–74
Hökfelt T (1969) Distribution of noradrenaline storing particles in peripheral adrenergic neurons as revealed by electron microscopy. Acta Physiol Scand 76:427–440
Hökfelt T (1979) Polypeptides: localization. Neurosci Res Prog Bull 17/3:425–443

Hökfelt T, Elfvin L-G, Elde R, Schultzberg M, Golstein M, Luft R (1977) Occurrence of somatostatinlike immunoreactivity in some peripheral sympathetic noradrenergic neurons. Proc Natl Acad Sci USA 74:3587–3591

Hökfelt T, Ljungdahl A, Steinbusch H et al. (1978) Immunohistochemical evidence of substance P-like immunoreactivity in some 5-hydroxytraptamine-containing neurons in the rat central nervous system. Neuroscience 3:517–538

Holstein AF, Orlandini GE, Baumgarten HG (1974) Morphological analysis of tissue components in the tunica dartos of man. Cell Tissue Res 154:329–344

Holzbauer M, Sharman DF (1972) The distribution of catecholamines in vertebrates. In: Blaschko H, Muscholl E (eds) Catecholamines. Springer, Berlin Heidelberg New York (Handbook of experimental pharmacology, vol XXXIII, pp 110–185)

Howard ER, Garrett JR (1973) The intrinsic myenteric innervation of the hindgut and accessory muscles of defaecation in the cat. Z Zellforsch Mikrosk Anat 136:31–44

Hoyes AD, Barber P (1980) Axonal terminal ultrastructure in the myenteric ganglia of the guinea-pig stomach. Cell Tissue Res 209:329–343

Hultén L (1969) Reflex control of colonic motility and blood flow. Acta Physiol Scand [Suppl] 335:77–93

Irvin JL, Irvin EM (1954) The interaction of quinacrine with adenine nucleotides. J Biol Chem 210:45–56

Jacobowitz D (1965) Histochemical studies of the autonomic innervation of the gut. J Pharmacol Exp Ther 149:358–364

Jacobowitz D (1974) The peripheral autonomic system. In: Hubbard JI (ed) The peripheral nervous system. Plenum, New York, London, pp 87–110

Jacobs JM (1980) Blood barriers in the nervous system studied with horse-radish peroxidase. Trends Neurosci 3/8: 187–189

Jansson G (1969) Extrinsic nervous control of gastric motility. Acta Physiol Scand [Suppl] 326:1–42

Jessen K, Mirsky R, Dennison M, Burnstock G (1979) Gaba in the vertebrate peripheral autonomic nervous system; cited from Burnstock G, Non-adrenergic, non-cholinergic neurotransmission. Neurosci Res Prog Bull 17/3:458

Jonakait GM, Tamir H, Rapport MM, Gershon MD (1977) Detection of a soluble serotonin binding protein in the mammalian myenteric plexus and other peripheral sites of serotonin storage. J Neurochem 28:277–284

Juorio AV, Gabella G (1974) Noradrenaline in the guinea-pig alimentary canal: regional distribution and sensitivity to denervation and reserpine, J Neurochem 22:851–858

Juorio AV, Killick SW (1972) Monoamines and their metabolism in some molluscs. Comp Gen Pharmacol 3:283–295

Kamijo K, Hiatt RB, Koelle GB (1953) Congenital megacolon, a comparison of the spastic and hypertrophied segments with respect to cholinesterase activities, and sensitivities to acetylcholine, D.F.P. and the barium ion. Gastroenterology 24:173–185

Koelle GB (1951) The elimination of enzymatic diffusion artifacts in the histochemical localization of cholinesterases and a survey of the cellular distributions. J Pharmacol Exp Ther 103:153–164

Kosterlitz HW (1968) Intrinsic and extrinsic nervous control of motility of the stomach and the intestines. In: Code CF (ed) Handbook of physiology, vol IV, sect 6. American Physiological Society, Washington, DC, pp 2147–2171

Kosterlitz HW, Lydon RJ, Watt AJ (1970) The effects of adrenaline, noradrenaline and isoprenaline on inhibitory α- and β-adrenoceptors in the longitudinal muscle of the guinea-pig ileum. Br J Pharmacol 39:398–413

Kreulen DL, Szurszewski JH (1979a) Reflex pathways in the abdominal prevertebral ganglia: evidence for a colo-colonic inhibitory reflex. J Physiol (Lond) 295:21–32

Kreulen DL, Szurszewski JH (1979b) Nerve pathways in celiac plexus of the guinea-pig. Am J Physiol 237:E90–E97

Langley JN (1921) The autonomic nervous system, part 1. Heffer, London

Larsson LI (1977) Ultrastructural localization of a new neuronal peptide (VIP) Histochemistry 54:173–176

Larsson LI, Fahrenkrug J, Schaffalitzky de Muckadell O, Sundler F, Häkanson R, Rehfeld JF (1976) Localization of vasoactive intestinal polypeptide (VIP) to central and peripheral neurons. Proc Nat Acad Sci USA 73:3197–3200
Leaming DB, Cauna N (1961) A qualitative and quantitative study of the myenteric plexus of the small intestine of the cat. J Anat 95:160–169
Leitner JW, Sussmann KE, Vatter AE, Schneider FH (1975) Adenine nucleotides in the secretory granule fraction of rat islets. Endocrinology 96:662–677
Lorén I, Björklund A, Falck B, Lindvall O (1976) An improved histofluorescence procedure for freeze-dried paraffin embedded tissue based on combined formaldehyde-glyoxylic acid perfusion with high magnesium content and acid pH. Histochemistry 49: 177–192
Lorén I, Björklund A, Falck B, Lindvall O (1980) The aluminum-formaldehyde (Alfa) histofluorescence method for improved visualization of catecholamines and indoleamines, 1. A detailed account of the methodology for central nervous tissue using paraffin, cryostat or vibratome sections, J Neurosci Methods 2:277–300
Lorez HP, da Prada M, Pletscher A (1975) Flashing phenomenon in blood platelets stained with fluorescent basic drugs. Experientia 31:593–595
Lundberg J, Ahlmann H, Dahlström A, Kewenter J (1976) Catecholamine-containing nerve fibres in the human-abdominal vagus. Gastroenterology 70:472–474
Lundberg JM, Hökfelt T, Schultzberg M, Uvnäs-Wallenstein K, Kohler C, Said SI (1979) Occurrence of vasoactive intestinal polypeptide (VIP)-like immunoreactivity in certain cholinergic neurons of the cat: evidence from combined immunohistochemistry and acetylcholinesterase staining. Neuroscience 4:1539–1559
Manber L, Gershon MD (1978) An axo-axonic synapse between adrenergic and cholinergic axons in the mammalian gut. Fed Proc 37:227
Meier-Ruge W, Morger R (1968) Neue Gesichtspunkte zur Pathogenese und Klinik des Morbus Hirschsprung. Schweiz med Wochenschr 98:209
Morita K, North RA, Katayama Y (1980) Evidence that substance P is a neurotransmitter in the myenteric plexus. Nature 287:151–152
Myhrberg HE (1972) Ultrastructural localization of monoamines in the central nervous system of Lumbricus terrestris (L.) with remarks on neurosecretory vesicles. Z Zellforsch 126:348–362
Nagasawa J, Suzuki T (1967) Electron microscopic study on the cellular interrelationships in the smooth muscle. Tohoku J Exp Med 91:299–313
Niemi M, Kouvalainen K, Hjelt L (1961) Cholinesterase and monoamine oxidase in congenital megacolon. J Pathol Bacteriol 82:363
Nishihara H (1970) Some observations on the fine structure of the guinea-pig taenia coli in hypertonic solution, with special reference to the nexus. J Anat 107:101–114
Norberg KA (1964) Adrenergic innervation of the intestinal wall studied by fluorescence microscopy. Int J Neuropharmacol 3:379–382
North R, Katayama Y, Williams JT (1979) On the mechanism and site of action of enkephalin on single myenteric neurons. Brain Res 165: 67–77
Ohga A, Nakazato Y, Saito K (1970) Considerations of the efferent nervous mechanism of the vago-vagal reflex relaxation of the stomach in the dog. Jpn J Pharmacol 20:116–130
Olson L, Ålund M, Norberg KA (1976) Fluorescence-microscopical demonstration of a population of gastro-intestinal nerve fibers with a selective affinity for quinacrine. Cell Tissue Res 171:407–423
Paton WDM, Vizi ES (1969) The inhibitory action of noradrenaline and adrenaline on acetylcholine output by guinea-pig ileum longitudinal muscle strips. Br J Pharmacol 35:10–28
Paton WDM, Zar MA (1968) The origin of acetylcholine released from guinea-pig intestine and longitudinal muscle strips. J Physiol (Lond) 194:13–33
Pearcy JF, Van Liere FJ (1926) Studies on the visceral nervous system. XVII. Reflexes from the colon. 1. Reflexes to the stomach. Am J Physiol 78:64–73
Pearse AGE (1977) The diffuse neuroendrocrine system and the APUD concept: related "endocrine" peptides in brain, intestine, pituitary, placenta and anuran cutaneous glands. Med Biol 55:115–125
Pletscher A (1975 Blood platelets as models for the study of neurohumoral transmission. Proc Sixth Int Pharmacol Helsinki

Prosser CL, Burnstock G, Kahn J (1960) Conduction in smooth muscle: comparative structural properties. Am J Physiol 199:545–552
Read JB, Burnstock G (1968) Fluorescent studies on the mucosa of the vertebrate gastrointestinal tract Histochemie 16:324–332
Richards JG, Da Prada M (1977) Uranaffin reaction: a new cytochemical technique for the localization of adenine nucleotides in organelles storing biogenic amines. J Histochem Cytochem 25:1322–1336
Richardson KC (1960) Studies on the structure of autonomic nerves in the small intestine correlating the silver-impregnated image in light microscopy with the permanganate-fixed ultrastructure in electron microscopy. J Anat 94:457–472
Richardson KC (1964a) The fine structure of autonomic nerve endings in smooth muscle with special reference to the vas deferens. Acta Neuroveg 26:373–376
Richardson KC (1964b) Fine structure of the albino rabbit iris with special reference to the identification of adrenergic and cholinergic nerves and nerve endings in its intrinsic muscles. Am J Anat 114:173–205
Richardson KC (1966) Electron microscopic identification of autonomic nerve endings. Nature 210:756
Roberts E, Chase TN, Tower DB (1976) GABA in nervous system function. Raven, New York
Robinson R, Gershon MD (1971) Synthesis and uptake of 5-hydroxytryptamine by the myenteric plexus of the small intestine of the guinea-pig. J Pharmacol Exp Ther 179:29–41
Rogers DC, Burnstock G (1966) Multiaxonal autonomic junctions in intestinal smooth muscle of the toad Bufo marinus. J Comp Neurol 126:625–652
Rothman TP, Ross LL, Gershon MD (1976) Separately developing axonal uptake of 5-hydroxytryptamine and norepinephrine in the fetal ileum of the rabbit. Brain Res 115:437–456
Schabadasch A (1930) Intramuturale Nervengeflechte des Darmrohrs. Z Zellforsch Mikrosk Anat 10:320–385
Schaumann W (1958) Zusammenhänge zwischen der Wirkung der Analgetica und Sympathicomimetica auf den Meerschweinchendünndarm. Arch Exp Pathol Pharmakol 233:112–124
Schofield GC (1968) The enteric plexus of mammals. In: Felts WJL, Harrison RJ (eds) International review of general and experimental zoology. Academic Press, New York
Schultzberg M, Dreyfus CF, Gershon MD et al. (1978) VIP, enkephalin-, substance P-, and somatostation-like immunoreactivity in neurons intrinsic to the intestine: immunohistochemical evidence from organotypic tissue cultures. Brain Res 155:239–248
Schultzberg M, Hökfelt T, Terenius L et al. (1979) Enkephalin immunoreactive nerve terminals and cell bodies in sympathetic ganglia of the guinea-pig and rat. Neuroscience 4:249–270
Schultzberg M, Hökfelt T, Nilsson L et al. (1980) Distribution of peptide- and catecholamine-containing neurons in the gastro-intestinal tract of rat and guinea-pig: immunohistochemical studies with antisera to substance P, vasoactive intestinal polypeptide, enkephalins, somatostatin, gastrin/cholecystokinin, neurotensin and dopamine β-hydroxylase Neuroscience 5:689–744
Segu L, Gaudin-Chazal G, Sevfritz N, Puizillout JJ (1981) A serotonergic system in the nodose ganglion of the cat. J Physiol (Paris) 77:187–190
Semba T, Fujii K, Kimuta N (1964) The vagal inhibitory responses of the stomach to stimulation of the dog's medulla oblongata. Jp J Physiol 14:319–327
Silva DG, Ross G, Osborne LW (1971) Adrenergic innervation of the ileum of the cat. Am J Physiol 220:347–352
Sjöstrand NO (1965) The adrenergic innervation of the vas deferens and the accessory male genital glands. Acta Physiol Scand [Suppl 257] 65:1–82
Sundler F, Håkanson R, Larsson LI, Brodin E, Nilsson G (1977) Substance P in the gut: an immunochemical and immunohistochemical study of its distribution and development. Nobel Symp 37:59–65
Szurszewski JH (1977) Modulation of smooth muscle by nervous activity: a review and a hypothesis. Fed Proc 2456–2461

Szurszewski JH (1979) Polypeptides: physiology. VIP and neurotensin in stomach. Neurosci Res Prog Bull 17/3:443–444
Taxi J (1958) Sur la structure du plexus d'Auerbach de la souris étudié au microscope électronique. CR Acad Sci [D] (Paris) 246:1922–1925
Taxi J (1959) Sur la structure des travées du plexus d'Auerbach: confrontation des données par le microscope ordinaire et par le microscope électronique. Ann Sci Nat Zool Biol Anim 1:571–593
Taxi J (1961) On the innervation of the smooth muscle fibers of the mouse intestine. C R Acad Sci [D] (Paris) 252:331–333
Taxi J (1965) Contribution a l'étude des connexions des neurones moteurs du système-nerveux autonome. Ann Sci Nat Zool Biol Anim 7:413–674
Taxi J, Droz B (1966) Etude de l'incorporation de noradrenaline-^{3}H (NA-^{3}H) et de 5-hydroxytryptophane-^{3}H(5-HTP-^{3}H) dans les fibres nerveuses du canal déférent et de l'intestin. CR Acad Sci [D] (Paris) 263:1237–1240
Tranzer JP, Richards JG (1976) Ultrastructural cytochemistry of biogenic amines in nervous tissue: methodologic improvements. J Histochem Cytochem 24:1178–1193
Trendelenburg P (1917) Physiologische und pharmakologische Versuche über die Dünndarmperistaltik. Naunyn Schmiedebergs Arch Exp Pathol Pharmakol 81:55–129
Van Driel C, Drukker J (1973) Contribution to the study of the architecture of the autonomic nervous system of the digestive truct of the rat. J Neural Transm 34:301–320
Watanabe H, Yamamoto TY (1974) Freeze-etch study of smooth muscle cells from vas deferens and taenia coli. J Anat 117:553–564
Watson AHD (1979) Fluorescent histochemistry of the teleost gut: evidence for the presence of serotonergic neurones. Cell Tissue Res 197:155–164
Weems WA, Szurszewski JH (1978) An intracellular analysis of some intrinsic factors controlling neural output from interior mesenteric ganglion of guinea-pigs. J Neurophysiol 41:305–321
Welsh JH (1972) Catecholamines in invertebrates. In: Blaschko H, Muscholl E (eds) Catecholamines. Springer, Berlin Heidelberg New York (Handbook of experimental pharmacology, vol XXXIII, pp 79–109)
Wilson AJ, Furness JB, Costa M (1979) A unique population of uranaffin-positive intrinsic nerve endings in the small intestine neurosci Lett 14:303–308
Wood JD, Mayer CJ (1979a) Intracellular study of tonic-type enteric neurons in guinea pig small intestine. J Neurophysiol 42:569–581
Wood JD, Mayer CJ (1979b) Serotonergic activation of tonic-type enteric neurons in guinea-pig small bowel. J Neurophysiol 42:582–593
Wood JD, Grafe P, Mayer CJ (1980) Comparison of the action of 5-hydroxytryptamine and substance P on intracellularly recorded electrical activity of myenteric neurons. In: Christensen J (ed) Gastrointestinal motility. Raven, New York, pp 131–138
Yamamoto M (1977) Electron microscopic studies on the innervation of the smooth muscle and the interstitial cell of Cajal in the small intestine of the mouse and bat. Arch Histol Jpn 40:171–201
Yamauchi A (1964) Electron microscopic studies on the autonomic neuromuscular junction in the taenia coli of the guinea-pig. Acta Anat Nippon 39:22–38

'CHAPTER 3

Morphological Basis of Gastrointestinal Motility: Ultrastructure and Histochemistry of Endocrine–Paracrine Cells in the Gut

E. SOLCIA, C. CAPELLA, R. BUFFA, L. USELLINI, and P. TENTI

A. General Cytology and Physiology

Endocrine–paracrine cells are specialized cells scattered in the epithelium lining the gastric glands, intestinal crypts, and villi. They are characterized by secretory granules, which, as a rule, are concentrated in the basal part of the cytoplasm, while the Golgi complex is supranuclear. In the pyloric and intestinal mucosa, most of such cells reach the lumen in a narrow, specialized area showing tufts of microvilli and a centriole; this area probably acts as a receptor surface facing the luminal contents (SOLCIA et al. 1967). Such a pattern suggests some functional polarity of the cell (Figs. 1, 2). In the fundic mucosa endocrine–paracrine cells lack luminal contacts and show less evident polarity (SOLCIA et al. 1975).

Secretory granules are released at the basal surface of the cell or along the lower part of its lateral surface (KOBAYASHI and SASAGAWA 1976), where intervening cells may form interstitial spaces and canaliculi (Figs. 1–3). In the upper (juxtaluminal) part of the epithelium these spaces are closed by junctional complexes with neighboring cells. Granule release at the luminal surface has never been observed. Smooth vesicles and elongated cisternae are often found just below the luminal surface of the cell or in the supranuclear cytoplasm between the Golgi complex and the luminal endings.

Microtubules, thin microfilaments, and intermediate (100 nm, cytoskeletal) filaments are found in gut endocrine cells. The latter filaments are particularly well developed in some cell types, such as gastric P or D_1 cells and intestinal motilin (Mo) cells, where they may form perinuclear bundles. A mechanical function is to be postulated for such filaments, possibly involved in the reception of mechanical stimuli from the lumen.

Nerves and neurons are widely distributed in the gastrointestinal wall, including the mucosa and submucosa. Nerve endings are lacking in the epithelium. They are only found in the lamina propria just below the base of endocrine-paracrine cells (Fig. 1). Although lacking direct reciprocal contacts, mucosal nerve endings and endocrine-paracrine cells might display some functional interaction through release and local diffusion of active amines and peptides.

Occasionally, endocrine-like cells are found in the lamina propria of the intestinal and gastric mucosa. These cells may be in contact with nerve endings and Schwann-like cells, thus displaying paraganglionic features. Such extraglandular cells are found more frequently in fetal life; they increase dramatically in the inflamed lamina propria of chronic atrophic gastritis.

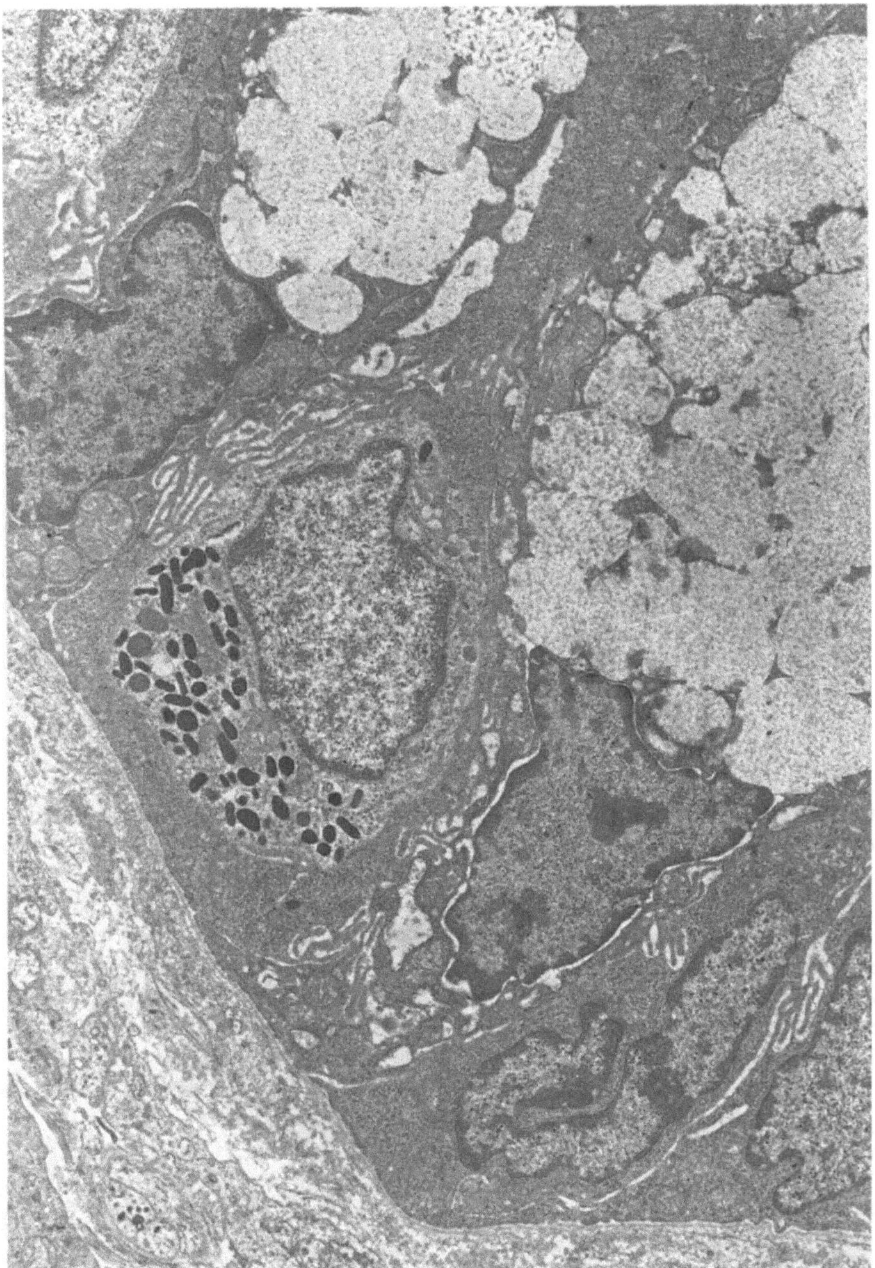

Fig. 1. EC cell with basal, pleomorphic granules in a crypt of the human rectum also showing two goblet cells and nerve endings in the lamina propria *(lower left corner)*. Note intercellular intraepithelial spaces with cell protrusions. × 7,500

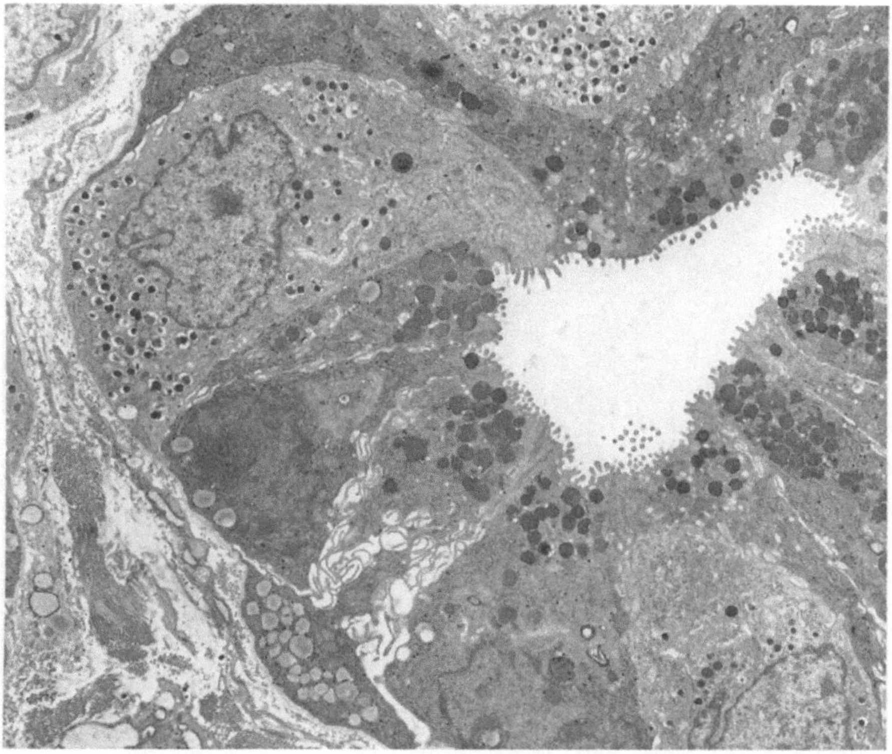

Fig. 2. Cat pyloric gland showing mucous neck cells and three G cells, one of which contacts both the basal lamina and the lumen. Note intercellular intraepithelial spaces. ×4,200

The endocrine-paracrine cells scattered in the gastrointestinal mucosa are now regarded as a diffuse modulatory system of digestive motor and secretory functions. They appear to be sensitive to chemical and mechanical stimuli acting from the lumen, to which they respond by releasing a series of extracellular mediators (SOLCIA et al. 1980b). Such mediators are mostly the same as those released by nerve endings disseminated in the gastrointestinal wall, certainly including substance P, somatostatin, enkephalin, gastrin-cholecystokinin COOH terminal sequences, bombesin, and catecholamines (FURNESS and COSTA 1980). Other mediators detected in endocrine-paracrine cells whose presence in gut nerves has been suggested, although not fully proven yet, are gastrin, neurotensin, pancreatic polypeptide (PP), motilin, and 5-hydroxytryptamine (5-HT). Among neural mediators possibly displaying some counterpart in gut endocrine-paracrine cells are vasoactive intestinal peptide (VIP), angiotensin, γ-aminobutyric acid (GABA), and acetylcholine.

It seems clear that in the gut the endocrine-paracrine cells and nerves function as two integrated regulatory systems, acting through the same extracellular mediators on targets largely in common. They differ mainly in the mechanism by which their mediators are released and transported at the receptor site. Nerves rep-

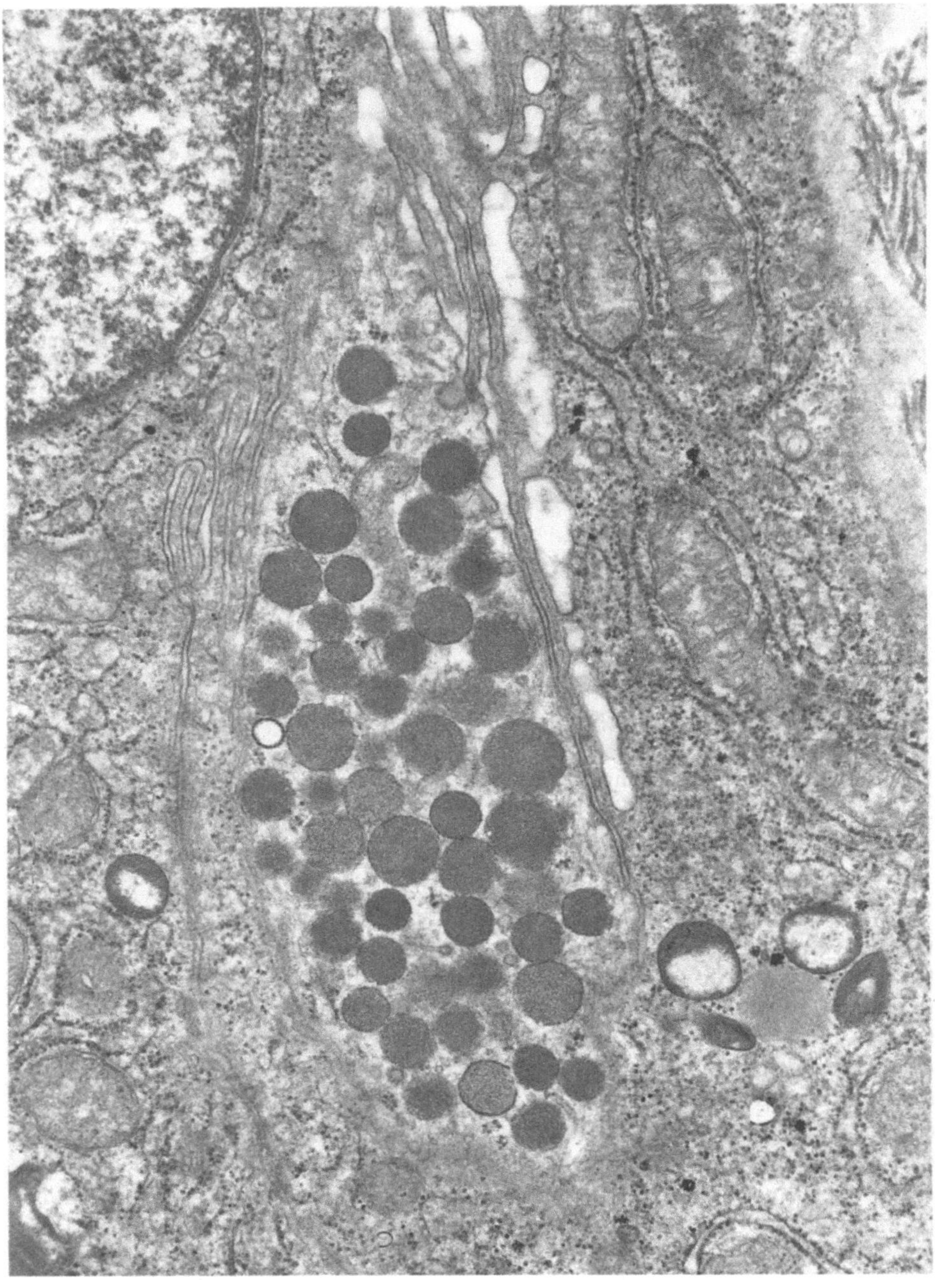

Fig. 3. D cell process filled with poorly dense granules in a fundic gland of the human stomach. Note intercellular spaces formed by cell protrusions of surrounding nonendocrine cells. ×28,000

resent a very efficient way for rapid and selective modulation of the receptor, requiring very small amounts of the mediator. The endocrine mechanism requires comparatively high amounts of the mediator to be released, diluted in the blood, and transported to the receptor site in sufficiently high concentration to effect stimulation. In the paracrine mechanism the mediator is transported to the receptor site either through direct cell–cell contacts between the paracrine cell and the target cell, sometimes facilitated by elongated cell processes of the paracrine cell (Fig. 3), or by diffusion in the extracellular–intraepithelial spaces and canaliculi surrounding endocrine–paracrine cells and other epithelial cells (SOLCIA et al. 1980b). The paracrine mechanism is somewhat intermediate between the endocrine and nerve mechanisms. By avoiding dilution in the blood, it provides high local concentrations of the mediator at the receptor site with relatively minor amounts of secretory product.

Besides acting on neighboring exocrine and endocrine cells of the epithelium, secretory products of endocrine–paracrine cells may cross the basal membrane of the glands and villi and act on structures present in the lamina propria of the mucosa, such as blood vessels, nerve endings, or even smooth muscle cells surrounding the gastric glands, entering the villi, or forming the muscularis mucosae. For instance, cholecystokinin (CCK) and cerulein are known to stimulate muscles and increase the motility of the villi (V. ERSPAMER 1976, personal communication). 5-HT, which is released from EC cells, has been shown to release VIP from perivascular vipergic nerves, thus dilating intramucosal blood vessels and increasing blood flow (FAHRENKRUG 1980).

A number of these mediators are known to affect gut motility; however the contribution of mediators coming from endocrine–paracrine cells to the function of the muscular wall of the gut is difficult to assess, given the presence of some mediators (such as enkephalin, VIP, etc.) in nerves ending in close contact with the muscle fibers. Anyway, by local extracellular diffusion and local or systemic intravascular transport, the possibility exists for a direct interaction of endocrine–paracrine cell products (motilin, CCK, gastrin, somatostatin, enkephalin, etc.) with muscle fibers of the proper musculature of the gut wall.

B. Classification and Description of Cell Types

The first attempt to classify endocrine cells of mammalian gut was made by SOLCIA, FORSSMANN, and PEARSE during the 1969 Wiesbaden Symposium, where seven cell types were considered. The Wiesbaden classification was revised during the Bologna meeting in 1973. Two more cell types were added to the gut endocrine cells and these were tentatively compared with pancreatic endocrine cells (SOLCIA et al. 1973). As many as 15 gastroenteropancreatic (GEP) endocrine cells have been considered in the 1977 Lausanne classification approved by 18 specialists working in the field (SOLCIA et al. 1978).

This classification, rearranged and improved according to a recent discussion held in Los Angeles (SOLCIA et al. 1981), is reported in Table 1. It is based on ultrastructural and immunohistochemical studies interpreted in the light of available biochemical data. A concise description of each cell type is added.

Table 1. Human gastroenteropancreatic endocrine–paracrine cells

Cell	Main product	Pancreas	Stomach		Intestine		
			Oxyntic	Antral	Small		Large
					Upper	Lower	
P	Peptides?	a	+	+	+		
D_1	Peptides?	f	+	f	+	f	f
EC	5-HT, Peptides	r	+	+	+	+	+
D	Somatostatin	+	+	+	+	f	r
PP(F)	P. P.	+				b	b
B	Insulin	+					
A	Glucagon	+	a				
X	Unknown		+				
ECL	Unknown (histamine, 5-HT)		+				
G	Gastrin			+	f		
S	Secretin				+	f	
I	CCK				+	f	
K	GIP				+	f	
M_0	Motilin				+	f	
N	Neurotensin				r	+	r
L	GLI				f	+	+

a = fetus or newborn; b = bovine pancreatic polypeptide serum, mostly GLI cells; + = cells present; f = few; r = rare

I. P and D_1 cells

Pulmonary-type P cells are small cells with very small (100–140 nm mean diameter), round granules, often with a thin halo surrounding a moderately electron-dense core (Fig. 4). Some vesicular granules may also be present; well-developed reticulum and Golgi, small mitochondria, and numerous microfilaments are regulary found (CAPELLA et al. 1978).

So called "enterocatecholamine" (ECT) cells described in the rat pyloric mucosa (FORSSMANN et al. 1969) are likely to be interpreted as a variant of P cells showing an abundance of vesicular, densely cored granules. Somewhat similar cells have been found in the human, cat, and rabbit stomach (CAPELLA et al. 1978). In the latter two species P and/or ECT cells might correspond to the dopamine-storing cells shown histochemically by HÅKANSON et al. (1970). Ultrastructurally, ECT cells may be difficult to distinguish from ECL cells. P cells are distributed in various tissues, especially the upper gut and lung; however in normal tissues they are never numerous at any site. In the human gut, they are more frequently found in the gastric and duodenal mucosa.

Fig. 4 a–c. Human duodenum. **a** P cell with small haloed granules; **b** D_1 cell likely to be interpreted as a motilin Mo cell, with homogeneous fairly dense granules; **c** D_1 cell likely to be interpreted as an intestinal gastrin IG cell, with poorly dense, somewhat inhomogeneous, sometimes haloed, granules. ×28,000

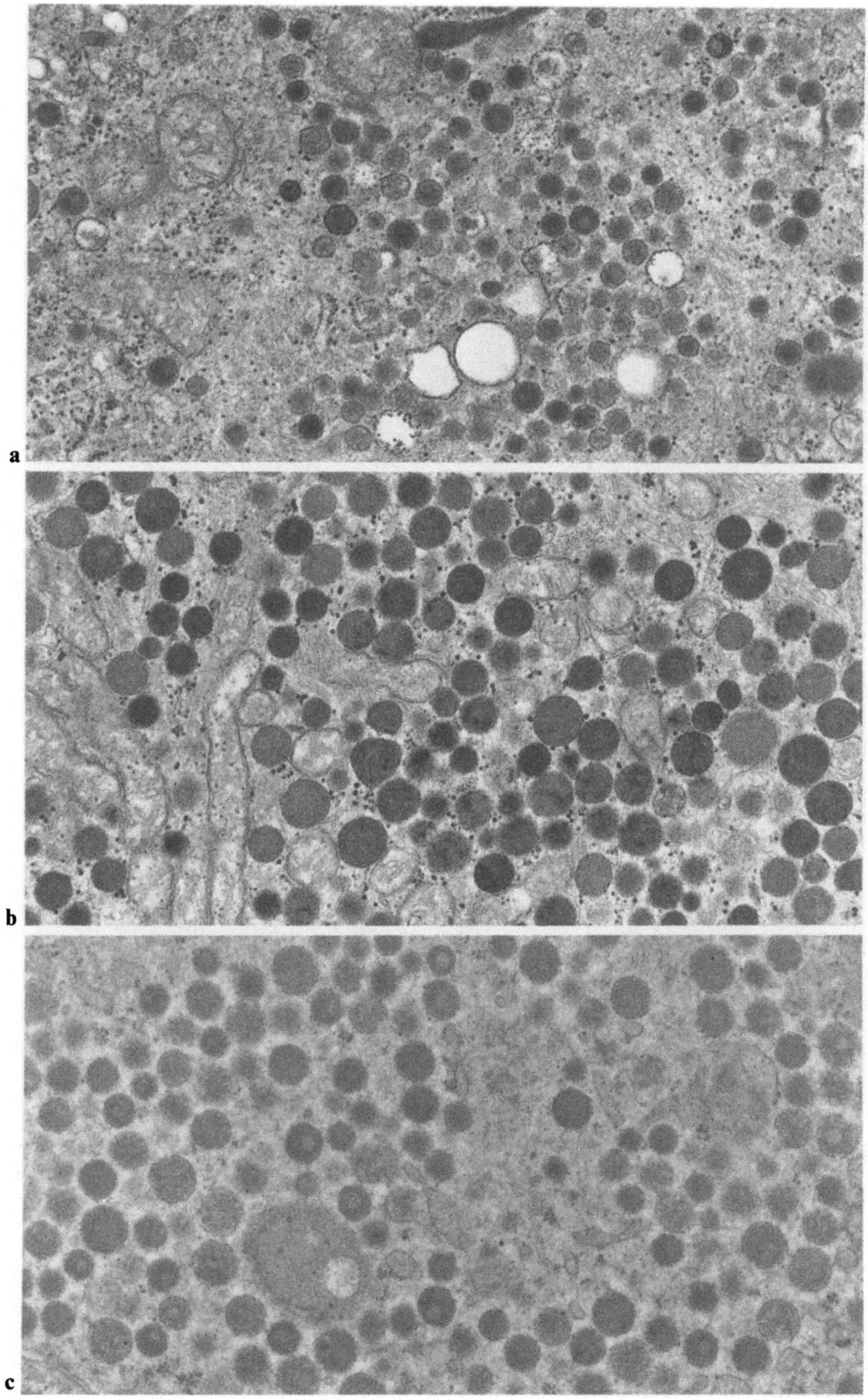
a
b
c

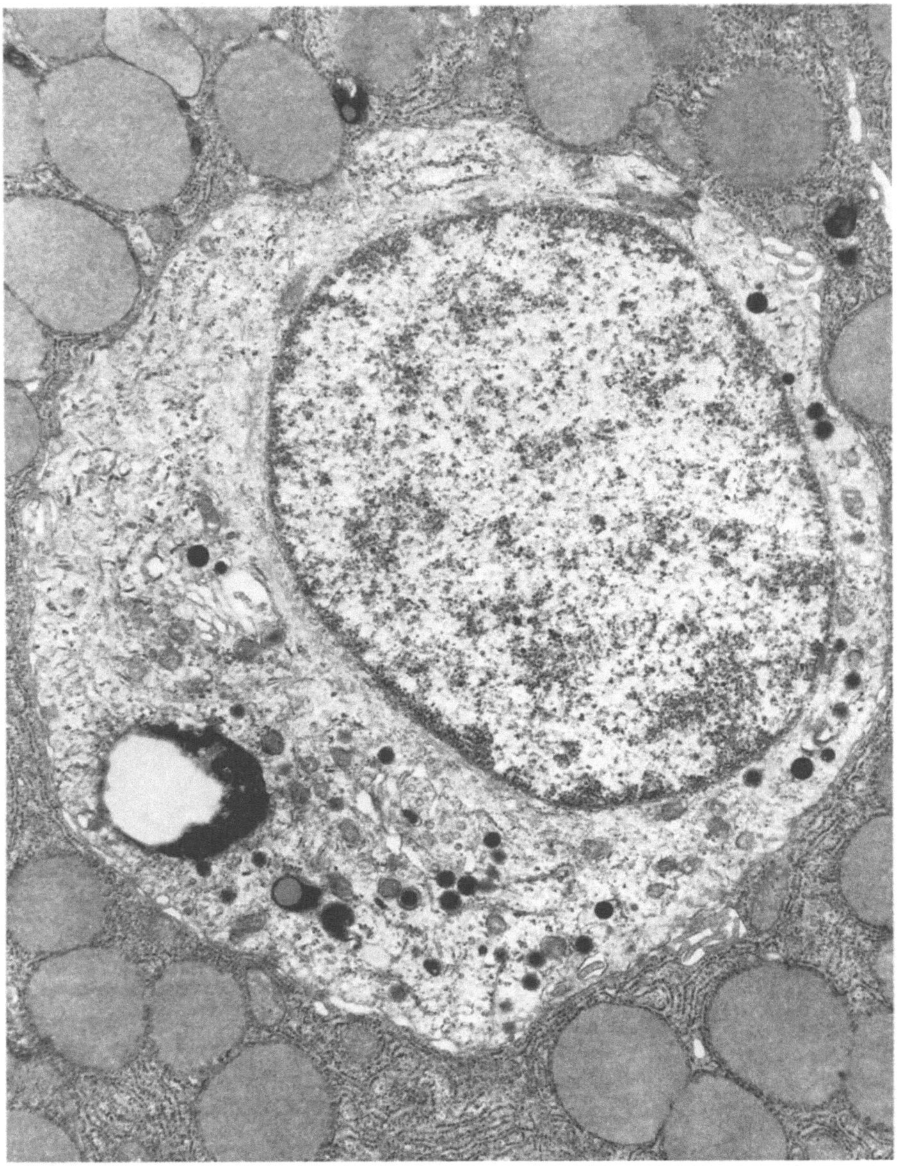

Fig. 5. Small D_1 cell of the human fundic mucosa "embedded" in a chief cell filled with pepsinogen granules. Note small intercellular spaces and canaliculi in between the two cells. ×13,800

D_1 cells are considered to be all the small cells with round granules of mean diameter 140–190 nm, showing a solid core of moderate osmiophilia with closely applied membranes or a very thin halo (CAPELLA et al. 1978). We are certainly dealing with a functionally heterogeneous population of cells, among which slight ultrastructural differences have been observed. D_1 cells of the human fundic mucosa are ovoid or dome-shaped, with round granules of homogeneous, fairly osmiophilic and argyrophilic core, small, thin, oblonged mitochondria, well-developed Golgi and reticulum, and abundant microfilaments (Fig. 5). They closely resemble part of the type 3 cells found in human fetal lung (CAPELLA et al. 1978). D_1 cells of the upper small intestine may be ovoid or elongated; their granules are often less osmiophilic and less argyrophilic than those of fundic cells. Hindgut D_1 cells are elongated cells showing relatively large, ovoid, or cuneiform mitochondria and granules with less dense, sometimes loose core, often with a thin, peripheral, clear space (BUFFA et al. 1978; SOLCIA et al. 1979). D_1 cells of the human pancreas are very scarce, if present at all, when distinguished from islet PP cells (CAPELLA et al. 1978).

The function of P and D_1 cells remains partly uncertain. The granules of some P cells closely resemble the neurosecretory granules of nerve endings in the external zone of hypothalamic median eminence; besides monoamines, some sort of neuropeptides might well be produced by such cells. In the gastric mucosa and lung, a fraction of these cells might be involved in the production of bombesin-like peptide or peptides (WHARTON et al. 1978). A relationship between some "D_1 cells" of the human duodenum and jejunum with motilin cells or with intestinal gastrin-immunoreactive cells and some cells present in pancreatic and duodenal gastrinomas is likely (BUCHAN et al. 1979; VASSALLO et al. 1972).

II. Enterochromaffin (EC) Cells

EC cells (Fig. 1) are cells with highly osmiophilic argentaffin and heavily argyrophil granules of pleomorphic shape: round, ovoid, oblong, pear-shaped, triangular, kidney-shaped, U-shaped, or frankly irregular. In appropriately fixed paraffin sections, EC cells can be stained with a number of methods known to react with 5-HT in vitro, such as chromaffin, diazonium, thioindoxyl, argentaffin, ferric ferricyanide, dimethylaminobenzaldehyde, and xanthydrol reactions. Formaldehyde-fixed EC cells, like formaldehyde-treated 5-HT in vitro, give yellow fluorescence under ultraviolet light, which is much more intense when hot formaldehyde vapor is allowed to react with EC cells in freeze-dried tissues (Falck–Hillarp technique). Direct application of the Masson argentaffin reaction to electron microscopy (SOLCIA et al. 1975) allows discrimination between EC cells and 5-HT-storing non-EC cells.

Earlier histochemical findings suggesting the presence of proteins or peptides in EC cell granules have been validated by recent immunohistochemical studies showing the presence of substance P (NILSSON et al. 1975; PEARSE and POLAK 1975) in cells likely to correspond to at least part of the "intestinal" EC cells described by SOLCIA et al. (1976). Recently, an EC cell fraction of various species, including some large granule EC cells of pig antroduodenal mucosa, but not human cells, has been found to react with anti-leu-enkephalin antibodies (ALUMETS et al. 1978b).

As producers and releasers of 5-HT, substance P, and enkephalin, EC cells seem well-qualified modulators of gastrointestinal motility. Increased numbers of EC cells have been reported in intestinal tracts near strictures or obstructions. Gut hypermotility is a well-known component of the so-called carcinoid syndrome associated with argentaffin EC cell carcinoids. An obvious increase of EC cells is also

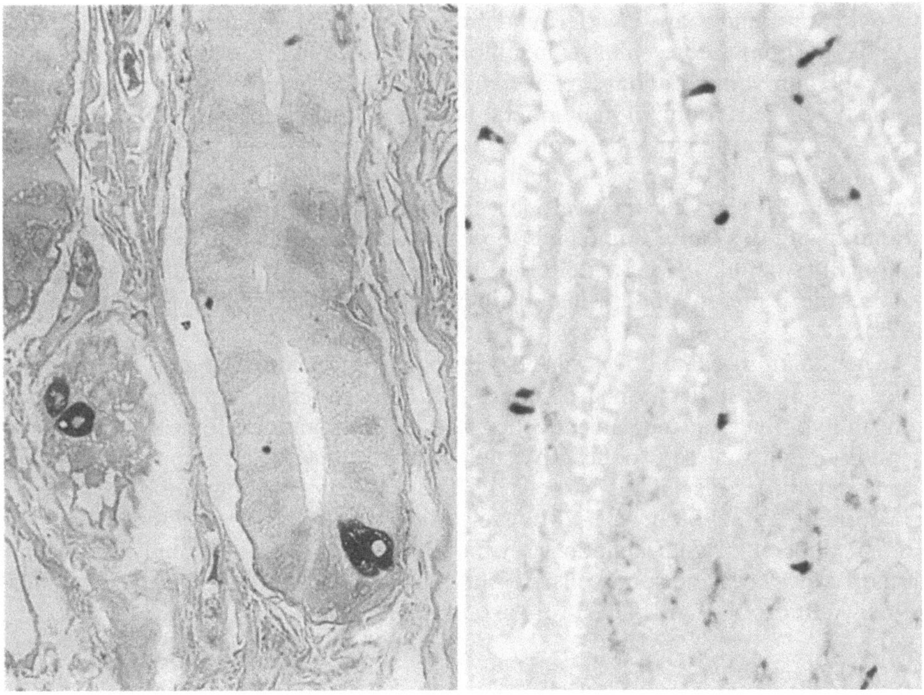

a b

Fig. 6a, b. Somatostatin cells in a 1 μm resin section of dog pyloric mucosa **a** and neurotensin cells in a paraffin section of dog ileum **b**. Immunoperoxidase. **a** × 1,200; **b** × 450

found in connection with chronic inflammatory diseases, with special reference to chronic gastritis (SOLCIA et al. 1970a), cholecystitis, appendicitis (FEYRTER, 1953), and celiac disease (CHALLACOMBE and ROBERTSON 1977).

III. Somatostatin (D) Cells

Somatostatin-immunoreactive cells (Fig. 6a) and ultrastructurally identified D cells have been observed along the whole gastrointestinal tract, from cardia to rectum, as well as in the pancreas (VASSALLO et al. 1971; SOLCIA et al. 1975; ALUMETS et al. 1977). Many of these cells show long cell processes (Fig. 3) contacting blood vessels (in the pancreatic islets) or other cell types, including G cells (LARSSON et al. 1979). D cell granules are usually round, homogeneous, of low electron density and with closely fitting membrane; they are large in most mammalian species and relatively small in the rat. D cells are blackened rather selectively with Davenport's alcoholic silver, while failing to react with a number of aqueous silver techniques.

The wide distribution of D cells in several tissues and in close contact with many cells of different function is in keeping with their proposed local modulatory (paracrine) function and the ability of somatostatin to inhibit many endocrine (gastrin, CCK, insulin, glucagon, etc.) and exocrine secretions (pancreatic enzymes, gastric hydrochloric acid, etc.). In particular, pyloric D cells, which have been shown to be stimulated by HCl infusion into the gastric lumen (KOBAYASHI and SASAGAWA 1976), may play an important role in the HCl-mediated feedback mechanism modulating gastrin and gastric acid secretion.

IV. Pancreatic Polypeptide (PP or F) Cells

F cells with irregular granules of highly variable, generally poor density, known for a long time to occur in the dog pancreas, have now been shown to store pancreatic polypeptide (PP) (LARSSON et al. 1976). Ultrastructurally equivalent cells have been found in all mammals so far investigated, although with some difficulty, owing to wide changes in the granular structure from one species to another (CAPELLA et al. 1977). With electron immunoperoxidase or the thin–semithin section technique, PP-immunoreactive cells of the human islets have been identified as cells with rather small granules that are difficult to distinguish from D_1 cells (LARSSON et al. 1976; GEPTS et al. 1978). The intense argyrophilia and occasional angularity of granules in PP(F) cells might be a distinguishing feature (CAPELLA et al. 1977; GEPTS et al. 1978). PP-immunoreactive cells have been observed in the pyloric mucosa and have been found to be very rare in the guinea pig, but relatively numerous in the dog. Cells storing a peptide (or peptides) reacting with antibodies directed against the COOH terminal sequence of PP are present in the human lower intestine (BUFFA et al. 1978). Many of the latter cells also react with anti-glucagon and anti-glicentin sera and are thus to be interpreted as L cells (FIOCCA et al. 1980).

V. Glucagon (A) Cells

The glucagon-producing A cell is now defined immunohistochemically as a cell reacting with antibodies directed against the COOH terminal part of the glucagon molecule, as well as with antibodies against the middle part of the glucagon or the non-glucagon (R-64) sequence of glicentin (BAETENS et al. 1976; GRIMELIUS et al. 1976; RAVAZZOLA et al. 1979). Ultrastructurally, the A cell is characterized by haloed or targetoid granules with a glucagon-storing core surrounded by an argyrophil material admixed with proglucagon molecules (BUSSOLATI et al. 1971; SOLCIA et al. 1975; RAVAZZOLA and ORCI 1980). The presence of glucagon A cells in the oxyntic and cardiac mucosa of the dog, cat, and monkey is now established (SOLCIA et al. 1970b, 1975; BAETENS et al. 1976; SUNDLER et al. 1976). Few A cells, as well as cells with intermediate features between A and L cells, occur in the human fetal stomach; however, in our experience glucagon A cells are not a regular component of nonpathologic human adult stomach. Cells immunohistochemically and ultrastructurally resembling L cells are well represented in the chicken proventriculus, where A cells are lacking (USELLINI et al. 1980, ultrastructural and immunohistochemical characterization of endocrine-paracrine cells in the chick proventriculus, unpublished work). The presence of very few, if any, A cells in the upper small intestine is suggested by some immunohistochemical findings; their ultrastructural identification remains uncertain because of the difficulty of distinguishing such cells from K and S cells.

VI. X Cells

The X cells are round to ovoid cells characterized ultrastructurally by their round to slightly irregular, compact, dense, fairly large granules of moderate and diffuse argyrophilia (SOLCIA et al. 1975; SOLCIA et al. 1979). In several species they react intensely with acidophilic dyes and are stained deep blue with phosphotungstic hematoxylin, thus suggesting their storage of some basic polypeptide. X cells also react heavily with lead–hematoxylin and the hydrochloric acid–basic dye or "masked metachromasia" technique (CAPELLA et al. 1969; SOLCIA et al. 1975). Despite some ultrastructural resemblance to pancreatic A cells (BAETENS et al. 1976), X cells fail to react with anti-glucagon and anti-glicentin sera (SUNDLER et al. 1976; RAVAZZOLA et al. 1979). Thus, they should not be confused with the A or A-like cells found in the fundic and cardiac mucosa of several species and now proven to store glucagon and related peptides.

In several species, such as the rat, mouse, and dog, X cells seem to be restricted to the fundic and cardial mucosa. In other species the presence of a few X cells in the pyloric mucosa cannot be excluded. In the pig cardia, X cells account for nearly 50% of the whole endocrine cell population; in the same tissue glucagon- and glicentin-immunoreactive cells are quite few. The function of the X cell remains unknown.

VII. ECL Cells

ECL cells are small, irregularly shaped, heavily argyrophilic cells (when stained by the Grimelius, Bodian, and Sevier–Munger methods) which are distributed in fairly large numbers in the fundic glands of the stomach. Ultrastructurally, they show either vesicular granules with an irregular argyrophilic core eccentrically located in a wide space, or a round , relatively compact (or coarsely granular), argyrophilic core surrounded by a membrane, often of wavy appearance and forming a thin, clear space (CAPELLA et al. 1971; SOLCIA et al. 1975). The argyrophilic histamine-storing cells peculiar to murine fundic mucosa are to be identified with ECL cells (RUBIN and SCHWARTZ 1979), although interference of X cells has also been suggested (HÅKANSON et al. 1970).

In many species, ECL cells of untreated animals are not argentaffin; like other argyrophilic cells, they become argentaffin only after treatment with amine precursors (HÅKANSON et al. 1970; CAPELLA et al. 1971; SOLCIA et al. 1975, 1976). However, ultrastructural findings suggest that some ECL cells of the cat and rabbit can display argentaffin granules even in untreated animals (SOLCIA et al. 1975). Occasional granules of some human ECL cells may also be argentaffin. ECL cells with argentaffin granules, probably storing 5-HT or 5-hydroxytryptophan, differ from argentaffin EC cells in being poorly reactive with the hydrochloric acid–basic dye technique and with lead–hematoxylin.

ECL cells are scattered in the fundic glands, especially in their deep half; very few ECL cells have been found in the neck of the glands, and none in the epithelium covering the luminal surface of the mucosa and the gastric pits. Like other endocrine cells of the fundic mucosa, ECL cells lack any contact with the lumen of the glands, usually being covered by oxyntic or peptic cells (CAPELLA et al. 1971). No ECL cells have been found in the well-developed cardiac mucosa of the pig.

ECL cells have been shown to be stimulated by gastrin (HÅKANSON et al. 1976) and to undergo hypertrophic and hyperplastic changes in hypergastrinemic patients suffering from Zollinger–Ellison syndrome (SOLCIA et al. 1975). It seems likely that they take part in the local control of acidopeptic secretion by releasing histamine, 5-hydroxyindoles and, possibly, active peptides.

VIII. Gastrin-Storing (G) Cells

The gastrin-storing cell of the pyloric mucosa or G cell is a medium-sized, ovoid to bottle or pear-shaped cell with abundant, slightly eosinophilic, faintly granular cytoplasm and a relatively large, round, clear nucleus. Selective staining of G cells without interference of nonendocrine cells can be obtained with Grimelius silver, lead–hematoxylin or amine precursor uptake and decarboxylation (APUD) fluorescence from injected precursors (SOLCIA et al. 1975). However, the only reliable technique for G cell detection and quantitation in the light microscope remains immunohistochemistry.

Most anti-gastrin sera contain antibody populations directed against the COOH terminal part of the molecule, which is in common with CCK (LARSSON and REHFELD 1977). By using these antisera in immunohistochemical tests, staining of both gastrin and CCK cells is obtained (BUFFA et al. 1976; LARSSON and REHFELD 1977). This cross-reactivity has little impact with studies of normal mammalian stomach, where no CCK cells have been detected so far; however, it is a major drawback when dealing with the small intestine or with pathologic stomachs showing intestinal (small bowel) mataplasia, where CCK cells occur (LARSSON et al. 1978; RUSSO et al. 1980). In the latter cases, the use of antibodies directed against the non-COOH terminal part of the gastrin molecule, a close comparison with results of CCK-specific antisera and parallel ultrastructural studies are mandatory.

In most species, pyloric G cells are characterized by the presence of vesicular granules with a floccular content, either finely dispersed as in the rat and rabbit, or aggregated in relatively dense cores of irregular contours, as in the cat (Fig. 2). However, haloed granules with homogeneous, dense core (chicken), relatively compact granules with closely applied membrane (pig, guinea pig), or dense, compact, angular granules (ox) are prevalent in G cells of some species. These various ultrastructural patterns of granules may coexist in the same species or even in the same cell. In humans, the large majority of pyloric G cells display at least a proportion of vesicular granules of high diagnostic value; however, compact granules of various density and round to angular shape are also present in many cells (SOLCIA et al. 1967, 1975; CAPELLA et al. 1969; FORSSMANN et al. 1969; VASSALLO et al. 1971). The functional state of the cell as well as the fixation procedures employed may also modify the ultrastructural appearance of G cell granules.

The G cells so far identified ultrastructurally in the human duodenal mucosa are very scarce and disproportionately few in comparison with the gastrin-immunoreactive cells (SOLCIA et al. 1975; LARSSON et al. 1977a) described in the small intestine of humans and several other mammals. This discrepancy has been now resolved by directly identifying gastrin-immunoreactive cells at the ultrastructural level and finding in many of them small (about 175 nm) solid, round granules differing from those of G cells (BUCHAN et al. 1979). It remains to be investigated whether these intestinal gastrin (IG) cells (Fig. 4c) represent a functional variant of G cells or a distinct population of cells producing somewhat different gastrin-related peptides – including a larger proportion of big gastrin – and responding to different stimuli.

IX. Cholecystokinin (I) Cells

Specific immunohistochemical staining of the cholecystokinin (CCK) cell has been achieved by using antibodies directed against the non-COOH terminal part of CCK molecule and lacking reactivity with mammalian G cells (BUFFA et al. 1976; BUCHAN et al. 1978). The identification of the CCK cell with the I cell of ultrastructural investigations (SOLCIA et al. 1975) is now confirmed immunocytochemically (BUCHAN et al. 1978b). Despite changes in size from one species to another, in all species so far investigated, the granules of the CCK (I) cell were solid, fairly dense, round, and unreactive with silver techniques (BUSSOLATI et al. 1971; SOLCIA et al. 1975). Human CCK cell granules were medium-sized (approximately 250 nm) and round to slightly irregular (CAPELLA et al. 1976; BUCHAN et al. 1978b). CCK cells are fairly represented in the duodenal and jejunal mucosa, occurring only occasionally in the ileum.

X. Secretin (S) Cells

The staining of secretin S cells in the small intestine of mammals by means of immunohistochemical techniques has been achieved by several groups (POLAK et al. 1971b; SOLCIA et al. 1972; LARSSON et al. 1977b). With some antisera, cross-reactivity may occur with cells producing chemically related peptides (SOLCIA et al. 1980a). In these cases addition of glucagon, glicentin, and gastric inhibitory polypeptide (GIP) to the anti-secretin serum is recommended to prevent nonspecific staining of L cells in the small and large intestine, of A cells in the pancreas or fundic type gastric mucosa, and of K (GIP) cells in the small intestine. When the specificity of the anti-secretin serum is ensured, secretin-immunoreactive cells are only detected in the small intestine, with special reference to the duodenum (Fig. 7) and jejunum (SOLCIA et al. 1972). They are more numerous in the dog, cat, and pig than in humans. In our experience secretin-immunoreactive cells of the rat and mouse are about as numerous in the upper and lower small intestine; moreover, their separation from the intestinal cells producing chemically related hormones seems less obvious than in other mammals.

Ultrastructurally, secretin S cells showed round to fairly irregular, slightly haloed, argyrophil, small to medium-sized granules (around 200 nm), mostly grouped at the base of the cell (BUSSOLATI et al. 1971; SOLCIA et al. 1972, 1975; LARSSON et al. 1977b).

XI. Gastric Inhibitory Polypeptide (K) Cells

This type of cell was first identified ultrastructurally thanks to its large granules, part of which show a target-like pattern with an argyrophil matrix surrounding the argyrophobe core, thus mimicking in part those of pancreatic A cells (BUFFA et al.

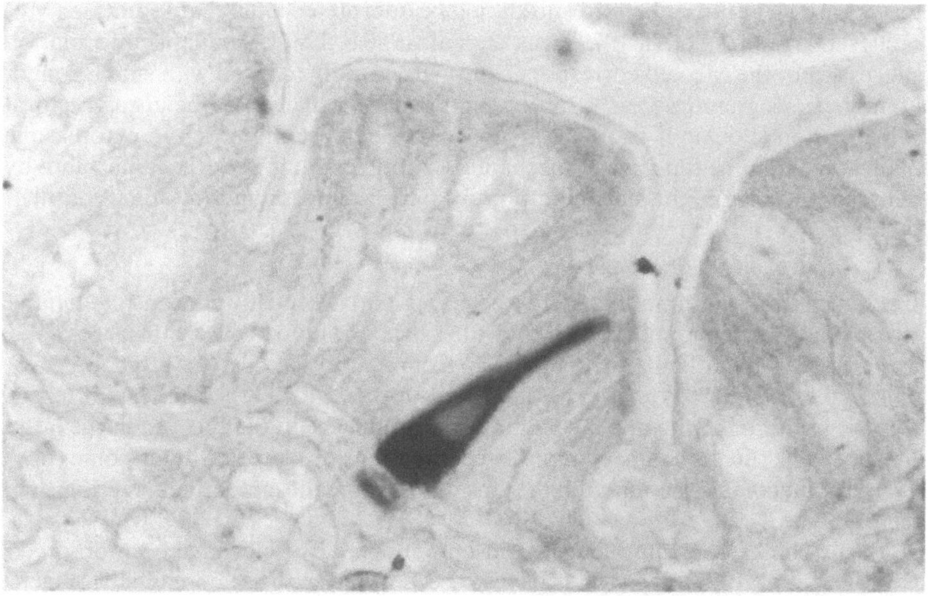

Fig. 7. Secretin cell in the dog duodenum stained with immunoperoxidase. ×1,200

1975; SOLCIA et al. 1975). However, K cell granules are often more variable than α granules and may lack any evidence of halo or double structure (CAPELLA et al. 1976; SOLCIA et al. 1979). These homogeneous granules are especially prevalent in pig K cells, which may be difficult to distinguish from L cells. In humans, K cell granules are usually larger than those of L cells; moreover, L cells are practically lacking in the human duodenum, where K cells are rather numerous (SOLCIA et al. 1975).

K cells of the dog and human lack reactivity with anti-glucagon, anti-glicentin, and anti-secretin sera, while reacting with anti-gastric inhibitory polypeptide (GIP) sera (BUCHAN et al. 1978a; SOLCIA et al. 1980a). However, some anti-GIP sera, besides K cells, also stain pancreatic or gastric A cells and intestinal L cells (ALUMETS et al. 1978a). Cross-reactivity with glicentin or some other component of A and L cells might be the cause of such behavior. Anyway, other anti-GIP sera selectively stain a population of cells peculiar to the upper small intestine, with special reference to deep crypts; such cells have been shown to correspond ultrastructurally to K cells (BUCHAN et al. 1978a). Thus, at least in some mammals, a specialized GIP-producing cell is present in the small intestine, where it is much better represented in adult than in fetal life. This cell may have a key role in the so-called enteroinsular axis.

XII. Motilin (Mo) Cells

Motilin immunoreactive cells have been found in the duodenal and jejunal mucosa of mammals (PEARSE et al. 1977). Although EC cells may be also stained with some anti-motilin sera (HEITZ et al. 1978), only a population of nonargentaffin cells reacts with all anti-motilin sera and should be interpreted as a reliable motilin (Mo) cell (HELMSTAEDTER et al. 1979; SOLCIA et al. 1980b, 1981). Available information suggests that the Mo cell corresponds ultrastructurally to a cell with small, round, homogeneous granules (Fig. 4b) fitting in the "D_1" cell class of the ultrastructural classification (POLAK et al. 1978a). Mo cells, which are often rich in cytoplasmic filaments, might be sensitive to mechanical stimuli coming from the lumen and, by releasing motilin, might play an important part in the modulation of gut motility.

XIII. Neurotensin (N) Cells

Specific neurotensin-immunoreactive cells (Fig. 6b) are mostly concentrated in the lower small intestine and correspond ultrastructurally to cells with large, solid granules (Fig. 8) difficult to distinguish from L (GLI) cells (ORCI et al. 1976; FRIGERIO et al. 1977; POLAK et al. 1977; SUNDLER et al. 1977). A few neurotensin cells are scattered in the upper small intestine and, exceptionally, even in the large intestine. The functional meaning of N cells remains obscure, although some (possibly pharmacologic) action of neurotensin on pancreatic endocrine secretions has been reported.

In the chicken antrum, neurotensin-immunoreactive cells have been observed (SUNDLER et al. 1977). No such cells have been found in the mammalian stomach.

XIV. Glucagon-Like Immunoreactive (L) Cells

Although a few cells reacting with glucagon-specific COOH terminal antibodies have been reported in the intestinal mucosa, ultrastructural evidence for the presence of true A cells at this site is lacking, apart, perhaps, from occasional cells in the upper small intestine. Instead, a large population of cells reacting with antibodies directed against the middle part of the glucagon molecule (POLAK et al. 1971a; GRIMELIUS et al. 1976; RAVAZZOLA et al. 1979) as well as antibodies against the non-glucagon part of the glicentin molecule (RAVAZZOLA et al. 1979; SOLCIA et al. 1980a) have been described and found to correspond ultrastructurally to a specific type of cell, the L cell, clearly differing from the A cell (BUSSOLATI et al. 1971; SOLCIA et al. 1975; GRIMELIUS et al. 1976; RAVAZZOLA et al. 1979). L cells show large (dog) to medium-sized (human), mostly round, solid granules of variable argyrophilia (Fig. 8). Only a few similar granules are found in adult A cells, while they may be numerous in fetal A cells.

It has been shown by MOODY et al. (1979) that the whole glucagon molecule is present in intestinal glucagon-like immunoreactive (GLI) peptides, although in a form leaving the COOH terminal part of the molecule inaccessible to specific antibodies. In fact, L cells become reactive with anti-COOH terminal glucagon-specific antibodies after treatment with some proteases known to split GLI molecules (RAVAZZOLA and ORCI 1980). A GLI molecule of 69 amino acid residues, called glicentin, has been extracted from pig intestine and purified (MOODY et al. 1978). A similar peptide has been extracted from tissues, like the pancreas and fundic mucosa, producing true glucagon; it has been identified with proglucagon. In keeping with this interpretation, antibodies reacting with the non-glucagon part of the glicentin molecule stained pancreatic and fundic A cells, besides intestinal L cells (RAVAZZOLA et al. 1979; SOLCIA et al. 1980a).

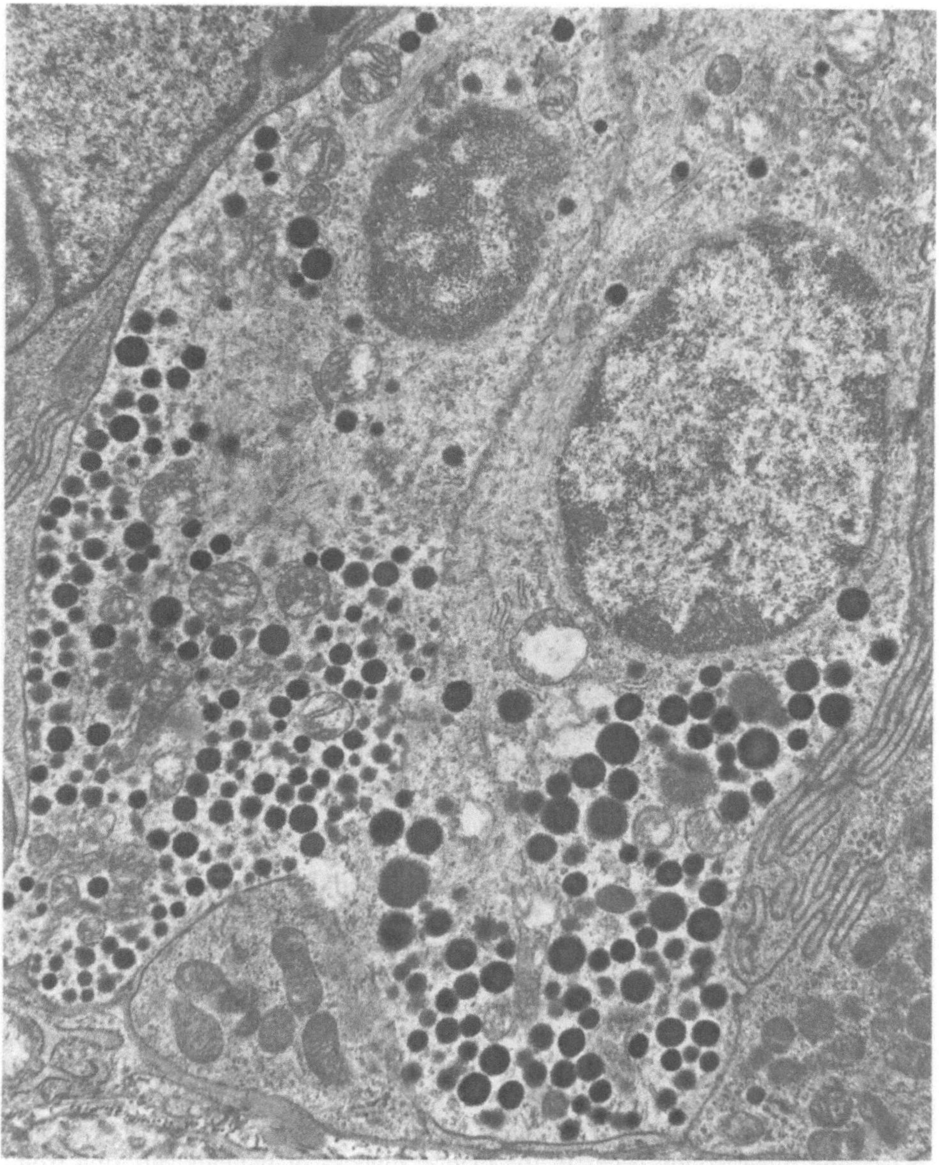

Fig. 8. Two adjacent endocrine cells in the human ileum, one with round, larger granules likely to be interpreted as an N cell, the other with round smaller granules, probably an L cell. × 13,800

Thus, present evidence suggests that the L cell should be interpreted as a somewhat primitive form of cell, producing GLI peptides of unknown function containing the glucagon sequence in an inactive, unreactive form. In the upper gut (stomach and possibly the duodenum) and its derivatives (pancreas) the A cell evoluted from the L cell by acquiring the proteases which split the GLI–proglucagon molecules up to release glucagon. Once released, glucagon molecules are concentrated in the core of the α granule – where they undergo conformational changes and even crystallization – leaving proglucagons, the argyrophil material, enzymes, and other components in the peripheral halo.

XV. Bombesin (BN) Cells

Bombesin-immunoreactive (BN) cells have been detected unequivocally in the chicken (TIMSON et al. 1979) and frog (LECHAGO et al. 1978) stomach. In both species the immunoreactive cells correspond ultrastructurally to cells with small, round to slightly irregular granules fitting in the P-D_1 cell class. Ultrastructurally comparable cells have been observed in the mammalian stomach and lung (CAPELLA et al. 1978). However, so far, immunohistochemical evidence of bombesin endocrine cells in the mammalian gastrointestinal mucosa is lacking or restricted to occasional antisera. Bombesin-immunoreactive nerves might account wholly or in part for the bombesin immunoreactivity detected by radioimmunoassay of mammalian gut extracts (POLAK et al. 1978b).

C. Pathology

Like other secretory cells, endocrine–paracrine cells of the gut may show morphological changes in different functional states. Increased Golgi complex, reticulum, ribosomes, and mitochondria as well as increased size of the whole cell, the nucleus, and/or the nucleolus are found in stimulated cells. Actively secreting cells may also show decreased numbers, increased margination, exocytosis, and structural changes of secretory granules. Dilation of endoplasmic reticulum is often a feature of cell overstimulation. Conversely, resting cells may show decreased organules and size, with various amounts of mature granules in the cytoplasm and increased granulolysis in the lyzosomes, which may be filled with granule debris and residues.

Besides cell hypertrophy, chronic overstimulation may lead to hyperplasia with or without micronodules, and sometimes to tumor growth. Moderate hyperplasia and hypertrophy of G cells have been found in the pyloric and duodenal mucosa of some patients with peptic ulcer disease (SOLCIA et al. 1970a), hyperparathyroidism or acromegaly (CREUTZFELDT et al. 1971). No changes of gastrin levels in blood have been shown in association with these G cell hyperplasias, which seem scarcely relevant to the pathogenesis of peptic ulcer disease. More obvious G cell hyperplasia and hypertrophy occur in the pyloric mucosa of patients suffering from chronic atrophic gastritis restricted to the body and fundus (as in pernicious anemia patients). In the latter case, hypergastrinemia and ultrastructural signs of G cell hyperstimulation and hyperfunction are found regularly, together with increased Golgi complex, reticulum and polyribosomes as well as changes in number and ultrastructural patterns of secretory granules (Fig. 9).

In rare cases of severe gastric hypersecretion with peptic ulcer disease and hypergastrinemia, extensive G cell hyperplasia and hypertrophy coupled with ultrastructural signs of hyperfunction have been observed in the pyloric mucosa (SOLCIA et al. 1970a, 1975; POLAK et al. 1972). Some of the latter patients have been cured

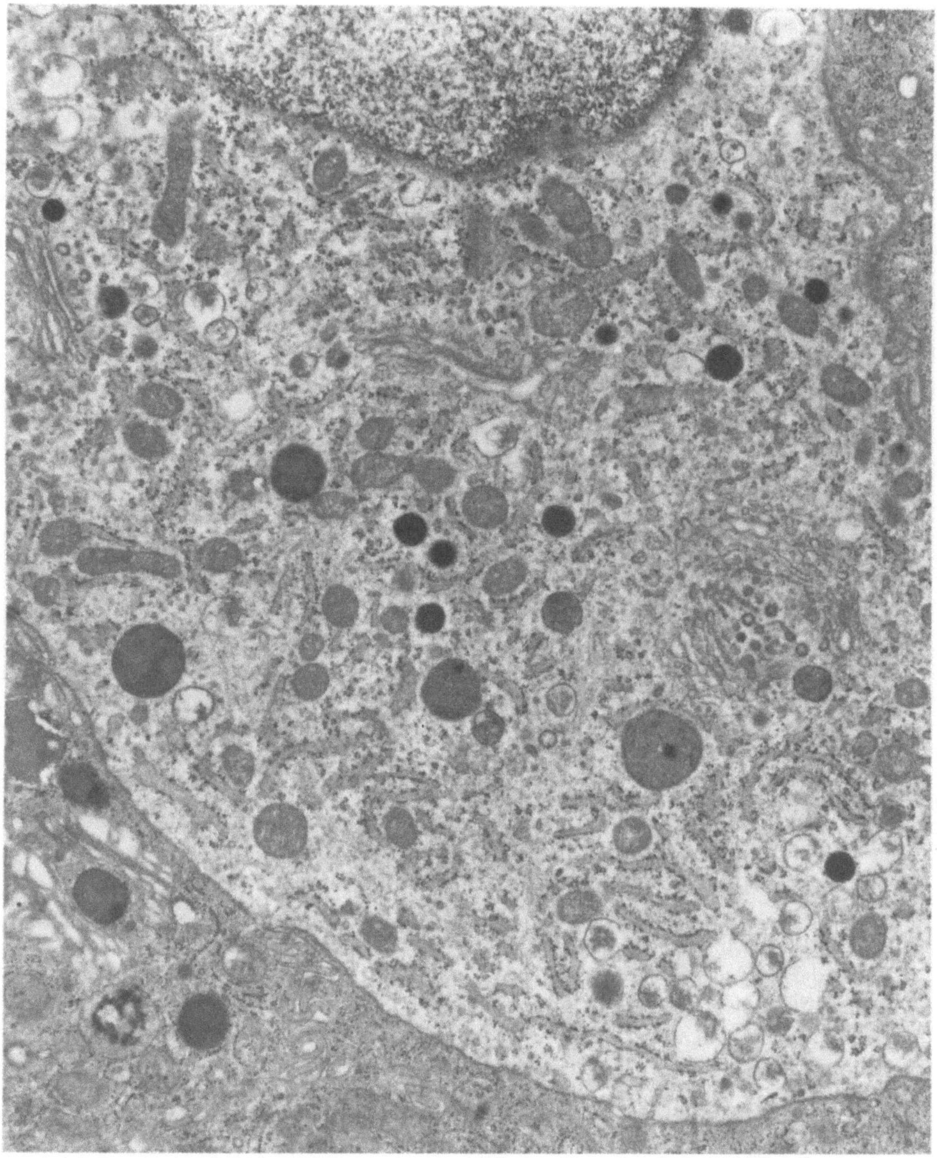

Fig. 9. Human pyloric gland of a pernicious anemia patient with atrophic gastritis and hypergastrinemia. Hypertrophic G cell with extensively developed Golgi and rough endoplasmic reticulum, whose cisternae are filled with proteinaceous material. Note relatively few secretory granules of vesicular or compact type. × 16,000

by surgical removal of the G cell area of the stomach, a behavior suggesting that their peptic ulcer disease was caused by the G cell hyperplasia and hyperfunction. Besides diffuse G cell hyperplasia, intraglandular G cell rows, microacini, or micronodules have sometimes been found in the pyloric mucosa of these patients. Multiple antroduodenal G cell "tumorlets" (micronodular G cell hyperplasias together with microgastrinomas) have been reported in a child with Zollinger–Ellison syndrome and normal pancreas (BHAGAVAN et al. 1974). True G cell tumors (gastrinomas) have been found only very rarely in the stomach, being more frequent in the duodenum (13% of total gastrinomas; HOFMANN et al. 1973) and pancreas (85%). The only gastric gastrinoma we have observed so far arose in association with G cell hyperplasia due to chronic atrophic gastritis (RUSSO et al. 1980).

Hyperplastic changes (rarely progressing up to tumor growth) have been found in G cells and in other endocrine cells of the gut, such as argyrophil ECL cells (especially of hypergastrinemic patients), gastric P or D_1 cells, gastric and intestinal EC cells, and intestinal endocrine cells in intestinal metaplasia of chronic gastritis or antritis (SOLCIA et al. 1975, 1979). Somatostatin, GIP, CCK, GLI, and motilin cells are increased in number, while secretin cells are decreased, in the duodenal mucosa of patients with celiac disease (SJÖLUND et al. 1979).

G cell loss may occur in chronic atrophic antritis, especially in areas of intestinal metaplasia. More subtle regressive changes have been also noted, with special reference to increased residual bodies or autophagosomes, often the site of lipid storage and peroxidation. Increased number, swelling or densification of mitochondria, excessive dilation of the reticulum, increased reticulum and/or cytoplasmic filaments, or even an increase in cytoplasmic density with cell shrinkage have all been noted in endocrine cells of damaged mucosa, more often during acute or chronic inflammation (SOLCIA et al. 1979).

So far, the following hormones have been found to be produced by gut endocrine tumors: 5-HT by argentaffin carcinoids of the appendix, small and large bowel, stomach, pancreas, and biliary tree; substance P by intestinal argentaffin carcinoids; gastrin by duodenal, gastric, and jejunal gastrinomas; GLI (including glicentin-like sequences) and PP-like sequences by rectal nonargentaffin carcinoids; histamine by some gastric argyrophil carcinoids, somatostatin by tumors of the upper small bowel and stomach (usually together with gastrin) or colorectal tumors (usually together with GLI and PP-immunoreactivity). Enkephalin, endorphin, adrenocorticotropic hormone (ACTH), calcitonin, and insulin immunoreactivities have also been detected in tumors. Insulin, glucagon, somatostatin, and PP are regularly produced by pancreatic insulinomas, glucagonomas, somatostatinomas, PP-omas, and related multicellular/multihormonal tumors arising from the four main types of pancreatic endocrine cells. Gastrin, VIP, and less frequently, calcitonin, ACTH, lipotropin, enkephalin, vasopressin, neurotensin, and parathyroid hormone have been also found in pancreatic tumors.

D. Summary

Up to 15 endocrine cell types have been identified ultrastructurally and/or histochemically in the gastroenteric mucosa. Of these, gastrin (G), somatostatin (D), glucagon (A), secretin (S), CCK (I), GIP (K), GLI (L), motilin (Mo), and neurotensin (N) cells have been confirmed cytochemically and seem to be considered as fully established cell types. EC cells, although well known ultrastructurally and as 5-HT producers, are still under investigation as producers of peptides (substance P, enkephalin, and perhaps motilin). P, D_1, X, and ECL cells, although ultrastructurally defined, are functionally unknown, with secretory products mostly to be identified. The individuality of PP-immunoreactive cells, well established in the pancreas, remains to be ascertained in the gut; at least most intestinal PP-immunoreactive cells are to be identified with GLI cells. So far, bombesin (BN) cells have been fully identified only in the stomach of the chicken and frog.

Functional and pathologic changes have been described in gut endocrine–paracrine cells, whose secretory products may have a role in the control of gut motility.

References

Alumets J, Sundler F, Håkanson R (1977) Distribution, ontogeny and ultrastructure of somatostatin immunoreactive cells in the pancreas and gut. Cell Tissue Res 185:465–479

Alumets J, Håkanson R, O'Dorisio T, Sjölund K, Sundler F (1978a) Is GIP a glucagon cell constituent? Histochemistry 58:253–257

Alumets J, Håkanson R, Sundler F, Chang K-J (1978b) Leu-enkephalin-like material in nerves and enterochromaffin cells in the gut. An immunohistochemical study. Histochemistry 56:187–196

Baetens D, Rufener C, Srikant C, Dobbs R, Unger R, Orci L (1976) Identification of glucagon-producing cells (A cells) in dog gastric mucosa. J Cell Biol 69:455–464

Bhagavan BS, Hofkin GA, Woel GM, Koss LG (1974) Zollinger-Ellison syndrome. Ultrastructural and histochemical observations in a child with endocrine tumorlets of gastric antrum. Arch Pathol 98:217–222

Buchan AMJ, Polak JM, Solcia E, Capella C, Pearse AGE (1978a) Electron immunocytochemical evidence for the K cell localization of gastric inhibitory polypeptide (GIP) in man. Histochemistry 56:37–44

Buchan AMJ, Polak JM, Solcia E, Capella C, Hudson D, Pearse AGE (1978b) Electron immunocytochemical evidence for the human intestinal I cell as the source of CCK. Gut 19:403–407

Buchan AMJ, Polak JM, Solcia E, Pearse AGE (1979) Localisation of intestinal gastrin in a distinct endocrine cell type. Nature 277:138–140

Buffa R, Polak JM, Pearse AGE, Solcia E, Grimelius L, Capella C (1975) Identification of the intestinal cell storing gastric inhibitory peptide. Histochemistry 43:249–255

Buffa R, Solcia E, GO VLW (1976) Immunohistochemical identification of the cholecystokinin cell in the intestinal mucosa. Gastroenterology 70:528–532

Buffa R, Capella C, Fontana P, Usellini L, Solcia E (1978) Types of endocrine cells in the human colon and rectum. Cell Tissue Res 192:227–240

Bussolati G, Capella C, Vassallo G, Solcia E (1971) Histochemical and ultrastructural studies on pancreatic A cells. Evidence for glucagon and non-glucagon components of the α granule. Diabetologia 7:181–188

Capella C, Solcia E, Vassallo G (1969) Identification of six types of endocrine cells in the gastrointestinal mucosa of the rabbit. Arch Histol Jp 30:479–495

Capella C, Vassallo G, Solcia E (1971) Light and electron microscopic identification of the histamine-storing argyrophil (ECL) cell in murine stomach and of its equivalent in other mammals. Z Zellforsch 118:68–84

Capella C, Solcia E, Frigerio B, Buffa R (1976) Endocrine cells of the human intestine. An ultrastructural study. In: Fujita T (ed) In: Endocrine gut and pancreas. Elsevier, Amsterdam, pp 42–59

Capella C, Solcia E, Frigerio B, Buffa R, Usellini L, Fontana P (1977) The endocrine cells of the pancreas and related tumors. Virchows Arch [Pathol Anat] 373:327–352

Capella C, Hage E, Solcia E, Usellini L (1978) Ultrastructural similarity of endocrine-like cells of the human lung and some related cells of the gut. Cell Tissue Res 186:25–37

Challacombe DN, Robertson K (1977) Enterochromaffin cells in the duodenal mucosa of children with coeliac disease. Gut 18:373–376

Creutzfeldt W, Arnold R, Creutzfeldt C, Feurle G, Ketterer H (1971) Gastrin and G-cells in the antral mucosa of patients with pernicious anaemia, acromegaly and hyperparathyroidism and in a Zollinger-Ellison tumour of the pancreas. Eur J Clin Invest 1:461–479

Fahrenkrug J (1980) Nervous release of VIP: physiological implications. In: Mihoshi A (ed) Gut peptides: secretion, function and clinicopathology. Kodansha Scientific, Tokyo, pp 73–78
Feyrter F (1953) Über die peripheren enkokrinen (parakrinen) Drüsen des Menschen. Maudrich, Vienna Düsseldorf, pp 1–231
Fiocca R, Capella C, Buffa R, et al. (1980) Glucagon-, glicentin- and pancreatic polypeptide-like immunoreactivities in rectal carcinoids and related colorectal cells. Am J Pathol 100:81–92
Forssmann WG, Orci L, Pictet R, Renold AE, Roullier C (1969) The endocrine cells in the epithelium of the gastrointestinal mucosa of the rat. J Cell Biol 40:692–715
Frigerio B, Ravazzola M, Ito S, Buffa R, Capella C, Solcia E, Orci L (1977) Histochemical and ultrastructural identification of neurotensin cells in the dog ileum. Histochemistry 54:123–131
Furness JB, Costa M (1980) Types of nerves in the enteric nervous system. Neuroscience 5:1–20
Gepts W, Baetens D, De Mey J (1978) The PP cell. In: Bloom SR (ed) Gut hormones. Churchill Livingstone, Edinburgh London, pp 229–233
Grimelius L, Capella C, Buffa R, Polak JM, Pearse AGE, Solcia E (1976) Cytochemical and ultrastructural differentiation of enteroglucagon and pancreatic-type glucagon cells of the gastrointestinal tract. Virchows Arch [Cell Pathol] 20:217–228
Håkanson R, Owman CH, Sjöberg NO, Sporrong B (1970) Amine mechanisms in enterochromaffin and enterochromaffin-like cells of gastric mucosa in various mammals. Histochemie 21:189–220
Håkanson R, Larsson LI, Liedberg G, Sundler F (1976) The histamine-storing enterochromaffin-like cells of the rat stomach. In: Coupland RE, Fujita T (eds) Chromaffin, enterochromaffin and related cells. Elsevier, Amsterdam, pp 243–263
Heitz PH, Kasper M, Krey G, Polak JM, Pearse AGE (1978) Immunoelectron cytochemical localization of motilin in human duodenal enterochromaffin cells. Gastroenterology 74:713–717
Helmstaedter V, Kreppein W, Domschke W, Mitznegg P, Yanaihara N, Wunsch E, Forssmann WG (1979) Immunohistochemical localization of motilin in endocrine non-enterochromaffin cells of the small intestine of humans and monkey. Gastroenterology 76:897–902
Hofmann, JW, Fox PS, Milwaukee SDW (1973) Duodenal wall tumors and the Zollinger-Ellison syndrome. Arch Surg 107:334–338
Kobayashi S, Sasagawa T (1976) Morphological aspects of the secretion of gastro-enteric hormones. In: Fujita T (ed) Endocrine gut and pancreas. Elsevier, Amsterdam, pp 255–271
Larsson LI, Rehfeld JF (1977) Characterization of antral gastrin cells with region-specific antisera. J Histochem Cytochem 25:1317
Larsson LI, Sundler F, Håkanson R (1976) Pancreatic polypeptide. A postulated new hormone: identification of its cellular storage site by light and electron microscopic immunocytochemistry. Diabetologia 12:211–226
Larsson LI, Rehfeld JF, Goltermann N (1977a) Gastrin in the human fetus. Distribution and molecular forms of gastrin in the antro-pyloric gland area, duodenum and pancreas. Scand J Gastroenterol 12:869–872
Larsson LI, Sundler F, Alumets J, Håkanson R, Schaffalitzky De Muckadell OB, Fahrenkrug J (1977b) Distribution ontogeny and ultrastructure of the mammalian secretin cell. Cell Tissue Res 181:361–368
Larsson LI, Rehfeld JF, Stockbrügger R, et al. (1978) Mixed endocrine gastric tumors associated with hypergastrinemia of antral origin. Am J Pathol 93:53–68
Larsson LI, Goltermann N, De Magistris L, Rehfeld JF, Schwartz TW (1979) Somatostatin cell processes as pathways for paracrine secretion. Science 205:1393–1395
Lechago J, Holmquist AL, Rosenquist GL, Walsh JH (1978) Localization of bombesin-like peptides in frog gastric mucosa. Gen Comp Endocrinol 36:553–558
Moody AT, Jacobsen H, Sundby S (1978) Gastric glucagon and gut glucagon-like immunoreactivity. In: Bloom SR (ed) Gut hormones. Churchill Livingstone, Edinburgh, London, pp 369–378

Moody AJ, Frandsen EK, Jacobsen H, Sundby F (1979) Speculations on the structure and function of gut GLIs. In: Rosselin G (ed) Second international symposium on hormonal receptors in digestive tract physiology. Elsevier, Amsterdam
Nilsson G, Larsson LI, Håkanson R, Brodin E, Pernow B, Sundler F (1975) Localization of substance P-like immunoreactivity in mouse gut. Histochemistry 43:97–99
Orci L, Baetens O, Rufener C, Brown M, Vale W, Guillemin R (1976) Evidence for immunoreactive neurotensin in dog intestinal mucosa. Life Sci 19:559–562
Pearse AGE, Polak JM (1975) Immunocytochemical localization of substance P in mammalian intestine. Histochemistry 41:373–375
Pearse AGE, Polak JM, Bloom SR (1977) The newer gut hormones. Cellular sources, physiology, pathology and clinical aspects. Gastroenterology 72:746–761
Polak JM, Bloom S, Coulling J, Pearse AGE (1971 a) Immunofluorescent localization of enteroglucagon cells in the gastrointestinal tract of the dog. Gut 12:311–318
Polak JM, Bloom S, Coulling J, Pearse AGE (1971 b) Immunofluorescent localization of secretin in the canine duodenum. Gut 12:605–610
Polak JM, Stagg B, Pearse AGE (1972) Two types of Zollinger-Ellison syndrome. Immunofluorescent, cytochemical and ultrastructural studies of the antral and pancreatic gastrin cells in different clinical states. Gut 13:501–512
Polak JM, Sullivan SN, Bloom SR, Buchan AMJ, Facer P, Brown MR, Pearse AGE (1977) Specific localisation of neurotensin to the N cell in human intestine by radioimmunoassay and immunocytochemistry. Nature 270:183–184
Polak JM, Buchan AMJ, Dryburgh JR, Christofides N, Bloom SR, Yanaihara N (1978 a) Immunoreactive motilins? Lancet 1:1364–1365
Polak JM, Buchan AMJ, Czykowska W, Solcia E, Bloom SR, Pearse AGE (1978 b) Bombesin in the gut. In: Bloom SR (ed) Gut hormones. Churchill Livingstone, Edinburgh, London, pp 541–543
Ravazzola M, Orci L (1980) Glucagon and glicentin immunoreactivity are topologically segregated in the α granule of the human pancreatic A cell. Nature 284:66–67
Ravazzola M, Siperstein A, Moody AJ, Sundby F, Jacobsen H, Orci L (1979) Glicentin immunoreactive cells. Their relationship to glucagon-producing cells. Endocrinology 105:499–508
Rubin W, Schwartz B (1979) Electron microscopic radioautographic identification of the ECL cell as the histamine-synthesizing endocrine cell in the rat stomach. Gastroenterology 77:458–467
Russo A, Buffa R, Grasso G, Giannone G, Sanfilippo G, Sessa F, Solcia E (1980) Gastric gastrinoma and diffuse G cell hyperplasia associated with chronic atrophic gastritis. Endoscopic detection and removal. Digestion 20:416–419
Sjölund K, Alumets J, Berg NO, Håkanson R, Sundler F (1979) Duodenal endocrine cells in adult coeliac disease. Gut 20:547–552
Solcia E, Vassallo G, Sampietro R (1967) Endocrine cells in the antro-pyloric mucosa of the stomach. Z Zellforsch 81:474–486
Solcia E, Capella C, Vassallo G (1970 a) Endocrine cells of the stomach and pancreas in states of gastric hypersecretion. Rend Gastroenterol 2:147–158
Solcia E, Vassallo G, Capella C (1970 b) Cytology and cytochemistry of hormone producing cells of the upper gastrointestinal tract. In: Creutzfeldt W (ed) Origin, chemistry, physiology and pathophysiology of the gastrointestinal hormones. Schattauer, Stuttgart, pp 3–29
Solcia E, Capella C, Vezzadini G, Barbara L, Bussolati G (1972) Immunohistochemical and ultrastructural detection of the secretin cell in the pig intestinal mucosa. Experientia 28:549–550
Solcia E, Pearse AGE, Grube D, Kobayashi S, Bussolati G, Creutzfeldt W, Gepts W (1973) Revised Wiesbaden classification of gut endocrine cells. Rend Gastroenterol 5:13–16
Solcia E, Capella C, Vassallo G, Buffa R (1975) Endocrine cells of the gastric mucosa. Int Rev Cytol 42:223–286
Solcia E, Capella C, Buffa R, Frigerio B (1976) Histochemical and ultrastructural studies on the argentaffin and argyrophil cells of the gut. In: Coupland RE, Fujita T (eds) Chromaffin, enterochromaffin and related cells. Elsevier, Amsterdam, pp 209–225

Solcia E, Polak JM, Pearse AGE, et al. (1978) Lausanne 1977 classification of gastroenteropancreatic endocrine cells. In: Bloom SR (ed) Gut hormones. Churchill Livinstone, Edinburgh London, pp 40–48
Solcia E, Capella C, Buffa R, Usellini L, Frigerio B, Fontana P (1979) Endocrine cells of the gastrointestinal tract and related tumors. Pathobiol Annu 9:163–203
Solcia E, Buffa R, Capella C, Fiocca R, Yanaiyara N, Go VLW (1980a) Immunohistochemical and ultrastructural characterization of gut cells producing GIP, GLI, glucagon, secretin, and PP-like peptides. Front Horm Res 7:7–12
Solcia E, Capella C, Buffa R, Usellini L, Fiocca R, Sessa F (1980b) Endocrine cells of the digestive system. In: Johnson L, Christensen J, Grossman M, Jacobson ED, Schultz SG (eds) Physiology of the digestive tract. Raven, New York
Solcia E, Creutzfeldt W, Falkmer S, et al. (1981) Human GEP endocrine-paracrine cells: Santa Monica 1980 classification. In: Lechago J, Grossman MI, Walsh JH (eds) Cellular basis of chemical messangers in the digestive system. Academic Press, New York, pp 159–165
Sundler F, Alumets J, Holst J, Larsson LI, Håkanson R (1976) Ultrastructural identification of cells storing pancreatic-type glucagon in dog stomach. Histochemistry 50:33–37
Sundler F, Alumets J, Håkanson R, Carraway R, Leeman SE (1977) Ultrastructure of the gut neurotensin cell. Histochemistry 53:25–34
Timson CM, Polak JM, Wharton J, et al. (1979) Bombesin-like immunoreactivity in the avian gut and its localisation to a distinct cell type. Histochemistry 61:213–221
Vassallo G, Capella C, Solcia E (1971) Endocrine cells of the human gastric mucosa. Z Zellforsch 118:49–67
Vassallo G, Solcia E, Bussolati G, Polak JM, Pearse AGE (1972) Non-G cell gastrin-producing tumours of the pancreas. Virchows Arch [Cell Pathol] 11:66–79
Wharton J, Polak JM, Bloom SR, Ghatei MA, Solcia E, Brown MR, Pearse AGE (1978) Bombesin-like immunoreactivity in the lung. Nature 273:769–770

CHAPTER 4

Ionic Basis of Smooth Muscle Action Potentials

T. TOMITA

A. Introduction

The action potentials of smooth muscles are important in triggering and synchronizing contraction. However, the contribution of action potentials to these functions varies greatly, depending on the type of smooth muscle. Although a typical action potential having an all-or-nothing property can be observed in the smooth muscle of some organs, such as gastrointestinal (GI) tract, vas deferens, or uterus, it is more common that the amplitude and the shape of action potentials vary from time to time even in the same tissue, and also from one tissue to another. Most tracheal smooth muscles and some vascular smooth muscles are inexcitable under normal conditions. There is a spectrum of smooth muscles between the highly excitable type and the inexcitable type (CREED 1979). Thus, in most smooth muscles, the action potential is graded, and does not obey the all-or-nothing law. Owing to this variability, a general description of the action potential is difficult.

Our knowledge of the action potential has been greatly advanced by the studies of HODGKIN and HUXLEY (1952a, b, c) on the squid giant axon. The action potential of most nerves and vertebrate skeletal muscles is generated by the influx of Na ions resulting from an increase in Na conductance, and this is followed by an increase in K conductance and efflux of K ions. This type of action potential is called the Na spike. From studies of the effects of altering the external ionic composition, and of some agents which affect the action potential, it is now clear that the action potential in smooth muscles differs from that of the Na spike.

The action potential in smooth muscles is less sensitive to Na ions but more sensitive to Ca ions than the Na spike. In this respect, the action potential of smooth muscles is similar to that in some invertebrate neurons or muscles, in which Ca influx is more dominant than Na influx during the action potential. However, the mechanism underlying the generation of action potentials has not been fully investigated in smooth muscle, mainly owing to technical difficulties. The voltage clamp method utilized for the analysis of ionic mechanism underlying the action potential in the squid giant axon, is not easy to apply to multicellular tissues such as smooth muscles (BOLTON et al. 1981).

Another unique property of the electrical activity in smooth muscles is the existence of slow rhythmic changes in the membrane potential, particularly in the GI tract. Each of these slow potential changes usually triggers an action potential or a train of action potentials, which is important for producing the contraction. Slow potentials are probably generated by a different mechanism from that for action potentials, but sufficient analyses have not been done to understand their generation (TOMITA 1981).

B. Methodologic Problems

It may be worth discussing the various problems concerning studies of the action potential in smooth muscles, before going into the mechanism. The smooth muscle fibers are small and appear to be electrically interconnected. This structural complexity introduces many technical difficulties in the analysis of mechanisms underlying excitation.

I. Electrophysiologic Analysis

Since the first attempt to record the membrane potential from the guinea-pig taenia coli intracellularly (BÜLBRING 1954), the effects of external ions have been studied mostly on the spontaneous action potentials in this preparation. Interpretation of such results is more difficult than interpretation of results obtained from electrically evoked action potentials, because the effect could easily be on the spontaneity or on the conduction rather than directly on the spike generation. The configuration of spikes is affected not only by the ionic fluxes during excitation but also by the spatial interaction of spikes (TOMITA 1967b, 1970).

The spike in the taenia coli conducts along a functional bundle, composed of many muscle fibers in parallel and in series, as expected from the cable-like properties (TOMITA 1966; ABE and TOMITA 1968). This is probably true in many other smooth muscles (TOMITA 1975), including small intestine of cat (NAGAI and PROSSER 1963), or vas deferens of guinea pig (TOMITA 1967a). However, this functional bundle is not a simple cable; it braches out and joins other bundles, forming a complex mesh, as shown by histologic studies (GABELLA 1976, 1977). Thus, the local circuit current of a spike which propagates along a bundle spreads into a larger area at each branching point. Therefore, at such a place the safety factor would be low, and the spike propagation would be more easily blocked. This may be one of the reasons why the conduction along smooth muscle is usually decremental and the degree of contraction can be graded. Similarly the shape of the spike near the point where different bundles join may be modified, depending on the time lag between the arrival of spikes conducted along the different bundles. This may be an explanation for notches on the spike, and also for the variability of spike amplitude. Thus, for analysis of the spike mechanism, the effect of ions should be investigated on the electrically evoked action potential, preferably under conditions in which conduction is prevented.

Another factor which should be considered is the effect of the geometry of the stimulating electrodes on the action potential, since the spread of current in the tissue greatly influences the apparent excitability of preparations. If the membrane is uniformly depolarized, the threshold is reached when the inward current activated by depolarization becomes larger than the outward current. This inward current is used to depolarize its own membrane by discharging the membrane capacity. However, if a part of the preparation is depolarized, the inward current generated at this part is used not only for depolarizing its own membrane but also for depolarizing the surrounding membrane. Owing to this effect, the threshold depolarization should be higher in nonuniform than in uniform depolarization, and some liminal area has to be depolarized above the threshold in order to generate enough current for a propagated action potential (RUSHTON 1937; NOBLE 1972; FOZZARD and SCHOENBERG 1972). Thus, intracellular stimulation often fails to generate an action potential or, even if it is effective, the response is usually graded and nonpropagated (KURIYAMA and TOMITA 1965; HASHIMOTO et al. 1966; BENNETT 1967a, b; TOMITA 1967a).

The double sucrose gap is a convenient method to use for electrophysiologic studies on many types of smooth muscle. By passing constant current pulses across one gap, the resultant changes of the membrane potential of the central portion of the preparation can be recorded across the other gap. The part of the preparation contained in the central pool is exposed to the test solution and it is often termed the node. For better understanding of the mechanism underlying the action potential, analysis of ionic currents flowing through the membrane is important. For this purpose, the membrane potential at the node can be clamped at a given level to record the membrane current, by combining the double sucrose gap method with a negative feedback amplifier. However, the results obtained by this method must be interpreted with some reservation.

Spatial uniformity of the membrane potential in the node is an important factor for obtaining a reliable estimate of membrane current during voltage clamping. However, this is very difficult to achieve in multicellular preparations, mainly owing to the series resistance (cleft resistance) of the extracellular space (ATTWELL and COHEN 1977; BOLTON et al. 1981; JOHNSON and LIEBERMAN 1971). The apparent membrane current recorded includes the current flowing through the membrane immersed in the extracellular space where mixing of the test solution and sucrose solution takes place. The contribution of the current flowing in the extracellular space to the true transmembrane current at the node increases as the node width is reduced. The center node is surrounded by muscle fibers immersed in sucrose solution, and lack of ions in this region probably alters the membrane properties. Since this region can be expected to interact with the nodal membrane through intercellular connections and the extracellular leakage, the electrical properties of the nodal membrane are apparently modified. Another complication is that strong currents during the voltage clamp may accumulate or deplete ions near the membrane. In this respect, the K concentration in the narrow extracellular space may be most affected.

II. Analysis of Ion Fluxes

Estimation of intracellular ionic concentration varies, depending on the assumptions made about the extracellular space and the distribution of the ions in various components of the tissue. The intracellular content of an ion is obtained from the total tissue content by subtracting the amount of the ion contained in the extracellular space and the amount of the ion bound to the cell membrane and extracellular structures. The extracellular space is estimated by using a substance, such as inulin or sorbitol, which is supposed to distribute itself only in the extracellular space. However, there is no assurance that these markers have the same distribution as that of the ions in the extracellular space.

Another way of estimating the ionic distribution in the tissue is to analyze the flux data of radioisotopes. However, the fluxes of all ions have a complex time course which is composed of multiexponential phases. Although it is generally assumed that these phases correspond to the fluxes from the different components in the tissue, there is no reliable basis for this assumption and the data are not necessarily interpreted in the same way by different investigators.

Even if these analyses approach the real evaluation, intracellular compartmentalization of ions makes estimation of ionic concentration in the cytoplasm very difficult. This is particularly serious when dealing with Ca ions. Furthermore, from the electrophysiologic point of view, the extracellular and intracellular ionic activities are important, rather than the ionic concentrations. The activity of ions is believed to be much lower in the cytoplasm than in the external medium. However, there is no way of obtaining this activity, except by using ion-selective microelectrodes, which have yet to be developed for smooth muscle fibers.

C. Effects of Na Ions

I. Changes in Electrical Properties

The action potential of most excitable membranes is dependent on the presence of Na ions, and is abolished in Na-free solution. In these membranes, the gates of the Na channels are opened when the membrane is depolarized beyond the threshold, but quickly closed by an inactivation process. The number of gates which can be opened depends on the membrane potential just before the depolarization and on the rate of depolarization, because the inactivation mechanism is a function of membrane potential and time (Hodgkin and Huxley 1952b). In the squid axon (Hodgkin and Katz 1949), or in the mammalian skeletal muscle fiber (Ferroni and Blanchi 1965), it has been shown that the amplitude of the action potential is roughly proportional to the logarithm of the external Na concentration, $[Na]_o$ and its rate of rise is directly proportional to $[Na]_o$. These results are expected from the fact that depolarizing current is carried by Na ions (Hodgkin and Huxley 1952a, c).

The action potential of many smooth muscles is rather insensitive to changes in $[Na]_o$. The action potential in the guinea-pig vas deferens is little affected in low (18 m*M*) Na (sucrose substitution) solution, when evoked by intracellular or external field stimulation (Bennett 1967a).

The action potential in the cat ureter is also insensitive to Na removal (Tris substitution) for the first 10–15 min, but after 30 min, the amplitude and the rate of rise are both reduced, and finally, stimulation fails to produce an action potential (Kobayashi 1969). The action potential of cat ureter is composed of a spike and a subsequent slow repolarization phase, while that in the guinea pig ureter usually has repetitive spikes appearing on a slow plateau phase. When $[Na]_o$ is reduced, the plateau of the guinea pig ureter potential becomes smaller, but its spike component larger (Kuriyama and Tomita 1970; Shuba 1977). Since the decrease of the plateau and the increase of the spike cancel each other, the peak of the spike remains more or less constant or even greater in Na-deficient or Na-free solutions (sucrose or Tris substitution). After a prolonged exposure to Na-free solution, the duration of the spike component is gradually increased (Shuba 1977). In the guinea pig urinary bladder, reduction of $[Na]_o$ to one-tenth of its normal valve (sucrose or Tris substitution) reduces the amplitude and the rate of rise of the action potential (CREED 1971).

The taenia coli of the guinea pig has a strong spontaneous activity, particularly when it is stretched. The spontaneous spike has a weak sensitivity to $[Na]_o$ between 15 and 137 m*M*, although the rate of rise is usually reduced in Na-deficient solution, and the spontaneous activity is lost in Na-free solution. When the Na substitute is choline, the spontaneous activity remains in 5 m*M* Na solution, but its amplitude and rate of rise are reduced. In 2 m*M* Na solution the spike is abolished after 35 min (HOLMAN 1958). When Tris is used for Na substitution, the activity stops after 45–60 min in 7 m*M* Na solution, and in 0 m*M* Na solution, the activity reappears after a transient suppression for 5–10 min, until it finally disappears after about 45 min accompanied by depolarization of the membrane (BÜLBRING and KURIYAMA 1963). The spikes observed in Na-free solution have more or less the normal amplitude for at least 20–30 min, while the rate of rise is reduced. These changes observed on Na removal may vary in different preparations, and precise evaluation is difficult because of spontaneous activity, as already discussed.

The spontaneous spike activity of the taenia coli disappears in 10 m*M* Na solution when Na is substituted with sucrose, but the spike can be easily evoked by depolarizing the membrane (BRADING et al. 1969). Under these conditions, the amplitude and rate of rise of the spike are greater than the control. In this solution the membrane is hyperpolarized by about 10 mV, but restoration of the membrane potential with conditioning depolarizing current does not significantly change the parameters of the action potential.

Applying the voltage clamp method to the guinea pig taenia coli, it has been shown that replacement of all Na with dimethyldiethanol ammonium (DDA) or choline has no significant effect on the early transient inward current and its reversal potential (INOMATA and KAO 1976).

In the pregnant rat myometrium, reduction of Na to 16.7 m*M* (sucrose substitute) reduces the amplitude and rate of rise of the spontaneous spikes, and stops the activity within 30 min (ABE 1971). However, a spike of larger amplitude and faster rate of rise than in normal solution can be evoked for a period of up to 4 h. In this solution, the membrane is reported to be hyperpolarized. In the pregnant mouse myometrium, there is little change in the spontaneous spike activity in 59 m*M* Na (choline, sucrose, or Tris substitute) solution, and in a solution containing about 15 m*M* Na, there is some depolarization, and the amplitude of the evoked spike is decreased (OSA 1971). However, this depolarization gradually recovers, and after 50 min, spikes having an overshoot larger than the control can be observed (OSA 1973). When $[Na]_o$ is reduced to less than 5 m*M*, the membrane is markedly depolarized to between -25 and -35 mV and contracture develops. The spike amplitude decreases as the membrane is depolarized. Under this condition, an action potential of 60 mV, having an overshoot of 10 mV, can be evoked provided that the membrane is polarized to a level between -40 and -50 mV by conditioning hyperpolarizing current. However, the evoked spike gradually decreases after 20 min, and then disappears after 40 min in Na-free solution.

The circular muscle of the guinea pig stomach is also depolarized by 20–30 mV (almost to the peak of the slow wave) when $[Na]_o$ is reduced to less than 5 m*M* (OH-

BA et al. 1977). The depolarization produced by removal of external Na ions is one of the unique properties of some smooth muscles. This is particularly prominent in the myometrium, but is generally observed to various degrees in many types of smooth muscle. The mechanism of the depolarization by Na removal is not known, but since the depolarization is antagonized by Mn, which reduces the Ca conductance, an increase in Ca conductance may be involved, at least in the myometrium (OSA 1973).

It seems that the presence of about 10 mM Na is necessary to maintain the normal membrane potential, and that the suppression of the spike in solutions containing less than 5 mM Na is mainly due to the depolarization of the membrane. This conclusion is supported by results obtained from the rat myometrium with intracellular recording. However, the results obtained from the sucrose gap are not necessarily in accord with this conclusion, particularly with reference to the change in membrane potential.

The membrane potential of the pregnant rabbit myometrium observed with the single sucrose gap method hyperpolarizes in Na-free solution (DDA substitution), even though the spontaneous spike frequency is increased during this period (KLEINHAUS and KAO 1969). The spontaneous activity lasts for variable periods (2.5–31.2 min; mean 11.9 min) in Na-free solution. With the double sucrose gap method, the estrogen-treated rat myometrium is hyperpolarized by about 10 mV in Na-free (Tris) solution and the evoked spike is abolished (ANDERSON et al. 1971). DDA or choline substitution have a similar effect on the spike, but the hyperpolarization is less marked. Similar hyperpolarization of the membrane in Na-free solution has been observed with pregnant rat myometrium also by means of sucrose gap recording (DDA substitute: KAO and MCCULLOUGH 1975; choline, sucrose, or Li substitute: MIRONNEAU 1976).

The difference in shift of membrane potential on removal of Na ions between intracellular recording and sucrose gap recording should be clarified. It is possible that, in the sucrose gap recording, some factor resulting from exposure to sucrose solution may cause an apparent shift in the recorded membrane potential.

The inward current obtained from the myometrium with the voltage clamp method is more sensitive to $[Na]_o$ than that of the taenia coli. The inward current in the estrogen-treated rat myometrium is reduced from 32 to 4 μA and its equilibrium potential is shifted from +30 to −15 mV after 11.5 min in Na-free solution (Tris substitute: ANDERSON 1969). Similar results are obtained with DDA substitution (ANDERSON et al. 1971). In the pregnant rat myometrium, a significant shift of the equilibrium is observed even by a 50% reduction of $[Na]_o$ and the inward current is abolished after 10 min in Na-free solution (KAO and MCCULLOUGH 1975).

On the other hand, rather less sensitivity to $[Na]_o$ has also been reported for the same tissue, i.e., the inward current is hardly affected by reducing $[Na]_o$ to 12.5 mM (MIRONNEAU 1973) or it is decreased by only 10–15% in Na-free solution when substituted with choline (MIRONNEAU 1974). The reason for this discrepancy is not clear, but from the results of intracellular recording, Na may not only act as a charge carrier for the inward current, but may also have rather complicated actions on the membrane permeability to other ions, particularly in the myometrium. Thus, the contribution of Na to the action potential as a charge carrier is still a matter of speculation. However, compared with other typical excitable membranes which utilize mainly Na for the action potential, the effect of removal of Na is very much weaker in smooth muscles. This is one of the strongest pieces of evidence suggesting that another ion, probably Ca, is acting as a main charge carrier.

II. Changes in Ionic Distribution

1. Extracellular Na

The fact that the spike activity can be maintained for some period of time in Na-deficient or Na-free solution may be argued to be due to very slow outward diffusion of Na from a deep region of a preparation (KAO 1967). It is certainly difficult to estimate the real ionic concentration just outside the membrane. However, it has been shown that the fast component of the Na efflux curve fits well with the washout curve of an extracellular marker, EDTA ^{60}Co or sorbitol ^{14}C, which follows a bulk diffusion process, and that reduction of ^{24}Na in the extracellular space is reasonably fast, both for the guinea pig taenia coli and the estrogen-dominated rabbit myometrium (BRADING and JONES 1969). Of course, the actual time of washout depends on the thickness of the preparation, the time to reduce an ion to a certain level being proportional to the square of the thickness. Although the diffusion coefficient of Na in the extracellular space is estimated to be about one-third of that in free water, it takes less than 3 min to wash out 90% of the extracellular space in a tissue 500 μm thick, and about 11 min in one 1,000 μm thick. A reduction to one-tenth of the normal concentration should have a dramatic effect on the action potential if Na is the main charge carrier.

Na loss from the narrow extracellular spaces in Na-free medium may be compensated by an intracellular source owing to a leak or to active transport (SOMMER and JOHNSON 1968). This factor probably contributes to some degree to slowing down the rate of decrease in concentration gradient across the membrane. However, since the membrane would be a greater rate limiting factor for Na efflux than extracellular diffusion, the absolute concentration of Na in the extracellular space should decrease rather quickly. Thus, the rate of rise of an action potential should also be reduced proportionally to the decrease of Na concentration in the extracellular space.

The strong argument against the idea that Na remaining outside the membrane is responsible for spike activity would be that readmission of Na after treatment with Na-free solution usually produces suppression of the action potential, i.e., an increase in Na concentration markedly raises the threshold, or makes the tissue inexcitable, before it recovers to normal in several smooth muscles (T. Tomita 1980, unpublished observation).

2. Intracellular Na

The intracellular Na concentration $[Na]_i$ is very difficult to estimate, as already described. When the $[Na]_i$ of the guinea pig taenia coli is calculated from the total Na content, water content (from dry weight: wet weight ratio) and extracellular space of the tissue using ethanesulfonate ^{35}S or sorbitol ^{14}C, it is 32–40 mmol/l cell water (CASTEELS and Kuriyama 1966). In another series of similar experiments, with sorbitol ^{14}C, $[Na]_i$ is reported to be 19 mM (CASTEELS 1969).

From the data on Na efflux, $[Na]_i$ can be estimated to be 13 mM (CASTEELS 1969) or 3.14 mM (BRADING 1971), depending on the interpretation of efflux curves. A very low value (3.14 mM) is obtained if it is assumed that the kinetics of Na efflux from the extracellular space fit with bulk diffusion of EDTA ^{60}Co. However, the amount of Na washed out by this efflux is larger than that which could be contained in the space occupied by the extracellular marker by 12–15 mequiv/kg wet weight (BRADING 1971; BRADING and JONES 1969). This discrepancy can be explained by the hypothesis that the extracellular space contains sites having a negative charge which can bind 12–15 mequiv./kg tissue, because La, which has a very strong binding ability, displaces this amount of Na from the tissue (BRADING and WIDDICOMBE 1977).

It is also possible that some Na ions are sequestered in the sarcoplasmic reticulum (BRADING 1975). Since this compartment is excluded from the distribution of extracellular markers, Na contained is apparently intracellular, but this Na seems to be rapidly transported outside through a Na–Na exchange process.

It has been reported (based on chemical analysis in the taenia coli) that when $[Na]_o$ is reduced from 126 to 10 mM, $[Na]_i$ falls from 35 to 24 mM after 1 h (BRADING et al. 1969). However, most of this Na seems to be bound and the actual intracellular concentration is probably very much lower than this, as already described. There has been no careful study of changes in $[Na]_i$ after removal of external Na in smooth muscle. $[Na]_i$ may fall rapidly as in crab muscle fiber, in which the intracellular Na activity has been measured with a Na-sensitive electrode (VAUGHAN-JONES 1977). However, it is unlikely that the reduction of $[Na]_o$ decreases $[Na]_i$ so much that the concentration gradient actually becomes larger than the control. Therefore, the cause of the increase in the overshoot of action potentials should be sought for in another mechanism.

D. Effects of Tetrodotoxin

Tetrodotoxin (TTX) obtained from pufferfish (Tetraodontoidea) or saxitoxin from Dinoflagellata is known to block most of the Na channels quickly (KAO 1966; NARAHASHI 1974), but to have no effect on the Ca spike (HAGIWARA and NAKAJIMA 1966a). Therefore, resistance to TTX is generally taken as evidence that the action potential is not Na dependent, and more likely Ca dependent.

However, there are some exceptions. The action potential in a snail neuron is dependent on $[Na]_o$ but this is not sensitive to TTX (WALD, 1972; KOSTYUK et al. 1974). Similar findings are reported for the muscle fiber of pufferfish (HAGIWARA and TAKAHASHI 1967b). The action potential in mammalian skeletal muscle is normally susceptible to TTX, but it becomes resistant after denervation, even though it is still Na dependent (REDFERN and THESLEFF 1971), suggesting the importance of some molecular structure of the Na channel for the susceptibility to TTX.

There are many studies on the effect of TTX on smooth muscles, but no action potentials are reported to be affected by TTX under normal conditions. The insensitivity to TTX is probably not due to a lack of accessibility of the membrane deep in the tissue, because junction potentials produced by stimulation of the intramural nerves can be easily abolished by TTX (BÜLBRING and TOMITA 1967; HASHIMOTO et al. 1967; TOMITA 1967a). The inability of TTX to affect the action potential of smooth muscles indicates either that the Na channel is not involved in the action potential or that the molecular organization of the smooth muscle Na channel is different from that of the typical Na channel.

In the guinea pig ureter, the plateau component of the action potential is probably due to an increase in the Na conductance of the membrane. However, this is not affected by TTX (KURIYAMA et al. 1967; BURY and SHUBA 1976). Therefore, the ineffectiveness of TTX is not a strong argument for the idea that Na contribution to the action potential is negligible. However, as with many other results, it is generally thought that Na is not the main charge carrier for the action potential in smooth muscles.

E. Na-Dependent Activity in Ca-Free Solution

Some spontaneous electrical activity can be observed in the guinea pig taenia coli exposed to Ca-free solution, when recorded as the extracellular potential field (GOLENHOFEN and PETRÁNYI 1969). To obtain this activity, the preparation was pretreated with Ca-free solution containing 5 m*M* NaF for more than 30 min, and then exposed to Ca- and Mg-free solution. The recorded potential field was about 100 μV, but no information is available about the transmembrane potential. The interesting finding is that TTX (10^{-5} g/ml) blocked this activity. Interpretation of this result needs further analysis.

A slow spike-like activity can be recorded with the double sucrose gap method when the taenia coli is immersed in Krebs solution containing no Ca with 0.1 m*M* EGTA and 0.5 m*M* Mg (BÜLBRING and TOMITA 1970a, b). Sometimes, a plateau type of potential appears. The presence of Na ion is necessary for these activities, but they are not affected by TTX (2×10^{-6} g/ml).

Similar potentials have been observed in the longitudinal and circular muscles of the cat small intestine and in the stomach muscles of the skate, frog, and toad, by means of pressure electrodes and intracellular microelectrodes (CURTIS and PROSSER 1977; PROSSER et al. 1977). Slow potentials of square wave type having a duration ranging from 4 to 44 s (mean 17.6 s) appear in Ca-free solution containing 3–5 m*M* EGTA. In the cat small intestine, the membrane is depolarized from −64.5 to −35 mV in Ca-free solution and the slow potential has an amplitude of 35 mV. As observed in the taenia coli, the potential is also dependent on $[Na]_o$, but not affected by TTX. This potential is abolished by Ca antagonists such as Mn, Co, La, or verapamil.

The common carotid artery of sheep, which normally does not show any spontaneous activity, becomes electrically active in Ca- and Mg-free solution (KEATINGE 1968). This spontaneous activity is suppressed by the addition of a small amount of Ca ($2.5–7.5 \times 10^{-5}$ *M*) or Mg ($0.25–7.5 \times 10^{-4}$*M*). This potential is also dependent on $[Na]_o$, but not influenced by TTX (10^{-5} g/ml).

Thus, in many smooth muscles exposed to Ca-free solution, a TTX-resistant channel through which Na ions can pass may be opened to produce regenerative activity. However, the properties of this channel are quite different from those responsible for the typical Na spike. Furthermore, it is unlikely that this mechanism is involved in the action potential observed in the normal solution containing appropriate Ca and Mg concentrations (BÜLBRING and TOMITA 1970a). It may be that a trace of Ca or Mg still remaining in the membrane controls this Na permeability mechanism so that it can be effective only when there is almost no Ca present in the external medium. In the *Helix* giant neuron, a chelating agent (EGTA or EDTA) is thought to modify the Ca channel so that it can pass inward Na current, with kinetics similar to that of the inward Ca current (KOSTYUK and KRISHTAL 1977). A similar change may also be produced in smooth muscle.

F. Effects of Ca Ions

I. Excitable Membranes and Ca

Most nerve fibers and mammalian skeletal muscle fibers produce an action potential by an influx of Na ions (Na spike) as a result of an increase in Na conductance. On the other hand, in crustacean muscle fibers and some invertebrate neuron somata, it is known that it is Ca influx that mainly contributes to the action potential (Ca spike) (REUTER 1973; HAGIWARA 1973, 1975). Thus, in some excitable membranes, Ca may carry charge to generate the Ca spike. The gate mechanism for the Ca channel differs from that for the Na channel. The amount of Ca conductance opened is much less and its rate of opening is much slower. Thus, the inward current carried by Ca is easily counteracted by a simultaneous flow of outward current carried by K ions. Ba, Sr, and probably Mn ions in some tissues, are known to be able to substitute for Ca as charge carriers through this channel.

Ca ions are also known to affect the excitability of the membrane. The effects of increasing the external Ca concentration are the same as those of membrane hyperpolarization in raising the threshold and also removing inactivation of the Na conductance. This is called the stabilizing action, and is observed not only in membranes which produce Na spikes (BRINK 1954; FRANKENHAEUSER and HODGKIN 1957; WEIDMANN 1955), but also in membranes which produce Ca spikes (HAGIWARA and TAKAHASHI 1967a). Many polyvalent cations, such as La, Mn, Co, Ni, and Tm are stronger than Ca in this action. Mg also has a stabilizing action, though weaker than Ca. Thus, the threshold is increased and the action potential can be suppressed by these cations. Opposite effects are expected when the external Ca is removed. It may increase excitability by lowering the threshold and by depolarizing the membrane. On the other hand, it may suppress the Na spike by causing inactivation of the Na conductance.

The third action of Ca ions is to control the membrane permeability to other ions. When the external Ca is removed, the membrane is gradually depolarized, accompanied by a reduction of the membrane resistance. This is thought to be due to an increase in Na conductance, but this mechanism is probably different from that involved in the generation of the Na-spike. The degree to which Na conductance increases varies greatly in different tissues. Under some conditions, particularly in smooth muscle, the depolarization can be controlled in an all-or-nothing manner, probably by a trace of divalent cations, as described in Sect. E. Many other polyvalent cations have a suppressing action on the membrane permeability similar to Ca.

The action of Ca on the passive membrane permeability to various ions and the action on the kinetics of excitation should be theoretically differentiated. However, it is difficult practically to distinguish these actions clearly. Furthermore, it is possible that divalent cations affect the permeability of Na and K differently. Some cations may decrease both permeabilities, but to different degrees, and other cations may decrease one and increase another permeability. For example, Ca may decrease Na permeability, probably from the outside and increase K permeability, probably from the inside of the membrane.

In erythrocytes (ROMERO 1976; ROMERO and WHITTAM 1971) and some neuron somata (ECKERT and TILLOTSON 1978; MEECH 1976, 1978; MEECH and STANDEN 1975), an increase in intracellular Ca leads to an increase in the K conductance of the membrane, which probably differs from that activated by depolarization (HEYER and LUX 1976; NEHER and LUX 1972). It may be that membranes having a large population of the Ca channels have a high density of the Ca-sensitive K channel (KLEINHAUS and PRICHARD 1977) and that the frequency of spike activity is limited by this mechanism.

II. Changes in Electrical Properties

Since Ca has complex effects on the membrane properties, interpretation of effects produced by changing $[Ca]_o$ is sometimes difficult. In general, changes of $[Ca]_o$ produce rather marked effects on the amplitude and the rate of rise of action potentials in smooth muscles. In the guinea pig vas deferens, the spike evoked by intracellular stimulation is increased in amplitude and rate of rise by increasing $[Ca]_o$ up to 10 m*M*, no further increase being found beyond this level (BENNETT 1967a). The overshoot of action potentials evoked by nerve stimulation increases by about 22 mV for each ten-fold increase in $[Ca]_o$ between 0.1 and 2.5 m*M*, but no further change is observed over 2.5 m*M*. In another series of experiments, the overshoot was increased from 14.6 to 18 mV and the maximum rate of rise from 14.5 to 32.2 V/s by increasing $[Ca]_o$ to 25 m*M* (KURIYAMA 1964).

In solutions containing normal $[Na]_o$, the amplitude of the action potential of cat ureter increases by about 12 mV with a ten-fold increase in $[Ca]_o$. The absolute values of amplitude and rate of rise at each concentration of $[Ca]_o$ and also the effects of $[Ca]_o$ on these parameters are greater in Na-containing than in Na-free solution (KOBAYASHI 1969). In the guinea pig ureter, excess Ca (12.5 m*M*) increases the amplitude and rate of rise of the spike component, but reduces the plateau component (BENNETT et al. 1962; KOBAYASHI 1971). On the other hand, the spike component is quickly suppressed by removal of Ca (BENNETT et al. 1962; GOLENHOFEN and LAMMEL 1972). The plateau and spike are both completely abolished by prolonged exposure to Ca-free solution containing 5 m*M* Mg (KURIYAMA and TOMITA 1970). The spontaneous activity of the guinea pig urinary bladder is blocked by increasing $[Ca]_o$ to 7.5 or 10 m*M*, but the amplitude and rate of rise of evoked spikes are increased by excess Ca without a change in membrane potential, while reduction of action potential and resting potential is observed in Ca-deficient solution (CREED 1971).

In the guinea pig taenia coli, excess Ca increases the amplitude and rate of rise of action potentials, both spontaneously generated (HOLMAN 1958; BÜLBRING and KURIYAMA 1963) and also electrically evoked (BRADING et al. 1969). The excess Ca also raises the critical firing level. When $[Ca]_o$ is reduced in Na-deficient (10 m*M*) solution, the action potential becomes smaller and slower, without much change in the membrane potential. The ability of Ca ions alone to produce an action potential is shown by the following experiments (BRADING et al. 1969), in which first the external Ca and then Na is removed. In this Ca- and Na-free solution, no action potential can be evoked, but the active response recovers on addition of only 0.2 m*M* Ca. When $[Ca]_o$ is increased to more than 1 m*M* in Na-free solution, the threshold is increased so that a stronger stimulating current pulse has to be applied to evoke the spike.

Inward currents of the taenia coli observed under voltage-clamp condition are abolished in Ca-free solution (KUMAMOTO and HORN 1970). In this experiment, the steady state current was little affected by removing Ca. Similarly, the inward current was reduced in Ca-deficient solution and increased in excess Ca solution (INOMATA and KAO 1976). Thus, as far as the guinea pig taenia coli is concerned, it is generally agreed that Ca is the main charge carrier for the action potential.

In the mouse myometrium, the membrane is depolarized from about -60 to -20 mV in Ca-free solution within 30 min, when measured intracellularly (OSA 1973). Irreversible depolarization has been observed in the rat myometrium when exposed to solutions containing less than 10^{-6} *M* Ca (KAO and MCCULLOUGH 1975). On the other hand, little change in the membrane potential is reported for the rat myometrium after 10 min exposure to Ca-free solution, when measured with the sucrose gap method, although the spike is abolished within 4–5 min and a small increase in leakage current is observed (ANDERSON et al. 1971). As described in the section on the effects of Na ions, the measurement of membrane potential with the double sucrose gap method may not be reliable. When the inward current is measured in the rat myometrium, it is very much reduced in Ca-free solution containing 0.2 m*M* EDTA, but some inward current remains even after 9 min (ANDERSON et al. 1971). Similar results on the inward current have been reported by MIRONNEAU (1973, 1974). When Ca is reduced from 1.9 to 0.5 m*M* the action potential and inward current of the rat myometrium are significantly reduced within 3 min (KAO and MCCULLOUGH 1975).

Although the effects of Ca are rather dramatic in the myometrium, these results have nevertheless been interpreted to indicate that the inward current is mainly carried by Na and that Ca ions have a regulatory role on the Na current. The easiest explanation is probably that the small inward current left in Ca-free solution is due to Na. However, since removal of bound Ca, particularly near the boundary between sucrose solution and test solution in the sucrose gap is likely to be difficult, some response in Ca-free solution may still be attributable to this Ca. The difficulty of removing Ca in Na-free sucrose solution has been shown in the frog stomach muscle (SPARROW et al. 1967).

Excess Ca increases the overshoot and the rate of rise of action potentials in the rat myometrium. At Ca concentrations of up to 25 m*M* this is accompanied by hyperpolarization of the membrane (CASTEELS and KURIYAMA 1965), whereas at 7.5 m*M* this occurs without a change in membrane potential (REINER and MAR-

SHALL 1975). On the other hand, in the mouse myometrium, little change in overshoot is observed between 2.5 and 66 mM $[Ca]_o$ (OSA 1973; OSA and TAGA 1973).

As already described, action potentials are abolished by a prolonged exposure to Na-free solution in the mouse myometrium. However, after a complete block of electrical activity in Na- and Ca-free solution, addition of Ca only (2.5 mM) causes recovery of the spontaneous spike activity, and this effect of Ca can be repeatedly observed for more than 2 h in Na-free solution (OSA 1971). These results suggest that Ca is the main ion producing the spike and that Na has some effect on Ca regulation, as will be described later.

III. Bound Ca and Action Potentials

When Ca is simply omitted without a chelating agent, the electrical activity and mechanical activity of the guinea pig taenia coli are gradually abolished. This occurs rather slowly, usually taking more than 10 min for complete abolition, and the speed varies greatly in different preparations (BÜLBRING and TOMITA 1970a, b). The activity may continue for more than 30 min in Ca-free solution, particularly at low temperature. When EGTA (0.1–0.5 mM) is added to Ca-free solution, the activity is blocked within 3 min in every preparation. On the other hand, when $[Na]_o$ is reduced by substituting with sucrose, the spike may be larger and the electrical activity in Ca-free solution is maintained longer. Reappearance of action potentials by addition of 0.2 mM Ca in Na-free, Ca-free solution also takes more than 10 min to reach a steady state. The action potential in the smooth muscle of cat intestine remains relatively constant when $[Ca]_o$ is altered by keeping the quotient $[Ca]_0/[Na]_0^2$ constant (CONNOR and PROSSER 1974; CURTIS and PROSSER 1977).

These results suggest that the action potential utilizes Ca bound at the outer surface of membrane, and that Na ions compete with Ca at this site. In addition to this, the slow disappearance of spike activity in Ca-free solution may be due to a continuous supply of Ca from a storage site inside the cell by a transport mechanism across the membrane.

In the mouse myometrium, the overshoot of evoked action potentials remains nearly the same between 2.5 and 10 mM Ca, although the intensity of depolarizing current has to be increased to evoke a spike in excess Ca solution (OSA and TAGA 1973). The overshoot is little affected even in 66 mM Ca solution (OSA 1973). The constancy of overshoot with varying $[Ca]_o$ may also be explained by assuming saturation of Ca bound at some sites in the membrane. However, the saturation concentration of Ca may vary greatly in different smooth muscles. The peak size of the action potential is very likely determined not only by Ca conductance but also by K conductance. Therefore, the difference in dependence of the overshoot on $[Ca]_o$ may be due to differences in the degree of increase in the K conductance during the action potential. Thus, the rate of rise, rather than the overshoot, may be a better indication for investigating the action potential mechanism. Furthermore, it would be desirable to study effects of $[Ca]_o$ on the spike parameters in the presence of tetraethyl ammonium (TEA), which reduces the K conductance.

According to studies on barnacle muscle fibers, which produce a typical Ca spike, the relationship between the rate of rise and $[Ca]_o$ and that between the overshoot and $[Ca]_o$ can be adequately explained if it is assumed that the adsorbed or bound form of Ca at the membrane, rather than the free Ca ion concentration, is utilized for excitation; for example the maximum rate of rise, representing the maximum inward current, is proportional to the amount of Ca occupying membrane sites of finite density (HAGIWARA and TAKAHASHI 1967a; HAGIWARA 1973). The suppressing effects of other polyvalent cations on the action potential is also explained by competition of the cations with Ca at the membrane sites.

From a chemical analysis of smooth muscle, a certain amount of cation is thought to be associated with negative sites in the extracellular space. The uptake of ^{28}Mg in the guinea pig taenia coli shows three different phases and only the intermediate phase is influenced by the presence of Ca or K ion in the external medium, suggesting a competition between Mg and K or Ca for fixed negative sites. The amount of ^{28}Mg in this phase in the absence of other cations is 6.1 mmol (12.2 mequiv.)/kg fresh weight (SPARROW 1969). Thus, this value may indicate the amount of fixed negative sites available for cation association. From the experiments with La, which displaces cations associated with the negative sites in the extracellular space, the negative sites in the taenia coli are estimated to retain 15–20 (12–15 Na, 2.74 Ca, and 0.96 Mg) mequiv/kg fresh weight (BRADING and WIDDICOMBE 1977). In the rabbit aorta, 8.1 mmol (16.2 mequiv.)/kg wet weight Ca can be displaced by La in Na-free sucrose solution (VAN BREEMEN et al. 1972). Thus, it is very likely that a large amount of Ca is able to bind with some sites in the extracellular space. However, the precise location of the sites and the kinetics of binding are not clear. The sites responsible for the action potential and those responsible for the control of resting membrane permeability should be at the plasma membrane, but they are likely to be independent of each other.

The negative charge of sialic acid may be responsible for one of Ca binding sites, at least in the cardiac muscle (LANGER 1978). Removal of sialic acid by an enzyme (neuraminidase) increases the rate of Ca uptake and washout, without affecting the K permeability of the membrane, and also reduces binding of La with the surface membrane. This kind of experiment in smooth muscle may help in clarifying the function of Ca associated with the negative sites.

G. Regulation of Intracellular Ca

I. Ca Distribution

The intracellular Ca concentration $[Ca]_i$ may be less than 10^{-7} *M* in the resting state and about 10^{-5} *M* during maximum contraction (ENDO et al. 1977). Therefore, the concentration gradient across the membrane will be 10^2–10^4, giving a positive equilibrium potential of 60–120 mV inside. However, when measurements of actual changes in $[Ca]_i$ are attempted, it is extremely difficult to obtain reliable data, because most of the Ca ions in the tissue are either contained in the extracellular space, of which some are bound, or sequestered in intracellular organelles. The fluxes of these Ca ions will be relatively large and complex, and will mask the functionally important component of transmembrane fluxes which may reflect changes in $[Ca]_i$. For example, in the rat myometrium, the depolarization and contraction produced by high external K is not associated with any significant increase in uptake or efflux of ^{45}Ca (KREJCI and DANIEL 1970). Similarly, carbachol fails to influence tissue Ca content or Ca flux in the longitudinal muscles of guinea pig small intestine (LÜLLMANN and SIEGFRIEDT 1968). In the guinea pig taenia coli, an increase of ^{45}Ca uptake associated with action potentials is not measurable, which leads to the assumption that Na ions play the predominant role in the generation of the action potential (LAMMEL and GOLENHOFEN 1971).

On the other hand, there is a report that the tissue Ca content of the longitudinal muscle of the guinea pig small intestine is increased from 1.7 to 2.8 mmol/kg wet weight by stimulation at 0.1 or 0.2 Hz for 30 min (LÜLLMANN and MOHNS 1969). However, this occurs without any change in influx or efflux kinetics. Further analyses seem necessary before these results can be properly interpreted in relation to the transmembrane flux of Ca ions.

The amount of Ca in the tissue, expressed in mmol/kg wet weight, is generally higher than the Ca concentration of the medium, indicating accumulation of Ca in some compartment (LÜLLMANN 1970). In an attempt to remove Ca bound in the extracellular space and to obtain more precise estimates of intracellular Ca, La ions have been used (VAN BREEMEN et al. 1972; VAN BREEMEN et al. 1973). La (1–10 m*M*) is believed to block transmembrane movement of Ca and displace extracellular Ca. With this method, it is now possible to detect the cellular uptake of ^{45}Ca caused by various stimulating agents. For example, in the rabbit aorta, a net Ca uptake of 102 µmol/kg smooth muscle cell occurred during the maximal contraction produced by excess K (VAN BREEMEN 1977). A comparable study on the relationship between action potential and Ca influx has not been reported.

Even though the measurement of intracellular Ca content is greatly improved by the La method, the sensitivity of the method for detecting small changes in $[Ca]_i$ does not seem to be high enough. It would be impossible, for instance, to measure small changes in the local concentration of free Ca ion near the membrane, which may be critical in determining the membrane permeability or membrane transport.

The intracellular Ca required for contraction or muscle tone may enter from the extracellular space during the action potential or during any process which increases Ca conductance. The same amount of Ca has then to be removed from the cell to maintain the equilibrium. The mechanism of Ca extrusion is still not clearly elucidated in smooth muscle. Two main mechanisms have been proposed, one is an ATP-dependent Ca pump and the other is Na–Ca exchange (VAN BREEMEN et al. 1979). It is likely that both mechanisms are involved, but that their relative contributions differ in various types of smooth muscles and also under various experimental conditions. The Na–Ca exchange process may be dominant in smooth muscles which produce a contracture on Na removal, while the Ca pump may be more important in smooth muscles which remain relaxed in Na-free solution. Interaction between Na and Ca for the regulation of $[Ca]_i$ or muscle tone seems very complicated, and thus, Na–Ca exchange may be proposed as one of several possible models (VAN BREEMEN et al. 1979).

II. Na–Ca Exchange

It has been shown that the inward movement of Na along its electrochemical gradient across the membrane drives Ca out of the cell in the mammalian cardiac muscle (REUTER and SEITZ 1968; GLITSCH et al. 1970), the squid giant axon (BAKER et al.

1969; BLAUSTEIN and HODGKIN 1969), and the barnacle muscle (RUSSEL and BLAUSTEIN 1974). In many other tissues, this Na–Ca-linked transport may play an important role in their function (BLAUSTEIN 1974).

Smooth muscles from rabbit aorta and pulmonary artery develop reversible contractures when the external Na is removed by replacing it with sucrose, choline, or Li (REUTER et al. 1973). The tension also increases when $[Na]_i$ is increased in K-free solution or ouabain-containing solution. When the Ca content in the rabbit aorta is measured, after reducing extracellular Ca by EGTA or La, it is found to increase from 650 μmol/kg wet weight in normal NaCl solution to 1,000 in Li, 1,140 in choline, and 1,570 μmol/kg wet weight in sucrose-substituted Na-free solutions. In La-containing solutions, the ^{45}Ca efflux (supposedly transmembrane) is decreased in Na-free (choline) solution by approximately 35%, and rapidly increased by readmission of Na. From these experiments, it has been concluded that there is Na–Ca exchange, that about 50% of total ^{45}Ca efflux is due to this process in arterial smooth muscle and that this mechanism is similar to that found in cardiac muscle and squid axon (REUTER et al. 1973).

Based on the effects of removing the external Na and of ouabain on the contractions produced by noradrenaline, the presence of a Na–Ca exchange mechanism has been proposed for the rabbit aorta and mesenteric artery (BOHR et al. 1969), and also for the dog mesenteric artery (SITRIN and BOHR 1971). The importance of the Na concentration gradient for regulation of $[Ca]_i$ and its effect on the muscle tone of blood vessels has been considered in relation to hypertension (BLAUSTEIN 1977).

However, there is a report that in the rabbit aorta, Na removal, ouabain, or K-free solution become ineffective in causing contracture after α-adrenergic blockade, reserpinization, or mechanical denervation (KARAKI and URAKAWA 1977). Thus, some precaution seems necessary to exclude the contribution of endogenous catecholamines in this type of experiment, although in the guinea pig aorta contracture produced by K-free solution and ouabain can be demonstrated in the presence of an α-blocker, phentolamine (OZAKI et al. 1978).

The giunea pig taenia coli does not develop a sustained contracture in Na-free or K-free solution, or in the presence of ouabain (KATASE and TOMITA 1972; RAEYMAEKERS et al. 1974). Thus, in contrast to what would be expected from Na–Ca exchange, an inhibitory action of intracellular Na on Ca uptake has been considered in order to explain the relaxed state in the presence of ouabain (BOSE 1975). However, once a contracture has been developed by increasing $[K]_o$, the external Na greatly affects the tension development. When $[K]_o$ is reduced to normal by replacing it with sucrose or Tris, instead of Na, the muscle remains contracted. The contracture recovers on adding more than 5–7 mM Na. However, since relaxation is also produced by high concentrations of Mn, Mg, La, or Ca ions, some other mechanism seems to be involved in this process in addition to a possible Na–Ca exchange (KATASE and TOMITA 1972). The suppressing action of Mn on the depolarization and contracture produced in Na-deficient solution has also been demonstrated in the mouse myometrium (OSA 1973). Na and polyvalent cations may compete with Ca at a membrane site and stabilize the membrane, thus reducing Ca influx. As a result, a metabolically driven Ca pump is now able to cope with reducing $[Ca]_i$ to cause relaxation.

Although the taenia coli maintains a relaxed state in Na-free solution, contracture is readily developed if the Ca conductance of the membrane is increased, for example by carbachol, and tissues which contain high intracellular Na after prolonged exposure to K-free solution produce contractures on reversing the Na concentration gradient by removing the external Na (BRADING 1977). Furthermore, it has been shown that the presence of external Na is necessary for these contractures to relax and that internal Na facilitates Ca influx to cause contracture in response to Ca readmission to preparations pretreated with Ca-free solution. It is also possible that the refilling of an intracellular Ca store is facilitated by the presence of Na.

The necessity for a small amount of external Na (more than 7 m*M*) for relaxation from K contracture has been confirmed by MA and BOSE (1977) for guinea pig taenia coli and also for rat myometrium. These authors considered the contracture to be maintained mainly by impairment of Ca efflux rather than continued Ca influx. Reduction of the tissue Ca was demonstrated during the Na-dependent relaxation, suggesting that Ca extrusion rather than intracellular sequestration was mainly responsible for the relaxation.

There are some arguments against a transmembrane Na–Ca exchange process in the guinea pig taenia coli (CASTEELS and VAN BREEMEN 1975; RAEYMAEKERS et al. 1974). The ^{45}Ca efflux decreases to 50% when Na is reduced to 15.5 m*M* by replacing with sucrose, and it is increased again by readmission of Na. However, similar changes can be observed even in dead preparations pretreated at 100 °C for 2 min, suggesting that Na–Ca exchange is a result of their physical competition (RAEYMAEKERS et al. 1974). Furthermore, Ca content remains nearly the same, when most of the internal K is replaced by Na after exposure to K-free solution. ^{45}Ca efflux is stimulated when metabolism is inhibited by dinitrophenol (DNP) and monoiodoacetic acid (IAA), probably as a result of Ca release from intracellular stores. This stimulated ^{45}Ca efflux is suppressed by replacing Na with choline. However, if the solution contains only 7 m*M* Na or if Na is substituted with Li, there is no such inhibition. ^{45}Ca loaded at 0 °C can be extruded on rewarming to 35 °C in the absence of Na. From these results, it was considered that the inhibition of ^{45}Ca was due to a nonspecific effect of choline, and that a Ca pump, but not Na–Ca exchange, is mainly responsible for the Ca extrusion in the guinea pig taenia coli under normal conditions (CASTEELS and VAN BREEMEN 1975).

III. Ca Pump

The guinea pig taenia coli cannot necessarily be made to increase its tension simply by reducing $[Na]_o$ or by increasing $[Na]_i$. Furthermore, Ca accumulated during cold storage can be extruded in the absence of Na when rewarmed (CASTEELS and VAN BREEMEN 1975). These indicate that Na–Ca exchange is not essential for the control of $[Ca]_i$, at least in the taenia coli. Thus, a Ca pump probably plays an important role in maintaining low $[Ca]_i$ under normal conditions.

In the smooth muscle of cat intestine, the rate of ^{45}Ca efflux is not affected when Na is reduced or Li is replaced with Tris or arginine after about 80 min exposure to Li solution (CURTIS and PROSSER 1977). This also suggests that there is no Na-sensitive Ca efflux. Bovine tracheal muscle remains relaxed when Na is replaced with sucrose, Li, or Mg and when the strip is treated with ouabain or K-free solution for 40 min (KIRKPATRICK and MCDANIEL 1976). Thus, contribution of Na–Ca exchange to the intracellular Ca regulation has also been questioned in this tissue. Although there is no direct evidence for a Ca pump, it is reasonable to assume that the intracellular Ca is mainly regulated by a Ca pump in these smooth muscles.

It has been demonstrated that the microsomal vesicles obtained from the guinea pig taenia coli can accumulate Ca in the presence of ATP (RAEYMAEKERS et al. 1977). These microsomal vesicles contain membrane fragments derived from both the plasma membrane and the sarcoplasmic reticulum. The lowest concentration at which an active uptake can take place is 4×10^{-8} *M* and the rate of Ca uptake is half-maximal at a concentration of 7×10^{-7} *M* Ca. The microsomal fraction is able to remove 150 μmol Ca/kg wet weight in 1 min at 25 °C, and this rate is considered to be high enough to account for physiologic relaxation.

ATP-dependent Ca uptake has also been shown in the microsomes from rabbit aorta (FITZPATRICK et al. 1972), and from the longitudinal muscle of the guinea pig ileum (HURWITZ et al. 1973). It has been suggested that the Ca sequestration may be associated primarily with the intracellular stores in the aorta and with the plasma membrane in the ileum.

H. Effects of Intracellular Ca

The active electrical response in crustacean muscle is due to inward current carried by Ca ions, but it is usually graded, depending on the stimulus intensity (FATT and GINSBORG 1958; FATT and KATZ 1953; WERMAN et al. 1961). The graded activity is likely to be related to the fact that the generation of the action potential is strongly influenced by $[Ca]_i$ in the barnacle muscle fiber (HAGIWARA and NAKA 1964; HAGIWARA and NAKAJIMA 1966b). When $[Ca]_i$ is reduced to less than 8×10^{-8} *M* by using a Ca buffer, the action potential can be produced in an all-or-nothing manner, but the active response becomes graded at a concentration between 8×10^{-8} and 5×10^{-7} *M*, and no active response is evoked above 5×10^{-7} *M*. Similar results have been obtained in the giant neuron of molluscs, *Helix* and *Lymnaea* spp. (KOSTYUK and KRISHTAL 1977). In this neuron, 5.8×10^{-8} *M* $[Ca]_i$ completely blocks the Ca channel, and 3.5×10^{-7} *M* $[Ca]_i$ also blocks the Na channel.

It is possible that internal Ca also affects the excitability of smooth muscles. Any condition which increases $[Ca]_i$ may reduce the amplitude and the rate of rise of the action potential. Removal of external Na tends to increase $[Ca]_i$ by suppression of a Na–Ca exchange mechanism and/or an increase in Ca conductance. However this could be counteracted by a simultaneous increase in Ca efflux due to activation of an active Ca pump, and one may speculate that in smooth muscles such as the taenia coli, which possess an effective Ca pump, the action potential is not much influenced by removal of Na. On the other hand, in smooth muscles such as the myometrium, which have a poorly developed active transport system for Ca, removal of Na causes contracture and suppresses the action potential.

Therefore, the possibility remains that the reduction of the action potential observed in many smooth muscles exposed to Na-deficient or Na-free solution is not simply due to there being less Na to carry current, but to a secondary effect of increasing $[Ca]_i$. This idea is supported by the observations in the mouse myometrium that Na-free contracture does not occur when the external Ca is simultaneously omitted and that readmission of Ca in the absence of Na initiates spike activity (OSA 1971). The most important condition for spike generation seems to be low $[Ca]_i$ and a large concentration gradient of Ca. The suppressing effects of internal Ca may be several. An increase in $[Ca]_i$ should reduce the concentration gradient, i.e., the driving force for Ca influx. An increase in $[Ca]_i$ may increase the K conductance, resulting in increase in outward K current, which counteracts the inward Ca current. Since an increase in $[Ca]_o$ cannot antagonize the suppressing effect of an increase in $[Ca]_i$ in the crustacean muscle, it is also possible that binding of Ca at the inner surface of the plasma membrane modifies some kinetics for excitation (HAGIWARA 1973).

J. Effects of Decreasing K Conductance

Excitable membranes may have complicated systems for K conductance. When such a membrane is depolarized, one type of K conductance is increased with a delay which is long compared with the conductance system carrying the inward current. Thus, this conductance has the property of delayed (or outward-going) rectification and is responsible for repolarization of the action potential. There is another conductance showing inward-going (or anomalous) rectification. This K conductance is present in skeletal muscle or cardiac muscle and is responsible for the negative afterpotential or the plateau potential. Another type of K conductance is sensitive to intracellular Ca (a Ca-activated K conductance). This type of K conductance is dominant in membranes which utilize Ca for the spike and is involved in the positive afterpotential.

It is possible that an excitable membrane may have all these forms of K conductance, but that the relative proportions may vary greatly from one type of membrane to another.The different types of K conductance can often be discriminated by their pharmacologic properties. For example, in some molluscan neurons, the delayed rectifying conductance is more susceptible to blockade by tetraethyl ammonium (TEA) than the Ca-sensitive conductance, and some polyvalent cations block the Ca-sensitive conductance by preventing Ca influx, but have less effect on other types of K conductance (HEYER and LUX 1976; NEHER and LUX 1972; HERMANN and GORMAN 1979; THOMPSON 1977). However, clear differentiation is not always achieved.

Although the properties of K conductance in smooth muscle are not fully understood, TEA and procaine are thought to reduce K conductance. In crustacean muscles, TEA is known to potentiate the action potential (FATT and GINSBORG 1958; FATT and KATZ 1953; WERMAN and GRUNDFEST 1961). Similarly, potentiation of the action potential by TEA has been observed in many different smooth muscles. In the guinea pig ureter, TEA (5 m*M*) gradually increases the amplitude and duration of the plateau and the first spike on top of the plateau, but it blocks repetitive discharges (SHUBA 1977). The resting potential and the membrane resistance remain nearly the same. Potentiation of the spike component can clearly be observed in Na-free (sucrose) solution.

The muscle of canine stomach antrum produces slow potentials which have a spike component at the beginning. When observed with the sucrose gap method, the peak amplitude of the spike is 23 mV and the slow wave is 17 mV on average, and the average duration of the slow wave is 7.1 s. TEA (20 m*M*) increases the spike peak to 35 mV, the slow wave amplitude to 24 mV, and the duration to 10.5 s (Szurszewski 1976). The frequency of spikes appearing on top of the slow wave is increased by TEA up to 30 m*M*, but beyond this concentration it decreases again. TEA (5 m*M*) can produce rhythmic spike activity on top of slow potentials which normally do not have such activity (El-Sharkawy et al. 1978). The effect of TEA is completely abolished in Ca-free solution. Potentiation and prolongation of evoked or spontaneous action potentials are also observed in the circular muscle of cat intestine (Connor et al. 1977).

In the circular muscle of the antrum (Ito et al. 1970), and the longitudinal muscle of the fundus (Osa and Kuriyama 1970) of the guinea pig stomach, the electrotonic potentials produced by hyperpolarizing current pulses are not much affected by TEA (3–5 m*M*). However, the membrane resistance is increased when outward current pulses are used, and the graded spike becomes all-or-nothing, accompanied by an increased amplitude. After TEA treatment, it becomes possible to evoke spikes, even in cells which do not normally produce active responses. The potentiation of the spike can be observed in the absence of Na.

Similar results are obtained in the mouse myometrium (Osa 1974). TEA increases the overshoot of action potentials by about 10 mV at 3 m*M*, and prolongs the duration of spikes at a higher concentration (10–30 m*M*), without significantly affecting the membrane potential or electrotonic potentials produced by inward currents.

When depolarization of more than 30 mV is imposed on the guinea pig myometrium under voltage clamp condition, the inward current is followed by a transient outward current, which appears as a hump, before the steady state is reached (Vassort 1975). This transient outward current is considered to be responsible for a transient undershoot of the action potential, although the afterpotential configuration varies in different preparations. Although the outward current in the steady state is not noticeably affected, TEA (15 m*M*) abolishes the undershoot of the spike within 10 min, and also blocks the transient outward current, increasing the inward current. The transient outward current apparently overlaps the inward current, thus it not only produces the undershoot of the spike, but also causes a reduction of spike amplitude. On the other hand, the possibility that this early outward current is an artifact caused by a technical failure to control the membrane potential has been considered (Kao 1977; Inomata and Kao 1976).

TEA is known to induce action potentials in some smooth muscles which are normally inexcitable (sheep carotid artery: Keatinge 1976; rabbit carotid artery: Mekata 1971; bovine trachea: Kirkpatrick 1975; rabbit ear artery: Droogmans et al. 1977; rabbit pulmonary artery: Casteels et al. 1977; Harder and Sperelakis 1978, 1979; rat anococcygeus muscle: Creed et al. 1975). The smooth muscle of the rabbit ear artery does not produce electrical activity in response to electrical stimulation. When TEA (10–15 m*M*) is applied, the membrane is depolarized from −63.3 mV to a level between −40 and −30 mV, and spontaneous action potentials appear at about −35 mV (Droogmans et al. 1977). Although ^{42}K efflux is accelerated by TEA in normal Krebs solution, probably owing to depolarization and generation of activity, TEA decreases the rate of ^{42}K efflux in excess K (138 mM) solution. Similarly, in sheep carotid artery, reduction of ^{42}K efflux is observed in K-excess solution, but not in normal solution (KEATINGE 1976). In rabbit ear artery in the presence of TEA, the amplitude and rate of rise of the action potentials are increased by increasing $[Ca]_o$ from 1.5 to 4.5 mM, and action potentials are abolished at a $[Ca]_o$ below 0.5 mM. On the other hand, action potentials can be observed after reduction of $[Na]_o$ to 3 mM (by replacing it with choline), although the amplitude and rate of rise are reduced. Similar results have been obtained in rabbit pulmonary artery, although only graded activity can be evoked in the presence of 10 mM TEA (CASTEELS et al. 1977).

The superior mesenteric artery of the guinea pig can also be made excitable by TEA (5–7.5 mM) treatment for 3–5 min (HARDER and SPERELAKIS 1978). The amplitude of this action potential increases by 29 mV for a ten-fold increase in $[Ca]_o$ (from 0.5 to 4.8 mM) and the maximum rate of rise increases from 2 to 15 V/s by increasing $[Ca]_o$ from 1 to 6 mM, but they are almost independent of $[Na]_o$ (HARDER and SPERELAKIS 1979).

Procaine (10^{-3} g/ml) converts graded responses into all-or-nothing action potentials in crayfish muscle fibers (TAKEDA 1967). Potentiation and prolongation of the Ca spike in barnacle muscles are also observed (HAGIWARA and NAKAJIMA 1966a).

The longitudinal muscle of guinea pig ileum is spontaneously active, while the circular muscle produces spikes only in response to depolarizing current pulses. Procaine (1 mM) increases the spike frequency in the longitudinal muscle and generates spontaneous activity in the circular muscle (SUZUKI and KURIYAMA 1975). Similar effects of procaine have been observed in the guinea pig urinary bladder (KURIHARA 1975). Procaine (1 mM) depolarizes the membrane from -41.2 to -36.8 mV, and increases spike frequency and overshoot (KURIHARA and SAKAI 1976). On the other hand, the maximum rates of rise and fall are reduced from 3.0 to 2.0 V/s and from 3.6 to 1.8 V/s, respectively. These reductions can be counteracted by increasing $[Ca]_o$ from 2.5 to 7.5 mM, but the overshoot is further increased.

As with TEA, smooth muscles which are normally inexcitable may initiate action potentials after treatment with procaine. The sheep carotid artery becomes spontaneously active after application of procaine (10–20 mM) (KEATINGE 1976). A decrease of ^{42}K efflux by procaine can be demonstrated in this preparation (KEATINGE 1976) and also in rabbit pulmonary artery (CASTEELS et al. 1977). In the presence of procaine (5 mM), rabbit pulmonary artery can produce a graded action potential and this change is accompanied by an increase in membrane resistance (CASTEELS et al. 1977; ITO et al. 1977).

The main effect of TEA or procaine is generally considered to be a reduction of the K current which overlaps with the inward current, so that in the presence of these agents, the action potential is potentiated by an increase of the net inward current. In the Na spike, the inward current carried by Na has a rapid onset and high density. Therefore, the overlap of slow and small outward currents is not serious. However, the inward current carried by Ca is low in density and slow in onset. Thus, the degree of overlap by the outward current can be relatively large. Furthermore, in membranes which have outward-going rectification, an increase in the inward current produced by depolarizing stimulation tends to be cencelled by a simultaneous increase in the outward current. In the normally inexcitable membranes, it may be that the inward current cannot increase enough to overcome the outward current. When the outward current is reduced by TEA or procaine, an inexcitable membrane can be converted into an excitable membrane, because the inward current can overcome the reduced outward current. However, the outward current affected by TEA or procaine has not been fully specified in smooth muscles.

K. Effects of Membrane Polarization

In the squid giant axon, the maximum Na conductance which can be activated by depolarization is a function of the resting membrane potential, being decreased with depolarization before excitation (HODGKIN and HUXLEY 1952b). The decrease in Na conductance by lowering the membrane potential is called inactivation. The Ca conductance which is responsible for the Ca spike is also inactivated by depolarization of the membrane. Thus, in many excitable membranes, it has been demonstrated that conditioning hyperpolarization increases the amplitude and rate of rise of action potentials, and conditioning depolarization produces the opposite effects.

In the guinea pig taenia coli, spontaneous spikes and evoked spikes are increased by conditioning hyperpolarization, while they are decreased by conditioning depolarization applied intracellularly using a bridge circuit (KURIYAMA and TOMITA 1965). The initial spike of the guinea pig ureter also increases its amplitude and rate of rise when conditioning hyperpolarization is applied with a suction electrode, and the plateau component is similarly affected, although less effectively (KOBAYASHI 1971). Conditioning depolarization has the opposite effect. The relationship between the amplitude or the rate of rise of the action potential and the membrane potential becomes steeper when $[Ca]_o$ is increased from 2.5 to 12.5 mM. In membranes which produce a Na spike, an increase in $[Ca]_o$ has the same effect as conditioning hyperpolarization (FRANKENHAEUSER and HODGKIN 1957; WEIDMANN 1955). This is probably due to neutralization of the surface negative charge. Therefore, in a Na-carrying system, an increase in $[Ca]_o$ is expected to shift the relationship between the rate of rise of the action potential and the membrane potential in a depolarizing direction. However, in the guinea pig ureter, excess Ca shifts the relationship upwards, i.e., the maximum rate of rise obtained during strong conditioning hyperpolarization is increased by excess Ca. This difference is probably due to the fact that Ca is not only affecting the channel control system, but is also utilized for carrying the inward current.

These observations that the action potential is increased with conditioning hyperpolarization are more or less in accord with removal of an inactivation process for the inward current system. However, there are some opposite findings. In the guinea pig vas deferens, it has been shown in the intracellularly evoked, mostly graded, spikes that conditioning hyperpolarization, applied intracellularly, reduces the amplitude of spikes while conditioning depolarization potentiates it (HASHIMOTO et al. 1966). Since this effect persists for some time after cessation of the conditioning pulse, it is not simply due to changes in stimulus strength. When the conditioning depolarization is strong enough to produce an active response by itself, the spikes produced by test pulses are depressed, corresponding to an ordinary type of inactivation.

In the longitudinal muscle of the guinea pig stomach fundus, the action potential is usually difficult to evoke, even with a large external electrode, particularly if a short current pulse is used (OSA and KURIYAMA 1970). However, when the duration of current pulses is increased to several seconds, action potentials can be generated with a long delay, and with a depolarization of more than 30 s some activity appears in most of the muscle fibers. The spike evoked by a 0.5 s pulse is increased with conditioning depolarization and decreased with conditioning hyperpolarization, and these changes are time dependent. The change in excitability is accompanied by alterations of the membrane resistance, i.e., the membrane resistance is gradually increased by a prolonged depolarization, taking several seconds to reach a steady state.

Similar results have been obtained with the mouse myometrium (OSA 1971). The peak of evoked action potentials is slightly increased with weak hyperpolarization, but it decreases markedly with further increase in the conditioning hyperpolarization. The maximum rate of rise is obtained when the resting potential is between −40 and −50 mV, although the overshoot is rather insensitive to membrane polarization between −30 and −50 mV. The maximum rate of rise and the overshoot both decrease sharply when the membrane potential becomes more negative than −50 mV.

With voltage clamp experiments on the rat myometrium, the inward current is shown to attain its maximum when the membrane potential, before the depolarizing test pulse is applied, is held at around −35 mV (KAO and MCCULLOUGH 1975). The maximum inward current decreases when the membrane potential is shifted to either depolarization or hyperpolarization, and it decreases to one-half of the maximum intensity at −26 (9 mV depolarization) and at −50 mV (15 mV hyperpolarization). Also with voltage clamp experiments, similar results have been obtained from the guinea pig myometrium (VASSORT 1975). The decrease in the inward current by hyperpolarization is interpreted as being due to an increase in a transient outward current which overlaps with the inward current. Since TEA reduces the transient outward current, the suppressing effect of conditioning hyperpolarization on the inward current is decreased, and the relationship between availability of inward current and conditioning depolarization is shifted towards membrane potentials less negative inside in the presence of TEA. Although the existence of a transient outward current in the myometrium may be questioned, mainly because of technical difficulties involved in clamping, the decrease in inward current or the reduction of action potential size with conditioning hyperpolarization is most likely to be due to an increase in outward current, the characteristics of which are not well defined yet.

The outward K current in marine gastropods (mainly *Anisidoris* spp.) is gradually inactivated with time during maintained depolarization, with a time constant of 220–600 ms at 5 °C. This transient outward current is different from the ordinary K current (delayed rectifying type) in its early onset and in showing clear inactivation (CONNOR and STEVENS 1971). This current is increased by conditioning hyperpolarization up to more than −100 mV, and decreased by reducing the membrane potential, complete inactivation being found in the range between −40 and −50 mV. TEA reduces the transient current by one-half in concentrations of 50–100 m*M*, which essentially blocks the steady state outward current (the delayed rectifying type). *Helix* spp. neurons also have two outward currents, fast and delayed, and the fast outward current masks the inward current, particularly when the membrane is hyperpolarized beyond −50 mV (NEHER 1971; KOSTYUK et al. 1975a). In these neurons, externally applied TEA depresses the delayed outward current more effectively, while internally applied TEA depresses the fast outward current more than the delayed one (KOSTYUK et al. 1975a). The decrease of inward current by conditioning hyperpolarization is almost completely prevented by intracellular application of TEA.

Helix spp. neurons can produce Na spikes in the absence of Ca, but the inward current observed under this condition is not reduced by conditioning hyperpolarization (STANDEN 1974). Some giant neurons of *Aplysia* spp. can utilize both Na and Ca for the spike, although the Ca current is slower and smaller than the Na current. When Ca is omitted from the solution, the inward current is not affected by conditioning hyperpolarization, while the inward current observed in Na-free solution containing Ca is decreased by conditioning hyperpolarization (GEDULDIG and GRUENER 1970). This apparent inactivation of the Ca-carrying system by conditioning hyperpolarization may be similarly explained by an activation of an early K current.

These results obtained from giant neurons suggest that an early transient outward current may also exist in smooth muscle, and that the suppressing effect of conditioning hyperpolarization can be similarly explained, both for giant neurons and smooth muscle. However, the contribution of the transient outward current to the electrical activity may vary greatly in different types of smooth muscle. The possibility that this outward current is closely related to the Ca influx during the action potential needs further investigation.

L. Effects of Ca Antagonists

The ions of some transition elements or rare earth elements, such as Mn, Co, Ni, or La, have a stabilizing action like Ca, but they cannot substitute for Ca in the spike mechanism as a current carrier. These ions are known to suppress Ca spikes in several tissues, without much effect on the Na-carrying mechanism (giant neuron: GEDULDIG and JUNGE 1968; KOSTYUK et al. 1974; crustacean muscle: HAGIWARA and NAKAJIMA 1966a; HAGIWARA and TAKAHASHI 1967a).

Sone organic agents, such as verapamil (iproveratril), D-600 (a methoxy derivative of verapamil), or nifedipine are also believed to block the slow channel responsible for the Ca spike (FLECKENSTEIN 1977). These agents are not so effective as membrane stabilizers, but are considered to act directly on the mechanism involved in spike generation, although their mode of action has not been analyzed enough.

I. Inorganic Antagonists

In many smooth muscles, it has been demonstrated that the action potential is suppressed by Mn or La. These results are thought to be strong evidence that smooth muscles utilize Ca for the spike. However, most experiments have been done on spontaneous spikes, and it is thus difficult to determine whether the suppressing effect is on the pacemaker mechanism or on the spike mechanism.

It has been reported in guinea pig taenia coli that 0.5 m*M* Mn blocks spontaneous spikes within 1 min when recorded extracellulary (NONOMURA et al. 1966; BÜLBRING and Tomita 1969). However, when recorded intracellularly, the block occurs gradually, taking more than 20 min for complete cessation (BRADING et al. 1969). The first effects are to increase the spike duration, to decrease the spike frequency, and to depolarize the membrane. Even after the block of spontaneous activity by 0.5–1 m*M* Mn, spikes of reduced amplitude can be evoked by depolarizing current, and the evoked spikes partially recover with conditioning hyperpolarization. To abolish the evoked spike, more than 2 m*M* Mn is necessary.

With voltage clamp experiments on guinea pig taenia coli, a marked suppression of both inward and the outward currents has been observed by Mn (1–5 m*M*) (INOMATA and KAO 1976). A concentration of 0.5 m*M* Mn reduces the inward current by about one-half in 14 min (KUMAMOTO and HORN 1970).

In the rat myometrium, similar results on the action potential to those in the taenia coli have been obtained (ABE 1971). Spikes can be evoked for about 30 min in the presence of 2 m*M* Mn, although their amplitude and rate of rise are depressed.

In the mouse myometrium, 0.1 m*M* Mn increases the intervals between the spontaneous bursts of activity, without affecting the activity itself (OSA 1973). At 0.3 m*M*, Mn reduces the overshoot, rate of rise, and also number of spikes in each burst. This effect develops slowly, taking 10–30 min to reach a steady state, when a single spike appears sporadically, or no activity is seen. It is found that 0.6 m*M* Mn hyperpolarizes the membrane by about 5 mV and completely blocks spontaneous activity within 3–10 min. However, a spike can be evoked, although the amplitude and rate of rise are reduced. Some graded activity can still be observed in 1.8 m*M* Mn. Excess Ca antagonizes the suppressing effect of Mn. When Ca is increased to 22 or 66 m*M* in the presence of 0.6 or 1.8 m*M* Mn, spikes having a rate of rise larger than the control can be evoked. La (2.4 m*M*), Mn (1–5 m*M*), and Co (1 m*M*) block or markedly reduce the inward current and also simultaneously reduce the steady state outward current in the rat myometrium, measured under voltage clamp conditions (ANDERSON et al. 1971; MIRONNEAU 1973, 1974).

In the guinea pig ureter, 2 m*M* Mn abolishes the spike component, and reduces the duration of the plateau component, without much change in resting membrane potential and resistance (SHUBA 1977). However, the amplitude of the plateau is markedly increased, even in Na-free solution. The underlying mechanism of this plateau-like potential observed in the presence of Mn is not known. However, since the potential is abolished in the absence of Ca, it is considered to be due to Ca entry through the slow Na channel, rather than due to Mn entry.

Although Mn can inhibit the action potential of smooth muscles, particularly if they are spontaneously generated, its action may be neither selective nor simple. Depolarization of the membrane and prolongation of the spike may suggest that Mn reduces the K conductance. This could partly counteract the blocking action on the Ca channel. Furthermore the possibility remains, at least at a low concentration and in some smooth muscle, that Mn may be able to carry charge through a Ca channel, as observed in the guinea pig ventricular papillary muscle (OCHI 1976).

II. Organic Antagonists

Verapamil (0.8–1.0×10^{-6} *M*) often slightly depolarizes the membrane in the guinea pig taenia coli, and reduces the frequency of spontaneous spikes and tension (GOLENHOFEN and LAMMEL 1972; RIEMER et al. 1974). The overshoot and the rates of rise and fall of spontaneous spikes are reduced, but these changes are antagonized by increasing $[Ca]_o$ to 7.5 m*M*. At a concentration of 10^{-5} *M*, the spontaneous activity is completely blocked. Since the spike frequency recovers earlier than tension after washing out verapamil, some impairment of the excitation–contraction coupling is suggested (RIEMER et al. 1974). However, the main factor which inhibits the mechanical activity is considered to be the decrease in spike frequency, i.e., suppression of the pacemaker potential by verapamil. The inward current recorded with the voltage clamp method is markedly reduced by 10^{-5} *M* D-600 without much effect on the outward current (INOMATA and KAO 1976).

From the competitive effect between Ca and verapamil on the phasic contraction produced by isotonic K solution, it has been shown that one verapamil molecule antagonizes 8,000 Ca ions in guinea pig taenia coli (RIEMER et al. 1974). However, since Ca has many different actions, the interpretation of this result is difficult. When the spontaneous activity is blocked by D-600 (2×10^{-5} *M*), Ca uptake by guinea pig taenia coli is significantly reduced, when estimated with the La method (MAYER et al. 1972). The uptake of Ca is 108 μ*M*/kg wet weight after 2 h in the control and 80 μ*M*/kg wet weight in the presence of D-600, the difference being 28 μ*M*/kg wet weight. However, the functional significance of this value is difficult to evaluate. It has also been shown that both the contracture and the increase in Ca uptake produced by 154 m*M* K solution are prevented by 2×10^{-5} *M* D-600.

The smooth muscle of guinea pig portal vein has spontaneous spike activity, similar to the taenia coli. Verapamil (10^{-5} *M*) depolarizes the membrane by about 15–20 mV and blocks the spontaneous activity (GOLENHOFEN et al. 1973; GOLENHOFEN and LAMMEL 1972).

The longitudinal muscle of rat myometrium shows a spontaneous slow depolarization which triggers a train of spikes. This slow depolarization is reduced by D-600 (10^{-8}–10^{-7} *M*), resulting in abolition of spontaneous spike activity (REINER and MARSHALL 1975). However, in the presence of 10^{-7} *M* D-600, a train of spikes can still be evoked by electrical stimulation. All electrical activity is abolished by 10^{-6} *M* D-600. At 10^{-7} *M*, D-600 reduces the amplitude of spikes from 48 to 40 mV, and the maximum rate of rise from 7.6 to 4.2 V/s without a change in resting potential, and excess Ca counteracts the reductions. An increase in $[Ca]_o$ from

2.5 to 7.5 m*M* increases these parameters to values larger than the controls, although the antagonistic effect of Ca on the tension and spike frequency is weaker than on the spike parameters. The increment of tension with each action potential remains nearly the same on treatment with D-600 (10^{-7} *M*). The inward current of rat myometrium recorded under voltage clamp conditions is reduced by D-600 (5×10^{-6} *M*) (MIRONNEAU 1973), and in guinea pig myometrium, D-600 (2×10^{-4} *M*) reduces both inward and outward currents (VASSORT 1975).

The slow wave in the stomach and the plateau component of the ureter in the guinea pig are rather resistant to 10^{-6} *M* verapamil, although the spike component which appears on top of them is markedly suppressed (GOLENHOFEN and LAMMEL 1972). At 10^{-5}–10^{-4} *M*, the plateau of the ureter is shortened, without much effect on the amplitude. Before suppression, verapamil transiently increases the spike amplitude (SHUBA 1977). The slow wave of the circular muscle of canine stomach antrum has a similar configuration to the cardiac action potential. D-600 (10^{-5} *M*) significantly depresses the amplitude and duration of the plateau phase, without much effect on the initial spike component and the resting potential (EL-SHARKAWY et al. 1978).

The organic Ca blockers have a suppressing effect on the action potentials of smooth muscles and this action is antagonized by Ca ions. However, their mode of action has not been fully investigated. Although the spontaneous activity is suppressed at a low concentration, much higher concentrations seem to necessary to eliminate the inward current when stimulated. At such a high concentration, the specificity of the Ca antagonists as blocking agents of the Ca channel may not be high. For example, it is known in the snail neuron that verapamil (5×10^{-6}–1.2×10^{-4} *M*) depresses not only the inward current but also the outward current (KOSTYUK et al. 1975b).

M. General Properties of Spikes and Their Implications

The action potential of smooth muscles seems to utilize mainly, if not exclusively, Ca ions. Ca influx, rather than Na influx during the spike is the physiologically significant influx in several respects, such as conduction and contraction. Besides these, the Ca-spike may be more suitable than the Na-spike for the excitations of muscles situated in the GI tract or the urogenitals organs, where the ionic composition of extracellular fluid may not be kept constant. This is because it is bound Ca, rather than free Ca, which plays a major role in the generation of Ca spikes. When the extracellular medium is diluted or concentrated, the electrical activity of the Ca spike would be more stable than that of the Na spike.

During conduction of the action potential, a local circuit current flows from the excited area to a preexcited area in front of the action potential owing to a potential gradient, and the preexcited area is stimulated by this current. The intensity of local circuit current is dependent on the potential gradient and the resistance of the circuit. The potential gradient can be calculated from the rate of rise of the action potential divided by the conduction velocity. In smooth muscle such as the guinea pig taenia coli, the maximum rate of rise is 5 V/s and the conduction velocity is 5 cm/s as a rough approximation. Thus, the maximum spatial gradient in front of the action potential is about 0.1 V/mm. The spatial gradient is more or less the same order of magnitude in all excitable tissues, as would be expected from a functional point of view. Thus, the slow rate of rise and slow conduction velocity are closely related to each other.

Since the increase in Ca conductance during the Ca spike is small and slow, compared with the Na conductance during the Na spike, the rate of rise is also slow and subsequently the conduction velocity is slow. Thus, a Ca spike would not be appropriate for nerve fibers or mammalian skeletal muscle fibers in which rapid transmission of information is required. However, slow conduction is teleologically important for proper control of synchronization of the activity and movement of visceral organs.

The contraction of smooth muscles is very much slower than that of mammalian skeletal muscles. Furthermore, in smooth muscles, owing to their small diameter, diffusion of Ca which enters through the plasma membrane during the action potential is fast enough for slow contraction, without having a special releasing mechanism of Ca from the sarcoplasmic reticulum, which is particularly well developed for excitation–contraction coupling in skeletal muscle fibers.

At the moment, it is difficult to estimate the amount of Ca which enters during the spike. If it is assumed that the membrane capacity is 1 $\mu F/cm^2$, the action potential is 60 mV in amplitude, and only Ca carries the inward current without simultaneous overlapping of an outward current, then the Ca influx would be 0.3×10^{-12} mol/cm^2. The Ca influx would be much larger for a typical large spike, if there was a significant overlap with an outward current. If the volume/surface quotient of a smooth muscle fiber is taken to be 0.39×10^{-4} cm (GABELLA 1976), a Ca influx of 0.3×10^{-12} mol/cm^2 corresponds to an increase of intracellular concentration of 7.7×10^{-6} *M*. Since Ca is bound to the contractile protein and the system responsible for Ca uptake or extrusion, the actual increase in $[Ca]_i$ should be less than 7.7×10^{-6} *M*, and it is probably around $1–2 \times 10^{-6}$ *M*. Nevertheless, Ca influx due to a spike would be expected to increase $[Ca]_i$ significantly even with calculations based on the least estimation of influx. It can thus be reasonably assumed from this calculation that the summation of contractions caused by a few spikes is enough to produce a nearly maximum tension development. Of course, smooth muscles may have a Ca-releasing system in addition to the Ca influx across the plasma membrane for initiating contraction, particularly those smooth muscles which do not rely on spike activity, such as most vascular smooth muscles.

N. Summary

It may be concluded that, in smooth muscles, the contribution of Ca ions to the action potential is larger and that of Na ions is smaller, than in other typical excitable tissues, such as nerve fibers or skeletal muscle fibers, which have well-developed Na channels.

Fig. 1 is a schematic illustration showing the possible ionic currents involved in action potential generation. The inward current responsible for the depolarizing phase of the action potential can be divided into fast and slow currents, which flow through separate fast and slow channels, respectively. The fast cannel is responsi-

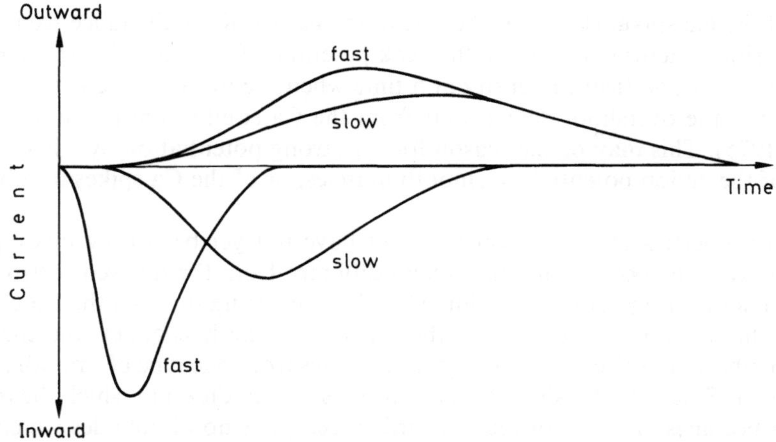

Fig. 1. Schematic drawing of possible ionic currents during an action potential. Their relationship is shown in a very qualitative way, and the magnitude and time course are not drawn to scale

ble for the typical Na spike. This channel is very selective for Na, but it can pass some Ca. All TTX-sensitive channels belong to this fast type.

The slow channel is thought to be responsible for the Ca spike and permeability of this channel is very high to Ca, Ba, and Sr ions, but generally very low to Na (HAGIWARA 1975; REUTER 1973). Nerve fibers possess mainly the fast channel but may have slow channels to various degrees. At the nerve terminal (the presynaptic membrane), the density of slow channels is very much increased (KATZ and MILEDI 1969, 1971). Cardiac muscle has well-developed fast and slow systems, and many molluscan giant neurons are also known to have both fast and slow channels, although their density varies in different neurons (REUTER 1973). Crustacean muscle is a typical example of a tissue which has the slow channel (HAGIWARA 1973, 1975).

Analyses are still far from satisfactory for precise description on the ionic channels in smooth muscles. However, considering the low sensitivity to $[Na]_o$, the slow rate of rise of the action potential, and the pharmacologic properties of the tissues, the existence of fast inward current channels is very doubtful in smooth muscle and the ionic channels are more likely to be of the slow type. The slow channel is inactivated at membrane potentials much lower than those nessessary to inactivate the fast channel (REUTER 1973). This accords with the fact that the resting membrane potential of smooth muscle is generally low (40–60 mV).

The problem is interpretation of the suppressing effect of Na deficiency on the action potential or the inward current. It may be that the selectivity of the slow channel is poor in smooth muscles, so that Na may carry significant inward current through this channel. Alternatively, Na as ions, or its concentration gradient across the membrane, may affect the kinetics of a slow channel which passes mainly Ca ions. The latter possibility should be further investigated, particularly in relation to the role of Na in regulation of $[Ca]_i$.

Since the Na inward current through the fast channel increases much faster than the outward K current, the overlap of the two currents is not serious at the peak of the Na spike. However, the rate of increase in K conductance seems to be a very critical factor determining the peak potential of the Ca spike. If the K current reaches a significant intensity at a time when the inward Ca current is at its maximum, the overshoot deviates far from the Ca equilibrium potential (HAGIWARA 1975). This may be the reason for the strong potentiation by TEA or procaine of the action potentials in smooth muscles, or of the Ca spikes in other tissues.

The properties of the outward current have not yet been fully investigated. There may be a special transient system different from the delayed rectifying K channel, and this system may be linked with Ca entering through the Ca channel during the spike. It is possible that the property of the K channel, and therefore the contribution of K current to excitation, varies from one type of smooth muscle to another. Since the threshold for excitation is the condition in which the inward current becomes in excess of the outward current, in smooth muscles which have high resting K conductance or in which depolarization rapidly increases K conductance, no excitation is possible unless the K conductance is suppressed by some means. A highly excitable smooth muscle may have low resting K conductance and only a slight overlap of K current with the inward current. There seems to be a spectrum of excitability in smooth muscle between these two extremes, and their properties are probably related to differences in intracellular Ca regulation.

References

Abe Y (1971) Effects of changing the ionic environment on passive and active membrane properties of pregnant rat uterus. J Physiol (Lond) 214:173–190
Abe Y, Tomita T (1968) Cable properties of smooth muscle. J Physiol (Lond) 196:87–100
Anderson NC (1969) Voltage-clamp studies on uterine smooth muscle. J Gen Physiol 54:145–165
Anderson NC, Ramon F, Snyder A (1971) Studies on calcium and sodium in uterine smooth muscle excitation under current-clamp and voltage-clamp conditions. J Gen Physiol 58:322–339
Attwell D, Cohen I (1977) The voltage clamp of multicellular preparations. Prog Biophys Mol Biol 31:201–245
Baker PF, Blaustein MP, Hodgkin AL, Steinhardt RA (1969) The influence of calcium on sodium efflux in squid axons. J Physiol (Lond) 200:431–458
Bennett MR (1967a) The effect of cations on the electrical porperties of the smooth muscle cells of the guinea-pig vas deferens. J Physiol (Lond) 190:465–479
Bennett MR (1967b) The effect of intracellular current pulses in smooth muscle cells of the guinea pig vas deferens at rest and during transmission. J Gen Physiol 50:2459–2475
Bennett MR, Burnstock G, Holman ME, Walker JW (1962) The effect of Ca^{2+} on plateau-type action potentials in smooth muscle. J Physiol (Lond) 161:47–48P
Blaustein MP (1974) The interrelationship between sodium and calcium fluxes across cell membranes. Rev Physiol Biochem Pharmacol 70:33–82
Blaustein MP (1977) Sodium ions, calcium ions, blood pressure regulation, and hypertension: a reassessment and a hypothesis. Am J Physiol 232:C165–173
Blaustein MP, Hodgkin AL (1969) The effect of cyanide on the efflux of calcium from squid axons. J Physiol (Lond) 200:497–527
Bohr DF, Seidel C, Sobieski J (1969) Possible role of sodium-calcium pumps in tension development of vascular smooth muscle. Microvasc Res 1:335–343
Bolton TB, Vassort G, Tomita T (1981) Voltage-clamp as applied to smooth muscle. In: Bülbring E, Brading AF, Jones AW, Tomita T (eds) Smooth muscle II. Arnold, London, pp 47–64

Bose D (1975) Mechanism of mechanical inhibition of smooth muscle by ouabain. Br J Pharmacol 55:111–116
Brading AF (1971) Analysis of the effluxes of sodium, potassium and chloride ions from smooth muscle in normal and hypertonic solutions. J Physiol (Lond) 214:393–416
Brading AF (1975) Sodium/sodium exchange in the smooth muscle of the guinea-pig taenia coli. J Physiol (Lond) 251:79–105
Brading AF (1977) Na, Ca and contraction in the smooth muscle of the guinea-pig taenia coli. In: Casteels R, Godfraind T, Rüegg JC (eds) Excitation-contraction coupling in smooth muscle. Elsevier/North-Holland, Amsterdam, pp 97–99
Brading AF, Jones AW (1969) Distribution and kinetics of CoEDTA in smooth muscle, and its use as an extracellular marker. J Physiol (Lond) 200:387–401
Brading AF, Widdicombe JH (1977) The use of lanthanum to estimate the numbers of extracellular cation-exchanging sites in the guinea-pig's taenia coli, and its effects on transmembrane monovalent ion movements. J Physiol (Lond) 266:255–273
Brading A, Bülbring E, Tomita T (1969) The effect of sodium and calcium on the action potential of the smooth muscle of the guinea-pig taenia coli. J Physiol (Lond) 200:637–654
Brink F (1954) The role of calcium ions in neural processes. Pharmacol Rev 6:243–298
Bülbring E (1954) Membrane potentials of smooth muscle fibres of the taenia coli of the guinea-pig. J Physiol (Lond) 125:302–315
Bülbring E, Kuriyama H (1963) Effects of changes in the external sodium and calcium concentrations on spontaneous electrical activity in smooth muscle of guinea-pig taenia coli. J Physiol (Lond) 166:29–58
Bülbring E, Tomita T (1967) Properties of the inhibitory potential of smooth muscle as observed in the response to field stimulation of the guinea-pig taenia coli. J Physiol (Lond) 189:299–315
Bülbring E, Tomita T (1969) Effect of calcium, barium and manganese on the action of adrenaline in the smooth muscle of the guinea-pig taenia coli. Proc R Soc Lond [Biol] 172:121–136
Bülbring E, Tomita T (1970 a) Effects of Ca removal on the smooth muscle of the guinea-pig taenia coli. J Physiol (Lond) 210:217–232
Bülbring E, Tomita T (1970 b) Calcium and the action potential in smooth muscle. In: Cathbert AW (ed) A symposium on calcium and cellular function. Macmillan, London, pp 249–260
Bury VA, Shuba MF (1976) Transmembrane ionic currents in smooth muscle cells of ureter during excitation. In: Bülbring E, Shuba MF (eds) Physiology of smooth muscle. Raven, New York, pp 65–75
Casteels R (1969) Calculation of the membrane potential in smooth muscle cells of the guinea-pig's taenia coli by the Goldman equation. J Physiol (Lond) 205:193–208
Casteels R, Kuriyama H (1965) Membrane potential and ionic content in pregnant and nonpregnant rat myometrium. J Physiol (Lond) 177:263–287
Casteels R, Kuriyama H (1966) Membrane potential and ion content in the smooth muscle of the guinea-pig's taenia coli at different external potassium concentrations. J Physiol (Lond) 184:120–130
Casteels R, Van Breemen C (1975) Active and passive Ca fluxes across cell membranes of the guinea-pig taenia coli. Pfluegers Arch 359:197–207
Casteels R, Kitamura K, Kuriyama H, Suzuki H (1977) The membrane properties of the smooth muscle cells of the rabbit main pulmonary artery. J Physiol (Lond) 271:41–61
Connor C, Prosser CL (1974) Comparison of ionic effects on longitudinal and circular muscle of cat jejunum. Am J Physiol 226:1212–1218
Connor JA, Stevens CF (1971) Voltage clamp studies of a transient outward membrane current in gastropod neural somata. J Physiol (Lond) 213:21–30
Connor JA, Kreulen D, Prosser CL, Weigel R (1977) Interaction between longitudinal and circular muscle in intestine of cat. J Physiol (Lond) 273:665–689
Creed KE (1971) Effects of ions and drugs on the smooth muscle cell membrane of the guinea-pig urinary bladder. Pfluegers Arch 326:127–141
Creed KE (1979) Functional diversity of smooth muscle. Br Med Bull 35:243–247

Creed KE, Gillespie JS, Muir TC (1975) The electrical basis of excitation and inhibition in the rat anococcygeus muscle. J Physiol (Lond) 245:33–47
Curtis BA, Prosser CL (1977) Calcium and cat intestinal smooth muscle. In: Casteels R, Godfraind T, Rüegg JC (eds) Excitation-contraction coupling in smooth muscle. Elsevier/North-Holland, Amsterdam, pp 123–129
Droogmans G, Raeymaekers L, Casteels R (1977) Electro- and pharmacomechanical coupling in the smooth muscle cells of the rabbit ear artery. J Gen Physiol 70:129–148
Eckert R, Tillotson D (1978) Potassium activation associated with intraneuronal free calcium. Science 200:437–439
El-Sharkawy TY, Morgan KG, Szurszewski JH (1978) Intracellular electrical activity of canine and human gastric smooth muscle. J Physiol (Lond) 279:291–307
Endo M, Kitazawa T, Yagi S, Iino M, Katuta Y (1977) Some properties of chemically skinned smooth muscle fibres. In: Casteels R, Godfraind T, Rüegg JC (eds), Excitation-contraction coupling in smooth muscle. Elsevier/North-Holland, Amsterdam, pp 199–209
Fatt P, Ginsborg BL (1958) The ionic requirements for the production of action potentials in crustacean muscle fibres. J Physiol (Lond) 142:516–543
Fatt P, Katz B (1953) The electrical properties of crustacean muscle fibres. J Physiol (Lond) 120:171–204
Ferroni A, Blanchi D (1965) Maximum rate of depolarization of single muscle fibre in normal and low sodium solutions. J Gen Physiol 49:17–25
Fitzpatrick DF, Landon EJ, Debbas G, Hurwitz L (1972) A calcium pump in vascular smooth muscle. Science 176:305–306
Fleckenstein A (1977) Specific pharmacology of calcium in myocardium, cardiac pacemakers, and vascular smooth muscle. Ann Rev Pharmacol Toxicol 17:149–166
Fozzard HA, Schoenberg M (1972) Strength-duration curves in cardiac Purkinje fibres: effects of liminal length and charge distribution. J Physiol (Lond) 226:593–618
Frankenhaeuser B, Hodgkin AL (1957) The action of calcium on the electrical properties of squid axons. J Physiol (Lond) 137:218–244
Gabella G (1976) Quantitative morphological study of smooth muscle cells of the guinea-pig taenia coli. Cell Tissue Res 170:161–186
Gabella G (1977) Arrangement of smooth muscle cells and intramuscular septa in the taenia coli. Cell Tissue Res 184:195–212
Geduldig D, Gruener R (1970) Voltage clamp of the *Aplysia* giant neurone: early sodium and calcium currents. J Physiol (Lond) 211:217–244
Geduldig D, Junge D (1968) Sodium and calcium components of action potentials in the *Aplysia* giant neurone. J Physiol (Lond) 199:347–365
Glitsch HG, Reuter H, Scholz H (1970) The effect of the internal sodium concentration on calcium fluxes in isolated guinea-pig auricles. J Physiol (Lond) 209:25–43
Golenhofen K, Lammel E (1972) Selective suppression of some components of spontaneous activity in various types of smooth muscle by iproveratril (verapamil) . Pfluegers Arch 331:233–243
Golenhofen K, Petrányi P (1969) Spikes of smooth muscle in calcium-free solution (Isolated taenia coli of the guinea-pig) . Experientia 25:271–273
Golenhofen K, Hermstein N, Lammel E (1973) Membrane potential and contraction of vascular smooth muscle (portal vein) during application of noradrenaline and high potassium, and selective inhibitory effects of iproveratril (verapamil) . Microvasc Res 5:73–80
Hagiwara S (1973) Ca spike. Adv Biophys (Tokyo) 4:71–102
Hagiwara S (1975) Ca-dependent action potential. In: Eisenman G (ed) , Membranes, a series of advances, vol 3. Dekker, New York, pp 359–381
Hagiwara S, Naka K (1964) The initiation of spike potential in barnacle muscle fibers under low intracellular $Ca.^{++}$ J Gen Physiol 48:141–162
Hagiwara S, Nakajima S (1966a) Differences in Na^{+} and Ca^{++} spikes as examined by application of tetrodotoxin, procaine, and manganese ions. J Gen Physiol 49:793–805
Hagiwara S, Nakajima S (1966b) Effect of the intracellular Ca-ion concentration upon the excitability of the muscle fiber membrane of a barnacle. J Gen Physiol 49:807–818
Hagiwara S, Takahashi K (1967a) Surface density of calcium ions and calcium spikes in the barnacle muscle fiber membrane. J Gen Physiol 50:583–601

Hagiwara S, Takahashi K (1967b) Resting and spike potentials of skeletal muscle fibres of salt-water elasmobranch and teleost fish. J Physiol (Lond) 190:499–518
Harder DR, Sperelakis N (1978) Membrane electrical properties of vascular smooth muscle from the guinea pig superior mesenteric artery. Pfluegers Arch 378:111–119
Harder DR, Sperelakis N (1979) Action potentials induced in guinea pig arterial smooth muscle by tetraethylammonium. Am J Physiol 237:C75–80
Hashimoto Y, Holman ME, Tille J (1966) Electrical properties of the smooth muscle membrane of the guinea-pig vas deferens. J Physiol (Lond) 186:27–41
Hashimoto Y, Holman ME, McLean AJ (1967) Effect of tetrodotoxin on the electrical activity of the smooth muscle of the vas deferens. Nature 215:430–432
Hermann A, Gorman ALF (1979) External and internal effects of tetraethylammonium on voltage-dependent and Ca-dependent K^+ currents components in molluscan pacemaker neurons. Neurosci Lett 12:87–92
Heyer CB, Lux HD (1976) Control of the delayed outward potassium currents in bursting pace-maker neurones of the snail, *Helix pomatia*. J Physiol (Lond) 262:349–382
Hodgkin AL, Huxley AF (1952a) Currents carried by sodium and potassium ions through the membrane of the giant axon of *Loligo*. J Physiol (Lond) 116:449–472
Hodgkin AL, Huxley AF (1952b) The dual effect of membrane potential on sodium conductance in the giant axon of *Loligo*. J Physiol (Lond) 116:497–506
Hodgkin AL, Huxley AF (1952c) A quantitative description of membrane current and its application to conductance and excitation in nerve. J Physiol (Lond) 117:500–544
Hodgkin AL, Katz B (1949) The effect of sodium ions on the electrical activity of the giant axon of the squid. J Physiol (Lond) 108:37–77
Holman ME (1958) Membrane potentials recorded with high-resistance micro-electrodes; and the effects of changes in ionic environment on the electrical and mechanical activity of the smooth muscle of the taenia coli of the guinea-pig. J Physiol (Lond) 141:464–488
Hurwitz L, Fitzpatrick DF, Debbas G, Landon EJ (1973) Localization of calcium pump activity in smooth muscle. Science 179:384–386
Inomata H, Kao CY (1976) Ionic currents in the guinea pig taenia coli. J Physiol (Lond) 255:347–378
Ito Y, Kuriyama H, Sakamoto Y (1970) Effects of tetraethylammonium chloride on the membrane activity of guinea-pig stomach smooth muscle. J Physiol (Lond) 211:455–460
Ito Y, Suzuki H, Kuriyama H (1977) Effects of caffeine and procaine on the membrane and mechanical properties of the smooth muscle cells of the rabbit main pulmonary artery. Jpn J Physiol 27:467–481
Johnson EA, Lieberman M (1971) Heart: excitation and contraction. Ann Rev Physiol 33:479–532
Kao CY (1966) Tetrodotoxin, saxitoxin and their significance in the study of excitation phenomena. Pharmacol Rev 18:997–1049
Kao CY (1967) Ionic basis of electrical activity in uterine smooth muscle. In: Wynn RM (ed) Cellular biology of the uterus. North-Holland, Amsterdam, pp 386–448
Kao CY (1977) Electrophysiological properties of the uterine smooth muscle. In: Wynn RM (ed) Biology of the uterus. Plenum, New York, pp 423–496
Kao CY, McCullough JR (1975) Ionic currents in the uterine smooth muscle. J Physiol (Lond) 246:1–36
Karaki H, Urakawa N (1977) Possible role of endogenous catecholamines in the contractions induced in rabbit aorta by ouabain, sodium-depletion and potassium-depletion. Eur J Pharmacol 43:65–72
Katase T, Tomita T (1972) Influences of sodium and calcium on the recovery process from potassium contracture in the guinea-pig taenia coli. J Physiol (Lond) 224:489–500
Katz B, Miledi R (1969) Tetrodotoxin-resistant electric activity in presynaptic terminals. J Physiol (Lond) 203:459–487
Katz B, Miledi R (1971) The effect of prolonged depolarization on synaptic transfer in the stellate ganglion of the squid. J Physiol (Lond) 216:503–512
Keatinge WR (1968) Ionic requirements for arterial action potential. J Physiol (Lond) 194:169–182
Keatinge WR (1976) Effect of local anaesthetics on electrical activity and voltage-dependent K permeability of arteries. INSERM Colloq 50:177–180

Kirkpatrick CT (1975) Excitation and contraction in bovine tracheal smooth muscle. J Physiol (Lond) 244:263–281
Kirkpatrick CT, McDaniel DG (1976) Tracheal smooth muscle – Failure to demonstrate Na-Ca exchange. Pfluegers Arch 361:301–302
Kleinhaus AL, Kao CY (1969) Electrophysiological actions of oxytocin on the rabbit myometrium. J Gen Physiol 53:758–780
Kleinhaus AL, Prichard JW (1977) Close relation between TEA responses and Ca-dependent membrane phenomena of four identified leech neurones. J Physiol (Lond) 270:181–194
Kobayashi M (1969) Effect of calcium on electrical activity in smooth muscle cells of cat ureter. Am J Physiol 216:1279–1285
Kobayashi M (1971) Relationship between membrane potential and spike configuration recorded by sucrose gap method in the ureter smooth muscle. Comp Biochem Physiol 38A:301–308
Kostyuk PG, Krishtal OA (1977) Effects of calcium and calcium-chelating agents on the inward and outward current in the membrane of mollusc neurones. J Physiol (Lond) 270:569–580
Kostyuk PG, Krishtal OA, Doroshenko PA (1974) Calcium currents in snail neurones. I. Identification of calcium current. Pfluegers Arch 348:83–93
Kostyuk PG, Krishtal OA, Doroshenko PA (1975a) Outward current in isolated snail neurones. II. Effects of TEA. Comp Biochem Physiol 51C:265–268
Kostyuk PG, Krishtal OA, Doroshenko PA (1975b) Outward currents in isolated snail neurones. III. Effect of verapamil. Comp Biochem Physiol 51C:269–274
Krejci I, Daniel EE (1970) Effect of contraction on movements of calcium 45 into and out of rat myometrium. Am J Physiol 219:256–262
Kumamoto M, Horn L (1970) Voltage clamping of smooth muscle from taenia coli. Microvasc Res 2:188–201
Kurihara S (1975) The effect of procaine on the mechanical and electrical activities of the smooth muscle cells of the guinea-pig urinary bladder. Jpn J Physiol 25:775–788
Kurihara S, Sakai T (1976) Relationship between effects of procaine and Ca on spontaneous electrical and mechanical activities of the smooth muscle cells of the guinea-pig urinary bladder. Jpn J Physiol 26:487–501
Kuriyama H (1964) Effect of calcium and magnesium on neuromuscular transmission in the hypogastric nerve-vas deferens preparation of the guinea-pig. J Physiol (Lond) 175:211–230
Kuriyama H, Tomita T (1965) The responses of single smooth muscle cells of guinea-pig taenia coli to intracellularly applied currents and their effect on the spontaneous electrical activity. J Physiol (Lond) 178:270–289
Kuriyama H, Tomita T (1970) The action potential in the smooth muscle of the guinea pig taenia coli and ureter studied by the double sucrose-gap method. J Gen Physiol 55:147–162
Kuriyama H, Osa T, Toida N (1967) Membrane properties of the smooth muscle of guinea-pig ureter. J Physiol (Lond) 191:225–238
Lammel E, Golenhofen K (1971) Messungen der ^{45}Ca-Aufnahme an intestinaler glatter Muskulatur zur Hypothese eines von Ca-Ionen getragenen Aktionsstromes. Pfluegers Arch 329:269–282
Langer GA (1978) The structure and function of the myocardial cell surface. Am J Physiol 235:H461–468
Lüllmann H (1970) Calcium fluxes and calcium distribution in smooth muscle. In: Bülbring E, Brading AF, Jones AW, Tomita T (eds) Smooth muscle. Arnold, London, pp 151–165
Lüllmann H, Mohns P (1969) The Ca metabolism of intestinal smooth muscle during forced electrical stimulation. Pfluegers Arch 308:214–224
Lüllmann H, Siegfriedt A (1968) Über den Calcium-Gehalt und den ^{45}Calcium-Austausch in der Längsmuskulatur des Meerschweinchendünndarms. Pfluegers Arch 300:108–119
Ma TS, Bose D (1977) Sodium in smooth muscle relaxation. Am J Physiol 232:C59–66

Mayer CJ, Van Breemen C, Casteels R (1972) The action of lanthanum and D-600 on calcium exchange in the smooth muscle cells of the guinea-pig taenia coli. Pfluegers Arch 337:333–350
Meech RW (1976) Intracellular calcium and the control of membrane permeability. Symp Soc Exp Biol 30:161–191
Meech RW (1978) Calcium-dependent potassium activation in nervous tissues. Ann Rev Biophys Bioeng 7:1–18
Meech RW, Standen NB (1975) Potassium activation in *Helix aspersa* neurones under voltage clamp: a component mediated by calcium influx. J Physiol (Lond) 249:211–239
Mekata F (1971) Electrophysiological studies of the smooth muscle cell membrane of the rabbit common carotid artery. J Gen Physiol 57:738–751
Mironneau J (1973) Excitation-contraction coupling in voltage clamped uterine smooth muscle. J Physiol (Lond) 233:127–141
Mironneau J (1974) Voltage clamp analysis of the ionic currents in uterine smooth muscle using the double sucrose gap method. Pfluegers Arch 352:197–210
Mironneau J (1976) Contraction and transmembrane currents of rat myometrium. Voltage clamp data: a new tool for pharmacological studies. INSERM Colloq 50:183–196
Nagai T, Prosser CL (1963) Electrical parameters of smooth muscle cells. Am J Physiol 204:915–924
Narahashi T (1974) Chemicals as tools in the study of excitable membranes. Physiol Rev 54:813–889
Neher E (1971) Two fast transient current components during voltage clamp on snail neurons. J Gen Physiol 58:36–53
Neher E, Lux HD (1972) Differential action of TEA^+ on two K^+-current components of a molluscan neurone. Pfluegers Arch 336:87–100
Noble D (1972) The relation of Rushton's 'liminal length' for excitation to the resting and active conductances of excitable cells. J Physiol (Lond) 226:573–591
Nonomura Y, Hotta Y, Ohashi H (1966) Tetrodotoxin and manganese ions: effects on electrical activity and tension in taenia coli of guinea pig. Science 152:97–99
Ochi R (1976) Manganese-dependent propagated action potentials and their depression by electrical stimulation in guinea-pig myocardium perfused by sodium-free media. J Physiol (Lond) 263:139–156
Ohba M, Sakamoto Y, Tomita T (1977) Effects of sodium, potassium and calcium ions on the slow wave in the circular muscle of the guinea-pig stomach. J Physiol (Lond) 267:167–180
Osa T (1971) Effect of removing the external sodium on the electrical and mechanical activities of the pregnant mouse myometrium. Jpn J Physiol 21:607–625
Osa T (1973) The effects of sodium, calcium and manganese on the electrical and mechanical activities of the myometrial smooth muscle of pregnant mice. Jpn J Physiol 23:113–133
Osa T (1974) Effects of tetraethylammonium on the electrical activity of pregnant mouse myometrium and the interaction with manganese and cadmium. Jpn J Physiol 24:119–133
Osa T, Kuriyama H (1970) The membrane properties and decremental conduction of excitation in the fundus of the guinea-pig stomach. Jpn J Physiol 20:626–639
Osa T, Taga F (1973) Effects of external Na and Ca on the mouse myometrium in relation to the effects of oxytocin and carbachol. Jpn J Physiol 23:97–112
Ozaki H, Karaki H, Urakawa N (1978) Possible role of Na-Ca exchange mechanism in contractions induced in guinea pig aorta by potassium free solution and ouabain. Arch Pharmacol 304:203–209
Prosser CL, Kreulen DL, Weigel RJ, Yau W (1977) Prolonged potentials in gastrointestinal muscles induced by calcium chelation. Am J Physiol 233:C19–24
Raeymaekers L, Wuytack F, Casteels R (1974) Na-Ca exchange in taenia coli of the guinea-pig. Pfluegers Arch 347:329–340
Raeymaekers L, Wuytack F, Batra S, Casteels R (1977) A comparative study of the calcium accumulation by mitochondria and microsomes isolated from the smooth muscle of the guinea-pig taenia coli. Pfluegers Arch 368:217–223

Redfern P, Thesleff S (1971) Action potential generation in denervated rat skeletal muscle. II. The action of tetrodotoxin. Acta Physiol Scand 82:70–78
Reiner O, Marshall JM (1975) Action of D-600 on spontaneous and electrically stimulated activity of the parturient rat uterus. Arch Pharmacol 290:21–28
Reuter H (1973) Divalent cations as charge carriers in excitable membranes. Prog Biophys Mol Biol 26:1–43
Reuter H, Seitz N (1968) The dependence of calcium efflux from cardiac muscle on temperature and external ion composition. J Physiol (Lond) 195:451–470
Reuter H, Blaustein MP, Haeusler G (1973) Na-Ca exchange and tension development in arterial smooth muscle. Philos Trans R Soc Lond [Biol] 265:87–94
Riemer J, Dörfler F, Mayer C-J, Ulbrecht G (1974) Calcium-antagonistic effects on the spontaneous activity of guinea-pig taenia coli. Pfluegers Arch 351:241–258
Romero PJ (1976) Role of membrane-bound Ca in ghost permeability to Na and K. J Membr Biol 29:329–343
Romero PJ, Whittam R (1971) The control by internal calcium of membrane permeability to sodium and potassium. J Physiol (Lond) 214:481–507
Rushton WAH (1937) Initiation of the propagated disturbance. Proc R Soc Lond [Biol] 124:210–243
Russell JM, Blaustein MP (1974) Calcium efflux from barnacle muscle fibers. Dependence on external cations. J Gen Physiol 63:144–167
Shuba MF (1977) The effect of sodium-free and potassium-free solutions, ionic current inhibitors and ouabain on electrophysiological properties of smooth muscle of guinea-pig ureter. J Physiol (Lond) 264:837–851
Sitrin MD, Bohr DF (1971) Ca and Na interaction in vascular smooth muscle contraction. Am J Physiol 220:1124–1128
Sommer JR, Johnson ES (1968) Cardiac muscle. A comparative study of Purkinje fibers and ventricular fibers. J Cell Biol 36:497–526
Sparrow MP (1969) Interaction of ^{28}Mg with Ca and K in the smooth muscle of guinea-pig taenia coli. J Physiol (Lond) 205:19–38
Sparrow MP, Mayrhofer G, Simmonds WJ (1967) Uptake and increased binding by smooth muscle in half isotonic sucrose and its relationship to contractility. Aust J Exp Biol Med Sci 45:469–484
Standen NB (1974) Properties of a calcium channel in snail neurones. Nature 250:340–342
Suzuki H, Kuriyama H (1975) Electrical and mechanical properties of longitudinal and circular muscles of the guinea-pig ileum. Jpn J Physiol 25:759:773
Szurszewski JH (1976) The effect of tetraethylammonium ion on the action potential of the longitudinal muscle of the canine antrum. INSERM Colloq 50:247–249
Takeda K (1967) Permeability changes associated with the action potential in procaine-treated crayfish abdominal muscle fibers. J Gen Physiol 50:1049–1074
Thompson SH (1977) Three pharmacologically distinct potassium channels in molluscan neurones. J Physiol (Lond) 265:465–488
Tomita T (1966) Electrical responses of smooth muscle to external stimulation in hypertonic solution. J Physiol (Lond) 183:450–468
Tomita T (1967a) Current spread in the smooth muscle of the guinea-pig vas deferens. J Physiol (Lond) 189:163–176
Tomita T (1967b) Spike propagation in the smooth muscle of the guinea-pig taenia coli. J Physiol (Lond) 191:517–527
Tomita T (1970) Electrical properties of mammalian smooth muscle. In: Bülbring E, Brading AF, Jones AW, Tomita T (eds) Smooth muscle. Arnold London, pp 197–243
Tomita T (1975) Electrophysiology of mammalian smooth muscle. Prog Biophys Mol Biol 30:185–203
Tomita T (1981) Electrical activities (spikes and slow waves) in gastrointestinal smooth muscles. In: Bülbring E, Brading AF, Jones AW, Tomita T (eds) Smooth muscle II. Arnold, London, pp 127–156
Van Breemen C (1977) Calcium requirement for activation of intact aortic smooth muscle. J Physiol (Lond) 272:317–329

Van Breemen C, Farinas BR, Gerba P, McNaughton ED (1972) Excitation-contraction coupling in rabbit aorta studied by the lanthanum method for measuring cellular calcium influx. Circ Res 30:44–54
Van Breemen C, Farinas BR, Casteels R, Gerba P, Wuytack F, Deth R (1973) Factors controlling cytoplasmic Ca^{++} concentration. Philos Trans R Soc Lond [Biol] 265:57–71
Van Breemen C, Aaronson P, Loutzenhiser R (1979) Sodium-calcium interactions in mammalian smooth muscle. Pharmacol Rev 30:167–208
Vassort G (1975) Voltage-clamp analysis of transmembrane ionic currents in guinea-pig myometrium: evidence for an initial potassium activation triggered by calcium influx. J Physiol (Lond) 252:713–734
Vaughan-Jones RD (1977) The effect of lowering external sodium on the intracellular sodium activity of crab muscle fibres. J Physiol (Lond) 264:239–265
Wald F (1972) Ionic differences between somatic and axonal action potentials in snail giant neurones. J Physiol (Lond) 220:267–281
Weidmann S (1955) Effects of calcium ions and local anaesthetics on electrical properties of Purkinje fibres. J Physiol (Lond) 129:568–582
Werman R, Grundfest H (1961) Graded and all-or-none electrogenesis in arthropod muscle. II. The effect of alkali-earth and onium ions on lobster muscle fibers. J Gen Physiol 44:997–1027
Werman R, McCann FV, Grundfest H (1961) Graded and all-or-none electrogenesis in arthropod muscle. J Gen Physiol 44:979–995

CHAPTER 5

Electrophysiology of Intestinal Smooth Muscle

R. CAPRILLI, G. FRIERI, and P. VERNIA

A. Introduction

The electrophysiology of smooth muscle, despite the continuous development of new techniques, is still one of the most confusing and controversial fields of digestive physiology. The marked differences in smooth muscle properties in different segments of the gastrointestinal tract, in the various layers, and in different species have greatly contributed to this confusion. Controversy has also arisen concerning the extrapolation of results obtained from in vitro experiments to those obtained in vivo.

This chapter deals with methods used in recording the electrical activity, the types and characteristics of potentials recorded from gastrointestinal muscle, the relationship between electrical and mechanical events, and the interdigestive migrating myoelectric complex. A brief account of some of the possible applications of electromyography in clinical practice is also included.

B. Methods Used in Electrical Recording

The physiologic activity of smooth muscle fibers is accompanied by electrical changes across the surface of their membranes. These may generate intracellular and extracellular electrical potentials which can be recorded by means of an electrode placed in the cytoplasm of the cell (intracellular technique) or in the extracellular fluid (extracellular technique). The electrode consists of a metal wire, usually platinum, silver, or steel which, with the exception of the tip, is completely insulated.

Intracellular and extracellular recordings can be performed in vitro, whereas for in vivo experiments only extracellular recordings are used. Electrical signals, after appropriate amplification, are recorded on paper for visual analysis or on magnetic tape for computer analysis.

I. Intracellular Recording Technique

With the intracellular recording technique introduced for smooth muscle by BÜLBRING and HOOTON (1954), it is possible to record the electrical activity of a single cell by penetration of a microelectrode within the cell. The electrode must be very thin to reduce damage when penetrating the biologic membranes and must have an exposed surface small enough to record only from a single cell. Microelectrodes pulled from glass capillary tubing and filled with 3 *M* KCl, may be used with a micromanipulator or suspended by very thin wire (floating electrode). Most microelectrodes are less than 0.5 μm in diameter and have a resistance between 20 and 50 MΩ. The main advantage of intracellular recording is the possibility of measuring the resting membrane potential (RMP), i.e., the potential beween the inside and outside of the cell in the absence of electrical activity. RMP of visceral muscle is fairly low and more variable than that of cardiac and skeletal muscle (HOLMAN 1968). The disadvantages of this method are mainly the difficulty in performing the technique itself and the impossibility of applying it to in vivo experiments. A further limitation is the lack of a direct method for evaluating the probability of injury occurring on impalation of the cell (injury potential).

II. Extracellular Recording Technique

With the extracellular technique, which consists of placing electrodes in the extracellular fluid surrounding the tissue, it is possible to record potentials generated by several cells. It can be employed in both in vivo and in vitro experiments. Potentials may be detected by the monopolar and bipolar methods. In the monopolar technique, the electrical activity of the tissue is recorded by a searching electrode and is referred to an area of low electrical activity by means of an indifferent electrode. In bipolar recordings, a pair of closely spaced electrodes are placed in the extracellular fluid and the potential difference between the two regions where the electrodes make contact is recorded.

The chief advantage of the extracellular technique is that it offers the possibility of performing in vivo studies in both acute and chronic experiments and of evaluating the effect of drugs. The most commonly used electrodes are described below.

1. Pressure Electrode

The pressure electrode is a glass tube with a tip diameter of less than 0.5 mm, filled with agar–Tyrode solution. The electrode is attached to a transducer which permits simultaneous recording of electrical and mechanical activity as well as of the pressure exerted by the electrode on the tissue (BORTOFF 1961).

2. Glass Pore Electrode

The glasspore electrode is a silver–silver chloride wire inserted into a glass capillary tube (diameter 0.05–0.1 mm). Bath fluid provides the electrical contact between muscle cell and the wire.

3. Wick Electrode

The wick electrode is similar to the glass pore electrode but contact with the tissue is made by a wick. Its diameter ranges from 0.1–2 mm. One advantage of the wick electrode consists in the absence of injury potential. This electrode has also been employed for recording electrical activity from the upper gastrointestinal tract in humans (CHRISTENSEN et al. 1964).

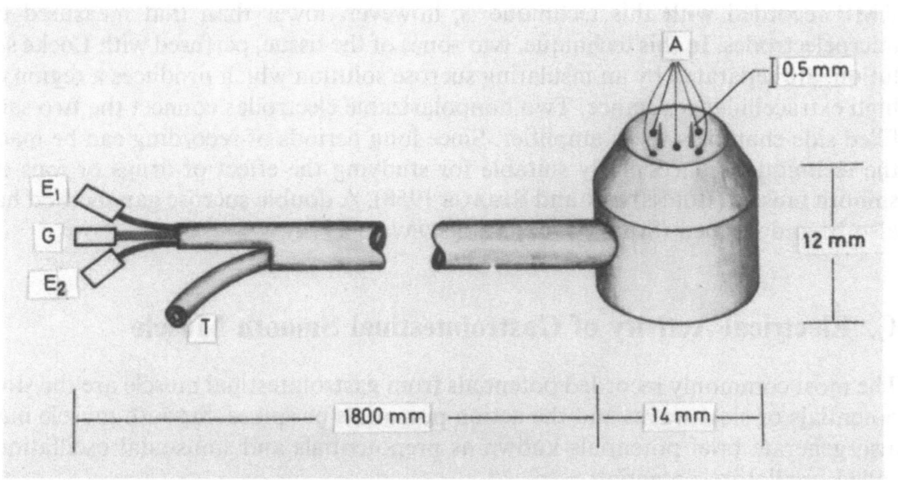

Fig. 1. Schematic representation of a probe for intraluminal recording of gastrointestinal myoelectric potentials. The distal end of the probe consists of a rubber cup containing a pair of silver needle electrodes and holes for aspiration (A). The probe also has a plastic tube connecting three shielded coaxial cables; two are connected to electrodes (E_1 and E_2) and one (G) grounds the electrodes. A polyethylene tube (T) enables suction to be applied near the electrodes

4. Needle Electrode

Needle or punctate electrodes consist of needles implanted in the smooth muscle and are particularly suitable for chronic experiments, but are affected by injury potential.

For chronic experiments the needles are embedded in an acrylic disk (McCoy and Bass 1973) from which the tip of the electrodes (0.5 mm in diameter) protrudes 2–3 mm and are then surgically implanted on the serosal surface of the bowel. Contact with the intestinal muscle layer is established by sutures passed through holes drilled into the disk. Insulated copper wires covered by polytetrafluoroethylene tubing connect the electrodes to a cannula implanted in the abdominal wall. Chronically implanted electrodes have been widely used in animals and occasionally also in humans (Provenzale and Pisano 1971; Taylor et al. 1975; Sarna et al. 1979). In humans, however, electrical activity is commonly recorded by means of intraluminal electrodes. A rubber suction cup (Fig. 1) (Monges et al. 1969; Caprilli et al. 1971; Taylor et al. 1974b) or clips (Snape et al. 1976) are used to ensure contact of the electrodes with the mucosal surface.

Several other types of electrodes have been described for recording electromyography in humans (Dorph et al. 1972; Taylor et al. 1975) and dogs (Thillier and Bertrand 1975; Smout et al. 1979) but are, as yet, little used.

5. Sucrose Gap

The sucrose gap method was first introduced by Stämpfli (1954) for recording resting membrane potentials with external electrodes in frog nerve fibers. The

RMP recorded with this technique is, however, lower than that measured by microelectrodes. In this technique, two zones of the tissue, perfused with Locke solution, are separated by an insulating sucrose solution which produces a region of high extracellular resistance. Two nonpolarizable electrodes connect the two salt-filled side chambers to an amplifier. Since long periods of recording can be made the technique is particularly suitable for studying the effect of drugs or ions on smooth muscle (BURNSTOCK and STRAUB 1958). A double sucrose gap method has also been described (SHUBA 1961; SZURSZEWSKI 1974).

C. Electrical Activity of Gastrointestinal Smooth Muscle

The most commonly recorded potentials from gastrointestinal muscle are the slow potentials or slow waves and the action potentials or spikes. Smooth muscle may also generate brief potentials known as prepotentials and sinusoidal oscillations called oscillatory potentials.

I. Slow Waves

Slow waves may be defined as spontaneous, slow, periodic fluctuations in transmembrane potential of smooth muscle cells. These electrical phenomena have also been termed basic electrical rhythm (BER), indicating omnipresent cyclically recurring electrical potentials (BASS et al. 1961), pacemaker potentials, since these waves set the maximal frequency of contractions (SZURSZEWSKI et a. 1970), and electrical control activity (ECA), suggested by the control function of slow waves on contraction (SARNA and DANIEL, 1973). The configuration of slow waves varies with the recording technique, but usually consists of a fast initial and a slower second component (Fig. 2). Spikes may be associated with the second component.

In experiments with intracellular microelectrodes, slow waves appear as monophasic waves consisting of an initial rapid depolarization followed by a plateau with a slow repolarization (Fig. 3). Spikes may be superimposed on the plateau. Duration varies (1–8 s), but is usually in the range 3–5 s. Amplitude also varies and is a function of RMP. A linear correlation exists between the amplitude of slow waves and RMP in circular muscle of cat colon (CHRISTENSEN et al. 1969). The mean RMP of smooth muscle in the gastrointestinal tract is about 50 mV and is low compared with that in other excitable tissues.

The configuration of slow waves, recorded by the extracellular technique, varies according to the type of the electrode employed, the pressure exerted by the electrodes on the tissue, the nature of the medium surrounding the tissue, and the site of the gastrointestinal tract from which the recording is made (Figs. 4, 7).

Three main patterns of slow waves may be recorded (BORTOFF 1961). The first has a monophasic configuration resembling slow waves recorded by intracellular microelectrodes and is obtained when the electrode exerts a pressure of about 1–2 g on the tissue. The second consists of a major upward positive deflection followed by a return to a less positive plateau ending with a small negative component. The third type of slow wave is similar to the second but the initial deflection is triphasic. Since the tissue in extracellular recording is usually surrounded by a large volume

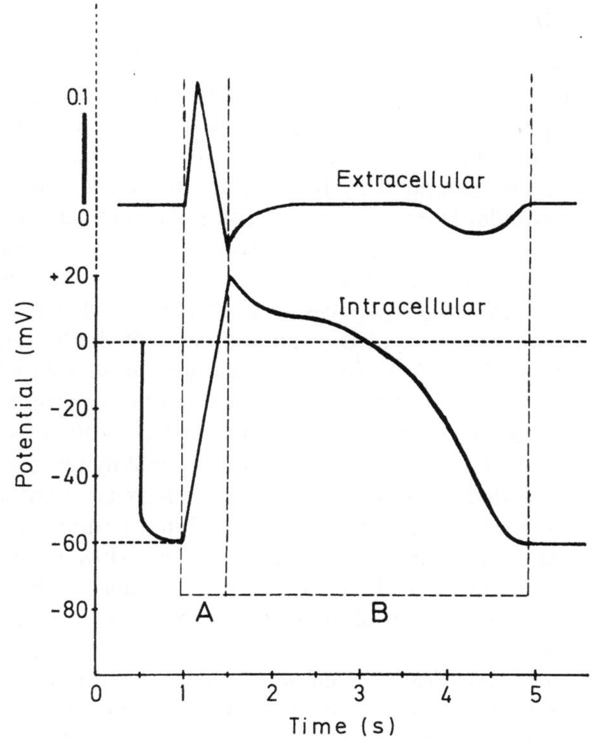

Fig. 2. Diagram of slow waves recorded with extracellular and intracellular techniques. The initial rapid component corresponds to depolarization (A) and second slow component to repolarization (B)

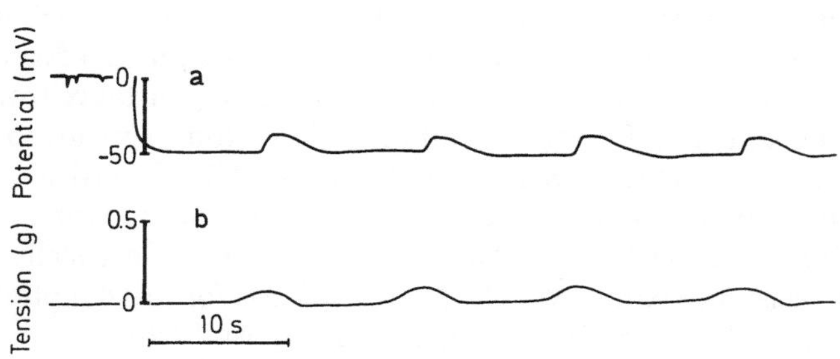

Fig. 3 a, b. Simultaneous recording of transmembrane potential (**a**) and tension (**b**) from circular muscle of cat colon in vitro. Penetration of microelectrode into the cell shows regular depolarizations corresponding to slow waves. Small amplitude contractions accompany each slow wave

of conducting medium, the potentials recorded under these conditions are called field potentials or volume conducted potentials (PROSSER and BORTOFF 1968). The electrical events of field potentials correspond to an outward–inward–outward flowing membrane current and are responsible for the triphasic component of slow waves.

1. Origin

Slow waves are myogenic since they remain unchanged after nerve blockage by tetrodotoxin, ganglionic blockage by hexamethonium, and cholinergic and adrenergic blockage by atropine, propanolol, and phentolamine, respectively. The site of origin of slow waves in the small intestine differs from that in the colon. In the small intestine, slow waves are generated by the longitudinal muscle and spread electrotonically into the circular layer, while in the colon they are generated by the circular muscle and spread into the longitudinal layer. The origin of slow waves from the longitudinal muscle in small intestine was demonstrated by BORTOFF (1961) and later confirmed by Kobayashi (KOBAYASHI et al. 1966). Intact segments of cat jejunum showed slow waves propagated in the long axis of the longitudinal fibers at an average velocity of 10 mm/s. Strips of pure circular muscle showed waves only when connected to the longitudinal layer (KOBAYASHI et al. 1966). Evidence of electrotonic spread into the circular muscle is supported by the observation that in flat preparations in which the longitudinal layer is partially removed, the amplitude of slow waves recorded from the circular muscle decreases exponentially with the distance from the longitudinal layer (BORTOFF 1965; KOBAYASHI et al. 1966). Furthermore, the transmembrane current associated with slow waves in the longitudinal layer is first outward and then inward, whereas that in the circular layer is always outward, indicating that slow waves in this layer are entirely passive (BORTOFF and SACHS 1970).

The origin of slow waves from the circular muscle in the colon was first suggested by CHRISTENSEN et al. (1969) and subsequently demonstrated by CAPRILLI and ONORI (1972). Strips of cat colon without mucosa exhibit slow waves from both the circular and longitudinal sides. Slow waves were recorded from the circular muscle (Fig. 4) in pure preparations of isolated muscle layers, but not from the longitudinal muscle. The longitudinal muscle showed slow waves only when it was attached to the underlying circular muscle. In preparations in which only part of the circular muscle was removed from the longitudinal layer, the decrease in slow wave amplitude in the longitudinal muscle with increasing distance from the circular fibers was exponential (Fig. 5), indicating an electrotonic spread from the circular to the longitudinal layer (CAPRILLI and ONORI 1972). Electrical interactions, therefore, appear to exist between the two muscle layers both in the small intestine and colon. The muscle cells of the two layers seem to be electrically coupled by means of low-resistance pathways. The anatomic basis for this coupling is thought to be provided by the nexus or gap junction (BORTOFF 1976).

Studies on muscle strips of resected human colon did not confirm the origin of slow waves from the circular layer, which in fact failed to show slow wave activity (KIRK and DUTHIE 1978). Of the 26 strips examined in this study, however, only 5 showed spontaneous activity. Conclusions derived from these experiments should therefore be considerd with some reserve.

2. Ionic Dependence

Slow wave generation is related to ionic fluxes through the cell membrane. The ionic dependence of gastric and small intestine slow waves, however, differs from that in the colon.

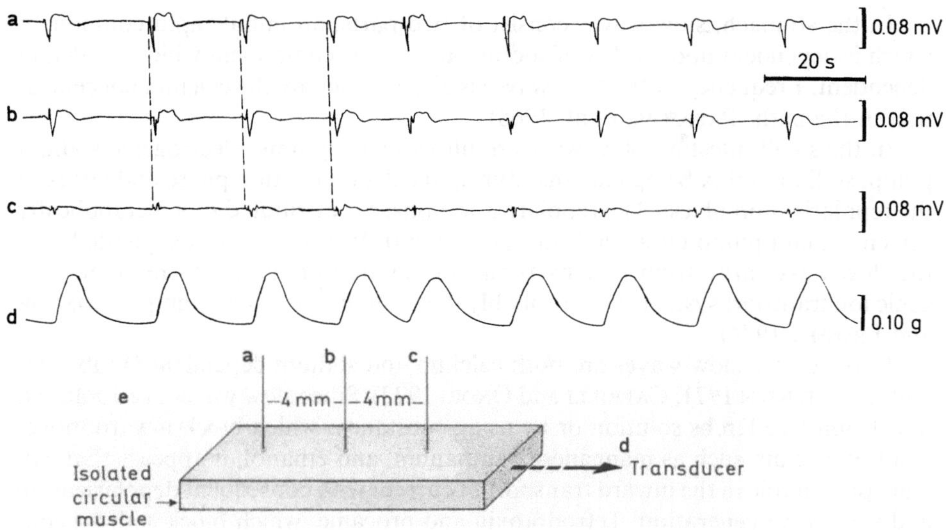

Fig. 4a–e. Electrical (**a–c**) and mechanical (**d**) activity recorded from a strip of isolated circular muscle of cat colon. Three glass pore electrodes are arranged as shown (**e**). Recordings show volume-conducted slow waves in regular sequence, propagated from *a* to *c* (phase-locked) and associated with phasic contractions. (CAPRILLI and ONORI 1972)

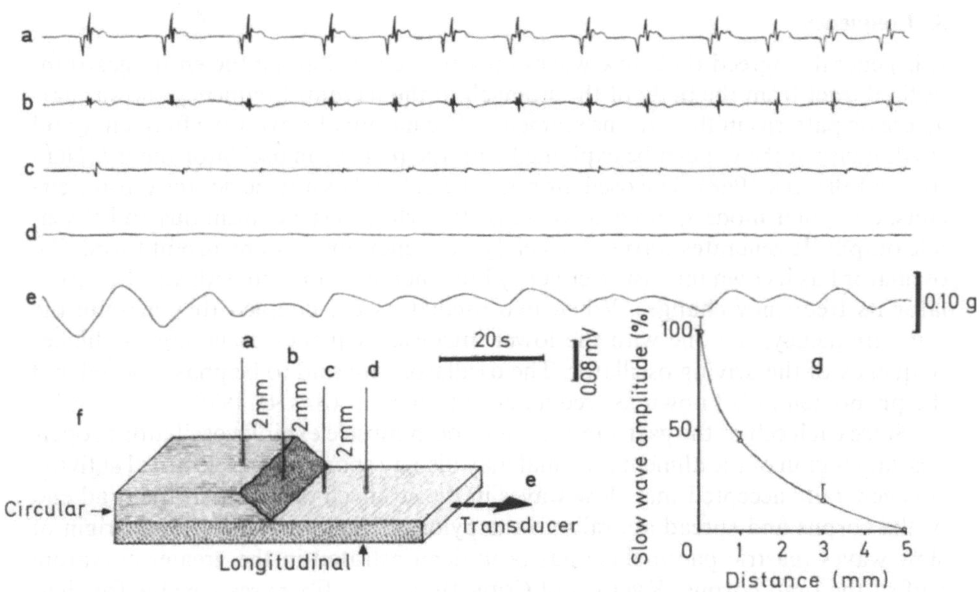

Fig. 5a–g. Electrical (**a–d**) and mechanical (**e**) activity recorded from a strip of cat colon in which the circular layer was partially removed. Electrode **a** records from circular muscle, electrode **b** from edge of circular layer, electrodes **c** and **d** from isolated longitudinal muscle. Electrodes are arranged as shown (**f**). Amplitude of slow waves gradually decreases with distance from the edge of the circular layer. Exponential spatial decrement of slow wave amplitude, expressed as a percentage of amplitude, is also shown (**g**). (CAPRILLI and ONORI 1972)

In the stomach, slow waves consist of two parts, an initial rapid component which is dependent upon influx of sodium and a slow component which is calcium dependent. Frequency of the slow waves is also affected by the calcium concentration in the bath (PAPASOVA et al. 1968).

In the small intestine, slow waves result from a rhythmic electrogenic sodium pump, sodium influx being maximal during the depolarization phase and efflux in the repolarization phase. The repolarization phase is attributed to a metabolically driven sodium pump (JOB 1969; LIU et al. 1969). It has also been suggested that the slow waves arise from "the rhythmic modulation of current from an electrogenic ion transport system, most probably the sodium–potassium pump" (CONNOR and PROSSER 1974).

In the colon, slow waves are both calcium and sodium dependent (WIENBECK and CHRISTENSEN 1971; CAPRILLI and ONORI 1972). Since slow waves are abolished in calcium-free Krebs solution or by using substances which block inward movement of calcium such as manganese, lanthanum, and ethanol, it appears that calcium plays a role in the inward transport of current with consequent depolarization and slow wave generation. Tetrodotoxin and procaine, which block sodium conductance, had no effect on slow waves, whereas dinitrophenol and ouabain, which block active sodium efflux, abolished slow waves. Evidence therefore exists that sodium plays an important role in the genesis of slow waves by its active cyclic extrusion (CAPRILLI and ONORI 1972).

3. Frequency

It is generally agreed that slow waves may be recorded along the entire gastrointestinal tract from the body of the stomach to the rectum. Frequency shows characteristic patterns in the various segments. The nature of slow wave frequency and rhythmicity in the gut can be explained with the relaxation oscillator theory (NELSON and BECKER 1968). The oscillator may be defined as any device (electronic circuits, computer models, biologic systems etc.) which has a continuous and rhythmic output. It generates waves at a steady frequency and with recurrent form. The oscillator has its own intrinsic frequency but when it is coupled with another oscillator its frequency changes. When two oscillators are coupled they acquire the same frequency, the one with the lower intrinsic frequency assuming the higher frequency of the driving oscillator. The oscillators are said to be phase-locked and the phenomenon is known as frequency entrainment (SARNA 1978).

Since each cell of the gastrointestinal smooth muscle exhibits oscillator properties, any region of the alimentary canal may display spontaneous electrical activity. It is generally accepted that slow waves in the stomach originate in the orad end of the corpus and spread aborally to the pylorus. The dominant site of origin of slow waves (gastric pacemaker) has been demonstrated in the greater curvature half of the orad corpus (KELLY and CODE 1971). The fibers responsible for their origin and spread have, however, not been identified. The cells of the orad corpus, in the greater curvature region, entrain those of the distal stomach as these have a slower intrinsic frequency and can be driven by the faster gastric pacemaker. Thus the frequency of slow waves is the same in all parts of the stomach (4–5 cycles/min in dogs, 3 cycles/min in humans).

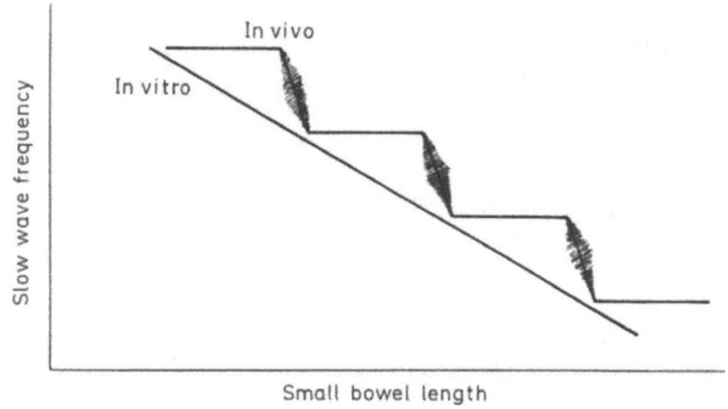

Fig. 6. Schematic representation of decrease of slow wave frequency along the intestine. Frequency decreases in stepwise fashion in vivo and linearly in vitro. Frequency plateaux are separated by areas of waxing and waning. (Adapted from DIAMANT and BORTOFF 1969 b)

No relationship exists between the frequency of the slow electrical rhythm in the stomach and duodenum (ALLEN et al. 1964). It has been shown, however, that antral slow waves spread by a myogenic transmission across the gastroduodenal junction into the duodenum where they periodically increase duodenal slow wave depolarizition and therefore the probability of duodenal spiking (BORTOFF and DAVIS 1968). The spread of antral slow waves may thus represent an integrative mechanism for the coordination of motility in the gastroduodenal junction during emptying of the stomach.

In the small intestine, the frequency of slow waves decreases caudally along the gut (ALVAREZ et al. 1922). Frequency is constant in the duodenum and the first few centimeters of the jejunum. Beyond this it decreases aborally with a characteristic pattern. In situ experiments in several animal species demonstrated a stepwise decrease in slow wave frequency along the intestine, with areas of waxing and waning between each frequency plateau, whereas in vitro experiments showed a linear decrease (Fig. 6) (DIAMANT and BORTOFF 1969 a). These results were explained by comparing the intestinal muscle to a series of loosely coupled oscillators with a decreasing intrinsic frequency in an aboral direction. In each frequency plateau, the oscillator with the highest intrinsic frequency drives the oscillators with lower intrinsic frequency. At the end of the plateau, corresponding to the waxing and waning zone, the frequency of the driving oscillator is too high for coupling maintenance of the next oscillator, so the distal oscillator becomes the driver of the next frequency plateau. The intrinsic frequency gradient determines the direction of slow wave propagation which proceeds from higher to lower frequency segments (DIAMANT and BORTOFF 1969 b). The shape and amplitude of the slow waves is more variable as the distance from the pylorus increases (Fig. 7), thus suggesting that oscillators are more loosely coupled (SZURSZEWSKI et al. 1970; CAPRILLI et al. 1975 a). Autonomic stimulation and blockade do not affect frequency, amplitude, or duration of slow waves.

Slow wave frequency in the duodenum is 17–22 cycles/min in dogs and 11.8 cycles/min in humans and in the ileum 9–13 cycles/min in dogs and 7–9 cycles/min in humans. Slow wave frequency in the colon has not yet been fully elucidated ow-

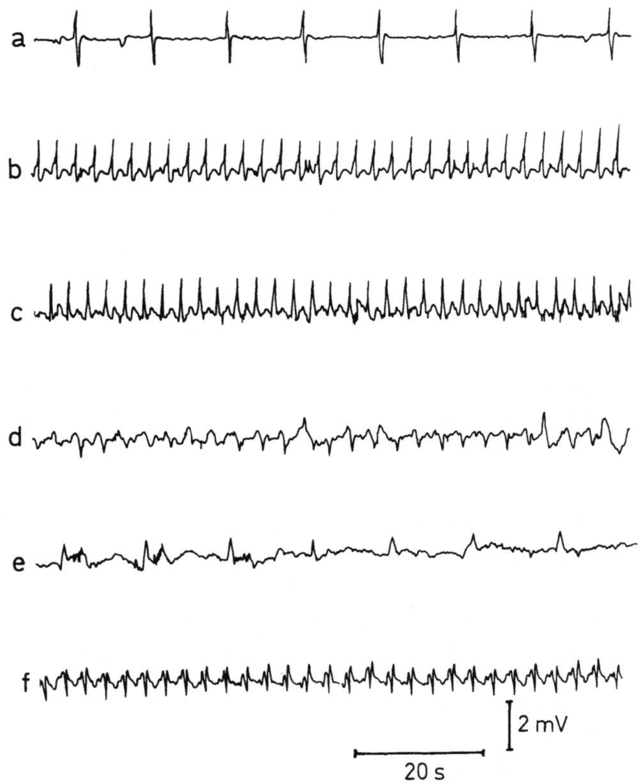

Fig. 7a–f. Electrical activity from six bipolar chronically implanted needle electrodes in an unanesthetized dog. Recordings from stomach (**a**), duodenum (**b**), upper jejunum (**c**), ileum (**d**), proximal colon (**e**), and distal colon (**f**). Different frequencies and configuration of slow waves in the various gastrointestinal segments are evident. In the stomach, slow waves show a triphasic configuration with a regular rhythm at 5 cycles/min. In the duodenum and jejunum, slow waves have similar characteristics at a frequency of 20 cycles/min. In the ileum, shape, amplitude, and rhythm of slow waves are variable, indicating that the recording is made from a region of uncoupled oscillators. The two colonic electrodes record two distinct rhythms, one at a frequency of about 5 cycles/min (proximal colon), the other at about 20 cycles/min (distal colon)

ing to conflicting results having been obtained from in vitro and in vivo recordings. Electrical activity in the in vitro experiments was found to be omnipresent and with only one frequency (CHRISTENSEN et al. 1969; CAPRILLI and ONORI 1972) (Figs. 3, 4), whereas in vivo slow waves have been detected only intermittently and at two or more frequencies, both in humans (TAYLOR et al. 1974a; SNAPE et al. 1976) and in dogs (CAPRILLI et al. 1975b) (Figs. 8, 9).

Most of the work on electrical activity of the colon in vitro has been done in the cat. Strips of proximal colon showed a frequency of 1–4 cycles/min, those of the distal colon a frequency of 5–7 cycles/min. Slow waves were shown to be propagated in the direction of the long axis of the circular muscle fibers at 1.6 cm/s, whereas propagation velocity in the long axis of the colon was 1–5 mm/s

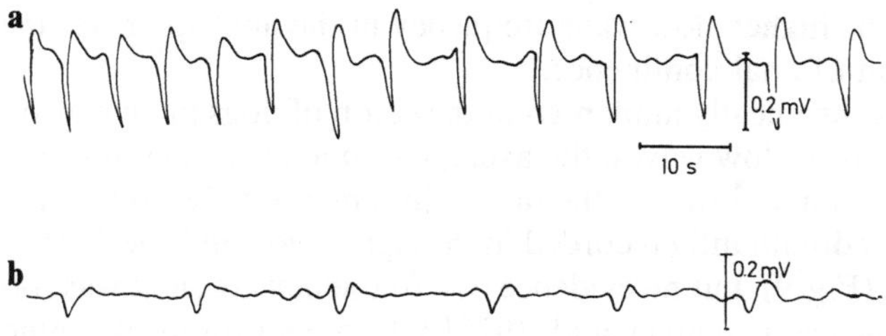

Fig. 8a, b. Electrical activity recorded from rectosigmoid region in humans. Two different rhythms at a frequency of 9 cycles/min (**a**) and 3 cycles/min (**b**) are detected at same site at different times

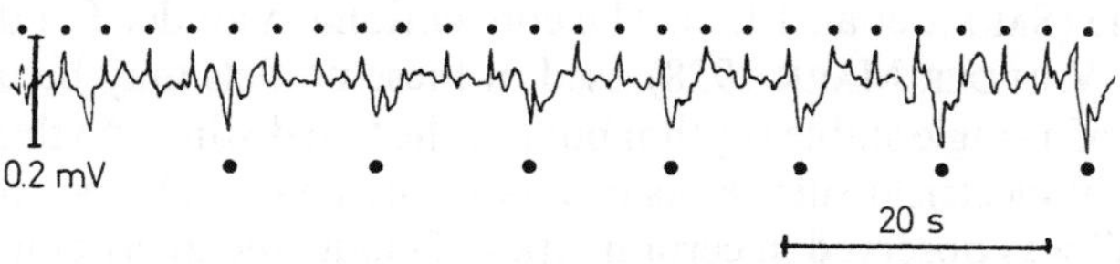

Fig. 9. Electrical activity recorded from dog colon in vivo. Recording shows simultaneous occurrence of a slower rhythm *(large circles)* with a frequency of 5.5 cycles/min and a faster rhythm *(small circles)* with a frequency of 19 cycles/min

(CHRISTENSEN et al. 1969). Coupling of the oscillator is tighter in the circumferential than in the longitudinal direction. In fact phase-locking occurred 95% of the time in the circumferential direction and only 67% in the longitudinal direction (CHRISTENSEN and HAUSER 1971 a, b). These results were recently confirmed in strips of dog colon (SHEARIN et al. 1979).

In experiments on whole cat colon in vitro the frequency of slow waves was found to rise from 4.5 cycles/min in the proximal colon to 6.0 cycles/min in the distal colon, with intermediate values at intermediate points (CHRISTENSEN et al. 1974). The frequency gradient in the colon would appear, therefore, to be the opposite to that in the small intestine but WIENBECK et al. (1972) failed to confirm these results in chronic experiments in vivo in the cat.

In humans, the frequency gradient of slow waves has not been extensively investigated as most studies have been performed with suction electrodes applied only to the mucosa of the rectosigmoid region. Difficulties arise in evaluating the frequency gradient in humans on account of the presence of more than one slow wave frequency at one recording site and periods of electrical silence in which slow waves are not recognizable. The possibility of recording two distinct electrical rhythms, a faster rhythm at 6–9 cycles/min and a slower rhythm at 3–5 cycles/min was first reported by TAYLOR et al. (1974 a, b) and later confirmed by others (SNAPE et al. 1976).

Electrodes implanted under the serosal coat along the colon of patients undergoing cholecystectomy revealed a similar frequency of the slower rhythm at all leves. The faster rhythm showed little change in frequency from the right to the left colon, but a definite gradient was found from the sigmoid into the rectum (TAYLOR et al. 1975). The gradient is reversed again from the rectum to the anal sphincter, where the frequency reaches values of 20 cycles/min (USTACH et al. 1970; MONGES et al. 1979). The higher slow wave frequency in the anal sphincter may be related to the function of anal continence.

Electrodes chronically implanted in the colon of dogs in vivo also revealed two distinct rhythms of slow waves, the average frequency of the slower rhythm being about 5 cycles/min and that of the faster rhythm about 18 cycles/min. The slower rhythm was predominantly recorded in the right colon and the faster in the middle and left colon (Fig. 9), but no evidence of a frequency gradient was found for other bands of frequency (CAPRILLI et al. 1975 b). In preparations of canine colon in vitro, however, only one frequency ranging from 3.5 to 6.5 cycles/min has been recorded (BOWES et al. 1978; SHEARIN et al. 1979; EL-SHARKAWY et al. 1979).

The relaxation oscillator theory, using chains and arrays of coupled nonlinear oscillators, has been widely used for modeling of slow waves in the stomach and small intestine (SARNA et al. 1972). The conventional Van der Pol dynamics (VAN DER POL and VAN DER MARK 1928) used in these models may be used to explain the existence of a single stable rhythm but not the two distinct rhythms interspersed with periods of electrical silence, as occurs in the colon. The occurrence of these "multimodes" was observed in certain types of chain and array connections, using a slight modification of the nonlinear conductance term in the Van der Pol equation and by coupling together oscillators with inductive or capacitive components (LINKENS and DARTARDINA 1978). These modified oscillators may generate two distinct stable frequencies, occurring separately or simultaneously and a pattern of zero activity. Another explanation for these findings has been suggested by BOWES et al. (1978) who, following experiments in dog colon in vivo, confirmed the visual existence of two frequencies of slow waves while computer analysis showed a single fundamental frequency (4–7 cycles/min). On the basis of these results and the widely demonstrated occurrence of only one rhythm in preparations in vitro, these authors suggested that a single frequency oscillator acts in the colon. Multiple frequencies and periods of "noise" activity may be due to "the simultaneous recording by each recording electrode of multiple oscillators having approximately the same frequency, with a variable degree of coupling between them." The main difference between in vivo and in vitro techniques consists in the respectively very small and large areas of contact between the electrode and the muscle cell. The small electrodes used in vitro are in contact with fewer oscillators and therefore record only one frequency rhythm, while large punctate electrodes used in chronic studies in vivo are in contact with multiple oscillators. If the oscillators are loosely coupled, several frequencies, or zero activity, or noise may be recorded.

II. Spikes

The spike or action potential may be defined as the response to depolarization during which the membrane may become polarized in the opposite direction to that of the RMP (HOLMAN 1968). Spikes are also referred to as fast activity, on account of their short duration, or electrical response activity, since they represent the response to the electrical control activity.

Spikes usually occur isolated or in bursts, superimposed on slow waves (Fig. 10). In contrast to the uniformity of action potentials in skeletal muscle, the spikes recorded in smooth muscle vary in shape, duration, and amplitude. Dura-

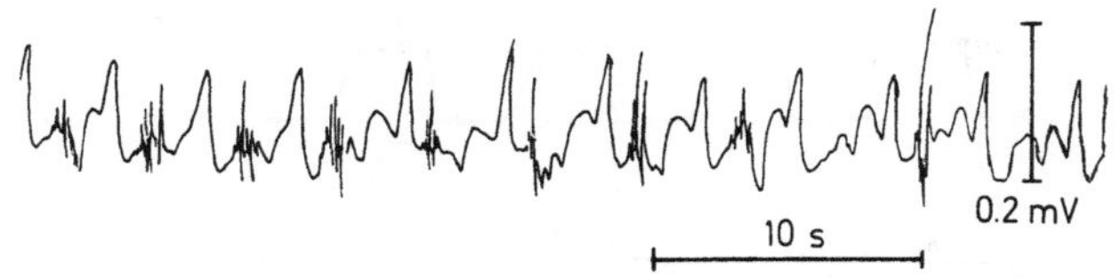

Fig. 10. Electrical activity recorded from the colon of an unanesthetized dog. A regular sequence of slow waves with bursts of spikes is superimposed on the second component

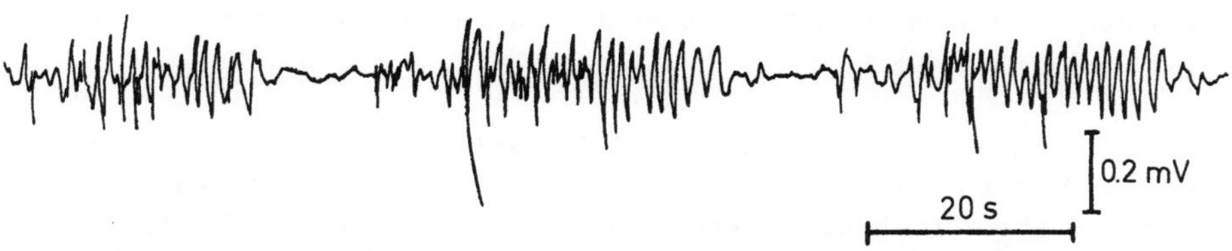

Fig. 11. Oscillatory potentials recorded in dog colon in vivo. They show a frequency of 45–55 cycles/min, a duration of 30–40 s, and are spindle-shaped. Single spikes are observed within the complex

tion ranges from 10 to 100 ms and amplitude from 5 to 70 mV (HOLMAN 1968; KOBAYASHI et al. 1967). Spikes may occur spontaneously or in response to electrical, chemical, pharmacologic, or mechanical stimulation. They propagate only over short distances and are associated with contraction, but do not necessarily represent a prerequisite for contraction. Acetylcholine and norepinephrine may initiate contractions in the absence of membrane potential changes.

Slow waves and spikes arise from two different membrane processes, but are strictly related. The slow waves may be considered as some form of continuous operating prepotential which acts to move the RMP to the spike firing level. When spikes occur they do not appear to change either the frequency or the amplitude of the slow waves. As far as the ionic dependence of spike potentials is concerned, a peak calcium influx has been shown to coincide with the occurrence of spikes (JOB 1969; BRADING et al. 1969). First, calcium controls the membrane permeability to other ions, especially sodium. Second, calcium ions carry the action current (BÜLBRING 1970). Since calcium is necessary for activating the contractile apparatus of the muscle cells, the coupling between spikes and contractions may be the direct consequence of calcium ion movement into the cell. Despite the importance of calcium, sodium also plays a role in spike generation. In calcium-free conditions, sodium-dependent spikes are seen to occur (GONELLA 1965).

III. Prepotentials

Prepotentials are brief potentials which precede the spikes. Spikes, however, can occur without prepotentials and prepotentials without spikes. Prepotentials occur particularly in the uterus and taenia coli. Their ionic nature is still to be elucidated.

IV. Oscillatory Potentials

Regular, sinusoidal oscillations at 30–40 cycles/min have been reported in cat colon (CHRISTENSEN et al. 1969). The oscillatory potentials do not influence the rhyth-

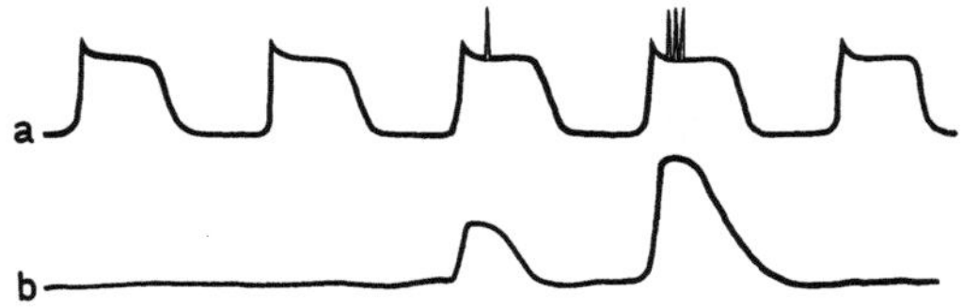

Fig. 12 a, b. Schematic representation of relationship between slow waves, spikes and contractions. Tracings of electrical activity recorded by intracellular microelectrodes showing slow waves with and without spikes (**a**) and muscle tension (**b**). Contraction occurs only when spikes are present. Bursts of spikes generate stronger contractions than single spikes

mic sequence of slow waves and are associated with sustained contractions. Spikes may precede or terminate a series of oscillatory potentials (WIENBECK et al. 1972). This electrical pattern has also been observed in dog colon (Fig. 11). Its meaning remains to be clarified.

D. Relationship Between Electrical and Mechanical Activity

The existence of a relationship between electrical events and smooth muscle contraction is generally accepted, but the mechanisms underlying excitation–contraction coupling have not yet been fully elucidated. Most investigations in this respect have been performed on muscle strips and concern primarily phasic contractions. The study of the relationship between electrical and mechanical activity is even more difficult in vivo, particularly in humans where elctrical activity is usually recorded from a small area of tissue by means of electrodes inserted into the muscle whereas mechanical activity is detected by intraluminal microballoons or open tips which record pressure from a large chamber.

Contraction in the smooth and skeletal muscles is related to depolarization of the cell membrane beyond a critical value. This critical value is invariably reached when spikes occur, whereas the almost omnipresent slow fluctuation of membrane potentials (slow wave) do not usually induce contraction (Fig. 12). In cat colon however, a regular 1:1 ratio between slow waves and contractions (Figs. 3, 4) has also been reported in the absence of spikes (CAPRILLI et al. 1970) and a correlation between slow waves and mechanical activity was confirmed by the simultaneous disappearance of both, in dinitrophenol, ouabain, sodium-free, and calcium-free solutions. Spikes, when present, corresponded to increased amplitude of contraction.

The main function of slow waves resides in increasing the excitability of intestinal muscle cells, enhancing the probability of spike discharge at predetermined time invervals, and therefore synchronizing spike activity. Spikes, when they do not appear in long bursts, are seen to coincide mainly with the peak depolarization of slow waves. The activation of myofilaments and the development of tension is attributed to calcium ions. An increase in intracellular calcium concentration is related to the appearance of spikes and is considered to be due both to its release by intracellular structures, not yet fully identified, and to a rapid calcium influx from outside the cell (JOB 1969; SYSON and HUDDART 1973). Since the absence of a definite mechanical threshold has been demonstrated in smooth muscle (SYSON and

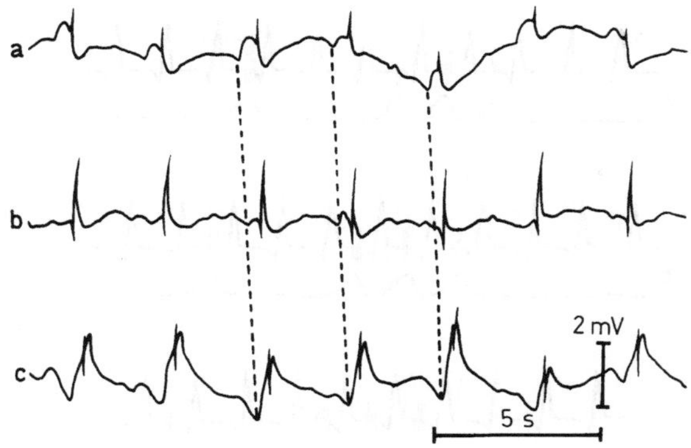

Fig. 13 a–c. Phase-locked slow waves recorded from three electrodes chronically implanted 2 cm apart in the long axis of dog duodenum. Tracings are ordered from the most orad electrode **a** to the most caudad electrode **c**. *Broken lines* show uniform aborad propagation of slow waves

HUDDART 1973) and the magnitude of muscle contraction is related to the number of spikes superimposed on each slow wave, the process of excitation–contraction coupling is probably not an all-or-nothing event but is related to the degree of depolarization of the cell. A variable calcium conductance related to the level of membrane depolarization has been proposed (SYSON 1974).

Slow waves in the stomach and small bowel have been shown to arise from longitudinal muscle and to spread electrotonically to circular muscle (BORTOFF 1965). Synchronous depolarization may therefore be recorded from the two layers, with both spikes and contractions occurring simultaneously (HUKUHARA and FUKADA 1965; BORTOFF and GHALIB 1972; MCKIRDY 1972). The small time lag described between longitudinal and circular muscle contraction (KOSTERLITZ et al. 1956) has not been confirmed in more recent experiments and should probably be interpreted as an artifact deriving from the Trendelenburg recording technique (HUKUHARA and FUKADA 1965; BORTOFF and GHALIB 1972). Also the assumption that the two layers are reciprocally innervated (KOTTEGODA 1969), which was used to explain the hypothesized slight temporal delay between the contraction of the two muscle layers, was not confirmed.

The occurrence of simultaneous contractions of both the circular and longitudinal muscle layers was used to interpret the segmenting contractions. Segmentation may be defined as a localized circumferential contraction of both layers, involving predominantly the circular muscle, and is a consequence of circumferentially synchronized slow waves (BORTOFF 1976). The electrotonic spread of slow waves from longitudinal to circular muscle in fact creates an almost simultaneous condition for spike firing all around the bowel. Since spikes are propagated for only very small distances, the contraction induced by this mechanism is self-limiting and usually involves an intestinal segment of no more than 1–2 cm.

When spike-generating slow waves propagate along the gut, a series of contractions are propagated in turn, resulting in content displacement. The occurrence of phase-locked slow waves (Fig. 13) is therefore necessary for coordinated motor activity. In fact, the uncoupling of slow waves leads to electrical "disorganization"

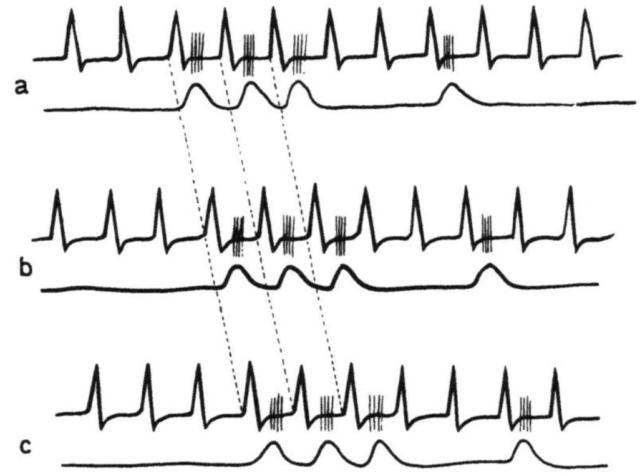

Fig. 14a–c. Schematic representation of electrical and mechanical activity recorded from three points equally spaced in the small bowel. Diagrams are ordered from the most orad electrode **a** to the most caudad electrode **c**. *Upper tracings* show electrical activity; *Lower tracings*, mechanical activity. Only spike-generating slow waves are accompanied by contractions. *Broken lines* show aborad propagation of slow waves and contractions

and prevents coordinated spike activity over large areas, with consequent failure of efficient muscle contraction and disappearance of intraluminal pressure (CAPRILLI et al. 1975a).

Peristalsis may be defined as coordinated motor activity of intestinal muscle layers, induced by intraluminal stimulation and self-propagating aborally. From a myoelectric viewpoint it may be considered the result of an aborad propagation of spike-generating slow waves (Fig. 14). No consistent difference seems to exist, therefore, between the electrical events which determine segmentation and peristalsis. It is only the distance over which spike-generating slow waves propagate that determines whether a peristaltic wave is likely to occur. It has in fact been demonstrated that contractions propagate at the same velocity as slow waves (BORTOFF and SACCO 1974). In experimental conditions, distension of the bowel produces contraction of the intestinal muscle on the oral side and relaxation on the anal side. During peristalsis in isolated rabbit colon, a marked spike activity was seen to correspond to mechanical activation above the bolus, whereas electrical silence corresponded to mechanical inhibition below the site of stimulation (FRIGO et al. 1972). The spike activity was referred to the "ascending excitatory" reflex, the electrical silence to the "descending inhibitory" reflex. Both reflexes are nerve mediated, the ascending pathway involving neurons releasing a 5-HT-like transmitter and cholinergic neurons, and the descending pathway involving noncholinergic inhibitory neurons (COSTA and FURNESS 1976). Examples of electrical and mechanical activities during peristaltic reflex from rabbit small intestine and colon are shown in Figs. 15 and 16.

In cat duodenum (WEISBRODT and CHRISTENSEN 1972) and small bowel (WEISBRODT and BURKS 1974) an orad spread of long bursts of action potentials was observed concomitantly with morphine-induced vomiting, possibly representing the electrical counterpart of so-called antiperistalsis.

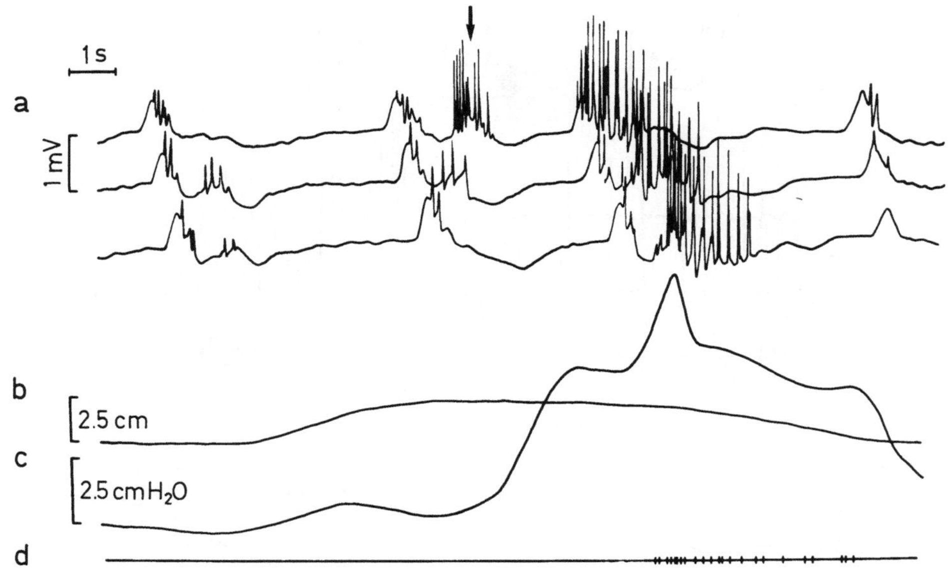

Fig. 15 a–d. Peristaltic reflex in rabbit ileum elicited by perfusing the intestinal lumen. Extracellular electrical activity recorded from three pressure electrodes on the serosal surface 2 cm apart along the longitudinal axis **a**; longitudinal movements recorded by an isotonic transducer **b**; intraluminal pressure **c**; ejection of fluid from the distal end of the intestine **d**. *Arrow* indicates segmental nonpropulsive activity preceeding the peristaltic firing. The propagation of peristalsis is preceded by increase in intraluminal pressure and contraction of longitudinal muscle end results in ejection of fluid. (Unpublished data, courtesy of G. M. FRIGO, M. TONINI, S. LECCHINI and A. CREMA, Institute of Medical Pharmacology, University of Pavia)

Slow waves are usually propagated strictly in an aborad direction. A reverse conduction has only rarely been observed (HIATT et al. 1971) and it is doubtful whether it should be considered a physiologic event. Reverse propagation of slow waves has been recorded in the stomach after vagotomy (KELLY and CODE 1969) and in the small bowel when the gut was mishandled or excessively probed with electrodes (BASS et al. 1961). This finding is probably more common in the proximal colon and it has been hypothesized that it might influence the polarity of movements and the average flow direction of colonic contents (CHRISTENSEN et al. 1974).

It is not known why slow waves and therefore peristalsis tend to propagate in the aborad direction. The frequency gradients in the stomach and small intestine are considered to be of prime importance. Since any frequency plateau invariably displays a higher frequency than the more caudad plateaux, and within each frequency plateau the driving oscillator is located mainly in a cranial position, the electrical entraining of slow waves and contraction propagation will most likely occur in an orocaudad direction. It is difficult to explain intestinal polarity merely on a myogenic basic since removal of mucosal and submucosal layers (FRIGO and LECCHINI 1970) as well as local anesthesia (HARDCASTLE and MANN 1968) abolish the peristaltic reflex elicited by an intraluminal stimulus. These and other findings such as the inhibitory effect of tetrodotoxin and ganglion-blocking agents on peristalsis suggest that autonomic innervation plays an important role in peristaltic reflex.

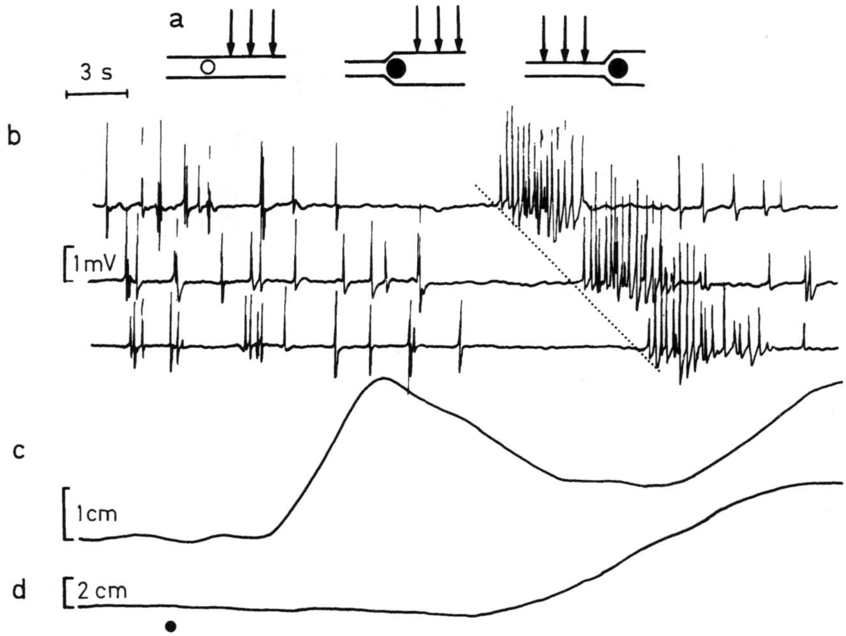

Fig. 16 a–d. Peristaltic reflex in the rabbit colon elicited by localized distension produced by filling an intraluminal balloon and recorded by means of three pressure electrodes on the serosal surface 2 cm apart along the longitudinal axis **a**; Tracings show: extracellular electrical activity **b**; longitudinal movements recorded by an isotonic transducer **c**; displacement of the intraluminal balloon **d**. Radial distension of the intestinal wall produces a contraction above and a relaxation below the site of stimulation. The propagation of peristaltic excitation is preceded by a wave of inhibition spreading in the aborad direction. Propagation is also preceded by a contraction of longitudinal muscle. The *solid circles* indicate distension of the balloon. (Unpublished data, courtesy of G. M. FRIGO, M. TONINI, S. LECCHINI and A. CREMA, Institute of Medical Pharmacology, University of Pavia)

The intestinal smooth muscle appears, therefore, to be responsible for rhythmicity and polarity in the alimentary canal. Both nerves and hormones interact, however, with the intrinsic properties of smooth muscle modulating and coordinating motor activity. The neural and hormonal factors affecting gastrointestinal motility are discussed in Chaps. 9, 10, and 11.

E. Migrating Myoelectric Complex

A front of large amplitude action potentials, migrating from the stomach or duodenum to the distal ileum, was first described in fasted dogs by SZURSZEWSKI (1969) and called the migrating myoelectric complex (MMC).

Four phases can be recognized in the MMC: phase I, in which only slow waves are present; phase II, characterized by spikes randomly superimposed on slow waves; phase III, in which spikes of larger amplitude are superimposed on every slow wave; phase IV, in which action potential activity subsides (CARLSON et al. 1972). A period without action potentials follows, corresponding to phase I of the

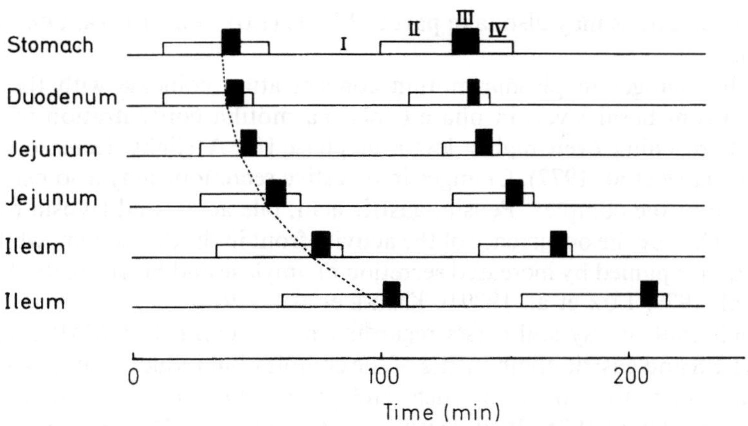

Fig. 17. Schematic representation of the migrating myoelectric complex. I, II, III, and IV indicate the four phases of the complex (see text). (Adapted from CODE and MARLETT 1975)

next complex. These four distinct phases can be detected, nearly unchanged, at each recording point during aborad propagation of the complex (Fig. 17). In the dog, the duration of each complete cycle is about 100 min while the period of action potentials (phase II, III, and IV) is more variable (10–50 min). The mean total time necessary for the electric complex to traverse the entire small bowel is about 110 min. Its propagation velocity is faster in the upper jejunum and increasingly slower as it migrates along the ileum (6.2–1.2 cm/min) (SZURSZEWSKI 1969).

The mechanical counterpart of the activity front is represented by a series of propagated contractions with a propulsive effect on intestinal contents. The function of the MMC has been compared to "one of a housekeeper which, between meals, periodically sweeps the stomach and the small intestine" (CODE and SCHLEGEL 1974).

The MMC in humans was first described as a cyclically recurring burst of rhythmic contractions moving caudad along the small intestine (VANTRAPPEN et al. 1977), and was considered to be the mechanical event of the activity front of the complex. A myoelectric complex was subsequently recognized by means of intraluminal (LUX et al. 1978; FLECKENSTEIN et al. 1978) and surgically implanted electrodes (STODDARD et al. 1978). The duration of each complete cycle was found to be about 40 min, the activity front ranging from 5 to 8 min. In humans, as in dogs, the complex migrates along the small bowel in a caudad direction and has a propulsive effect on the contents. Its propagation velocity decreases as the complex moves from the duodenum (12 cm/min) to the distal small bowel (6 cm/min) (VANTRAPPEN et al. 1977).

In the early studies, MMC was describes as starting from the duodenum (SZURSZEWSKI 1969) or from the gastric antrum (MARLETT and CODE 1971). Recently however, cyclic pressure fluctuations of the gastric fundus, wide variations of pressure of the lower esophageal sphincter (LES), and occasional motor activity of the esophagus have been reported to correlate with the development of MMC from the gastric antrum. These findings indicate that the esophageal body, LES, and gastric fundus may also take part in MMC (LUX et al. 1979a; DIAMANT et al. 1979a).

Cyclic changes in plasma motilin concentration coincide with the onset of MMC. From basal levels in phase I, plasma motilin concentration increases in phase II, reaching even higher levels in phase III. A slight decrease is seen in phase IV (LEE et al. 1977). Changes in digestive secretions may also occur during migration of the complex. Pepsin, gastric acid, bile acids, and trypsin output are enhanced before the occurrence of the activity front in the duodenum, while its passage is accompanied by increased secretion of amylase and bicarbonate (VANTRAPPEN et al. 1979; LUX et al. 1979b; KEANE et al. 1979).

Much controversy still exists regarding neural control of MMC. Vagotomy blocks LES and gastric components of the complex but induces only slight changes in its duration, time of occurrence, and propagation along the small intestine (MARIK and CODE 1975; RUCKEBUSH and BUENO 1977; DIAMANT et al. 1979b). Atropine, however, blocks the migrating complex, prevents the onset of a new MMC, and inhibits the cyclic fluctuation of plasma motilin concentration (LEE et al. 1977). α-Blocking agents also affect MMC, suppressing its coordinate activity and replacing it by irregular spike activity (EL-SHARKAWY et al. 1978).

In humans and dogs, but not in sheeps and rats, feeding promptly intenupts MMC (MARLETT and CODE 1971; GRIVEL and RUCKEBUSH 1972). The prandial electric pattern consists of a nonpropagated irregular spike activity uniformly distributed along the entire small bowel. A prolonged and continuous elevation of LES pressure has recently been reported in the prandial pattern (DIAMANT et al. 1979a). The duration of MMC prandial disruption depends upon the amount of food and its chemical composition. A small amount of oil has been shown to induce a longer disruption than sucrose and proteins (EECKHOUT et al. 1978). Total parenteral nutrition does not affect MMC, indicating that the presence of nutrients in the blood is not responsible for MMC disruption (WEISBRODT et al. 1976). An electric pattern closely resembling the feeding pattern is observed after infusion of pentagastrin in both intact and vagotomized dogs (MARIK and CODE 1975). In humans, doses of pentagastrin sufficient to induce more than half the maximal acid output did not affect MMC. In addition, normal complexes were recorded in patients with pernicious anemia and Zollinger–Ellison syndrome. At pharmacologic doses, however, pentagastrin invariably stopped MMC (HELLEMANS et al. 1978).

In vagotomized dogs, feeding fails to disrupt MMC (MARIK and CODE 1975; REVERDIN et al. 1979; DIAMANT et al. 1979b). Sometimes, however, a delay in the onset of the feeding pattern, a significant reduction in its duration, and a mixture of fasting and fed pattern have been reported (RUCKEBUSH and BUENO 1977; REVERDIN et al. 1979). These observations emphasize the importance of vagal control mechanisms in the feeding MMC response.

MMC may be experimentally induced by parenteral infusion of 13-norleucine-motilin (WINGATE et al. 1976; LUX et al. 1978) and by acidification of the duodenum (COLLINS et al. 1978; LEWIS et al. 1978).

The motilin-induced complex initiates from the stomach and propagates along the small intestine like the naturally occurring MMC. The acid-induced complex, invariably starts from the duodenum and is never preceded by the LES and gastric component of MMC, even in the presence of high plasma motilin concentration. Also the naturally occurring MMC never initiates from the stomach when duodenal pH is low. Emptying of acid from the stomach may then be one of the mechanisms regulating MMC propagation from the stomach to duodenum.

The absence of MMC has been reported in association with bacterial overgrowth in the human small intestine (VANTRAPPEN et al. 1977). It has been suggested that the lack of the cyclically sweeping function leads to accumulation of residual digestion products, desquamated cells, and secretions, thus favoring bacterial overgrowth. Infusion of motilin, even at high doses, failed to initiate MMC, in these patients (VANTRAPPEN et al. 1978).

F. Electromyography in Clinical Practice

I. Vagotomy

In patients submitted to vagotomy for duodenal ulcer electrical activity in the antrum is disorganized for up to 3 weeks, but the shape of slow waves remains changed for years. The disturbances in slow wave activity depend upon the extent of vagus denervation, an increase being observed from highly selective vagotomy to selective vagotomy and even more to truncal vagotomy (STODDARD et al. 1973). Experiments in animals indicate that vagotomy does not significantly alter interdigestive MMC, but does abolish the feeding response (MARIK and CODE 1975; RUCKEBUSH and BUENO 1977). The vagus does not appear, therefore, to be essential for the organization and propagation of MMC, but plays an important role in prandial response.

II. Idiopathic Intestinal Pseudoobstruction

Conflicting results have been obtained with the use of electromyography (EMG) in the diagnosis of idiopathic intestinal pseudoobstruction (IIP), a condition characterized by symptoms of obstruction with no demonstrable organic occlusion of the intestinal lumen. The intestinal slow wave activity and the spike motor response to exogenous neurohormonal stimulation was found to be normal in the duodenum and colon of four patients with IIP (SULLIVAN et al. 1977). Using electrodes implanted at laparatomy in the jejunum of a patient with IIP, SARNA et al. (1978) observed a complete absence of slow waves, except during the passage of MMC or after cholinergic stimulation. A possible explanation for these findings may be that slow waves are generated only when cells are depolarized by stimulant substances. Whether EMG will be of value in the pathophysiology and diagnosis of IIP remains to be established.

III. Diverticular Disease of the Colon

A rapid slow wave rhythm (frequency 12–18 cycles/min) has been reported in the rectosigmoid region of patients with symptomatic diverticular disease (TAYLOR et al. 1974a). The abnormally rapid electric rhythm present in 80% of the patients

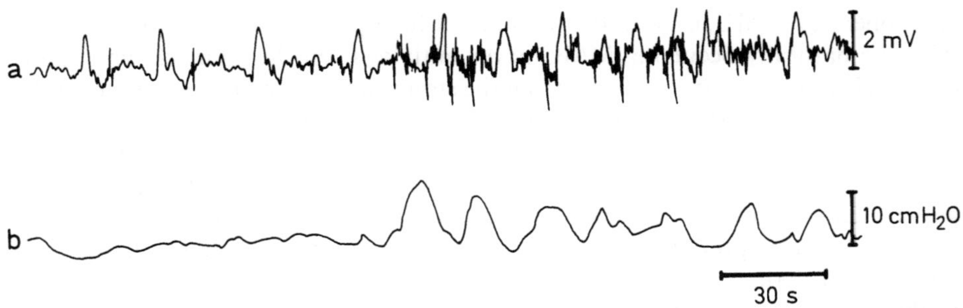

Fig. 18a, b. Electrical (**a**) and mechanical (**b**) activity recorded in an aganglionic segment of rectum in a patient with Hirschsprung's disease. Intense spike activity is accompanied by enhanced motor activity. A regular sequence of waves at 2.5 cycles/min is also present

was restored to normal frequency in 50% of cases after 1 month treatment with bran. These finding were accompanied by a reduction of high basal intracolonic pressure to within normal limits (TAYLOR and DUTHIE 1976).

IV. Irritable Colon Syndrome

A specific abnormality in myoelectric activity of the colon consisting of an increased incidence of 3 cycles/min slow wave activity (slower rhythm) has recently been reported in irritable colon syndrome (ICS) (SNAPE et al. 1976, 1977). These observations seem to be in keeping with the finding in ICS of nonpropulsive contractions at a frequency of 2–3 cycles/min. The myoelectric disorder did not appear to be related to altered bowel habit (TAYLOR et al. 1978).

Basal spike and motor activities in patients with ICS and in normal subjects were similar, but were abnormally prolonged after eating in ICS (SULLIVAN et al. 1978). Anticholinergic drugs reduced the duration and magnitude of the abnormal postprandial colonic response. The increased percentage of slower rhythm was not affected, however, by eating or by anticholinergic drugs.

V. Hirschsprung's Disease

Motility studies on the aganglionic segment of patients with Hirschsprung's disease (HD) showed increased motor activity in basal conditions and a heightened response to cholinergic stimulation (SUSTER et al. 1970), suggesting that in this segment a spastic contraction without propulsion occurs preventing the fecal column from moving down. These findings are in agreement with histochemical evidence of increased acetylcholinesterase content and the presence of large, nonmyelinated fibers in the aganglionic segment.

Increased electrical activity in the narrow tract has also been found in the colonic muscle strips from patients with HD (LYNEN 1974). In agreement with these results, increased electrical activity, mainly represented by discharge of large action potentials, may be recorded in vivo from the aganglionic segment in patients with HD (Fig. 18). The absence of spikes in the aganglionic segment of children with HD has, however, also been reported (MARIN et al. 1976).

References

Allen GL, Poole EW, Code CF (1964) Relationship between electrical activities of antrum and duodenum. Am J Physiol 207:906–910
Alvarez WC, Mahoney LJ (1922) Action currents in stomach and intestine. Am J Physiol 58:476–493
Bass P, Code CF, Lambert EH (1961) Motor and electric activity of the duodenum. Am J Physiol 201:287–291
Bortoff A (1961) Slow potential variation of small intestine. Am J Physiol 201:203–208
Bortoff A (1965) Electrical transmission of slow waves from longitudinal to circular intestinal muscle. Am J Physiol 209:1524–1560
Bortoff A (1976) Myogenic control of intestinal motility. Physiol Rev 56:419–434
Bortoff A, Davis RS (1968) Myogenic transmission of antral slow waves across the gastroduodenal junction in situ. Am J Physiol 215:889–897
Bortoff A, Ghalib E (1972) Temporal relationship between electrical and mechanical activity of longitudinal and circular muscle during intestinal peristalsis. Am J Dig Dis 17:317–325
Bortoff A, Sacco J (1974) Myogenic control of intestinal peristalsis. In: Daniel EE (ed) Proceedings of the 4th International Symposium on Gastrointestinal Motility. Mitchell, Vancouver, pp 53–60
Bortoff A, Sachs F (1970) Electrotonic spread of slow waves in circular muscle of small intestine. Am J Physiol 218:576–581
Bowes KL, Shearin NL, Kingma YJ, Koles ZJ (1978) Frequency analysis of electrical activity in dog colon. In: Duthie HL (ed) Gastrointestinal motility in health and disease. MTP Press, Lancaster, pp 251–269
Brading AF, Bülbring E, Tomita T (1969) The effect of sodium and calcium on the action potentials of smooth muscle of the guinea-pig taenia coli. J Physiol (Lond) 200:637–654
Bülbring E (1970) The role of electrophysiology in the investigation of factors controlling intestinal motility. Rend Gastroenterol 2:197–207
Bülbring E, Hooton IN (1954) Membrane potentials of smooth muscle fibres in the rabbit's sphincter pupillae. J Physiol (Lond) 125:292–301
Burnstock G, Straub RW (1958) A method for studying the effect of ions and drugs on the resting membrane potentials in a smooth muscle with external electrodes. J Physiol (Lond) 104:156–167
Caprilli R, Onori L (1972) Origin, transmission and ionic dependence of colonic electrical slow waves. Scand J Gastroenterol 7:65–74
Caprilli R, Onori L, Tonini M, Zapponi G (1970) Slow waves and mechanical activity of cat colon circular muscle. Rend Gastroenterol 2:65–66
Caprilli R, Onori L, Torsoli A (1971) Etudes électro-physiologiques sur la musculature du colon chez l'homme et chez le chat. Biol Gastroenterol 2:175–178
Caprilli R, Melchiorri P, Improta G, Vernia P, Frieri G (1975a) Effects of bombesin and bombesin-like peptides on gastrointestinal myoelectric activity. Gastroenterology 68:1228–1235
Caprilli R, Vernia P, Frieri G, Melchiorri P (1975b) Two electrical rhythms in the colon. Rend Gastroenterol 7:65–66
Carlson GM, Bedi BS, Code CF (1972) Mechanism of propagation of intestinal interdigestive myoelectric complex. Am J Physiol 222:1027–1030
Christensen J, Hauser RL (1971a) Circumferential coupling of electrical slow waves in circular muscle of cat colon. Am J Physiol 221:1033–1037
Christensen J, Hauser RL (1971b) Longitudinal axial coupling of slow waves in proximal cat colon. Am J Physiol 221:246–250
Christensen J, Schedl HP, Clifton JA (1964) The basic electrical rhythm of the duodenum in normal human subjects and in patients with thyroid disease. J Clin Invest 43:1659–1667
Christensen J, Caprilli R, Lund GF (1969) Electrical slow waves in circular muscle in cat colon. Am J Physiol 222:771–776
Christensen J, Anuras S, Hauser RL (1974) Migrating spike bursts and electrical slow waves in the cat colong: effect of sectioning. Gastroenterology 66:240–247

Code CF, Schlegel JF (1974) The gastrointestinal interdigestive housekeeper: motor correlates of the interdigestive myoelectric complex of the dog. In: Daniel EE (ed) Proceedings of the 4th International Symposium on Gastrointestinal Motility. Mitchel, Vancouver, pp 631–634
Collins SM, Lewis TD, Track N, Fox J, Daniel EE (1978) Release of motilin. Gastroenterology 74:1020
Connor JA, Prosser CL (1974) A studdy of pacemaker activity of intestinal smooth muscle. J Physiol (Lond) 240:671–701
Costa M, Furness B (1976) The peristaltic reflex: an analysis of the nerve pathways and their pharmacology. Naunyn-Schmiedberg Arch Pharmacol 294:47–60
Diamant NE, Bortoff A (1969 a) Effects of transection on the intestinal slow wave frequency gradient. Am J Physiol 216:734–743
Diamant NE, Bortoff A (1969 b) Nature of intestinal slow wave frequency gradient. Am J Physiol 216:301–307
Diamant NE, Hall K, Mui H, El-Sharkawy TY (1979 a) Vagal control of the feeding motor pattern in the lower esofageal sphincter, stomach and upper intestine of dog. In: Proceedings of the 7th International Symposium on Gastrointestinal Motility, Iowa City
Diamant NE, Mui H, El-Sharkawy TY, Hall K (1979 b) Dog lower esophageal sphincter pressure changes during fasting and after feeding. Gastroenterology 76:1121
Dorph S, Øigaard A, Kragsholm M (1972) A new disposable probe for recording intraluminal pressure and electric potentials in the human gastrointestinal tract. Gut 13:732–734
Eeckhout C De Wever I, Hellemans J, Vantrappen G (1978) The effect of different test meals on the interdigestive myoelectric complex (MMC) in dogs. In: Duthie HL (ed) Gastrointestinal motility in health and disease. MTP Press, Lancaster, pp 43–46
El-Sharkawy TY, Markus H, Diamant NE (1978) Integrity of the autonomic nervous system is required for the interdigestive myoelectric complex (IDMC). Gastroenterology 74:1121
El-Sharkawy TY, MacDonald WM, Diamant NE (1979) Characteristic of the slow wave activity of the canine colon. In: Proceedings of the 7th International Symposium on Gastrointestinal Motility, Iowa City
Fleckenstein P, Krogh F, Øigaard A (1978) The interdigestive myoelectric complex and other migrating electric phenomena in the human small intestine. In: Duthie HL (ed) Gastrointestinal motility in health and disease. MTP Press, Lancaster
Frigo GM, Lecchini S (1970) An improved method for studying the peristaltic reflex in the isolated colon. Br J Pharmacol 39:346–356
Frigo GM, Torsoli A, Lecchini S, Falaschi CF, Crema A (1972) Recent advances in the pharmacology of peristalsis. Arch Int Pharmacodyn Ther 196:9–24
Gonella J (1965) Variation de l'activité électrique spontanée du duodenum de lapin avec le lieu de dérivation. CR Acad Sci [D] (Paris) 260:5362–5365
Grivel ML, Ruckebusch Y (1972) The propagation of segmental contractions along the small intestine. J Physiol (Lond) 227:611–625
Hardcastle JD, Mann CV (1968) Study of the large bowel peristalsis. Gut 9:512–520
Hellemans J, Vantrappen G, Janssens J, Peters T (1978) Effect of feeding and of gastrin on the interdigestive myoelectric complex in man. In: Duthie HL (ed) Gastrointestinal motility in health and disease. MTP Press, Lancaster, pp 23–31
Hiatt RB, Goodman I, Overweg NIA (1971) Intestinal motility. 1. Control mechanisms in the basic electric rhythm of canine small bowel. J Surg Res 11:454–463
Holman ME (1968) Introduction to electrophysiology of visceral smooth muscle. In: Code CF (ed) Alimentary canal. American Physiological Society, Washington, DC (Handbook of physiology, vol IV, sect 6, pp 1665–1708)
Hukuhara T, Fukada H (1965) The motility of the isolated guinea pig small intestine. Jpn J Physiol 15:125–139
Job DD (1969) Ionic basis of intestinal electrical activity. Am J Physiol 217:1534–1541
Keane FB, Di Magno EP, Dozois RR, Go VLW (1979) Fasting canine pancreatic endocrine and exocrine function related to duodenal motor activity. In: Proceedings of 7th International Symposium on Gastrointestinal Motility, Iowa City

Kelly KA, Code DF (1969) Effect of transthoracic vagotomy on canine gastro-electric activity. Gastroenterology 57:51–58
Kelly KA, Code CF (1971) Canine gastric pacemaker. Am J Physiol 220:112–118
Kirk D, Duthie HL (1978) In vitro studies of the electrical activity of the longitudinal and circular muscle layers of the human colon. In: Duthie HL (ed) Gastrointestinal motility in health and disease. MTP press, Lancaster, pp 327–333
Kobayashi M, Nagai T, Prosser CL (1966) Electrical interaction between muscle layers of cat intestine. Am J Physiol 211:1281–1291
Kobayashi M, Prosser CL, Nagai T (1967) The electric properties of intestinal muscle as measured intracellularly and extracellularly. Am J Physiol 213:275–286
Kosterlitz HW, Pirie VW, Robinson JA (1956) The mechanism of the peristaltic reflex in the isolated guinea pig ileum. J Physiol (Lond) 133:681–689
Kottegoda SR (1969) An analysis of possible nervous mechanisms involved in the peristaltic reflex. J Physiol (Lond) 200:687–712
Lee KY, Chey WY, Tai HH, Wagner D, Yajima H (1977) Cyclic changes in plasma motilin levels and interdigestive myoelectric activity of canine antrum and duodenum. Gastroenterology 72:1162
Lewis TD, Collins SM, Fox JE, Daniel EE (1978) Initiation of migrating myoelectric complex (MMC) by intraduodenal acid. Gastroenterology 74:1055
Linkens DA, Dartardina SP (1978) Human colonic modelling and multiple solutions in non linear oscillators. In: Duthie HL (ed) Gastrointestinal motility in health and disease. MTP Press, Lancaster, pp 669–675
Liu J, Prosser CL, Job DD (1969) Ionic dependence of slow waves and spikes in intestinal muscle. Am J Physiol 217:1542–1547
Lux G, Strunz V, Domschke S, Femppel J, Rosch W, Domschke W (1978) 13-NLE motilin and interdigestive motor and electrical activity of human small intestine. Gastroenterology 74:1058
Lux G, Lederer P, Femppel J, Rosch W, Domschke W (1979a) Spontaneous and 13-NLE motilin induced interdigestive motor activity of esofagus, stomach and small intestine in man. In: Proceedings of the 7th International Symposium on Gastrointestinal Motility, Iowa City
Lux G, Lederer P, Femppel J, Rosch W, Domschke W (1979b) Motor and secretory activity of the duodenal interdigestive complex: an integrated function. In: Proceedings of the 7th International Symposium on Gastrointestinal Motility, Iowa City
Lynen FK (1974) Motor function of the large bowel in megacolon congenitum. In: Daniel EE (ed) Proceeding of the 4th International Symposium on Gastrointestinal Motility. Mitchell, Vancouver, pp 189–196
Marik F, Code CF (1975) Control of the interdigestive myoelectric activity in dogs by the vagus nerves and pentagastrin. Gastroenterology 69:387–395
Marin AM, Rivarola A, Garcia H (1976) Electromyography of the rectum and colon in Hirschsprung's disease. J Pediatr Surg 11:547–552
Marlett JA, Code CF (1971) The interdigestive gastrointestinal electric complex. Fed Proc 30:609
McCoy EJ, Bass P (1963) Chronic electrical activity of gastroduodenal area: effects of food and certain catecolamines. Am J Physiol 205:439–445
McKirdy HC (1972) Functional relationship of longitudinal and circular layers of muscularis externa of the rabbit large intestine. J Physiol (Lond) 227:839–853
Monges H, Salducci J, Roman C (1969) Etude électromyographique de la motricité gastrique chez l'homme normal. Arch Fr Mal Appar Dig 58:517–530
Monges H, Salducci J, Naudy B, Ranieri F, Gonella J, Bouvier M (1979) The electrical activity of the internal anal sphincter: a comparative study in man and cat. In: Proceedings of the 7th International Symposium on Gastrointestinal Motility, Iowa City
Nelsen TS, Becker TC (1968) Simulation of electrical and mechanical gradient of the small intestine. Am J Physiol 214:749–757
Papasova MP, Nagai T, Prosser CL (1968) Two component slow waves in smooth muscle of cat stomach. Am J Physiol 214:695–702

Prosser CL, Bortoff A (1968) Electrical activity of intestinal muscle under in vitro conditions. In Code CF (ed) Altimentary canal. American Physiological Society, Washington, DC (Handbook of physiology, vol IV, sect 6, pp 2025–2050)
Provenzale L, Pisano M (1971) Methods for recording electrical activity of the human colon in vivo. Am J Dig Dis 16:712–722
Reverdin N, Hutton M, Ling A et al. (1979) Vagotomy and the motor response to feeding. In: Proceedings of the 7th International Symposium on Gastrointestinal Motility, Iowa City
Ruckebush Y, Bueno L (1977) Migrating myoelectrical complex of the small intestine. Gastroenterology 73:1309–1314
Sarna SK, (1978) Relaxation oscillators. In: Duthie HL (ed) Gastrointestinal motility in health and disease. MTP Press, Lancaster, pp 659–668
Sarna SK, Daniel EE (1973) Electrical stimulation of gastric electrical control activity. Am J Physiol 225:125–131
Sarna SK, Daniel EE, , Kingma YJ (1972) Simulation of the electric control activity of the stomach by an array of relaxation oscillators. Dig Dis 17:299–310
Sarna SK, Daniel EE, Waterfall WE, Lewis TD, Marzio L (1978) Postoperative gastrointestinal electrical and mechanical activities in a patient with idiopathic intestinal pseudoobstruction. Gastroenterologoy 74:112–120
Sarna SK, Bardakjian BL, Waterfall WE, Lind JF, Daniel EE (1979) Human colonic electrical activity. In: Proceedings of the 7th International Symposium on Gastrointestinal Motility, Iowa City; pp 101–102
Shearin NL, Bowes KL, Kingma YJ (1979) In vitro electrical activity in canine colon. Gut 20:780–786
Shuba MF (1961) Physical electrotonus in smooth muscle. Biophysics 6:34–38
Smout AJPM, Schel EJ, Grashnis JL (1979) Postprandial and interdigestive gastric electrical activity in the dog recorded by means of cutaneous electrodes. In: Proceedings of the 7th International Symposium on Gastrointestinal Motility, Iowa City
Snape WJ, Carlson GM, Cohen S (1976) Colonic myoelectric activity in the irritable bowel syndrome. Gastroenterology 70:326–330
Snape WJ, Carlson GM, Matarazzo SA, Cohen S (1977) Evidence that abnormal myoelectric activity produces colonic motor disfunction in the irritable bowel syndrome. Gastroenterology 72:383–387
Stämpfli R (1954) A new method for measuring membrane potential with external electrodes. Experientia 10:508–509
Stoddard CJ, Waterfall WE, Brown BH, Duthie HL (1973) The effect of varying the extent of the vagotomy on the myoelectric and motor activity of the stomach. Gut 14:657–664
Stoddard CJ, Smallwood RH, Duthie HL (1978) Migrating myoelectric complex in man. In: Duthie HL (ed) Gastrointestinal motility in health and disease. MTP Pres, Lancaster, pp 6–16
Sullivan MA, Snape WJ, Matarazzo SA, Petrokubi RJ, Jeffries G, Cotten S (1977) Gastrointestinal myoelectrical activity in idiopathic intestinal pseudoobstruction. N Engl J Med 297:233–238
Sullivan MA, Cohen S, Snape WJ (1978) Colonic electric activity in irritable bowel syndrome. Effect of eating and anticholinergics. N Engl J Med 298:878–883
Suster G, Chin KI, Barbero GJ (1970) Rectal motility pattern in infants and children with agangionic megacolon. Am J Dis Child 119:494–497
Syson AJ (1974) Studies on the excitation-contraction coupling mechanism of mammalian smooth muscle. PhD thesis, University of Lancaster
Syson AJ, Huddart H (1973) Contraction tension in rat vas deferens and ileal smooth muscle and its modification by external calcium and the tonicity of the medium. Comp Biochem Physiol 45A:345–362
Szurszewski JH (1969) A migrating electric complex of the canine small intestine. Am J Physiol 217:1757–1763
Szurszewski JH, (1974) Recording of electrical activity of smooth muscle by means of the sucrose gap. In: Daniel EE (ed) Proceedings of the 4th International Symposium on Gastrointestinal Motility. Mitchell, Vancouver, pp 102–108

Szurszewski JH, Elveback LR, Code CF (1970) Configuration and frequency gradient of electric slow wave over canine small bowel. Am J Physiol 218:1468–1473

Taylor I, Duthie HL (1976) Bran tablets and diverticular disease. Br Med J 1:988–990

Taylor I, Smallwood R, Duthie HL (1974a) Myoelectric activity in the rectosigmoid in man. In: Daniel EE (ed) Proceedings of the 4th International Symposium on Gastrointestinal Motility. Mitchell, Vancouver, pp 109–118

Taylor I, Duthie HL, Smallwood R, Brown BH, Linkens D (1974b) The effect of stimulation on the myoelectrical activity of the rectosigmoid in man. Gut 15:599–607

Taylor I, Duthie HL, Smallwood R, Linkens D (1975) Large bowel myoelectrical activity in man. Gut 16:808–814

Taylor I, Darby C, Hammond P, Basu P (1978) Is there a myoelectric abnormality in irritable bowel syndrome? Gut 19:391–395

Thillier JL, Bertrand J (1975) External digestive electromyography: theory and technique. Rend Gastroenterol 7:133–134

Ustach TJ, Tobon F, Hambrect T, Bass DD, Schuster MM (1970) Electrophysiological aspects of human sphincter function. J Clin Invest 49:41–48

Van der Pol B, Van der Mark (1928) The heart beat considered as a relaxation oscillator and electrical model of the heart. Philos Mag [Suppl] 6:763

Vantrappen G, Janssens J, Hellemans J, Ghoos Y (1977) The interdigestive motor complex of normal subjects and patients with bacterial overgrowth of the small intestine. J Clin Invest 59:1158–1166

Vantrappen G, Janssens J, Peeters TL, Bloom S, Van Tongeren J, Hellemans J (1978) Does motilin have a role in eliciting the interdigestive migrating motor complex (MMC) in man? Gastroenterology 74:1149

Vantrappen G, Peeters TL, Janssens J (1979) The secretory component of the interdigestive complex. In: Proceedings of the 7th International Symposium on Gastrointestinal Motility, Iowa City

Weisbrodt NW, Burks TF (1974) Central nervous control of intestinal motility. In: Daniel EE (ed) Proceedings of the 4th International Symposium on Gastrointestinal Motility. Mitchell, Vancouver, pp 649–656

Weisbrodt NW, Christensen J (1972) Electrical activity of the cat duodenum in fasting and vomiting. Gastroenterology 63: 1004–1010

Weisbrodt NW, Copeland EM, Thor PJ, Dudrick SJ (1976) The myoelectric activity of the small intestine of the dog during total parenteral nutrition. Proc Soc Exp Biol Med 153:121–124

Wienbeck M, Christensen J (1971) Cationic requirements of colon slow waves in the cat. Am J Physiol 220:513–519

Wienbeck M, Christensen J, Weisbrodt NW (1972) Electromyography of the colon in unanesthetized cat. Dig Dis 17:356–362

Wingate DL, Rupping H, Green WER et al. (1976) Motilin induced electrical activity in the canine gastrointestinal tract. Scand J Gastroenterol [Suppl] 11:111–118

CHAPTER 6

Electrophysiology of the Enteric Neurons

R. A. NORTH

A. Introduction

Early investigations into the function of the neurons of the enteric nervous system were made by studying the peristaltic reflex of isolated tissue (TRENDELENBURG 1971). That this reflex involved intramural neurons was first demonstrated by FELDBERG and LIN (1949) and PATON and ZAIMIS (1949). Their demonstration of the presence within the gut wall of cholinergic neurons was followed by the finding of a powerful group of nonadrenergic inhibitory neurons in the intestine (BENNETT et al. 1966a). The functional roles of these neuronal classes will be discussed in Chaps. 4, 8; the present chapter deals exclusively with studies of the electrical properties of the enteric neurons themselves. An indication will be given, so far as is possible, of how knowledge of single neuron properties has been helpful in working out the ways in which the enteric nervous system functions, and of how this function can be interpreted in terms of gastrointestinal motility.

Most knowledge regarding the electrophysiology of the enteric nervous system has come from single cell recordings – with either extracellular or intracellular recording techniques. However, KOSTERLITZ and LYDON (1971) provided useful information by recording the population response of the nerves and the smooth muscle evoked by nerve stimulation. This indicated that transmission along enteric neurons continued for up to 15 mm.

B. Extracellular Recording Techniques

I. Neuronal Types and Properties

YOKOYAMA (1966) was the first to record the electrical activity of single neurons in the intestinal wall. He inserted insulated platinum-iridium electrodes (10–20 μm tip size) through the serosal and longitudinal muscle layers onto nodes of the myenteric plexus of the rabbit jejunum. Characteristic bursts of action potentials were recorded at 10–30-s intervals, and these were rapidly abolished by hexamethonium (100 μM). This may indicate that, in the intact isolated intestine, the spontaneous activity was largely the results of continuous nicotinic synaptic input to the myenteric neurons. YOKOYAMA (1971) was able to record evoked but not spontaneous activity in stripped preparations, and made the important observation that this was conducted for up to 15 mm longitudinally and 3 mm circumferentially. Hexamethonium much reduced the distance of longitudinal conduction as did longitudinal incisions across the preparation 3–5 mm apart. There is good agreement between this value of 15 mm and that found by KOSTERLITZ and LYDON (1971) although those workers found that the distance was not reduced by hexamethonium (140 μ*M*). In both cases, the amplitude of the evoked multiunit response was considerably reduced when the distance between stimulating and recording electrodes exceeded 1 mm.

The next effort to record activity in single myenteric neurons was made by WOOD (1970). Recordings were made with insulated stainless steel electrodes (1–5 μm tip size) or iridium-filled glass capillaries (5–30 μm tip size). In the intact isolated jejunum of the cat, neurons were found which fired action potentials in one of three somewhat arbitrary patterns: "burst" units, "mechanoreceptor" (sometimes referred to as "mechanosensitive") and "single-spike" units. The burst units were frequently encountered (50% of total cells) and discharged about 8 (range 2–34) spikes at a frequency of 4–30 Hz, the burst being repeated after a silent interval of about 3 s (range 0.4–40). This pattern of discharge was not obviously related to longitudinal or circular muscle activity. It was stated that burst units were encountered only after the electrode tip was advanced into the ganglion by 75–150 μm. This must cause very marked deformation of the neuronal membrane within the ganglion; a clean penetration of this distance would put the electrode through the ganglion and far out the other side.

The mechanoreceptor units were most easily observed with large tipped (10–30 μm) electrodes which dimpled the ganglion surface. It was concluded that this activity probably arose through mechanical distortion of the cell membrane – although two types of response to such mechanical stimulation were apparent. Some cells (tonic mechanoreceptors) continued to discharge action potentials at a slowly decreasing frequency for 8–30 s after the termination of the mechanical deformation. Other cells (phasic mechanoreceptors) showed larger spike amplitudes and fired action potentials for only as long as the stimulation (mechanical deformation) was maintained. These cells also fired single spikes unrelated to mechanical stimuli. Phasic mechanoreceptor activity was said to correlate with circular muscle activity; tonic mechanoreceptor activity did not. Single spike units fired at random intervals (0.1–1 s, i.e., 1–10 Hz), were infrequently observed (about 10% of total), and had activity which was not related to that of the muscle layers. WOOD (1970) found essentially similar results whether the jejunum was intact, or the ganglia adhered to the stripped longitudinal muscle layer.

This initial classification may be somewhat misleading. It is possible that different neuronal types chiefly differed in the degree of mechanical deformation caused by the stimulating electrode. Burst units may be sufficiently depolarized by the trauma of the electrode to fire spikes, but may then progressively recover (perhaps due to active sodium extrusion). Mechanoreceptors may be a misnomer because almost all neurons will fire action potentials when their membrane is distorted or punctured by a large metal electrode – whether they are phasic or tonic would depend on the accommodative properties of the cell membrane (which were later shown by intracellular recording, see Sect. C). Single-spike units were probably the least damaged by the recording electrode, and it is noteworthy that these cells were very sensitive to excitation by acetylcholine. The burst units, which probably reflect greater or lesser degrees of neuronal damage, were excited by acetylcholine and nicotine in only some experiments, and not affected by catecholamines, 5-hydroxytryptamine (5-HT), glycine, glutamate, or γ-aminobutyric acid (GABA). It seems unlikely that the activity of any cells was synaptically driven as it was not affected by (+)-tubocurarine, hexamethonium, or pentolinium; nicotine blocked cell discharge, presumably by profound depolarization. The mechanoreceptors were little affected by any drugs.

Later studies (OHKAWA and PROSSER 1972a, b) essentially confirmed the findings of WOOD (1970). As in the original experiments of WOOD (1970), methylene blue was used to visualize ganglia in intact jejunal segments; methylene blue has subsequently been shown to cause an irreversible suppression of firing in myenteric neurons (NOZDRACHEV et al. 1977). OHKAWA and PROSSER (1972a, b) showed that raising the intraluminal pressure excited myenteric neurons, but could not show any relation between nerve activity and slow wave activity in the muscle. No activity could be evoked by electrical stimulation of the preparation at a distance from the site of recording. These papers were also the first to describe electrical activity from neurons in the submucous plexus of the intestine, reporting that the cell types and properties were almost identical to those in the myenteric plexus. Periarterial nerve stimulation had no effect on the firing rate of myenteric neurons; an analogous finding was later made with intracellular recording (HIRST and MCKIRDY 1974a).

Acetylcholine (ACh) in rather high concentrations (about 5 μ*M*) excited the burst cells in both plexuses; this was blocked by atropine and/or (+)-tubocurarine. The physiologic significance of these observations is unclear; the acetylcholine perhaps excited the underlying muscle and thereby increased the firing rate by moving the cell with respect to the electrode, this being blocked by atropine. Adrenaline and noradrenaline showed only excitatory effects, and those were blocked by phenoxybenzamine and phentolamine but not by sotalol. However, the α-blockers were used in too high concentration to discriminate among several receptors types. 5-HT (about 2 μ*M*) was reported to have no effect on neurons in the submucous plexus. All in all, the results from these early extracellular studies did not allow many useful conclusions to be drawn about enteric neuron electrophysiology. The activity recorded was certainly neuronal (blocked by TTX and lidocaine), but may have arisen from mechanical deformation of the neuron by the recording electrode. In the cat, it was apparently not due to continuous nicotinic synaptic activity because it was not reduced by nicotinic antagonists, but in the rabbit, hexamethonium blocked both evoked and spontaneous activity (YOKOYAMA 1966, 1971). The findings with other drugs were either negative (e.g., 5-HT, GABA, glycine) or in a direction opposite to that expected (noradrenaline).

These studies on cat intestine by WOOD (1970) and OHKAWA and PROSSER (1972a, b), and the rabbit by YOKOYAMA (1966, 1971), were extended to the guinea pig by WOOD (1973). He stripped off the longitudinal muscle layer, allowing the plexus to remain attached to the circular muscle, and stained the preparation with methylene blue before recording with electrodes of 5–25 μm tip size. WOOD (1973) again distinguished between burst units, mechanoreceptor units, and single-spike units – although he questioned the usefulness of such a distinction when he found that most single-spike units were excited by mechanical deformation. Most of the cells which responded to electrode movement discharged action potentials only at the onset of the movement – the so-called phasic mechanoreceptors. These may be the rapidly accommodating type 2 (AH) cells which were stimulated by the mechanical distortion of the recording electrode (see Sect. C.I). The single-spike units were identical to the phasic mechanoreceptors, except that they were not excited by moving the electrode tip.

The next group of studies using extracellular techniques came from the laboratory of Takagi (Sato et al. 1973, 1974). The major difference between these studies and those of earlier workers was the failure to report mechanoreceptor units. This is presumably because these workers realized that all units were mechanoreceptors in that the activity which was recorded was a consequence of the suction applied to the recording electrode. The electrodes used by Sato et al. (1973), and later groups (Sato et al. 1974; Takayanagi et al. 1974, 1977; Ehrenpreis et al. 1976), recorded neuronal action potentials occurring either in discrete bursts, or apparently at random intervals. The single-spike, random units had very low amplitudes (20 μV) and a resting frequency of about 3–15 Hz. The pharmacologic experiments on these units are discussed in detail in a later section. Briefly, morphine inhibited spike firing, nicotine and acetylcholine excited cells through nicotinic sites, and (in contrast to the findings of Ohkawa and Prosser 1972a, b), adrenaline and noradrenaline inhibited spike firing. It is unlikely that continuous synaptic activity contributed to the spike activity recorded because hexamethonium, while blocking the excitatory effects of ACh, had no effects of its own on the firing rate. This contrasts with the original findings of Yokoyama (1966) in the rabbit; this discrepancy is discussed further below. The burst unit described by Sato et al. (1973) fired groups (3–7) of action potentials at a frequency of about 10 Hz, and at intervals of 1–2 s. These cells are probably type 2 (AH) cells as described by the intracellular recordings, and similar to the class of slowly firing units observed with suction electrodes by Dingledine and Goldstein (1976) and North and Williams (1977) (see Sect. C.I).

Dingledine et al. (1974) and Dingledine and Goldstein (1975, 1976) made a detailed study of the action of several agents on the myenteric neurons. On the basis of spike discharge patterns they concluded that only two types of cell could be distinguished by an extracellular suction electrode, and these were analogous to those described by Sato et al. (1973). The majority were single-spike units, firing randomly at frequencies of 2–6 Hz. A small proportion were units of higher amplitude and a slower, more regular discharge rate. Dingledine et al. (1974) stated in their original report that the activity which they recorded may be "at least ... in part ... the result of mechanical stimulation provided to the ganglion by the suction electrode." Activity of single cells could be recorded for up to 7 h, a period which far exceeded that described by the earlier workers who did not use suction electrodes. This group made the further important observation (Dingledine and Goldstein 1976) that the neuronal activity recorded with a suction electrode persisted in calcium-free solutions containing high Mg concentrations. This was direct evidence that the activity was not due to continuous synaptic bombardment of the neuron under study. Neurons continued to fire for up to 3–5 h in such calcium-free solutions.

North and Williams (1977) also used extracellular suction electrodes to record activity of myenteric neurons in guinea pig ileum. The activity recorded (Fig. 1) was very similar to that reported by Sato et al. (1973, 1974) and

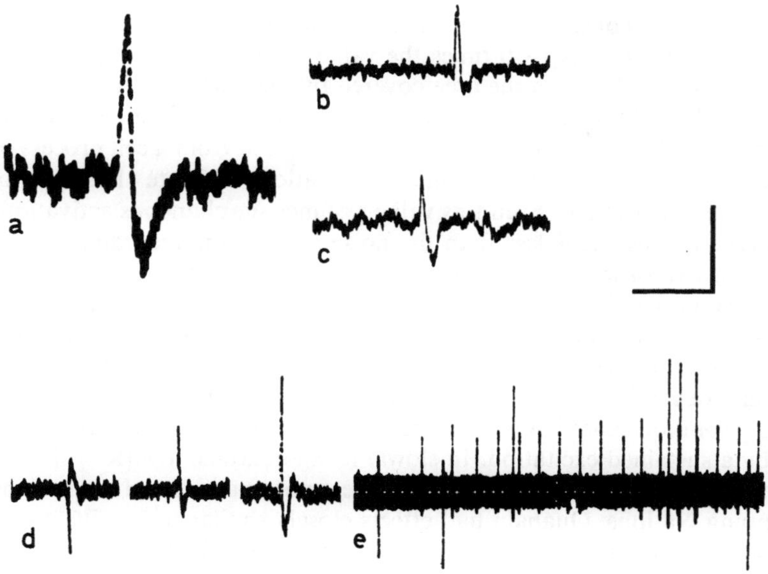

Fig. 1 a–e. Spike waveforms recorded with extracellular suction electrodes from myenteric neurons. **a–c** typical biphasic spikes recorded from different neurons with different electrodes; **d** different spike waveforms; **e** waveforms as in **d** photographed at a slow sweep speed. In **d** and **e** the suction electrode was recording activity from three neurons at the same time. Calibrations: vertical **a** 100 μV, **b–e** 200 μV; horizontal **a** 2 ms, **b–c** 10 ms, **d** 20 ms, **e** 2 s. Negativity upwards. (NORTH and WILLIAMS 1977)

DINGLEDINE and GOLDSTEIN (1975, 1976). Several features of this study indicated that the activity recorded was a consequence of the mechanical deformation of the cell membrane. First, as described by DINGLEDINE and GOLDSTEIN (1976), the recorded activity persisted in calcium-free solutions, suggesting that it is not generated by continuous excitatory synaptic input. Spontaneous spike activity is only rarely recorded with intracellular electrodes under otherwise identical experimental conditions (see Sect. C.I), and this implies that the extracellular electrode is inducing the neuronal action potentials. Second, no spike activity was ever recorded without suction being applied to the recording electrode. An increase in the amount of suction increased the cell firing rate. Third, although the activity of a neuron often disappeared rapidly after placing the suction electrode on the ganglion, such cells would usually discharge action potentials again by enhancing their excitability with a calcium-free solution.

NORTH and WILIAMS (1977) argued that the activity recorded was not likely to arise in a cell process, but that the waveform of the action potential suggested a direct capacitative coupling through the cell membrane with the cell interior. A cell soma of diameter 30 μm might be expected to present a cross-sectional area of 14×10^{-6} cm^2 (a hemisphere) to a favorably positioned extracellular electrode of the same tip diameter (30 μm). In such a cell, a single action potential of amplitude 100 mV would result in a charge movement across the membrane beneath the electrode of 1.4 pC (assuming a membrane capacitance of 1 μF/cm^2). If this occurs within 1 ms, and the electrode (resistance 500 kΩ) has pure resistive coupling onto the cell, then the recorded potential change would be approximately 700 μV for a duration of 1 ms. This is 1–10 times the value usually recorded, presumably because of disparities between the area covered by the electrode tip and the surface of a single cell, and because of resistive losses at the point of contact between electrode and cell membrane. If the activity were recorded from a cell process of 2 μm diameter and 30 μm in length, similar calculations indicate that the potential change would be 0.25 μV. Even this value assumes simultaneous activation of the entire nerve process. It is less than the noise level of the recording systems employed, and two orders of magnitude lower than the spikes actually recorded. It is therefore most unlikely that the activity recorded is not somatic.

Under the conditions of recording with the extracellular suction electrodes, in which the cells are firing spontaneously at rates of 1–5 Hz, the neurons appeared to be quite dependent on active sodium extrusion (NORTH and WILLIAMS 1977). Low $[K^+]$ solution caused only a transient reduction in firing rate, and this was followed by a sustained excitation. In slowly firing neurons, low $[K^+]$ slowed spike discharge, as might be expected if the cells were not so active and not so dependent on pumping Na ions. Ouabain had effects closely similar to low $[K^+]$ solutions. These observations are compatible with a significant sodium entry and active extrusion in the conditions of these experiments.

II. Neuronal Connections

1. Connections Within Ganglia

WOOD (1970) and OHKAWA and PROSSER (1972a, b) described examples of cells which they stated to be "coupled" in some way in their discharge patterns. However, in view of the likelihood that the same mechanical deformation is causing the discharge of both the neurons recorded with a single electrode, it is to be expected that their discharges would appear to be coupled in some circumstances. They failed to perform the crucial experiment to show that such coupling was abolished in solutions known to block synaptic connections, e.g., low calcium and high magnesium. Similar criticisms may be applied to the studies of neuronal interactions between cells whose activity was recorded with a single electrode (WOOD 1976). Slight muscle movements would be sufficient to change the rate of discharge of the myenteric neurons, by moving them against the large recording electrode; useful conclusions cannot be drawn from such experiments with multiunit recordings from a single extracellular microelectrode unless the cells can be clearly shown to be synaptically connected by pharmacologic tests.

2. Connections Within the Plexus

YOKOYAMA (1971) reported that hexamethonium blocked the excitatory responses of single myenteric neurons to stimulation of a distant part of the myenteric plexus. He also studied the spread of excitation throughout the myenteric and submucous plexus of the rabbit small intestine by recording the compound action potential, evoked by stimulating a single node. YOKOYAMA et al. (1977a) corroborated their earlier report that excitation speads longitudinally in the myenteric plexus for up to 15 mm, but circumferentially for only 3 mm. The conduction velocities from one node of the plexus to another were 3–50 cm/s. This velocity was the same in the oral and aboral directions. At greater conduction distances the latency of the compound evoked response was much prolonged, and the amplitude was reduced, until at 10–15 mm no activity could be evoked. The amplitude of the evoked potential recorded aboral to the site of stimulation was always greater than the amplitude of the potential evoked by stimulating aboral to the recording electrode. As transmission over long distances (>1 mm) was reduced by hexamethonium, this implies that nicotinic synapses are interposed. The greater amplitude of the aborally recorded compound action potential therefore implies an orientation of the synaptic connections which favors transmission in the descending direction. Longitudinal sections of the plexus greatly shortened the distance over which evoked activity could be recorded, indicating that the nerve processes did not necessarily travel aborally by the shortest direct pathway in the interconnecting strands.

YOKOYAMA (1971) also reported that the size of the compound extracellular action potential recorded at a distance of less than 1.2 mm from the stimulating site was not reduced when it was evoked at a frequency of 20 Hz. This is compatible with excitation of a cell process which propagated to the soma (see DINGLEDINE and GOLDSTEIN 1976; NORTH and WILLIAMS 1977; discussed further in Sect. C.II), and also agrees with the finding that hexamethonium did not affect the responses at short distances (due to direct excitation of a cellular process) but reduced the responses recorded at longer distances. It is particularly significant that the long-distance (>2 mm) responses to nerve stimulation were not completely abolished by hexamethonium, indicating either a few long nerve processes or a noncholinergic interneuron (YOKOYAMA et al. 1977a; see also DINGLEDINE and GOLDSTEIN 1976). These studies were important in showing that excitation propagated more readily in the longitudinal than in the oblique or circumferential directions, and that the number and size of the evoked potentials elicited aboral to the stimulus site were greater than that recorded oral to such site. This indicates a preferential propagation in the aboral direction, in agreement with the earlier findings of HIRST et al. (1975) using intracellular recording (see Sect. C.I). The results on propagation agreed well with those of KOSTERLITZ and LYDON (1971), and the findings with hexamethonium were compatible with the known findings that the agent blocks the peristaltic reflex in the isolated ileum (HUKUHARA et al. 1958).

In a companion paper, YOKOYAMA et al. (1977b) described experiments on submucous neurons. Chronaxie and conduction velocity were similar, but the maximum distance for transmission of evoked activity was 4 mm. When the distance was short between stimulating and recording electrodes (<2 mm) hexamethonium had no effect, but at longer distances hexamethonium completely abolished the evoked activity. As in the myenteric plexus, hypoxia abolished the late evoked activity. The conclusion was that nicotinic synapses usually occur within 2 mm in the submucous plexus.

Evoked activity was also recorded by DINGLEDINE and GOLDSTEIN (1976) and similar findings were made by NORTH and WILLIAMS (1977). Pulses were delivered from a glass stimulating electrode (tip diameter 30 μm) positioned on the same or an adjacent ganglion. A single pulse stimulus evoked a graded compound action potential, with usually a predominant component from one unit (Fig. 2). An inflection on the rising phase of the spike suggested an initial segment blockade (see

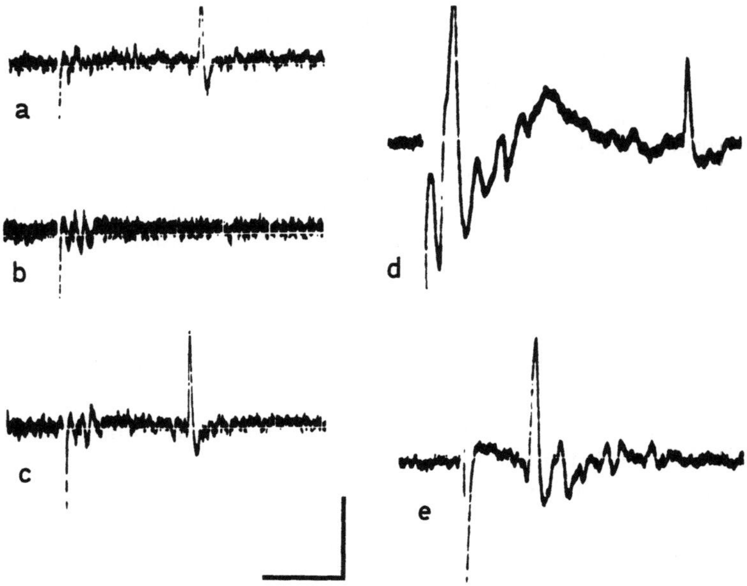

Fig. 2a–e. Responses of myenteric neurons to focal stimulation of the ganglion with single pulses. **a** a single stimulus evoked a compound "antidromic" response followed by a later single spike – the low-amplitude "antidromic" spikes may represent axonal activity; **b** the same unit as in **a** after perfusion for 2 min with hexamethonium (100 μ*M*); **c** the same unit as in **a** and **b** 4 min after washing out the hexamethonium, **d** compound "antidromic" spike arising from the stimulus artifact followed by a single biphasic spike – the later single spike disappeared on repetitive stimulation and was judged to be synaptic; the first "antidromic" spike has a prominent inflexion on its rising phase, analogous to that observed with intracellular recording (see Fig. 3b); **e** compound "antidromic" spike recorded from a different unit at a greater distance from the stimulating electrode. Calibrations: vertical **a–e** 200 μV; horizontal **a–c** 10 ms, **d–e** 4 ms. (NORTH and WILLIAMS 1977)

Sect. C.I). The second response was a unitary, all-or-nothing spike, occurring at a longer latency (4–20 ms) and failing to follow the stimulus at frequencies exceeding 0.5 Hz. This conforms to the intracellular recording (NISHI and NORTH 1973a) that the excitatory postsynaptic potential (e.p.s.p.) declines in amplitude at frequencies over 0.5 Hz. This second late response disappeared in calcium-free solutions, again indicating that it was due to release of a synaptic transmitter. The late synaptic response to a single stimulus was abolished by 100 μ*M* hexamethonium (NORTH and WILLIAMS 1977). However, the response to a train of pulses was only slightly affected by 100 μ*M* hexamethonium, indicating that it may be due to a noncholinergic excitatory transmitter. It was abolished by 10 μ*M* 5-HT (DINGLEDINE and GOLDSTEIN 1976); this may represent a presynaptic action of 5-HT to inhibit the release of a noncholinergic interneuronal transmitter released by repetitive nerve stimulation (see Sect. D.III).

3. Connections with Extrinsic Nerves

In the original studies of OHKAWA and PROSSER (1972a, b) stimulation of periarterial nerves had no effect on the activity of single neurons. However, that group reported only excitatory effects of α-agonists on neuronal firing, so that the interpretation of the work is difficult. TAKAYANAGI et al. (1977) presented evidence for three types of response of myenteric neurons to sympathetic nerve stimulation. In the majority of cells (70%), periarterial nerve stimulation was without effect even on neurons whose firing was inhibited by exogenous noradrenaline, and even when the periarterial nerve stimulation inhibited muscle activity. This finding is compatible with the demonstrations by HIRST and MCKIRDY (1974a) and NISHI and NORTH (1973b) that the principal action of periarterial nerve stimulation or exogenous catecholamines is presynaptic inhibition of acetylcholine release at interneuronal ganglionic synapses. The second type of cell (20%) was inhibited by periarterial nerve stimulation, a finding which may relate to the finding with intracellular recordings that a proportion of neurons [especially type 2 (AH) cells] are hyperpolarized by noradrenaline (see Sect. D.II). The third type of cell (10%) showed a long-lasting excitation of myenteric neurons following periarterial stimulation. This was not blocked by hexamethonium, from which the authors concluded that it was probably not due to excitation of vagal fibers. However, noncholinergic vagal transmitters may be involved. The conclusion of these authors that the sympathetic innervation does not influence activity within the myenteric plexus is incorrect – HIRST and MCKIRDY (1974a) had already provided direct evidence from intracellular recording that it does inhibit transmission within the plexus.

C. Intracellular Recording Techniques

I. Neuronal Types and Properties

There were several attempts made to record activity from myenteric neurons prior to 1972 – some reported in the literature (OHKAWA and PROSSER 1972a) and others not (S. YOKOYAMA 1970 unpublished work; M. E. HOLMAN 1971 unpublished work). The principal difficulties are three-fold: the delicacy of the single cells (which are devoid of a protective Schwann cell); the constant movement of the underlying smooth muscle; and the scattered nature of the cell bodies in the plexus which necessitates visual placement of recording electrodes. The first successful recordings from myenteric neurons were reported in abstract form at a meeting of The Physiological Society in Oxford in 1972 (HIRST et al. 1972). The first papers describing the cellular properties appeared soon after (NISHI and NORTH 1973a; HIRST et al. 1974). The essential feature of these earlier intracellular recordings was that there existed two types of neurons in the myenteric plexus of the guinea pig. The first type was called type 1 by NISHI and NORTH (1973a) and S (synaptic) cells by HIRST et al. (1974). These cells had high input resistances (40–200 MΩ), relatively low thresholds (5–20 mV), fired action potentials repetitively in response to depolarizing current injection, and responded to focal stimulation of the ganglion surface with excitatory postsynaptic potentials (e.p.s.p.; Fig. 3) similar in all respects to those recorded in other autonomic ganglia (NISHI 1974). The voltage–current relationship for these cells was generally linear, with anomalous rectification occurring for larger outward currents. The synaptic potentials were of short duration (~20 ms), reached threshold for spike initiation (Fig. 3c), were associated with an increased membrane conductance, were mimicked by acetylcholine ionto-

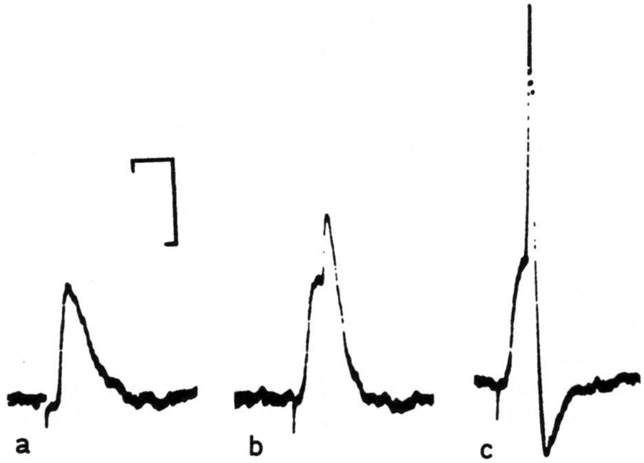

Fig. 3 a–c. Intracellular recording from a myenteric neuron. Excitatory postsynaptic potentials in response to a single stimulus applied 100 μm from the cell soma. **a** stimulus strength 9 V; **b** stimulus strength 10 V – a local response appears which does not reach threshold for the full soma membrane spike; **c** stimulus strength 11 V – a full spike is evoked. Calibrations; vertical 20 mV; horizontal 10 ms. (Nishi and North 1973 a)

phoresis, became larger with hyperpolarizing currents, became smaller with depolarizing currents, and reversed in polarity at a membrane potential close to zero (Nishi and North 1973 a; Hirst et al. 1974; North and Nishi 1976; North et al. 1980 b). These fast e.p.s.p. were abolished by either hexamethonium (10–200 μ*M*) (Nishi and North 1973 b) or (+)-tubocurarine (0.14–1.4 μ*M*) (Hirst et al. 1974) and greatly prolonged by eserine (North and Nishi 1974; North 1974). The fast e.p.s.p. appears to be generated predominantly on the soma membrane – its amplitude is readily altered by changing the soma membrane potential and the amplitude of the orthodromic spike is less than that of the spike evoked by direct depolarization of the cell membrane (Nishi and North 1973 a). These excitatory synapses mediated by nicotinic receptors appear to be responsible for the major connections between neurons within the myenteric plexus and their discovery explained the findings of Paton and Zaimis (1949) and Hukuhara et al. (1958) that nicotinic antagonists blocked the peristaltic reflex. The e.p.s.p. showed a marked decline in amplitude when it was evoked at frequencies greater than 0.1 Hz (Nishi and North 1973 a, b). This "rundown" appeared to be due to presynaptic factors–either a reduction in transmitter release for each action potential, or a reduction in the number of fibers excited by the focal stimulating electrode, perhaps owing to a prolonged postspike hyperpolarization of nerve terminals.

The other type of neuron within the enteric ganglion was of considerable interest because similar cells had not been previously described in autonomic ganglia. These cells (type 2, Nishi and North, 1973 a; AH cells, Hirst et al. 1974; Hirst and Spence 1973; North 1973) are characterized by an absence of fast e.p.s.p. and by the presence of a long-lasting (1–30 s) afterhyperpolarization which follows the action potential (Fig. 4). Type 2 (AH) cells have somewhat higher resting membrane potentials, a lower input resistance, and a higher threshold. However, considerable caution is necessary in making comparisons between the neuronal types

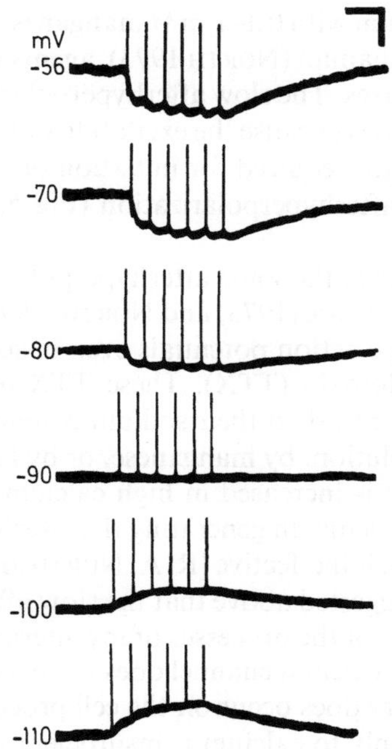

Fig. 4. Reversal of the slow afterhyperpolarization in a single myenteric neuron. The membrane potential (indicated beside each record) was changed by passing a constant hyperpolarizing current through the recording microelectrode. At each membrane potential, six action potentials were elicited by passing sufficiently strong depolarizing current pulses (10 ms) through the recording microelectrode. At membrane potentials above 90 mV the slow afterhyperpolarization became a slow afterdepolarization. The resting potential of this cell, determined by withdrawal of the microelectrode, was − 56 mV. Calibrations: vertical 20 mV; horizontal 2 s. (NORTH and NISHI 1974)

on the basis of passive membrane properties recorded with high resistance electrodes. It is not unusual to find type 2 (AH) cells which progressively become more and more excitable during a 2–3-h period, until their membrane properties are similar to those of a type 1 (S) cell (G.D.S. HIRST 1974 unpublished work; R.A. NORTH 1974 unpublished work). A recent finding (GRAFE et al. 1979) that type 2 (AH) cells show fast e.p.s.p. may be assumed to lack physiologic relevance in view of the fact that the amplitude of the potential changes was 1–2 mV and the membrane potential was artificially hyperpolarized to 90 mV – these potentials may however represent electrical coupling from synapses onto other neurons. One type 2 (AH) cell which was subsequently identified ultrastructurally showed no synaptic specializations in 200 sections of its soma (GABELLA and NORTH 1974).

The afterhyperpolarization of the type 2 (AH) cells is due to calcium entering across the soma membrane during the spike and causing an increase in the potassium conductance of the soma membrane. This is evidenced by the findings that the slow afterhyperpolarization is associated with an increase in membrane conductance, reverses its polarity at the potassium equilibrium potential (Fig. 4), and disappears in calcium-free solutions (NISHI and NORTH 1973a; HIRST et al. 1974).

It is also blocked by perfusion with 0.1–1 m*M* manganese (HIRST and SPENCE 1973; NORTH 1973) or 1 m*M* lanthanum (NORTH 1973), agents which are known to block transmembrane calcium fluxes. The slow afterhyperpolarization appears not to develop on the cellular processes because the excitability of the process, judged by the extracellular current strength required for initiation of a propagated spike, is not reduced during the soma afterhyperpolarization (Fig. 5; NORTH and NISHI 1974, 1976).

The calcium dependence of the soma afterhyperpolarization is compatible with the findings by HIRST and SPENCE (1973) and NORTH (1973) that the soma of type 2 (AH) neurons can discharge action potentials even in solutions devoid of sodium ions, or containing tetrodotoxin (TTX). These TTX-resistant action potentials have a much lower rate of rise than their sodium counterparts and are reversibly blocked by calcium-free solution, by manganese, or by lanthanum. The amplitude of the TTX-resistant spike is increased in high calcium solutions (NORTH 1976). Strontium substitutes for calcium in generating the afterhyperpolarization (NORTH and NISHI 1976) but barium is ineffective (R. A. NORTH and Y. KATAYAMA 1979 unpublished work). It was suggested above that the slow afterhyperpolarization does not occur on the membrane of the processes of myenteric neurons. This may imply that the voltage-dependent calcium channel does not exist on the membrane of the cell process; if calcium entry does occur on the cell process then the density of the inward current flow due only to calcium is insufficient to cause propagation, because the spike initiated in a cell process fails rapidly in TTX (NORTH and NISHI 1976). The inability of the cellular process to propagate a calcium spike, contrasting with the ability of the soma membrane to generate a calcium spike, is analogous to the situation reported in dorsal root ganglion cells in culture (DICHTER and FISCHBACH 1977). The similarities between type 2 (AH) cells and primary afferent neurons in the dorsal root ganglia did not pass unnoticed by the first investigators. The absence of demonstrable synaptic input as well as the voltage-dependent calcium permeability of the soma membrane prompted the suggestion that the type 2 (AH) cells may themselves be the afferent cells of the enteric nervous system. Direct evidence for the suggestion remains lacking, but the proposal has not been substantially refuted.

Intracellular recordings have shown no evidence of pacemaker activity in any myenteric neuron. Spontaneous somatic spikes can occasionally be recorded; these appear to be due to invasion of the ganglion cell body by an action potential arising in one of its processes (HIRST et al. 1974; NISHI and NORTH 1973a; R. A. NORTH 1972 unpublished work). It is not clear whether such spikes arise from physiologic input to a cell process (i.e., an e.p.s.p. or generator potential) or from damage during the preparation of the tissue. These propagated action potentials, whether spontaneous or evoked by stimulation, readily fractionate at the soma, leaving a small, all-or-nothing, rapid potential change analogous to the initial segment (IS) spike of motoneurons (Fig. 5; NISHI and NORTH 1973a). Spontaneous synaptic potentials are rarely recorded in isolated myenteric plexus longitudinal muscle strips. However, when the preparation is made with an attached segment of intact bowel, spontaneous e.p.s.p. can be recorded. Hyperpolarizing synaptic potentials (i.p.s.p.) do not occur in myenteric neurons except, in a few cells, in response to repetitive presynaptic nerve stimulation.

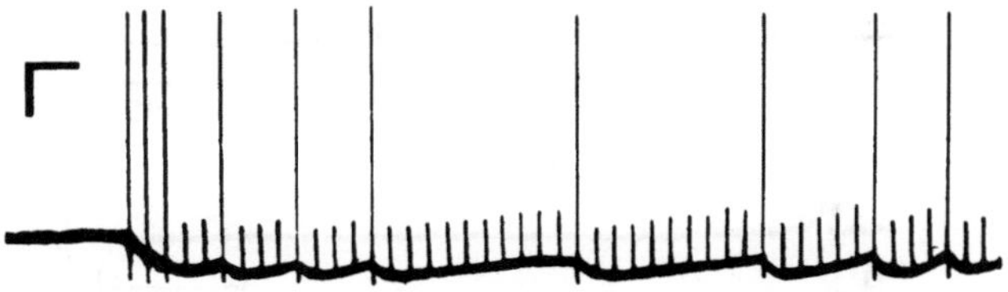

Fig. 5. Intracellular recording from a myenteric neuron. Fractionation of "antidromic" input by the slow afterhyperpolarization. A cell process was stimulated at 1 Hz. The spikes were followed by a slow afterhyperpolarization which summated to a value sufficient to block incoming action potentials – only proximal process potentials were recorded in the soma. When the hyperpolarization declined, incoming spikes again invaded the soma and the slow afterhyperpolarization followed. The soma fired at a mean frequency of 0.2 Hz when the neuronal process was stimulated at 1 Hz. Calibrations: vertical 20 mV; horizontal 3 s. (NORTH and NISHI 1974)

Slow synaptic potentials of time courses up to one thousand times longer than the fast e.p.s.p. also occur in both types of myenteric neurons when the presynaptic nerves are stimulated repetitively. The contention that these potential changes occur only in type 2 (AH) cells (WOOD and MAYER 1979a) is incorrect (see JOHNSON et al. 1980a). These potential changes require repetitive presynaptic nerve stimulation except in a small proportion of type 1 (S) cells where single pulse stimuli are effective. Depolarizing potentials can be elicited in 42% of type 1 (S) and 19% of type 2 (AH) cells. They are long latency and up to 90 s in duration. They are not affected by atropine or hexamethonium and their ionic mechanism has been shown to be an inactivation of the resting membrane potassium conductance (JOHNSON et al. 1980a). These synaptic potentials are abolished in calcium-free, high magnesium (WOOD and MAYER 1979a), or calcium-free solutions (JOHNSON et al. 1980a).

A slow hyperpolarizing response was observed in a small proportion (less than 10%) of myenteric neurons, following repetitive presynaptic nerve stimulation. This response, with latency 0.5–2 s and duration 2–40 s, was never evoked by single pulse stimuli. This contrasts with the i.p.s.p. in submucous neurons (see subsequent discussion). The slow i.p.s.p. appeared to be due to an increase in the potassium permeability of the soma membrane (JOHNSON et al. 1980a). It is possible that the high frequency focal stimulation excites some sympathetic terminals, and an overflow of noradrenaline hyperpolarizes the cell or that one of the peptides which hyperpolarize myenteric neurons and which are found in the plexus may be responsible for the potential (see Sect. D).

The physiologic significance, if any, of these slow potential changes which follow repetitive nerve stimulation remains to be clarified. It seems that simultaneous activation of many presynaptic fibers is necessary to elicit the responses – and this does not represent the kind of electrical activity which may normally occur in vivo. Despite this uncertainty the presence of the slow potential has fueled considerable speculation that it plays a fundamental role in controlling the activity of the myenteric neurons (WOOD 1979a). The action of exogenous substances on the slow potential and their possible chemical mediators will be discusses in Sect. D.

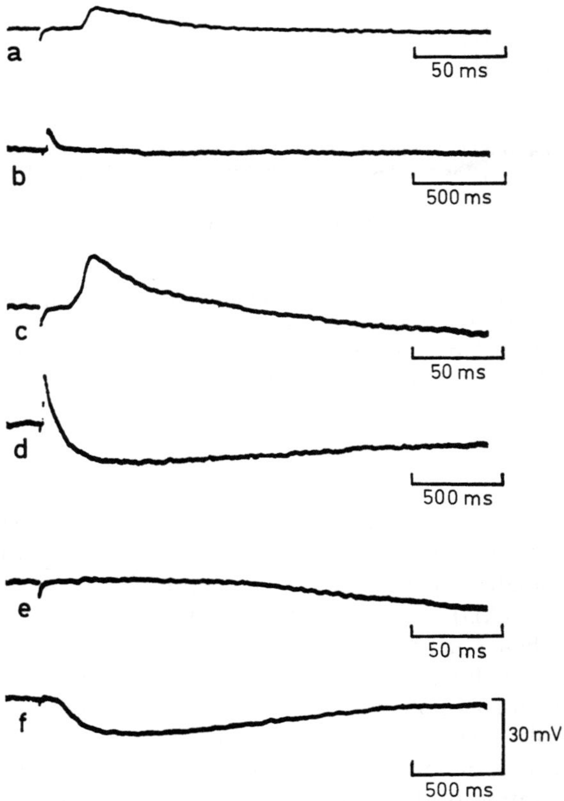

Fig. 6 a–f. Synaptic responses recorded from a submucous neuron at two different stimulus strengths. **a** and **b** show the threshold response to transmural stimulation at two different recording speeds. In this pair of records and in each of the other pairs the upper trace was taken at the faster scan speed. When the stimulus strength was increased a second e.p.s.p. was also evoked and this complex was now followed by a long hyperpolarization. Tubocurarine abolished e.p.s.p. (1×10^{-4} g/ml), **e** and **f** and an i.p.s.p. was recorded. Note the long latency and the slow time course of the i.p.s.p. The voltage calibration applies to each trace. (HIRST and MCKIRDY 1975)

Neurons in the submucous plexus resemble type 1 (S) cells of the myenteric plexus (HIRST and MCKIRDY 1975). As in the myenteric neurons, the fast e.p.s.p. is graded with the strength of presynaptic nerve stimulation, has a duration of 20–60 ms, and is reversibly abolished by nicotinic blocking drugs (Fig. 6). The main distinguishing feature of the submucous neurons is the finding that 40% of the cells show long-lasting inhibitory synaptic potentials in addition to the nicotinic e.p.s.p. This hyperpolarizing response was usually not graded with the stimulus strength – indicating perhaps that a single presynaptic nerve impulse underlay it. The i.p.s.p. lasted from 1 to 5 s, was associated with a marked increase in conductance, and reversed its polarity at a potential close to the potassium equilibrium potential (Fig. 7). The i.p.s.p. persisted after extrinsic denervation, and is not likely therefore to be mediated by release of catecholamines from nerve terminals. In any case, it appears that significant noradrenaline release can only be elicited with repetitive

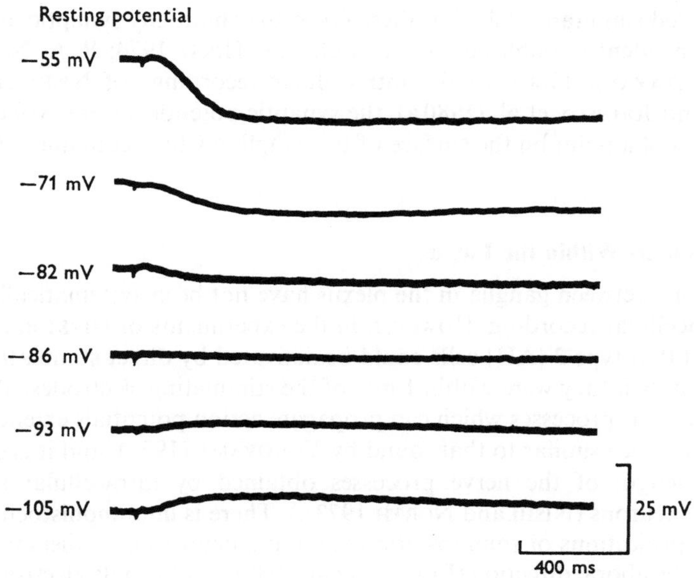

Fig. 7. Intracellular recording from a submucous neuron. Effect of passing inward current through the recording electrode on the amplitude and polarity of i.p.s.p. The upper trace is a record of an i.p.s.p. at cell resting potential. Between each trace the intensity of steady hyperpolarizing current was increased. It can be seen that the amplitude of the i.p.s.p. was gradually reduced and finally the polarity of the i.p.s.p. was reversed. The values of membrane potential are listed beside each trace. Voltage and time calibrations apply to each trace. (HIRST and MCKIRDY 1975)

stimulation, whereas the i.p.s.p. follows a single pulse stimulus (see for example BENNETT et al. 1966 b). Atropine did not block the slow i.p.s.p., thereby distinguishing it from the slow i.p.s.p. of sympathetic ganglion cells (NISHI 1974) and the slow i.p.s.p. of amphibian parasympathetic cardiac ganglion cells (HARTZELL et al. 1977). Bicuculline was also ineffective, but guanethidine and 2-bromolysergic acid both reversibly abolished the i.p.s.p. without blocking the fast e.p.s.p. Unfortunately, our poor understanding of the mechanism and specificity of action of these agents makes it difficult to conclude anything about the nature of the transmitter. It is not likely to be 5-HT, because iontophoretic application of 5-HT depolarizes submucous neurons (NEILD 1978). Dopamine and noradrenaline both increase the potassium conductance of the submucous neurons, but their virtual absence in the denervated preparation makes their participation unlikely (HIRST and SILINSKY 1975). The possibility that a peptide mediates this i.p.s.p. deserves active exploration.

II. Neuronal Connections

1. Connections Within Ganglia

Intracellular recordings have failed to provide any evidence for connections between neurons within the same myenteric ganglia. Many pairs of neurons have been impaled simultaneously, but these failed to exhibit any synaptic interconnection (independent unpublished work of G. D. S. Hirst 1974; R. A. North 1975; and Y. Katayama 1980). In the intracellular recordings of Nishi and North (1973 a) and Johnson et al. (1980 a), the synaptic potentials were evoked by focal stimulation of a point on the surface of the ganglion which contained the impaled cell.

2. Connections Within the Plexus

Connections between ganglia in the plexus have not been systematically explored with intracellular recording. However, in the experiments of Hirst et al. (1974) it was noted that type 2 (AH) cells could be activated by direct stimulation of their processes only if they were within 1 mm of the stimulating electrodes. This implies that the cellular processes which can propagate action potentials extend for about 1 mm, a distance similar to that found by Yokoyama (1971), and it is compatible with the length of the nerve processes obtained by intracellular staining of myenteric neurons (Nishi and North 1973 a). There is immunohistochemical evidence for projections of somatostatin-containing neurons over distances of up to 10 mm in the aboral direction (Furness et al. 1980), but the only electrophysiologic evidence for cell processes which conduct for distances greater than 1 mm is the work of Kosterlitz and Lydon (1971). This subject merits further careful study.

The influence of the extrinsic nerves on the properties of the myenteric neurons has also been little studied. Nishi and North (1973 b) showed that noradrenaline caused presynaptic inhibition of acetylcholine release within myenteric ganglia, but it was Hirst and McKirdy (1974 a) who showed that this effect could also be observed during and following repetitive stimulation of the periarterial nerves. Stimulation of the periarterial nerves (at frequencies greater than 20 Hz) hyperpolarized the circular but not the longitudinal muscle layer. Such stimulation was without effect on the resting membrane potential of myenteric neurons, but it caused a depression of both spontaneous e.p.s.p., and also e.p.s.p. evoked by close transmural stimulation. The depression of the first e.p.s.p. in a series was more marked than the depression of later e.p.s.p. – an exactly similar finding was reported for the action of exogenous noradrenaline (Nishi and North 1973 b; North and Nishi 1974). In only 2 of 40 experiments did periarterial nerve stimulation result in e.p.s.p. in myenteric neurons, presumably owing to excitation of vagal fibers.

III. Comparison of Results of Intracellular Recording with Results of Extracellular Recording

Extracellular recording from enteric neurons is technically simpler than intracellular recording; for this reason, it was the first technique to be successfully applied. Before any comparison can be made between the results obtained, it is important to consider the nature of the preparation from which the recordings were made. The separation of the layers of the intestinal wall during the preparation of the tissue for electrophysiologic recording might be expected to influence the type of recordings obtained.

1. Recording from Enteric Neurons in Intact Segments of Gut Wall

In these studies, the intestinal wall was intact, or else small pieces of longitudinal muscle had been removed to expose the myenteric plexus lying on the circular muscle layer. Only extracellular recordings have been made in these circumstances (YOKOYAMA 1966; OHKAWA and PROSSER 1972a, b; WOOD 1970, 1973). These studies used insulated metal electrodes and described spontaneously firing neurons of three types – single-spike, mechanoreceptor, and burst. It is possible that this spontaneous activity was: (a) induced by contact with the large tipped recording electrode (either as an artifact or as a representation of physiologically relevant mechanoreceptor units); or (b) due to inherent "pacemaker" properties of the neurons, similar to that observed in certain cardiac and smooth muscle tissue; or (c) due to continuous synaptic bombardment by excitatory transmitters. Evidence favoring the first interpretation came from later extracellular recordings from stripped myenteric plexus longitudinal muscle preparations, and evidence in favor of the third interpretation came from YOKOYAMA's (1966) finding that hexamethonium blocked the spontaneous spike activity. Strong evidence against the second interpretation came from later intracellular recordings.

2. Recording from Enteric Neurons in Separate Layers of Gut Wall Close to Intact Segments of Intestine

Intracellular recordings have been made from myenteric neurons on the surface of a flap of stripped longitudinal muscle yet remaining neurally connected to an intact segment of ileum (HIRST and MCKIRDY 1974b; HIRST et al. 1975). These recordings revealed a low rate (0.2–2 Hz) of spontaneous e.p.s.p. in the myenteric neurons, a few of which gave rise to spontaneous action potentials. Distension of a balloon within the ileal lumen, or transmural stimulation of the ileal wall, increased the frequency of e.p.s.p. recorded from myenteric neurons, so long as the flap was situated aboral to the point of distension or stimulation (Fig. 8). Electrical activity has been recorded with extracellular electrodes from neurons of the rabbit small intestine lying on a small flap of stripped longitudinal muscle (YOKOYAMA and OZAKI 1980). In these circumstances, spontaneous action potentials occurred, and their frequency was usually increased following gut distension. This was accompanied by a contraction of the longitudinal muscle layer. These neurons were probably primary afferent mechanoreceptor cells, because they continued to respond to distension, even in the presence of hexamethonium and manganese, which would be expected to block the cholinergic transmission between neurons in the plexus (YOKOYAMA and OZAKI 1980; and personal communications). No pacemaker potentials were observed, and nor were any spontaneous action potentials observed which did not arise from e.p.s.p.

3. Recording from Enteric Neurons in Isolated Layers of Gut Wall

This is the simplest and most common experimental procedure. Extracellular recordings in these circumstances have been made with insulated metal, or glass suction electrodes. The tip diameter of the former was 5–30 μm, and these were used to record spontaneous activity from rabbit myenteric neurons in the studies by

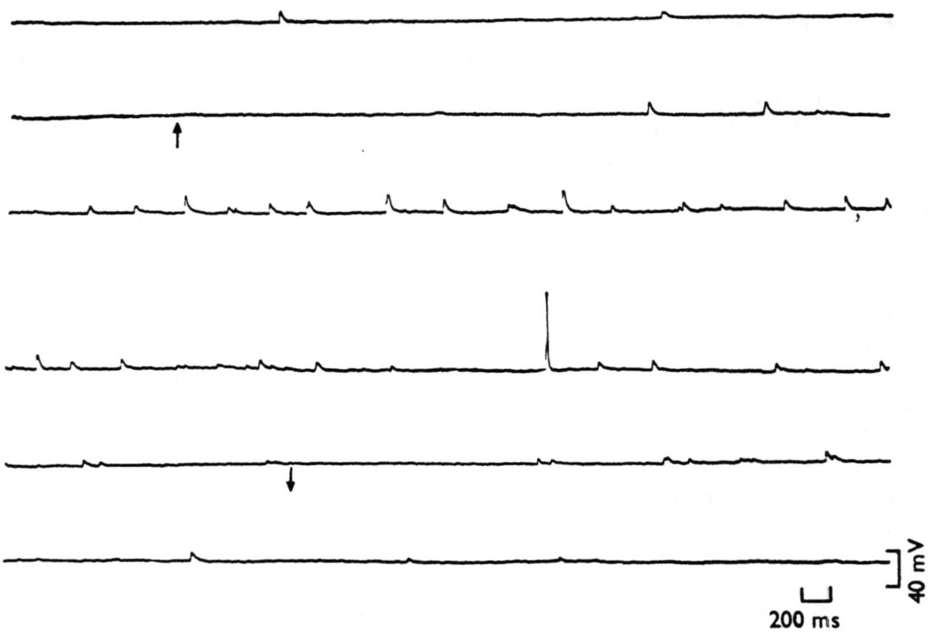

Fig. 8. Effect of distension on the rate of discharge of synaptic potentials in a myenteric neuron in a flap attached to the anal side of an intestinal segment. Onset and termination of distension (0.2 ml) are indicated by upward and downward arrows respectively. Note that increased frequency of e.p.s.p. begins about 3 s after inflation of the balloon and continues for some time after deflation (action potential retouched). (HIRST and MCKIRDY 1974b)

YOKOYAMA (1971) and YOKOYAMA et al. (1977a). There are no reports of spontaneous activity recorded with such electrodes from the isolated myenteric plexus–longitudinal muscle preparation of the guinea pig. Evoked activity can be readily induced by electrical stimulation of another node of the plexus (YOKOYAMA 1971). Glass suction electrodes have been widely used to record activity from isolated strip preparations (SATO et al. 1973, 1974; TAKAYANAGI et al. 1974, 1977; EHRENPREIS et al. 1976; DINGLEDINE et al. 1974; DINGLEDINE and GOLDSTEIN 1975, 1976; NORTH and WILLIAMS 1977; NORTH and KARRAS 1978a; NORTH and ZIEGLGANSBERGER 1978; KATAYAMA et al. 1979; WILLIAMS et al. 1979; WILLIAMS and NORTH 1979a, b). The glass suction electrode not only records but also induces the electrical activity recorded in these circumstances. The evidence for this conclusion is two-fold. First, no spike activity could be recorded without the application of suction, even though the electrode was making good contact with the ganglion surface. Increasing the suction pressure increased the firing rate of the cells. Second, units which stopped firing with no change or an increase in spike amplitude started firing again if the perfusion solution was changed to one which contained an excitatory substance, or one which was devoid of calcium ions (NORTH and WILLIAMS 1977; J. T. WILLIAMS 1979 unpublished work). Evidence has been discussed above that the neuronal activity recorded with suction electrodes is the action potential occurring on the entire soma membrane. In view of the evidence that most myenteric neurons have a small localized part of the membrane (the axon hillock region) which is more excitable than the bulk soma membrane (NISHI and NORTH 1973a; Fig. 3), it is likely that this is the site of generation of the spike recorded. It may be presumed that the initiating factor is the mechanical deformation of the axon hillock membrane against the edge of the glass at the tip of the suction electrode.

Intracellular recordings have been made from neurons in the stripped preparation (NISHI and NORTH 1973a; HIRST et al. 1974). A most important feature of these studies is that no evidence was found for inherent pacemaker activity in myenteric or submucous neurons and spontaneous synaptic potentials were only rarely observed. Spontaneous proximal process potentials – caused by the action potential entering the soma electrotonically from the proximal process (or axon hillock region) – are occasionally observed (NISHI and NORTH 1973; WOOD and MAYER 1978a). It is likely that the action potentials are initiated in more distal parts of the cell processes which are damaged during the preparation, but the possibility that these result from continuous excitatory synaptic input or on spontaneously occurring mechanoreceptor input to distal processes has not been satisfactorily excluded.

The major critique of the extracellular technique is three-fold. First, only somatic activity is likely to be recorded with the electrodes used in studies to date. Second, it is clear in at least some of the studies that the electrode (suction) actually induces the activity which is recorded. Third, when suction electrodes were not used (which was largely in situations in which the gut wall was relatively intact and spontaneous activity might be anticipated) only one study (YOKOYAMA 1966) presents any evidence that the activity recorded was the result of continuous synaptic input to the neurons under study.

Criticisms of the intracellular recording experiments are also in order. First, it is possible that larger cell bodies are preferentially penetrated, although the same consideration pertains to the extracellular recordings. Second, only somatic events can be directly recorded. Third, and most significant, all the experiments with intracellular recording have been made on stripped preparations, either totally isolated from, or with only partial connections to the cells in the intact gut wall. A major need exists for intracellular recordings to be made from neurons lying in the intact wall of the intestine.

The studies with intracellular recording, with the notable exception of those by HIRST and MCKIRDY (1974b) and HIRST et al. (1975), help to elucidate the basic forms of interneuronal communication which are available within myenteric ganglia, but do not as yet offer much insight as to which of these are important in the continuous activity of intestine in vivo.

The studies using intact segments of intestine with extracellular recording have so far failed to distinguish unequivocally between activity induced by the elecrode (artifactual or due to relevant mechanoreceptors) and activity due to continuous synaptic activity. It is therefore not possible to use with confidence the division of neuronal types into "burst", "single-spike", and "mechanoreceptor", nor to interpret studies purporting to show interaction between such cell types. It seems likely that the single-spike units represent relatively undamaged type 1 (S) cells in view of their firing pattern and sensitivity to various agents (see Sect. I). The burst units may represent type 2 (AH) cells whose discharge is limited to bursts by the after-hyperpolarization of the soma membrane. The mechanoreceptor units are probably either type 1 (S) cells (tonic mechanoreceptors) or type 2 (AH) cells (phasic mechanoreceptors) which are deformed by contact with the recording electrode. The studies using suction electrodes are completeley artifactual in the physiologic sense and provide no information relevant to our understanding of gastrointestinal motility; they do, however, offer a useful technique for studying drug action on membrane excitability. Only two types of unit are recorded with suction electrodes; the randomly firing (1–5 Hz), lower amplitude (usually 40–70 μV) units which are sensitive to many pharmacologic agents, and the more slowly firing (0.2–2 Hz) units of generally higher amplitude (5–200 μV) which discharge regularly. These two types almost certainly represent the type 1 (S) cells and type 2 (AH) cells of the intracellular records – the preponderance (85%) of extracellular recordings from the first type is presumably a reflection of the greater excitability of type 1 (S) cells compared with type 2 (AH) cells, so that they can more readily be brought to threshold by the suction electrode.

D. Actions of Endogenous and Exogenous Substances on Enteric Neurons

I. Acetylcholine and Related Substances

Although there was substantial indirect evidence that acetylcholine (ACh) and other nicotinic agonists excited myenteric neurons, the first direct demonstration of this effect was provided by WOOD (1970) and OHKAWA and PROSSER (1972b). The single-spike units, now considered to represent relatively undamaged type 1 (S) cells, were excited by quite low concentrations of ACh, dimethylphenylpiperazinium, and nicotine, and these effects were unchanged by atropine. The burst units were less frequently affected by acetylcholine – which is also compatible with their probable identification as type 2 (AH) cells.

Further study of the acetylcholine excitation was made by SATO et al. (1973) using glass suction electrodes. The single-spike units were excited by acetylcholine and nicotine, and these effects were blocked by hexamethonium (10 μ*M*) but not by a similar concentration of atropine. Studies with intracellular recording have indicated that ACh depolarizes the type 1 (S) cells of the myenteric plexus; the depolarization caused by iontophoresis of ACh has properties closely similar to the fast e.p.s.p., reversing at about 0 mV and being blocked by hexamethonium. Type 2 (AH) cells, which do not receive nicotinic fast e.p.s.p., seem to be less sensitive to iontophoretic application of ACh (NORTH and NISHI 1976). The action of ACh has not been tested directly on submucous neurons, but the block of their fast e.p.s.p. by nicotinic antagonists implies that it probably excites all cells in this plexus. OHKAWA and PROSSER (1972b) stated that ACh excited all of the three submucous neurons to which it was applied. These findings with ACh establish it unequivocally as a major interneuronal transmitter substance within the enteric nervous system.

II. Noradrenaline and Other Catecholamines

The initial experiments with catecholamines using extracellular recording showed either no effect (WOOD 1970) or excitations of myenteric and submucous neurons (OHKAWA and PROSSER 1972b). The threshold concentration for noradrenaline was 300 nM, and for adrenaline 3 μM. These excitatory effects have not been reported by later workers using either suction electrodes or intracellular electrodes, and their significance remains a mystery. However, there are other notable differences between the results of these early studies and those of later work with regard to the effects of 5-HT (see Sect. D.III). SATO et al. (1973) reported inhibition of neuronal activity by noradrenaline (3 μM). This was later confirmed by TAKAYANAGI et al. (1977) and by DINGLEDINE and GOLDSTEIN (1975) both of whom used suction electrodes. No units were excited by noradrenaline (3 μM) in the study of DINGLEDINE and GOLDSTEIN (1975).

Intracellular recordings indicate that adrenaline (0.1–1 μM) and noradrenaline (1 μM) generally have no effect on the membrane potential and input resistance of type 1 (S) cells. Their most prominent effect is rapid and reversible depression of the fast e.p.s.p. (NISHI and NORTH 1973b; NORTH and NISHI 1974; WOOD and MAYER 1979c). The first e.p.s.p. in a train is much more depressed than later e.p.s.p. indeed, the second and subsequent e.p.s.p. evoked at frequencies greater than 1 Hz are not inhibited by noradrenaline. This action of noradrenaline and adrenaline is blocked by phenoxybenzamine (3 μM) (NISHI and NORTH 1973b) and phentolamine (1 μM) (WOOD and MAYER 1979c). Noradrenaline does not reduce the amplitude of the depolarization evoked by iontophoretic application of ACh to type 1 (S) cells (NISHI and NORTH 1973b) and it may therefore be concluded to be acting presynaptically.

Some type 1 (S) cells and many type 2 (AH) cells of the myenteric plexus are hyperpolarized by perfusion with a solution containing noradrenaline (NORTH and HENDERSON 1975). The probability of this effect being observed is greater when the perfusion solution contains a reduced calcium ion content (MORITA and NORTH 1981). This hyperpolarization is associated with a fall in membrane resistance (NORTH and HENDERSON 1975; see also WOOD and MAYER 1979c).

A similar hyperpolarization by noradrenaline (and also by dopamine) has been observed in neurons of the guinea pig submucous plexus. The equilibrium potential for the actions of noradrenaline and dopamine were the same as that of the inhibitory synaptic potential in these cells (HIRST and SILINSKY 1975). Furthermore, methysergide, which blocked the inhibitory synaptic potential, also blocked the effects of dopamine and noradrenaline. Methysergide also had small hyperpolarizing effects of its own in some cells. In the submucous neurons, noradrenaline depressed the amplitude of the cholinergic fast e.p.s.p., even in cells whose membrane potential was not affected by noradrenaline (cells without inhibitory synaptic potentials) (HIRST and SILINSKY 1975). This indicates a presynaptic action of noradrenaline on acetylcholine release similar to that found in the myenteric plexus (HIRST et al. 1972; NISHI and NORTH 1973b) and other autonomic ganglia (CHRIST and NISHI 1971).

The slow e.p.s.p. which can be recorded from both types of myenteric neurons was also reversibly blocked by noradrenaline (0.2–2 μM) (WOOD and MAYER 1979c) without change in resting potential of the postsynaptic cell.

III. 5-Hydroxytryptamine and Related Substances

5-HT has been of considerable interest to students of gastrointestinal motility since the findings of Gaddum and Picarelli (1975) that it could excite the gut both by direct action on the muscle (blocked by dibenzyline, the D receptor) and stimulatory action on the neurons (causing action potentials and release of acetylcholine, blocked therefore by atropine and morphine, the M receptor) (see Costa and Furness 1979a, b). The direct actions of 5-HT on enteric neurons were therefore of some interest, but early reports with extracellular electrodes (Wood 1970; Ohkawa and Prosser 1972b) showed a lack of any effect. Later studies with suction electrodes (Sato et al. 1973; Takayanagi et al. 1974; Dingledine et al. 1974) left no doubt that 5-HT had excitatory effects on some, but not all, myenteric neurons. In the fullest study Dingledine et al. (1974) found that 5-HT (1 μ*M*) excited about one-half of the single-spike myenteric neurons, inhibited one tenth, and had no effect on the remainder. 5-HT (1 μ*M*) increased the firing rate by a mean of four-fold. This excitation was dose dependent (0.1–10 μ*M*) and tended to pass off during the exposure to 5-HT.

Intracellular studies of the action of 5-HT are compatible with these findings. In the study by Johnson et al. (1980b) almost one-half of type 1 (S) cells were excited by perfusion with 5-HT, which is compatible with the proposal made earlier that these cells are the ones most readily induced to fire and be recorded with the extracellular suction electrodes. The depolarizing responses often passed off after 2–3 min perfusion with 5-HT. In contrast, the predominant response of type 2 (AH) cells to bath application of 5-HT was a membrane hyperpolarization similar in all respects to that caused by noradrenaline. This membrane hyperpolarization was often profound, and it persisted for as long as the 5-HT was present in the perfusing solution. More detailed studies of the response to 5-HT were made by employing iontophoretic application (Wood and Mayer 1979b; Johnson et al. 1980b). About one-half of type 1 (S) cells were depolarized, as was the case with 5-HT perfusion. Type 2 (AH) cells were more often hyperpolarized than depolarized with low iontophoretic charges (less than 100 nC), but higher charge ejections led to a higher probability of observing a depolarization of type 2 (AH) cells (Johnson et al. 1980b).

The ionic mechanism of the depolarizing response in both cell types may be predominantly potassium inactivation. The hyperpolarizing response appears to be due to potassium activation (Johnson et al. 1980b). The 5-HT depolarization, whether induced by iontophoretic or bath application, augments the excitability of the myenteric neurons, presumably because of the concomitant increase in membrane resistance. Other studies (Wood and Mayer 1979b) which have used iontophoretic application of 5-HT have used exceedingly high iontophoretic currents (typically 200–400 nA, 10–15 ms, 2,000–4,500 nC), and these applications caused potential changes associated with a reversible fall in input resistance (e.g., Fig. 2b of Wood and Mayer 1978b).

A third action of 5-HT on myenteric neurons is quite distinct from the depolarization and hyperpolarization described. This is presynaptic inhibition of transmitter release very similar to that described for noradrenaline (Henderson and North 1975; North and Henderson 1975; North et al. 1980b). Low concentrations (0.01–1 μ*M*) of 5-HT cause reversible and readily repeatable depressions of the fast e.p.s.p. in cells in which there is no change in postsynaptic membrane potential or resistance. This action of 5-HT is observed on all e.p.s.p. (NORTH et al. 1980b). It is considered presynaptic because the postsynaptic response to ACh iontophoresis in unchanged, and because the first e.p.s.p. in a train is much more strongly depressed than the second and subsequent e.p.s.p. (NORTH et al. 1980b). This presynaptic action would be effective in blocking transmission through the myenteric ganglia and may underlie the inhibition of the peristaltic reflex caused by serosal application of 5-HT (BÜLBRING and CREMA 1958; see KOSTERLITZ and LEES 1964).

5-HT (1 μ*M*) also depresses the noncholinergic slow e.p.s.p. in myenteric neurons with a time course similar to that by which it depresses the fast e.p.s.p. (NORTH et al. 1980b; WOOD and MAYER 1979b; JOHNSON et al. 1981). WOOD and MAYER (1978b, 1979b) contended that the blockade by exogenous 5-HT of the slow e.p.s.p. was due to a desensitization of the postsynaptic receptors to 5-HT released by nerve stimulation (i.e., that 5-HT is the transmitter mediating the slow e.p.s.p.); this in inferential. In view of the findings that 5-HT blocks the fast e.p.s.p. and the slow e.p.s.p. (and the slow i.p.s.p., JOHNSON et al. 1981), it would most likely be due to a presynaptic action of 5-HT to inhibit the release of whatever transmitter underlies the slow e.p.s.p. The rapid reversibility of the inhibition by 5-HT of the fast e.p.s.p. and slow e.p.s.p. is also compatible with a presynaptic action – the blockade caused by postsynaptic desensitization might not be expected to be rapidly reversible on washing out the 5-HT. It is of interest that the excitatory effect of 5-HT in myenteric neurons usually passed off during the application, but no such tachyphylaxis was observed to the inhibitory effects of 5-HT (membrane hyperpolarization and presynaptic inhibition).

5-HT has been applied by iontophoresis to submucous neurons (HIRST and SILINSKY 1975). It caused only depolarizing responses, which were not blocked by methysergide or lysergic acid diethylamide, but which were blocked by (+)-tubocurarine (HIRST and SILINSKY 1975) and partially by cyproheptadine (NEILD 1978).

The use of agents as 5-HT antagonists in the intestine is plagued with confusion. This is particularly acute in studies on single neurons, where it is seldom possible to test several concentrations and show competition, or to test several agonists and show specificity. Much of the work with putative 5-HT-antagonists therefore resides on inference from effects describes elsewhere. COSTA and FURNESS (1979a) studied the responses of the intact guinea pig intestine to 5-HT and distinguished carefully betweeen the nerve-mediated contractions (due to release of ACh, M receptor) and the direct muscle action (D receptor; GADDUM and PICARELLI 1957). Methysergide had no effect on the excitation of nerves, but blocked the action of 5-HT on the muscle. Phenylguanide and 2-bromolysergic acid were also ineffective in blocking the nerve stimulatory action of 5-HT on the guinea pig ileum.

The findings of HIRST and SILINSKY (1975) on submucous neurons, and COSTA and FURNESS (1979a) on myenteric cholinergic neurons, that methysergide does *not* block the action of 5-HT should be considered when attempting to interpret the results of WOOD and MAYER (1978b, 1979b), that methysergide (3–30 μ*M*) blocked the depolarizing responses to exogenous 5-HT. Details of the concentrations of 5-

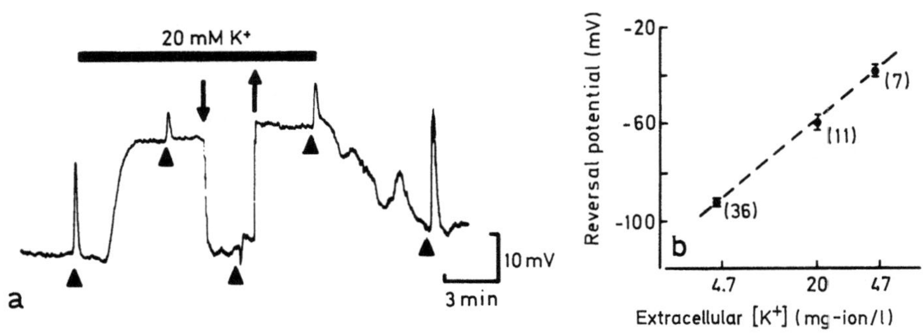

Fig. 9. a Intracellular recording from a myenteric neuron. Substance P (40 nA, 5 s) was applied, by electrophoresis, to the soma membrane *(triangles)*. High concentration potassium solution (20 m*M*; *solid bar*) depolarized the membrane and much reduced the amplitude of the substance P potential. The membrane potential was shifted back to the control level *(downward arrow)* by hyperpolarizing current. The substance P potential was now reversed. These changes were reversible when the hyperpolarizing current was terminated *(upward arrow)* and when perfusion with normal Krebs solution was restarted. **b** Relation between substance P reversal potential and extracellular potassium ion concentration. Reversal potentials are mean values (with standard errors indicated) for the number of cells shown. The slope of the line (fitted by eye) is 54 mV per ten-fold change in potassium ion concentration. (KATAYAMA et al. 1979)

HT or the number of experiments are not given. Methysergide does reduce or abolish the slow e.p.s.p. (WOOD and MAYER 1979b; JOHNSON et al. 1981), an observation argued by WOOD and MAYER (1979b) to support the identity of 5-HT as the underlying transmitter. But methysergide is quite nonspecific for it also blocks the fast e.p.s.p. and the slow i.p.s.p. (NORTH et al. 1980b; JOHNSON et al. 1981). This is quite possibly the result of agonist actions of methysergide itself (see HIRST and SILINSKY 1975), perhaps similar to those described for lysergic acid diethylamide (HUGHES 1973). It is therefore of little value in identifying the transmitter responsible for the slow e.p.s.p.

IV. Substance P

Substance P was isolated from the intestine by VON EULER and GADDUM (1931), and has recently been shown to be present within a large number of varicose nerve processes of the myenteric and submucous plexuses (FURNESS et al. 1980; COSTA et al. 1981). The possibility exists that it may function as an excitatory transmitter from nerve to muscle (FRANCO et al. 1979). Experiments on single neurons indicate that it may also function as a transmitter between neurons of the myenteric plexus. Low concentrations (0.1–100 n*M*) of substance P increase the firing rate of neurons recorded with an extracellular suction electrode; this excitation persists in calcium-free solutions, and is not shared (except at much higher concentrations) by the pentapeptide analog of substance P, or substance P from which the COOH terminal amide group has been removed (KATAYAMA and NORTH 1978; KATAYAMA et al. 1979).

Intracellular recordings indicate that substance P depolarizes myenteric neurons by inactivating their resting potassium conductance (KATAYAMA et al. 1979; Fig. 9).

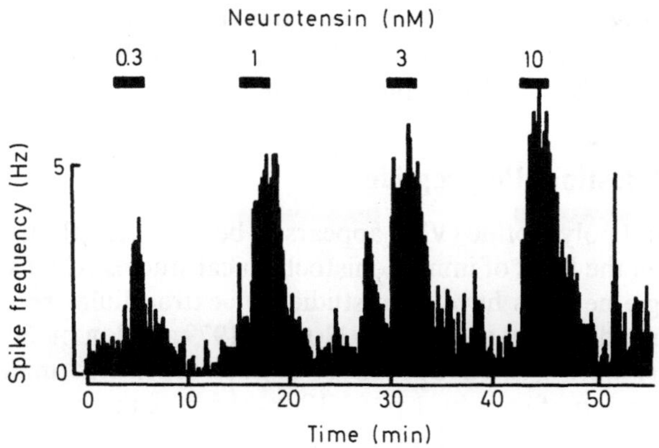

Fig. 10. Concentration-dependent excitation of a myenteric neuron by neurotensin. During the periods indicated by the *solid bars* the perfusing solution was changed to one which contained neurotensin in the concentrations indicated. (WILLIAMS et al. 1979)

This ionic mechanism is therefore common to the slow e.p.s.p. and the 5-HT depolarization, and is not particularly helpful in ascribing a transmitter role to one or other substance. This ionic mechanism is also shared by other peptides in the plexus. The high density of substance P terminals in the myenteric plexus and the high proportion of cells which respond to it suggest that it may play a role in intercellular communication such as the slow e.p.s.p.; on the other hand, the variability of the responses to 5-HT and the difficulty in demonstrating its presence within myenteric neurons argue against such a role for 5-HT. Actually many cells in which a depolarizing slow e.p.s.p. can be recorded are hyperpolarized by exogenous 5-HT (JOHNSON et al. 1981).

A further indication that substance P can be released from presynaptic nerves to act upon myenteric neurons is the finding that chymotrypsin will block the slow e.p.s.p. Chymotrypsin (200 μg/ml) also reversibly blocks the response to iontophoretic application of substance P, indicating that it can destroy substance P between the time of its release from the iontophoresis electrode and its subsequent action on the neuronal membrane. Chymotrypsin did not reduce the fast e.p.s.p. or the response to iontophoretic application of 5-HT, arguing against a nonspecific action to reduce the sensitivity of the postsynaptic membrane (MORITA et al. 1980).

V. Neurotensin

Extracellular recordings with glass suction electrodes indicated that neurotensin (0.1–300 n*M*) caused a dose-dependent excitation of about 50% of myenteric neurons in the class of randomly firing single-spike units, not the more regularly firing units of higher amplitude and slower frequency (Fig. 10; WILLIAMS et al. 1979). In agreement with this, exactly one-half of type 1 (S) cells identified by intracellular recording were depolarized by neurotensin (100 n*M*). This depolarization was accompanied by an increase in neuronal membrane resistance, perhaps indicating potassium inactivation. Type 2 (AH) cells were either depolarized

(25%), hyperpolarized (15%), or not affected (60%) by neurotensin (100 n*M*). The depolarization was similar to that observed in type 1 (S) cells; the hyperpolarization was accompanied by a fall in input resistance.

VI. Vasoactive Intestinal Polypeptide

Vasoactive intestinal polypeptide (VIP) appears to be relatively abundant with the myenteric plexus on the basis of immunohistochemical studies (COSTA et al. 1981). Its actions on single neurons have been studied by extracellular recordings with glass suction electrodes (WILLIAMS and NORTH 1979a; NORTH et al. 1980a). Myenteric neurons were excited by VIP in a dose-dependent manner (0.001–300 n*M*). The excitation of cells by 1 p*M* VIP marks it as the most potent of the peptides studied to date. The populations of cells which were excited by VIP, substance P, and neurotensin are not identical, indicating a degree of biologic specificity in the responses. Repeated applications of VIP at intervals of 15 min caused reproducible excitations.

VII. Morphine

The inhibitory action of morphine on the firing rate of neurons in the guinea pig ileum has been shown repeatedly (SATO et al. 1973; SATO et al. 1974; DINGLEDINE et al. 1974; DINGLEDINE and GOLDSTEIN 1975, 1976; NORTH and WILLIAMS 1977; EHRENPREIS et al. 1976; NORTH and KARRAS 1978a; NORTH and ZIEGLGANSBERGER 1978). It is concentration dependent; the proportion of neurons inhibited by morphine increases as the concentration is increased, and in a given neuron the degree of inhibition increases with increasing concentration of morphine. A concentration of 100 n*M* morphine is sufficient to inhibit the firing of most cells by at least 50% (DINGLEDINE and GOLDSTEIN 1975) and several neurons are inhibited by concentrations as low as 1 n*M* (see NORTH and KARRAS 1978a). This action of morphine is shared by normorphine, ketocyclazocine, cyclazocine, and levorphanol, but not dextrorphan (KARRAS and NORTH, 1981). It is also shared by enkephalin and β-endorphin (see Sect. D.VIII). The action of morphine is reversed or prevented by (−)-naloxone but not by (+)-naloxone, by the antagonist (−)5,9,α-diethyl-2-(3-furylmethyl)-2′-hydroxy-6,7-benzomorphan (Mr 2266) but not by the (+)-isomer (Mr 2267), and by a morphine antibody (KARRAS 1980).

It has been reported that a higher concentration of morphine is required to inhibit neuronal firing than to depress the contractile response of the longitudinal muscle–myenteric plexus preparation (EHRENPREIS et al. 1976). For this reason, and because of evidence that the inhibition of firing was less readily antagonized by naloxone than the inhibition of the contractile response, EHRENPREIS et al. (1976) proposed that the receptors mediating the two effects may be different. However, DINGLEDINE and GOLDSTEIN (1976) showed that the concentration of morphine required to cause 50% depression of the contractile response (about 300 n*M*) was actually higher than that required to inhibit cell firing by 50% (about 60 n*M*). In any event, the amount of morphine required to inhibit the contractile response varies markedly from laboratory to laboratory, significantly depending on whether maximal or submaximal electrical stimulation is used (see NORTH and

TONINI 1977) and also is quite different between the stripped longitudinal muscle myenteric plexus preparation and the intact ileum. This substantially undermines the conclusions of EHRENPREIS et al. (1976) who compared their own data on cell firing rates with published reports of others of the inhibition of the contractile response.

In addition to depressing activity of spontaneously firing neurons in the guinea pig ileum, morphine inhibits neuronal firing induced by 5-HT (SATO et al. 1973, 1974; DINGLEDINE et a. 1974; DINGLEDINE and GOLDSTEIN 1975, 1976), sodium picrate, cerulein, pentagastrin, and acetylcholine (SATO et al. 1974). The action of morphine is not much affected by high or low potassium concentrations, or by low chloride concentrations (NORTH and WILLIAMS 1977). The inhibition by morphine is also not affected in low or zero calcium solutions, indicating both that it is a direct action on the neuron whose activity is being recorded, and also that the action itself is not dependent on extracellular calcium ions (DINGLEDINE and GOLDSTEIN 1976; NORTH and WILLIAMS 1977). In calcium-free solutions, type 2 (AH) cells begin to discharge spontaneously (NISHI and NORTH 1973a) and their firing is less sensitive to inhibition by morphine (DINGLEDINE and GOLDSTEIN 1976; NORTH and WILLIAMS 1977). However, the type 1 (S) cells, whose firing rate is increased in Ca-free solutions, retain a normal sensitivity to morphine (DINGLEDINE and GOLDSTEIN 1976).

The inhibition of firing by morphine is demonstrable, therefore, under almost all experimental circumstances in which it is sought. DINGLEDINE and GOLDSTEIN (1976) proposed that it may be due to a membrane hyperpolarization. Morphine, normorphine, and levorphanol, but not dextrorphan, were later shown to hyperpolarize the membrane of a proportion of myenteric neurons. These effects were blocked by naloxone in concentrations equal to (0.01–1 μM) or ten times lower than the concentration of agonist (NORTH 1976; NORTH and TONINI 1977); they occurred without change in the cell action potential.

Such a hyperpolarizing action of morphine is likely to contribute to the inhibition of firing observed with the suction electrodes. Furthermore, the hyperpolarization probably contributes to the inhibition of acetylcholine release by morphine when the ileum (or stripped muscle–plexus preparation) is stimulated by an electric field. In the latter case, the action of morphine is to hyperpolarize neurons sufficiently to prevent their excitation altogether, or to prevent action potential propagation into the nerve varicosities (for full discussion see NORTH and TONINI 1977). Strong independent support for this hypothesis is provided by the experiments of DOWN and SZERB (1980), who showed that morphine inhibits ACh release, not by reducing the rate of efflux from an ACh pool of constant size, but by reducing the size of the pool (i.e., reducing the number of neurons excited). Morphine also hyperpolarizes myenteric neurons in the cat small intestine (WOOD 1979b).

Morphine has not only acute inhibitory actions in single myenteric neurons, but following long-term application (either by injecting live animals with morphine, or by exposing the ileum to morphine for 24 h after its removal from the animal) the neurons become quite insensitive to the usual inhibitory effect of the morphine. In these circumstances, application of the antagonist(−)-naloxone [but not its inactive isomer (+)-naloxone] causes a profound excitation of the single cells (NORTH and KARRAS 1978a; NORTH and ZIEGLGANSBERGER 1978). This excitation causes massive release of acetylcholine, which leads to a muscle contracture. These effects of naloxone (similar effects can also be observed by simple withdrawal of morphine) may be interpreted as signs of physical dependence of the neurons on morphine. It is of particular interest that the development of this long-term change (dependence) does not require continuous neuronal activity as it is unchanged by concomitant exposure to lidocaine (NORTH and KARRAS 1978b). The membrane potential of myenteric neurons is not affected by incubation for 24 h with morphine; however, such incubated neurons are depolarized by naloxone, whereas naloxone has no effect on neurons incubated for 24 h without morphine (JOHNSON and NORTH 1980).

VIII. Enkephalin

Met^5-enkephalin, Leu^5-enkephalin, and β-endorphin have the same effect on the firing of myenteric neurons as do morphine and normorphine; these actions are also antagonized by naloxone (NORTH and WILLIAMS 1976; WILLIAMS and NORTH 1979b). Intracellular recordings also show a hyperpolarizing action of enkephalin in many type 1 (S) neurons. This hyperpolarization appears to be generated, not on the soma membrane itself, but on a region of the cellular process. It may or may not be associated with a resistance change detectable with an intrasomatic electrode, probably depending on the geometrical and passive electrical properties of the neuron. It cannot be mimicked by iontophoretic application of enkephalin to the cell soma, but only by application of enkephalin to the cell processes (NORTH et al. 1979). Moreover, the excitability of the cell axon (or axon hillock) is reduced during enkephalin application, even without change in the soma membrane potential, which also reflects a hyperpolarization and/or conductance increase at the point of stimulation on the axonal membrane (MORITA and NORTH 1981).

The difficulty in observing hyperpolarizing inhibitory synaptic potentials in myenteric neurons (see Sect. C.II), except by repetitive and simultaneous excitation of a number of nerve fibers, makes it unlikely that enkephalin normally functions as an inhibitory transmitter between myenteric neurons. Occasionally the slow i.p.s.p. is reversibly blocked by naloxone (JOHNSON et al. 1981) but this is not a consistent finding. It is possible that the principal action of enkephalin released from myenteric neurons may occur at a site distant from the cell soma, perhaps acting to inhibit release of acetylcholine by hyperpolarizing block of varicose fibers.

IX. Somatostatin

Somatostatin is present in about 5% of myenteric neurons in the guinea pig ileum (COSTA et al. 1977; FURNESS et al. 1980). Extracellular recordings indicated that somatostatin (300 pM–1 μM) inhibited the firing of most myenteric neurons – both the randomly firing (single-spike cells) and the more slowly firing units of higher amplitude [type 2 (AH) cells; WILLIAMS and NORTH 1978]. Tachyphylaxis to the depressant effect of somatostatin was apparent with concentrations greater than 100 nM. Somatostatin also inhibited firing in calcium-free solutions. Intracellular recordings showed both depolarizing and hyperpolarizing responses to somatostatin, associated with an increase and a decrease in cell input resisistance respectively (similar to the effects of 5-HT) (KATAYAMA and NORTH 1980). With bath application of somatostatin (100 nM) about one-third of type 1 (S) cells and one-fifth of type 2 (AH) cells were hyperpolarized. In each respective group, one-sixth and one-tenth of cells were depolarized. Iontophoretic application of somatostatin was more likely to produce membrane depolarization (also similar to 5-HT), but the ionic mechanism of the responses was the same whether produced by bath application (known concentration) or by iontophoresis (unknown concentration). It should be borne in mind that bath application applies somatostatin to the entire cell surface, whereas iontophoretic application puts the somatostatin on the soma. As in the case of enkephalin (NORTH et al. 1979) several cells were observed which were depolarized by iontophoretic application to the soma and hyperpolarized by bath application.

The responses are closely similar to those observed with enkephalin – the predominant effect of application to the soma is depolarization (due to potassium inactivation) whereas the predominant effect of application to the entire cells is hyperpolarization. The hyperpolarization is probably generated at a nonsomatic site. Like the enkephalin responses, the hyperpolarization caused by somatostatin persisted in a calcium-free solution. The hyperpolarization caused by somatostatin is not blocked by naloxone (1 μM).

X. Adenosine and Cuclic Nucleotides

Adenosine and related nucleotides depress the nerve-mediated contractile response of the longitudinal muscle of the guinea pig ileum by inhibiting the release of acetylcholine (HAYASHI et al. 1978). The firing of substantial proportion of myenteric neurons is inhibited by cyclic AMP and dibutyryl cyclic AMP. These agents appear not to affect the inhibition of firing induced by morphine (KARRAS and NORTH 1979).

E. Functional Roles of Enteric Neurons in the Control of Gastrointestinal Motility

In discussing the role of enteric neurons in controlling the activity of the intestinal musculature, the experimental circumstances must be carefully described. When the longitudinal muscle of guinea pig small intestine is stripped from the circular muscle so that the myenteric plexus adheres to its surface, and this piece of tissue is completely separated from the intestinal wall, no spontaneous activity can be recorded from myenteric neurons. That is to say, no evidence exists for "pacemaker" cells or reverberating synaptic circuits involving one or a few ganglia. The "spontaneous" activity recorded from neurons in these circumstances is a consequence of mechanical deformation with a large extracellular suction electrode. As the physiologic significance of this deformation is not known, it is not appropriate to call such cells mechanoreceptors. The primary afferent neurons which respond to radial stretch of the intestinal wall have so far not been identified unequivocally by electrophysiologic methods, except in the recent study of YOKOYAMA and OZAKI (1980). Some of the properties of the type 2 (AH) neuron are compatible with its function as an afferent cell (see Sect. C).

When the recordings are made from myenteric ganglia which remain on the circular muscle (e.g. WOOD 1970, 1973), spontaneous spike activity can be recorded extracellularly from neurons in the guinea pig or cat small intestine. In these circumstances, prevention of neuronal firing by any of several means increases the contractile activity of the circula muscle layer (WOOD 1970, 1972). WOOD (1972) suggested that it was the burst neurons which exerted the tonic inhibitory input to the circular muscle. What might be the underlying reason for their burst activity remains to be tested by intracellular recording in intact intestinal wall.

Intracellular recordings have been made from myenteric neurons lying on a flap of stripped longitudinal muscle, connected to an intact piece of intestine within 1 cm of the recording site (HIRST et al. 1975). In these circumstances, spontaneous fast e.p.s.p. were recorded from about 50% of type 1 (S) myenteric neurons and occasionally spontaneous action potentials – these were recorded only in neurons situated anally to the intact segment of ileum. No slow e.p.s.p. were observed. Transmural stimulation of the intact piece of ileum, or distending a balloon within its lumen, always initiated a burst of e.p.s.p. in the neurons on anally situated flaps (Fig. 8). These findings constitute an elegant demonstration of descending orientation of physiologically relevant pathways within the myenteric plexus. Of particular interest was the finding that the balloon distension led to inhibitory junction potentials (i.j.p.) only in the anally located circular muscle layer. In the longitudinal muscle i.j.p. were not observed. This descending inhibitory pathway which was activated by balloon distension or transmural stimulation and led to an i.j.p. in the circular muscle was completely and reversibly blocked by (+)-tubocurarine. This is in agreement with many observations which indicate that peristalsis is blocked by (+)-tubocurarine. This finding, as well as the absence of any slow potential changes in these experiments, raises the question whether slow e.p.s.p. or slow i.p.s.p. is of functional significance in the intact gut wall.

Intracellular recordings have not yet been made from myenteric neurons in an intact segment of bowel which is distended. The technical difficulties are substantial. However, only such an experiment can succeed in identifying the afferent cells themselves [presumably type 2 (AH) cells] and in elucidating the detailed nature of the descending inhibitory pathway. Extracellular recordings from the longitudinal muscle-myenteric plexus preparation of the rabbit small intestine show that repetitive electrical stimulation of a node of the plexus causes excitation of the orally situated longitudinal muscle and inhibition of the aborally situated longitudinal muscle. This is further evidence that a polarity of the pathways exists within the myenteric plexus itself (YOKOYAMA and OZAKI 1978a).

HIRST et al. (1975) described further important observations on the descending pathway in the guinea pig ileum. Some neurons responded to balloon distension with a burst of e.p.s.p. within 1 s. Another group of neurons responded to balloon distension with a burst of e.p.s.p. occurring only after a longer latency (3–7 s). Intracellular recordings from the muscle layers indicated that balloon distension was followed by an i.j.p. in circular muscle with a delay of about 1 s. This was followed by excitatory junction potentials (e.j.p.) in both muscle layers with a latency of 3–7 s. These values agreed well with the latencies to the appearance of e.p.s.p. The inference was that balloon distension activated two descending pathways – the first was rapidly conducting, initiated fast e.p.s.p. in anally situated neurons (at least the bulk of which were inhibitory), the fast e.p.s.p. brought these cells of threshold for an action potential and this led to the i.j.p. in the circular muscle. At the same time, a descending pathway was activated which was much slower and this led to a burst of e.p.s.p. in another separate group of anally situated neurons (most of which were excitatory); action potentials arising from the e.p.s.p. in these cells propagated anally, released ACh and caused e.j.p. in both muscle layers.

These experiments provided the framework for understanding of the descending inhibition described by BAYLISS and STARLING (1899, 1900). The reason for the long delay in the descending excitatory pathway may be transmission to and from the submucous plexus (see HIRST 1979) on the basis of the finding that the e.j.p. in the muscle layers could not be evoked by radial distension in preparations from which the submucosa had been removed.

The overall picture which emerges is of a tonic inhibitory control on the circular musculature, and of little tonic control of the longitudinal muscle. This leads to a spontaneously active longitudinal muscle – which has no significant propulsive capacity – and a circular muscle which is not spontaneously active. In the resting state, with an empty lumen, what constitutes the input to the intrinsic inhibitory neurons? This could be "pacemaker cells" or "reverberating circuits" as proposed by WOOD (1970), but none have ever been observed with electrophysiologic techniques which leave beyond doubt the origin and nature of the activity recorded. Or could it be the continuous activation of the type 2 (AH) cells, or whatever are the real mechanoreceptors, by a slightly moving, tonically active, longitudinal muscle layer? Or it could be other continuous afferent inputs to the plexus – chemoreceptors in the mucosa – as yet undetected? Upon distension of the lumen, type 2 (AH) cells would be more strongly activated and they would excite two pathways – descending inhibition followed by descending excitation. The cause of the interval between them is yet not understood, but may involve the submucosa.

One final caveat is in order. Much of the work described above has been carried out on the small intestine of the guinea pig, rabbit, and cat. There remains a relative ignorance of the electrophysiologic properties of enteric neurons in other parts of the intestinal tract and from other animal species.

Acknowledgements. The author is grateful to Professor HANS KOSTERLITZ for initiating and fostering his interest in this subject, to Dr. SYOMATU YOKOYAMA and Dr. DAVID HIRST for valuable discussions and to the various funding agencies which supported the work carried out in his own laboratory (USPHS NIH Grant NSO6672, NIDA Grant DA01730, and the Schweppe Foundation).

References

Bayliss WM, Starling EH (1899) The movements and innervation of the small intestine. J Physiol (Lond) 24:99–143

Bayliss WM, Starling EH (1900) The movements and innervation of the large intestine. J Physiol (Lond) 26:107–118

Bennett MR, Burnstock G, Holman ME (1966a) Transmission from intramural inhibitory nerves to the smooth muscle of the guinea-pig taenia coli. J Physiol (Lond) 182:541–558

Bennett MR, Burnstock G, Holman ME (1966b) Transmission from perivascular inhibitory nerves to the smooth muscle of the guinea-pig taenia coli. J Physiol (Lond) 182:527–540

Bülbring E, Crema A (1958) Observations concerning the action of 5-hydroxytryptamine on the peristaltic reflex. Br J Pharmacol 13:444–457

Christ DD, Nishi S (1971) Effects of adrenaline on nerve terminals in the superior cervical ganglion of the rabbit. Br J Pharmacol 41:331–338
Costa M, Furness JB (1979a) The sites of action of 5-hydroxytryptamine in nerve-muscle preparations from the guinea-pig small intestine and colon. Br J Pharmacol 65:237–248
Costa M, Furness JB (1979b) On the possibility that an indoleamine is a neurotransmitter in the gastrointestinal tract. Biochem Pharmacol 28:565–571
Costa M, Patel Y, Furness JB, Arimura A (1977) Evidence that some intrinsic neurons of the intestine contain somatostatin. Neurosci Lett 6:215–222 (1977)
Costa M, Furness JB, Buffa R, Said SI (1981) Distribution of enteric nerve cell bodies and axons showing immunoreactivity for vasoactive intestinal polypeptide (VIP) in the guinea-pig intestine. Neuroscience 5:587–596
Costa M, Cuello AC, Furness JB, Franco R (1980) Distribution of enteric neurons showing immunoreactivity for substance P in the guinea-pig ileum. Neuroscience 5:323–331
Dichter MA, Fischbach GD (1977) The action potential of chick dorsal root ganglion neurones maintained in cell culture. J Physiol (Lond) 267:281–298
Dingledine R, Goldstein A (1975) Single neuron studies of opiate action in the guinea-pig myenteric plexus. Life Sci 17:57–62
Dingledine, R, Goldstein A (1976) Effect of synaptic transmission blockade on morphine action in the guinea-pig myenteric plexus. J Pharmacol Exp Ther 196:97–106
Dingledine R, Goldstein A, Kendig J (1974) Effects of narcotic opiates and serotonin on the electrical behavior of neurons in the guinea pig myenteric plexus. Life Sci 14:2299–2309
Down JA, Szerb JC (1980) Kinetics of morphine-sensitive [^{3}H]-acetylcholine release from the guinea-pig myenteric plexus. Br J Pharmacol 68:47–55
Ehrenpreis S, Sato T, Takayanagi I, Comaty JE, Takagi K (1976) Mechanism of morphine block of electrical activity in ganglia of Auerbach's plexus. Eur J Pharmacol 40:303–309
Feldberg W, Lin RCY (1949) The action of local anaesthetics and d-tubocurarine on the isolated intestine of the rabbit and guinea-pig. Br J Pharmacol Chemother 4:33–44
Franco R, Costa M, Furness JB (1979) Evidence for the release of endogenous substance P from intestinal nerves. Naunyn-Schmiedeberg Arch Pharmakol 306:195–201
Furness JB, Costa M, Franco R, Llewellyn-Smith IJ (1980) Neuronal peptides in the intestine: distribution and possible functions. Adv Biochem Psychopharmacol 22:601–617
Gabella G, North RA (1974) Intracellular recording and electron microscopoy of the same myenteric plexus neurone. J Physiol (Lond) 240:28–30P
Gaddum JH, Picarelli ZP (1957) Two kinds of tryptamine receptor. Br J Pharmacol 9:240–248
Grafe P, Wood JD, Mayer CJ (1979) Fast excitatory postsynaptic potentials in AH (Type 2) neurons of guinea-pig myenteric plexus. Brain Res 163:349–352
Hartzell HC, Kuffler SW, Stickgold R, Yoshikami D (1977) Synaptic excitation and inhibition resulting from direct action of acetylcholine on two types of chemoreceptors on individual amphibian parasympathetic neurones. J Physiol (Lond) 271:817–846
Hayashi E, Mori M, Yamada W, Kunimoto M (1978) Effects of purine compounds on acetylcholine release in electrically stimulated guinea-pig ileum. Eur J Pharmacol 48:297–307
Henderson G, North RA (1975) Presynaptic action of 5-hydroxytryptamine in the myenteric plexus of the guinea-pig ileum. Br J Pharmacol 54:265P
Hirst GDS (1979) Mechanisms of peristalsis. Br Med Bull 35:263–268
Hirst GDS, McKirdy HC (1974a) Presynaptic inhibition at a mammalian peripheral synapse? Nature 250:430–431
Hirst GDS, McKirdy HC (1974b) A nervous mechanism for descending inhibition in guinea-pig small intestine. J Physiol (Lond) 238:129–143
Hirst GDS, McKirdy HC (1975) Synaptic potentials recorded from some neurones of the submucous plexus of guinea-pig small intestine. J Physiol (Lond) 249:369–385
Hirst GDS, Silinsky EM (1975) Some effects of 5-hydroxytryptamine, dopamine and noradrenaline on neurones in the submucous plexus of the guinea-pig small intestine. J Physiol (Lond) 251:817–832
Hirst GDS, Spence I (1973) Calcium action potentials in mammalian peripheral neurones. Nature 243:54–56

Hirst GDS, Holman ME, Prosser CL, Spence I (1972) Some properties of the neurones of Auerbach's plexus. J Physiol (Lond) 225:60P
Hirst GDS, Holman ME, Spence I (1974) Two types of neurones in the myenteric plexus of duodenum in the guinea-pig. J Physiol (Lond) 236:303–326
Hirst GDS, Holman ME, McKirdy HC (1975) Two descending nerve pathways activated by distension of guinea-pig small intestine. J Physiol (Lond) 244:113–127
Hughes J (1973) Inhibition of noradrenaline release by lysergic acid diethylamide. Br J Pharmacol 49:706–708
Hukuhara T, Yamagami M, Nakayama S (1958) On the intestinal intrinsic reflexes. Jpn J Physiol 8:9–20
Johnson SM, North RA (1980) Membrane potential changes in neurones undergoing withdrawal from opiates. Brain Res 190:559–563
Johnson SM, Katayama Y, North RA (1980a) Slow synaptic potentials in neurones of the myenteric plexus. J Physiol (Lond) 301:505–516
Johnson SM, Katayama Y, North RA (1980b) Multiple actions of 5-HT on myenteric neurones of the guinea-pig ileum. J Physiol (Lond) 304:459–470
Johnson SM, Katayama Y, Morita K, North RA (1981) Mediators of slow synaptic potentials in the myenteric plexus of the guinea-pig ileum. J Physiol (Lond) 320:175–186
Karras PJ (1980) A study of the acute and chronic effects of narcotics on single myenteric neurons. PhD thesis, Loyola University of Chicago
Karras PJ, North RA (1979) Inhibition of neuronal firing by opiates: evidence against the involvement of cyclic nucleotides. Br J Pharmacol 65:647–652
Karras PJ, North RA (1981) Acute and chronic effects of morphine on single myenteric neurons. J Pharmacol Exp Ther 217:70–80
Katayama Y, North RA (1978) Does substance P mediate slow synaptic excitation within the myenteric plexus? Nature 274:387–388
Katayama Y, North RA (1980) The action of somatostatin on neurones of the myenteric plexus of the guinea-pig ileum. J Physiol (Lond) 303:315–323
Katayama Y, North RA, Williams JT (1979) The action of substance P on neurons of the myenteric plexus of the guinea-pig small intestine. Proc R Soc Lond [Biol] 206:191–208
Kosterlitz HW, Lees GM (1964) Pharmacological analysis of intrinsic intestinal reflexes. Pharmacol Rev 16:301–339
Kosterlitz HW, Lydon RJ (1971) Impulse transmission in the myenteric plexus – longitudinal muscle preparation of the guinea-pig ileum. Br J Pharmacol 43:74–85
Morita K, North RA (1981) Enkephalin reduces excitability of cellular processes of myenteric neurones. Neuroscience 6:1943–1951
Morita K, Katayama Y, North RA (1980) Chymotrypsin attenuates synaptic potentials: evidence that substance P is a neurotransmitter within the myenteric plexus. Nature 287:151–152
Neild TO (1978) On the response of autonomic neurones to 5-hydroxytryptamine. Proc Aust Physiol Pharmacol Soc 9:120P
Nishi S (1974) Ganglionic transmission. In: Hubbard JI (ed) The peripheral nervous system. Plenum, London, pp 225– 256
Nishi S, North RA (1973a) Intracellular recording from the myenteric plexus of the guinea-pig ileum. J Physiol (Lond) 231:471–491
Nishi S, North RA (1973b) Presynaptic action of noradrenaline in the myenteric plexus. J Physiol (Lond) 231:29–30P
North RA (1973) The calcium-dependent slow, after-hyperpolarization in myenteric plexus neurone with tetrodotoxin-resistant action potentials. Br J Pharmacol 49:709–711
North RA (1974) Action of anticholinesterase on synaptic transmission in the myenteric plexus. Proc Int Union Physiol Sci XI:394
North RA (1976) Effects of morphine on myenteric plexus neurons. Neuropharmacology 15:719–721
North RA, Henderson G (1975) Action of morphine on guinea-pig myenteric plexus and mouse vas deferens studied by intracellular recording. Life Sci 17:63–66
North RA, Karras PJ (1978a) Opiate tolerance and dependence induced in vitro in single myenteric neurons. Nature 272:73–75

North RA, Karras PJ, (1978b) Tolerance and dependence in vitro. In: Van Ree JM, Terenius L (eds) Characteristics and functions of opioids. Developments in neuroscience 4. Elsevier, Amsterdam, pp 25–36

North RA, Nishi S (1974) Properties of ganglion cells of the myenteric plexus of the guinea-pig ileum determined by intracellular recording. In: Daniel EE (ed) Proceedings of the IVth International Symposium on Gastrointestinal Motility. Mitchell, Vancouver, pp 667–676

North RA, Nishi S (1976) The slow after-hyperpolarization of myenteric neurons. In: Bülbring E, Kostyuk PG, Subha MF (eds) Physiology of smooth muscles. Raven, New York, pp 303–307

North RA, Tonini M (1977) The mechanism of action of narcotic analgesics in the guinea-pig ileum. Br J Pharmacol 61:541–549

North RA, Williams JT (1976) Enkephalin inhibits firing of myenteric neurones. Nature 264:460–461

North RA, Williams JT (1977) Extracellular recording from the guinea-pig myenteric plexus and the action of morphine. Eur J Pharmacol 45:23–33

North RA, Zieglgansberger W (1978) Opiate withdrawal signs in single myenteric neurones. Brain Res 144:208–211

North RA, Katayama Y, Williams JT (1979) On the site and mechanism of action of enkephalin on myenteric neurons. Brain Res 165:67–77

North RA, Katayama Y, Williams JT (1980a) Actions of peptides on enteric neurons. In: Trabucchi M, Costa E (eds) Regulation and function of neural peptides. Raven, New York, pp 83–92

North RA, Henderson G, Katayama Y, Johnson SM (1980b) Electrophysiological evidence for presynaptic inhibition of acetylcholine release by 5-hydroxytryptamine in the enteric nervous system. Neuroscience 5:581–586

Nozdrachev AD, Gnetov AV, Kachalov YP, Fedorova LD (1977) Effects of methylene blue on transmission in autonomic ganglia. Neurosci Lett 5:205–208

Ohkawa H, Prosser CL (1972a) Electrical activity in myenteric and submucous plexuses of cat intestine. Am J Physiol 222:1412–1419

Ohkawa H, Prosser CL (1972b) Functions of neurons in enteric plexuses of cat intestine. Am J Physiol (Lond) 222:1420–1426

Paton WDM, Zaimis EJ (1949) The pharmacological actions of polymethylene bistrimethylammonium salts. Br J Pharmacol Chemother 4:381–400

Sato T, Takayanagi I, Takagi K (1973) Pharmacological properties of electrical activities obtained from neurons in Auerbach's plexus. Jpn J Pharmacol 23:665–671

Sato T, Takayanagi I, Takagi K (1974) Effects of acetylcholine releasing drugs on electrical activities obtained from Auerbach's plexus in the guinea-pig ileum. Jpn J Pharmacol 24:447–451

Takayanagi I, Sato T, Takagi K (1974) Action of 5-hydroxytryptamine on electrical activity of Auerbach's plexus in the ileum of the morphine dependent guinea-pig. Eur J Pharmacol 27:252–254

Takayanagi I, Sato T, Takagi K (1977) Effects of sympathetic nerve stimulation on electrical activity of Auerbach's plexus and intestinal smooth muscle tone. J Pharm Pharmacol 29:376–377

Trendelenburg P (1979) Physiologische und pharmakologische Versuche über die Dünndarmperistaltik. Arch Exp Pathol Pharmakol 81:55–129

Von Euler US, Gaddum JH (1931) An unidentified pressor substance in certain tissue extracts. J Physiol (Lond) 72:74–87

Williams JT, Katayama Y, North RA (1979) The action of neurotensin on single myenteric neurones. Eur J Pharmacol 59:181–186

Williams JT, North RA (1978) Inhibition of firing of myenteric neurones by somatostatin. Brain Res 155:165–168

Wiliams JT, North RA (1979a) Vasoactive intestinal polypeptide excites neurones of the myenteric plexus. Brain Res 175:174–177

Williams JT, North RA (1979b) Effects of endorphins on single myenteric neurons. Brain Res 165:57–65

Wood JD (1970) Electrical activity from single neurons in Auerbach's plexus. Am J Physiol 219:159–169
Wood JD (1972) Excitation of intestinal muscle by atropine, tetrodotoxin and xylocaine. Am J Physiol 222:118–125
Wood JD (1973) Electrical discharge of single enteric neurons of guinea pig small intestine. Am J Physiol 225:1107–1113
Wood JD (1979a) Neurophysiology of the enteric nervous system. In: Brooks CMcC, Koizumi K, Sato A (eds) Integrative functions of the autonomic nervous system. Elsevier/North-Holland Biomedical, Amsterdam, pp 177–193
Wood JC (1979b) Intracellular study of effects of morphine on electrical activity of myenteric neurons in cat small intestine. Fed Proc 38:959
Wood JD (1976) Neuronal interactions within ganglia of Auerbach's plexus of the small intestine. In: Bülbring E, Shuba MF (eds) Physiology of smooth muscle. Raven, New York, pp 321–330
Wood JD, Mayer CJ (1978a) Intracellular study of electrical activity of Auerbach's plexus in guinea-pig small intestine. Pfluegers Arch 374:265–275
Wood JD, Mayer CJ (1978b) Slow synaptic excitation mediated by serotonin in Auerbach's plexus. Nature 276:836–837
Wood JD, Mayer CJ (1979a) Intracellular study of tonic-type enteric neurons in guinea-pig small intestine. J Neurophysiol 42:569–581
Wood JD, Mayer CJ (1979b) Serotonergic activation of tonic-typer enteric neurons in guinea-pig small bowel. J Neurophysiol 42:582–593
Wood JD, Mayer CJ (1979c) Adrenergic inhibition of serotonin release from neurons in guinea-pig Auerbach's plexus. J Neurophysiol 42:594–603
Yokoyama S (1966) Aktionspotentiale der Ganglienzelle des Auerbachschen plexus im Kaninchendünndarm. Pfluegers Arch 288:95–102
Yokoyama S (1971) Evoked potentials of the Auerbach's plexus in the rabbit small intestine. Proc Int Union Physiol Sci IX:613
Yokoyama S, Ozaki T (1978a) Polarity of effects of stimulation of Auerbach's plexus longitudinal muscle. Am J Physiol 235:E345–E353
Yokoyama S, Ozaki T (1978b) Functions of Auerbach's plexus. Jpn J Smooth Muscle Res 14:173–187
Yokoyama S, Ozaki T (1980) Effects of gut distension on Auerbach's plexus and intestinal muscle. Jpn J Physiol 30:143–160
Yokoyama S, Ozaki T, Kajitsuka T (1977a) Excitation conduction in Auerbach's plexus of rabbit small intestine. Am J Physiol 232:E100–E108
Yokoyama S, Ozaki T, Kajitsuka T (1977b) Excitation conduction in Meissner's plexus of rabbit small intestine. Am J Physiol 232:E109–E113

CHAPTER 7

In Vivo Techniques for the Study of Gastrointestinal Motility

E. CORAZZIARI

A. Introduction

Myogenic activity, contractions of the walls, and transit of contents represent the main aspects of motor activity in the alimentary canal. Myoelectric properties of the muscle layers regulate contractions of the walls. Wall contractions cause variations in the intraluminal pressure and thus pressure gradients, which may either stimulate or inhibit the transit of the contents. Knowledge of these motor aspects is due to, and, at the same time limited by, a particular technique; these limitations are even more evident in human investigations.

Techniques for the study of myoelectric properties will not be discussed as they have been already described in Chap. 5. In this chapter a description is given of the techniques employed in the study of gastrointestinal mechanical activity, i.e., contractions of the walls, intraluminal pressure variations, and transit of contents. No attempt has been made to enumerate all techniques, but methods are discussed which, in our opinion, have greatly contributed to our knowledge of gastrointestinal motor activity, are less invasive, and have less effect on physiologic conditions during investigation.

B. Techniques for the Study of Wall Movements

I. Direct Inspection

The method of opening an animal's abdomen in a warm saline bath in order to observe the contractions of the gastrointestinal tract directly made a great contribution to the knowledge of small intestinal motor activity (SANDERS 1871, cited in HIGHTOWER 1968, ENGELMANN 1871; HOUCKGEEST 1872). Less invasive techniques have replaced this method which is mentioned here for historical purposes.

II. Radiologic Techniques

Radiology offers qualitative information on the motor behavior of the alimentary tract. Following the administration of a carefully selected contrast material, this method can be used indirectly to evaluate wall contractions, as well as displacement of contents. In modern radiologic techniques, image intensifier systems (WOLF and KHILNANI 1966) are employed. These offer a better resolution of this images, significantly reduce radiation, and make telefluorography, photofluorography, cinefluorography, and magnetic tape recording possible.

Fluoroscopic observation performed with an X-ray image intensifier is, at least in clinical practice, the oldest and the most useful test for investigation of motor activity of the gastrointestinal tract. Fluoroscopic images can be recorded by means of photofluorography or cinefluorography. When cinefluorographic systems synchronized with the X-ray generator are used, subjects are exposed to irradiation only while the shutter of the camera is open (HENDEE 1970).

Cinefluorography has proved to be very useful in the evaluation of motor activity of the pharyngoesophageal (COHEN and WOLF 1968) and the rectoanal tracts (BROWN 1965), where the rapid sequence of motor events cannot be fully analyzed by telefluoroscopy. Cinefluorographic films are then evaluated with equipment which permits slow motion or still pictures to be analyzed frame by frame (COHEN and WOLF 1968). Alternatively, fluoroscopic images can be converted into video signals in the television camera, and recorded on magnetic tape or disk (GEBAUER et al. 1967). Unlike cinefluorography, images of the motor events can be immediately displayed by playback of the magnetic tape or disk recorder. However, unlike the disk recorder, slow motion and still picture replay of the magnetic tape system results in a considerable quality loss of the images and it is therefore less accurate than cinefluorography for the analysis of rapid motor events.

Cinefluorography may be useful to observe wall movements caused by contraction of the circular muscle layer, but is inadequate to investigate motor events of the longitudinal muscle layer. Movements of the longitudinal layer are better evaluated by cinefluorography when small radioopaque metal markers, implanted on the external side of the viscus, can be observed. Simultaneous analysis of displacement of the metal marker and intraluminal pressure recordings may be used to study the role of the two muscle layers during gastrointestinal motor events. The most widely used metal markers are made of tantalum and have been employed to evaluate motor activity of the longitudinal muscle layer of the esophagus (NAUTA 1956; DODDS et al. 1974), stomach (GIANTURCO 1934), and small bowel (TASAKA and FARRAR 1969). However, this method can be used only in experimental animals and analysis of data is rather difficult. It is used to measure changes in the length of the muscle, but offers no information on variation in the wall tension.

III. External Transducers

The force displacement transducer is a device which changes its electrical properties in response to a deformation in its shape. Force displacement tranducers usually have a wire or sheet strain gauge which induces variations in the electrical resistance proportional to the deformation. If the strain gauge is placed in an appropriate electrical circuit such as a Wheatstone bridge, the change in resistance can be measured (JACOBY et al. 1963; WEISBRODT 1974).

Transducers sewn to the extraluminal side of the viscus evaluate motor activity of muscle fibers in the direction of the long axis of the sensor. They can therefore monitor contractile activity of the longitudinal and circular layers separately without interfering with the flow of intraluminal contents and without stimulating mucosal receptors.

As the transducers are implanted, they can record repeatedly, and therefore compare motor activity at the same site in the viscus at different times and under

Fig. 1. Extraluminal strain gauge pressure transducer

different experimental conditions. With the use of several transducers, contractile activity can be evaluated simultaneosly in several areas (BASS and WILEY 1972). Mechanical events of the longitudinal and circular muscle layer as well as myoelectric activity can be monitored at the same time and in the same area by means of a device made up of two sensors with their long axes placed perpendicularly. This transducer can also avoid artifacts secondary to extraluminal pressure variations which may deform the sensor (LAMBERT et al. 1976).

Another type of external transducer is made up of silicone rubber tubing filled with mercury under pressure. Resistance of the electric circuit, in which the transducer is placed, varies according to its length. This device, which is about 4 cm long, is located around the circumference of the viscus and, unlike the force displacement transducer, is affected by changes in length and measures variations in the diameter of the gastrointestinal tract (KELLY 1974).

For a long time, external transducers were used exclusively in experimental animals in which gastrointestinal motor activity could be evaluated under physiologic conditions without surgery and anesthesia. The only disadvantage of this method is the relative immobility of the animals during the investigation on account of the electrical connections to the transducers.

The early transducers were too large to be used in humans or to investigate motor activity of small gastrointestinal structures in animals. Miniaturized transducers have recently been produced and used in small animals (PASCAUD et al. 1978). Similarly, microminiaturized transducers made of material which does not provoke an inflammatory reaction and which can be detached from the serosal surface by simply pulling the electrical wire connections, are employed in humans whenever implantations is possible during abdominal surgery (NELSEN and ANGELL 1979).

C. Techniques for the Study of Intraluminal Pressure

Modern manometric methods are characterized by strain gauge pressure transducers in the recording system (Fig. 1). These instruments have a very low pressure–

volume coefficient and transform intraluminal pressure variations into electrical signals. Transducers can be located directly within the lumen of the gastrointestinal tract or outside the lumen. When the transducer is located within the lumen, the frequency response depends exclusively on the sensitivity of the transducer which sends the electrical signals to an amplifier and then to a polygraph; outside the lumen, the frequency response depends also, and essentially, on the system used to transmit intraluminal pressure variations from the gut to the transducer. Most of the manometric techniques currently used depend upon catheters which, introduced into the alimentary canal, transmit intraluminal pressures to external transducers.

Modern recording polygraphs and strain gauge pressure transducers display a flat frequency response to pressure variations exceeding any present in the alimentary canal; therefore, for practical purposes, frequency responses depend exclusively on the characteristics of the catheters. The shorter and the wider the catheter, the better the frequency response.

I. Intraluminal Balloons

1. Large Balloons

The intraluminal balloon was the first system to be used in the investigation of gastrointestinal motor activity in vivo (KRONECKER and MELTZER 1883; ALVAREZ 1929). In the early methods, large rubber balloons filled with air or water were connected by the catheter to a manometer in order to measure volume variation. Even now, despite the connection to strain gauge pressure transducers, the system is not suitable for recording intraluminal pressures. The system measures the ballon volume variation secondary to intraluminal pressure and to wall tension, related to its inherent elasticity. Furthermore, the balloon has been shown to modify the mechanical activity of the viscus by stimulating wall contractions and by obstructing the lumen (QUIGLEY and BRODY 1952). Nevertheless with some variations, large balloons can be used to evaluate the intraluminal pressure–volume relationship and have been usefully employed in the detection of wall tension variations and contraction responses during distension of the alimentary canal. This approach has been used to investigate the adaptation response of the proximal part of the stomach (STAADAS 1975 a, b), the colon (PARKS 1970), and the rectal ampulla (IRHE 1974; ARHAN et al. 1976).

A different use of the balloon system has been suggested by JAHNBERG and coworkers; the balloon is placed in the proximal part of the stomach and connected through a wide-bore catheter to a water-filled tank where the water is kept 0–30 cm above the fundus (JAHNBERG et al. 1977). Using this system, measurement of water flow from the tank to the intragastric balloon evaluates the adaptation mechanism of the proximal stomach in various pharmacologic and pathologic conditions.

2. Small Balloons

Compared with large balloons, balloons of smaller dimensions, e.g., 7 × 10 mm (ATKINSON et al. 1957) have less compliance, stimulate wall contractions to a lesser degree, and do not prevent transit of contents. Recordings, using small balloons,

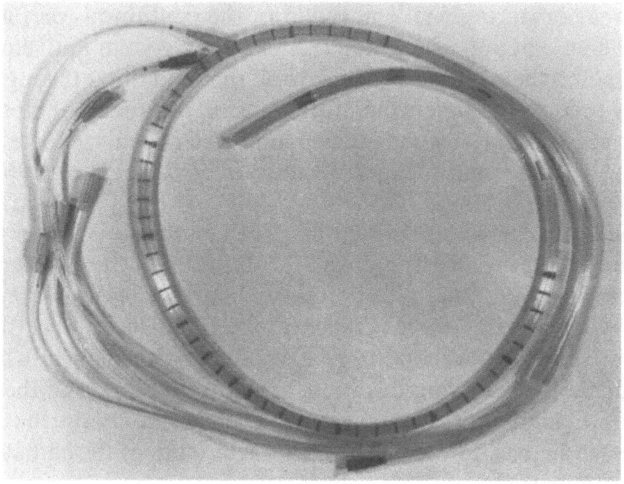

Fig. 2. Manometric catheter with recording side holes

however, represent both pressure and volume variations. This systems was in fact widely used before continuously perfused catheters were employed, particularly in investigations of large bowel motor activity since semisolid contents easily occluded the openings of the catheters. Small balloons are no longer used for pressure recordings, but may be usefully employed to evaluate motor activity from gastrointestinal segments with a wide lumen. Furthermore, even if small balloons overestimate pressure values in sphincteric areas (POPE 1967), the system can be satisfactorily employed to investigate motor activity of the anal canal (HANCOCK 1976). If a balloon is located in the proximal part of the anal canal and a second balloon in the distal part, it is possible to evaluate the reflex response of the internal and external anal sphincters to intrarectal distension (GASTON 1948; SCHUSTER et al. 1963).

II. Catheters with Distal Opening

Catheters with a distal opening (Fig. 2) transmit only intraluminal pressure variations to the external transducers; furthermore, they interfere less with the motor activity of the digestive tract than the balloon system. In the past, catheters were filled with air or water but results were inaccurate. The air column is highly compressible and the intraluminal pressure is dispersed during transmission; moreover the openings are easily occluded by the contents and/or the gastrointestinal mucosa. Recording openings are occluded by the mucosa when these are localized in narrow luminal tracts such as sphincteric areas and segments where wall annular contractions squeeze the catheter. Under these conditions the catheter does not transmit the true intraluminal pressure, but rather the last pressure present within the recording system before occlusion of the opening (HARRIS et al. 1966). To avoid these drawbacks catheters are perfused with a continuous flow of distilled water (HARRIS and POPE 1964; HARRIS et al. 1966). When manometric recording is from sphincteric areas or during wall contractions, the continuous perfusion induces a progressive increase in pressure within the recording system until equilibrium with the force occluding the opening is reached and the fluid flows through the opening. Continuously perfused catheters therefore transmit intraluminal pressure, when recording from segments with a wide lumen, and wall occluding force, when recording from areas with a narrow lumen (HARRIS et al. 1966; POPE 1967; COHEN and HARRIS 1970).

Recording accuracy with this method depends on the infusion rate, the compliance of the system, the amplitude, and the duration of the intraluminal pressure variation. The more rapidly equilibrium between pressure within the recording system and the wall occluding force is reached, the more accurate is the manometric measurement. This optimal condition is obtained with high infusion rates, low compliance of the system, long duration, and low amplitude of the motor events (DODDS 1976). Compliance of the system depends mainly on the infusion technique, to a lesser degree, on the physical characteristics of the catheters and, negligibly, on the transducers.

1. Infusion Systems

a) Syringe Pump

Compliance of the syringe pump infusion system increases during intraluminal pressure transients because of fluid leakage along the syringe barrel and there is some play in the pump gear train (STEF et al. 1974). This disadvantage can be reduced by greasing the syringes and using heavy duty pumps with low gear train tolerance; despite these improvements, accurate recordings of motor events characterized by short duration and elevated pressure, such as may occur at the level of the upper esophageal sphincter and of the cervical tract of the esophagus, can be obtained only with high infusion rates. Infusion rates, however, cannot be increased beyond certain limits without affecting manometric recording. Excessively high infusion rates raise the baseline on the manometric tracing and perfusion of large amounts of fluid in the lumen of the alimentary canals stimulates motor activity. Elevated perfusion rates cannot be employed in investigations on intraluminal pressures at the level of the upper esophageal sphincter because of the cough reflex secondary to fluid aspiration in the airway.

b) Hydraulic Capillary

The hydraulic capillary infusion system has a negligible compliance compared with the pump infusion technique. Using this method, a highly pressurized fluid reservoir is connected to the manometric catheters by means of inextensible tubes of capillary caliber. When pressure within the reservoir is raised to 1,000 mmHg or above, the resistance of the capillary tubes to flow allows fluid to reach the catheters at atmospheric pressure and at an infusion rate proportional to the pressure gradients existing along the capillary. The vast difference which exists between the pressure within the reservoir and the maximal pressure in the alimentary canal, even during motor events of short duration and high force, prevents variations in the flow rate. With this method the compliance of the system depends essentially upon the physical characteristics of the manometric catheters; furthermore, because of the low perfusion rates of the hydraulic capillary technique, catheters with a reduced inner diameter can be employed and thus reduce another source of inaccuracy (see Sect. II.2; ARNDORFER et al. 1977).

c) Intraflow

In the Intraflow infusion system fluid is perfused from a pressurized reservoir to the transducers passing through a catheter partially occluded by a short obstacle which causes, distally, a sudden decrease in the infusion pressure and flow rate. The Intraflow system, therefore, like the hydraulic capillary technique, but with cheaper and more easily available material, determines, at a low compliance, very low perfusion rates. However the Intraflow system displays flow rates which vary irregularly with time as the elastic material of the device is affected by pressure, temperature, and rapid deterioration. Furthermore this technique has a high compliance compared with the hydraulic capillary system (A. BIANCO and R. ISABELLI, unpublished work 1980).

2. Physical Characteristics

The internal diameter, length, elasticity and thickness of the catheter walls affect accuracy of manometric recording. The compliance of the catheters is reduced by decreasing the internal diameter and the length of the tube as well as by using less elastic material (STEF et al. 1974). Choice of the inner diameter of the catheter depends on the infusion system. In the infusion pump technique, catheters with an internal diameter of 1.6–2.0 mm should be used to facilitate high infusion rates without altering the baseline pressure value (STEF et al. 1974); with the hydraulic capillary system accurate manometric recordings can be made with a catheter with an internal diameter of 0.8–1.0 mm (ARNDORFER et al. 1977).

Several catheters are generally used in manometric recordings in order to get a more comprehensive evaluation of motor activity. The total number of catheters depends on the number of available transducers and recording channels as well as on the maximal external diameter of the probe which can be comfortably passed into the alimentary canal. The external diameter of the probe also affects pressure values recorded during wall contractions or in sphincteric areas (KAYE and SHOWALTER 1974; LYDON et al. 1975). The larger the external diameter, the greater the contraction force and thus this relation can be usefully employed to evaluate tension of the sphincteric muscle (BIANCANI et al. 1975).

Recording orifices, with a size equal, or very close, to the internal diameter, of the catheter, can open either at the distal tip or the side of the catheter. Side holes, however, are more widely used as open ended tips underestimate contracting forces of the wall when several catheters are assembled together (POPE 1967).

In a manometric probe comprising several catheters, the recording orifices are orientated according to intraluminal pressure characteristics of the segment to be studied. Pressure values, in the sphincteric zones, are not uniform on the radial axis (KAYE and SHOWALTER 1971; WINANS 1972) and vary at the different levels on the longitudinal axis; furthermore, sphincteric areas can move orally and aborally, causing pressure variations in the manometric recordings on account of the un-

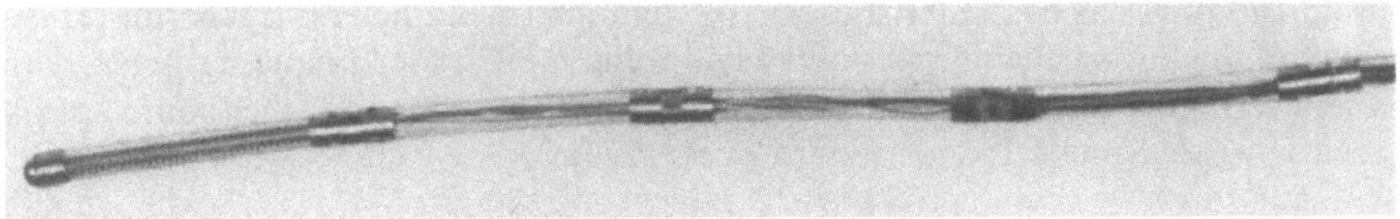

Fig. 3. Manometric catheter with three intraluminal miniaturizes pressure transducers placed 5 cm apart

stable spatial relationship between side holes and the segment under investigation (DODDS et al. 1974). The pressure profile of the sphincteric areas can be thoroughly evaluated by locating some recording orifices at the same level, radially and others at short intervals along the longitudinal axis (DODDS 1976). Thus the asymmetry of radial pressure can be recorded with the former, and pressure variations on the longitudinal axis, with the latter, However, the longitudinal pressure profile is better evaluated by pulling the catheters, and therefore the recording orifices, through the sphincter. Withdrawal of the probe through the high pressure zones can be performed either in a stepwise fashion (station or slow pull-through), thus evaluating intraluminal pressure at the different levels at regular time intervals, or in a constant and rapid sequence during apnea (rapid pull-through) (BOTHA et al. 1957; WALDECK et al. 1973; DODDS et al. 1975). With the first method, phasic responses such as inhibition and contraction are recorded in addition to the resting pressure, but the manometric data are affected by pressure variations owing to mobility of the sphincter and respiration (DODDS et al. 1975); while the second method does not present these limitations it can only be used to evaluate the pressure profile in resting conditions. At the level of the lower esophageal sphincter, however, pressures were more variable when measured by the rapid than by the station pull-through technique (WELCH and DRAKE 1980). Intraluminal pressure studies of nonsphincteral areas can be performed with recording orifices located at intervals of about 3–5 cm (DODDS et al. 1978).

The maximal closing force of the sphincters can be accurately measured with the *sleeve method*. The sleeve, made up of thin silicone rubber sheets, is 5–6 cm long and is located at the distal end of the catheter. Distilled water is infused at a constant rate into the sleeve at the sealed proximal edge and flows into the lumen from the open distal end (DENT 1976; WHEATLEY et al. 1977). This sensor transmits the maximal closing force applied at its circumference to the transducer. On account of its length, recordings obtained with the sleeve are not affected by pressure variations secondary to mobility of the sphincter, but are affected by pressure variations occurring in the adjacent luminal tracts. This method appears to be particularly useful in prolonged recording of the maximal closing force of the sphincters.

III. Miniaturized Intraluminal Transducers

Factors unfavorably affecting the recording fidelity of manometric systems with external transducers can be avoided by using intraluminal transducers (Fig. 3). Several strain gauge pressure microtransducers can be assembled together on the same

Fig. 4. Telemetering capsule with piezotransistor

manometric probe and located at different levels (MILLHON et al. 1968; Hollis and CASTELL 1972). Evaluation of intraluminal pressure in nonsphincteral areas with this system is as accurate or more accurate than with the low compliance perfused system, but may be lower in the sphincteral areas (HAY et al. 1979).

Microtransducers, however, have never been widely used because technical drawbacks, related to higher costs and shorter lifetimes than external transducers, make them less practical than systems with perfused catheters. Commercially available probes have only three microtransducers located at fixed intervals of 5 cm and all orientated in the same direction. Furthermore, microtransducers must be calibrated at body temperature, requiring a long heating period; during manometric recordings, the baseline may drift unexpectedly (HAY et al. 1979). In the sphincteral areas, variability of pressure secondary to radial asymmetric and mobility of the sphincter cannot be evaluated with these instruments; however, a microtransducer has been devised which can detect pressure applied to the entire circumference of the probe and, therefore, evaluate the maximal closing force (KAYE et al. 1973).

IV. Radiotelemetering Capsules

Radiotelemetering capsules (Fig. 4) are minute instruments which contain a pressure transducer and which can be easily swallowed. They transmit radio signals at a frequency modulated by variation in intraluminal pressure (FARRAR et al. 1957; MACKAYE and JACOBSON 1957). The recording system is made up of a radio transmitter, a frequency modulation receiver connected to an antenna, and a polygraph. Movements of the transducer membrane modulate the oscillating frequency from the transmitter; radio signals are then taken up by the antenna and transmitted to the receiver. Signals are then demodulated and sent to the polygraph. The radio transmitter is transistorized and has a 15-h battery-powered energy supply within the capsule (FARRAR and BERNSTEIN 1958; CONNELL and ROWLANDS 1960). With a piezotransistor the operating lifetime is increased to about 100 h (BARBARO et al. 1971).

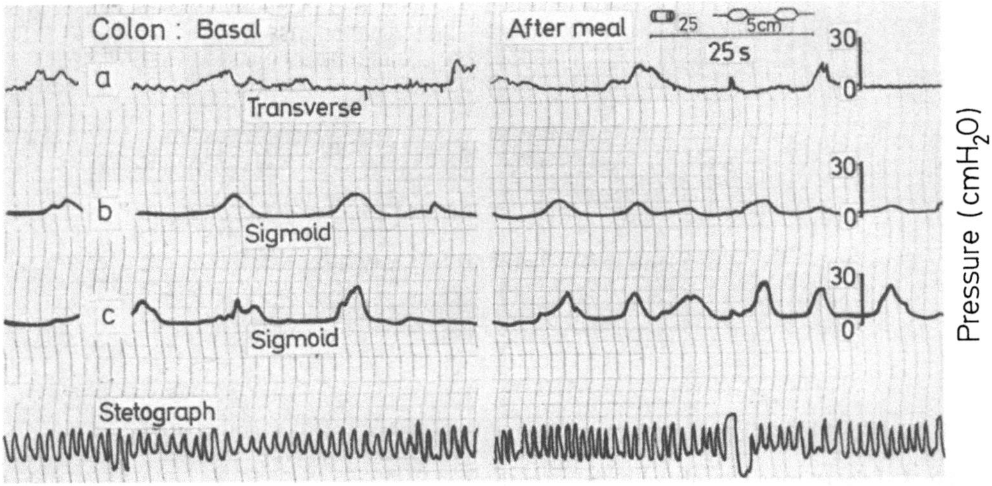

Fig. 5 a–c. Intraluminal pressure recordings from the transverse colon **a** by means of a telemetering capsule and from the sigmoid tract **b, c** by means of small balloons

Radiotelemetering capsules have been used to record intraluminal pressures in the ileum and proximal colon (Fig. 5) which are relatively inaccessible to probes (BARANY and JACOBSON 1964; RAMORINO and COLAGRANDE 1964). As no catheters are needed with this technique, capsules are more acceptable in human studies. However, mobilitiy of the capsule represents the main disadvantage of this instrument as it is not possible to prolong the length of manometric recording in a segment of the alimentary canal. Furthermore, as the capsule moves with the contents in accordance with the intraluminal pressure gradient, pressure variations during its movement are not recorded.

D. Techniques for the Study of Transit of Contents

Transit is the passage of contents through the alimentary canal or a segment of it and is the result of mixing, propulsive, and retropulsive processes present to different degrees in the various gastrointestinal tracts. Flow in the digestive tract depends upon intraluminal pressure gradients, resistance caused by the wall, and on viscosity of the contents. The rate of transit, in the different segments of the alimentary canal, therefore, varies according to the anatomic structure and to the specific motor activity of each tract as well as, in a particular tract, the physical characteristics of the contents.

No single technique exists which can evaluate all the various aspects of transit at each level of the alimentary canal, but various techniques, radiologic, chemical, radioisotopic, and particulate have to be employed, each of which can evaluate transit in only one ore more segments and/or for a single characteristic of the contents.

To measure total or segmental gastrointestinal flow accurately, these methods all use a *marker* which is chosen not only according to the technique employed but also according to the characteristics of the contents and the segment investigated. The ideal marker to measure transit, without affecting it, should be chemically, physiologically, and pharmacologically inert; it should not be absorbed, it should mix thoroughly with the contents, be easily counted, and completely recovered. Since no substance has all these requisites; our knowledge of gastrointestinal transit is the result of, and at the same time is limited by, the measuring technique and the marker used. Techniques for the study of total or segmental transit offer an overall evaluation of the flow rate of the contents, but do not provide any information on how motor activity displaces the contents.

I. External Collection of Contents

Following the administration of one or more markers, contents flowing from the anus or from natural or surgical fistulas which may be located at any level of the alimentary canal, are collected. All the markers can thus be recovered, but only the time of appearance and disappearance can be estimated. These methods therefore evaluate transit time without providing any information on the modality of flow. Use of these techniques has so far been limited on account of difficulties in collecting and storing the gastrointestinal outflow, detection and measurement of the markers and, whenever segmental transit time is investigated, the need to operate surgically. Furthermore, external diversion of contents may affect the physiologic characteristics of transit.

Various markers have been used in these techniques: insoluble dye powders such as carmine and charcoal (ROTHMAN and KATZ 1964), different colored glass beads (ALVAREZ and FREEDLANDER 1924), millet seeds (BURNETT 1923), etc. More accurate data are obtained with chemical markers, but they require more complex systems of analysis. Moreover, data derived from the use of chemical markers are affected by the homogeneity of contents investigated; markers such as, for example, chromium oxide and polyethylene glycol reflect transit of the solid and liquid phase, respectively (WILKINSON 1971). Chemical markers tagged with radioisotopic substances are widely used to simplify measurements in the material collected (HANSKY and CONNELL 1962; WINGATE et al. 1972). If carefully chosen, these indicators offer numerous advantages, but, like the particulate and chemical markers, do not provide information on the modality of transit. Furthermore these markers should be used with caution since toxicity of some of these markers may be enhanced in pathologic conditions, e.g., sodium chromate in achloridric patients (DONALDSON and BARRERAS 1966).

II. Radiologic Techniques

1. Barium Transit

In the radiologic technique, bismut subnitrate and more recently barium sulfate have been employed as markers to measure transit of gastrointestinal contents (HURST 1919; WALLACE et al. 1938; MANOUSOS et al. 1967). With X-ray films of the

a 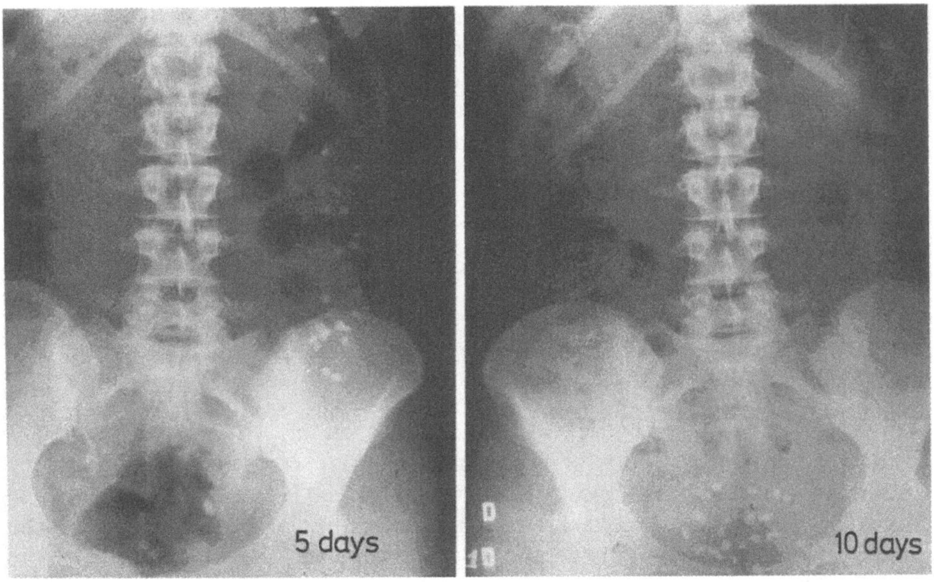 b

Fig. 6 a, b. Radioopaque markers retained at the level of left colon and rectum 5 and 10 days after ingestion

abdomen taken at constant time intervals it is possible to evaluate progression of the radioopaque bolus at different levels of the alimentary canal; it is also possible by radiography of feces, to measure the oroanal transit time withouth risk of radiation. However this method, which is very simple to perform, gives only a semiquantitative evaluation as it detects the head, but less precisely, the tail of the radioopaque bolus. Furthermore the great variability in viscosity and stability of barium sulfate suspensions may have different effects upon the flow rate of the contents (MILLER 1967) and it is likely that a dose of more than 60 g barium sulfate accelerates transit (ALVAREZ and FREEDLANDER 1924).

2. Particulate Radioopaque Markers

Particulate radioopaque markers are more useful as they allow quantitative measurement of transit. These markers, made up of polyethylene mixed with a 20% (W/W) barium sulfate, are shaped as disks 3–5 mm in diameter, 3 mm thick, and with a relative density of 1.05–1.5; similar markers can be made by cutting 3 mm segments of polyethylene radioopaque tube with a 1–5 mm external diameter. Total gastrointestinal transit time is measured by taking radiographs of feces for several days after a single ingestion of these markers (HINTON et al. 1969). It is also possible to measure total gastrointestinal transit time by radiography of a single stool specimen evacuated on the 4th day after administration, for three consecutive days, of different shaped radioopaque markers; but this method can only be used in subjects with a normal transit time not exceeding 4 days (CUMMINGS and WIGGINS 1976).

The physical properties of particulate radioopaque markers make them suitable for measuring transit of semisolid, solid contents through the large bowel. Following ingestion of markers, X-ray films of the abdomen taken at regular time intervals (CASSANO et al. 1967; HINTON et al. 1969) can be used to detect their location within the large bowel.

Distribution of markers in the various segments of the colon and in the rectum can be employed to identify modalities of transit which differ markedly in patients with chronic constipation (CASSANO et al. 1967; CORAZZIARI et al. 1975; Fig. 6). Measuring markers present at the various levels of the large bowel in a series of successive X-ray films is a simple method of obtaining disappearance curves of the markers from any single segment of the large bowel (MARTELLI et al. 1978). This method, however, does not allow evaluation of transit of the contents in any one segment of the large bowel independently from that of the adjacent segments; moreover several X-ray films, and thus a substantial number of observations, are necessary to plot disappearance curves. Evaluation of the "mean percentage transit time of the markers" which have entered the different segments of the large bowel or the measurement of the transit time in intestinal segments characterized by outflow of contents only (CORAZZIARI et al. 1975; ZAPPONI et al. 1979) enable flow rate to be measured in each large bowel segment and the number of X-ray films can be markedly reduced.

3. Time-Lapse Cinematography

This technique has been used to investigate movements of the contents in the large bowel. With cinefluorography one photogram is made every minute over a 20–60-min period following the arrival of barium sulfate in the colon (RITCHIE 1968). The sequence of photograms can then be run at a faster rate and the hardly perceptible movements of the colon appear accelerated and therefore easier to analyze visually. The method offers only qualitative data on the movements of the contents in the large bowel without providing information on the wall contractions which caused the displacement; it exposes the patient to a rather large dose of radiation.

III. Intubation Techniques

In the intubation techniques, transit is evaluated by means of catheters which are located at various levels of the alimentary canal to collect liquids marked with a nonabsorbable indicator introduced into the gastrointestinal segment proximal to the sampling area. The main disadvantages of these methods are the difficulty in introducing and positioning the catheter as well as the possibility of investigating transit of liquid contents only; intubation techniques, on the other hand, can be used to evaluate transit and absorption as well as gastrointestinal secretions simultaneously.

1. Fractional Test Meal

Sampling of gastric contents at regular time intervals after ingestion of a meal was the first method in which a catheter was used to measure gastric emptying (LEUBE

1883; Ehrenreich 1912). With this test only the total emptying time of a liquid is measured and no additional information on the transit pattern is provided (Hollander and Penner 1939).

2. Serial Test Meal

The serial test meal is a more accurate method for evaluating the rate of gastric emptying. A fixed amount of liquid containing a marker is administered on different days and gastric contents are sampled at various time intervals. The rate and pattern of gastric emptying can therefore be evaluated by measuring the volume of the marker and its concentration in each gastric sample. At the same time this test can be used to measure the volume of gastric secretion (De Salamanca 1949; Hunt and Spurrell 1951). Although the serial test meal has proved to be an accurate method, it has never been extensively used on account of the need to repeat nasogastric intubation on consecutive days.

3. Marker Dilution Tests

a) Double Sampling Test

The double sampling method can be used for accurate studies of gastric emptying withouth the need to repeat intubation. In this test, following introduction of a liquid meal containing a marker, a small fraction of gastric contents is aspirated at regular time intervals, then 20 ml of a solution with a higher concentration of marker is instilled into the stomach, the gastric contents are thoroughly mixed, and finally a second sample is aspirated (George 1968). The volume of contents in the stomach at different times, and therefore the pattern of gastric emptying during the period of the test, can be estimated from the concentration of the marker in the meal and in the two samples.

The method, as described by George, cannot distinguish changes in volume secondary to gastric emptying or secretion, but the latter can be evaluated by measuring the Cl^- ion concentration in the gastric samples (Hunt 1974). Gastric emptying and oroanal transit time can be evaluated simultaneously if ^{51}Cr is used as marker with the double sampling method before collection of feces (McKelvey 1970).

b) Double Marker Test

An accurate test which avoids aspiration of contents from the stomach and evaluates gastric emptying time during digestion and absorption of alimentary substancces in the duodenum has been described by Meerof et al. (1973). In this test, a nasoduodenal tube is introduced, a first marker (polyethylene glycol 4000) is constantly perfused into the proximal part of the duodenum, and a liquid meal with a second marker is introduced into the stomach while contents are continuously aspirated at the level of the duodenojejunal junction. This method is used to evaluate the pattern of gastric emptying as it is affected by the entrance of various alimentary substances into the duodenum. If a fairly sophisticated method of analysis is used, this technique can be employed for simultaneous measurement of gastric secretion of acid and pepsin as well as gastric emptying time during ingestion of a meal composed of both liquids and solids (MALAGELADA et al. 1976). Continuous emptying of the duodenal contents might however affect gastric emptying.

c) Marker Dilution Test

The marker dilution test with a triple lumen tube is used to measure transit and absorption of alimentary substances in the intestine. In this technique, a triple lumen tube is passed into the alimentary tract; a solution to be tested containing a first nonabsorbable marker is infused from the proximal opening, and a solution with a second marker is intermittently flushed from the opening positioned 10–15 cm distally as contents are sampled, at regular time intervals, from the third opening which is located distally. Dilution of the second marker is used as a measure of the mean transit time while the concentrations of the first marker and of the substance under study are used as a measure of absorption in the selected intestinal segment (COOPER et al. 1966).

If the triple lumen catheter is located with the proximal opening in the stomach, the middle opening in the upper part of the duodenum and the distal opening in the small bowel, it is possible, with some technical variations, to measure gastric, biliary, and pancreatic secretion as well as gastric emptying, transit time, and the absorption pattern in the selected intestinal segment (JOHANSSON et al. 1972).

d) Multiple Indicator Test

All these variables can be estimated if several markers are administered consecutively through the middle opening of the triple lumen catheter and a dilution curve of each marker in the collected sample is plotted. With a sophisticated system of analysis (LAGERLOF et al. 1972), the multiple indicator dilution technique can be used for continuous measurement of transit time, mixing with the various secretions, and absorption of the different components of a meal without affecting physiologic conditions.

Measurement of intestinal transit by means of the indicator dilution curves, however, has some disadvantages. Flow of the contents is, in fact, dependent on the infusion rate of the solution (CHAUVE et al. 1976); furthermore the length of the segment under study can vary as intestinal walls may either shorten or lengthen around the catheter. Transit measured by means of this method is not, therefore, an absolute value but is related to the technique employed. This approach to the evaluation of transit may, however, be useful in comparative studies in which the same techniques are used.

IV. Radioisotopic Techniques

Radioisotopic techniques have been extensively used to evaluate transit of contents because markers are easily measured, the method is not invasive and therefore well accepted. Furthermore it does not affect physiology of the motor events. Radioactivity counts can be measured externally or directly over the contents, collected after partial or total passage through the alimentary canal, by employing radioactive markers together with one of the previously described techniques.

Transit of radioactive markers through the gastrointestinal tract can be followed from the outside of the body by rectilinear scintigraphy and a computerized gamma camera. The computerized gamma camera has been extensively employed to study gastric emptying of liquid, and unlike intubation techniques, also of solid food. Gastric emptying, however, can be accurately measured by means of a single scintillation counter (OSTICK et al. 1975). Measurements made be rectilinear and gamma camera scintigraphy are affected by variation in depth in the radiation source during passage trough the alimentary canal as well as body thickness. These sources of error, which appeared irrelevant when evaluated in an in vitro experimental model (HARVEY et al. 1970) may be significant in studies of gastric emptying in vivo (TOTHILL et al. 1978).

Various isotopes can be used as markers, nevertheless, substances should be chosen which are well absorbed by the meal (HEADING et al. 1971), limit the dose of radiation, and have an elevated γ-emission.

1. Rectilinear Scintigraphy

Rectilinear scintigraphy offers a fairly accurate image of the organs but the techniques is limited for dynamic events by the long time required to record a single image. This method has been used mainly to study gastric emptying, but errors are likely if the stomach empties rapidly (HARVEY et al. 1970).

2. Gamma Camera Scintigraphy

Gamma camera scintigraphy can detect and simultaneously record the distribution of the radioactive substance over the entire area of interest (HARVEY et al. 1970). Rapid processing of the results, which are then expressed as quantitative data can be achieved by connecting a computer to the gamma camera. As soon as the marker is administered, this technique, which at the same time measures distribution and activity of the radioisotope, can also evaluate events characterized by rapid transit of the contents as well as sudden changes in the modality of gastrointestinal flow. The gamma camera with an on-line computer has been used to investigate displacement of contents in the upper gastrointestinal tract.

a) Esophageal Transit

Esophageal transit can be observed and quantitatively measured by recording the scintillation count rate in the esophageal area with a computerized gamma camera after the subject has swallowed a bolus containing a radioactive marker (KAZEM 1972; TOLIN et al. 1979b).

b) Gastroesophageal Reflux

Reflux of radioactive substance from the stomach to the esophagus can also be observed and measured. Detection of reflux during abdominal compression has been suggested as a useful test to differentiate normal subjects from patients with a pathologic gastroesophageal reflux (FISHER et al. 1976). The method, however, is not sufficiently sensitive to detect reflux in resting conditions (SCOPINARO et al. 1979).

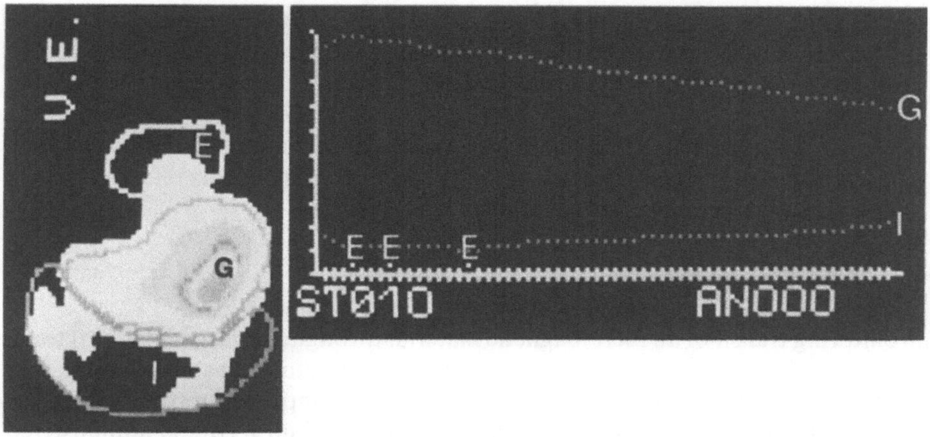

a b

Fig. 7 a, b. Gastroesophageal scintigraphy obtained by means of computerized gamma camera. In the scintiscan **a** areas of interest are as follows: esophageal *E*, gastric *G*, and intestinal *I* areas. Spontaneous gastroesophageal reflux is visualized. In the diagram **b**, upper, middle, and lower curves: count recorded from gastric *G*, intestinal *I*, and esophageal *E* areas. Emptying rate of stomach evaluated by counts of gastric area. Three episodes of gastroesophageal reflux detected by counts of esophageal area

c) Gastric Emptying

The gamma camera with an on-line computer can accurately evaluate gastric emptying (DELIN et al. 1978; Fig. 7) and, unlike other techniques, it measures the disappearance rate of liquid and/or solid food from stomachs with a normal or accelerated emptying time. This method is therefore useful to study emptying of the stomach after vagotomy and/or gastric resection. Furthermore, evaluation of the time–activity curve by means of images recorded at 1-s intervals from small, closely spaced areas at the level of the antrum can be used to measure the frequency and propagation rate of the antral peristaltic contractions (AKKERMANS et al. 1979). It is also possible to measure the emptying pattern of the liquid and solid components of a meal simultaneously, if the two phases are marked with different isotopes (HEADING et al. 1976; MEYER et al. 1976). After administration of a radioactive substance which is excreted with the bile into the duodenum and another radioisotope with a meal, this technique can detect duodenogastric reflux as well as the emptying rate of the stomach (TOLIN et al. 1979 a).

3. Radioactive Capsules

a) Oroanal Transit

Capsules made of the epoxy resin, Araldite (ROSSWICK et al. 1969; WALLER 1975) or polymethyl methacrylate (KIRWAN and SMITH 1974) and containing a small dose of radioactive material can be easily localized during transit through the alimentary canal by an external scintillation counter. This method has been proposed to evaluate the oroanal transit time and progression of the contents through the various segments of the gastrointestinal tract. The capsule (4 × 10 mm) cannot be used to measure gastric emptying without altering physiologic conditions and it probably affects motor activity of the intestine. Furthermore, as this capsule cannot mix with the contents, it may be separated from the main flow and the transit time may not correspond to that of the contents; finally, localization of the capsule site within the gastrointestinal tract may not be accurate inasmuch as reference points are based on fixed abdominal areas.

V. Magnetic Techniques

Monitoring transit of an inert magnetic material through the gastrointestinal tract, by means of an extraabdominal sensor placed over the stomach and cecal area, has been employed to measure gastric and small bowel emptying, respectively (Benmair et al. 1977a, b). With this method, 50 g magnesium ferrite is administered with the meal and a magnetic probe, which magnetizes the ferrite compound and at the same time measures the magnetic field produced in the gut, is placed over the area of interest. The output signal is then amplified, filtered, and recorded on a chart recorder. With the magnetic technique, which is easy to perform and not expensive, no radiation is involved and no intubation is required; however the ferromagnetic tracer, on account of its physical properties, does not intimately mix with the contents and its flow in the alimentary canal may not represent movement of the substance or substances contained in the meal under investigation. Furthermore, this method has not yet been validated by directly comparing its results with those of a more standardized technique.

VI. Breath Test

The orocecal transit time can be measured by administering an oral dose of lactulose, which is not absorbed and, as it reaches the colon, is metabolized by the bacterial flora. The fermentation process releases hydrogen ions which are rapidly absorbed by the mucosa and appear in the expired air, where they can be detected by a hydrogen electrode. Time elapsing from ingestion of lactulose to appearance of hydrogen in the expired air can be used to evaluate, with a simple and noninvasive method, the rate of orocecal transit (Bond and Levitt 1975). The use of a nonabsorbable carbohydrate, however, may effect the intestinal flow rate.

E. Use of Combined Techniques

Motor events of the gastrointestinal tract are characterized by contraction of the walls, intraluminal pressure variations, and displacements of contents, but no constant relationship links these three aspects. Knowledge of one or two of these motor characteristics does not necessarily mean the third can be precisely defined. The methods previously described, with the exception of radiology, analyze only one aspect of motility and therefore, when applied alone, cannot investigate the various aspects of motility. This limitation is reflected in the numerous attempts to classify motor events according to the results obtained by a single technique, e.g., the different types of wall contractions and displacement of contents described by radiology taking into account only movements of the radioopaque contrast material, or the different types of contractions described by manometric techniques taking into account only duration, amplitude, and morphology of waves (TEMPLETON and LAWSON 1931; ADLER et al. 1941).

Better knowledge of motor events can be obtained by combining several techniques and some of these combinations have already been mentioned earlier in this chapter. Increased invasiveness and difficulty have, however, limited these studies in vivo, particularly so in humans. The combined use of manometric and radiologic techniques has provided a large amount of information on motor characteristics of the intestinal tract since the combination of these two methods, which are suitable for use in humans, can be employed to evaluate intraluminal pressure variations, contractions of the walls, and displacement of contents, simultaneously (INGELFINGER and ABBOTT 1940; CHAPMAN and PALAZZO 1949; FRIEDMAN et al. 1965; DELLER and WANGEL 1965; TURANO 1957; TORSOLI et al. 1971).

Synchronization of radiologic images with manometric recording is an essential part of this method. Synchronization between cinefluorography and manometry is obtained by means of a mechanical system which is connected to the motor axis of the cinecamera and sends an electrical signal to the manometric recording (RAMORINO and COLAGRANDE 1964). Cinefluorographic and manometric recording can also be synchronized by simultaneously recording on a magnetic tape recorder (BORGSTRÖM and ARBORELIUS 1971). Analysis of radiologic frames is then performed with slow motion or still picture equipment in order to correlate images with the manometric counterpart. Although this combined method is affected by limitations of both the radiologic and manometric techniques, it has proved to be valuable in the analysis of mechanical events and particularly of motor events associated with propulsive effects on the contents (TURANO 1957; TORSOLI et al. 1971).

References

Adler HF, Atkinson AJ, Ivy AC (1941) A study of the motility of the human colon: an explanation of dysynergia of the colon, or of the "unstable colon". Am J Dig Dis 8:197–202

Akkermans LMA, Jacobs F, Hong-Yoe O (1979) A non-invasive method to quantify antral contractile activity in man and dog. Abstracts 7 th International Symposium on Gastrointestinal Motility, Iowa City, University of Iowa, Iowa City, p 45

Alvarez WC (1929) Physiologic studies on the motor activities of the stomach and bowel of man. Am J Physiol 88:650–662

Alvarez WC, Freedlander BL (1924) The rate of progress of food residues through the large bowel. JAMA 83:576–580

Arhan P, Faverdin C, Persoz B, Devroede G, Dubois F, Dornic C, Pellerin D (1976) Relationship between viscoelastic properties of the rectum and anal pressure in man. J Appl Physiol 41:677–682

Arndorfer RC, Stef JJ, Dodds WJ, Linchan JH, Hogan WJ (1977) Improved infusion system for intraluminal esphageal manometry. Gastroenterology 73:23–27

Atkinson M, Edwards DAW, Honour AJ, Rowlands EN (1957) Comparison of cardiac and pyloric sphincters. A manometric study. Lancet 2:918–922

Barany F, Jacobson B (1964) Endoradiosonde study of propulsion and pressure activity induced by test meals, prostigmine, and diphenoxylate in the small intestine. Gut 5:90–95

Barbaro V, Daniele Sargentini A, Frank M, Macellari V, Neroni M (1971) Use of a piezotransistor for a pressure-sensitive radiotelemetering capsule. Rend Gastroenterol 3:34–37
Bass P, Wiley JN (1972) Contractile force transducers for recording muscle activity in unasthesized animals. J Appl Physiol 32:567–570
Benmair Y, Dreyfuss F, Fishel B, Frei EH, Gilat T (1977a) Study of gastric emptying using a ferromagnetic tracer. Gastroenterology 73:1041–1045
Benmair Y, Fishel B, Frei EH, Gilat T (1977b) Evaluation of a magnetic method for the measurement of small intestinal transit time. Am J Gastroenterol 68:470–475
Biancani P, Zabinski MP, Behar J (1975) Pressure, tension, and forces of closure at the human lower esophageal sphincter and esophagus. J Clin Invest 56:476–483
Bond JH, Levitt MD (1975) Investigation of small bowel transit time in man utilising pulmonary hydrogen (H_2) measurement. J Lab Clin Med 85:546–555
Borgström S, Arborelius M Jr (1971) A technique for studying propulsion and the displacement of contents in the duodenum and proximal jejunum. Rend Gastroenterol 3:174–177
Botha GSM, Astley R, Carre IJ (1957) A combined cineradiographic and manometric study of the gastro-esophageal junction. Lancet 1:659–662
Brown BSJ (1965) Defecography or anorectal studies in children including cinefluorographic observations. J Assoc Radiol 16:66–76
Burnett FL (1923) The intestinal rate and form of the feces. Am J Roentgenol 10:599–604
Cassano CC, Amoruso M, Arullani P, Badalamenti G (1967) Ricerche sui tempi di transito gastrointestinal totali parziali. Arch Ital Mal Appar Dig 34:290–293
Chapman WP, Palazzo WL (1949) Multiple-balloon-kymograph recording of intestinal motility in man with observations on the correlation of the tracing patterns with barium movements. J Clin Invest 28:1517–1525
Chauve A, Devroede G, Bastin E (1976) Intraluminal pressures during perfusion of the human colon in situ. Gastroenterology 70:336–340
Cohen S, Harris LD (1970) Lower esophageal sphincter pressure as an index of lower esophageal sphincter strength. Gastroenterology 58:157–162
Cohen BR, Wolf BS (1968) Cineradiographic and intraluminal pressure correlations in the pharynx and esophagus. In: Code CF (ed) Alimentary canal. American Physiological Society, Washington, DC (Handbook of physiology, vol IV, sect 6, pp 1841–1860)
Connell AM, Rowlands EN (1960) Wireless telemetering from the digestive tract. Gut 1:266–272
Cooper H, Levitan R, Fordtran JS, Ingelfinger FJ (1976) A method of studying absorption of water and solute from the human small intestine. Gastroenterology 50:1–7
Corazziari E, Dani S, Pozzessere C, Anzini F, Torsoli A (1975) Colonic segmental transit times in non-organic constipation. Rend Gastroenterol 7:67–69
Cummings JH, Wiggins HS (1976) Transit through the gut measured by analysis of a single stool. Gut 17:219–223
Delin NA, Axelsson B, Johansson C, Poppen B (1978) Comparison of gamma-camera and withdrawal method for the measurement of gastric emptying. Scand J Gastroenterol 13:867–872
Deller DJ, Wangel AG (1965) Intestinal motility in man. I. A study combining the use of intraluminal pressure recording and cineradiography. Gastroenterology 48:45–57
Dent J (1976) A new technique for continuous sphincter pressure measurement. Gastroenterology 71:263–267
De Salamanca FE Jr (1949) Estudio de fisiologiá gástrica en el perro. Trab Inst Nac Cienc Med 12:2–64
Dodds WJ (1976) Instrumentation and methods for intraluminal esophageal manometry. Arch Intern Med 136:515–523
Dodds WJ (1978) Efficient manometric techniques for accurate regional measurement of esophageal body motor activity. Am J Gastroenterol 70:21–24
Dodds WJ, Stewart ET, Hogan WJ, Stef JJ, Arndorfer RC (1974) Effect of esophageal movement on intraluminal esophageal pressure recording. Gastroenterology 67:592–600

Dodds WJ, Hogan WJ, Stef JJ, Lydon SB, Arndorfer RC (1975) A rapid pull-through technique for measuring lower esophageal sphincter pressure. Gastroenterology 68:437–443
Donaldson RM Jr, Barreras RF (1966) Intestinal absorption of trace quantities of chromium. J Lab Clin Med 68:484–493
Ehrenreich M (1912) Über die kontinuierliche Untersuchung des Verdauungsablaufs mittels der Magenverweilsonde. Z Klin Med 75:231–252
Engelmann TW, Van Brakel G (1871) Über die peristaltische Bewegung, insbesondere des Darms. Arch Physiol 4:33–50
Farrar JT, Bernstein JS (1958) Recording of intraluminal gastrointestinal pressures by a rediotelemetering capsule. Gastroenterology 35:603–612
Farrar JT, Zworkin VR, Baum J (1957) Pressure-sensitive telemetering capsules for study of gastrointestinal motility. Science 162:975–976
Fisher RS, Malmud LS, Roberts CS, Lobis IF (1976) Gastroesophageal (GE) scintiscanning to detect and quantitate GE reflux. Gastroenterology 70:301–308
Friedman G, Wolf BS, Waye JD, Janowitz HD (1965) Correlation of cineradiográphic and intraluminal pressure changes in the human duodenum: an analysis of the functional significance of monophasic wave. Gastroenterology 49:37–49
Gaston EA (1948) The physiology of fecal continence. Surg Gynecol Obstet 87:280–290
Gebauer A, Lissner J, Schott O (1967) Roentgen television. Grune & Stratton, New York London
George JD (1968) New clinical method for measuring the rate of gastric emptying: the double sampling test method. Gut 9:237–242
Gianturco C (1934) Some mechanical factors of gastric physiology, study I. Am J Roentgenol Radium Ther Nucl Med 31:735–750
Hancock BD (1976) Measurement of anal pressure and motility. Gut 17:645–651
Hansky YJ, Connell AM (1962) Measurement of gastrointestinal transit using radioactive chromium. Gut 3:187–188
Harris LD, Pope CE II (1964) "Squeeze" vs. resistance, an evaluation of the mechanism of sphincter competence. J Clin Invest 43:2272–2278
Harris LD, Winans CS, Pope CE II (1966) Determination of yield pressures; a method for measuring anal sphincter competence. Gastroenterology 50:754–760
Harvey RF, MacKie DB, Brown NJG, Keeling DH (1970) Measurement of gastric emptying time with a gamma camera. Lancet 1:16–18
Hay DJ, Goodall RJ, Temple JG (1979) The reproducibility of the station pullthrough technique for measuring lower oesophageal sphincter pressure. Br J Surg 66:93–97
Heading RC, Tothill P, Laidlaw AJ, Shearman DJC (1971) An evaluation of ^{131}indium DTPA chelate in the measurement of gastric emptying by scintiscanning. Gut 12:611–615
Heading RC, Tothill P, McLoughlin GP, Shearman DJC (1976) Gastric emptying rate measurement in man. A double isotope scanning technique for simultaneous study of liquid and solid components of a meal. Gastroenterology 71:45–50
Hendee WR (1970) Medical radiation physics. Year Book Medical Publisher, Chicago
Hightower NC (1968) Motor action fo the small bowel. In: Code CF (ed) Alimentary canal. American Physiological Society, Washington, DC (Handbook of physiology, vol IV, sect 6, pp 2001–2024)
Hinton JM, Lennard-Jones JE, Young AW (1969) A new method for studying gut transit time using radioopaque markers. Gut 10:842–847
Hollander F, Penner A (1939) History and development of gastric analysis procedures. Am J Dig Dis 5:22–25, 739–743, 786–791
Hollis JB, Castell DO (1972) Amplitude of esophageal peristalsis as determined by rapid infusion. Gastroenterology 63:417–422
Houckgeest JP, Van Braam (1872) Untersuchungen über Peristaltic des Magens und Darmkanals. Arch Ges Physiol 6:266–302
Hunt JN (1974) A modification to the method of George for studying gastric emptying. Gut 15:812–813
Hunt JN, Spurrell WR (1951) The pattern of emptying of the human stomach. J Physiol (Lond) 113:157–168

Hurst AF (1919) Constipation and allied disorders, 2nd edn. Frowde, London
Ingelfinger FJ, Abbott WD (1940) Intubation studies of the small intestine. XX. The diagnostic significance of motor disturbance. Am J Dig Dis 7:468–474
Irhe T (1974) Studies on anal function in continent and incontinent patients. Scand J Gastroenterol [Suppl 25] 9:1–64
Jacoby HI, Bass P, Bennett DR (1963) In vivo extraluminal contractile force transducer for gastrointestinal muscle. J Appl Physiol 18:658–665
Jahnberg T, Abrahmsson H, Jansson G, Martinson J (1977) Gastric relaxatory response to feeding before and after vagotomy. Scand J Gastroenterol 12:225–228
Johansson C, Legerlof HO, Ekelund K, Kulsdom N, Larsson I, Nylond B (1972) Determination of gastric secretion and evacuation, biliary and pancreatic secretion, intestinal absorption, intestinal transit time and flow of water in man. Scand J Gastroenterol 7:489–499
Kaye MD, Showalter JP (1971) Manometric configuration of the lower esophageal sphincter in normal human subjects. Gastroenterology 61:213–223
Kaye MD, Showalter JP (1974) Measurement of pressure in the lower esophageal sphincter. The influence of catheter diameter. Am J Dig Dis 19:860–863
Kaye MD, Showalter JP, Rock KC, Johnson E (1973) A circumferentially-sensitive miniature transducer for study of human esophageal motility. Gastroenterology 64:752
Kazem I (1972) A new scintigraphic technique for the study of the esophagus. Am J Roentgenol Radium Ther Nucl Med 115:681–688
Kelly MA (1974) The use of miniaturized mercury strain gauges to record gastric motility. In: Daniel EE (ed) Proceedings of the 5th International Symposium on Gastrointestinal Motility. Mitchell, Vancouver, pp 323–329
Kirwan WD, Smith AN (1974) Gastrointestinal transit estimated by an isotope capsule. Scand J Gastroenterol 9:763–766
Kronecker H, Meltzer SJ (1883) Der Schluckmechanismus, seine Erregung und seine Hemmung. Arch Ges Anat Physiol [Suppl] 7:328–332
Lagerlof HO, Ekelund K, Johansson C (1972) A mathematical analysis of jejunal indicator concentrations used to calculate jejunal flow and mean transit time. Scand J Gastroenterol 7:379–389
Lambert A, Eloy R, Grenier JF (1976) Transducer for recording electrical and mechanical chronic intestinal activity. J Appl Physiol 41:942–945
Leube W (1883) Beiträge zur Diagnostik der Magenkrankheiten. Dtsch Arch Klin Med 33:1–21
Lydon SB, Dodds WJ, Hogan WJ, Arndorfer RC (1975) Effect of manometric assembly diameter on intraluminal esophageal pressure. Am J Dig Dis 20:968–970
McKay RS, Jacobson B (1957) Endoradiosonde. Nature 179:1239–1240
Malagelada JR, Longstreth GF, Summerskill WHJ, Go VLW (1976) Measurement of gastric functions during digestion of ordinary solid meals in man. Gastroenterology 70:203–210
Manousos ON, Truelove SC, Lumsden K (1967) Transit times of food in patients with diverticulosis or irritable colon syndrome and normal subjects. Br Med J 3:760–762
Martelli H, Devroede G, Arhan P, Duguay C, Dornic C, Faverdin C (1978) Some parameters of large bowel motility in normal man. Gastroenterology 75:612–618
McKelvey STD (1970) Gastric incontinence and postvagotomy diarrhea. Br J Surg 57:741–747
Meerof JC, Go VL, Phillips SF (1973) Gastric emptying of liquids in man. Quantification by duodenal recovery marker. Mayo Clin Proc 48:728–732
Meyer JH, MacGregor IL, Gueller R, Martin P, Cavalieri R (1976) ^{99}mTc tagged chicken liver as a marker of solid food in the human stomach. Am J Dig Dis 21:296–304
Miller RE (1967) Barium sulphate as a contrast medium: In: Margulis AR, Burheune HJ (eds) Alimentary tract roentgenology. Mosby, Saint-Louis, pp 25–36
Millhon WA, Hoffman DE, Jarvis P, Cross CJ, Millhon JS, Crites NA (1968) Preliminary report on Millhon-Crites intraesophageal motility probe. Am J Dig Dis 13:929–934
Nauta J (1956) The closing mechanism between the esophagus and the stomach. Gastroenterologia 86:219–241

Nelsen TS, Angell JB (1979) Microminiature force transducers for chronic in vivo use. Abstracts 7th International Symposium on Gastrointestinal Motility, Iowa City. University of Iowa, Iowa City, p 55
Ostick DG, Green G, Howe K, Dymach IW, Cowley DJ (1975) A simple clinical method for measurement of gastric emptying of solid meals. Br J Surg 62:663
Parks TG (1970) Motor responses in unresected and resected diverticular disease of the colon. Proc Roy Soc Med [Suppl] 63:3–6
Pascaud XB, Genton JH, Bass P (1978) Gastroduodenal contractile activity in fed and fasted unrestrained rats. In: Duthie HL (ed) Gastrointestinal motility in health and disease. MTP, Lancaster England, pp 637–645
Pope CE II (1967) A dynamic test of sphincter strength: its application to the esophageal sphincter. Gastroenterology 52:779–786
Quigley JP, Brody DA (1952) A physiological and clinical consideration of the pressure developed in the digestive tract. Am J Med 13:73–81
Ramorino MA, Colagrande C (1964) Intestinal motility: preliminary studies with telemetering capsules and synchronized fluorocinematography. Am J Dig Dis 9:64–71
Ritchie JA (1968) Colonic motor activity and bowel function. Part I: Normal movement of contents. Gut 9:442–456
Rosswick RP, Stedford RD, Brooke BN (1969) New methods of studying intestinal transit time. Gut 8:195–196
Rothman MM, Katz AB (1964) Analysis of feces. Carmine or charcoal bowel motility test. In: Bockus HL (ed) Gastroenterology, 2nd ed, vol 2. Philadelphia, pp 695–696
Schuster MM, Hendrix TR, Mendeloff AI (1963) The internal anal sphincter response. J Clin Invest 42:196–207
Scopinaro F, Vignoni A, Gatti V, Pozzessere C, Anzini F, Corazziari E (1979) Role of scintigraphy in the detection of spontaneous gastroesophageal (GE) reflux. Eur J Nucl Med 4:140
Staadas JO (1975a) Intragastric pressure/volume relationship before and after proximal gastric vagotomy. Scand J Gastroenterol 10:129–134
Staadas JO (1975b) Intragastric pressure/volume relationship in the normal human stomach. Scand J Gastroenterol 10:135–140
Stef JJ, Dodds WJ, Hogan WJ, Linehan JH (1974) Esophageal manometry component analysis of systems used to record intraluminal pressure. In: Daniel EE (ed) Proceedings of the 5th International Symposium on Gastrointestinal Motility. Mitchell, Vancouver, pp 337–345
Stef JJ, Dodds WJ, Hogan WJ, Linehan JJ, Stewart ET (1977) Intraluminal esophageal manometry: an analysis of variables affecting recording fidelity of peristaltic pressures. Gastroenterology 73:23–27
Tasaka K, Farrar JT (1969) Mechanics of small intestinal muscle function in the dog. Am J Physiol 217:1224–1229
Templeton RD, Lawson H (1931) Studies in the motor activity of the large intestine. Am J Physiol 96:667–676
Tolin RD, Malmud LS, Menin R, Reilley J, Maxler PT, Fisher RS (1979a) Detection and quantitation of enterogastric reflux following Billroth II surgery. J Nucl Med 19:739
Tolin RD, Malmud LS, Reilley J, Fisher RS (1979b) Esophageal scintigraphy to quantitate esophageal transit (quantitation of esophageal transit). Gastroenterology 76:1402–1408
Torsoli A, Ramorino ML, Ammaturo MV, Capurso L, Arcangeli G, Paoluzi P (1971) Mass movement and intracolonic pressures. Am J Dig Dis 16:693–696
Tothill P, McLoughlin GP, Heading RC (1978) Techniques and errors in scintigraphic measurement of gastric emptying. J Nucl Med 19:256–261
Turano L (1957) Malattie non neoplastiche dell'esofago. In: Atti del Congr Ital Med Int, September 1957. Pozzi, Rome
Waldeck F, Jennewein HM, Siewert R (1973) The continuous withdrawal method for the quantitative analysis of the lower esophageal sphincter (LES) in humans. Eur J Clin Invest 3:331–337
Wallace RP, Ehrenfeld I, Cowett HP, Joliffe N, Shapiro LL, Sturtevant M (1938) Motility of the gastrointestinal tract. Am J Roentgenol 39:64–66

Waller SL (1975) Differential measurements of small and large bowel transit times in constipation and diarrhea. A new approach. Gut 16:372–378
Weisbrodt NW (1974) Application of extraluminal force transducers to motility. In: Daniel EE (ed) Proceedings of the 5th International Symposium on Gastrointestinal Motility. Mitchell, Vancouver, pp 331–335
Welch RW, Drake ST (1980) Normal lower esophageal sphincter pressure: a comparison of rapid vs. slow pull-through techniques. Gastroenterology 78:1446–1451
Wheatley IC, Hardy KJ, Dent J (1977) Anal pressure studies in spinal patients. Gut 18:488–490
Wilkinson R (1971) Polyethylene glycol 4000 as a continuously administered non-absorbable fecal marker for metabolic balance studies in human subjects. Gut 12:654–660
Winans CS (1972) The pharyngoesophageal closure mechanism: a manometric study. Gastroenterology 63:768–777
Wingate DL, Sondberg RJ, Phillips SF (1972) A comparison of stable and ^{14}C labelled polyethylene glycol or volume indicators in the human jejunum. Gut 3:812–815
Wolf BS, Khilnani MT (1966) Progress in gastroenterological radiology. Gastroenterology 51:542–559
Zapponi GA, Bausano G, Corazziari E, Pozzessere C, Anzini F, Torsoli A (1979) Progressione del contenuto attraverso l'intestino crasso in soggetti con tempo di transito normale. Ital J Gastroenterol [Suppl] 11:132

CHAPTER 8

In Vitro Techniques for the Study of Gastrointestinal Motility

G.J. SANGER and A. BENNETT

A. Introduction

Gastrointestinal motilitiy in vitro can be studied by many different techniques, each having advantages over the others for particular investigations. Experiments can be carried out on responses of muscle strips or whole segments of gut to added substances, to stretch and to nerve stimulation. These systems can be controlled to give good precision and reproducibility, and this can help elucidate the physiological control of motility and the ways that it can be modified for therapeutic purposes.

Although isolation of a tissue overcomes obvious ethical and technical problems associated with studies in vivo, consideration must be given to how well the in vitro situation reflects in vivo function. BENNETT (1968) concluded that in many cases in vitro and in vivo results with human gastrointestinal muscle are similar, and some discussion of this aspect is included in the following descriptions of in vitro techniques for measuring gastrointestinal motility. With methods described well in the references quoted, only brief technical details are given here.

B. General Methodology

The basic techniques for maintaining viable isolated gastrointestinal preparations have been described previously (BENNETT 1973). Details of isolated tissue baths, bathing solutions, oxygenation and temperature are important, since small changes can greatly influence smooth muscle reponses. These details can be found in the reference cited. Movements of isolated gastrointestinal muscle are usually registered by means of transducers and pen recorders. The types of transducer available (isotonic, isometric or auxotonic) generally give similar results, although one type may have a particular advantage for certain uses (see PATON 1975a).

Methods of killing animals prior to removal of tissues may be important. For example, anaesthesia can influence muscle responses (JOHNSON 1976). Preparations are usually used soon after their removal, to avoid degenerative changes. Storage of animal intestine at 4°–6 °C for 48 h in bathing solution equilibrated with the appropriate oxygen mixture, prevents peristalsis (AMBACHE 1946; INNES et al. 1957) and reduces responses to postganglionic nerve stimulation and α-adrenoceptor stimulants in rabbit preparation (GILLESPIE and WISHART 1957; LUM and KERMANI 1963; BUCKNELL 1966). However, human tissue resected at operation late in the day may have to be stored overnight in bathing solution at 4 °C. Such specimens do not seem to change their sensitivity to drugs, and they do not lose their ability to respond to nerve stimulation (BENNETT and WHITNEY 1966), even when stored for up to 3 days (BUCKNELL 1966).

C. Muscle Segments and Strips

Segments of gut, usually 2–3 cm long, are often used to study the movements of the longitudinal muscle. These segments are usually suspended in a tissue bath under a load of 0.5–1.5 g, depending on the tissue, with the lumen open (see BENNETT 1973). Light loads are usually necessary for sensitivity and for delicate preparations; the heavier loads can be used on more robust preparations to aid muscle relaxation between contractions or to reduce spontaneous activity.

Instead of muscle segments, strips of muscle can be cut parallel to either the longitudinal or circular muscle fibres. This is the most common method of studying smooth muscle from large preparations such as human gut, and has the added advantage of minimising mechanical interaction between longitudinal and circular muscle layers (see BORTOFF 1976). Responses of intestinal circular muscle from small animals may also be studied by cutting spirals (e.g. guinea-pig intestine, BENNETT and FLESHLER 1970), or a series of transverse cuts (HARRY 1963). In addition, longitudinal muscle strips can be separated from the circular muscle in thick tissues such as human alimentary muscle. In certain smaller tissues the longitudinal muscle may be prepared as a flat sheet (AMBACHE 1954), or gently rolled off the circular muscle after tangential stroking through the longitudinal muscle at its mesenteric attachment (AMBACHE and FREEMAN 1968).

MACKENNA and MCKIRDY (1970) investigated the functional relationship between the longitudinal and circular muscle layers of the rabbit intestine using a flat preparation of muscle. The mucosa was stripped off and a 1 cm square section of muscle cut so that the longitudinal and circular muscle fibres were parallel to the respective cut edges. Two adjacent cut edges were fixed, while the other cut edges were suspended from separate isometric transducers which recorded longitudinal and circular muscle activity. The simultaneous isometric responses of the longitudinal and circular muscle layers can therefore be measured with little mechanical interference from each other. In a similar type of preparation, OZAKI (1979) cut an L-shaped section from the rabbit small intestine, with one length cut parallel to the longitudinal fibres and the other length parallel to the circular muscle fibres. The junction of the strips was fixed and each free end connected to an isotonic transducer. To prevent any circular muscle interference with the longitudinal muscle, OZAKI separated the two layers and removed the circular muscle from the strip cut parallel to the longitudinal axis. This also exposed the myenteric plexus on the longitudinal muscle, so that a stimulating microelectrode could be placed on a node of the plexus (see Sect. G).

In such preparations, drugs added to the bathing solution reach both the serosal and mucosal surfaces. Diffusion into the muscle may more closely mimic local endogenous release of that substance, than does diffusion from the bloodstream following intravascular injection in vivo (BENNETT 1968). Studies on muscle strips generally seem to reflect in vivo gastrointestinal motility (BENNETT 1968, 1974) but in some cases the effects in vivo and in vitro are different. Thus morphine does not affect muscle strips but stimulates motility in vivo or when perfused in vitro through the vasculature (DANIEL et al. 1959). In humans, 5-hydroxytryptamine has comparable effects in vitro and in vivo (BENNETT 1968) but this substance, histamine or ATP have different effects in blood-perfused rat ileum than in isolated muscle preparations (SAKAI 1979; SAKAI et al. 1979a, b).

Techniques for perfusing the vascular system of isolated gut with blood usually rely on the use of a donor animal (SAKAI et al. 1979a, b), although in one technique heparinized blood diluted with Ringer solution has been used to maintain a viable preparation (SALERNO et al. 1966).

D. Gastric Motility

Although in vivo methods are often used to study gastric motility, some authors have used in vitro methods. The following descriptions refer to the original apparatus, but appropriate transducers and pen recorders would now be used in preference. PATON and VANE (1963) removed stomachs from guinea-pigs, kittens, rats and mice. After ligating the pylorus and oesophagus, a glass cannula was tied into a hole approximately 2 cm long, cut in the proximal part of the stomach near the greater curvature. The stomach was suspended by this cannula in bathing solution, and fluid was instilled intragastrically in a volume depending on the size of the stomach, the method of recording and the nature of the experiment.

Gastric motility was measured either as volume changes at a constant pressure, or as pressure changes at a constant volume, by attaching appropriate recorders to the glass cannula. In experiments where measurements of changes in volume were required, PATON and VANE (1963) fitted two float recorders in parallel (Fig. 1). An 80 ml capacity volume recorder measured absolute changes in stomach volume, while a 10 ml capacity recorder, with a small side arm through which the large variations in baseline could be controlled, gave a more sensitive measure of volume changes. The intraluminal pressure at a constant volume can be measured with a water manometer, and the constant pressure opposes muscle shortening and so eliminates large variations in baseline. The stomachs were also divided into "pyloric" and "fundal" sections by cutting the stomach transversely along a line which left the oesophagus on the "fundal" section. T-shaped cannulas were fitted to each section and motility was measured by means of water manometers.

Another method for measuring whole-stomach motility (ARMITAGE and DEAN 1962, 1966) allows simultaneous measurement of pressures in different parts of the stomach, and propulsion of fluid through the pylorus (Fig. 2). These authors suspended rat stomach in a tissue bath by two cannulae, one tied into the fundus, another in the duodenum. Pressure in the body of the stomach was measured by a transducer connected through a side arm of the fundic cannula; pressure in the antrum was measured by a fine cannula passed through the antral wall.

Intraluminal pressure can be increased by raising a fluid-filled Marriotte bottle connected to the fundic cannula, and the column of fluid propelled through the pylorus measured. By passing a fine tube into the body of the stomach, and preventing outflow, fluid can be removed from the stomach by suction after periods of stimulated activity (BENNETT et al. 1966). Substances released by the stomach can therefore accumulate intraluminally, giving higher concentrations which aid detection and measurement.

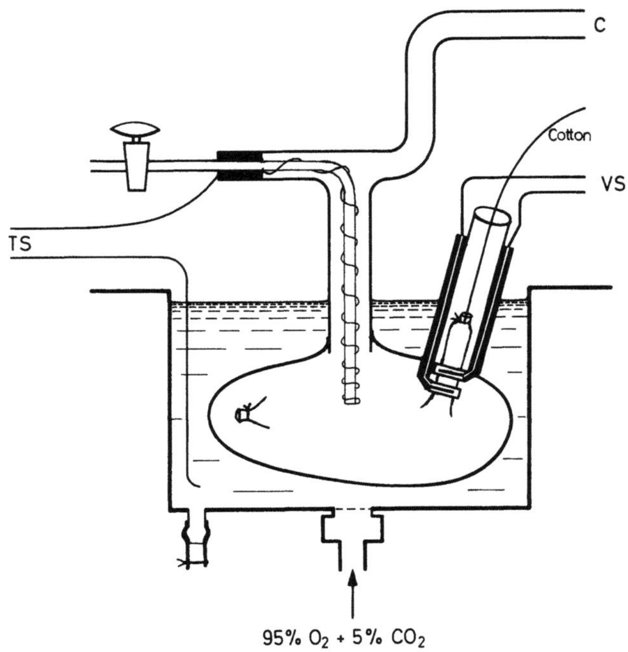

Fig. 1. Apparatus for measuring gastric motility. A cannula *C* was tied into the stomach near the greater curvature and fitted to a volume or pressure recorder. A polyethylene tube was threaded through a side arm in this cannula, so that fluid could be removed. An electrode coiled around the polyethylene tube was used for transmural stimulation *TS*. For vagal stimulation the oesophagus and the attached vagal nerves were ligated proximally and pulled into a polyethylene tube containing two platinum rings connected to a stimulator *VS*. (Adapted from PATON and VANE 1963)

Simple increases of intraluminal pressure alone did not satisfactorily stimulate gastric peristalsis in either technique (ARMITAGE and DEAN 1962, 1966; PATON and VANE 1963), although a "receptive relaxation" response could be detected (PATON and VANE 1963). Motility was therefore usually induced by vagal or transmural stimulation (PATON 1955; see Sect. G) or with drugs.

CAMPBELL (1966) described a method for guinea-pig stomach similar to that of ARMITAGE and DEAN, except that the oesophagus was ligated. The stomach was filled with about 25 ml saline, and motility was measured with either a float recorder or a water manometer connected to a pyloric cannula.

E. Intestinal Peristalsis

Peristalsis is the coordinated movement of alimentary muscle resulting in the propulsion of contents. However, this activity is only part of the gut motility that occurs in vivo. The peristaltic reflex is mediated through the intrinsic nerves, but can be modulated by extrinsic nerves and hormones (BAYLISS and STARLING 1899; KOS-

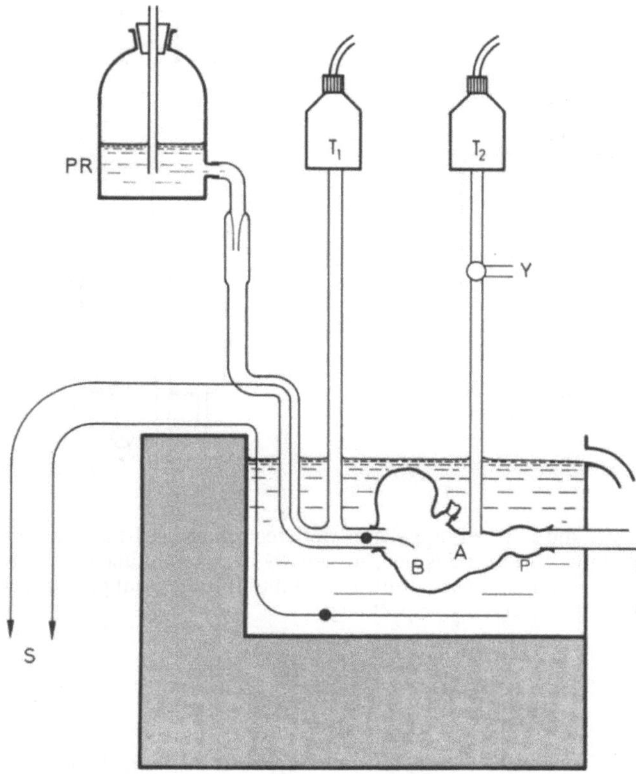

Fig. 2. Apparatus for measuring rat gastric motility and propulsion. The polyrus *P*, gastric antrum *A* and body *B* were cannulated. Pressures in the body and antrum were measured separately with transducers attached to the respective cannulas via air-filled T_1 and saline-filled tubes T_2. Gastric pressure was increased by raising the pressure reservoir *PR*. Fluid propelled through the duodenum was collected and measured. For gastric transmural stimulation coaxial platinum wires were connected to a stimulator *S*. Slow injection of saline through a three-way tap *Y* prevented blockage of the antral cannula by mucus produced during stimulation. (Adapted from ARMITAGE and DEAN 1966)

TERLITZ et al. 1956). It is this intrinsic nature of the reflex that enables us to investigate peristalsis in vitro.

TRENDELENBURG (1917) described the first in vitro method for studying peristalsis, using guinea-pig ileum. The reflex is evoked by a radial distension of the gut wall (KOSTERLITZ and ROBINSON 1959), which stimulates sensory receptors lying in either the mucosal or submucosal layer (BÜLBRING et al. 1958; GINZEL 1959; DIAMENT et al. 1961; FRIGO and LECCHINI 1970). The most appropriate type of stimulus to elicit peristalsis may vary in different parts of the gut. FRIGO and LECCHINI (1970) concluded that unlike the ileum where the intraluminal contents are relatively fluid, a physiological stimulus for muscle activity in the greater part of the distal colon is a solid bolus. Methods for eliciting peristalsis in ileum and colon are therefore described separately.

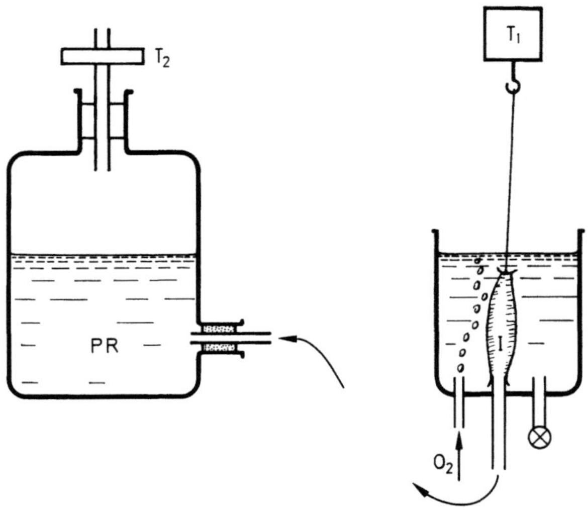

Fig. 3. Simplified and modernised apparatus for recording peristaltic activity by the Trendelenburg technique. Intestine *I* is tied proximally over a cannula leading to the pressure reservoir *PR*. Longitudinal muscle contractions and intraluminal pressure changes are registered by transducers T_1 and T_2 respectively

I. Ileum

For the method of TRENDELENBURG (1917), a segment of ileum 3–5 cm long is tied at its distal end over a glass tube of about 0.5 cm diameter connected via tubing to a reservoir containing bathing solution. The proximal end of the fluid-filled ileum is then closed by tying with thread (Fig. 3), the preparation suspended in bathing solution, and the proximal end tied to an isotonic recording system under a 0.5–2 g load. In the original method a sensitive volume recorder was connected to the reservoir, but now a pressure transducer is usually used to measure intraluminal pressure changes in the ileum. Raising the reservoir so that its fluid level is up to a few centimetres above that of the bathing solution in the tissue bath increases the intraluminal pressure; a rise of 0.5 cm H_2O may be enough to elicit a peristaltic response, but rises of 1–5 cm H_2O are often used. The response is recorded as an initial longitudinal muscle contraction (usually registered by an isotonic or an isometric transducer which give similar responses; KOSTERLITZ et al. 1956), followed by peristaltic longitudinal and circular muscle movements. Increased intraluminal pressure is thought to correspond to the circular muscle contractions which begin at the oral end and move along the intestine. As a result fluid is expelled aborally in the direction of the pressure reservoir (for a detailed description see KOSTERLITZ et al. 1956). The reservoir is only partly filled with bathing solution, so that there is a large air space which can be compressed; consequently the rise in pressure is small and does not greatly impede muscle activity (KOSTERLITZ et al. 1956). Calibration of the pressure transducer is achieved by clamping the tube between the ileum and the reservoir, and injecting known volumes of fluid into the tube leading to the reservoir.

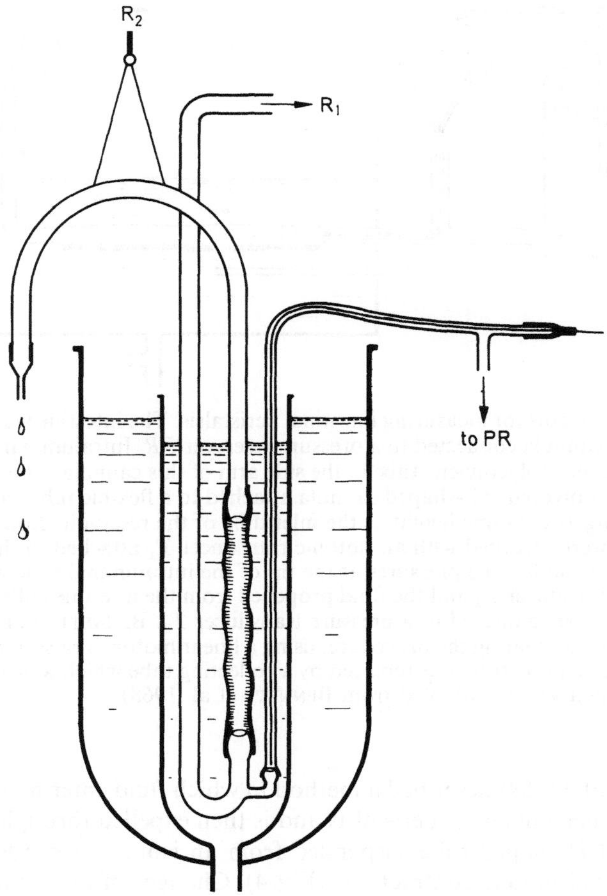

Fig. 4. Apparatus for measuring intestinal peristalsis. Intestine was tied proximally over a J-shaped tube partly filled with bathing solution and attached to a float recorder R_1. The distal end of the intestine was tied over an inverted U-shaped tube and fluid was propelled through the valve (flat rubber tubing) during peristalsis. Longitudinal muscle contractions were registered by a lever R_2 fixed to the U-shaped tube. The pressure reservoir *PR* was connected to the J-shaped tube by a side arm, which contained a fine polyethylene tube for intraluminal injections. (Adapted from BÜLBRING et al. 1958)

Peristalsis is usually elicited for 1 min periods, each followed by 2–10 min rest. Frequent stimulation or the maintenance of a raised intraluminal pressure causes fatigue and the responses become irregular. Changes in the peristaltic response are often assessed by eye. Attempts to quantify peristalsis in the Trendelenburg preparation by measuring various parameters, such as the number and amplitude of contractions, have been described (VAN NEUTEN et al. 1973; SANGER and WATT 1978), but these are usually laborious and do not greatly improve on visual assessment.

Peristalsis can better be quantified in other methods by measuring the fluid propelled by the peristaltic contractions, and this is both simple and satisfactory. BÜL-

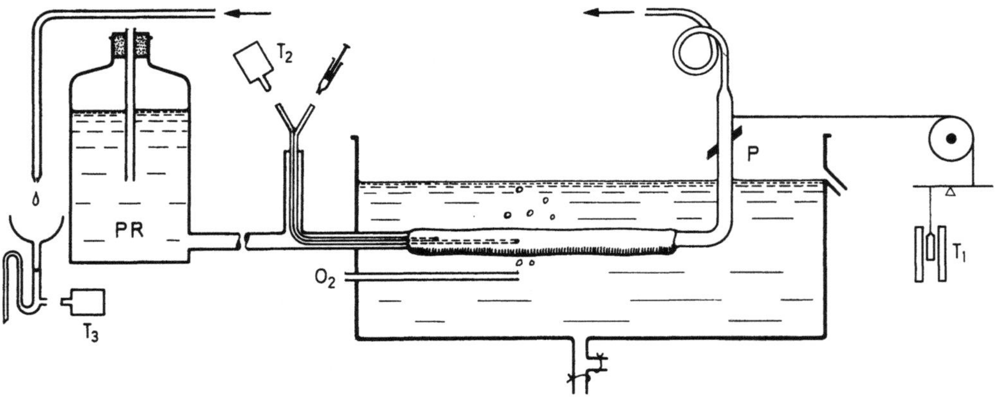

Fig. 5. Apparatus for measuring intestinal peristalsis. The intestine was attached proximally to a fixed cannula connected to a pressure reservoir *PR*. Intraluminal injections were made through a fine polyethylene tube in the side arm of this cannula. The intestine was tied distally over a pivoted *P* L-shaped cannula attached to a flexible tube with a non-return valve terminating at the same height as the inlet tube of the reservoir. Longitudinal muscle contractions were recorded with an isotonic transducer T_1 attached to the vertical limb of the L-shaped cannula. The pressures at the tip of the intraluminal tube were measured with a pressure transducer T_2 and the fluid propelled from the intestine collected and measured in a syphon tube connected to a pressure transducer T_3. BENNETT et al. (1976) modified this method by automating the procedure, using a linear motor to raise and lower the Marriotte bottle. The syphon tube was replaced by a collecting tube which was emptied automatically by opening a valve. (Adapted from BENNETT et al. 1968)

BRING et al. (1958) described a method in which fluid entering the proximal end of the ileal segment elicits peristalsis and is then expelled through the distal end into a vertical U-shaped tube suspended from an isotonic recorder used to measure longitudinal muscle contractions (Fig. 4). Changes of intraluminal pressure measured from the proximal end of the ileum give an indication of circular muscle activity (Fig. 4). BENNETT et al. (1968) adapted the Trendelenburg technique so that as well as measuring longitudinal and circular muscle movements, fluid was expelled aborally from the ileum, collected and measured (Fig. 5). Fluid propulsion in a modified Trendelenburg preparation has also been measured by following the movement of an air bubble (VAN NEUTEN et al. 1973) in the tubing leading from the distal end of the ileum.

In vivo intestinal activity probably combines the components of both "open" and "closed" peristaltic systems. The open systems of BÜLBRING et al. (1958) and BENNETT et al. (1968) have certain advantages over closed systems other than ease of assessment of peristaltic activity. With an open system the intraluminal fluid is renewed so that substances do not accumulate, and these methods are more suitable for the intraluminal administration of drugs, achieved by injecting drugs through fine polyethylene tubing inserted through the inflow tube at the proximal end of the gut segment. In the Trendelenburg method, intraluminally administered drugs might not reach all of the mucosa; they would tend to be expelled from the gut and diluted in the reservoir fluid, and once administered the drugs would be difficult to remove from the closed system.

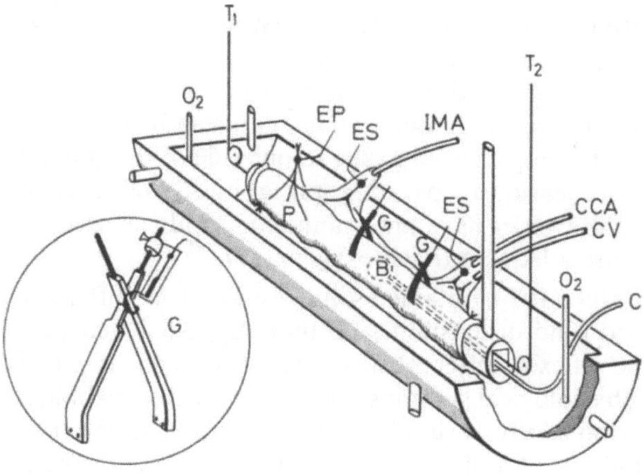

Fig. 6. Apparatus for measuring cat colonic segmental activity. The common colonic artery *CCA*, inferior mesenteric artery *IMA* and colonic vein *CV* were cannulated before removal of the tissue. The colon was fixed proximally and connected distally to an isotonic transducer T_1. Oxygenated Tyrode solution containing polyvinylpyrrolidone 35 mg/ml was perfused through the arteries (1.5–3 ml/min; 65–80 mmHg) and drained through the veins. The reflex was elicted by rapidly distending a proximally placed thin rubber balloon *B* with warm water passed through cannula *C*. Displacement of the balloon was recorded by the isotonic transducer T_2. Circular muscle activity was recorded with two auxotonic force transducers *G* attacched to the serosal surface of the colon. The effects of electrical stimulation of extrinsic nerves were studied using three pairs of electrodes placed around the pelvic nerves *P* and the arteries (*EP* and *ES* respectively). (Adapted from FRIGO and LECCHINI 1970)

Finally, since in an open system muscle contraction might not produce such great changes of pressure and there is no sustained increase of intraluminal pressure which might otherwise affect muscle activity, ELEY et al. (1977) suggested that differences with open and closed peristaltic systems might affect the response of guinea-pig ileum to prostaglandins.

II. Colon

The method originally described by TRENDELENBURG (1917) and adaptations of this method have been used for eliciting and measuring colonic peristalsis in guinea-pigs and rabbits (LEE 1960; MACKENNA and MCKIRDY 1972; ELEY et al. 1977). Results are generally satisfactory, although coordinated and regular muscle activity is often difficult to obtain in guinea-pig colon (K. G. ELEY, personal communication).

FRIGO and LECCHINI (1970) concluded that for the greater part of the colon a solid bolus is a more physiological stimulus than a diffuse distension with fluid. They mounted a segment of guinea-pig colon horizontally, fixed its proximal end, and connected the distal end to an isotonic transducer which registered longitudinal muscle movements. Peristalsis was elicited by pushing a thin rubber balloon proximally into the lumen, and rapidly inflating it with warm water. This induced a reflex distal propulsion of the balloon, as measured by an isotonic transducer connected to the balloon through the proximal part of the colon. With the more robust cat colon, circular muscle activity can also be measured (FRIGO and LECCHINI 1970) by attaching a strain gauge to one side of the colon and fixing the other side to a support bearing a separating spring (Fig. 6). Because of the thickness of cat colonic wall, the arteries have to be cannulated and perfused with physiological solution to ensure adequate tissue oxygenation.

FRIGO and LECCHINI (1970) obtained some results which differed from those of LEE (1960) who used fluid distenstion, and they suggested that the two types of stimuli triggered the reflex in different ways. Apart from eliciting activity by what is possibly a more physiological stimulus, the use of a semisolid bolus makes quantification of propulsion relatively simple. ISHIZAWA and MIYAZAKI (1973 a, b, 1975) measured the propulsion of a plastic ball through guinea-pig isolated colon. The ball was connected to a thread which passed through the distal end of the colon to an isotonic transducer.

F. Effects of Stretch on Muscle Strips

Radial distension of the intact small intestine activates stretch receptors and initiates coordinated peristaltic activity which depends largely on the intrinsic nerves. Muscle strips which are stretched can also show an increase in tension, due at least partly to an intrinsic property of the muscle. Although stretch usually depolarises the membrane and causes muscle contraction (BÜLBRING 1955; BURNSTOCK and PROSSER 1960; GILLESPIE 1962), certain preparations and types of stretch, such as the application of a quick stretch along the longitudinal axis of guinea-pig ileum, do not elicit a response (KOSTERLITZ and ROBINSON 1959). MEISS (1971) showed that tissues may respond only to certain types of stretch.

The effects of stretch on gastrointestinal muscle strips can be modified by intrinsic nerve activity. In experiments on guinea-pig colonic circular muscle (FURNESS and COSTA 1977, DAVISON and PEARSON 1979), segments of distal colon were placed in bathing solution and a stainless steel rod passed through the lumen and fixed in position. A metal hook was passed through the colon wall and fastened to an isotonic transducer to register circular muscle contractions. Application of a series of weights (4–24 g) to the colon induced distension which, by addition of various blocking drugs, was found to be partially facilitated by a nonadrenergic inhibitory nerve mechanism. Such inhibitory pathways may account for relaxation of intestinal circular muscle around and distal to a distending bolus.

Essentially the methods for applying stretch to strips of smooth muscle can be divided into those applying quick stretch (GORDON and SIEGMAN 1971), a linearly increasing stretch force (ROSSBERG and KIESSLING 1969) or a rhythmic sinusoidal stretch (GOLENHOFEN 1964), since contractions in response to stretch may vary according to the rate of stretching (BURNSTOCK and PROSSER 1960). Details of these techniques have been well reviewed by KOSTERLITZ and WATT (1975) and STEPHENS (1975).

G. Electrical Stimulation of Autonomic Nerves

Electrical stimulation of autonomic nerves is an important technique for studying nerve functions and mechanisms of action. Controlled and consistent responses can be produced which facilitate the quantitation of drug action or neurotransmitter release. However, since a nerve bundle or plexus may contain various different types of neurone, it can be difficult to activate one particular nerve pathway selectively. Besides cholinergic and adrenergic fibres, there is increasing evidence for the existence of nonadrenergic, noncholinergic inhibitory (BURNSTOCK 1972; FURNESS and COSTA 1973) and excitatory (AMBACHE and FREEMAN 1968; BENNETT and FLESHLER 1969; FURNESS and COSTA 1973) fibres, and of nerves which may release peptides such as vasoactive intestinal peptide (BRYANT et al. 1976). The effect of electrical stimulation of autonomic nerves therefore depends on the types and relative importance of the neurones present, in addition to other factors such as the state of the muscle and its type of response and sensitivity to the neurotransmitters. Distinction between the different nerve types may be achieved by using drugs which selectively inhibit responses to a particular nerve type and/or by varying the intensity of stimulation. In general, noradrenergic nerves are excited at frequencies of approximately 2 Hz or above (pulse width 0.1–0.5 ms), whereas lower frequencies preferentially stimulate cholinergic nerves (GARRY and GILLESPIE 1955).

In human circular and longitudinal oesophageal, ileal and colonic muscle strips, contractions elicited at frequencies below 1 Hz (pulse width 1 ms) involve excitation of noncholinergic nerves, whereas contractions above this frequency seem due only to cholinergic nerve stimulation (STOCKLEY and BENNETT 1974; BENNETT and STOCKLEY 1975).

BENNETT and STOCKLEY (1975) also reported that fast aftercontractions following inhibitory nerve stimulation in human longitudinal and circular gastrointestinal muscle strips were due mainly to cholinergic nerve stimulation, although part of the response seemed resistant to hyoscine. However in ascending colonic circular muscle, a slow aftercontraction induced by noncholinergic nerves was observed at frequencies of stimulation of 8–32 Hz. FURNESS and COSTA (1973) reported similar slow aftercontractions in guinea-pig colon. These were maximal at 20–50 Hz (pulse width 0.2–0.5 ms) but could be detected at frequencies as low as 5 Hz. Unlike the experiments of BENNETT and STOCKLEY (1975) the aftercontractions in guinea-pig colon could be blocked with the 5-hydroxytryptamine antagonist methysergide.

In tissues such as human taenia coli, which are densely innervated with nonadrenergic inhibitory fibres, the adrenergic response may be masked by the nonadrenergic inhibitory response (STOCKLEY and BENNETT 1977), even though substantial amounts of noradrenaline may be released (HOUGHTON and BENNETT 1977). Frequencies of 1–10 Hz caused nonadrenergic inhibition of human taenia coli, whereas frequencies greater than 10 Hz elicited adrenergic responses (see STOCKLEY and BENNETT 1977, for references). In addition, an increased pulse duration may reduce the adrenergic contribution to the overall inhibitory response (STOCKLEY and BENNETT 1977). One explanation of these results may be that adrenergic nerves synapse with nonadrenergic inhibitory nerves.

The gut contains intramural plexuses and also receives extrinsic innervation which goes mainly to the plexuses. Techniques for stimulating gastrointestinal nerves are therefore divided into two categories: those for stimulating intrinsic nerves, and those for stimulating extrinsic nerves.

I. Intrinsic Nerves

PATON (1955) described a method of coaxial stimulation which produces a voltage gradient across the wall of guinea-pig ileum. Appropriate stimuli preferentially stimulate the intrinsic nerves. The technique has since been used mostly on intestinal tissues (see PATON 1975b) but ARMITAGE and DEAN (1962, 1966) and PATON and VANE (1963) adapted the method for use with whole-stomach preparations. The technique described below is that most commonly used for intestinal preparations.

A segment of intestine about 3 cm long is threaded over a platinum wire electrode partly insulated by polyethylene tubing at its upper end. The proximal end of this tissue is tied over the insulated area, and the distal end is tied over a glass tube into which the lower end of the electrode protrudes and moves freely with muscle contractions. The glass tube is fixed in a tissue bath, and the upper end of the electrode is attached to a device for measuring longitudinal muscle activity.

A stimulator supplies current via a fine wire connected to the intraluminal electrode; stimulation is more effective when this electrode is the anode. A second platinum electrode suspended parallel to the intestine in the external bathing solution is the cathode. Stimulation with single square wave pulses elicits a single muscle contraction which, at a constant pulse duration (usually 0.5–1 ms), is dependent on voltage strength. With pulses of 0.5 ms duration the threshold voltage for stimulation is approximately 1 V, and about 5–25 V are required for a maximum contraction (PATON 1955).

There is considerable evidence that these parameters of transmural stimulation contract guinea-pig ileum by stimulating postganglionic cholinergic nerves (see PATON 1975b). However, particularly with frequencies of electrical stimulation at or above 2 Hz, other types of neurone can be stimulated, and locally produced substances can be released. For the guinea-pig ileum, these include stimulation or noradrenergic nerve terminals (KADLEC et al. 1978; SANGER and WATT 1980), the release of prostaglandin-like material (BOTTING and SALZMANN 1974; SANGER 1977) and enkephalin (PUIG et al. 1977). Direct muscle stimulation also seems to occur at these frequencies since part of the contraction is resistant to tetrodotoxin (BENNETT and STOCKLEY 1973, 1974). These authors examined various parameters of field stimulation for their ability to induce tetrodotoxin-resistant contractions (indicative of direct muscle stimulation) in guinea-pig ileum and human gastrointestinal muscle strips. In general, frequencies of stimulation greater than 2 Hz, high pulse durations (1 ms) or high voltages (60 V) caused "unacceptable" stimulation of tetrodotoxin-resistant contractions. The train lengths of stimulation did not seem to influence the selectivity of nerve stimulation.

The method of PATON (1955) produces consistent and controlled twitch or tetanic responses but luminal substances and products of electrolysis tend to accumulate; large pulse durations or tetanic stimulation cause large changes in pH

(PATON 1975b). Flushing the lumen with bathing solution alleviates this problem but is rather awkward. However, the use of two linked stimulators which deliver alternate pulses of opposite polarity (BENNETT and STOCKLEY 1973) largely circumvents the problems of electrolysis.

A disadvantage of the Paton technique is that only changes in longitudinal muscle tension can be satisfactorily recorded. If the lumen is filled with fluid, similar to the Trendelenburg method, and the intestine distended at a pressure of approximately 1–3 cm H_2O, an electrically induced emptying reaction can be demonstrated although the twitch response is somewhat irregular (PATON 1955; KOTTEGODA 1969). The changes of intraluminal pressure are assumed to correspond to circular muscle activity, but the restrictions by the intraluminal electrode may interfere mechanically with the response of both muscle layers.

Electrical stimulation of intrinsic neurones in strips or whole segments of intestine can be achieved with field stimulation, usually by passing an electric current across two platinum electrodes positioned near to the muscle at either side. Current is therefore short-circuited through the tissue and bathing solution. This method was originally designed for direct electrical stimulation of muscle cells (SPERELAKIS 1962), but the lower voltages and frequencies of stimulation described can selectively stimulate the intrinsic neurones. It may be an advantage for the electrodes to be as large as the tissue itself, so that all the nerve cells are stimulated simultaneously and more equally. If the electrodes are opposite only a part of the tissue, a localized area of excitation is produced, and the response may involve propagation of the induced stimulus (BURNSTOCK et al. 1966).

Because current is applied across a bathing solution of low resistance, power requirements of the stimulator can be high and could heat the solution sufficiently to interfere with muscle contraction (SPERELAKIS 1962). High voltage can also present problems due to electrolysis, including oxidation of compounds such as noradrenaline (WYSE 1977). Periods of high-voltage stimulation should therefore be kept to a minimum.

Various electrode arrangements have been examined for their ability to stimulate nerves selectively (BENNETT and STOCKLEY 1974). The methods compared were: an electrode above and below the tissue (BUCKNELL 1965); uninsulated platinum wires either side of the intenstine (BIRMINGHAM and WILSON 1963; AMBACHE and FREEMAN 1968); wires placed either side of the tissue but insulated on entry to the bath (CREMA et al. 1968); the method of PATON (1955); and ring electrodes 2 mm apart around one end of the muscle strip (BURNSTOCK et al. 1966; FURNESS 1970; SMALL 1971). Threshold frequencies for tetrodotoxin-resistant contractions indicated that the method of CREMA et al. (1968) was the most selective for nerve stimulation, even though with other stimulus parameters this arrangement most effectively stimulates the muscle (SPERELAKIS 1962). Electrodes above and below the tissue gave least selectivity for nerve stimulation.

Finally the intrinsic nerves may be stimulated directly, after peeling off the circular muscle layer and exposing the myenteric plexus attached to the longitudinal muscle. OZAKI (1979) used a glass microelectrode through which Ag–AgCl wire was passed and placed on a node of the plexus; an indifferent electrode was placed in the bathing solution. With the dissection previously described in Sect. C, the effects of this electrode's stimulation of a fixed area could be recorded on longitudinal and circular muscle activity. Similarly, a tissue holder which stretches and fixes muscle in one position can be used to record electrical activity from intestinal muscle and also simultaenous mechanical activity of muscle outside the fixed area (SMALL and WESTON 1977).

II. Extrinsic Nerves

A technique for stimulating nerves extrinsic to the gut was first described by FINKLEMAN (1930) for the sympathetic nerve fibres running alongside the mesenteric blood vessels of rabbit isolated ileum. A mesenteric blood vessel and its perivascular nerves are threaded through two platinum ring electrodes, and the segment is set up to record longitudinal muscle activity.

Some techniques for stimulation of extrinsic neurones require a specialised dissection and preparation of the tissue. These include a periarterial nerve–longitudinal muscle (taenia) or circular muscle preparation from the caecum of the guinea-pig (AKUBE 1966, 1977), rabbit rectum–pelvic nerve preparations (GARRY and GILLESPIE 1954, 1955), and rat anococcygeus muscle (GILLESPIE 1972).

Techniques for studying nerves supplying the rat isolated whole stomach have also been described (PATON and VANE 1963; ARMITAGE and DEAN 1966). The dissection and preparation of the rat stomach have previously been described in Sect. D. For vagal stimulation the distal end of the oesophagus and its associated vagal trunks can be threaded through two platinum ring electrodes (PATON and VANE 1963). Alternatively, the vagal trunks themselves may be dissected away from the oesophagus (ARMITAGE and DEAN 1966). Stimulation of the sympathetic periarterial nerves can similarly be produced by means of the dissected coeliac artery.

H. Conclusions

In vitro techniques are valuable in studying gastrointestinal motility because they allow approaches that are difficult with anaesthetised animals and impossible in conscious animals in vivo. Many questions can be answered with precision, because interference from other controlling mechanisms can often be avoided. However, it follows that in vitro methods give at best only an approximation of motility which in vivo is affected by the interplay of numerous factors such as the hormones and other substances reaching the gut through the bloodstream. Furthermore, isolation of the tissue causes damage, and some results may more closely reflect changes seen in vivo with diseased rather than with normal tissues. Our understanding of motility must therefore include studies of intact anaesthetised and unanaesthetised animals, as indicated in Chap. 7.

References

Akubue PI (1966) A periarterial nerve-circular muscle preparation from the caecum of the guinea-pig. J Pharm Pharmacol 18:390–395

Akubue PI (1977) A periaterial nerve-longitudinal muscle (taenia) preparation from the guinea-pig caecum. J Pharm Pharmacol 29:122–124

Ambache N (1946) Interaction of drugs and the effect of cooling on the isolated mammalian tissue. J Physiol (Lond) 104:266
Ambache N (1954) Separation of the longitudinal muscle of the rabbits ileum as a broad sheet. J Physiol (Lond) 125:53–55P
Ambache N, Freeman MA (1968) Atropine-resistant longitudinal muscle spasm due to excitation of non-cholinergic neurones in Auerbach's plexus. J Physiol (Lond) 199:705–727
Armitage AK, Dean ACB (1962) A new technique for studying gastric peristalsis in small animals. World Med Electron 1:17–19
Armitage AK, Dean ACB (1966) The effects of pressure and pharmacologically active substances on gastric peristalsis in transmurally stimulated rat stomach – duodenum preparation. J Physiol (Lond) 182:42–56
Bayliss WM, Starling EH (1899) The movements and innervation of the small intestine. J Physiol (Lond) 24:99–143
Bennett A (1968) Relationship between in vitro studies of gastrointestinal muscle and motility of the alimentary tract in vivo. Am J Dig Dis 13:410–414
Bennett A (1973) The pharmacology of isolated gastrointestinal muscle. In: Holton P (ed) Encylcopedia of pharmacology, vol 39A. Pergamon, Oxford New York, pp 399–432
Bennett A (1974) Relation between gut motility and innervation in man. Digestion 11:392–396
Bennett A, Bucknell A, Dean ACB (1966) The release of 5-hydroxytryptamine from the rat stomach in vitro. J Physiol (Lond) 182:57–65
Bennett A, Eley KG, Scholes GB (1968) Effect of prostaglandins E_1 and E_2 on intestinal motility in the guinea-pig and rat. Br J Pharmacol 34:639–647
Bennett A, Eley KG, Stockley HL (1976) Inhibition of peristalsis in guinea-pig isolated ileum and colon by drugs that block prostaglandin synthesis. Br J Pharmacol 57:335–340
Bennett A, Fleshler B (1969) A hyoscine resistant excitatory nerve pathway in guinea-pig colon. J Physiol (Lond) 203:62P–63P
Bennett A, Fleshler B (1970) Prostaglandins and the gastrointestinal tract. Gastroenterology 59:790–800
Bennett A, Stockley HL (1973) Electrically-induced contractions of guinea-pig isolated ileum resistant to tetrodotoxin. Br J Pharmacol 48:357–360P
Bennett A, Stockley HL (1974) Effect of electrode positions on contractions of guinea-pig isolated ileum to electrical stimulation. Br J Pharmacol 50:453P
Bennett A, Stockley HL (1975) the intrinsic innervation of the human alimentary tract and its relation to function. Gut 16:443–453
Bennett A, Whitney B (1966) A pharmacological investigation of the human isolated stomach. Br J Pharmacol Chemother 27:286–298
Birmingham AT, Wilson AB (1963) Preganglionic and postganglionic stimulation of the guinea-pig isolated vas deferens preparation. Br J Pharmacol Chemother 21:569–580
Bortoff A (1976) Myogenic control of intestinal motility. Physiol Rev 56:418–434
Botting JH, Salzmann R (1974) The effect of indomethacin on the release of PGE_2 and acetylcholine from guinea-pig isolated ileum at rest and during field stimulation. Br J Pharmacol 50:119–124
Bryant MD, Polak JM, Modlin I, Bloom SR, Albuquero RH, Pearse AGE (1976) Possible dual role for vasoactive intestinal peptide as gastrointestinal hormone and neurotransmitter substance. Lancet 1:991
Bucknell A (1965) Effects of direct and indirect stimulation on isolated colon. J Physiol (Lond) 177:58–59P
Bucknell A (1966) Studies on the physiology and pharmacology of the colon of man and other animals. PhD thesis, University of London
Bülbring E (1955) Correlation between membrane potential spike discharge and tension in smooth muscle. J Physiol (Lond) 128:200–221
Bülbring E, Crema A, Saxby OB (1958) A method for recording peristalsis in isolated intestine. Br J Pharmacol Chemother 13:440–443
Burnstock G (1972) Purinergic nerves. Pharmacol Rev 24:509–581

Burnstock G, Campbell G, Rand MT (1966) The inhibitory innervation of the taenia of the guinea-pig caecum. J Physiol (Lond) 182:504–526
Burnstock G, Prosser CL (1960) Responses of smooth muscle to quick stretch: relation of stretch to conduction. Am J Physiol 198:921–925
Campbell G (1966) The inhibitory nerve fibres in the vagal supply to the guinea-pig stomach. J Physiol (Lond) 185:600–612
Crema A, del Tacca M, Frigo GM, Lecchini S (1968) Presence of a nonadrenergic inhibitory system in the human colon. Gut 9:633–637
Daniel EE, Sutherland WH, Bogoch A (1959) Effects of morphine and other drugs on motility of the terminal ileum. Gastroenterology 36:510
Davison JS, Pearson GT (1979) The role of intrinsic non-adrenergic non-cholinergic inhibitory nerves in the regulation of distensibility of the guinea-pig colon. Pflügers Arch 381:75–77
Diament ML, Kosterlitz HW, McKenzie J (1961) Role of the mucous membrane in the peristaltic reflex in the isolated ileum of the guinea-pig. Nature 190:1205–1206
Eley KG, Bennett A, Stockley HL (1977) The effect of prostaglandins E_1, E_2, $F_{1\alpha}$ and $F_{2\alpha}$ on guinea-pig ileal and colonic peristalsis. J Pharm Pharmacol 29:276–280
Finkleman B (1930) On the nature of inhibition of the intestine. J Physiol (Lond) 70:185–202
Frigo GM, Lecchini S (1970) An improved method for studying the peristaltic reflex in the isolated colon. Br J Pharmacol 39:346–356
Furness JB (1970) An examination of nerve mediated, hyoscine-resistant excitation of the guinea-pig colon. J Physiol (Lond) 207:803–821
Furness JB, Costa M (1973) The nervous release and the action of substances which affect intestinal muscle through neither adrenoceptors nor cholinoreceptors. Philos Trans R Soc Lond [Biol] 265:123–133
Furness JB, Costa M (1977) The participation of enteric inhibitory nerves in accommodation of the intestine to distension. Clin Exp Pharmacol Physiol 4:37–41
Garry RC, Gillespie JS (1954) An in vitro preparation of the distal colon of the rabbit with orthosympathetic and parasympathetic innervation. J Physiol (Lond) 123:60–61P
Garry RC, Gillespie JS (1955) The response of the musculature of the colon of the rabbit to nerve stimulation in vitro of the parasympathetic and the sympathetic outflows. J Physiol (Lond) 128:557–576
Gillespie JS (1962) Spontaneous mechanical and electrical activity of stretched and unstretched intestinal smooth muscle cells and their response to sympathetic-nerve stimulation. J Physiol (Lond) 162:54–75
Gillespie JS (1972) The rat anococcygeus muscle and its response to nerve stimulation and some drugs. Br J Pharmacol 45:404–416
Gillespie JS, Wishart M (1957) The effect of cooling on the response of rabbit colon to nerve and drug stimulation. J Physiol (Lond) 135:45P
Ginzel KH (1959) Are mucosal nerve fibres essential for the peristaltic reflex? Nature 184:1235–1236
Golenhofen K (1964) "Resonance" in the tension response of smooth muscle of guinea-pig's taenia coli to rhythmic stretch. J Physiol (Lond) 173:13–15P
Gordon AR, Siegman MJ (1971) Mechanical properties of smooth muscle. II. Active state. Am J Physiol 221:1250–1259
Harry J (1963) The action of drugs on the circular muscle strip from the guinea-pig isolated ileum. Br J Pharmacol Chemother 20:399–417
Houghton J, Bennett A (1977) Release of (^{3}H) noradrenaline by electrical stimulation of human isolated taenia coli. In: Gastrointestinal motility in health and disease. Proceedings of the 6th International Symposium on Gastrointestinal Motility. MTP Press, Lancaster, pp 137–142
Innes IR, Kosterlitz HW, Robinson JA (1957) The effects of lowering the bath temperature on the responses of the isolated guinea-pig ileum. J Physiol (Lond) 137:396
Ishizawa M, Miyazaki E (1973a) Action of prostaglandins on gastrointestinal motility. Sapporo Med J 42:366–373
Ishizawa M, Miyazaki E (1973b) Effect of prostaglandins on the movement of guinea-pig isolated intestine. Jpn J Smooth Muscle Res 9:235–237

Ishizawa M, Miyazaki E (1975) Effect of prostaglandin $F_{2\alpha}$ on propulsive activity of the isolated segmental colon of the guinea-pig. Prostaglandins 10:759–768
Johnson IT (1976) Alternative methods of animal sacrifice: the effect on intestinal function in vitro. Experientia 32:347–348
Kadlec O, Măsek K, Šeferna I (1978) Modulation by prostaglandins of the release of acetylcholine and noradrenaline in guinea-pig isolated ileum. J Pharmacol Exp Ther 205:635–645
Kosterlitz HW, Robinson JA (1959) Reflex contractions of the longitudinal muscle coat of the isolated guinea-pig ileum. J Physiol (Lond) 146:369–379
Kosterlitz HW, Watt AJ (1975) Stimulation by stretch. In: Daniel EE, Paton DM (eds) Methods in pharmacology, vol 3. Plenum, New York, pp 347–358
Kosterlitz HW, Pirie VW, Robinson JA (1956) The mechanism of the peristaltic reflex in the isolated guinea-pig ileum. J Physiol (Lond) 133:681–694
Kottegoda SR (1969) An analysis of possible nervous mechanisms involved in the peristaltic reflex. J Physiol (Lond) 200:687–712
Lee CY (1960) The effect of stimulation of extrinsic nerves on peristalsis and on the release of 5-hydroxytryptamine in the large intestine of the guinea-pig and of the rabbit. J Physiol (Lond) 152:405–418
Lum BKB, Kermani MH (1963) Selective loss of response to alpha adrenergic agents following cold storage of the rabbit jejunum. Fed Proc 22:449
MacKenna BR, McKirdy HC (1970) A simple method for investigating the functional relationship of the longitudinal and circular layers of the muscularis externa of the rabbit bowel using the "flat" preparation. J Physiol (Lond) 211:18P
MacKenna BR, McKirdy HC (1972) Peristalsis in the rabbit distal colon. J Physiol (Lond) 220:33–54
Meiss RA (1971) Some mechanical properties of cat intestinal muscle. Am J Physiol 220:2000–2007
Ozaki T (1979) Effects of stimulation of Auerbach's plexus on both longitudinal and circular muscles. Jpn J Physiol 29:195–209
Paton WDM (1955) The response of the guinea-pig ileum to electrical stimulation by coaxial electrodes. J Physiol (Lond) 127:40–41P
Paton WDM (1975a) The recording of mechanical responses of smooth muscle. In: Daniel EE, Paton DM (eds) Methods in pharmacology, vol 3. Plenum, New York, pp 261–264
Paton WDM (1975b) Transmural and field stimulation of nerve-smooth muscle preparations. In: Daniel EE, Paton DM (eds) Methods in pharmacology, vol 3. Plenum, New York, pp 313–320
Paton WDM, Vane JR (1963) An analysis of the responses of the isolated stomach to electrical stimulation and to drugs. J Physiol (Lond) 165:10–46
Puig MM, Gascon P, Craviso GL, Musacchio JM (1977) Endogenous opiate receptor ligand: electrically induced release in the guinea-pig ileum. Science 195:419–420
Rossberg F, Kiessling A (1969) Die aktive Reaktion der Taenia coli des Meerschweinchens auf Dehnung und Entdehnung. Z Biol 116:220–234
Sakai K (1979) A pharmacological analysis of the contractile action of histamine upon the ileal region of the isolated blood-perfused small intestine of the rat. Br J Pharmacol 67:587–590
Sakai K, Akima M, Matsushita H (1979a) Analysis of the contractile responses of the ileal segment of the isolated blood-perfused small intestine of rats to adenosine triphosphate and related compounds. Eur J Pharmacol 58:157–162
Sakai K, Akima M, Shiraki Y (1979b) Comparative studies with 5-hydroxytryptamine and its derivatives in isolated blood-perfused small intestine and ileum strip of the rat. Jpn J Pharmacol 29:223–233
Salerno RA, Iijima K, Healey WV (1966) Extra-corporeal circulation of excised sigmoid colon segments. Surg Gynecol Obstet 122:767–772
Sanger GJ (1977) Modulation by prostaglandins of the autonomic control of motility in guinea-pig isolated ileum. PhD thesis, University of Manchester
Sanger GJ, Watt AJ (1978) The effect of PGE_1 on peristalsis and on perivascular nerve inhibition of peristaltic activity in guinea-pig isolated ileum. J Pharm Pharmacol 30:762–765

Sanger GJ, Watt AJ (1980) Some mechanisms which may modulate noradrenaline release in guinea-pig isolated ileum. J Pharm Pharmacol 32:188–191
Small RC (1971) Transmission from cholinergic neurones to circular smooth muscle obtained from the rabbit caecum. Br J Pharmacol 42:656–657P
Small RC, Weston AH (1977) Simultaneous recording of electrical and mechanical activity from intestinal and vascular smooth muscle. Br J Pharmacol 61:491P
Sperelakis N (1962) Contraction of depolarized smooth muscle by electric fields. Am J Physiol 202:731–742
Stephens NJ (1975) Physical properties of contractile systems. In: Daniel EE, Paton DM (eds) Methods in pharmacology, vol 3. Plenum, New York, pp 265–296
Stockley HL, Bennett A (1974) The intrinsic innervation of human sigmoid colonic muscle. In: Daniel EE (ed) Proceedings of the 4th International Symposium on Gastrointestinal Motility. Mitchell, Vancouver pp 165–176
Stockley HL, Bennett A (1977) Relaxations mediated by adrenergic and nonadrenergic nerves in human isolated taenia coli. J Pharm Pharmacol 29:533–537
Trendelenburg P (1917) Physiologische und pharmakologische Versuche über die Dünndarmperistaltik. Naunyn-Schmiedeberg's Arch Exp Pathol Pharmakol 81:55–129
Van Neuten JM, Geivers H, Fontaine J, Janssen PAJ (1973) An improved method for studying peristalsis in the isolated guinea-pig ileum. Arch Int Pharmacodyn Ther 203:411–414
Wyse DG (1977) Alteration of exogenous noradrenaline caused by electrical "field" stimulation and its role in poststimulant relaxation. Can J Physiol Pharmacol 55:990–1000

CHAPTER 9

Nervous Control of Esophageal and Gastric Motility

C. ROMAN

A. Nervous Control of the Esophagus

The esophagus may be regarded as a tube extending from pharynx to stomach and serving for transport of material between these two organs. At rest (i.e., between swallows and regurgitations), sphincter mechanisms at either end of the tube prevent easy access of air from above and gastric contents from below. For purposes of clarity the control of the tubular esophagus (or esophageal body) and that of the sphincter areas will be dealt with separately.

I. Esophageal Body

1. General Survey of Motility Patterns

The only transport which will be considered in this chapter is that occurring in the aboral direction, which is due to a propagated contraction generally recognized as peristalsis. According to its method of elicitation, esophageal peristalsis was divided by MELTZER (1899, 1907) into primary and secondary types: primary peristalsis is initiated by swallowing; secondary peristalsis is independent of swallowing and starts in the esophagus in response to a local stimulus. Such a local stimulus can be provided either by a bolus of food stuck in the gullet or by a rubber balloon suddenly inflated.

In manometric studies, the dominant feature of the esophageal swallowing patterns is a positive forceful wave called the third or peristaltic wave which moves down more rapidly in the proximal part of the esophagus than in the distal part. This difference of velocity is not only observed in species in which the muscular coat of the distal esophagus is composed of smooth muscle (e.g., opossum, cat, monkey, human) but also in species with a muscular coat composed entirely of striated muscle (e.g., dog, sheep). For instance, the progression velocity of primary peristalsis in the dog is 7 cm/s in the cervical esophagus and only 3.7 cm/s in the distal thoracic esophagus (JANSSENS 1978).

Electromyographic (EMG) recordings show that the peristaltic contraction constists of a burst of spike potentials whose characteristics vary according to the type of muscular coat. Thus, smooth muscle spikes have a greater amplitude, a longer duration, and a lower frequency than spikes originating from the striated muscle. At rest, EMG examinations have revealed no activity, i.e., no tonic contraction of the muscular walls. When swallows are taken in rapid succession, no peristaltic contraction appears until after the last swallow, because each new swallow inhibits the esophageal contraction which should be triggered by the previous one. This phenomenon has been called deglutitive inhibition. Other inhibitory phenomena can be elicited from the esophagus itself. Thus, distension of an esophageal balloon inhibits the contractions (either primary or secondary) below the level of the balloon.

For a more detailed description of the motility patterns and for the extensive literature on the subject, the interested reader should refer to the following reviews: INGELFINGER (1958), CODE and SCHLEGEL (1968), HELLEMANS and VANTRAPPEN (1974), HELLEMANS et al. (1974).

2. Anatomic and Histologic Data on Esophageal Innervation

a) Vagal Innervation

The motor innervation of the esophagus is supplied by the vagus nerves. Classically, the cervical esophagus is innervated by the recurrent laryngeal nerves and the rest of tube by branches coming from the thoracic vagal trunks (INGELFINGER 1958). However, HWANG et al. (1948) have demonstrated that the cervical portion of the esophagus, at least its upper end, receives its innervation either from the pharyngoesophageal nerve (e.g., dog, cat, rabbit) or from the external branch of the superior laryngeal nerve (e.g., monkey, guinea pig, rat). The pharyngoesophageal nerve (leaving the vagus nerve above the superior laryngeal nerve) also supplies the cervical esophagus of sheep (DOUGHERTY et al. 1958; ROMAN 1967). In humans, nerves of similar anatomic distribution have not been described (HWANG and GROSSMAN 1953).

The myenteric plexus of Auerbach lies between the longitudinal and circular muscle layers of the esophagus. These intramural neurons are generally considered as relay neurons between the vagal fibers and the smooth muscle cells (KUNTZ 1947). But a myenteric plexus also exists in the striated muscle coat, i.e., in the proximal part of the human esophagus and throughout the esophagus of various species whose esophageal muscle coat is composed entirely of striated muscle (INGELFINGER 1958; ABE 1959; MANN and SHORTER 1964; JACOBOWITZ and NEMIR 1969). The function of this intramural plexus within the striated muscle esophagus remains obscure since many histologic studies have suggested that the efferent vagal fibers do not synapse on cells of the myenteric plexus but end directly on the striated muscle cells at neuromuscular junctions, similar to those of skeletal muscle fibers elsewhere (JABONERO 1958; ABE 1959; GRUBER 1968; SAMARASINGHE 1972; FLOYD 1973).

The vagus nerves also contain sensitive fibers originating from the esophagus. These fibers have their cell bodies in the nodose ganglion. As indicated by various experiments involving section, stimulation, and reception, the sensitive fibers from the upper cervical esophagus of dog, rat, cat, and sheep join the vagus high in the neck through the superior laryngeal nerve (HWANG 1954; ANDREW 1956; MEI 1970; ROMAN 1967; ROMAN and CAR 1970). Those innervating the rest of the esophagus join the vagus within the thorax, either through the recurrent laryngeal nerves (lower cervical and upper thoracic esophagus), or through esophageal branches of the vagus (lower thoracic esophagus). Histologic studies have shown the existence of many sensory endings in the mucosa, submucosa, and muscular layers (JA-

BONERO 1958; ABE 1959; SPASSOVA 1959; YAMAMOTO 1960; JABONERO 1962). These endings are generally simple, i.e., free endings. But a few encapsulated structures resembling neuromuscular spindles have also been described in the human striated muscle esophagus (SPASSOVA 1959) and that of various animals (JABONERO 1962). Most of these endings probably belong to vagal sensory fibers, although solid support for such a contention is lacking.

b) Sympathetic Innervation

The esophagus seems to receive an abundant sympathetic innervation, but the origin and course of these fibers remain unclear. As for their termination within the esophageal wall, more precise information has been obtained with the Falck–Hillarp histofluorescence technique by BAUMGARTEN and LANGE (1969). According to these investigators, the sympathetic fibers which enter the cat and rhesus monkey esophagus not only supply the vessels but also mainly innervate ganglia of the myenteric plexus and the muscularis mucosae. Only a few fibers are confined to the muscular layers of the distal esophagus (smooth muscle). Surprisingly, densely innervated myenteric ganglia are also observed in the upper and middle portion of the esophagus, despite the fact that the muscular wall is exclusively or mainly composed of striated muscle.

3. Nervous Control of the Striated Muscle Esophagus

a) Effects of Stimulation and Section of the Sympathetic Supply

Many investigators have found no change in esophageal function of various animals following section or stimulation of the possible components of the sympathetic supply to the esophagus (INGELFINGER 1958; GREENWOOD et al. 1962; BURGESS et al. 1972). In humans, extensive sympathectomy also failed to alter the esophageal motility (INGELFINGER 1958). KNIGHT (1934) reported somewhat different results in the cat but his findings have not been duplicated by others.

b) Effects of Stimulation of Vagal Efferent Fibers

Stimulation of vagal efferent fibers by a single shock or a train of pulses triggers, after a short latency, a twitch or a tetanus which are typical of striated muscle responses. If the stimulation is delivered on the cervical vagus below the nodose ganglion, the response observed in most mammals is restricted to the middle and lower esophagus, because the upper esophagus receives its motor fibers either from the pharyngoesophageal nerve or from the external branch of the superior laryngeal nerve (see Sect. A.I.2.a).

The motor effects of vagal stimulation on the striated muscle esophagus are suppressed by neuromuscular blocking agents such as curare or succinylcholine (INGELFINGER 1958; ROMAN 1967; BARTLET 1968; TOYAMA et al. 1975; Dodds et al. 1978). In contrast, atropine has no effect (INGELFINGER 1958; TOYAMA et al. 1975). With ganglion-blocking agents like hexamethonium or nicotine controversial results have been reported. According to BARTLET (1968), the esophageal responses are unaffected by these drugs. On the contrary, TOYAMA et al. (1975) observed that hexamethonium, at doses which did not affect the neuromuscular transmission in the diaphragm, significantly reduced the vagally induced EMG response of the canine esophagus. Nicotine at high dosage had a similar effect. In view of these results, TOYAMA et al. (1975) have suggested that a certain proportion of the esophageal striated muscle fibers are innervated via myenteric cells of the myenteric (Auerbach's) plexus. This assumption might explain why the esophageal striated muscle does not degenerate after total extrinsic denervation (JURICA 1926).

c) Effects of Vagotomy

Bilateral vagotomy is followed by a paralysis of the striated muscle esophagus (see INGELFINGER 1958; CARVETH et al. 1962; GREENWOOD et al. 1962; HIGGS and ELLIS 1965). When the section is performed below the nodose ganglion, the cervical esophagus of the dog and many other animals is not affected. But this part of the gullet is also paralyzed if the pharyngoesophageal nerves or the external branch of the superior laryngeal nerves are severed. Surprisingly, the striated muscle does not degenerate (JURICA 1926). Moreover, it seems able to exhibit a motility consisting of weak general tonic contractions when food or a balloon is placed in the esophagus (JURICA 1926; HWANG et al. 1947). However, peristalsis probably never returns (INGELFINGER 1958; CARVETH et al. 1962; HIGGS and ELLIS 1965).

d) Recording of Vagal Efferent Fiber Activity

After microdissection of the superior laryngeal nerve in anesthetized rats, ANDREW (1956) recorded the swallowing discharge of a few vagal motor fibers controlling the upper cervical esophagus. Following ANDREW's pioneer work, more extensive studies have been performed in conscious sheep (ROMAN 1966, 1967) and monkeys (ROMAN and TIEFENBACH 1972) by means of a nerve suture technique. In these animals, the central end of the left vagus had been sutured to the peripheral end of the left spinal accessory nerve. When reinnervation was effective, the activity of motor units of the sternocleidomastoideus and trapezius muscle indicated vagal motor fibers activity. By means of electromyographic techniques for unit detection (by a Bronk needle) in the *unasthetized* sheep or monkey, it was possible to study the discharge of vagal fibers which previously supplied the esophagus. The main results emerging from these experiments may be summarized as follows:

1) The vagal motor fibers have no spontaneous discharge at rest. This means that there is no tonic contraction of the esophagus, a conclusion which is in keeping with data obtained by direct EMG recording of esophageal muscle activity (MONGES et al. 1968; ARIMORI et al. 1970; HELLEMANS et al. 1974). They fire only during esophageal peristalsis either primary or secondary with a burst of spikes. Their discharge frequency is fairly high and typical of a striated muscle command (mean frequency between 15 and 30 spikes/s; instantaneous frequency between 20 and 70 spikes/s).

2) During peristalsis, the various vagal motor fibers controlling the different portions of the esophagus discharge in succession (Fig. 1).

3) During primary peristalsis, the vagal motor discharges are reinforced by stimulation of afferent fibers from the esophagus: deglutition of a bolus elicits a more powerful discharge than that occurring during a "dry" swallow (Fig. 1).

4) During secondary peristalsis, the vagal motor fiber activity is usually weaker than that exhibited during primary peristalsis (Fig. 1).

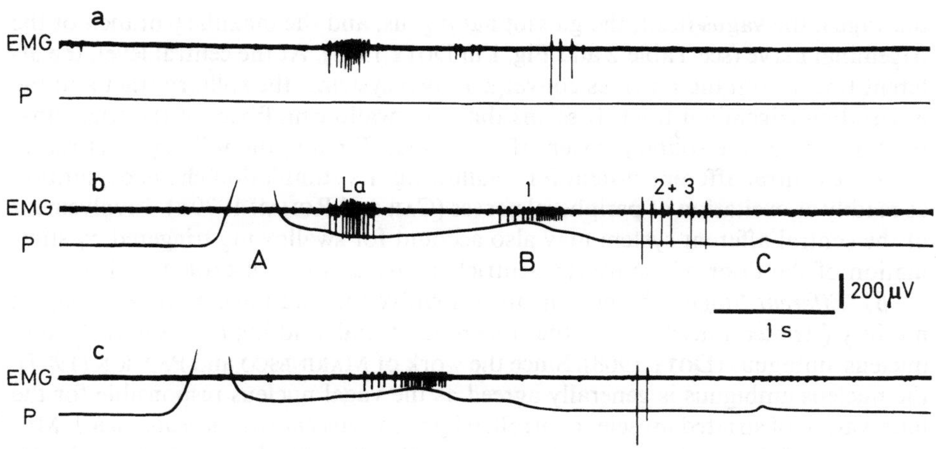

Fig. 1 a–c. Discharge of vagal efferent fibers during primary and secondary peristalsis. Recording made in sheep (striated muscle esophagus) with nerve suture technique (see text). EMG: electromyograms from neck muscles innervated by laryngeal and esophageal efferent fibers. La: "laryngeal" unit innervated by motoneuron which probably supplied the larynx originally; 1, 2, 3: "esophageal" units probably innervated by fibers originally destined for lower cervical (1) and midthoracic (2, 3) esophagus. P: pressure variations recorded by an intraesophageal balloon propelled by the peristaltic wave ("dynamic" pressure recording). A: balloon inflation; B: passage through the narrow area of the beginning of the thoracic esophagus; C: balloon entering the stomach. **a** "dry" swallow, balloon empty; **b** deglutition of a bolus (balloon inflated with 20 ml air), the discharge of all the esophageal units is markedly increased; **c** secondary peristalsis with a bolus of the same size as in **b**, the activity of unit 1 is similar to **b** whereas that of unit 2 and 3 is less. (ROMAN 1966)

5) Firing of vagal motor fibers is inhibited during the buccopharyngeal stage of deglutition or after distension of an esophageal portion above their innervation area. This inhibition has a central origin (see Sect. A.I.3.e.γ); its effects on esophageal motility have also been observed directly, either by manometric (for review see INGELFINGER 1958) or by EMG techniques (HELLEMANS et al. 1974; JANSSENS 1978).

e) Central Mechanisms Responsible for Esophageal Peristalsis

The organization of esophageal motility takes place within the swallowing center, which in fact is responsible for all the phenomena observed during deglutition, i.e., the buccopharyngeal component and the esophageal peristalsis. A large body of evidence resulting from the destruction of various nervous structures or section of the brain stem at different levels (for references see DOTY 1968) indicates that the circuits necessary to the swallowing motor performance are contained within the rhombencephalon (medulla and pons). The swallowing center, or more exactly its two halves (left and right) may be divided into three stages or subsystems (see ROMAN 1967; DOTY 1968; JEAN 1972a): afferent (inputs), efferent (outputs, i.e., motoneurons), and organizing stage (internuncial system).

α) *Afferent Stage.* The sensitive fibers responsible for elicitation of swallowing have been demonstrated by nerve stimulation in the superior laryngeal branch of the vagus, the vagus itself, the glossopharyngeus, and the maxillary branch of the trigeminal nerve (see Table 2 and Fig. 1 in DOTY 1968). At the central level, the afferent fibers from these nerves converge in two systems: the solitary tract and the descending trigeminal tract. It seems that the swallowing fibers of the trigeminal tract also enter the solitary system (DOTY 1968). Finally, the solitary tract represents the central afferent system for swallowing. Its stimulation elicits deglutition as readily as excitation of peripheral nerves (CAR and ROMAN 1970 a). Involvement of the central afferent system may also account for swallowing triggered by stimulation of the floor of the fourth ventricle (MILLER and SHERRINGTON 1916).

β) Efferent Stage. The motoneurons involved in swallowing and esophageal motility (striated muscle) lie in the trigeminal, facial, and hypoglossal nuclei and nucleus ambiguus (DOTY 1968). Since the work of MARINESCO and PARHON (1907), the nucleus ambiguus is generally agreed as the vagal nucleus responsible for the innervation of striated muscles controlled by the vagus *(noyau musculo-strié)*. Motoneurons of this nucleus certainly innervate the pharynx, larynx, and esophagus (see DOTY 1968). Histologic studies of muscle representation within the nucleus have provided evidence that esophageal motoneurons are concentrated in the rostral portion (DOTY 1968). This point has been fully confirmed by electrical stimulation of the medulla, in both rabbits (LAWN 1964) and sheep (ROMAN and CAR 1967). In addition, microelectrode recordings in the nucleus ambiguus have evidenced the swallowing discharge of motoneurons occurring during either the buccopharyngeal (SUMI 1964; JEAN 1972 a) or the esophageal stage of deglutition (JEAN 1978 b). By the way, it may be noticed that motor innervation of the smooth muscle esophagus is probably provided by the dorsal motor nucleus of the vagus instead of the nucleus ambiguus. Such a statement is supported by histologic data (MARINESCO and PARHON 1907). Besides, in cases of lesions of the vagal dorsal motor nucleus in the cat, the motility of the distal esophagus (smooth muscle) is markedly impaired (HIGGS et al. 1965). Finally, when horseradish peroxidase is injected into the last few centimeters of the cat esophagus, labeled cell bodies are observed in the dorsal motor nucleus of the vagus (NIEL et al. 1980), although some are also seen in the nucleus ambiguus.

γ) Organizing stage. This stage consists of an internuncial system which programs the successive excitation of motoneurons, thereby organizing the whole motor sequence of deglutition. These programming interneurons are obviously placed between the afferent fibers and the motoneurons, but their exact location is still disputed. On the basis of central destructions in acute experiments, DOTY et al. (1967) localized the organizing stage of the swallowing center in the reticular substance at the level between the posterior pole of the facial nucleus and the rostral pole of the inferior olive, 1–3 mm dorsal to these structures and about 1.5 mm off the midline. Subsequent experiments based on stimulation of the medulla failed to confirm this interpretation (ROMAN 1967; CAR and ROMAN 1970 a). In addition, microelectrode soundings indicated that the medullary neurons exhibiting a swallowing activity, except the motoneurons, were located in the solitary tract nucleus (STN) and the underlaying reticular substance, about 2–4 mm in front of the obex (SUMI 1964; JEAN 1972 a; NEYA et al. 1974; JEAN 1978 a).

These STN or reticular neurons present either a phasic discharge (only when swallowing occurs) or a spontaneous activity which is altered during deglutition (increased or inhibited). Depending upon the temporal relationship of their activity with the mylohyoideus contraction (onset of deglutition), the phasic swallowing neurons have been classified into three categories by JEAN (1972a, 1978a): "early" neurons, firing before or during the buccopharyngeal stage; "late" and "very late" neurons, discharging during the esophageal stage (Fig. 2). After lesions of the STN area by means of electrocoagulation, either extensive (JEAN1978a) or restricted (JEAN 1972b), the deglutitions elicited by peripheral nerve stimulation disappear either totally or in part (suppression of esophageal stage). This emphasizes that interneurons of the STN area play an important role in programming the deglutitive motor sequence.

Current concepts about the nervous mechanisms of esophageal peristalsis in particular, and deglutition in general, have been deeply influenced by the pioneer work of MOSSO (1876) and MELTZER (1899). From these fundamental studies, especially those by MELTZER, emerged the concept of central programming of deglutition. According to this concept, the motor sequences of deglutition would depend on a central mechanism (connections between neurons) which, after starting, could run its entire course without any new afferent support. This opinion was based on a fact previously noticed by MOSSO (1876) that in the dog, primary peristalsis could jump a gap produced by esophageal transection. According to MELTZER, the mechanism of secondary peristalsis, quite different from that of primary peristalsis, would consist of a succession of reflexes originating in esophageal receptors sequentially stimulated by a moving bolus. In support of this difference, MELTZER argued that in the dog, secondary peristalsis, unlike primary peristalsis, cannot jump a gap produced by esophageal transection (MELTZER 1899, 1907).

More recent studies of the effects of esophageal transection with or without bolus deviation have confirmed the theory of the central programming of primary peristalsis in the striated muscle esophagus (CARVETH et al. 1962; GREENWOOD et al. 1962; JANSSENS 1978). In addition, microelectrode recordings have shown that, in sheep, the swallowing discharge of medullary neurons, either motoneurons or interneurons, persisted after curarization of the animals (Fig. 3; JEAN 1972a, 1978a, b). This means that the swallowing program triggered by peripheral nerve stimulation can run its full course within the center even if muscles are paralyzed (striated muscle) and thus no bolus is transported.

However, if afferent feedback is not necessary for primary peristalsis, under normal circumstances when a bolus is swallowed, esophageal receptors are stimulated. This afferent stimulation increases the discharge of the various esophageal motoneurons (see Sect. A.I.3.d). A similar facilitation is also observed at the level of interneurons which program the motor sequence of peristalsis (Fig. 4; JEAN 1972a, 1978a). Thus, the central program may function in the absence of afferent support, but under normal circumstances this program is permanently modified by esophageal afferent fibers which adjust the force and progression velocity of the peristaltic contraction to the esophageal contents.

Concerning secondary peristalsis, various investigators (HWANG 1954; CREAMER and SCHLEGEL 1957; FLESHLER et al. 1959; SIEGEL and HENDRIX 1961; JANSSENS et al. 1974) have produced evidence that its central mechanism should not be markedly different from that of primary peristalsis. The motoneurons and interneurons involved in both cases are the same (ROMAN 1966, 1967; JEAN 1972a,

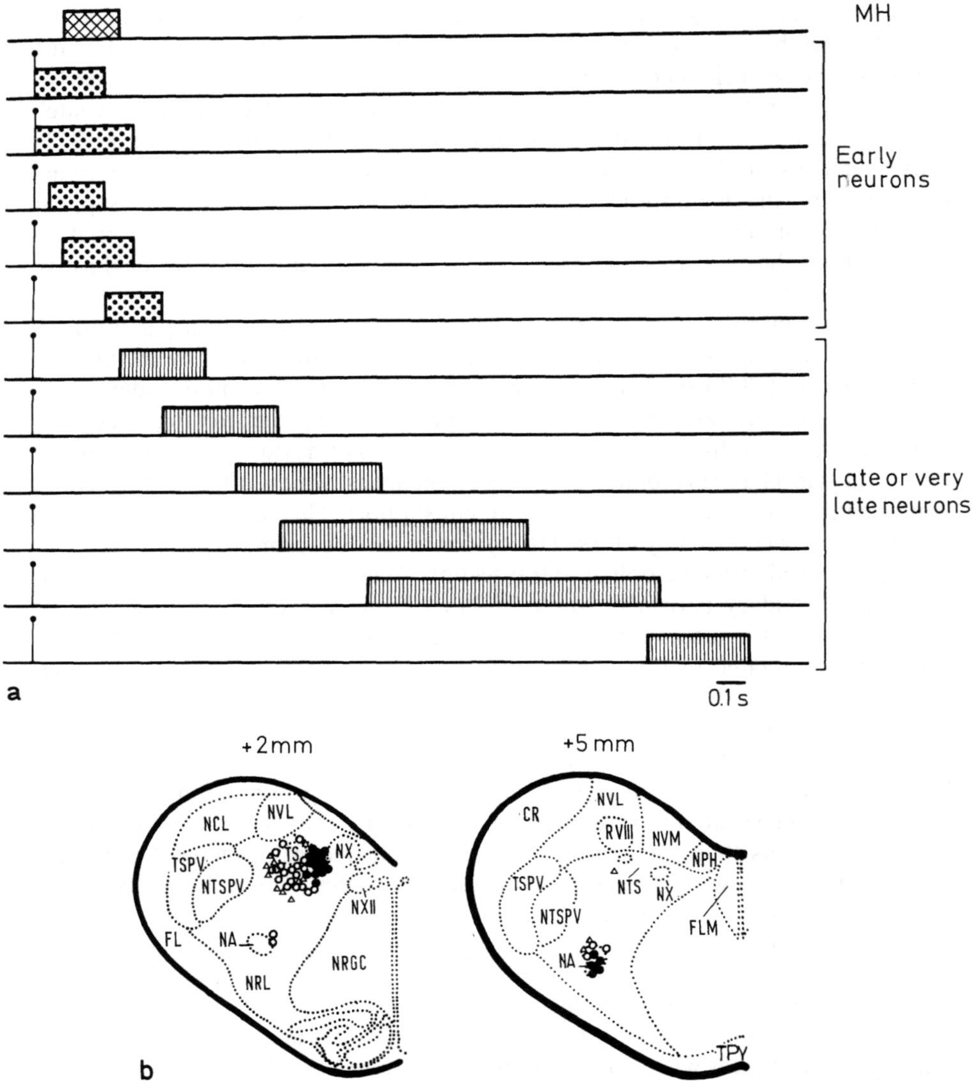

Fig. 2. a Diagram showing the swallowing discharge sequence of different medullary neurons. The vertical *lines with dots* indicate the time of superior laryngeal nerve stimulation (induction of swallowing). The *cross-hatched rectangle* shows the mylohyoideus contraction (MH). The duration of the swallowing discharges of various interneurons and motoneurons involved either in the buccopharyngeal (early neurons) or the esophageal (late and very late neurons) stages of deglutition are shown below. **b** Two maps corresponding to transverse sections of hemimedulla, 2 and 5 mm in front of the obex. The swallowing neurons are located either around the solitary tract (interneurons) or in the nucleus ambiguus (motoneurons). *Open circles* indicate early neurons; *full circles* late and very late neurons; *triangles* neurons with spontaneous activity. *CR*, corpus restiformis; *FL*, fasciculus lateralis; *FLM*, fasciculus longitudinalis medialis; *NA*, nucleus ambiguus; *NCL*, nucleus cuneatus lateralis (Von Monakov); *NPH*, nucleus praepositus hypoglossi; *NRGC*, neucleus reticularis gigantocellularis; *NRL*, nucleus reticularis lateralis; *NTS*, nucleus tractus solitarius; *NTSPV*, nucleus tractus spinalis nervi trigemini; *NVL*, nucleus vestibularis lateralis; *NVM*, nucleus vestibularis medialis; *NX*, nucleus dorsalis nervi vagi; *NXII.*, nucleus nervi hypoglossi; *RVIII.*, radices descendentes nervi vestibuli; *TPY*, tractus pyramidalis; *TS*, tractus solitarius; *TSPV*, tractus spinalis nervi trigemini. (Jean 1972a, 1978a)

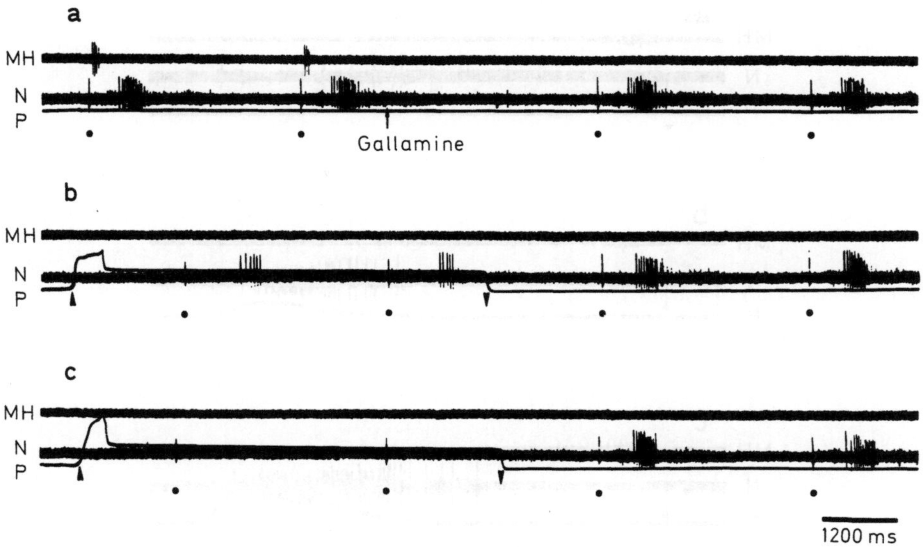

Fig. 3 a–c. Effect of curarization and distension of a proximal esophageal region on the swallowing discharge of a very late interneuron. Anesthetized sheep; microelectrode recording from the medulla. **a** Swallowing activity of a very late interneuron *N* corresponding to the beginning of the thoracic esophagus during several deglutitions (superior laryngeal nerve stimulation at the *dots*) before and after curarization (gallamine). The intraesophageal balloon *P* located in the midcervical esophagus is deflated. Note that after curarization the mylohyoideus EMG (MH) disappears, while the swallowing activity of the interneuron persists. **b** First the balloon is inflated with 10 ml air *(upward arrow)*: the swallowing activity of the interneuron is then weaker and delayed. After balloon deflation *(downward arrow)* the swallowing activity returns to normal *(end of tracing)*. **c** The distension is stronger (20 ml air). The neuronal discharge is absent as long as the balloon is inflated. (JEAN 1972a, 1978a)

1978a). However, during primary peristalsis the central chain of neurons which programs deglutition (organizing system) is excited from its beginning whereas during secondary peristalsis the excitation starts at the level of one or other link of the chain. According to ROMAN (1967) and JEAN (1978a), an effect of this difference would be that, in sheep at least, excitation of the central organizing system is weaker during secondary than during primary peristalsis. Then, if this central excitation was not permanently reinforced by afferent feedback during secondary peristalsis, it would not proceed to the end of the central neuronal chain, contrary to what is observed during primary peristalsis.

The central mechanism responsible for the successive excitation of interneurons during deglutition is not yet understood. However, it may be assumed that it is probably different from a simple transmission of excitation between neurons. In this connection, attention must be paid to inhibitory phenomena which have already been mentioned about motoneuron activity (see Sect. A.I.3.d). Inhibition is also observed at the level of interneurons, i.e., within the central system subserving the programming of deglutition. Indeed, JEAN (1972a, 1978a) has shown that in sheep, all the esophageal interneurons were strongly inhibited during the buccopharyngeal stage of deglutition. Besides, the interneurons controlling a distal esophageal segment were also inhibited when interneurons controlling more proximal

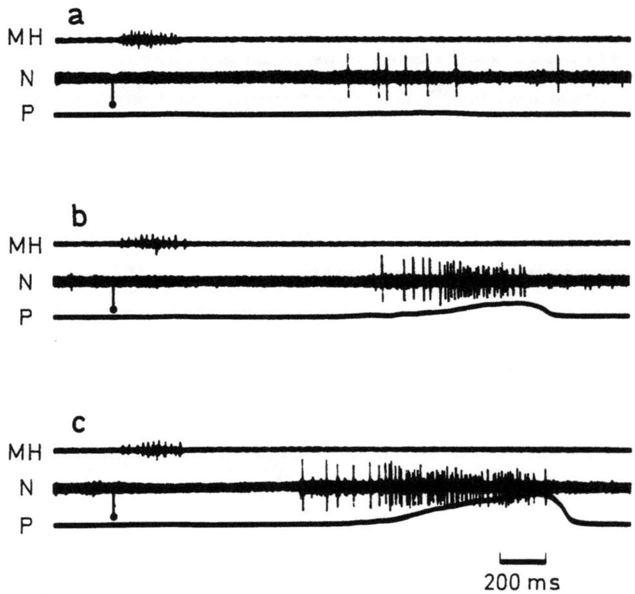

Fig. 4a–c. Swallowing discharge of a late interneuron; effects of esophageal distension. Anesthetized sheep; microelectrode recording from the medulla. *MH*, mylohyoideus EMG; *N*, discharge of the medullary interneuron; *P*, pressure recorded by an intraesophageal balloon. The recordings were obtained during swallowing induced by superior laryngeal nerve stimulation *(dots)*. The balloon is deflated **a**, inflated with 5 ml air **b**, and with 10 ml **c**. Note the facilitation of neuronal discharge by esophageal distension. (JEAN 1972A, 1978A)

segments were called into play (Fig. 3). JEAN (1978a) put forward arguments supporting the assumption that the successive activation of the swallowing interneurons might result, at least in part, from a succession of postinhibitory rebounds.

4. Nervous Control of the Smooth Muscle Esophagus

a) Effects of Severing the Extrinsic Nerves: "Autonomous" Peristalsis

In contrast with the striated muscle esophagus, bilateral vagotomy does not result in paralysis of the smooth muscle portion of esophagus (CANNON 1907; JURICA 1926; BINDER et al. 1968; ROMAN and TIEFFENBACH 1971; BURGESS et al. 1972; UEDA et al. 1972; DIAMANT and EL SHARKAWY 1977). Indeed, the smooth muscle esophagus remains able to exhibit peristaltic contractions in response to liquid injection or balloon distension. Similar reactions can be observed in vitro on the isolated organ (CHRISTENSEN and LUND 1969), indicating that the responses observed after vagotomy alone were not due to involvement of the sympathetic nervous system. In this connection it may be recalled that, according to BURGESS et al. (1972), the sympathectomy alone performed in vivo does not change the function of the feline smooth muscle esophagus.

The peristalsis occurring in the smooth muscle esophagus deprived of its extrinsic innervation was termed "tertiary peristalsis" by CANNON (1907) and JURICA (1926). Unfortunately, this therm is now confusing since clinicians use it to describe nonpropulsive contractions occuring in elderly patients. Thus, for purposes of clarity, "autonomous" will be employed instead of "tertiary" in the following text.

The autonomous motility of the smooth muscle esophagus has been studied either in vivo after bivagotomy or in vitro on the isolated organ. Usually, esophageal distension produced by inflating a balloon elicits a peristaltic conctraction which propels the balloon towards the stomach (ROMAN and TIEFFENBACH 1971). By recording the EMG at various sites along the cat esophagus, it can be observed that the autonomous peristalsis results from a propagated cocontraction of the two muscular layers (Fig. 5) and that the EMG activity of the circular layer increases as a function of the size of the transported bolus (ROMAN and TIEFFENBACH 1971).

When distension remains localized to a portion of esophagus, three types of responses are generally recorded:

The "on" response, which is a brief circumferential contraction above the point of distension, following inflation of the distending balloon.
The "off" response, occurring after deflation of the balloon and consisting of a simultaneous or propagated circumferential contraction of the entire esophagus below the point of stretch.
The "duration" response, which is an active shortening of the esophagus while the distending balloon is filled.

These three types of response, first described in the opossum by CHRISTENSEN's group (CHRISTENSEN and LUND 1969; CHRISTENSEN 1970 a, 1971) have also been observed in the cat by ROMAN and TIEFFENBACH (1971) who showed in addition that the circular muscle below the point of stretch was strongly inhibited throughout the distension.

Electrical stimulation of the serosal surface of the esophagus in vitro can also elicit the three types of response (CHRISTENSEN and LUND 1969; CHRISTENSEN 1970 a). Electrical stimulation of isolated muscle strips has shown that the "on" and "off" responses involve only circular muscle; the "duration" response involves only the longitudinal muscle and the muscularis mucosae (CHRISTENSEN 1971). The "on" response, at least in the opossum, could be due to a direct stimulation of muscle, since it is not antagonized by tetrodotoxin (TTX) (CHRISTENSEN 1970 a; LUND and CHRISTENSEN 1969; CHRISTENSEN 1971). However, in the cat, the "on" response of circular smooth muscle strips probably results from excitation of cholinergic neurons, since it is suppressed by atropine (DIAMANT and EL SHARKAWY 1975; EL SHARKAWY and DIAMANT 1976). Because the "duration" response is abolished by TTX and also by atropine, it is probably due to activation of intrinsic cholinergic nerves. The "off" response is antagonized by TTX, but not by antagonists of adrenergic, histaminergic, or serotoninergic receptors (LUND and CHRISTENSEN 1969; CHRISTENSEN 1970 a; DE CARLE et al. 1976). Whether or not the "off" response has a cholinergic component is still the subject of debate. Studies by CHRISTENSEN's group in the opossum showed that atropine had no effect, whereas other experiments in the cat (EL SHARKAWY and DIAMANT 1976; DIAMANT and EL SHARKAWY 1977; DE CARLE et al. 1977) indicated that the "off" response could be at least partially sensitive to atropine. Be that as it may, the "off" response is, at least in part, a rebound phenomenon, i.e., a membrane depolarization which follows active membrane hyperpolarization induced by noncholinergic, nonadrener-

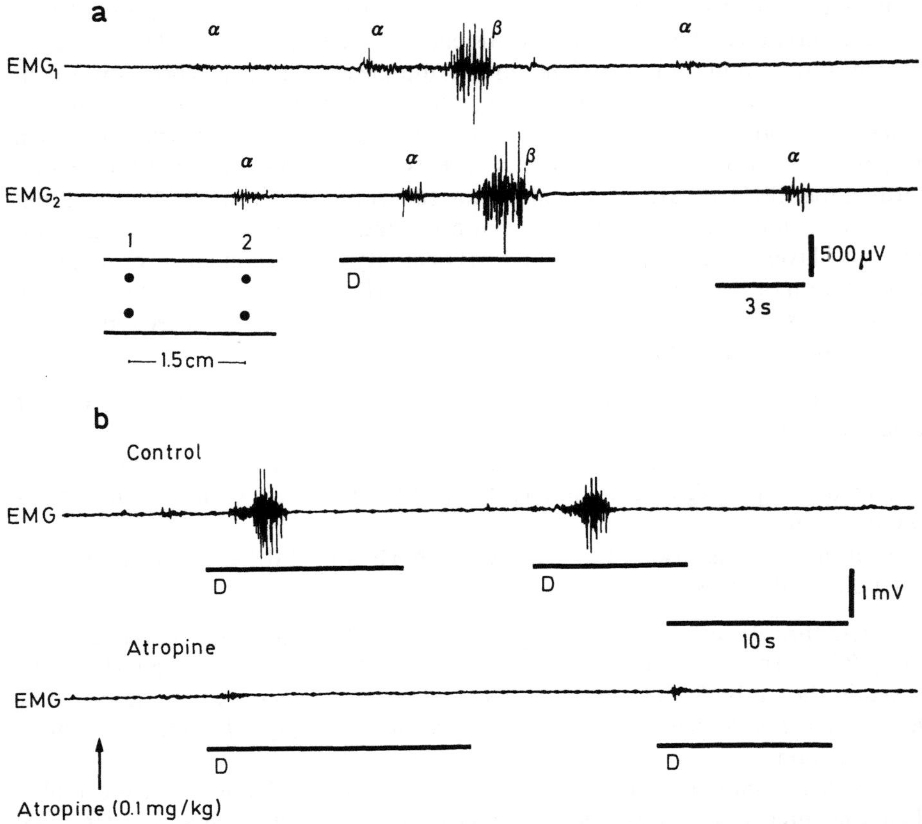

Fig. 5 a, b. EMG of smooth muscle esophagus during "autonomous" peristalsis. Anesthetized and bivagotomized cat. **a** EMG recording at two sites (EMG_1, EMG_2); location and spacing of the pressure electrodes indicated on the diagram; alternating current coupled amplification (time constant 0.1 s). The autonomous peristalsis is induced by inflation *D* of a small balloon above the recording site 1. During the propulsion of the balloon toward the stomach, a complex EMG activity is recorded successively at level 1, then 2. This complex activity consists of α-potentials identical to those recorded spontaneously and corresponding to longitudinal muscle contraction as well as β-potentials which are specific for peristalsis and which indicate circular muscle contraction. **b** EMG recording at only one site. Control tracing: the complex EMG activities typical of peristalsis are recorded after balloon inflation *D*. After administration of atropine *(lower tracing)* esophageal distensions no longer elicit peristalsis; only weak longitudinal contractions are observed. (C. ROMAN 1977, unpublished work)

gic inhibitory myenteric neurons (see DIAMANT 1974; DIAMANT and EL SHARKAWY 1977). These intramural inhibitory neurons are probably identical to those described for other parts of the gut and named by BURNSTOCK (1972, 1975) "purinergic" neurons on the basis of experimental data suggesting that their transmitter might be a purine nucleotide. This question, however, is still a matter of dispute.

The "off" response is often, but not always, propagated in a peristaltic manner. Furthermore, studies on smooth muscle strips from opossum esophagus (WEISBRODT and CHRISTENSEN 1972) showed that the time interval between the end of stimulus and the onset of the "off" response was longer for strips taken from more distally located segments. WEISBRODT and CHRISTENSEN (1972) proposed that these regional differences could be the basic mechanism responsible for the aboral propagation of the peristaltic contraction in the smooth muscle esophagus. However, as stressed by DIAMANT and EL SHARKAWY (1977): *There are large differences in the observed or calculated velocity at which a peristaltic wave can or would pass through the smooth muscle esophagus, depending on whether peristaltic velocity was observed after swallowing, after a variable period of balloon distension or vagal stimulation, or was calculated from the graded off-responses delays seen in isolated muscle strips.*

In addition, as already pointed out, the "off" responses below the stimulated segment are not always propagated but rather simultaneous.

In summary, although the mechanism of the "off" response may take part in the production of esophageal peristalsis, it is likely that other mechanisms are also involved. In the cat, for instance, autonomous peristalsis, i.e., the effective transport of a bolus (inflated balloon) is abolished by atropine (C. ROMAN unpublished work 1977), which indicates that cholinergic mechanisms also are important (Fig. 5).

b) Effects of Efferent Fiber Stimulation

Available data concern only the effects of vagal stimulation. With pressure or suction electrodes on cat smooth muscle esophagus during in vivo or in vitro (isolated organ) experiments, it is possible to record two kinds of EMG response induced by vagal efferent fiber stimulation with a single shock or a short train of pulses.

Excitatory responses consist of excitatory junction potentials (e.j.p., i.e., slow depolarizations which may give rise to a burst of spikes inducing a contraction of the muscle (TIEFFENBACH and ROMAN 1972; DIAMANT 1974; DIAMANT and EL SHARKAWY 1977; GONELLA et al. 1977). The e.j.p. are recorded from the two muscle layers, the circular e.j.p. having a higher threshold than the longitudinal e.j.p. (GONELLA et al. 1977). These excitatory responses are suppressed by atropine or hexamethonium, which indicates that the excitatory pathway is entirely cholinergic, i.e., composed of cholinergic preganglionic fibers exciting cholinergic intramural neurons which in turn activate smooth muscle.

Inhibitory responses consist of inhibitory junction potentials (i.j.p.) corresponding to slow hyperpolarizations of smooth muscle fibers often followed by a transient depolarization which may initiate spikes (postinhibitory rebounds). The i.j.p. are more easily recorded on the circular muscle (DIAMANT 1974; GONELLA et al. 1977) and after atropine treatment (GONELLA et al. 1977). They are not affected by antiadrenergic drugs but they are suppressed by hexamethonium (GONELLA et al. 1977). In view of the preceding results, i.j.p. might result from stimulation of cholinergic preganglionic fibers which excite intramural inhibitory neurons whose mediator is neither adrenaline nor acetylcholine ("purinergic" nerves, described by BURNSTOCK 1972, 1975). As in other parts of the digestive tract (MARTINSON and MUREN 1963; JULE 1975), it is likely that preganglionic vagal fibers leading to inhibition have a smaller diameter (higher excitation threshold; lower conduction velocity) than those leading to excitation (GONELLA et al. 1977).

c) Possible Roles of the Vagal Extrinsic Innervation

The existence of an autonomous peristalsis in the smooth muscle portion of the esophagus raises questions like: what may be the role of the vagal fibers supplying this region? Is there any central programming of peristalsis during deglutition? One might speculate that, in opossum, cat, monkey, and humans, the proximal esophagus (striated muscle) is under central control, while peristalsis of the distal esophagus (smooth muscle) would only depend on a peripheral mechanism (myenteric neurons) triggered by arrival of the bolus in this area. This assumption does not fit in with several experimental data. First, there is an extrinsic vagal supply whose stimulation induces excitatory and inhibitory effects, even with single pulse stimulation, which underlines the powerful influence exerted by the vagal fibers upon esophageal smooth muscle. On the other hand, various experiments strongly suggest the existence of an extrinsic command during deglutition. For instance, deviation of the swallowed bolus at the level of the cervical esophagus in the opossum or rhesus monkey does not eliminate the primary peristalsis in the thoracic esophagus (JANSSENS et al. 1976; JANSSENS 1978). In other words, the smooth muscle esophagus does not need the presence of an intraluminal bolus for the progression of primary peristalsis. An identical conclusion can be drawn from curarization experiments. In conscious baboons, primary peristalsis of the smooth muscle esophagus occurring during swallowing induced by superior laryngeal nerve stimulation is still elicited after curarization of the animal, i.e., after paralysis of the oropharynx and cervical esophagus (TIEFFENBACH and ROMAN 1972). Since no bolus can reach the thoracic esophagus after curarization, the persistent contractions of the smooth muscle are due to central commands. Similar observations have also been done by JANSSENS (1978) on three patients with severe cerebral trauma but a preserved function of the swallowing center and who needed curarization for artificial respiration. Besides, RYAN et al. (1977) have observed in the anesthetized opossum, that bilateral vagal cooling or bivagotomy significantly decreased the primary peristalsis occuring in the smooth muscle segment of the esophagus in response to pharyngeal stimulation.

Using a nerve suture technique previously described (see Sect. A.I.3.d), ROMAN and TIEFFENBACH (1972) recorded in the conscious monkey the spontaneous discharge of vagal efferent fibers controlling the smooth muscle esophagus (Fig. 6). These fibers fired in succession during either primary or secondary peristalsis. Their discharge was weak: only a few spikes at a frequency lower than 5 spikes/s. According to ROMAN and TIEFFENBACH, the main effect of these discharges would be to facilitate the intramural neurons which are by themselves able to organize esophageal motility (autonomous peristalsis). In other words, the vagal fibers whose activity was recorded could belong to the vagal excitatory pathway which would be called into play during primary or secondary peristalsis. As already mentioned, the vagal excitatory effects on smooth muscle are blocked by atropine. Thus, primary peristalsis should also be suppressed by atropine. This seems to be true in the cat (DODDS et al. 1978) but not in the opossum or humans (KANTROWITZ et al. 1966; MUKHOPADYAY and WEISBRODT 1975; DODDS et al. 1977). However as far as humans are concerned, intravenous injection of anticholinergic drugs is followed by disturbances of esophageal motility resembling those of diffuse spasm or presbyesophagus (HELLEMANS et al. 1974).

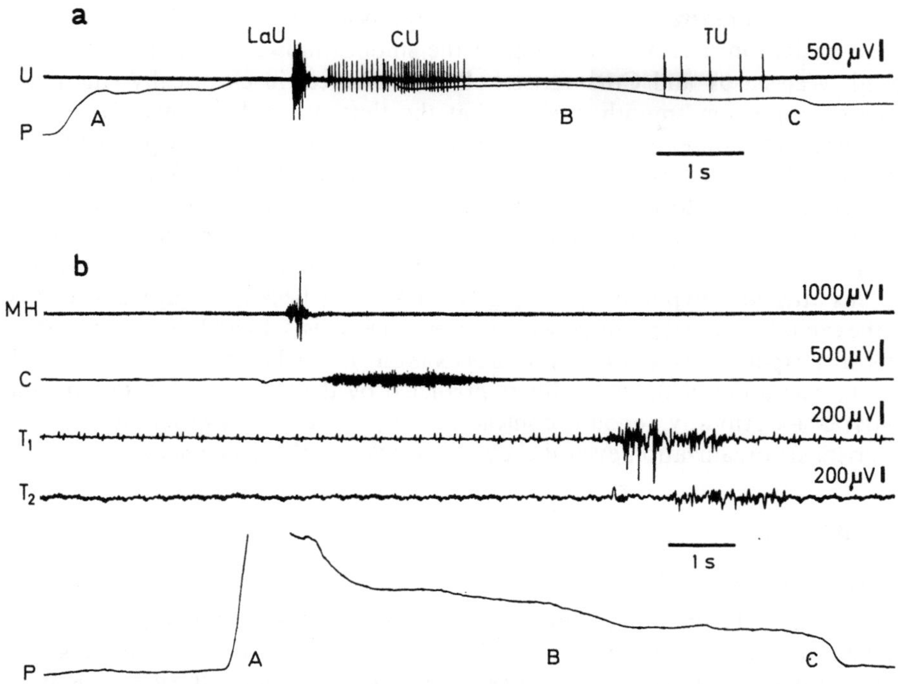

Fig. 6 a, b. Recording of vagal efferent fiber activity and esophageal EMG during deglutition in conscious baboons. **a** Recording of vagal activity by the nerve suture technique (see text). *U* unit discharge of three different vagal efferent fibers which originally supplied the larynx *LaU*, the cervical esophagus *CU* and the distal thoracic esophagus *TU*. P pressure variations recorded by an intraesophageal balloon (10 ml air) during its propulsion towards the stomach ("dynamic" recording). For the significance of the A, B, and C waves, see Fig. 1. **b** Another baboon; *MH*, EMG of mylohyoideus recorded by a Bronk needle and indicating the buccopharyngeal stage of deglutition; *C*, T_1, T_2, bipolar recording of esophageal EMG by electrodes implanted in the muscular wall at the diaphragm level T_2, 2 cm above T_1 and at the level of lower cervical esophagus *C; P*, "dynamic" pressure recording. Note (1) the difference of EMG in the cervical (striated muscle) and thoracic esophagus (smooth muscle); (2) the corresponding difference in the discharge pattern of vagal fibers supplying these two regions *CU; TU*. (Adapted from ROMAN and TIEFFENBACH 1972)

Nerve stimulation experiments have also indicated the existence of vagal preganglionic fibers whose excitation leads to an inhibition of esophageal motility (vagal inhibitory pathway). At the present time, it is only possible to speculate about the physiologic role of such fibers. For instance, they might be involved in the deglutitive inhibition of esophageal peristalsis. In monkeys and humans, EMG studies have shown that the distal progression of primary peristalsis in the thoracic esophagus could be stopped by a second swallow (HELLEMANS and VANTRAPPEN 1974; HELLEMANS et al. 1974; JANSSENS 1978). It is fairly possible that this inhibition results, at least in part, from an active inhibition of esophageal smooth muscle, for suppression of vagal excitatory discharges alone should not impede the progression of an autonomous peristalsis initiated by the arrival of the swallowed bolus in the distal esophagus. On the other hand, it is known that inhibition induced by nonadrenergic, noncholinergic intramural neurons is followed by a rebound excitation, i.e., a contraction of the smooth muscle. As previously mentioned, WEISBRODT and CHRISTENSEN (1972) observed, in electrically stimulated strips of oppossum smooth muscle, that the time interval between the end of stimulus and the onset of postinhibitory rebound was longer for strips taken from more distally located esophageal segments. This observation led WEISBRODT and CHRISTENSEN (1972) to assume that primary peristalsis in the smooth muscle esophagus could simply result from a rebound excitation following from a simultaneous excitation of all the vagal fibers synapsing with intramural inhibitory neurons.

This attractive hypothesis is not fully satisfactory because it provides no role for the excitatory vagal pathway which nevertheless exists. In addition, during their recording experiments, ROMAN and TIEFFENBACH (1972) did not observe the simultaneous excitation of the vagal fibers predicted by the theory of WEISBRODT and CHRISTENSEN. Anyway, rebound contractions do exist and they probably take part in peristalsis organization even if they are coupled with other phenomena.

II. Sphincters

1. Upper Esophageal Sphincter

The closure of the pharyngoesophageal junction at rest is probably mainly due to a tonic contraction of the cricopharyngeus (striated muscle). EMG recordings have demonstrated this tonic activity either in conscious animals (CAR and ROMAN 1970 b; HELLEMANS and VANTRAPPEN 1974; JANSSENS 1978) or in conscious humans (MONGES et al. 1968; HELLEMANS et al. 1974). Upon swallowing, the continuous spiking activity is immediately and completely inhibited while intraluminal pressure falls. This inhibition is followed by an intense burst of spikes which corresponds to the deglutitive contraction of the pharyngoesophageal sphincter (peak of intraluminal pressure). Then the EMG activity and the intraluminal pressure return to their resting level.

The behavior of the striated muscle of the pharyngoesophageal sphincter is necessarily the direct consequence of motor fiber discharge. Indeed ANDREW (1956) demonstrated that the vagal motor fibers supplying the upper sphincter had a spontaneous discharge which was inhibited during the buccopharyngeal stage of deglutition and which then resumed at a transiently raised level.

2. Lower Esophageal Sphincter

At the present time, most authors agree that the closure of the lower esophageal sphincter (LES) is due to a tonic contraction of the sphincter muscle (circular layer) rather than to factors outside the sphincter itself (see INGELFINGER 1958; CODE and SCHLEGEL 1968; HELLEMANS and VANTRAPPEN 1974). A surprising feature in view of this tonic closure is the fact that several EMG studies have failed to reveal any permanent spiking activity in the resting sphincter muscle of various animals with electrodes either chronically implanted (HELLEMANS and VANTRAPPEN 1967; HELLEMANS et al. 1968) or placed on muscle only during acute experiments (GONELLA et al. 1977). This seems to be true in humans as well (MONGES et al. 1968; HELLEMANS et al. 1974). However some investigators (ARIMORI et al. 1970, MIOLAN and

ROMAN 1973) reported a weak and irregular tonic discharge of sphincter smooth muscle in dogs with chronically implanted electrodes. A more permanent spiking activity was recently recorded from the LES of anesthetized opossums (ASOH and GOYAL 1978). But obviously, this activity was not the only determinant of the resting sphincter pressure since, during spontaneous absence of spikes, over 55% of the maximal sphincter pressure was retained. Then, a part of the LES closure might result from a contraction caused by a permanent muscle depolarization without spike potential, something resembling a physiologic contracture similar to that described by JOHANSSON (1971) and BOHR (1973) in vascular smooth muscle. This view is consistent with the finding of DANIEL et al. (1976) showing that in the opossum, the membrane potential is significantly lower in the sphincter than in esophageal smooth muscle. It is also in agreement with the fact that in sucrose gap experiments on circular smooth muscle strips from cat LES, GONELLA et al. (1977, 1979) usually recorded only slow depolarizations without spikes during the contraction induced by acetylcholine or noradrenaline.

a) LES Deprived of Extrinsic Nerve Supply

During in vitro experiments on the terminal esophagus of guinea pig and kitten (MANN et al. 1968) or dog (THOMAS and EARLAM 1974), it was demonstrated that some resting tone persisted in the LES totally deprived of its extrinsic innervation. In addition, MANN et al. (1968) showed that transient distension of esophageal body produced a sphincter relaxation immediately followed by a phasic contraction which closed the gastroesophageal junction.

The origin of the resting tone in vitro may be myogenic and/or neurogenic. In support of a myogenic origin is the fact that the ionic composition of the intracellular fluid in opossum LES is different from that of adjacent nonsphincteral area (SCHULTZE et al. 1977), i.e., there is a lower concentration of potassium ions, which might explain a lower membrane resting potential and hence a permanent contraction. Among the nervous influences likely to be involved in sphincter tone maintenance in vitro, the most plausible might result from a tonic activity of cholinergic intramural neurons. These neurons were demonstrated in cat sphincter walls by vagal stimulation (see Sect. A.II.2.c.α). In addition, acetylcholine has a motor effect on the LES (CHRISTENSEN 1975; CASTELL 1975) and causes the circular muscle to depolarize (GONELLA et al. 1977; Fig. 7). Finally, the sphincter tone in vitro is decreased (but not abolished) by atropine (THOMAS and EARLAM 1974). The LES is also sensitive to noradrenaline (Fig. 7) which can exert excitatory effects mediated by α-receptors (ELLIS et al. 1960; CHRISTENSEN and DONS 1968; CHRISTENSEN 1970b, 1975; DI MARINO and COHEN 1974; GOYAL and RATTAN 1978; GONELLA et al. 1979) and also inhibitory effects due to β-receptors (ELLIS et al. 1960; CHRISTENSEN and DONS 1968; ZFASS et al. 1970; DI MARINO and COHEN 1974; CHRISTENSEN 1975; GOYAL and RATTAN 1978). However, on the basis of many data obtained by the histofluorescence technique, it is currently agreed that gut walls contain adrenergic fibers but no adrenergic cell bodies, except in the proximal colon of the guinea pig (FURNESS and COSTA 1974; GABELLA 1979). Therefore, it is unlike that an adrenergic mechanism might play a role in LES contraction (or relaxation) in vitro. In the opossum, RATTAN and GOYAL (1977) showed that exogenous serotonin

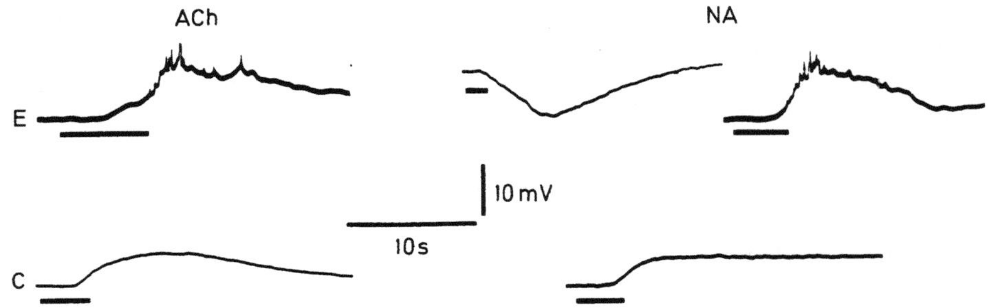

Fig. 7. Action of acetylcholine *ACh* and noradrenaline *NA* on circular muscle strips from the cat lower esophagus. *C*, LES strips; *E* strips from esophageal body, 4–5 cm orally to the LES. Sucrose gap experiment. *Bars* under the tracings indicate the time of ACh and NA application. ACh (10^{-5} g/ml) depolarizes both types of strips. NA (10^{-6} g/ml) depolarizes LES strips but may either hyperpolarize or depolarize esophageal strips, depending on the animal. Notice that LES strip depolarizations do not trigger spike potentials. (Adapted from GONELLA et al. 1977, 1979)

caused contraction of the LES by acting on two different receptors, i.e., a muscle receptor and a neural receptor probably located on the intramural cholinergic neurons, but so far there is no available evidence of the presence of serotoninergic neurons within the sphincter walls.

LES relaxation observed in vitro during distension of the esophageal body probably involves inhibitory noncholinergic, nonadrenergic neurons whose existence was previously mentioned in connection with contol of the smooth muscle esophagus (see Sects. A.I.4.a, b). Identical neurons were demonstrated in the LES by direct electrical stimulation of circular sphincter strips (TUCH and COHEN 1973; EL SHARKAWY et al. 1975). These stimulations induced a muscle relaxation which was suppressed by TTX but not by adrenergic and serotoninergic blockers. In addition, it is now well documented that these inhibitory neurons act by hyperpolarizing the circular muscle. Just such, a hyperpolarization of the circular muscle of the LES was recorded in vitro during esophageal distensions which relax the sphincter (DIAMANT 1974). Obviously, there is in the walls of the smooth muscle esophagus an intramural nervous mechanism which provides descending inhibition similar to that described for the intestine (BAYLISS and STARLING 1899; HIRST et al. 1975). Dopamine receptors whose stimulation leads to a reduction in basal LES muscle tension have been described in the opossum (DE CARLE and CHRISTENSEN 1976; RATTAN and GOYAL 1976; MUKHOPADHYAY and WEISBRODT 1977). However, several observations indicate that dopamine is not the transmitter of intramural inhibitory neurons. For example, haloperidol antagonizes the effect of exogenous applied dopamine but does not alter the muscle inhibition caused by direct electrical stimulation. Thus, the physiologic role of dopamine receptors remains to be determined.

b) Effects of Extrinsic Nerve Section and Pharmacologic Blockade

Since the sphincter function is preserved at least in part after extrinsic denervation, what is then changed in comparison with the normal animal? Many investigators have attempted to answer this question. For detailed bibliography prior to 1958;

the reader is refered to the review of INGELFINGER (1958); only work published since 1958 will be individually quoted in the following text.

α) Effects of Vagal Section or Atropine Injection. The results of vagotomy can be summarized as follows:

Difficulty or impossibility for the LES to relax during swallowing, which constitutes an experimental achalasia (INGELFINGER 1958; CARVETH et al. 1962; HIGGS and ELLIS 1965; BINDER et al. 1968; BURGESS et al. 1972). These data have been more recently confirmed by vagal cooling experiments (RYAN et al. 1977; COHEN et al. 1977).
Increased resting tone, i.e., a "spasm" which may be permanent, transient, or even of late occurrence (INGELFINGER 1958; BURGESS et al. 1972; RATTAN and GOYAL 1974).
Unaltered resting tone (INGELFINGER 1958; COHEN et al. 1977).
Decreased resting tone (INGELFINGER 1958; CARVETH et al. 1962; GREENWOOD et al. 1962; HIGGS and ELLIS 1965; JENNEWEIN et al. 1976).

These conflicting results certainly reflect differences in species and experimental conditions. As indicated below, the vagus may exert both excitatory and inhibitory influences on LES tone. The relative importance of the two types of influences is probably different according to the species. On the other hand, data obtained on anesthetized animals are questionable because under anesthesia vagal activity is markedly altered (see MIOLAN and ROMAN 1978b).

Concerning the effects of atropine, the results are less variable than for vagotomy since most of data include some fall in resting tone (LIND et al. 1968; SKINNER and CAMP 1968; PEDERSEN et al. 1971; JENNEWEIN et al. 1976; FISCHER et al. 1977). This is not surprising because atropine only blocks the vagal excitatory pathway while vagotomy also suppresses the inhibitory one.

β) Effects of Section or Pharmacologic Blockade of the Sympathetic Nerves. Here again, the results are rather conflicting. Very often, no clear-cut change of the sphincter resting tone has been observed after various sympathetic denervations, but sometimes slight decreases have been reported (see INGELFINGER 1958). More recent work in the opossum has shown that adrenergic denervation by 6-hydroxydopamine resulted in decreased basal pressure (DI MARINO and COHEN 1974) and that phentolamine (α-blocker) prevented the transient sphincter hypertension caused by vagotomy (RATTAN and GOYAL 1974). On the other hand, sphincter relaxation during swallowing was found to be unaffected by sympathectomy (DI MARINO and COHEN 1974).

In sum, the control exerted on LES tone by the adrenergic supply, if there is one, appears to be mainly excitatory. This fact has been confirmed by nerve stimulation experiments (see Sect. A.II.2.c).

c) Effects of Efferent Nerve Stimulation

α) Vagal Supply. Literature prior to 1958 (see INGELFINGER 1958) indicates that repetitive stimulation of the distal cut end of the cervical vagus in various animals (e.g., dog, rabbit, cat) produces conflicting results: either a sphincter contraction or a relaxation often followed by a contraction after the end of stimulation. It is also reported that the type of response might depend on the stimulation parameters or on the initial state of sphincter tone.

Subsequent work has somewhat clarified the problem. Thus, it is established that vagal stimulation in the cat and the opossum can relax the LES (CLARK and VANE 1961; RATTAN and GOYAL 1974; DIAMANT 1974; GONELLA et al. 1977). EMG recordings show that relaxation follows from a hyperpolarization of the circular muscle (DIAMANT 1974; GONELLA et al. 1977; Fig. 8). When stimulation is performed with long trains of pulses, the hyperpolarization is permanent, but when single pulses or a short train of pulses are used, the hyperpolarization consists of brief inhibitory junction potentials (i.j.p.). Usually these inhibitory responses are followed by a rebound excitation, i.e., a membrane depolarization which gives rise to spikes; thereby inducing a muscle contraction (Fig. 8). All these data easily explain an early and often reported observation (see INGELFINGER 1958) according to which vagal stimulation produced a sequence of both responses, i.e., sphincter relaxation followed by contraction.

In the opossum, it seems that vagal stimulation induces only inhibitory effects, i.e., sphincter relaxations (RATTAN and GOYAL 1974), but in the cat, excitatory responses are also elicited on both muscular layers (GONELLA et al. 1977). With single-pulse stimulation, these excitatory responses consist of brief muscle depolarizations or excitatory junction potentials (e.j.p.) which may give rise to spike potentials triggering muscle contraction (Fig. 9). The e.j.p. are suppressed, either by atropine or by hexamethonium, indicating that these responses are due to stimulation of preganglionic cholinergic fibers which excite intramural postganglionic neurons whose mediator responsible for muscle excitation is also acetylcholine.

As for inhibitory responses, they would involve preganglionic fibers synapsing with intramural postganglionic neurons whose mediator is neither acetylcholine nor adrenaline (GOYAL and RATTAN 1975; GONELLA et al. 1977). These intramural inhibitory neurons probably correspond to the so-called purinergic neurons (BURNSTOCK 1972, 1975). The synaptic transmission between vagal preganglionic fibers and intramural inhibitory neurons is predominantly cholinergic and utilizes nicotinic as well as muscarinic receptors (GOYAL and RATTAN 1975). Nevertheless, according to RATTAN and GOYAL (1978), the vagal inhibitory effects are not completely suppressed under hexamethonium and atropine. The persistent relaxation would involve serotonine (5-HT), for it is no longer observed after pretreatment of animals with *p*-chlorophenylalanine (PCPA). On the other hand, under PCPA together with atropine and hexamethonium direct electrical stimulations of the esophagus still induce maximal sphincter relaxations, which demonstrates that 5-HT is not the mediator of intramural inhibitory neurons. Then 5-HT together with acetylcholine might participate in transmission between pre and postganglionic fibers.

β) Sympathetic Supply. According to many investigators, electrical stimulation of the various components of the sympathetic supply has no effect on LES motility; for others, however, a contraction resembling a "spasm" can be obtained (INGELFINGER 1958). Recent experimental data in cats are in keeping with this last conclusion (ROMAN et al. 1975; GONELLA et al. 1979). From these data, the following points emerge. Sympathetic fibers supplying the cat LES come from the stellate ganglion or run along the splanchnic nerve. Repetitive stimulation of these fibers

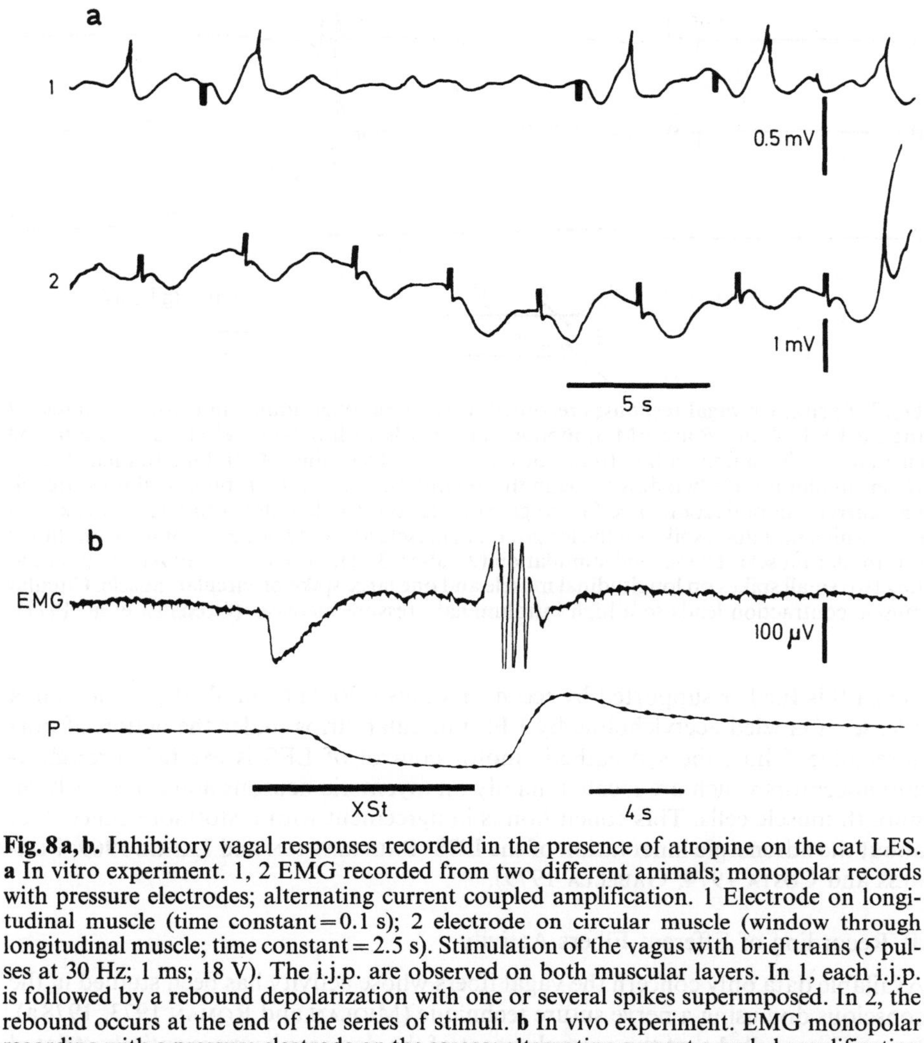

Fig. 8 a, b. Inhibitory vagal responses recorded in the presence of atropine on the cat LES. **a** In vitro experiment. 1, 2 EMG recorded from two different animals; monopolar records with pressure electrodes; alternating current coupled amplification. 1 Electrode on longitudinal muscle (time constant = 0.1 s); 2 electrode on circular muscle (window through longitudinal muscle; time constant = 2.5 s). Stimulation of the vagus with brief trains (5 pulses at 30 Hz; 1 ms; 18 V). The i.j.p. are observed on both muscular layers. In 1, each i.j.p. is followed by a rebound depolarization with one or several spikes superimposed. In 2, the rebound occurs at the end of the series of stimuli. **b** In vivo experiment. EMG monopolar recording with a pressure electrode on the serosa; alternating current coupled amplification (time constant = 2.5 s). *P*, intraluminal sphincter pressure; direct current amplification. A long-lasting vagal stimulation (*straight line XSt* under the tracings: pulses 0.5 ms, 20 Hz, 19 V) induces a hyperpolarization and a sphincter opening (the hyperpolarization seems to be transient because of the alternating current coupled amplification). 1 s after the end of stimulation a postinhibitory rebound appears, characterized by a depolarization with spike potentials superimposed and an exagerated closing of the LES. The tops of the spikes are cut off because of the high amplification. (Adapted from GONELLA et al. 1977)

induces a long latency (5–8 s) sustained or rhythmic contraction of the sphincter, which is suppressed by adrenergic α-blockers (phentolamine; dihydroergotamine) and greatly reduced by atropine. This surprising effect of atropin suggests that noradrenaline released by sympathetic endings might act chiefly (but not exclusively) on the myenteric cholinergic neurons to release acetylcholine which in turn excites sphincter muscle. The existence of a cholinergic link in the sympathetic control of

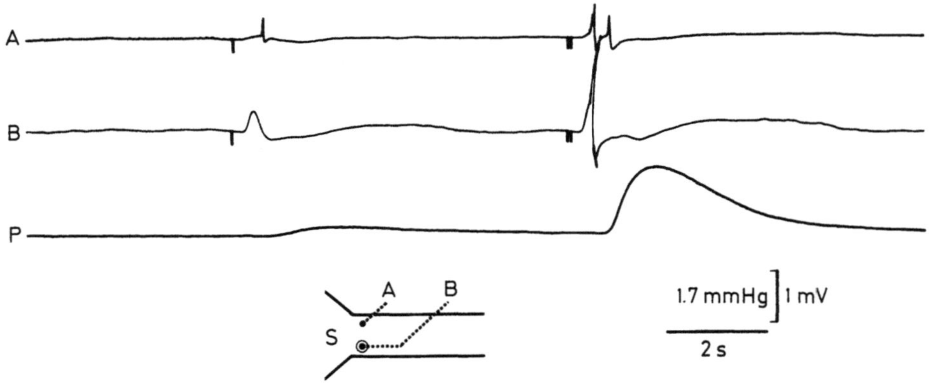

Fig. 9. Excitatory vagal responses recorded in vitro on longitudinal and circular muscle of the cat LES. *A* and *B* are EMG; monopolar records with pressure electrodes; alternating current coupled amplification (time constant 0.1 s). Recording: *A*, on longitudinal muscle; *B*, on circular muscle (window through the longitudinal layer). *P*, intraluminal pressure; direct current amplification. The first vagal stimulation (single pulse; 1 ms; 15 V) evokes an e.j.p. which initiates a spike on the longitudinal muscle and only one e.j.p. of high amplitude on circular muscle. The second stimulation (2 pulses, 30 Hz, 1 ms, 15 V) evokes e.j.p. initiating two small spikes on longitudinal muscle and one large spike on circular muscle. Circular muscle contraction leads to a high intraluminal pressure increase. (GONELLA et al. 1977)

the LES is further supported by recent findings (GONELLA et al. 1980) showing a release of labeled acetylcholine by LES muscular strips under the action of noradrenaline. Thus, the sympathetic motor control of LES is exerted through α-adrenoceptors which are located mainly on myenteric neurons and accessorily on smooth muscle cells. This conclusion is in agreement with histofluorescence data about the adrenergic innervation of the LES (BAUMGARTEN and LANGE 1969; FURNESS and COSTA 1974; GABELLA 1979).

d) Recording of Efferent Fibers Activity

Available data only concern the vagal fibers whose activity has been studied in the conscious dog using a nerve suture technique (MIOLAN and ROMAN 1973, 1978a). It must be recalled that the muscular coat of the dog esophagus is entirely striated except in the LES area where a band of circularly disposed smooth muscle becomes visible within the striated coat. Vagal fibers which might control the LES smooth muscle have a spontaneous discharge which is altered during swallowing. Two types of pattern can be distinguished (Fig. 10). For some fibers, the spontaneous firing rate (1.5–4.5 spikes/s) is suddenly enhanced just after the buccopharyngeal stage of swallowing (12–16 spikes/s). Then, this discharge stops abruptly just before the end of esophageal peristalsis and starts again 2–3 s later at a low frequency (1.5–4.5 spikes/s). Other fibers which also have a low discharge frequency (1–3 spikes/s) stop firing soon after the onset of swallowing and remain silent until the end of esophageal peristalsis. At this time, i.e., when the bolus enters the stomach, the discharge resumes with a transient increased frequency (5–9 spikes/s).

According to MIOLAN and ROMAN (1973, 1978a), the former fibers, termed VIC (vagal inhibitory fibers of the cardia) are preganglionic fibers which probably ac-

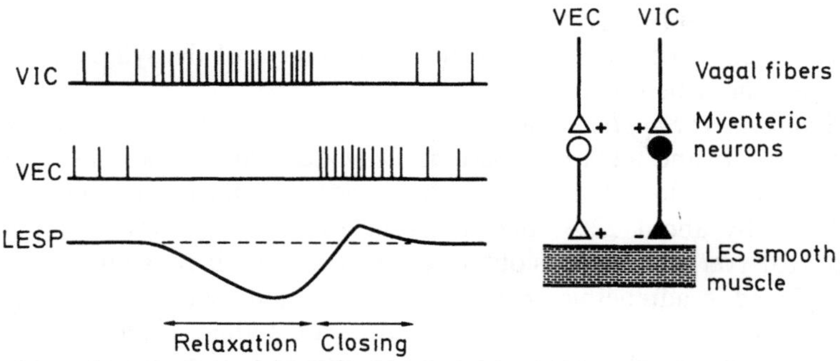

Fig. 10. Diagrammatic representation of the vagal control of LES (cardia). VEC, vagal excitatory fiber of the cardia; VIC vagal inhibitory fiber of the cardia; LESP, endoluminal pressure recorded in the LES during esophageal peristalsis. The two types of fiber exhibit a spontaneous tonic discharge. During esophageal peristalsis (either primary or secondary), the firing of VIC fibers is increased while the LES relaxes; this sustained actitivity is interrupted at the end of peristalsis, i.e., when the LES closes again. The behavior of VEC fibers is exactly the opposite; cessation of discharge during LES relaxation and transient activation during sphincter closing. (MIOLAN and ROMAN 1978a)

tivate intramural inhibitory neurons, while the latter called VEC (vagal excitatory fibers of the cardia) are preganglionic fibers which probably synapse with intramural excitatory neurons.

e) Possible Roles of Extrinsic Fibers

α) Vagal Supply. Since the two types of vagal fibers (VEC, VIC) have a spontaneous resting discharge, it is difficult to predict the net effect of the global vagal output on the LES tone. As mentioned above, vagotomy leads to very conflicting results which do not allow firm conclusions. Moreover, the respective importance of excitatory and inhibitory pathways may differ from one species to another.

If it is difficult to assess the role of vagal fibers in resting tone maintenance, on the contrary, it seems clear that these fibers are involved in the changes of sphincter tone occuring during swallowing or during gastric contractions. The LES opening during swallowing probably depends on an increased discharge of VIC fibers and on an inhibition of VEC fibers, at least in the dog. The closure observed afterwards results from an opposite change in the two discharge types. The cessation of VIC fiber discharge must be followed by a rebound excitation which is probably reinforced by the transient acceleration of VEC fiber discharge.

On the other hand, several investigators (CARLSON et al. 1922; CLARK and VANE 1961; LIND et al. 1961; DIAMANT and AKIN 1972; MIOLAN and ROMAN 1973) have reported an increase of LES tone occurring with increased gastric pressure due to gastric contractions. This response, obviously designed to prevent a possible gastroesophageal reflux, appears to be a reflex effect mainly mediated by the vagus nerves since it is reduced or even abolished by vagotomy (DIAMANT and AKIN 1972). In this connection, MIOLAN and ROMAN (1978a) have reported that the discharge of VIC fibers was decreased when the intragastric pressure increased. At the same time, the activity of VEC fibers is enhanced (MIOLAN 1980). Afferent fibers for such a reflex are probably sensitive fibers from the stomach which are carried by the vagus nerves (see Sect. B.II.2.a). As for the efferent pathway, besides the vagal fiber already mentioned, a possible involvement of the sympathetic supply of the LES cannot be ruled out.

β) Sympathetic Supply. Data on the possible role of sympathetic supply of the LES are scarce. In anesthetized opossums, DI MARINO and COHEN (1974) observed that, after pharmacologic blockade of adrenergic nerves the resting tone of the LES was reduced by about 22%, but that relaxation during swallowing was not changed. However, these results obtained under anesthesia may not reflect the actual part played by adrenergic fibers, especially in maintenance of the normal resting tone. Anyhow, sympathetic fibers may also, and perhaps chiefly, participate in reflex increase of the sphincter closure in response to various stimuli. In this connection, RATTAN and GOYAL (1974) showed in the opossum that a reflex increase of LES pressure ensued stimulation of the central cut end of the vagus (stimulation of sensitive fibers) even if the other vagus was severed. This reflex response was antagonized by phentolamine as well as by atropine, which can be easily interpreted, taking into account what is known about the mechanism of action of sympathetic fibers upon the LES (see Sect. A.II.2.c.β).

B. Nervous Control of the Stomach

I. General Survey of Motility Patterns and Control Mechanisms

The stomach receives and stores food, mixes it with gastric secretions, and then delivers it to the duodenum gradually. As far as motility is concerned, it is convenient to divide the stomach into two areas which have different functions (KELLY 1974; COOKE 1975; LAPLACE and ROMAN 1979): a proximal area, the fundus and the main part of the body, which has a reservoir function regulating, by variations in its volume, the mean intragastric pressure; and a distal area, the antrum and the lower body, which is mainly a pump, churning food and expelling it into the duodenum through the pylorus. This latter is a variable-geometry orifice which probably plays a part in gastric emptying and also in preventing duodenogastric reflux. But these two points are not yet very clear.

The motility of the distal gastric area is characterized by peristaltic contractions. The frequency, direction, and progression velocity of peristaltic contraction is predetermined by a complex EMG activity which, in dogs and humans, consists of two successive potentials. The initial potential or electrical control activity (ECA) (DANIEL and IRWIN 1971) is a rhythmic depolarization or upstroke potential (EL SHARKAWY et al. 1978) which is omnipresent and which constitutes the so-called basic electrical rhythm (BER). The initial potential is usually followed by a second potential or electrical response activity (ERA) (DANIEL and IRWIN 1971) which on microelectrode recordings consists of a plateau with superimposed oscillations or true spike potentials depending on the explored region (see EL SHARKAWY et al. 1978). Muscle contraction occurs only when the second potential has a certain amplitude and duration. It must be stressed that when the EMG is recorded with chronically implanted electrodes and alternating current coupled amplification, the plateau potential is not observed and then the ERA corresponds to oscillations or spike potentials of the distal antrum (Fig. 11).

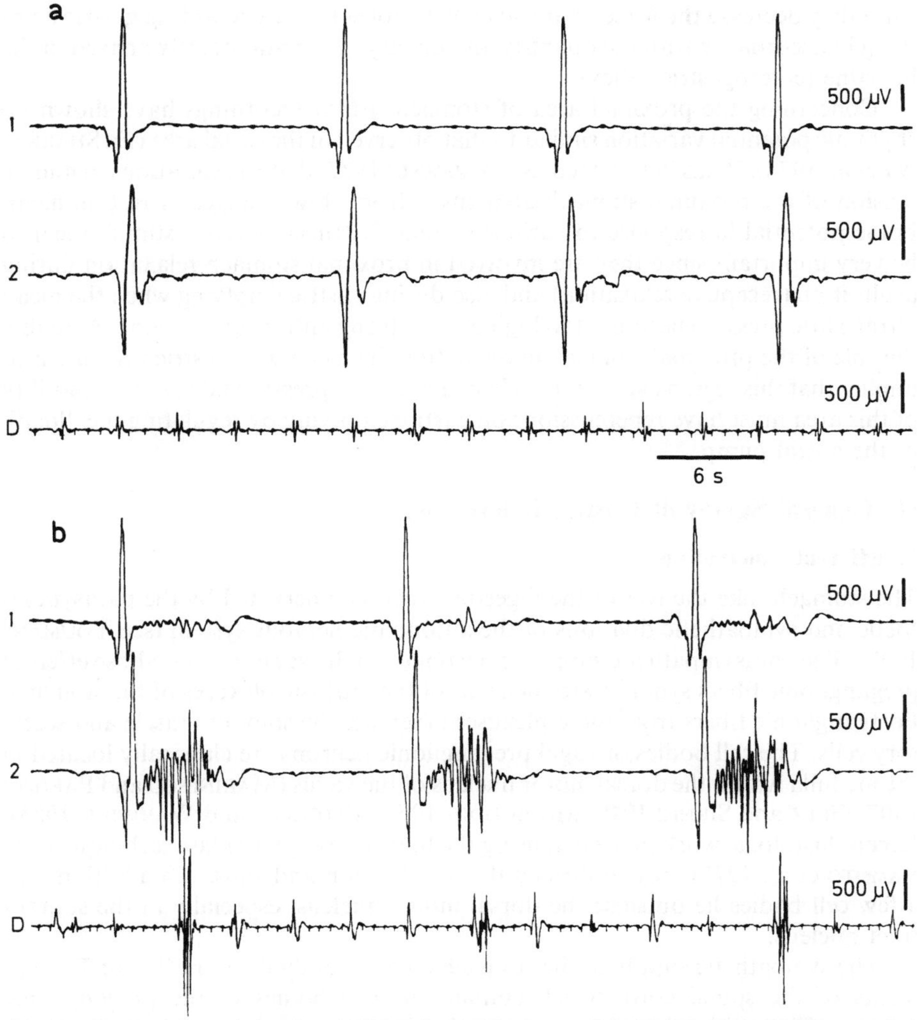

Fig. 11 a, b. Gastric and duodenal EMG during fasting and feeding in dog. EMG recorded by chronically implanted bipolar electrodes. Alternating current coupled amplification (time constant 0.15 s). Electrodes 1 and 2 in the antrum, 4 and 2 cm above the pylorus, respectively. Electrodes D in the duodenum 3 cm below the pylorus. **a** Fasted animal. Only ECA (antrum) and slow waves (duodenum) are recorded. **b** Fed animal. ERA, i.e., spikes or fast activity, follow each gastric ECA. Duodenal spiking occurs only after each antral contraction; typical pattern of antroduodenal coordination. (Miolan 1980)

Since TTX and atropine alone or in combination have no effect on the frequency or configuration of the spontaneous potential complex of the canine or human stomach, El Sharkawy et al. (1978) concluded that this activity was totally myogenic in origin and did not depend on neurogenic influences. However under normal circumstances neurogenic influences can alter gastric activity either by intrinsic (intramural) or extrinsic reflexes. Among the extrinsic reflexes, those arising from the proximal part of small intestine play a prominent role in gastric emptying since they decrease the force of antral contractions, thereby retarding gastric emptying in accordance with the quantity and quality of ingesta recently arrived in the intestine (enterogastric reflex).

Concerning the proximal area of stomach, EMG recordings have shown no rhythmic potential variation similar to that observed in the distal area (EL SHARKAWY et al. 1978). It has been stated (SZURSZEWSKI 1977) that the variations in muscle tension of the proximal stomach area ensue from slow changes of resting membrane potential in response to various stimuli. Extrinsic nervous stimuli seem to be very important since they are involved in proximal stomach relaxation during deglutition (receptive relaxation) and also during gastric emptying when the mean intragastric pressure becomes too high due to strong antral contractions. Actually, the role of the proximal stomach in regulating the mean intragastric pressure also implies that this region can contact when intragastric pressure falls, so that motility of this area must have repercussions on gastric emptying by regulating the "load" of the antral pump.

II. General Survey of Gastric Innervation

1. Efferent Innervation

The stomach, like the rest of the digestive tract, is innervated by the parasympathetic and sympathetic divisions of the autonomic nervous system (see YOUMANS 1968). The parasympathetic supply corresponds to the vagus nerves whose efferent preganglionic fibers synapse with neurons of the intrinsic plexuses of the stomach. Postganglionic fibers from these plexuses innervate the smooth muscle and secretory cells. The cell bodies of vagal preganglionic neurons are classically located in the medulla within the dorsal motor nucleus of the vagus (MARINESCO and PARHON 1907; GETZ and SIRNES 1949; MOHIUDDIN 1953; MITCHELL and WARWICK 1955). Recent histologic work in the cat, using the horseradish peroxidase technique (YAMAMOTO et al. 1977), has confirmed this localization and shown in addition that a few cell bodies lie outsidse the dorsal motor nucleus, especially in the solitary tract nucleus.

The sympathetic supply to the stomach comes mainly from the T-6 to T-9 segments of the spinal cord, which contain the cell bodies of the preganglionic neurons. The cell bodies of the postganglionic neurons are located mostly in the celiac ganglion, as far as the stomach and duodenum are concerned. Then the postganglionic fibers pass from the celiac ganglion to the stomach and duodenum along the branches of the celiac artery. Since the postganglionic fibers are adrenergic, the distribution of their terminals within the gastric wall can be visualized by the histofluorescence method. The main feature emerging from histofluorescence data is that most of the terminals supply the intramural ganglia and blood vessels, but that few fibers innervate the muscular coats or the muscularis mucosae (see FURNESS and COSTA 1974).

2. Sensitive Innervation

a) Vagal Fibers

Two sorts of gastric vagal receptor have been described: stretch receptors and chemoreceptors. Stretch receptors, presumably located within the muscular coat, are activated by gastric passive distension (PAINTAL 1954; IGGO 1955) and also by contractions of the walls (MEI 1970). Most of these receptors are found in the antrum, but some of them lie in the proximal stomach near the LES. According to TAKESHIMA (1971), antral receptors respond to distension and peristalsis while those of the cardiac portion are activated only by changes in gastric volume. Chemoreceptors correspond to mucosal endings. Actually they are rapidly adapting mechanoreceptors which function also as slowly adapting chemoreceptors (IGGO 1957b; DAVISON 1972; CLARKE and DAVISON 1978); the respond to balloon distension of stomach by an "on–off" discharge. Their chemosensitivity is limited to acid or alkali in the cat (IGGO 1957b; DAVISON 1972) but it seems rather nonspecific in the rat, since in this animal the same nerve ending can be excited by several of the following stimuli: organic and inorganic acids, water, alcohol, hypertonic saline, NaOH, NH_4Cl, $CuSO_4$, casein hydrolysate, mustard powder, and cayenne pepper (CLARKE and DAVISON 1978).

b) Splanchnic Fibers

According to RANIERI et al. (1973), splanchnic mechanosensitive endings are present in the gastric walls, especially those of antrum and lower body. They are activated by spontaneous or vagally induced contractions and also by distension or compression of the stomach. Most of them exhibit a slowly adapting discharge, except a few "on–off" receptors. The cell bodies of these sensitive fibers lie in spinal dorsal root ganglia between T-7 and T-11.

III. Effects of Severing the Extrinsic Nerves

1. Vagotomy

MCSWINEY, who surveyed the data available in 1931, concluded that section of both vagus nerves resulted in decreased tone, dilatation of the stomach, weakened peristalsis, and delayed emptying. However, subsequent studies have shown that gastric tone, instead of being lowered, was in fact increased after vagotomy. More precisely, vagotomy impairs the reservoir function of stomach. The adaptative relaxation of the proximal somach in response to a gastric distension is no longer observed and therefore the intragastric pressure of the distended stomach is higher than normal (MARTINSON 1965; JANSSON 1969a; AUNE 1969; KOSTER and MADSEN 1970; STADAAS and AUNE 1970; CARTER et al. 1972; WILBUR and KELLY 1973; JAHNBERG 1977). On the other hand, the gastric relaxation observed during swallowing or after esophageal distension (receptive relaxation) is also suppressed by vagotomy (CANNON and LIEB 1911, JANSSON 1969a; JAHNBERG 1977). Proximal gastric vagotomy, which denervates only the fundus and body, is as effective as truncal vagotomy in producing these alterations of the gastric reservoir function (WILBUR and KELLY 1973).

Concerning gastric emptying, all investigators agree that truncal vagotomy slows down emptying of solids (MCSWINEY 1931; MEEK and HERRIN 1934; INTERONE et al. 1971; WILBUR and KELLY 1973; COOKE 1975) and in contrast increases that of liquids (MEEK and HERRIN 1934; WILBUR and KELLY 1973; COOKE 1975). Proximal gastric vagotomy also results in an increased emptying of liquids (WIL-

BUR and KELLY 1973; BERGER et al. 1976) but does not affect emptying of solids (INTERONE et al. 1971; WILBUR and KELLY 1973; COOKE 1975). These data indicate that emptying of liquids depends to a great extent on the proximal stomach, which adjusts by its tension variations the intragastric transmural pressure, while emptying of solids is chiefly under the control of the distal stomach.

The fact that emptying of solids is delayed after truncal vagotomy or total gastric vagotomy probably results from a decreased force of antral peristalsis (MCSWINEY 1931; WILBUR and KELLY 1973) and not from a possible spasm of pylorus since this sphincter has been shown to be patulous and open following vagotomy (MCSWINEY 1931; QUIGLEY and LOUCKES 1951).

As far as gastric EMG is concerned, obvious disturbances of BER can be observed for the few days immediately following a transthoracic vagotomy. The antral electrical activity shows periods of total disorganization, with groups of slow potentials, i.e., ECA exhibiting various amplitudes, frequencies, and configurations (KELLY and CODE 1969; PAPAZOVA and ATANASSOVA 1972; MIOLAN 1980; Fig. 12). At the same time, the prevagotomy BER pattern is preserved on the body so that ECA of antrum and body are no longer synchronized (PAPAZOVA and ATANASSOVA 1972). The abnormal antral ECA are not followed by ERA and therefore they do not trigger muscular contractions. Usually, the normal pattern of antral BER is restored within a week, although some dogs continue to have occasional periods of disorganized rhythm for several months after vagotomy (KELLY and CODE 1969; PAPAZOVA and ATANASSOVA 1972). After restoration of the normal antral BER, perturbations similar to those observed in the early postvagotomy period can be provoked by intravenous injection of atropine (PAPAZOVA and ATANASSOVA 1972). Hence, EMG disturbance ensuing from vagotomy probably expresses a decrease of the cholinergic tone which is then able to recover, probably because of an adaptative function of intraparietal cholinergic structures.

Usually during emptying of a meal there is coordination between the duodenal and antral motor activities (THOMAS and CRIDER 1935; THOMAS 1957). Each antral ECA is accompanied by an ERA while, at the duodenal level, bursts of spike potentials are superimposed on the first or second slow waves which immediately follow the antral contraction (ALLEN et al. 1964; ATANASSOVA 1970; BEDI and CODE 1972; MIOLAN 1980; Fig. 11). According to AEBERHARD and BEDI (1977), 4 weeks after truncal vagotomy antroduodenal coordination could be demonstrated in the same way as before in two out of three dogs; in the third dog restitution of previously impaired coordination occurred during the second month after vagotomy. However, when EMG recordings are analyzed soon after vagotomy (MIOLAN 1980) deficient coordination is observed as a general rule. Namely, duodenal spike potentials occur with each slow wave and not only with those following the gastric contractions (Fig. 13). In addition, and contrary to what occurs in the normal dog, there are postprandial periods during which motility is totally absent since neither gastric ERA nor duodenal spike potential are recorded (Fig. 13). These disturbances (deficient coordination; periodic absence of motility) may also account for emptying deficiency following vatogomy. The recovery of normal motility patterns and of antroduodenal coordination some time after vagotomy once more confirms the adaptive function of the visceral effector, presumably owing to the presence of intramural ganglia.

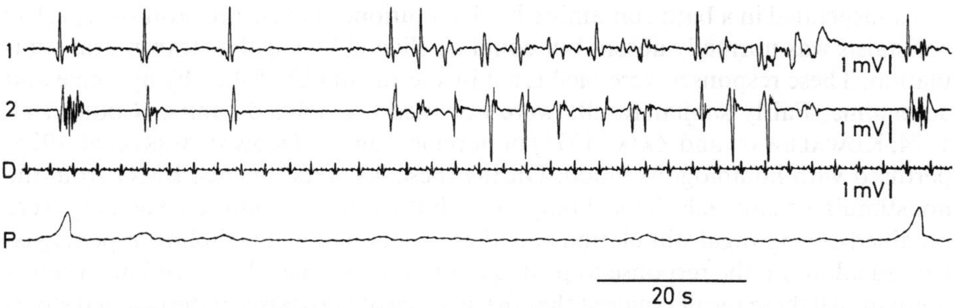

Fig. 12. Disturbances of dog gastric EMG following vagotomy. Recordings made two days after vagotomy. EMG recorded by chonically implanted bipolar electrodes. Alternating current coupled amplification (time constant 0.15 s). Electrodes 1 and 2 in the antrum, 4 and 2 cm above the pylorus, respectively. Electrodes D in the duodenum, 3 cm below the pylorus. P intraluminal pressure recorded in the antrum with an open tip catheter. Notice the complete disorganization of antral electrical activity and the related absence of contraction for more than 1 min. (MIOLAN 1980)

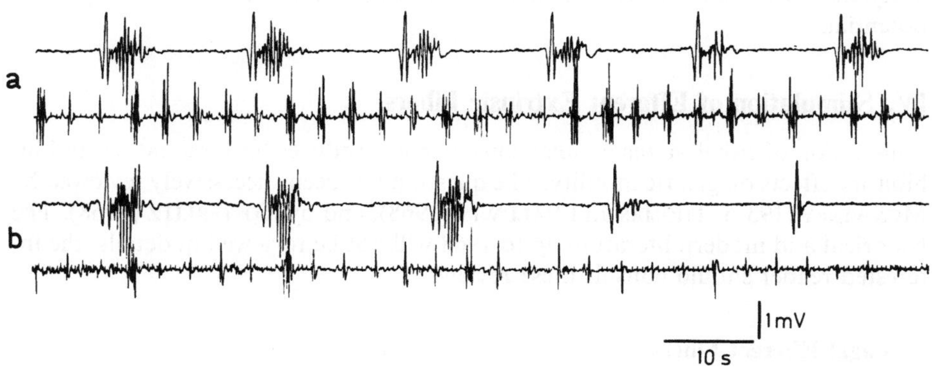

Fig. 13 a, b. Influence of feeding on gastric and duodenal EMG in the vagotomized dog. Recordings made 7 days after vagotomy. EMG recorded by bipolar electrodes chronically implanted in the antrum *(upper trace)* and duodenum *(lower trace)*. Alternating current coupled amplification (time constant 0.15 s). Two examples of EMG patterns frequently observed after a meal. **a** Disruption of antroduodenal coordination; almost all the duodenal slow waves trigger spike potentials (compare with Fig. 11); **b** sudden disappearence of antral contractions. (J.P. MIOLAN 1980 unpublished work)

2. Sympathectomy

Data on sympathectomy are less numerous than those on vagotomy. However, it is generally stated that section of the splanchnic nerves results in accelerated gastric emptying, probably owing to increased tone and more powerful peristalsis (MCSWINEY 1931; ALVAREZ 1948; KOSTERLITZ 1968). It is also agreed that splanchnic section has less disturbing effects on the stomach than has bivagotomy.

3. Stomach Totally Deprived of Extrinsic Nerves

The motility of the stomach totally deprived of extrinsic innervation was first studied by ARMITAGE and DEAN (1966) on rat isolated stomach–duodenum preparations suspended in a bath containing Krebs solution. In these preparations, gastric peristalsis was normally absent but could be elicited by vagal or transmural stimulation. These responses were facilitated by eserine but abolished by hyoscine and adrenaline. Fairly similar results were obtained on isolated canine (COOK et al. 1974; KOWALEWSKI and ZAJAC 1975) or porcine stomach (KOWALEWSKI et al. 1975) perfused with homologous blood. During these last experiments, EMG from the nonstimulated stomach showed only ECA, but ERA and contractile activity were obtained during vagal stimulation and after injection of methacholine or pentagastrin. In addition, the response to pentagastrin were abolished by atropine or tetrodotoxin. All these results suggest that the absence of peristalsis in the isolated stomach results from a cholinergic tone deficiency, probably due to a depressed activity of myenteric neurons. In this connection, one must recall the observations of SZURSZEWSKI's group (SZURSZEWSKI 1975; EL SHARKAWY and SZURSZEWSKI 1978) on dog gastric muscle strips in vitro. The spontaneous complex action potential (upstroke and plateau) recorded from longitudinal or circular muscle strips did not trigger contraction. Muscle shortening occured only in the presence of either acetylcholine or pentagastrin which increased the ERA, i.e., the size of the plateau potential.

IV. Stimulation of Efferent Extrinsic Fibers

Stimulation of either vagus or splanchnic nerves produces both excitatory and inhibitory effects on gastric motility. The question has been successively reviewed by McSWINEY (1931), THOMAS and BALDWIN (1968), and by KOSTERLITZ (1968). The historical and modern literature up to 1968 will not be reviewed in details; the interested reader should consult these reviews.

1. Vagal Efferent Fibers

The vagus nerves carry two kinds of preganglionic fibers: (1) fibers whose stimulation leads to excitatory effects; (2) others whose stimulation leads to inhibitory effects. The former have a lower excitation threshold (hence a larger diameter) than the latter (MARTINSON and MUREN 1963; MARTINSON 1964, 1965).

a) Excitatory Effects

When recorded by manometry, the excitatory effects first consist of an increase of gastric tone which may chiefly reflect the contraction of the proximal stomach, i.e., of the fundus and body (McSWINEY 1931; MARTINSON and MUREN 1963; PATON and VANE 1963). At the antral level, vagal stimulation can trigger peristaltic contractions in rat isolated stomach (ARMITAGE and DEAN 1966) and in anesthetized opossum (SARNA and DANIEL 1975).

When the EMG of the distal stomach is recorded, the vagal excitatory effects are revealed by the occurrence or the facilitation of the ERA in those species like the dog or pig which present complex gastric action potentials. Depending on the recording technique, one observes either an increase of the amplitude and duration of the plateau potential (MIOLAN and ROMAN 1971; Fig. 14) or the occurrence of spikes after each ECA (SARNA and DANIEL 1975; KOWALEWSKI and ZAJAC 1975;

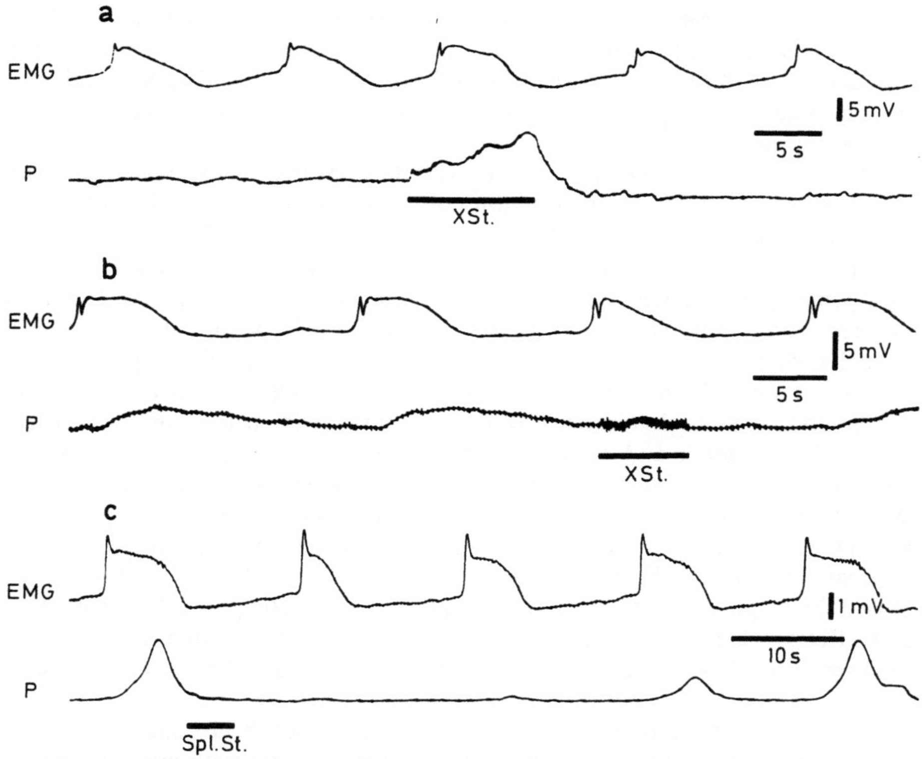

Fig. 14a–c. Effects of vagal and splanchnic stimulations on dog gastric EMG. In vivo experiments; dogs anesthetized with pentobarbital. EMG monopolar records made with a suction electrode; alternating current coupled amplification in **a** (time constant 5 s) and direct current amplification in **b** and **c**. P intraluminal pressure recorded by a balloon located in the distal antrum. **a** In the absence of atropine, stimulation of the vagus (XSt; 10 V, 2 ms, 10 Hz) causes a gastric contraction and increases the amplitude and duration of the gastric complex action potential. **b** In the presence of atropine vagal stimulation (XSt) with identical parameters decreases the amplitude and duration of the plateau potential (inhibitory effect also recorded on the pressure tracing). **c** Another animal, nonatropinized; stimulation of splanchnic nerve (Spl/St; 5 V, 0.5 ms, 30 Hz) results in gastric motility inhibition which is obvious on both EMG and pressure tracings. (Miolan and Roman 1971)

Kowalewski et al. 1975). Sarna and Daniel (1975) have also reported that intense vagal stimulation could cause premature control potentials (i.e., ECA) as did localized intra-arterial injections of acetylcholine (Daniel and Irwin 1968). In addition, Daniel and Sarna (1976) have shown that there is a segmental distribution of the excitatory vagal fibers which reach the stomach through various branches of the two nerves of Latarjet. Thus, if these branches are cut sequentially from proximal to distal or vice versa, the production of ERA and contraction drops out sequentially. Furthermore, each nerve of Latarjet sends what seems to be segmentally arranged preganglionic fibers to the opposite side of the stomach so that stimulation of one nerve of Latarjet causes the contraction of both sides of the stomach. By the way, it may be noticed that stimulation of the nerves of Latarjet, which causes gastric contractions, at the same time produces an inhibition of duodenal spontaneous motility (DANIEL 1977). If stimulation is continued, there is a gradual recovery to a pattern in which one or two duodenal contractions follow each gastric contraction, i.e., a return to a pattern resembling the one observed during emptying of a meal.

The EMG of the guinea pig stomach differs from that of the canine stomach since it does not exhibit complex action potentials but only true spike potentials, occurring more or less continously or grouped in bursts (especially at the antral level). Vagal excitatory responses recorded by intracellular microelectrodes from isolated guinea pig stomach (BEANI et al. 1971) consist of e.j.p. either pure or with superimposed spikes. These e.j.p. follow each vagal pulse within 200 ms and last about 400–500 ms. Then, the vagal excitatory responses of the guinea pig stomach resemble very closely those of the cat smooth muscle esophagus (see Sect. A.I.4.b).

The concensus is that neuroeffector transmission from the excitatory vagal pathway is cholinergic and that consequently it is blocked by atropine or hyoscine (KOSTERLITZ 1968; BEANI et al. 1971; MIOLAN and ROMAN 1971; SARNA and DANIEL 1975). Acetylcholine is present in gastric walls and is released by vagal stimulation (KOSTERLITZ 1968). In addition, during in vitro experiments on muscular strips from canine stomach (EL SHARKAWY and SZURSZEWSKI 1978), acetylcholine increased the amplitude and duration of the plateau of the complex action potential; these effects, which were blocked by atropine but not by hexamethonium, very closely resembled those triggered by vagal stimulation in vivo (MIOLAN and ROMAN 1971).

The contractor response induced by vagal stimulation is also abolished or at least markedly reduced by ganglion blocking drugs like hexamethonium (KOSTERLITZ 1968; SARNA and DANIEL 1975; KOWALEWSKI et al. 1975). Under the effect of hexamethonium, transmural stimulations are still able to trigger a contraction which in turn is abolished by atropine or hyoscine (PATON and VANE 1963). Thus, the excitatory vagal path appears to be composed of preganglionic cholinergic fibers synapsing with myenteric cholinergic neurons which excite smooth muscle. However, as previously noticed by LANGLEY (1922) the vagal efferent fibers are far less numerous than intramural neurons; it is likely that not all intramural neurons receive impulses from vagal fibers. This might explain the fact that the motor response to maximal vagal stimulation is always smaller than the response to transmural stimulation (PATON and VANE 1963).

b) Inhibitory Effects

These effects are easily observed after administration of atropine, which blocks the vagal excitatory effects. The inhibitory effects correspond to a relaxation of the gastric wall especially evident in the proximal stomach (fundus and body) and well recorded by manometric techniques (MARTINSON and MUREN 1963; PATON and VANE 1963; MARTINSON 1964, 1965; CAMPBELL 1966). On electrical recordings from the body of guinea pig stomach, vagal inhibitory responses consist of i.j.p., hyperpolarizations lasting 1,000–1,800 ms and following each stimulation pulse with a short latency period (200–300 ms). When trains of pulses (20 Hz) are used, hyperpolarization becomes permanent, but a rebound excitation (train of spikes) occurs after the cessation of stimulation (BEANI et al. 1971). EMG recording of antral activity in anesthetized dogs has shown that vagal inhibition also affects the antrum, but in a different manner from that observed in guinea pig stomach, since there is a decrease of amplitude and duration of the plateau of the complex action potential (MIOLAN and ROMAN 1971; Fig. 14).

Vagal inhibitory effects were first interpreted as being adrenergic in nature. Indeed, a large body of evidence indicates that the vagus of various species carries adrenergic fibers, originating chiefly from the superior cervical and stellate ganglia (MURYOBAYASHI et al. 1968; LIEDBERG et al. 1973; LUNDBERG et al. 1976). However, it is now widely accepted that the vagal inhibitory responses differ from those induced by sympathetic stimulation or catecholamine injection. The vagal responses are generally stronger, they develop more rapidly, and they are obtained at a lower stimulation frequency than adrenergic responses (MARTINSON 1965; CAMPBELL 1966). Furthermore, they are not blocked by antiadrenergic drugs such as guanethidine, bretylium, or α- and β-blockers (MARTINSON 1965; CAMPBELL 1966; BEANI et al. 1971). They are still present in isolated stomachs of animals treated with reserpine in concentrations sufficient to deplete tissue stores of noradrenaline (HEAZELL 1977). The ability of ganglion-blocking agents to reduce the vagal inhibitory responses to a significant extent suggests the presence of cholinergic (nicotinic) synapses between extrinsic vagal fibers and nonadrenergic, noncholinergic inhibitory intramural neurons (PATON and VANE 1963; BULBRING and GERSHON 1967; BEANI et al. 1971). The existence of such inhibitory intramural neurons in the gastric wall is suggested by the fact that transmural stimulation of the whole stomach or gastric strips from various animal species leads to a relaxation (PATON and VANE 1963; CHRISTENSEN and TORRES 1975; HEAZELL 1977) and to the production of i.j.p. in guinea pig preparations (ATANASSOVA et al. 1972; ITO and KURIYAMA 1975). These responses, obtained after administration of atropine, are not affected by hexamethonium; they persist under various antiadrenergic drugs, although often diminished. Thus it is beyond doubt that relaxation of strips is mainly due to stimulation of nonadrenergic nerves, even if a part of the response may result from stimulation of some intramural sympathetic nerve endings.

Concerning the neurotransmitter of the inhibitory intramural neurons, many experimental data gathered by BURNSTOCK (1972, 1975, 1978) suggest that a purine nucleotide, probably ATP, might be an excellent candidate. VALENZUELA (1976) proposed dopamine on the basis that this substance is able to relax the stomach. But dopamine antagonists (methoclopramide and pimozide) which block the dopamine-induced gastric relaxation, only partially antagonize the reflexly induced receptive relaxation which is known to be under the control of the vagal inhibitory pathway. Besides, the presence of dopamine in intramural neurons and the release of this substance during vagal stimulation have not been demonstrated. Recent work by FAHRENKRUG and co-workers (FAHRENKRUG et al. 1978a, b) has drawn attention to a possible involvement of vasoactive intestinal peptide (VIP) in vagal inhibition. VIP is released in the venous blood from the stomach during vagal stimulation. This release is not affected by atropine but is blocked by hexamethonium. In atropinized animals, electrical stimulation of the high threshold vagal fibers produces gastric relaxation and also induces a significant increase of venous plasma VIP concentration. When a similar vagal relaxation of the stomach is elicited by distending a balloon in the esophagus, an increase of venous plasma VIP concentration is also observed. In addition, close intra-arterial infusion of VIP results in a marked relaxation of the stomach. Taking into account all the preceding results together with the data of various histochemical studies showing an abundant localization of VIP in the intramural neurons of the alimentary canal (particularly those of the stomach) FAHRENKRUG et al. (1978 b) suggested that VIP might be a neurotransmitter in the gastrointestinal tract, and in this case, the transmitter of noncholinergic, nonadrenergic intramural nerves. However, this attractive hypothesis does not explain why, in the canine stomach, the plateau potential of the complex action potential is altered during vagal inhibition (MIOLAN and ROMAN 1971) but not at all during VIP-induced relaxation (MORGAN et al. 1978). It might be that several transmitters are simultaneously involved in vagal relaxation.

Part of the vagal gastric inhibition remains in the presence of ganglion-blocking agents such nicotine or hexamethonium (MARTINSON 1965; BULBRING and GERSHON 1967; BEANI et al. 1971). BULBRING and GERSHON (1967) suggested that serotonin (5-HT) might be a transmitter, participating with acetylcholine in the vagal inhibitory path to the stomach. This suggestion was based on the following observations: (1) 5-HT activated inhibitory ganglion cells of the myenteric plexus through specific receptors distinct from nicotinic receptors; (2) antagonism or desensitization of these neural receptors inhibited vagal relaxation of the stomach; (3) stimulation of the mouse stomach led to a neurally mediated 5-HT release (TTX sensitive). More recently, GERSHON and DREYFUS (1977) have produced evidence that the myenteric plexus contains intrinsic serotonergic neurons. Thus, it is conceivable that 5-HT intramural neurons serve as interneurons between some preganglionic fibers and some postganglionic inhibitory neurons (the same possibility might also exist for the excitatory vagal pathway).

2. Sympathetic Efferent Fibers

Both inhibition and facilitation of gastric motility have been described in response to splanchnic nerve stimulation (MCSWINEY 1931; THOMAS and BALDWIN 1968; KOSTERLITZ 1968). However, the chief and in any case the best known effect is the inhibitory one.

a) Relaxant Responses

These responses correspond to a decrease in the tone of the proximal stomach and to a weakened peristalsis of the antrum. These effects are adrenergic in nature since they are blocked by guanethidine (MARTINSON 1965; BEANI et al. 1971) or by bretylium (CAMPBELL 1966). In addition, when sympathetic stimulation is performed at the preganglionic level (splanchnic nerve) the inhibitory effects are also abolished by ganglion-blocking agents (hexamethonium, nicotine) either applied on the ganglia (celiac and superior mesenteric) or injected intravenously (SEMBA and HIRAOKA 1957; NAKAZATO et al. 1970).

It is now widely accepted that adrenergic nerves may reduce gastric motility by acting at two different sites, i.e., the myenteric neurons and the muscle itself. The action on myenteric neurons is strongly supported by the anatomic observation that adrenergic fibers ramify extensively in the myenteric (Auerbach's) plexus (see

FURNESS and COSTA 1974). Besides, experimental data gathered by JANSSON and co-workers (JANSSON and MARTINSON 1966; JANSSON and LISANDER 1969; JANSSON 1969 a) indicate that the more powerful the cholinergic tone (activity of myenteric neurons), the stronger the splanchnic inhibition of stomach. Thus, the sympathetic inhibition is prompt and pronounced, even for weak stimulation frequencies when the two vagi are intact. After acute vagotomy or atropine injection, the inhibitory responses are much weaker, retarded, and requiring higher stimulation frequencies. Since the gastric contractions caused by exogenous acetylcholine were unaffected by adrenergic nerve activations that promptly inhibited vagally induced contractions, JANSSON (1969 a) concluded that "the major, possibly only, site of inhibitory action of the adrenergic fibers of stomach ist localized to intramural cholinergic neurons at least with respect to the nervous control of reservoir function of the stomach."

Nevertheless, a direct action of adrenergic fibers on smooth muscle does exist. From histologic data based on the histofluorescence method, it can be stated that there is some adrenergic innervation of muscular coats (see FURNESS and COSTA 1974). In addition, nerve stimulation in vivo indicates that some sympathetic inhibition remains after administration of atropine (JANSSON 1969 a). In vitro, sympathetic relaxation of muscular strips taken from guinea pig stomach does not appear to be changed under hyoscine or atropine (GERSHON 1967; BEANI et al. 1971). Intracellular recording of the activity of muscular strips from guinea pig stomach (BEANI et al. 1971) showed that sympathetic stimulation at 10–20 Hz caused a hyperpolarization of the cell membrane and subsequently a rebound excitation. BEANI et al. (1971) did not mention whether such hyperpolarizations were also obtained in the presence of atropine. But it seems likely that they are not due to a decreased activity of intramural cholinergic neurons, since it was observed on other strips that atropine per se did not change the resting membrane potential of muscular cells. In studies of dog stomach in vivo, MIOLAN and ROMAN (1971) showed that splanchnic stimulation at 30 Hz reduced the amplitude and duration of the plateau of the complex action potential (Fig. 14). In vitro, EL SHARKAWY and SZURSZEWSKI (1978) observed similar modifications of the complex action potential of the dog stomach under the action of noradrenaline. According to EL SHARKAWY and SZURSZEWSKI (1978), noradrenaline could act directly on smooth muscle since the gastric action potential is not altered by atropine or TTX. However, it is fairly possible that a part of the inhibitory responses observed in vivo by MIOLAN and ROMAN (1971) result from an action on myenteric ganglion cells because the most obvious inhibitory effects were observed when gastric peristalsis was powerful, i.e., when the cholinergic tone was important.

Concerning the receptors involved in the effects of noradrenaline released from sympathetic nerves in mammalian gastrointestinal tissues, it was suggested that inhibition of intramural cholinergic neurons is mediated by α-adrenoceptors, while direct muscle inhibition is chiefly achieved through β-adrenoceptors (FURNESS and COSTA 1974; VIZZI 1976; GILLESPIE and KHOYI 1971).

Concerning the α-adrenoceptors of the myenteric neurons, data obtained from guinea pig intestine suggest that they are located at the presynaptic level, i.e., on the terminals of the vagal preganglionic fibers (VIZI 1976; GILLESPIE and KHOYI 1977). This seems also to be true for the chick stomach (SENO et al. 1978).

As for the adrenoceptors located on gastric muscle it is now firmly established that β-adrenoceptors, when present, always mediate inhibitory effects. This was checked in the guinea pig (GUIMARAES 1969; HAFFNER 1973; OHKAWA 1976), dog (NAKAZATO et al. 1970), rat (HAFFNER 1973), rabbit (HAFFNER and STADAAS 1972), and chick (SENO et al. 1978). However α-adrenoceptors mediating inhibition and probably situated in gastric muscle were also demonstrated in some of the preceding species, i.g., dog (EL SHARKAWY and SZURSZEWSKI 1978) and guinea pig (GUIMARAES 1969).

Taking into account the fact that adrenergic innervation of muscular coats is scarce in comparison with that of the myenteric plexus, it has often been claimed that part of the direct action of noradrenaline on smooth muscle during sympathetic stimulation resulted from an overflow of transmitter from nerves supplying the myenteric ganglia (FURNESS and COSTA 1974). This would explain why higher rates of stimulation are needed to produce inhibition in the absence of cholinergic tone, i.e., when myenteric ganglia inhibition has practically no more effect. Still the physiologic importance of a possible overflow remains obscure and, in any case, difficult to assess.

b) Motor Responses

According to MCSWINEY and co-workers (for references see MCSWINEY 1931), gastric motor effects can be elicited by stimulating the splanchnic nerves or the periarterial nerves running to the stomach. The same authors reported that high frequency stimulations led to inhibition whereas low frequency ones favored contraction of the body and fundus in particular. Splanchnic motor responses were more recently confirmed by SEMBA and HIRAOKA (1957), JANSSON (1969a), NAKAZATO et al. (1970), and SEMBA and MIZONISHI (1978). These motor responses, or part of them, are suppressed by atropine. Therefore, they probably result from cholinergic fibers stimulation. According to SEMBA and HIRAOKA (1975), the splanchnic cholinergic fibers would leave the spinal cord by the dorsal roots of rachidian nerves. Such cholinergic fibers emerging from the spinal cord in the dorsal roots of the spinal nerves have been extensively described in amphibians (see CAMPBELL and BURNSTOCK 1968).

However, NAKAZATO et al. (1970) have reported that in the atropinized dog, stimulation of splanchnic or periarterial nerves was still effective in producing gastric contractions. A similar motor effect was also obtained by close intra-arterial injection of noradrenaline. In this connection, early observations of MUREN (1957) must be recalled. MUREN noticed that intravenous injection of adrenaline or noradrenaline (2 μg/kg) in the dog produced gastric inhibition when the two vagi were intact and therefore the gastric tone was important. The response was reversed to excitation when gastric tone was decreased by various means, i.e., bivagotomy, deep anesthesia, or atropine injection. A similar inversion of adrenaline effects in the presence of atropine was also reported by MCSWINEY and co-workers in the rabbit (see MCSWINEY 1931). These truly sympathetic motor effects are very likely mediated by α-adrenoceptors localized on smooth muscle because in the atropinized dog they are blocked by phenoxybenzamine (α-blocker) and are replaced by a relaxation which is in turn antagonized by β-blockers like pronethalol

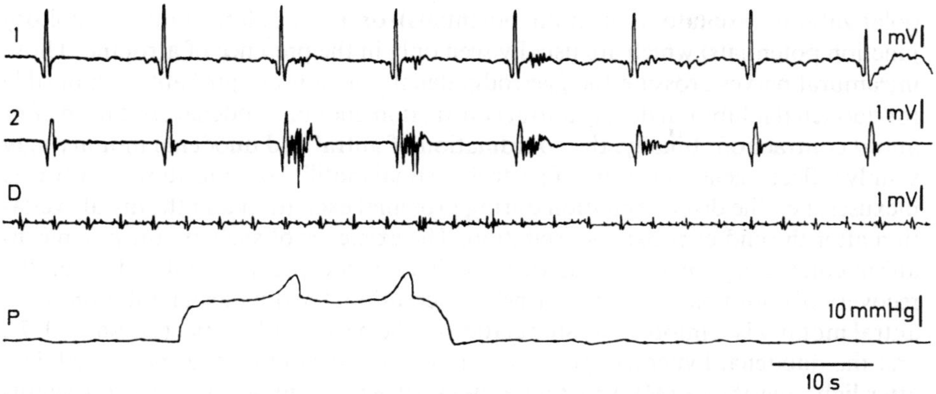

Fig. 15. Effect of antral distension on antral EMG activity in the vagotomized dog. EMG recorded by bipolar electrodes chronically implanted in the antrum *1*, *2* and duodenum *D*. Alternating current coupled amplification (time constant 0.15 ms). P pressure recorded by an intragastric balloon located in the antrum. Inflation of the balloon (10 ml air) triggers antral contractions visible on both EMG and pressure recordings. (J.P. MIOLAN unpublished work 1980)

(NAKAZATO et al. 1970). Moreover, the existence of gastric motor effects mediated by α-adrenoceptors is attested by several reports concerning two other species: guinea pig (GUIMARAES 1969; OHKAWA 1976) and rabbit (HAFFNER and STADAAS 1972).

Finally, it seems clear that motor effects can be induced by splanchnic stimulation. They result in fine from involvement of muscarinic or α-adrenoceptors, but no information is available for the assessment of their physiologic significance.

V. Possible Roles of Intrinsic Innervation

Data on the role of the intrinsic nerves of the stomach are scarce. However, it seems reasonable to assume that intrinsic circuits are responsible for peristalsis reappearing shortly after total gastric denervation in chronic animals. In rat isolated stomach–duodenum preparations, ARMITAGE and DEAN (1966) observed no spontaneous motility, but peristalsis could be elicited by vagal or by transmural stimulation; surprisingly, gastric distension by itself seemed ineffective. Nevertheless, in chronically vagotomized dogs, J. P. MIOLAN (1970 unpublished work) showed that antral distension promptly elicited strong peristaltic contractions (Fig. 15) which indicated that stretch receptors in the gastric wall are likely to play an important role in activation of the intrinsic mechanism responsible for gastric movements.

On the other hand, intrinsic nerves are probably involved in the coordination of antral and duodenal motility. Transection of the gastroduodenal junction (ATANASSOVA 1970) disrupts the coordination and leads to an increased duodenal motility which probably expresses a release from a tonic inhibition normally exerted by the active stomach upon the duodenum. Transmural stimulation of the gastric end of cat gastroduodenal strips induces changes in resting membrane potential of duodenal cells (ATANASSOVA 1976). These changes consist of either depolarizations (excitatory junction potentials) or hyperpolarizations (inhibitory junction potentials) which are usually seen only in the presence of atropine. Those intramural nerves crossing the gastroduodenal junction are probably responsible for duodenal inhibition during antral contraction and for duodenal excitation after antral contraction. Finally, the coordination of antral and duodenal motility may simply reflect a general feature of gastrointestinal motility organization by intrinsic plexuses, i.e., the distal inhibition during proximal excitation and the distal excitation after the end of proximal excitation. The existence of such an intrinsic mechanism correlating antral and duodenal activity may explain several facts, e.g., the recovery of coordination after extrinsic denervation, the duodenal inhibition when antral motility is reinforced by stimulation of the nerves of Latarjet (Daniel 1977), and the duodenal hyperactivation each time the antral motility is depressed, i.e., after light anesthesia (Miolan and Roman 1978 b) or immediately after a vagotomy (Miolan 1980).

VI. Possible Roles of Extrinsic Innervation

1. Receptive Relaxation and Related Phenomena

The term "receptive relaxation" was proposed by Cannon and Lieb (1911) to designate the relaxation which affects chiefly the fundus and the body during deglutition. As already mentioned by Cannon and Lieb (1911), receptive relaxation is suppressed by vagotomy and hence it results from an extrinsic reflex. The fact that the vagal inhibitory pathway is the efferent link of this reflex has been demonstrated by extensive studies by Martinson and co-workers (see Martinson 1965; Jansson 1969 a; Abrahamsson 1973; Jahnberg 1977). The same investigators have also shown that similar gastric relaxations occurred in several different circumstances. Thus, apart from deglutition, another effective stimulus is provided by esophageal distension which usually stimulates secondary peristalsis (Jansson 1969 a). Proximal stomach relaxations are also induced by distensions of the whole stomach, by distensions restricted to the antrum, or even by close intra-arterial injections of acetylcholine which increase antral motility (Abrahamsson 1973). It seems reasonable to assume that slowly adapting antral stretch receptors, activated either by distension or by forceful contractions, are responsible for such relaxations of the corpus–fundus area. In addition to the vagally mediated reflex relaxation, antral distensions can also elicit a sympathetic gastrogastric inhibitory reflex with both afferent and efferent links conveyed by the splanchnic nerves; the efferent adrenergic link of the reflex acts essentially by inhibiting the excitatory cholinergic neurons in the gastric walls (Abrahamsson 1973). Finally, gastric relaxations have also been produced by artifical stimulation, i.e., by electrical stimulation of sensitive fibers carried by the vagus nerve and originating from various receptors located in the esophagus, stomach, and probably also in the intestine (Jansson 1969 a; Ohga et al. 1970; Abrahamsson 1973). The physiologic significance of all these reflex relaxations seems obvious, i.e., they prevent the gastric pressure becoming too high during food ingestion and afterwards, during the strong antral contractions of the emptying period. Besides, vagally mediated relaxations of the stomach are also observed during vomiting (Abrahamsson 1973). They correspond to a sort of recep-

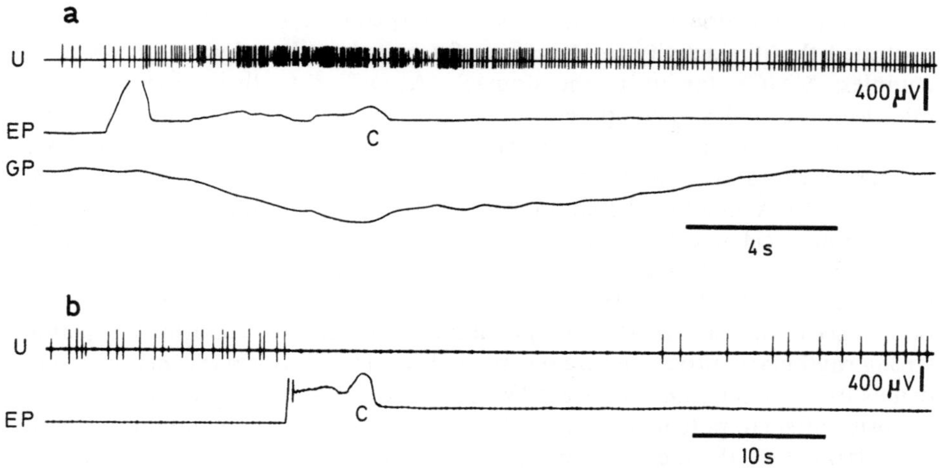

Fig. 16a, b. Discharge pattern of vagal efferent fibers probably supplying the fundus in conscious dog. *U*, unit activity of a vagal efferent fiber recorded using the nerve suture technique; *EP*, esophageal pressure recorded by a freely moving balloon propelled by a secondary peristaltic contraction ("dynamic" pressure tracing). The last wave *C* of the tracing is recorded when the balloon enters the stomach. *GP*, gastric pressure recorded by an inflated balloon (100 ml air) located in the proximal stomach. **a** During a secondary peristalsis the proximal stomach relaxes (receptive relaxation). At the same time, the vagal fiber is strongly excited (vagal inhibitory fiber of fundus). **b** During the receptive relaxation induced by a secondary peristalsis this other type of vagal fiber is inhibited (vagal excitatory fiber of fundus). (MIOLAN and ROMAN 1974; MIOLAN 1980)

tive relaxation since during vomiting the stomach is usually refilled with ingesta returning from the upper intestine.

The discharge pattern of vagal efferent fibers (preganglionic fibers) suspected of being involved in the corpus–fundus relaxation has been studied in conscious dogs by MIOLAN and ROMAN (1974) using a nerve suture technique. These vagal fibers fired spontaneously at low frequency (from 0–3 spikes/s). During the proximal stomach relaxation (swallowing, esophageal distension, strong antral contractions) two opposite patterns were observed (Fig. 16). For many fibers, the discharge frequency was increased to 10–40 spikes/s; for some others, the rate of firing was decreased or even abolished. From these results, MIOLAN and ROMAN (1974) concluded that the proximal stomach relaxation resulted from a decreased activity of vagal preganglionic fibers synapsing with excitatory myenteric neurons, but also, and chiefly, from an activation of preganglionic fibers synapsing with inhibitory myenteric neurons. In the following text, these fibers will be termed vagal excitatory fibers of the fundus (VEF) and vagal inhibitory fibers of the fundus (VIF) respectively.

2. Gastric Motility During Fasting and in the Fed State

a) Motility Patterns

By recording the EMG simultaneously at multiple sites along the small bowel of fasted conscious dogs, SZURSZEWSKI (1969) noticed a caudad moving band of large amplitude spike potentials (the activity front) sweeping in recurring cycles every

90–180 min along the entire small bowel. Later, CODE and MARLETT (1975) demonstrated that the activity front is part of a complex (the migrating myoelectric complex, MMC) starting in the stomach. According to their description, the myoelectric complex is composed of four distinct phases that recur regularly at each point of detection. As far as the gastric antrum is concerned, these four phases can be described as follows. Phase I is characterized by the absence of spike potentials or ERA. Only ECA (pacemaker potentials) are recorded. This means that there is no mechanical activity. Persistent but random spike potential activity marks the onset of phase II. During this phase, spike potentials increase in incidence and intensity. Phase III is characterized by the sudden onset and continuous occurrence of bursts of large action potentials with every ECA. During this period, strong rhythmic contractions are recorded. The end of this phase is nearly as sudden as its beginning. During phase IV, there is a rapid decrease in incidence and intensity of spike potentials and a return to phase I within few minutes. Recent data (ITOH et al. 1978) confirming early reports have shown that the fundus and body of the stomach also contract in association with the antrum during the MMC.

After feeding, electrical activity in the antrum consists mainly of bursts of large spike potentials of long duration following every ECA. This intense contractile activity lasts several hours and is responsible for the gastric emptying. At the same time, there is also a close functional coordination between stomach and duodenum, each antral contraction being followed by a proximal duodenum contraction. The behavior of the proximal stomach during the postprandial period is not well known. However it is generally stated that this region plays the role of a pressure regulator so that, if intragastric pressure increases too much during the forceful antral contractions, the fundus and body relax, and on the contrary, when intragastric pressure falls because of emptying, the fundus and body contract in order to maintain a certain load on the antral pump.

b) Control

Concerning the control of MMC, numerous data point out the importance of humoral factors and particularly of motilin in the initiation or at least the regulation ("modulation", etc.) of the interdigestive motor activity (ITOH et al. 1976; WINGATE et al. 1976; THOMAS et al. 1980; PEETERS et al. 1980). It is widely accepted that the extrinsic nerves are of minor importance since vagotomy and splanchnicotomy do not suppress the MMC (WEISBRODT et al. 1975; MARIK and CODE 1975; RUCKEBUSCH and BUENO 1975, 1977). However, one cannot take for granted that section is the best experimental approach to get reliable information about the actual role of the extrinsic nerves. In this connection, it is noteworthy that atropine suppresses the MMC, which indicates the prominent part taken by the cholinergic intramural nerves in interdigestive motor phenomena. It would be very surprising if the activity of cholinergic intramural neurons during MMC were controlled only by hormonal factors and not at all by extrinsic fibers.

Indeed, MIOLAN and ROMAN (1978b) have shown that the spontaneous discharge of vagal preganglionic fibers fluctuated with the various phases of the MMC occurring on the gastric antrum: mean frequencies of about 0.1–1.5 spikes/s during phase I; 1.5–5 spikes/s in phase III; intermediate values for phases II and

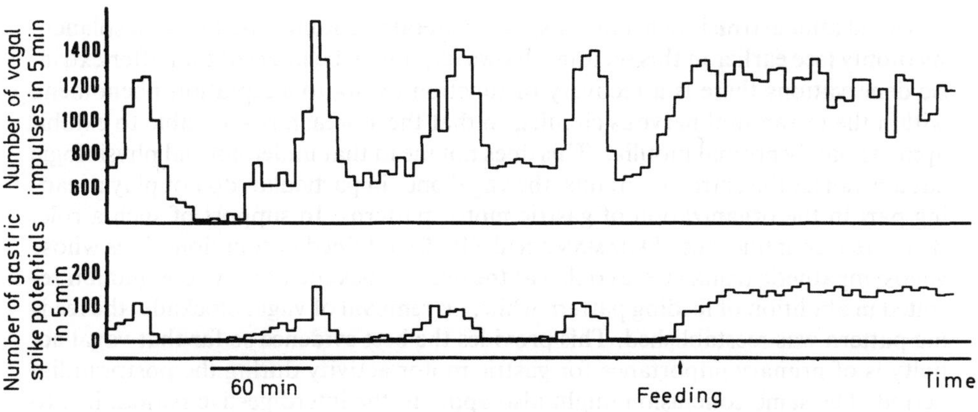

Fig. 17. Discharge pattern of vagal efferent fiber probably supplying the antrum (VEA type = vagal excitatory fiber of antrum) during fasting and feeding in conscious dog. Vagal impulses recorded by the nerve suture technique; gastric spike potentials recorded with electrodes chronically implanted within the antrum walls. In the fasted dog, there are cyclic variations of vagal discharge with parallel variations in gastric motility (MMC). Feeding increases both vagal discharge and gastric motilitiy. (MIOLAN and ROMAN 1978b)

IV (Fig. 17). These fibers were believed to supply the gastric antrum for the following reasons:

1) Discharge frequency very closely related to the spontaneous motility of the antrum.
2) Firing practically unaffected by the receptive relaxation of stomach.
3) Discharge facilitated by a gastric distension which increased antral motility and relaxed the proximal stomach.

Taking into account that their discharge increased each time the gastric motility was enhanced, one could conclude that these fibers belonged to the vagal excitatory pathway. Then, they will be termed VEA (vagal excitatory fibers of the antrum) in the rest of the text. More recent data (MIOLAN 1980) have revealed that vagal fibers controlling the proximal stomach (VEF and VIF types, see Sect. B.VI.1) were also concerned in MMC: the discharge of VEF fibers increased during phase III, while that of VIF fibers decreased at the same time. After feeding, the discharge pattern of VEF and VIF fibers is not known. As for the VEA fibers, their discharge frequency increased as much as during phase III of a MMC. This sustained activity, which started as soon as the food was offered, lasted several hours. If the food was only shown, but not given, the vagal hyperactivity was of short duration (only a few minutes).

Concerning the physiologic significance of the changes in vagal fiber activity, two interpretations may be proposed. First, these changes represent the extrinsic inputs which cause gastric motor events. Second, vagal firing changes do not program the gastric motor events but provide an additional mechanism which facilitates the functioning of the intramural network responsible for motor programming. In support of the second hypothesis and against the first, one might put forward that in conscious dogs, the MMC and the postprandial hypermotility are still observed after extrinsic denervations, i.e., bilateral vagotomy and bilateral splanchnicotomy (see earlier in this section). However, it may be inferred that after extrinsic denervations there is a recovery of function owing to adaptation phenomena within the intramural nervous circuits, so that the stomach is soon able to exhibit approximately normal motility. This does not mean that under normal physiologic circumstances the extrinsic circuits, the vagal ones in particular, do not play a leading part in the organization of gastric motor patterns. In support of such a role, there are recent findings of DIAMANT et al. (1980) obtained in conscious dogs whose vagosympathetic trunks were cooled at the neck. Blockade of nerve conduction resulted in abolition of feeding pattern while, on removal of vagal blockade, the feeding pattern was reestablished. This provides the best evidence so far that vagal activity is of primary importance for gastric motor activity during the postprandial period. The same conclusion might also apply to the interdigestive period, i.e., to MMC since the discharge of VEA fibers for instance is the same during phase III of MMC and after feeding. But additional data are obviously needed to answer this question. In particular, it would be very useful to know the effect of vagal cooling on MMC.

Another important question concerning the vagal firing changes during MMC or after feeding is that of their origin. One possibility is a reflex origin, i.e., changes resulting from stimulation of gastric stretch receptors which are known to be responsive to distension and also to contraction of the stomach (see Sect. B.II.2.a). This interpretation seems very plausible for changes observed after feeding. In this connection, DAVISON and GRUNDY (1978) have shown that in anesthetized rats, the discharge of single vagal efferent fibers was markedly altered (excitation or inhibition) by gastric inflation, gastric contraction, or compression of the stomach. According to these investigators, the opposite changes of vagal discharges would correspond to a reciprocal reflex control of antagonist excitatory and inhibitory vagal pathways. Thus, it is likely that during the postprandial period, vagovagal gastrogastric reflexes play a prominent role, in addition to local reflexes which can substitute for them if necessary (vagotomy). However, it is difficult to assume that during MMC the vagal firing changes also result from feedback phenomena (stimulation of gastric receptors by hypermotility), since atropine, which suppresses motor expression of MMC, does not affect the periodic changes of the vagal discharges (MIOLAN 1980).

Since the PAVLOV's demonstration of a cephalic phase of gastric secretion, it has been several times suggested that the sight, smell, and taste of food could stimulate the gastric motility. Some early observations are in keeping with this suggestion (for references, see THOMAS and BALWIN 1968). The finding of MIOLAN and ROMAN (1978b), that the discharge of VEA fibers is enhanced as soon as food is offered also strongly supports the hypothesis of a cephalic phase of gastric motility. However, other data (THOMAS and BALWIN 1968) have suggested otherwise by showing a decreased gastric motility at the sight or smell of food. These conflicting results may be ascribed to gastric receptive relaxation or to the cephalic phase of acid secretion and the subsequent reflex gastric inhibition triggered by arrival of acid in the upper intestine (see Sect. B.VI.3.a). It is possible that, depending on experimental conditions, the cephalic facilitation of gastric motility may be overridden by inhibitory influences.

3. Enterogastric Reflex and Gastric Emptying

Appropriate chemical or mechanical stimulations of the mucosa of the upper intestine produce an inhibition of gastric peristalsis and hence a slowing of gastric emptying. This has been called the "enterogastric inhibitory reflex" by THOMAS et al. (1934). Many stimuli have been found to initiate this reflex, e.g., mineral acids, fats and fatty acids, hypertonic solutions of various substances (salt, glucose, etc.), duodenal distension, and more recently duodenal thermal stimulation.

a) Response to Acids

Slowing of gastric emptying by mineral acids (in particular hydrochloric acid) has been known for a long time (see references in THOMAS 1957; HUNT and KNOX 1968). Extensive studies of the question have been done by HUNT and co-workers who have clarified many aspects of acid effectiveness, the influence of concentration, molecular wheight, pH. etc. (HUNT and KNOX 1968, 1969, 1972). Judging by the effects of restricted intestinal perfusions, acid receptors appear to be located in the duodenum (first 5 cm) and also in the jejunum (COOKE 1975; COOKE and CLARK 1976; COOKE 1977) on the other hand, electrophysiologic studies have indicated that vagal sensitive fibers could be activated by acid perfusion of the proximal intestine (PAINTAL 1957; ANDREWS and ANDREWS 1971; DAVISON 1972; PAINTAL 1973; CLARKE and DAVISON 1978). These vagal receptors are probably mucosal endings which function as rapidly adapting mechanoreceptors and also as slowly adapting chemoreceptors.

Concerning the reflex circuits involved in acid inhibition of gastric motility, investigators differ in opinion as to the importance of vagal pathways. According to THOMAS et al. (1934) and QUIGLEY and MESHAN (1938), the acid reflex in dogs was abolished, or at least greatly diminished, by vagotomy. This result has been fully confirmed by ROZE et al. (1977) in pigs. However, SCHAPIRO and WOODWARD (1959) reported that only celiac ganglionectomy was able to suppress acid-induced gastric inhibition in dogs, suggesting that the reflex circuit would consist of afferent fibers arising in the upper small intestine and synapsing with cells bodies of efferent adrenergic neurons lying in the celiac ganglion. Such connections have been demonstrated by recent electrophysiologic work (KREULEN and SZURZEWSKI 1979). Anyway, it is quite possible that several different reflex mechanisms act in association. The fact that the sympathetic circuit is involved in the acid reflex has been confirmed by COOKE and CLARK (1976). But, in addition, MIOLAN (1980) has recently observed that the discharge of the vagal efferent fibers controlling the dog stomach (recording made by the nerve suture technique previously mentioned) was modified when hydrochloric acid was introduced into the duodenum (Fig. 18). The activity of the excitatory fibers (VEA and VEF types, see Sect. B.VI.1, 2.b) was decreased while that of inhibitory fibers (VIF) was enhanced.

b) Response to Fat and Fatty Acids

Slowing of gastric emptying by fat and fatty acids is also well documented (see HUNT and KNOX 1968; COOKE 1975). The receptors responsible for such an effect are more sensitive to fatty acids than to triglycerides. They appear to be located mainly in the jejunum, at least in the dog (COOKE 1977); but there is no available

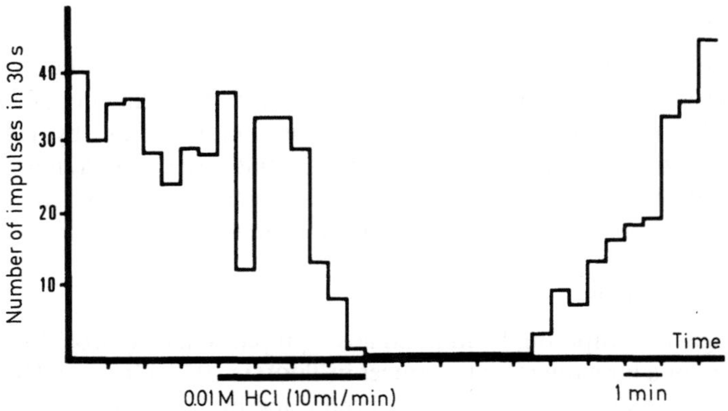

Fig. 18. Effect of acid stimulation of duodenum upon the discharge of a vagal efferent fiber probably supplying the antrum (VEA type). Conscious dog; recording of vagal activity made using the nerve suture technique. Duodenal infusion with hydrochloric acid (intraluminal catheter) produces a very pronounced and long-lasting inhibition of the vagal discharge. (MIOLAN 1980)

information about their electrophysiologic properties. Moreover, the mechanism of fat inhibition is still a controversial subject. Since the observations of FARRELL and IVY (1926), indicating that introduction of fats into the duodenum inhibited the motility of a transplanted fundic pouch, it is generally believed that a hormone, enterogastrone, is involved in the mediation of this inhibitory effect. However, according to HUNT and KNOX (1968), fat inhibition has a short latency and persists only for one or two minutes after withdrawal of the stimulus, which makes a hormonal mechanism hardly credible. In addition, it was been reported that vagotomy either reduced the slowing of gastric emptying by fast in humans (WADDEL and WANG 1953) or suppressed fat-induced gastric inhibition of dogs (KELLY and CODE 1969) and pigs (ROZE et al. 1977).

c) Response to Osmotic Factors

In order to explain that various hypertonic solutions (e.g., salts, sugars, amino-acids) can produce a slowing of gastric emptying, it has been postulated that osmoreceptors were present in intestinal walls (HUNT and KNOX 1968; COOKE 1975). Such receptors seem localized in the duodenum in humans (HUNT and KNOX 1968, MEEROFF et al. 1975) and the jejunum in dogs (COOKE 1977). Nevertheless, electrophysiologic evidence of their existence, i.e., of the existence of specific receptors responding to the osmotic pressure per se is still lacking. Vagal intestinal receptors described by CLARKE and DAVISON (1978) in rats were excited by hypertonic saline, but also by water, and in addition they did not respond to hypertonic glucose. In contrast, the vagal glucoreceptors reported by MEI (1978) responded only to glucose and various carbohydrates. Their discharge frequency increased when sugar concentration (and therefore osmolarity) was raised. But substances other than carbohydrates, e.g., NaCl, KCl of the same osmolarity, were ineffective, which indicated that here again the osmotic pressure per se was not directly related to the receptor activation. Anyway, it is quite possible that activation of such vagal glucoreceptors instead of "true" osmoreceptors may account for the slowing of gastric emptying by sugar solutions. In this connection, it is most significant that vagotomy in the pig abolishes the inhibition of gastric motility induced by intestinal infusion of glucose (ROZE et al. 1977).

d) Response to Intestinal Distension

Several studies have shown that duodenal distension (either by balloon inflation, or as a consequence of gastric evacuation) exerts an inhibitory influence on the mechanical activity of the stomach (QUIGLEY 1943; THOMAS et al. 1934; QUIGLEY and LOUCKES 1962). A similar effect can be observed on the electrical activity of the gastric musculature (DANIEL and WIEBE 1966). Electrophysiologic investigations have revealed the existence of many mechanoreceptors connected with vagal or splanchnic sensitive fibers. The vagal mechanoreceptors are either stretch receptors located within the muscular layers or mucosal endings which are responsive to both chemical and mechanical stimuli (IGGO 1957a; PAINTAL 1957; MEI 1970; DAVISON 1972; PAINTAL 1973; CLARKE and DAVISON 1978). As for splanchnic mechanoreceptors, the only ones described so far respond to distension or contraction of the intestine, thus resembling very much the vagal stretch receptors (RANIERI et al. 1973).

The inhibition of gastric motility in response to duodenal distension probably involves a reflex mechanism mediated by extrinsic nerves, since it is not affected by the transection of the gastroduodenal junction, while it is markedly altered by vagotomy or by pharmacologic blockade of the sympathetic system (DANIEL and WIEBE 1966). The existence of a vagal reflex is further supported by the finding of MIOLAN and ROMAN (1978b) that the discharge of vagal excitatory fibers supplying the dog antrum (VEA type) was inhibited by duodenal distension. On the other hand, it is well known that intestinal distension is able to elicit a sympathetic inhibition of gastric motility which is part of the intestinogastrointestinal sympathetic reflex (see Sect. B.VI.4; FURNESS and COSTA 1974). Contrary to what it is generally believed this reflex is not necessarily noxious in character since it can be evoked for moderate intestinal distensions which are probably within the physiologic range (JANSSON and MARTINSON 1966).

e) Response to Thermal Stimuli

Recently, EL OUAZZANI and MEI (1979) discovered vagal thermoreceptors located in the walls of cat antrum and duodenum. These receptors discharged with an optimum frequency when warm (46°–49 °C) or cold (12°–10 °C) solutions were infused into the gastroduodenal area. They did not respond to mechanical stimuli (compression and distension of the gut) nor to chemical ones (glucose or acid solutions). Thus, they must be considered as true thermoreceptors, responding specifically to temperature variations. By means of EMG recording, EL OUAZZANI and MEI (1979) have observed that cold and warm stimulation of the duodenum (sufficient to elicit the optimum discharge of thermoreceptors) induced an inhibition of antral activity. This inhibition persisted after bisplanchnicotomy, but disappeared after bivagotomy. Thus, it is likely that besides the other enterogastric reflexes, duodenal thermal stimulation also elicits a reflex inhibition of gastric motility, thereby providing an additional mechanism for the gastric emptying regulation.

4. Pain Reflexes

It is well known that distension or mucosal irritation of one part of the gastrointestinal tract can inhibit the movements of other parts, including the stomach (FURNESS and COSTA 1974). Such gastric inhibitions induced by distension of the intestine have been repeatedly reported (JANSSON and MARTINSON 1966; JANSSON and LISANDER 1969; JANSSON 1969a). The efferent link of the reflex involves sympathetic, i.e., adrenergic, fibers which act mainly on the intramural cholinergic ganglion cells. Generally speaking, the sympathetic inhibitory reflexes are considered as noxious because they are elicited by overdistension of the intestine. But it is now established that they are also induced by moderate distensions of the gut (JANSSON and MARTINSON 1966; JANSSON and LISANDER 1969). Therefore, they certainly play a part in the normal regulation of gastrointestinal flow rate. Besides, gastric motility can also be inhibited by noxious cutaneous stimuli (THOMAS and BALDWIN 1968; SATO et al. 1975) or by stimulation of somatic nerves (BABKIN and KITE 1950; JANSSON 1969b). Most of these inhibitory reflexes induced from skin or somatic nerves involve a sympathetic efferent pathway (SATO et al. 1975).

References

Abe S (1959) On the histology and the innervation of the esophagus and the first forestomach of the goat. Arch Histol Jpn 16:109–129
Abrahamsson H (1973) Studies on the inhibitory nervous control of gastric motility. Acta Physiol Scand [Suppl 390] 88:1–38
Aeberhard P, Bedi BS (1977) Effects of proximal gastric vagotomy (PGV) followed by total vagotomy (TV) on post prandial and fasting myoelectrical activity of the canine stomach and duodenum. Gut 18:515–523
Allen GL, Poole EW, Code CF (1964) Relationship between electrical activities of antrum and duodenum. Am J Physiol 207:906–910
Alvarez WC (1948) An introduction to gastroenterology, 4th edn. Hoeber, New York
Andrew BJ (1956) The nervous control of cervical oesophagus of the rat. J Physiol (Lond) 134:729–740
Andrews CJH, Andrews wHH (1971) Receptors activated by acid in the duodenal wall of rabbits. Q J Exp Physiol 56:221–230
Arimori M, Code CF, Schlegel JF, Sturm RE (1970) Electrical activity of the canine esophagus and gastroesophageal sphincter: its relation to intraluminal pressure and movement of material. Am J Dig Dis 15:191–208
Armitage AK, Dean ACB (1966) The effects of pressure and pharmacologically active substances on gastric peristalsis in a transmurally stimulated rat stomach-duodenum preparation. J Physiol (Lond) 182:42–56
Asoh R, Goyal RK (1978) Electrical activity of the opossum lower esophageal spincter in vivo. Gastroenterology 74:835–840
Atanassova E (1970) On the mechanism of correlation between the spike activities of the stomach and duodenum. Bull Inst Physiol Acad Sci Bulg 13:229–242
Atanassova E (1976) The role of the intrinsic nervous system in the correlation between the spike activites of the stomach and duodenum. In: Bülbring E, Shuba MF (eds) Physiology of smooth muscle. Raven, New York, pp 127–135
Atanassova ES, Vladimorava IA, Shuba MF (1972) Non adrenergic inhibitory, post-synaptic potentials of stomach smooth muscle cells (in Russian). Neirofiziologiia 4:216–222
Aune S (1969) Intragastric pressure after vagotomy in man. Scand J Gastroenterol 4:447–452
Babkin BP, Kite WC Jr (1950) Central and reflex regulation of motility of pyloric antrum. J Neurophysiol 13:321–334

Bartlet AL (1968) The effect of vagal stimulation and eserine on isolated guinea-pig oesophagus. Q J Exp Physiol 53:170–174
Baumgarten HG, Lange W (1969) Adrenergic innervation of the oesophagus in the cat (Felis domestica) and rhesus monkey. Z Zellforsch 95:529–545
Bayliss WB, Starling EH (1899) The movements and innervation of the small intestine. J Physiol (Lond) 24:99–143
Beani L, Bianchi C, Crema AM (1971) Vagal non-adrenergic inhibition of guinea-pig stomach. J Physiol (Lond) 217:259–279
Bedi BS, Code CF 81972) Pathway of coordination of postprandial antral and duodenal action potentials. Am J Physiol 222:1295–1298
Berger T, Ceder L, Hampelt A, Meurling S (1976) Effect of highly selective vagotomy on gastric emptying. Scand J Gastroenterol 11:829–832
Binder HJ, Bloom DL, Stern H, Solitaire GB, Thayer WR, Spiro HM (1968) The effect of cervical vagectomy on esophageal function in the monkey. Surgery 64:1075–1083
Bohr DF (1973) Vascular smooth muscle updated. Circ Res 32:665–678
Bülbring E, Gershon MD (1967) 5 Hydroxytryptamine participation in the vagal inhibitory innervation of the stomach. J Physiol (Lond) 192:823–846
Burgess JN, Schlegel JF, Ellis FH Jr (1972) Effect of denervation on feline esophageal function and morphology. J Surg Res 12:24–33
Burnstock G (1972) Purinergic nerves. Pharmacol Rev 24:509–581
Burnstock G (1975) Purinergic transmission. In: Iverson LL, Iverson SD, Snyder SH (eds) Handbook of psychopharmacology, vol 5. Plenum, London, pp 131–194
Burnstock G, Cocks T, Kasakov L, Wong HK (1978) Direct evidence for ATP release from non-adrenergic, non-cholinergic ("purinergic") nerves in the guinea-pig taenia coli and bladder. Eur J Pharmacol 49:145–149
Campbell G (1966) The inhibitory nerve fibres in the vagal supply to the guinea-pig stomach. J Physiol (Lond) 185:600–612
Campbell G, Burnstock G (1968) Comparative physiology of gastrointestinal motility. In: Code CI (ed) Alimentary canal. American Physiological Society, Washington, DC (Handbook of physiology, vol IV, sect 6, pp 2213–2226)
Cannon WB (1907) Esophageal peristalsis after bilateral vagotomy. Am J Physiol 19:436–444
Cannon WB, Lieb CW (1911) The receptive relaxation of the stomach. Am J Physiol 29:267–273
Car A, Roman C (1970a) Déglutitions et contractions oesophagiennes réflexes produites par la stimulation du bulbe rachidien. Exp Brain Res 11:75–92
Car A, Roman C (1970b) L'activité spontanée du sphincter oesophagien supérieur; ses variations au cours de la déglutition et de la rumination. J Physiol (Paris) 62:505–511
Carlson AJ, Boyd TE, Pearcy JF (1922) Studies on the visceral sensory nervous system. XII. The innervation of the cardia and the lower end of the esophagus in mammals. Am J Physiol 61:14–41
Carter DC, Whitfield HN, MacLeod IB (1972) The effect of vagotomy on gastric adaptation. Gut 13:874–879
Carveth SW, Schlegel JF, Code CF, Ellis FH (1962) Esophageal motility after vagotomy phrenicotomy, myotomy and myomectomy in dogs. Surg Gynecol Obstet 114:31–42
Castell DO (1975) The lower esophageal sphincter: physiologic and clinical aspects. Ann Intern Med 83:390–401
Christensen J (1970a) Patterns of some esophageal responses to stretch and electrical stimulation. Gastroenterology 59:909–916
Christensen J (1970b) Pharmacological identification of the lower esophageal sphincter. J Clin Invest 49:681–690
Christensen J (1971) Responses of the smooth muscle segment of the opossum esophagus to distension and electrical stimulation and their modification by antagonists. In: Demling L, Ottenjann R (eds) Gastrointestinal motility. Academic Press, New York London, pp 167–174
Christensen J (1975) Pharmacology of the esophageal motor function. Annu Rev Pharmacol Toxicol 15:243–258

Christensen J, Dons RF (1968) Regional variations in response of cat esophageal muscle to stimulation with drugs. J Pharmacol Exp Ther 161:55–58
Christensen J, Lund GF (1969) Esophageal responses to distension and electrical stimulation. J Clin Invest 48:408–419
Christensen J, Torres EI (1975) Three layers of the opossum stomach: responses to nerve stimulation. Gastroenterology 69:641–648
Clark GC, Vane JR (1961) The cardiac sphincter in the cat. Gut 2:252–262
Clarke GD, Davison JS (1978) Mucosal receptors in the gastric antrum and small intestine of the rat with afferent fibres in the cervical vagus. J Physiol (Lond) 284:55–67
Code CF, Marlett JA (1975) The interdigestive myoelectric complex of the stomach and small bowel of dogs. J Physiol (Lond) 246:289–309
Code CF, Schlegel JF (1968) Motor action of the esophagus and its sphincters. In: Code CF (ed) Alimentary canal. American Physiological Society, Washington DC (Handbook of physiology, vol IV, sect 6, pp 1821–1839)
Cohen S, Ryan J, Matarazzo S, Snape WJ Jr (1977) Nervous control of esophageal motor activity. In: Brooks FP, Evers PW (eds) Nerves and the gut. Slack, Thorofare, pp 207–222
Cook MA, Kowalewski K, Daniel EE (1974) Electrical and mechanical activity recorded from the isolated perfused canine stomach. The effects of some G1 polypeptides. In: Daniel EE (ed) Proceedings of the 4th International Symposium on Gastrointestinal Motility. Mitchell, Vancouver, pp 233–242
Cooke AR (1975) Control of gastric emptying and motility. Gastroenterology 68:804–816
Cooke A (1977) Localization of receptors inhibiting gastric emptying in the gut. Gastroenterology 72:875–880
Cooke A, Clark E (1976) Effect of first part of duodenum on gastric emptying in dogs: response to acid, fat, glucose and neural blockade. Gastroenterology 70:550–555
Creamer B, Schlegel J (1957) Motor responses of esophagus to distension. J Appl Physiol 10:498–504
Daniel EE (1977) Nerves and motor activity of the gut. In: Brooks FP, Evers PW (eds) Nerves and the gut. Slack, Thorofare, pp 154–196
Daniel EE, Irwin J (1968) Electrical activity of gastric musculature. In: Code CF (ed) Alimentary canal. American Physiological Society, Washington, DC (Handbook of physiology, vol IV, sect 6, pp 1969–1984)
Daniel EE, Irwin J (1971) Electrical activity of the stomach and upper intestine. Am J Dig Dis 16:602–610
Daniel EE, Sarna SK (1976) Distribution of excitatory vagal fibers in canine gastric wall to control motility. Gastroenterology 71:608–613
Daniel EE, Taylor GS, Holman ME (1976) The myogenic basis of active tension in the lower esophageal sphincter. Gastroenterology 70:874 A
Daniel EE, Wiebe GE (1966) Transmission of reflexes arising on both sides of the gastroduodenal junction. Am J Physiol 211:634–642
Davison JS (1972) Response of single vagal afferent fibres to mechanical and chemical stimulation of the gastric and duodenal mucosa in cats. J Exp Physiol 57:405–416
Davison JS, Grundy D (1978) Modulation of single vagal efferent fibre discharge by gastrointestinal afferents in the rat. J Physiol (Lond) 284:69–82
De Carle DJ, Christensen J (1976) A dopamine receptor in esophageal smooth muscle of the opossum. Gastroenterology 70:216–219
De Carle DJ, Brody MJ, Christensen J (1976) Histamine receptors in esophageal smooth muscle of the opossum. Gastroenterology 70:1071–1075
De Carle DJ, Tampleman DL, Christensen J (1977) The cat esophagus: responses of the circular layer of smooth muscle from the body to electrical field stimulation. In: Duthie HL (ed) Proceedings of the VIth International Symposium on Gastrointestinal Motility. MTP Press, Lancaster, pp 513–521
Diamant NE (1974) Electrical activity of the cat smooth muscle esophagus: a study of hyperpolarizing responses. In: Daniel EE (ed) Proceedings of the 4th International Symposium on Gastrointestinal Motility. Mitchell, Vancouver, pp 593–605
Diamant NE, Akin A (1972) Effect of gastric contraction of the lower esophageal sphincter. Gastroenterology 63 (1):38–44

Diamant NE, El Sharkawy TY (1975) The "on response" contraction in esophageal smooth muscle. Clin Res 23:620 A
Diamant NE, El Sharkawy TY (1977) Neural control of the esophageal peristalsis, a conceptual analysis. Gastroenterology 72:546–556
Diamant NE, Hall K, Mui H, El Sharkawy TY (1980) Vagal control of the feeding motor pattern in the lower esophageal sphincter, stomach and small intestine of dog. In: Christensen J (ed) Proceedings of the VII th International Symposium on Gastrointestinal Motility. Raven Press, New York
Di Marino AJ, Cohen S (1974) The adrenergic control of lower esophageal function. In: Daniel EE (ed) Proceedings of the 4 th International Symposium on Gastrointestinal Motility. Mitchell, Vancouver, pp 623–630
Dodds WJ, Christensen J, Wood JD, Arndorfer RC (1977) Effect of pharmacologic agents on primary esophageal peristalsis in the opossum (Abstr). Gastroenterology 72:1050
Dodds WJ, Stef JJ, Stewart ET, Hogan WJ, Arndorfer RC, Cohen EB (1978) Responses of feline esophagus to cervical vagal stimulation. Am J Physiol 235:E63–E73
Doty RW (1968) Neural organization of deglutition. In: Code CF (ed) Alimentary canal. American Physiological Society, Washington, DC (Handbook of physiology, vol IV, sect 6, pp 1861–1902)
Doty RW, Richmond WH, Storey AT (1967) Effect of medullary lesions on coordination of deglutition. Exp Neurol 17:91–106
Dougherty RW, Habel RE, Bond HE (1958) Esophageal innervation and the eructation reflex in sheep. Am J Vet Res 19:115–128
Ellis FG, Kauntze R, Trounce JR (1960) The innervation of the cardia and lower esophagus in man. Br J Surg 47:466–472
El Ouazzani T, Mei N (1979) Mise en évidence électrophysiologique des thermorécepteurs vagaux dans la région gastro-intestinale. Leur rôle dans la régulation de la motricité digestive. Exp Brain Res 34:419–434
El Sharkawy TY, Diamant NE (1976) Contraction patterns of esophageal circular smooth muscle induced by cholinergic excitation. Gastroenterology 70:969 A
El Sharkawy TH, Szurzewsky JH (1978) Modulation of canine antral circular smooth muscle by acetylcholine, noradrenaline and pentagrastrine. J Physiol (Lond) 279:309–320
El Sharkawy TY, Chan WWL, Diamant NE (1975) Neural mechanism of lower esophageal sphincter relaxation: a pharmacological analysis. In: Vantrappen G (ed) Proceedings of the V th International Symposium on Gastrointestinal Motility. Typoff, Herentals, pp 176–180
El Sharkawy TY, Morgan KG, Szurzewski JH (1978) Intracellular electrical activity of canine and human gastric smooth muscle. J Physiol (Lond) 279:291–307
Fahrenkrug J, Galbo H, Holst JJ, Schaffalitzky de Muckadell OB (1978 a) Influence of the autonomic nervous system on the release of vasoactive intestinal polypeptide from the porcine gastrointestinal tract. J Physiol (Lond) 280:405–422
Fahrenkrug J, Haglund U, Jodal M, Lundgren O, Olbe L, Shaffaltitzky de Muckadell OB (1978 b) Nervous release of vasoactive intestinal polypeptide in the gastrointestinal tract of cats: possible physiological implications. J Physiol (Lond) 284:291–305
Farrell JI, Ivy AC (1926) Studies on the motility of the transplanted gastric pouch. Am J Physiol 76:227–228
Fisher RS, Malmud LS, Roberts GS, Lobis IF (1977) The lower esophageal sphincter as a barrier to gastroesophageal reflex. Gastroenterology 72:19–22
Fleshler B, Hendrix IR, Kramer P, Ingelfinger FJ (1959) The characteristics and similarity of primary and secondary peristalsis in the esophagus. J Clin Invest 38:110–116
Floyd K (1973) Cholinesterase activity in sheep oesophageal muscle. J Anat 116:357–373
Furness JB, Costa M (1974) The adrenergic innervation of the gastrointestinal tract. Ergeb Physiol Biol Chem Exp Pharmacol 69:1–51
Gabella G (1979) Innervation of the gastrointestinal tract. Int Rev Cytol 59:129–193
Gershon MD (1967) Inhibition of gastrointestinal movement by sympathetic nerve stimulation: the site of action. J Physiol (Lond). 189:317–327
Gershon MD, Dreyfus CF (1977) Serotonergic neurons in the mammalian gut: In: Brooks FP, Evers PW (eds) Nerves and the gut. Slack, Thorofare, pp 197–206

Getz B, Sirnes T (1949) The localization within the dorsal motor vagal nucleus. J Comp Neurol 90:95–110

Gillespie JS, Khoyi MA (1977) The site and receptors responsible for the inhibition by sympathetic nerves of intestinal smooth muscle and its parasympathetic motor nerve. J Physiol (Lond). 267:767–789

Gonella J, Niel JP, Roman C (1977) Vagal control of lower oesophageal sphincter motility in the cat. J Physiol (Lond) 273:647–664

Gonella J, Niel JP, Roman C (1979) Sympathetic control of lower oesophageal sphincter motility in the cat. J Physiol (Lond) 287:177–190

Gonella J, Niel JP, Roman C (1980) Mechanism of the noradrenergic motor control on the lower oesophageal sphincter in the cat. J Physiol (Lond) 306:251–260

Goyal RK, Rattan S (1975) Nature of the vagal inhibitory innervation to the lower esophageal sphincter. J Clin Invest 55:1119–1126

Goyal RK, Rattan S (1978) Neurohumoral, hormonal, and drug receptors for the lower esophageal sphincter. Gastroenterology 74:598–619

Greenwood RK, Schlegel JF, Code CF, Ellis FH Jr (1962) The effect of sympathectomy, vagotomy and esophageal interruption on the canine gastroesophageal sphincter. Thorax 17:310–319

Gruber H (1968) Über Struktur und Innervation der quergestreiften Muskulatur des Oesophagus der Ratte. Z Zellforsch 91:236–247

Guimarães S (1969) Alpha excitatory, alpha inhibitory and beta inhibitory adrenergic receptors in the guinea-pig stomach. Arch Int Pharmacody Ther 179:188–201

Haffner JFW (1973) Pressure responses to cholinergic and adrenergic agents in the fundus, corpus and antrum of isolated rat and guinea pig stomachs. Acta Chir Scand 139:650–655

Haffner JFW, Stadaas J (1972) Pressure responses to cholinergic and adrenergic agents in the fundus, corpus and antrum of the isolated rabbit stomachs. Acta Chir Scand 138:713–719

Heazell MA (1977) A non-adrenergic inhibitory innervation in the rat stomach. Arch Int Phamacodyn Ther 226:109–117

Hellemans J, Vantrappen G (1967) Electromyographic studies on canine esophageal motility. Am J Dig Dis 12:1240–1255

Hellemans J, Vantrappen G (1974) Physiology. In: Schwiegk H (ed) Diseases of the esophagus. Springer, Berlin Heidelberg New York (Handbuch der inneren Medizin, vol III/1, pp 40–102)

Hellemans J, Vantrappen G, Valembois P, Janssens J, Vandenbroncke J (1968) Electrical activity of striated and smooth muscle of the esophagus: Am J Dig Dis 13:320–334

Hellemans J, Vantrappen G, Janssens J (1974) Electromyography of the esophagus. In: Schwiegk H (ed) Diseases of the esophagus. Springer, Berlin Heidelberg New York (Handbuch der inneren Medizin, vol III/I, pp 270–285)

Higgs B, Ellis FH (1965) The effect of bilateral supranodosal vagotomy on canine esophageal function. Surgery 58:828–834

Higgs B, Kerr FWL, Ellis FH (1965) The experimental production of esophageal achalasia by electrolytic lesions in the medulla. J Thor Cardiovasc Surg 50:613–625

Hirst GDS, Holman ME, McKirky HC (1975) Two descending nerve pathways activated by distension of guinea pig small intestine. J Physiol (Lond) 244:113–127

Hunt JN, Knox HT (1968) Regulation of gastric emptying. In: Code CF (ed) Alimentary canal. American Physiological Society, Washington, DC (Handbook of physiology, vol IV, sect 6, pp 1917–1935

Hunt JN, Knox JT (1969) The slowing of gastric emptying by nine acids. J Physiol (Lond) 201:161–179

Hunt JN, Knox HT (1972) The slowing of gastric emptying by four strong acids and three weak acids. J Physiol (Lond) 222:187–208

Hwang K (1954) Mechanism of transportation of the content of the esophagus. J Appl Physiol 6:781–796

Hwang K, Grossman MI (1953) A note on the innervation of the cervical portion of the human esophagus. Gastroenterology 25:375–377

Hwang K, Essex H, Mann FC (1947) A study of certain problems resulting from vagotomy in dogs with special reference to emesis. Am J Physiol 149:429–448

Hwang K, Grossman MI, Ivy AC (1948) Nervous control of cervical esophagus. Am J Physiol 154:343–357

Iggo A (1955) Tension receptors in the stomach and urinary bladder. J Physiol (Lond) 128:593–607

Iggo A (1957a) Gastrointestinal tension receptors with unmyelinated afferent fibres in the vagus of the cat. Q J Exp Physiol 42:130–143

Iggo A (1957b) Gastric mucosal chemoreceptors with vagal afferent fibres in the cat. Q J Exp Physiol 42:398–409

Ingelfinger FJ (1958) Esophageal motility. Physiol Rev 38:533–584

Interone CV, Del Finado JE, Miller B, Bombeck CT, Nyhus LM (1971) Parietal cell vagotomy: studies of gastric emptying and observations of protection from histamine induced ulcer. Arch Surg 102:43–44

Ito Y, Kuriyama H (1975) Responses to field stimulation of the smooth muscle cell membrane of the guinea pig stomach. Jpn J Physiol 25:333–344

Itoh Z, Honda R, Hiwatashi K, Takeuchi S, Aizawa R, Takayanagi R, Couch EF (1976) Motilin induced mechanical activity in the canine alimentary tract. Scand J Gastroenterol, [Suppl 39] 11:93–110

Itoh Z, Takayanagi R, Takeuchi S, Isshiki S (1978) Interdigestive motor activity of Heindenhain pouches in relation to main stomach in conscious dog. Am J Physiol 234:E333–E338

Jabonero V (1958) Mikroskopische Studien über die Innervation des Verdauungstraktes. Teil 1 Oesophagus. Acta Neuroveget 17:308–353

Jabonero V (1962) Nuevas observaciones sobre la fina inervacion del esofago. Trab Inst Cajal Invest Biol 54:37–92

Jacobowitz D, Nemir P (1969) The autonomic innervation of the esophagus of the dog. J Cardiovasc Surg 58:678–684

Jahnberg T (1977) Gastric adaptative relaxation. Effects of vagal activation and vagotomy. An experimental study in dogs and in man. Scand J Gastroenterol [Suppl 46] 12:5–32

Janssens J (1978) The peristaltic mechanism of the esophagus. Acco, Leuven

Janssens J, Valembois P, Hellemans J, Pelemans W, Vantrappen G (1974) Studies on the necessity of a bolus for the progression of secondary peristalsis in the canine esophagus. Gastroenterology 67:245–251

Janssens J, De Wever I, Vantrappen G, Hellemans J (1976) Peristalsis in smooth muscle esophagus after transection and bolus deviation. Gastroenterology 71:1004–1009

Jansson G (1969a) Extrinsic nervous control of gastric motility. An experimental study in the cat. Acta Physiol Scand [Suppl] 326:5–42

Jansson G (1969b) Effect of reflexes of somatic afferents on the adrenergic outflow to the stomach in the cat. Acta Physiol Scand 77:17–22

Jansson G, Lisander B (1968) On adrenergic influence on gastric motility in chronically vagotomized cats. Acta Physiol Scand 76:463–471

Jansson G, Martinson J (1966) Studies on the ganglionic site of action of sympathetic outflow to the stomach. Acta Physiol Scand 68:184–192

Jean A (1972a) Localisation et activité des neurones déglutiteurs bulbaires. J Physiol (Paris) 64:227–268

Jean A (1972b) Effet de lésions localisées du bulbe rachidien sur le stade oesophagien de la déglutition. J Physiol (Paris) 64:507–516

Jean A (1978a) Contrôle bulbaire de la déglutition et de la motricité oesophagienne. PhD Thesis, University of Marseille

Jean A (1978b) Localisation et activité des motoneurones oesophagiens chez le mouton. J Physiol (Paris) 74:737–742

Jennewein HM, Hummelt H, Meyer U, Siewert R, Koch A, Waldeck F (1976) The effect of vagotomy on the resting pressure and reactivity of the L.E.S. in man and dog. In: Vantrappen G (ed) Proceedings of the 5th International Symposium on Gastrointestinal Motility. Typoff, Herentals, pp 186–189

Johansson B (1971) Electromechanical and mechanoelectrical coupling in vascular smooth muscle. Angiologica 8:129–143

Jule Y (1975) Modifications de l'activité électrique du côlon proximal de lapin in vivo par stimulation des nerfs vagues et splanchniques. J Physiol (Paris) 70:5–26
Jurica EJ (1926) Motility of denervated mammalian esophagus. Am J Physiol 77:371–384
Kantrowitz PA, Siegel CI, Hendrix TR (1966) Differences in motility of the upper and lower esophagus in man and its alteration by atropine. Bull Johns Hopkins Hosp 118:476–491
Kelly KA (1974) Canine gastric motility and emptying: electric, neural and hormonal controls. In: Daniel EE (ed) Proceedings of the 4th International Symposium on Gastrointestinal Motility. Mitchell, Vancouver, pp 463–470
Kelly KA, Code CF (1969) Effect of transthoracic vagotomy on canine gastric electrical activity. Gastroenterology 57 (1):51–58
Knight GC (1934) Relation of the extrinsic nerves to the functional activity of the esophagus. Br J Surg 22:155–168
Koster N, Madsen P (1970) The intragastric pressure before and immediately after truncal vagotomy. Scand J Gastroenterol 5:381–383
Kosterlitz HW (1968) Intrinsic and extrinsic nervous control of motility of the stomach and the intestines. In: Code CF (ed) Alimentary canal. American Physiological Society. Washington, DC (Handbook of physiology, vol IV, sect 6, pp 2147–2171)
Kowalewski K, Zajac S (1975) Electrical and mechanical activity of the isolated canine stomach perfused with homologous in vitro oxygenated blood. Pharmacology 13:448–457
Kowalewski K, Zajac S, Kolodef A (1975) The effect of drugs on the electrical and mechanical activity of the isolated porcine stomach. Pharmacology 13:86–95
Kreulen DL, Szurszewski JH (1979) Nerve pathways in celiac plexus of the guinea pig. Am J Physiol 237:E90–E97
Kuntz A (1947) The autonomic nervous system. Lea & Febiger, Philadelphia
Langley JN (1922) Connexions of the enteric nerve cells. J Physiol (Lond) 56:393
Laplace JP, Roman C (1979) Activité de la musculeuse gastrointestinale et mouvements des contenus digestifs. Ann Biol Anim Biochim Biophys 19:841–879
Lawn AM (1974) The localization, by means of electrical stimulation, of the origin and path in the medulla oblongata of the motor nerve fibres of the rabbit oesophagus. J Physiol (Lond) 174:232–244
Liedberg G, Nielsen KG, Owman C, Sjoberg NO (1973) Adrenergic contribution to the abdominal vagus nerves in the cat. Scand J Gastroenterol 8:177–180
Lind JF, Duthie HC, Schlegel JF, Code CF (1961) Motility of the gastric fundus. Am J Physiol 201:197–202
Lind JF, Crispin JS, McIver DK (1968) The effect of atropine on the gastroesophageal sphincter. Can J Physiol 46:233–238
Lund GF, Christensen J (1969) Electrical stimulation of esophageal smooth muscle and effects on antagonists. Am J Physiol 217:1369–1374
Lundberg J, Ahlman H, Dahlström A, Kewenter J (1976) Catecholamine-containing nerve fibres in the human abdominal vagus. Gastroenterology 70:472–474
McSwiney BA (1931) Innervation of the stomach. Physiol Rev 11:478–514
Mann CV, Shorter RG (1964) Structure of the canine esophagus and its sphincters. J Surg Res 4:160–164
Mann CV, Code CF, Schlegel JF, Ellis FH Jr (1968) Intrinsic mechanisms controlling the mammalian gastroesophageal sphincter deprived of extrinsic nerve supply. Thorax 23 (6):634–639
Marik F, Code CF (1975) Control of the interdigestive myoelectrical activity in dogs by the vagus nerves and pentagastrin. Gastroenterology 69:387–395
Marinesco G, Parhon C (1907) Recherches sur les noyaux d'origine du nerf pneumogastrique et sur les localisations dans les noyaux. J Neurol 13:61–67
Martinson J (1964) The effect of graded stimulation of efferent vagal nerve fibres on gastric motility. Acta Physiol Scand 62:256–262
Martinson J (1965) Studies on the efferent vagal control of the stomach. Acta Physiol Scand [Suppl 225] 65:1–23
Martinson J, Muren A (1963) Excitatory and inhibitory effects of vagus stimulation on gastric motility in the cat. Acta Physiol Scand 57:309–316

Meek WJ, Herrin RC (1934) The effect of vagotomy on gastric emptying time. Am J Physiol 109:221–231

Meeroff JC, Go VLW, Phillips SF (1975) Control of gastric emptying by osmolality of duodenal contents in man. Gastroenterology 68:1144–1151

Mei N (1970) Mécanorérecepteurs digestifs chez le chat. Exp Brain Res 11:502–514

Mei N (1978) Vagal glucoreceptors in the small intestine of the cat. J Physiol (Lond) 282:485–506

Meltzers SY (1899) On the causes of the orderly progress of the peristaltic movements in the esophagus. Am J Physiol 2 (3):266–272

Meltzer SY (1907) Secondary peristalsis of the oesophagus. A demonstration on a dog with a permanent fistula. Proc Soc Exp Biol Med 4:35–37

Miller FR, Sherrington CS (1916) Some observations on the buccopharyngeal stage of reflex deglutition in the cat. Q J Exp Physiol 9:147–186

Miolan JP (1980) La motricité de l'estomac et du sphincter oesophagien inférieur: étude électromyographique; rôle de l'innervation extrinsèque. PhD thesis, University of Marseille

Miolan JP, Roman C (1971) Modification de l'électromyogramme gastrique du chien par stimulation des nerfs extrinsèques. J Physiol (Paris) 63:561–576

Miolan JP, Roman C (1973) Décharge des fibres vagales efférentes destinées au cardia du chien. J Physiol (Paris) 66:171–198

Miolan JP, Roman C (1974) Décharge unitaire des fibres vagales efférentes lors de la relaxation réceptive de l'estomac du chien. J Physiol (Paris) 68:693–704

Miolan JP, Roman C (1978 a) Activité des fibres vagales efférentes destinées à la musculature lisse du cardia du chien. J Physiol (Paris) 74:709–723

Miolan JP, Roman C (1978 b) Discharge of efferent vagal fibers supplying the gastric antrum: indirect study by nerve suture technique. Am J Physiol 235:E366–E373

Mitchell GAG, Warwick R (1955) The dorsal vagal nucleus. Acta Anat 25:371–395

Mohiuddin A (1953) Vagal preganglionic fibres to the alimentary canal. J Comp Neurol 99:289–318

Monges H, Salducci J, Roman C (1968) Etude EMG de la contraction oesophagienne chez l'homme normal. Arch Fr Mal Appar Dig 57:545–560

Morgan KG, Schmalz PF, Szurszewski JM (1978) The inhibitory effects of vasoactive intestinal polypeptide on the mechanical and electrical activity of canine antral smooth muscle. J Physiol (Lond) 282:437–450

Mosso A (1876) Über die Bewegungen der Speiseröhre. Untersuch Z Natur 11:327–349

Mukhopadhyay AK, Weisbrodt NW (1975) Neural organization of esophageal peristalsis: role of vagus nerve. Gastroenterology 68:444–447

Mukhopadhyay AK, Weisbrodt NW (1977) Effect of dopamine on esophageal motor function. Am J Physiol 232:E19–E24

Muren A (1957) Influence of the vagal innervation on gastric motor responses to adrenaline and noradrenaline. Acta Physiol Scand 39:195–202

Muryobayashi T, Mori J, Fujiwara M, Shimamoto K (1968) Fluorescence histochemical demonstration of adrenergic nerve fibres in the vagus nerve of cats and dogs. Jpn J Pharmacol 18:285–293

Nakazato Y, Saito K, Ohga A (1970) Gastric motor and inhibitor response to stimulation of the sympathetic nerve in the dog. Jpn J Pharmacol 20:131–141

Neya T, Watanabe K, Yamasoto T (1974) Localization of potentials in medullary reticular formation relevant to swallowing. Rend Gastroenterol 6:107–110

Niel JP, Gonella J, Roman C (1980) Localisation par la technique de marquage à la peroxydase des corps cellulaires des neurones ortho et parasympatiques innervant le sphincter oesophagien inférieur du chat. J Physiol (Paris) 76:591–599

Ohga A, Nakazato Y, Saito K (1970) Considerations on the efferent nervous mechanism of the vago-vagal reflex relaxation of the stomach in the dog. Jpn J Pharmacol 20:116–130

Ohkawa H (1976) Evidence for alpha excitatory action of catecholamines on the electrical activity of the guinea pig stomach. Jpn J Physiol 26:41–52

Paintal AS (1954) A study of gastric stretch receptors, their role in the peripheral mechanism of satiation of hunger and thirst. J Physiol (Lond) 126:255–270

Paintal AS (1957) Responses from mucosal mecanoreceptors in the small intestine of the cat. J Physiol (Lond) 139:353–368
Paintal AS (1973) Vagal sensory receptors and their reflex effects. Physiol Rev 53:159–227
Papazova M, Atanassova E (1972) Changes in the bioelectric activity of the stomach after bilateral transthoracal vagotomy. Bull Inst Physiol 14:121–133
Paton WDM, Vane JR (1963) An analysis of the responses of the isolated stomach to electrical stimulation and to drugs. J Physiol (Lond) 165:10–46
Pedersen SA, Nielsen PA, Sorensen HR (1971) The effect of atropine and hexamethonium in combination on the lower esophageal sphincter. Scand J Gastroenterol [Suppl] 9:43–47
Peeters TL, Vantrappen G, Janssens J (1980) Fluctuations of motilin and gastrin levels in relation to the interdigestive motility complex in man. In: Christensen J (ed) Proceedings of the 7th International Symposium on Gastrointestinal Motility. Raven, New York
Quigley JP (1943) A modern explanation of the gastric emptying mechanism. Am J Dig Dis 10:418–421
Quigley JP, Louckes HS (1951) The effects of complete vagotomy on the pyloric sphincter and the gastric evacuation mechanism. Gastroenterology 19:533–537
Quigley JP, Louckes HS (1962) Gastric emptying. Am J Dig Dis 7:672–676
Quigley JP, Meschan I (1938) The role of the vagus in the regulation of the pyloric sphincter and adjacent portions of the gut with special reference to the process of gastric evacuation. Am J Physiol 123:166
Ranieri F, Mei N, Crousillat J (1973) Les afférences splanchniques provenant des mécanorécepteurs gastrointestinaux et péritonéaux. Exp Brain Res 16:276–290
Rattan S, Goyal RK (1974) Neural control of the lower esophageal sphincter. J Clin Invest 54:899–906
Rattan S, Goyal RK (1976) Effect of dopamine on esophageal smooth muscle in vivo. Gastroenterology 70:377–381
Rattan S, Goyal RK (1977) Effects of 5-hydroxytryptamine on the lower esophageal sphincter in vivo. J Clin Invest 59:125–133
Rattan S, Goyal RK (1978) Evidence for 5-HT participation in vagal inhibitory pathways to opossum lower esophageal sphincter. Am J Physiol 234:E273–E276
Roman C (1966) Contrôle nerveux du péristaltisme oesophagien. J Physiol (Paris) 58:79–108
Roman C (1967) La commande de la motricité oesophagienne et sa régulation. PhD thesis, University of Marseille
Roman C, Car A (1967) Contractions oesophagiennes produites par la stimulation du vague ou du bulbe rachidien. J Physiol (Paris) 59:377–398
Roman C, Car A (1970) Déglutitions et contractions oesophagiennes réflexes obtenues par la stimulation des nerfs vague et laryngé supérieur. Exp Brain Res 11::48–74
Roman C, Tieffenbach L (1971) Motricité de l'oesophage à musculeuse lisse après bivagotomie: étude électromyographique (EMG). J Physiol (Paris) 63:733–762
Roman C, Tieffenbach L (1972) Enregistrement de l'activité unitaire des fibres motrices vagales destinées à l'oesophage du babouin. J Physiol (Paris) 64:479–506
Roman C, Gonella J, Niel JP, Condamin M, Miolan JP (1975) Effets de la stimulation vagale et de l'adrénaline sur la musculeuse lisse du bas oesophage de chat. INSERM Colloq, vol 50, pp 415–422
Roze C, Couturier D, Chariot J, Debray C (1977) Inhibition of gastric electrical and mechanical activity by intraduodenal agents in pigs and effects of vagotomy. Digestion 15:526–539
Ruckebusch Y, Bueno L (1975) Electrical activity of the ovine jejunum and changes due to disturbances. Am J Dig Dis 20:1027–1035
Ruckebusch Y, Bueno L (1977) Migrating myoelectric complex of the small intestine. An intrinsic activity mediated by the vagus. Gastroenterology 73:1309–1314
Ryan JP, Snape WJ, Cohen S (1977) Influence of vagal cooling on esophageal function. Am J Physiol 232:E159–E164
Samarasinghe DD (1972) Some observations on the innervation of the striated muscle in the mouse oesophagus. An electron microscope study. J Anat 112:173–184

Sarna SK, Daniel EE (1975) Vagal control of gastric electrical control activity and motility. Gastroenterology 68:301-308
Sato A, Sato Y, Shimada F, Torigata Y (1975) Changes in gastric motility produced by nociceptive stimulation of the skin in rats. Brain Res 87:151–159
Schapiro H, Woodward ER (1959) Pathway of the enterogastric reflex. Proc Soc Exp Biol Med 101:407-409
Schulze K, Conklin JL, Christensen J (1977) A potassium gradient in smooth muscle segment of the opossum esophagus. Am J Physiol 232:E270–E273
Semba T, Hiraoka T (1957) Motor response of the stomach and small intestine caused by stimulation of the peripheral end of the splanchnic nerve, thoracic sympathetic trunk and spinal roots. Jpn J Physiol 7:64–71
Semba T, Mizonishi T (1978) Atropine resistant excitation and colon induced by stimulation of the extrinsic nerves and their centers. Jpn J Physiol 28:239–248
Seno N, Nakazato Y, Ohga A (1978) Presynaptic inhibitory effects of catecholamines on cholinergic transmission in the smooth muscle of the chick stomach. Eur J Pharmacol 51:229–237
Siegel CI, Hendrix TR (1961) Evidence for the central mediation of secondary peristalsis in the esophagus. Bull Johns Hopkins Hosp 108:297–307
Skinner DB, Camp TF (1968) Relation of esophageal reflux to lower esophageal pressures decreased by atropine. Gastroenterology 54:543–551
Spassova J (1959) Über die afferent Innervation der Speiseröhre des Menschen. Anat Forsch Dtsch 65:327
Stadaas J, Aune S (1970) Intragastric pressure/volume relationship before and after vagotomy. Acta Chir Scand 136:611–615
Sumi T (1964) Neuronal mechanisms in swallowing. Arch Ges Physiol 278:467–477
Szurszewski JH (1969) A migrating electric complex of the canine small intestine. Am J Physiol 217:1757–1763
Szurszewski JH (1975) Mechanism of action of pentagastrin and acetylcholine on the longitudinal muscle of the canine antrum. J Physiol (Lond) 252:335–361
Szurszewski JH (1977) Modulation of smooth muscle by nervous activity: a review and hypothesis. Fed Proc 36:2456–2461
Takeshima T (1971) Functional classification of the vagal afferent discharges in the dog's stomach. Jpn J Smooth Muscle Res 7:19–27
Thomas JE (1957) Mechanics and regulations of gastric emptying. Physiol Rev 37:453–474
Thomas JE, Baldwin MV (1968) Pathways and mechanisms of regulation of gastric motility. In: Code CF (ed) Alimentary canal. American Physiological Society, Washington, DC (Handbook of physiology, vol IV, sect 6, pp 1937–1968)
Thomas JE, Crider JO (1935) Rhythmic changes in duodenal motility associated with gastric peristalsis. Am J Physiol 111:124–129
Thomas PA, Earlam RJ (1974) the effect of the gastrointestinal polypeptide hormones on the electrical activity and pressure of the isolated perfused canine gastro-oesophagal sphincter. In: Daniel EE (ed) Proceedings of the 4th International Symposium on Gastrointestinal Motility. Mitchell, Vancouver, pp 243–250
Thomas JE, Crider JO, Morgan CJ (1934) A study of reflexes involving the pyloric sphincter and antrum and their role in gastric evacuation. Am J Physiol 108:683–700
Thomas PA, Kelly KA, Go VLW (1980) Hormonal regulation of gastrointestinal interdigestive motor cycles: In: Christensen J (ed) Proceedings of the 7th International Symposium on Gastrointestinal Motility. Raven, New York
Tieffenbach L, Roman C (1972) Rôle de l'innervation extrinsèque vagale dans la motricité de l'oesophage à musculeuse lisse: étude électromyographique chez le chat et le babouin. J Physiol (Paris) 64:193–226
Toyama T, Yokoyama I, Nishi K (1975) Effects of hexamethonium and other ganglionic blocking agents on electrical activity of the esophagus induced by vagal stimulation in the dog. Eur J Pharmacol 31:63–71
Tuch A, Cohen S (1973) Lower esophageal sphincter relaxation: studies on the neurogenic inhibitory mechanism. J Clin Invest 52:14–20
Ueda M, Schlegel JF, Code CF (1972) Electric and motor activity of innervated and vagally denervated feline esophagus. Am J Dig Dis 17:1075–1088

Valenzuela JE (1976) Dopamine as a possible neurotransmitter in gastric relaxation. Gastroenterology 71:1019–1022
Vizi ES (1976) The role of α-adrenoreceptors situated in Auerbach's plexus in the inhibition of gastrointestinal motility. In: Bülbring E, Shuba MF (eds) Physiology of smooth muscle. Raven, New York, pp 357–367
Waddel WR, Wang CC (1953) Effect of vagotomy on gastric evacuation of high fat meals. J Appl Physiol 5:705–711
Weisbrodt NW, Christensen J (1972) Gradients of contractions in the opossum esophagus. Gastroenterology 62:1159–1166
Weisbrodt NW, Copeland EM, Moore EP, Kearly RW, Johnson LR (1975) Effect of vagotomy on electrical activity of the small intestine of the dog. Am J Physiol 228:650–654
Wilbur BG, Kelly KA (1973) Effects of proximal gastric, complete gastric and truncal vagotomy on canine gastric electric activity motility and emptying. Ann Surg 178:295–303
Wingate DL, Ruppin H, Green WER et al. (1976) Motilin-induced electrical activity in the canine gastrointestinal tract. Scand J Gastroenterol [Suppl 39] 11:111–118
Yamamoto T (1960) Histological studies on the innervation of the esophagus in formosan macaque. Arch Histol Jpn 18:545–564
Yamamoto T, Satomi H, Hiromi I, Takahashi K (1977) Evidence of the dual innervation of the cat stomach by the vagal dorsal motor and medial solitary nuclei as demonstrated by the horseradish peroxydase method. Brain Res 122:125–131
Youmans WB (1968) Innervation of the gastrointestinal tract. In: Code CF (ed) Alimentary canal. American Physiological Society, Washington, DC (Handbook of physiology, vol IV, sect 6, pp 1655–1663)
Zfass AM, Prince M, Allen FN, Farrar JT (1970) Inhibitory beta adrenergic receptors in the human esophagus. Am J Dig Dis 15:303–310

CHAPTER 10

Nervous Control of Intestinal Motility

M. COSTA and J. B. FURNESS

A. Introduction

The most important functions of the intestinal musculature are to transport the contents through the long tubular digestive tract and to ensure their adequate mixing and exposure to absorptive surfaces. The behaviour of the muscle associated with these functions is complex and highly adaptable to ensure the efficient utilisation of nutrients, even though feeding habits and diet may vary. The movements of the muscle depend on three primary factors: the intrinsic properties of the musculature itself, which undergoes rhythmic changes in excitability that vary along the intestine; the influence of circulating hormones; and the influence of nerves. The systems of nerves associated with the intestine are remarkable. Within the wall of the intestine is a complex network of intrinsic neurones, the enteric nervous system, which has connections with the muscle, mucosal cells and blood vessels of the gut wall, as well as with sympathetic ganglia and the central nervous system. LANGLEY, in his book on the autonomic nervous system (1921), had already realised the relative autonomy of the enteric nervous system, and classified it as one of the three divisions of the autonomic nervous system, the others being the sympathetic and parasympathetic divisions. It is at first surprising that the total number of enteric neurones is similar to the total number of neurones in the spinal cord, but the enteric nervous system, like the spinal cord, controls an extensive system of muscle with a varied repertoire of movements. The enteric nervous system is also involved in the modulation of secretion, intestinal blood flow and probably absorption. Like the spinal cord, the enteric ganglia are reflex centres, receiving sensory information, integrating this information and influences from other centers and providing an output to effectors.

Because an understanding of the organization and properties of the muscle is needed to discuss its control by nerves, these subjects are discussed briefly in Sect. B. Fuller descriptions are given in Chaps. 4 and 5. Likewise, there is a small section on transmission in enteric ganglia which is dealt with more thoroughly in Chap. 6. Some topics, such as the innervation of the muscularis mucosa and of the accessory muscles to the anal sphincter, have not been included. Analysis of the control of motility by extrinsic nerves has been limited to reflexes or originating from the digestive tract.

B. Neuromuscular Transmission in the Intestine

I. Arrangement of Muscle and Nerves in the Intestine

The external smooth muscle of the intestine is organised in two layers, an outer thinner longitudinal and an inner thicker circular layer. Another thin layer of smooth muscle, the muscularis mucosa separates the submucosa from the mucosa. It is the contraction of the layers of the external musculature that is mainly responsible for the progression of the intestinal contents from the stomach to the rectum. This basic arrangement of muscle shows variations in some parts of the intestine. The circular muscle layer is thicker in the ileocaecal region and in the anal region where it forms the ileocaecal sphincter and the internal anal sphincter, respectively. In some species the longitudinal muscle of the caecum and parts of the proximal colon is restricted to narrow bands (taeniae) leaving the peritoneum in direct apposition to the circular muscle between the taeniae.

Two ganglionated nerve plexuses are embedded in the intestinal wall, the myenteric (Auerbach) plexus, located between the longitudinal and the circular muscle layers, and the submucous (Meissner) plexus in the submucosa (see FURNESS and COSTA 1980; Chaps. 2, 11). Nerve processes arise from the nerve cells of these plexus and form nonganglionated nerve plexuses in the longitudinal muscle (the longitudinal muscle plexus) and throughout the circular muscle with a denser plexus being formed in its inner layers (the deep muscular plexus). There are also plexuses of nerve fibres in the mucosa, i.e. in the muscularis mucosa, around the intestinal glands, and in the villi forming the subepithelial plexus. Nerves accompany the submucosal blood vessels and form a network in the adventitia. Some of the nerve fibres of these plexuses are of extrinsic origin; they reach the intestine via the mesenteric nerves, the vagus and the pelvic nerves.

II. The Neuromuscular Junction in the Intestine

The muscle cells in the intestine are interconnected. They form bundles in which adjacent muscle cells are connected both electrically and mechanically and the bundles themselves are interconnected. The nerves which run in the muscle layers are formed by bundles of axons partly enveloped by Schwann cells. Within these bundles, individual axons are varicose for distances of several millimetres, although the muscle cells are 0.5 mm or less in length. It is generally believed, although not proved, that transmitter is released from varicosities along the length of the axon. Thus, because the axons run in spaces between several muscle cells and their varicose portions are considerably longer than the muscle cells, each axon probably influences several muscle cells directly and, because of the interdependence of the muscle cells, influences many muscle cells indirectly. There are no structurally identifiable featurs (synaptic specialisations) to identify points of transmission. Our present concept of the intestinal neuromuscular junction is that the prejunctional element is a branching, varicose axon which releases transmitters at many points adjacent to different muscle cells and that the postjunctional elements are bundles of muscle cells, mechanically and electrically coupled, whose influence on each other decreases with distance (see Sect. B.III). It is likely that axons which lie at the surface of the myenteric plexus and face the muscle with no intervening cells are in fact part of the innervation of the muscle.

III. Basic Properties of Intestinal Smooth Muscle

Within the layers of the intestine the individual muscle fibres (2–5 μm in diameter by about 500 μm long) are loosely arranged in muscle bundles of different sizes incompletely separated by connective septa into larger units (BURNSTOCK 1970). Since these bundles are interconnected in most cases, it is unlikely that they represent independent functional units (GABELLA 1979a). Transmission of the force generated by the smooth muscle to elicit contractions of the whole layer of muscle involves junctional structures between individual muscle cells and junctions between muscle cells and the stroma interspersed between the muscle cells (GABELLA 1979a). The force generated by the intestinal muscle per unit area is of the same order as that of striated muscles (GABELLA 1979a).

As described in Chap. 5, intestinal smooth muscle shows two types of electrical activity; *slow waves*, which are cyclical variations of the membrane potential not accompanied by mechanical activity, and *action potentials*, which initiate contractions.

1. Slow Waves

A very important feature of intestinal smooth muscle is that it generates cyclic changes in membrane potential. These slow waves have been also called "basic electrical rhythm", "pacesetter potentials" and "electrical control potentials". They do not initiate contraction of the muscle and represent only cyclical changes in the excitability of smooth muscle cells. Slow waves have been recorded in vivo from all parts of the small intestine and from most parts of the large intestine in both longitudinal and circular muscle layers in all species studied (DANIEL 1968, 1975; PROSSER and BORTOFF 1968; BASS 1968; PROSSER 1974; BORTOFF 1976. These slow waves play an important role in determining patterns of intestinal movements.

a) Origin

Connection between the two muscle layers is important for the appearance of normal slow waves. It was originally thought that slow waves were initiated in the longitudinal muscle and spread to the circular muscle (see BORTOFF 1976; DANIEL and SARNA 1978) but recent evidence suggests that they are probably generated by pacemaker cells in the circular muscle (TAYLOR et al. 1975a). A more complex theory suggests the initiation of a slow wave of small amplitude in the longitudinal muscle which spreads to the circular muscle where it is amplified and spreads back to reinforce the longitudinal muscle slow waves (KOBAYASHI et al. 1966; CONNOR et al. 1977). The result of the spread of slow waves between the longitudinal and circular muscle layers is that the two layers in the same segment of intestine depolarise simultaneously. The consensus is that not all individual muscle cells initiate slow waves but that a few specialised cells act as pacemakers that transmit the wave to neighbouring cells. The nature of the cells that generate slow waves and the way in which they drive other cells have not yet been established (DANIEL and SARNA 1978).

b) Frequency

The frequency of the slow waves varies in different species and in different parts of the intestine (PROSSER and BORTOFF 1968). In the human duodenum the frequency is 11–13 cycles/min (STODDART et al. 1978), in sheep it is 17 cycles/min (RUCKEBUSCH and BUENO 1975), in dogs 18–19 cycles/min (PROSSER and BORTOFF 1968; MCCOY and BAKER 1969) while in small mammals such as rats the frequency is much higher, 36–40 cycles/min (RUCKEBUSCH and FIORAMONTI 1975). In the small intestine in vivo the frequency of slow waves decreases from the duodenum to the ileocaecal region in a stepwise fashion so that lengths of intestine show the same frequency. For instance, in dogs the frequency decreases from 18–19 cycles/min in the duodenum to 10–12 cycles/min in the distal ileum (PROSSER and BORTOFF 1968). Each frequency plateau is separated from the next by an area in which the amplitudes of the slow waves periodically wax and wane (DIAMANT and BORTOFF 1969; BORTOFF 1976). The lengths of the plateaus and of the areas of waxing and waning vary along the intestine and also fluctuate with time. In the species studied so far (dog, cat and human) a long plateau is present in the upper intestine and comprises the whole duodenum and part of the jejunum with short plateaus of variable length in the ileum (ARMSTRONG et al. 1956; CHRISTENSEN et al. 1966; BUNKER et al. 1967; DIAMANT and BORTOFF 1969; SARNA et al. 1971; DUTHIE et al. 1972).

The stepwise nature of the gradient of slow wave frequency in vivo contrasts with a linear decrease of slow wave frequency observed in isolated segments. The usual explanation is that every segment in vivo acts as an individual oscillator which is incompletely coupled with adjacent oscillators. The oscillator with the highest frequency drives the slower caudal oscillators. At a certain distance the frequency differences is too great for the next oscillator to follow and a new plateau at a lower frequency is formed. The waxing and waning patterns may be a reflection of the interaction between the terminal oscillators of each plateau (BORTOFF 1976).

c) Propagation

There is a propagation of the slow waves around the circumference of the intestine through the circular muscle at a speed of 5.9 cm/s (BASS et al. 1961; KOBAYASHI et al. 1966; BORTOFF 1976). The almost simultaneous spread of the slow waves around the circumference of the intestine, involving a large number of individual smooth muscle cells, represents a mechanism for the synchronisation of the muscle fibres into a mechanically effective constriction of the lumen.

The slow waves are propagated anally along the longitudinal axis of the small intestine. The speed of propagation decreases along the intestine in the dog, being 14 cm/s in the duodenum and 3 cm/s in the distal jejunum (MCCOY and BAKER 1969). The slow waves do not propagate beyond the end of each frequency plateau.

Anal propagation of the slow waves does not always occur; reverse propagation may occur and slow waves propagate orally for short distances and for short periods. The reversal of the direction of propagation has been attributed to the appearance of ectopic pacemakers (KOBAYASHI et al. 1966; GONELLA 1970; HIATT et al. 1976). Acetylcholine has been shown to evoke premature slow waves (DANIEL 1977) and to initiate them if they are absent (PROSSER 1974; BOLTON 1979 a, b). The release of acetylcholine during activation of the ascending excitatory reflex is probably responsible for the appearance of an area of higher frequency of slow waves which can temporarily propagate orally for a short distance (see Sect. J.I.e).

d) Slow Waves in the Large Intestine

Slow waves have also been recorded from the large intestine of several species both in vitro (GILLESPIE 1962; CHRISTENSEN et al. 1969; CAPRILLI and ONORI 1972; JULÉ 1974; SHEARIN et al. 1978; VAN MERWYK and DUTHIE 1979; EL-SHARKAWY 1978) and in vivo (KERREMANS 1968; COUTURIER et al. 1969; PROVENZALE and PISANO 1971; JULÉ and GONELLA 1972; WIENBECK 1972; JULÉ 1975; TAYLOR et al. 1975b; SNAPE et al. 1976; DANIEL 1975; STODDART et al. 1979; LORD et al. 1979; SARNA et al. 1980). The waves in the colon resemble the gastric slow waves (see Chap. 9) and, unlike those in the small intestine, are not always regular. Furthermore, two rhythms appear to be superimposed, a faster (6–13 cycles/min) and a slower (2–6 cycles/min) rhythm, which vary in their relative strength along the large intestine and may interact to form frequency gradients (STODDART et al. 1979; CHRISTENSEN et al. 1969; CHRISTENSEN and HAUSER 1971; TAYLOR et al. 1975b; SARNA et al. 1980; SNAPE et al. 1976, 1978; CHRISTENSEN 1978). The electrical activity of the large intestine is independent of that of the small intestine and, in fact, an area of electrical silence has been demonstrated at the ileocaecal junction (BALFOUR and HARDCASTLE 1978). There is a general consensus that slow waves in the large intestine are poorly coupled and thus there is little or no propagation of this electrical activity along the large intestine (CHRISTENSEN et al. 1969; CHRISTENSEN and HAUSER 1971; TAYLOR et al. 1975b; SHEARIN et al. 1978, SARNA et al. 1980). The mid-colon appears to act as an area for initiation of slow waves which travel, for variable distances, orally and anally from this area (CHRISTENSEN 1978). At the level of the anal sphincter a high frequency of slow waves has been reported in humans (14–30 cycles/min) and in dogs (13–24 cycles/min; HOWARD and NIXON 1968; KERREMANS 1968; WANKLING et al. 1968; USTACH et al. 1970; MONGES et al. 1979; WIENBECK and ALTAPARMACOV 1979). CHRISTENSEN (1978) and SARNA et al. (1980) have suggested that the stationary nature of the slow waves, particularly in the distal colon, favours the appearance of nodal points of contraction which may be related to the formation of the haustrations in humans (Sect. G.II).

2. Action Potentials

The action potentials in intestinal smooth muscle, which are superimposed on the slow waves, have also been referred to as spike potentials, fast activity, spike bursts or electrical control activity. There is no doubt that the action potentials cause contraction of smooth muscle in the intestine (BASS et al. 1961; DANIEL and CHAPMAN 1963; DANIEL 1968; BASS 1968).

Intracellular recordings show that action potentials are elicited only when a slow wave or another depolarisation reaches a threshold level of depolarisation (HOLADAY et al. 1968; BASS et al. 1961; GONELLA 1970) and numerous observations show that the action potentials occur during a limited period of the slow wave, usually at the crest or just after it (DANIEL et al. 1960; BASS et al. 1961; BORTOFF 1961; BASS and WILEY 1965; BASS 1968; SANCHOLUZ et al. 1975). Calcium is the main ion involved in the generation of action potentials (BRADING et al. 1969; BÜLBRING and TOMITA 1969, 1970; LIU et al. 1969; KURIYAMA and TOMITA 1970; BORTOFF 1972; CONNOR and PROSSER 1974; CONNOR et al. 1977; BOLTON 1979b).

Action potentials, unlike the slow waves, are not propagated for more than about 5 mm in intestinal muscle (BASS 1968; GONELLA 1970; DANIEL 1968). This means that any propagated contraction can only result from a sequential initiation of action potentials along the intestine.

a) Phasic and Tonic Contractions

The intestine undergoes both phasic and tonic changes of length and tension. Phasic contractions, which depend on action potentials for their generation, can fuse to produce a tonic contraction. Some tonic contractions, however, appear to be due to activation of the contractile apparatus without action potentials occurring or even without apparent changes in muscle membrane potentials.

α) Contractions Dependent on Action Potentials. If action potentials appear on the crest of a slow wave, a single contraction occurs, its strength being proportional to the number and the frequency of the action potentials. If every slow wave reaches the threshold for the initiation of action potentials, rhythmic conctractions are generated at the same frequency as the slow waves and with a force proportional to the frequency and number of action potentials superimposed on the wave crest (BASS et al. 1961; BASS and WILEY 1965; Gonella 1970; Grivel and RUCKEBUSCH 1972; BORTOFF 1976; SZURSZEWSKI 1977b; BOLTON 1979a, b).

Further depolarisation of the smooth muscle membrane increases the number and the frequency of action potentials and they produce contractions that begin to fuse and appear as a more complex wave of localised contraction (see records in GRIVEL and RUCKEBUSCH 1972; RUCKEBUSCH and BUENO 1975). Further depolarisation results in an almost continuous firing of action potentials throughout the slow wave cycle with a maintained spasm of the muscle (BOZLER 1939; BASS et al. 1961; GRIVEL and RUCKEBUSCH 1972). It would appear therefore that tonic contractions on which phasic contractions are superimposed can be explained by fusion of contractions initiated by action potentials.

β) Contractions Independent of Action Potentials. It has been suggested that in the intestinal muscle, as in other smooth muscles, prolonged tonic contractions may be initiated by graded maintained depolarisations and in some cases even without any change in membrane potential (SZURSZEWSKI 1977b; BOLTON 1979a, b; CREED 1979). There have even been suggestions that the tonic and phasic contractions are mediated by two different types of smooth muscle (MCKIRDY 1972; GOLENHOFEN 1976; KAJIRSUKA 1979). In unanaesthetised dogs, BASS and WILEY (1965) demonstrated that changes in muscle tone occur spontaneously and are not accompanied by any electrical activity of the muscle. GRIVEL and RUCKEBUSCH (1972) could not record action potentials from rabbit longitudinal muscle during prolonged contractions. MENDEL et al. (1978) did not record action potentials in vivo from longitudinal muscle contracting independently from the circular muscle. During the initial nerve-mediated shortening of the longitudinal muscle of the guinea-pig ileum in the peristaltic reflex, NAKAYAMA (1962) recorded a graded depolarisation but no action potentials. Other authors observed spontaneous tonic contractions in cat and rabbit longitudinal muscle which could also be induced by electrical stimulation as a graded response (YOKOYAMA 1966; WOOD and PERKINS 1970; KAIRSUKA 1979). Slow wave potentials but not action potentials are associated with contractions and relaxations of the cat internal anal sphincter (KERREMANS 1968; USTACH et al. 1970; MONGES et al. 1979; WIENBECK and ALTAPARMACOV 1979). It cannot be excluded, however, that in all these experiments the mechanical events are still dependent on action potentials which occur in a limited number of smooth muscle cells and are not detected by the recording electrodes. Moreover, in some of these experiments the shortening and narrowing of the intestine could be due to the contraction of the muscularis mucosae. It has been shown that contractions of the muscularis mucosa can determine the radiographic appearance of the intestinal lumen (GOLDEN 1945) and that contraction of the muscularis mucosa may be responsible for contractions in a whole segment of intestine that were attributed to a contraction of the external musculature (KING et al. 1947).

b) Pacemaker Cells and Resting Tension

Some intestinal smooth muscle cells can initiate action potentials spontaneously and show a prepotential typical of cardiac pacemaker cells (HOLMAN 1968; GONELLA 1970; BENNETT 1972). The action potential generated by pacemaker cells does not excite all the other smooth muscle cells since the amplitudes of the spontaneous contractions are not usually maximal (MASHIMA et al. 1966).

It is likely that the small irregular contractions elicited by those smooth muscle cells which are coupled to pacemaker cells are partially responsible for the maintenance of a certain degree of resting tension of the intestinal muscle. It is on this resting tension that excitatory and inhibitory nerves, acting on the muscle, will increase the number of action potentials or will reduce the existing action potentials leading to contraction or relaxation of the muscle.

3. Interaction Between Muscle Layers

Since slow waves are propagated between the longitudinal and the circular muscle layers generating a synchronous depolarisation of the two layers, it is not surprising that many authors have recorded a simultaneous increase in tension of the two muscle layers (BAYLISS and STARLING 1899; BASS and WILEY 1965; BORTOFF and SACHS 1970; MCKIRDY 1972; KRANTIS et al. 1970). However, there is convincing evidence that the two muscle layers can also act independently (COWIE and LASHMET 1928; GONELLA 1970, 1971; GRIVEL and RUCKEBUSCH 1972; MCKIRDY 1972). It has been shown, for instance, that during the preparatory phase of the peristaltic reflex of the guinea-pig ileum the longitudinal muscle contracts well before the circular muscle (BAUR 1928; SCHAUMANN et al. 1953; KOSTERLITZ et al. 1956; NAKAYAMA 1962). That the conduction of nerve-mediated junction potentials between layers is not very effective has been pointed out by HIRST et al. (1975). These results taken together indicate that, despite the propagation of slow waves between the two muscle layers, action potentials do not propagate effectively from one layer to the other. Thus, the synchronous contraction of the two muscle layers is coordinated by the slow waves propagated between the layers and depends on both layers being excited to a sufficient level to initiate contraction. There is, however, convincing evidence that the contractions of one layer influences the other layer mechanically. When the circular muscle contracts vigorously and is not opposed by contraction of the longitudinal muscle, elongation of the intestine occurs; conversely, strong contractions of the longitudinal muscle produce an apparent dilation of the intestine (MALL 1896; COWIE and LASHMET 1928; ALVAREZ and BENNET 1931; KRISHNAN 1932; HUKUHARA and FUKUDA 1965; GREGORY and BENTLEY 1968; WOOD and PERKINS 1970). Even the maximal response of the longitudinal muscle to drugs is reduced if the circular muscle is also contracted (POMEROY and RAPER 1971).

The mechanical elongation elicited by the strong circular muscle contraction is the most likely explanation of the apparent relaxation of the longitudinal muscle during the peristaltic reflex in the guinea-pig ileum (SCHAUMANN et al. 1953; KOSTERLITZ et al. 1956; HUKUHARA and FUKUDA 1965; KOTTEGODA 1969). Thus, the net shortening or elongation of the intestine depends on the balance of the strengths of contraction of each muscle layer and the mechanical interactions between the two layers.

IV. Transmission from Enteric Neurones to Intestinal Muscle

Three methods of stimulation can be used to study transmission from nerves embedded in the intestinal wall: they can be stimulated by transmural electrodes or drugs or they can be activated by evoking reflexes, for example by distending the intestine. Transmural stimulation has the disadvantage of being nonselective. Nevertheless, by using different conditions of stimulation and antagonist drugs it has been possible to distinguish transmission via cholinergic nerves, from transmission via enteric inhibitory nerves and noncholinergic excitatory nerves. Evidence for the presence of these nerve types has also come from experiments with drug stimulation and evoked reflexes. Nevertheless, there is histochemical and other evidence for the existance of ten or more nerve types in the in the intestine (see Chap. 11) and it is possible that transmission from some types of nerve has not been detected because their effects were obscured by responses to other nerves for which there are no known specific antagonists.

1. Transmission from Cholinergic Nerves

Acetylcholine has been well established as a neurotransmitter in the intestine (see Chap. 11) and antimuscarinic drugs appear to be reliable tools for identification and study of cholinergic transmission to intestinal muscle (AMBACHE 1955; and see Chap. 11). Cholinergic nerves supply both the longitudinal and the circular muscle coats. PATON (1955) first described the highly effective transmission from myenteric cholinergic neurones to the longitudinal muscle of the guinea-pig small intestine. His technique of coaxial stimulation and the modifications which followed have become the most used techniques to investigate the pharmacology of cholinergic transmission to smooth muscle. Single electrical pulses elicit brief contractions which have been shown to be abolished by atropine, hyoscine and tetrodotoxin but not by antinicotinic drugs and have been shown to be potentiated by anticholinesterase drugs. The prompt contraction elicited even by just one pulse indicates that this transmission is very effective and since there are no nerve fibres within the longitudinal muscle layer in the preparation (see GABELLA 1979 b) transmission must occur from axons located at the surface of the myenteric plexus facing the muscle. This nerve transmission is regarded as a good example of "pure" transmission from cholinergic nerves. The lack of relaxation elicited by stimulation, the complete abolition of all response by antimuscarinic drugs and the lack of any detectable effect by antinicotinic drugs has been taken as evidence that his nerve-mediated response is due to the activation of excitatory cholinergic neurones without detectable involvement of cholinergic interneurones or other nerve types. However, when duodenal segments are used and the longitudinal muscle shows some tone, inhibitory transmission is also observed (FURNESS and COSTA 1973). Thus, after administration of hyoscine, single pulses elicit prompt nerve-mediated relaxations which are often followed by a quick brief contraction. Both inhibitory and excitatory junction potentials (i.j.p. and e.j.p.) can be recorded electrically (KURIYAMA et al. 1967).

Cholinergic transmission to the longitudinal muscle layer occurs when more physiological stimuli are applied; distension of a middle segment of guinea-pig small intestine with a small balloon resulted in the appearance of a delayed cholinergic e.j.p. in the longitudinal muscle below the distension (HIRST et al. 1975). Since the work of PATON (1955), other workers have described cholinergic excitatory effects in other muscle layers and in different preparations. There is good evidence for the presence of excitatory cholinergic transmission to the circular muscle of the guinea-pig small intestine. Cholinergic e.j.p. were also recorded from this muscle layer following distension (HIRST et al. 1975). Cholinergic transmission to the longitudinal muscle of several mammalian intestinal preparations has been reported by a number of authors: in guinea-pig taenia coli (BENNETT 1966a; CAMPBELL 1966a); guinea-pig and rabbit distal colon (FURNESS 1969a; GILLESPIE 1968); guinea-pig proximal and distal ileum (MUNRO 1953; DAY and VANE 1963); rabbit and kitten ileum (DAY and WARREN 1968); opossum and cat duodenum (ANURAS et al. 1977, 1979); and human ileum and colon (BENNETT and STOCKLEY 1975). Cholinergic responses were also recorded in several circular muscle preparations from rabbit caecum (SMALL 1971, 1972); dog jejunum (DANIEL 1968); guinea-pig distal colon (FURNESS 1969a); guinea-pig anal sphincter (COSTA and FURNESS 1974); human ileum and ascending colon (BENNETT and STOCKLEY 1975); opossum and cat duodenum (ANURAS et al. 1977, 1979); and rabbit ileum (YOKOYAMA and OZAKI 1978). Most authors confirmed that the enteric inhibitory neurones that are stimulated simultaneously often mask the excitatory responses.

As it will be shown in Sects. H and J, enteric excitatory cholinergic neurones probably represent the most important class of enteric neurones, being the final pathway involved in contractions of the intestinal muscle. Although in isolated segments of intestine there is a spontaneous nerve-mediated release of acetylcholine (PATON and ZAR 1968; and see Chap. 11), there is no evidence that the cholinergic motor neurones are spontaneously active; this indicated that they must be driven by other enteric interneurones (OHKAWA and PROSSER 1972a, b; NISHI and NORTH 1973a; HIRST et al. 1974; WOOD and MAYER 1978a). The enteric cholinergic neurones receive excitatory inputs from other cholinergic enteric interneurones and from preganglionic cholinergic neurones, either from the vagus or from the pelvic nerves. It is possible that some of the cholinergic neurones also receive inputs from enteric noncholinergic interneurones (Chap. 6). Inhibitory inputs which result in a decrease in acetylcholine release are provided by the entrinsic noradrenergic fibres although it has bee suggested that their main action is on presynaptic cholinergic interneurones (Sect. E.II.1).

2. Transmission from Enteric Inhibitory Neurones

The existence of intrinsic inhibitory neurones in the intestine was suggested at the end of the nineteenth century when the classic studies of MALL (1896), BAYLISS and STARLING (1899) and LANGLEY and MAGNUS (1905) showed that the intestine possesses intrinsic nerve reflexes which involve relaxation of the muscle. That such neurones were not sympathetic postganglionic neurones was established by LANGLEY and DICKINSON (1889) who showed, with the use of nicotine, that the cell bodies of the postganglionic sympathetic neurone were located in prevertebral ganglia (see FURNESS and COSTA 1974a).

It is rather surprising that the nature of such neurones remained unstudied for many decades until AMBACHE (1951) found that a relaxation was revealed after blocking the excitatory responses to nicotine with antimuscaranic drugs. This was interpreted at that time as due to stimulation of short intrinsic adrenergic neurones. It was only in the early 1960s, with the advent of adrenergic neurone-blocking drugs (guanethidine and bretylium) and the development of β-adrenoceptor antagonist drugs; that the inhibitory responses elicited by nicotine (AMBACHE 1951) or by transmural stimulation (MUNRO 1953) were recognised by scientists in Melbourne not to be mediated by adrenergic neurones (BURNSTOCK et al. 1964). Although it is now well established that these nerves are not adrenergic, their transmitter has not yet been determined (see Chap. 11). Compelling evidence that they are not adrenergic is the persistence of responses to their stimulation after degeneration of noradrenergic nerves supplying the guinea-pig distal colon (FURNESS 1969b; CREMA 1970). Furthermore, transmission from enteric inhibitory neurones was observed in the taenia coli after 3 weeks in tissue culture when noradrenergic nerves which ar of extrinsic origin had degenerated (RIKIMARU 1971a). These nerves were found by PATON and VANE (1963) in the stomachs of various species and it soon became apparent that the inhibitory neurones were involved in the relaxation recorded by LANGLEY (1898) in the lower oesophagus and stomach following stimulation of the vagus nerve (MARTINSON 1965; CAMPBELL 1966b) and in the inhibitory responses of the small and large intestine to vagus and pelvic nerve stimulation (Sects. E.I, E.IV). The term *enteric inhibitory neurone* is used here to describe this type of neurone (see also FURNESS and COSTA 1980).

Probably the first evidence that electrical stimulation activates intrinsic inhibitory intestinal neurones was the observation made by CLAUDE BERNARD and annotated in his *Cahier Rouge* (1850–1860) as follows

> ... (tonus des intestines) sur un chien et sur un lapin à jeun depuis 3 jours les intestine étaient aplatis. En les galvanisant, il s'élargissent, ce qui est inverse pour les artères qui se rétrecissent quand on le galvanise. Il semblerait donc en résulter que l'état de dilatation est un état actif de l'intestin. Comment comprendre d'autre part le galvanisme porté sur la fibre musculaire fasse dilater l'intestin? [1]

---

1 ... (on the tone of the intestine) in a cat and a rabbit fasted for 3 days the intestine was collapsed. When a galvanic current is passed the intestine dilates which is the opposite of what happens to the arteries which constrict in response to galvanic current. It appears from the results that the dilation is an active state of the intestine. How is it to be otherwise understood then that the galvanic current applied to the muscle fibre dilates the intestine?

Table 1. Evidence of transmission from enteric inhibitory nerves in mammalian intestine

Species	Preparation	Method of recording	Reference
Mouse	Duodenum LM	M	BURNSTOCK et al. (1972)
			HOLMAN and HUGHES (1965)
	Ileum LM	M	BURNSTOCK et al. (1972)
			HOLMAN and HUGHES (1965)
	Colon LM	M	HOLMAN and HUGHES (1965)
Rat	Duodenum LM	M	BURNSTOCK et al. (1972)
			GARCIA-RODRIGUEZ et al. (1971)
			HOLMAN and HUGHES (1965)
	Ileum LM	M	BARLETT et al. (1979)
			HOLMAN and HUGHES (1965)
	Colon LM	M	HOLMAN and HUGHES (1965)
Guinea-pig	Small intestine LM	M, E	FURNESS and COSTA (1973)
			KOSTERLITZ and LYDON (1968)
			KURIYAMA et al. (1967)
	Small intestine CM	M, E	BYWATER and TAYLOR (1979)
			MIR et al. (1977)
			VERMILLION et al. (1979)
	Taenia coli	M, E	BENNET et al. (1966)
			BENNET (1966a)
			BÜLBRING and TOMITA (1967)
			BURNSTOCK et al. (1966)
			SMALL (1972)
			VERMILLION et al. (1979)
	Caecum CM	E	LANG (1979)
	Proximal colon LM	M	COSTA and FURNESS (1972)
			FURNESS and COSTA (1973)
	Distal colon LM	M, E	BIANCHI et al. (1968)
			BURNSTOCK et al. (1972)
			FURNESS (1969b)
			FURNESS (1970)
	Distal colon CM		FURNESS (1969b)
	Internal anal sphincter	M, E	COSTA and FURNESS (1974)
			FURNESS and COSTA (1973)
Rabbit	Small intestine LM	M	DAY and WARREN (1968)
			GERSHON and THOMPSON (1973)
			SMALL and WESTON (1979)
			VERMILLION et al. (1979)
			HOLMAN and HUGHES (1965)
			WESTON (1973a, b)
	Caecum CM	M, E	SMALL (1972)
	Colon LM	M	BUCKNELL (1965)
			FURNESS (1969a)
			HOLMAN and HUGHES (1965)
Cat	Small intestine LM		
	Small intestine CM	E	WOOD (1972)
	Sphincter of Oddi	M	PERSSON (1976)
	Ileocaecal sphincter	M	PERSSON (1976)
Echidna	Small intestine LM	M	FURNESS and COSTA (1973)

Table 1 (continued)

Species	Preparation	Method of recording	Reference
Opossum	Duodenum LM	M	ANURAS et al. (1977)
	Duodenum CM	M	ANURAS et al. (1977)
	Ileocaecal sphincter	M	CONKLIN and CHRISTENSEN (1975)
Dog	Pylorus	M	MIR et al. (1977)
	Taenia coli	M	RIKIMARU (1971 b)
	Distal colon LM	E	FURNESS and COSTA (1973)
Sheep	Distal colon LM	E	FURNESS and COSTA (1973)
Monkey	Taenia coli	M	RIKIMARU (1971 b)
	Internal anal sphincter LM	M	RAYNER (1979)
Human	Small intestine LM	M	BENNET and STOCKLEY (1975)
	Small intestine CM	M	BENNET and STOCKLEY (1975)
	Colon LM	M	BENNET and STOCKLEY (1975) BURNSTOCK et al. (1972) CREMA et al. (1970)
	Colon CM	M	BENNET and STOCKLEY (1975)
	Taenia coli	M	STOCKLEY and BENNET (1977)
	Internal anal sphincter	M	BURLEIGH et al. (1979)

LM, longitudinal muscle; CM, circular muscle; M, mechanical; E, electrophysiological

Responses to the electrical stimulation of these enteric inhibitory nerves have been recorded in many preparations tested of mammalian (including human) intestine (see CAMPBELL and BURNSTOCK 1968; CAMPBELL 1970; BURNSTOCK 1972, FURNESS and COSTA 1973; Table 1); these nerves supply both longitudinal and circular muscle layers of small and large intestine, including the ileocaecal and internal anal sphincters and accessory muscles. The cell bodies are located in the myenteric plexus and form the efferent link in a cascade of descending reflexes which extend from the lower oesophageal sphincter to the internal anal sphincter and which ensure the polarity of the propulsion of the intestinal contents (FURNESS and COSTA 1973; COSTA and FURNESS 1976; and Sect. J). These neurones receive inputs not only from intrinsic cholinergic interneurones in the descending inhibitory pathways but also from cholinergic preganglionic axons from the vagus and the pelvic nerves (Sects. E.I, E.III). The extrinsic noradrenergic nerves do not appear to affect the enteric inhibitory nerves (JULÉ 1975).

The enteric inhibitory neurones are very effective and single pulses elicit a transient hyperpolarisation of the muscle (inhibitory junction potential i.j.p.) which lasts 800–1,200 ms (BENNETT and BURNSTOCK 1968). The i.j.p. elicited by reflex activation of the descending inhibitory pathways have been recorded (HIRST et al. 1975; JULÉ 1980). As a result of their long time course the i.j.p. summate at low frequencies of stimulation (BENNETT and BURNSTOCK 1968; SMALL 1972), with 80% of the maximal relaxation of the muscle being achieved at about 5 Hz (BURNSTOCK et al. 1966). The hyperpolarisation is due to an increase in the K^+ conductance of the membrane (BENNETT 1966a; TOMITA 1972).

Although i.j.p. can be elicited by transmural stimulation in most intestinal smooth muscle cells (BENNETT 1966a; BENNETT et al. 1966; FURNESS 1969a), in the stomach they were observed following stimulation of the vagus nerve in only 18% of impalements (BEANI et al. 1971). However, the spontaneous appearance of action potentials was blocked by stimulation of the vagal inhibitory pathways, even in cells in which i.j.p. were not observed (BEANI et al. 1971), indicating that the enteric inhibitory neurones could selectively innervate gastric pacemaker cells (FURNESS and COSTA 1973). In muscles with little or no tone, activation of these nerves fails to elicit a relaxation although the i.j.p. can be recorded. In such cases the presence of inhibitory nerves is often revealed by the rebound excitation which occurs following inhibition of the smooth muscle (CAMPBELL and BURNSTOCK 1968). This rebound excitation is myogenic (BENNETT 1966b; CAMPBELL 1966a; FURNESS 1971), although excitation occurring after stimulation can also be due to the release of other substances from nerves (e.g. FURNESS 1971; BYWATER and TAYLOR 1979; and see Sect. B.IV.3).

It is difficult to ascertain whether the enteric inhibitory neurones are spontaneously active since there are no drugs which specifically antagonise their transmission. Blockade of nerve activity with tetrodotoxin increased the spontaneous firing of smooth muscle action potentials (WOOD 1972; TONINI et al. 1974; BIBER and FARA 1973). This suggests that there was continuous activity in enteric inhibitory nerves; whether this normally occurs in vivo is not known.

3. Transmission from Noncholinergic Excitatory Neurones

There is very good evidence that a number of transmitter other than acetylcholine, noradrenaline and the unidentified transmitter of the enteric inhibitory neurones are released in the intestinal tract (see FURNESS and COSTA 1980; and Chap. 11).

The first clear evidence for the release of excitatory substances other than acetylcholine from enteric nerves cam from the work of AMBACHE and his collaborators (AMBACHE and FREEMAN 1968; AMBACHE et al. 1970). These authors showed that when the guinea-pig small intestine was stimulated electrically, atropine-resistant, nerve-mediated contractions were elicited. These contractions are due, at least in part, to the release of substance P (FRANCO et al. 1979a). Such responses mediated by substance P are not affected by chronic extrinsic denervation of the intestine (FRANCO et al. 1979b), indicating that these neurones are intrinsic to the intestine. The neurones that release substance P are capable of acting both on the longitudinal and on the circular muscle in vitro (FRANCO et al. 1979a; R. FRANCO unpublished work 1980) and project both orally and anally within the intestinal wall. Furthermore, they receive an excitatory input from excitatory cholinergic interneurones which form pathways within the myenteric plexus that are up to 5 cm long (FRANCO et al. 1979c). Noncholinergic excitatory junction potentials have been recorded from the circular muscle of the guinea-pig small intestine (BYWATER and TAYLOR 1979). The role of these neurones is at present unknown.

Another example of excitatory noncholinergic transmission has been found in the proximal colon of the guinea-pig. Transmural stimulation produced a prolonged contraction of the longitudinal muscle which was insensitive to hyoscine

(COSTA and FURNESS 1972; FURNESS and COSTA 1973). Such contractions, which last for up to 5 min, can be abruptly interrupted by tetrodotoxin, indicating that they are due to the prolonged firing of noncholinergic neurones. The blockade of these responses by methysergide and by exposure to 5-hydroxytryptamine (5-HT) led to the suggestion that an indoleamine might be released from the nerves (see Chap. 11). In addition, antinicotinic drugs and even guanethidine antagonised the contractions although there is compelling evidence that these neurones are not noradrenergic. In unpublished observations in the authors' laboratory, it was shown that the contractions migrate slowly along the proximal colon. It is quite possible that this represents an example of the intrinsic migration of electrical activity in the enteric plexuses discusses in Sect. H.V.3.b). Responses with similar pharmacological characteristics were obtained in the circular muscle of the distal colon in response to distension as part of the *ascending excitatory reflex* (COSTA and FURNESS 1976; see Chap. 11). Neither of these noncholinergic responses in the guinea-pig large intestine is likely to be mediated by substance P since desensitisation of its receptors failed to affect the nerve-mediated responses (M. COSTA and J. B. FURNESS unpublished results 1980). The possibility that such responses may be mediated by an indoleamine has been received (COSTA and FURNESS 1979; Chap. 11); it seems unlikely that the nerves in the distal colon release 5-HT.

Transmural stimulation, following blockade of cholinergic muscarinic receptors, usually elicits a transient relaxation mediated by enteric inhibitory nerves, which is followed by a rebound contraction; it is often not possible to distinguish this myogenic rebound (BENNETT 1966b; CAMPBELL 1966a; FURNESS 1971) from the action of a noncholinergic spasmogen released from other nerve types. Noncholinergic excitatory nerves are possibly involved in the contraction obtained by transmural electrical stimulation in the guinea-pig distal colon by BENNETT and FLESCHLER (1969) and in human colon by STOCKLEY and BENNETT (1974).

C. Transmission and Pathways in Enteric Ganglia

Conclusions about transmission in the enteric ganglia can be reached from pharmacological studies of mechanical responses to nerve stimulation or more directly from studies using intracellular microelectrodes (see Chap. 6). From such studies it can be concluded that there are excitatory cholinergic inputs to many neurones in both the submucous and myenteric plexuses, and noncholinergic excitatory inputs to neurones in the myenteric plexus. In both plexuses there are also inhibitory nonadrenergic inputs to some neurones. It thus appears that both excitatory and inhibitory interneurones are present in the enteric plexuses. The extrinsic nerves act on the intestine mainly by modifying the activity of the intrinsic neurones. The nerve pathways within the enteric plexuses have been studied by eliciting intrinsic reflexes, by stimulating nerve pathways within the intestinal wall electrically or by demonstrating specific nerve pathways by histochemical techniques.

As will be discussed in Sect. J, responses of the muscle to the activation of intrinsic reflexes can be recorded several diameters of the intestine away from the point of initiation. Electrical stimulation has also demonstrated the presence of long nerve pathways within the intestine, although evoked potentials recorded extracellularly from nerve bundles could not be recorded from more than 10–15 mm in the longitudinal direction and 3 mm in the circular direction (YOKOYAMA et al. 1977; KOSTERLITZ and LYDON 1971).

Recent morphological studies have demonstrated pathways of enteric neurones which contain neuropeptides in the enteric plexuses (FURNESS and COSTA 1980). These studies show that populations of enteric neurones identified by immunohistochemistry have specific projection and polarity. Thus, those neurones in the myenteric ganglia which contain vasoactive intestinal polypeptide project anally (FURNESS and COSTA 1979). Neurones that contain somatostatin project anally to other myenteric neurones (COSTA et al. 1980a), while those which contain substance P project both anally and orally (COSTA et al. 1980b) and so do the fibres that contain enkephalin (FURNESS et al. 1980). These types of studies are likely to provide the morphological basis for understanding the reflex pathways within the intestinal wall.

D. Initiation of Sensory Inputs in the Intestine

The existence of intrinsic and extrinsic reflexes in the intestine (see Sects. J, K) and the detection of visceral pain from the abdominal cavity imply that sensory nerves are activated, either directly or indirectly, by mechanical and chemical stimuli in the intestine. However, receptors and sensory nerve fibres have not been identified structurally in the intestine, although some speculations about the structure of sensory detectors and the cellular basis for the activation of sensory nerves have been made. For example, it has been suggested that the unusual small dark smooth muscle cells and closely related interstitial cells and nerves of the inner layer of circular smooth muscle may form a mechanoreceptive system (GABELLA 1979b; DANIEL 1977). These muscle cells would thus be analogous to intrafusal fibres of skeletal muscle. It has also been suggested that 5-HT released from enterochromaffin cells by mechanical stimuli might stimulate sensory nerve endings (BÜLBRING and CREMA 1959). Motilin released from mucosal endocrine cells has also been proposed as an intermediate in stimulating sensory endings (LEWIS et al. 1979). It has been suggested that modified epithelial cells of the mucosa might be specific sensory receptor cells for different stimuli which activate specific classes of sensory neurones (FUJITA and KOBAYASHI 1978; MATSUO and SEKI 1978).

I. Intrinsic Sensory Neurones

The existence of sensory neurones within the enteric ganglia is implied by the demonstration of reflexes in segments of intestine disconnected from the central nervous system (see Sect. J.I). Several authors have reported that some neurones in the myenteric plexus are activated by mechanical distortion of the ganglia (OHKAWA and PROSSER 1972a; WOOD 1970, 1973, 1975; MAYER and WOOD 1975). Distension or stretching of the intestinal wall also initiates or increases the firing of some myenteric and submucous neurones (OHKAWA and PROSSER 1972b; YOKOYAMA and OZAKI 1978) and the contraction of the muscle elicited by acetylcholine increases the firing of myenteric neurones (OHKAWA and PROSSER 1972b). However, since many of the myenteric neurones begin to fire action potentials when they are mechanically distorted by a suction electrode used for extracellular recording (NORTH and WILLIAMS 1977), it is difficult to establish whether neurones themselves are mechanoreceptors or whether their activation by mechanical distortion is unnatural. No reports are available on nonsynaptically activated enteric neurones during mechanical or chemical stimulation. However, there is little doubt that cell bodies of sensory neurones of the intrinsic nerve pathways involved in peristalsis are present in the myenteric plexus and the mechanoreceptor nerve endings occur within the external musculature (Sect. J.I.2). Perfusion of the lumen with 0.1 *M* HCl increased the firing rate of some myenteric neurones which did not respond to distension (YOKAYAMA and OZAKI 1978), indicating that in the enteric plexuses there are cell bodies of chemosensitive neurones, as would be expected from investigations on chemically induced intrinsic reflexes (Sect. J.I.5).

The original proposal by NISHI and NORTH (1973a) and by HIRST et al. (1974) that the myenteric neurones which show a prolonged afterhypopolarisation following a directly evoked action potential (type 2, AH cell) are sensory was based on the apparent lack of synaptic input on these cells. This extrapolation is no longer justified since the discovery that these neurones receive synaptic inputs (GRAFE et al. 1979).

II. Activation of Extrinsic Sensory Nerves

There is ample evidence that a variety of stimuli applied to the intestine initiate intestinofugal nerve discharges in the extrinsic nerves supplying the intestine (PAINTAL 1963, 1973; IGGO 1966; SHARMA 1967; LEEK 1972, 1977). The activity could be located in the peripheral process of spinal sensory neurones or centrally directed processes of intrinsic enteric neurones which project outside the intestinal wall. The experiments do not always distinguish between these two possibilities.

Distension of the intestine has been shown to initiate intestinofugal discharge in mesenteric nerves (BESSOU and PERL 1966; MEI 1970; NINOMIYA et al. 1974), in the splanchnic nerves (GERNANDT and ZOTTERMAN 1946; RANIERI et al. 1975) and in the vagus nerve (IGGO 1957; CLARKE and DAVISON 1978). Some of the mesenteric fibres activated by distension arise from intrinsic cholinergic neurones which project to the postganglionic sympathetic neurones in the sympathetic ganglia forming the afferent limb of a peripheral sympathetic reflex (see Sect. K.I.1). Other fibres activated by distension have their cell bodies either in spinal ganglia or in the sensory ganglia of the vagus (LEEK 1972) and supply the central nervous system with a continuous monitoring of the state of the intestine. Some of these are involved in extrinsic intestinal reflexes (see Sect. K). The exact location of the peripheral endings of centrally directed neurones activated by distension is not well established but it is almost certain that some are located in the external musculature, some in the serosa and some in the mesentery (LEEK 1972, 1977).

Not only distension bus also contraction of the intestinal muscle is capable of initiating afferent activity. Thus, in addition to evidence that spasm of the intestine causes pain (see IRVING et al. 1937; LEEK 1972), increased afferent activity in splanchnic nerves occurs when a strong peristaltic wave passes down the intestine (GERNANDT and ZOTTERMAN 1946). PAINTAL (1957) also reported that, during what he described as peristaltic rushes in cat small intestine, the afferent discharge in the mesenteric nerves increased. Afferent activity was also recorded by BESSOU and PERL (1966) during peristaltic contractions but not with segmental contractions. Recently IANTSEV (1979) found that spontaneous muscle contractions and their related muscle action potentials in conscious dogs were correlated with the frequency of spike potentials in the mesenteric nerves. Contractions caused by acetylcholine or other muscle stimulants initiate afferent nerve discharge and prevention of the contraction by antagonist drugs abolished the discharge (GERNANDT and ZOTTERMAN 1946; BROWN and GRAY 1948; BENELLI and SANTINI 1974). Some of the afferent nerves which respond to distension also respond to contractions and are likely to represent tension receptors in series with the smooth muscle while those which respond to distension alone are most likely arranged in parallel with the muscle (length receptors; LEEK 1972).

Some authors have attributed the afferent activity which occurs during propagated contractions in the intestine to mucosal mechanoreceptors sensitive to stroking but not to distension and proposed that these mechanoreceptors are located in the smooth muscle of the muscularis mucosae or of the villi (PAINTAL 1957, 1963, 1973). Afferents fibres with similar properties were also found in the vagus (CLARKE and DAVISON 1978; MEI 1970). Recent evidence indicates that there are also thermoreceptors sensitive to both cooling and warming in the intestine (EL QUAZZANI and MEI 1979).

There is also ample evidence for the existence of extrinsic afferent nerves sensitive to the presence of various chemical stimuli in the lumen of the intestine. Thus, SHARMA and NASSET (1962) perfused the cat intestinal lumen and a Thiry–Vella loop in dogs with glucose and various amino acids, evoking afferent discharges in the mesenteric nerves. Similar receptors that are stimulated by glucose and other carbohydrates but not by distension or contraction were reported in the vagus nerve of cats (MEI 1978). Afferent fibres that are stimulated by glucose in the intestinal lumen are also found in the vagus, in splanchnic and in mesenteric nerves (SANFORD 1976; HARDCASTLE et al. 1978; ITINA and SERGEEV 1978). Some of these intestinal chemoreceptor fibres that are capable of detecting small changes in the composition in the lumen have been suggested to signal the composition of the intestinal content to the central nervous system continuously (LEEK 1972; MADDISON and HORRELL 1979).

The effect of osmotic and acid stimuli and fats in slowing gastric emptying (enterogastric reflex) indicates that there must be receptors in the duodenal mucosa activated by these stimuli (THOMAS 1957; HUNT and KNOX 1968). The indirect evidence for mechanoreceptors in the rectum involved in the extrinsic reflexes of defecation will be discussed in Sect. K.II.

E. Electrical Stimulation of Extrinsic Nerves

The intestine receives extrinsic nerves from the central nervous system via three routes: the vagus; the splanchnic and mesenteric nerves; and the pelvic nerves (KUNTZ 1953; MITCHELL 1953; PICK 1970). The vagus provides the oesophagus and the stomach directly and the small and large intestines indirectly by joining the mesenteric nerves. Pathways from the thoracic splanchnic nerve, which form the greater and the lesser splanchnic nerves, run through the coeliac and the superior mesenteric ganglia and from these pass to the intestine via the superior mesenteric nerves. The lumbar splanchnic pathways run through the inferior mesenteric ganglion and from there reach the large intestine via the inferior mesenteric (colonic) nerves and hypogastric (presacral) nerves. Pelvic outflows follow the pelvic nerves through the mixed ganglia of the pelvic plexus and from there to the large intestine; they also follow the pudendal nerves which supply the internal and the external anal sphincter.

The problem encountered with the simultaneous electrical stimulation of different types of intrinsic intestinal nerves is also found with the electrical stimulation of extrinsic nerves. Mixed responses are often obtained because excitatory and inhibitory pathways are activated simultaneously and, moreover, responses obtained can be due to orthodromic activation of efferent fibres or to antidromic activation of afferent fibres.

I. Vagus Nerve

Vagus nerve fibres supply the whole small intestine and the proximal part of the large intestine (MITCHELL 1953; SCHOFIELD 1968; and Chap. 2). The dominant effect of the stimulation of the vagus nerve on the muscle of the small intestine and proximal colon is excitatory. If the intestine is quiescent, rhythmic contractions are initiated by stimulating the vagus or existing rhythmic contractions increase in amplitude with partial fusion to give a sustained contraction. Excitatory responses were recorded in the small intestine and proximal colon in dogs, cats, rabbits and guinea-pigs and in the proximal colon in pigs and monkeys (BUNCH 1898; BAYLISS and STARLING 1899, 1901; MELTZER and AUER 1907; THOMAS and KUNTZ 1926a; CARLSON 1930; KEN KURE et al. 1931; RENTZ 1938; WELLS et al. 1942; GRAY et al. 1955; BLAIR et al. 1959; HARPER et al. 1959; FÜLLGRAFF and SCHMIDT 1963; STAVNEY et al. 1963; VAN HARN 1963; GONELLA 1964; NAKAYAMA 1965; HULTEN 1969; ROSTADT 1973; JULE 1975; ORMSBEE et al. 1976; TANSY and KENDALL 1977; MIR et al. 1978). These excitatory responses appear to be mediated by cholinergic neurones since, at least in dogs, cats and rabbits, atropine abolished them (BAYLISS and STARLING 1899, 1901; FÜLGRAFF and SCHMIDT 1963; STAVNEY et al. 1963; Van Harn 1963; Nakayama 1965; Hulten 1969; Rostad 1973; Rubin et al. 1979; TANSY et al. 1979; TELFORD et al. 1979).

Many of the authors who studied the effects of vagal stimulation on intestinal motility also observed inhibitory effects which usually occurred transiently before the excitation or were revealed after blockade of the excitatory response by atropine (BUNCH 1898; BAYLISS and STARLING 1899, 1901; KLEE 1912; THOMAS and KUNTZ 1926a; KURU and SUGIHARA 1955; NAKAYAMA 1965; FUKUDA 1968a; JULÉ 1975; ORMSBEE et al. 1976; VAN HARN 1963; TELFORD et al. 1979; RUBIN et al. 1979). These inhibitory effects were not blocked by guanethidine (TELFORD et al. 1979; JULÉ 1975) and it is most likely that they are mediated by the enteric inhibitory neurones also involved in the vagal relaxation of the lower oesophageal sphincter and stomach (Sect. B.IV.2).

Nicotinic blocking agents antagonized both excitatory and inhibitory responses of the intestine (THOMAS and KUNTZ 1926a; MULINOS 1927; NAKAYAMA 1965;

JULÉ 1975; TELFORD et al. 1979). This indicates that, for the stomach, there are preganglionic cholinergic nerves in the vagus nerve which synapse within the intestinal wall with enteric excitatory cholinergic neurones thus forming the *vagal excitatory pathway to the intestine*, and that there are also preganglionic cholinergic nerves which synapse, through nicotinic receptors, with enteric inhibitory neurones, forming the *vagal inhibitory pathway to the intestine*. These two pathways appear to be distinct. JULÉ (1975) found that in the rabbit, the preganglionic fibres of the vagal excitatory pathway to the proximal colon conduct at 1.0 m/s while the preganglionic fibres of the inhibitory pathway conduct at 0.5 m/s. On the basis of the different excitability to electrical stimulation MARTINSON and MUREN (1963) also reached the conclusion that the vagal fibres of the excitatory pathways to the stomach are larger in diameter (and these would conduct faster) than the fibres of the inhibitory pathway. In cats, KURU and SUGIHARA (1955) found that stimulation of the region of the dorsal vagal nucleus elicited excitatory effects in the stomach and jejunum while stimulation of the nucleus ambiguous and its surroundings elicited inhibitory responses.

The fact that these parasympathetic pathways have their excitatory and inhibitory neurones embedded within the intestinal wall led many writers of textbooks to regard the enteric neurones as postganglionic parasympathetic neurones. It is not possible at present to establish whether the postganglionic cholinergic neurones of the excitatory vagal pathway are the same neurones involved in the intrinsic reflex pathway (see Sect. J). If this is the case, then some enteric excitatory cholinergic neurones would represent a common final pathway for extrinsic and intrinsic nerve pathways just as the α-motor neurones are the common final pathway in the somatic motor system. Similarly, the enteric inhibitory neurones may represent the common final neurones of vagal inhibitory pathways and of intrinsic inhibitory reflexes. Although the role of these two vagal pathways to the intestine is not fully defined, they probably have independent functions as do the two pathways to the stomach (see Chap. 9). The excitatory pathways appear to be activated during the cephalic and gastric phases of feeding (see Sect. K.II) while there is no clearly identified role for the inhibitory pathways to the intestine. Both excitatory and inhibitory pathways to the small intestine are activated during nausea and vomiting (see Sect. K.II).

The path that the vagal fibres follow to reach the intestine has not been fully clarified. The early work of M. H. TAYLOR (in BAYLISS and STARLING 1899) and subsequent studies have shown that the vagus reaches the small intestine via the mesenteric nerves (SCHOFIELD 1968). More recent work has shown that the inhibition of the upper part of the duodenum appears to be mediated by vagal pathways to the antrum which reach the duodenum by running within the enteric plexus (DANIEL 1977; MIR et al. 1978).

II. Splanchnic and Mesenteric Nerves

The intestine receives most of its sympathetic nerves via the thoracic and lumbar splanchnic nerves and the superior mesenteric and inferior mesenteric (colonic) nerves. Stimulation of these nerves, however, also activates afferent and efferent nonsympathetic nerve fibres whose effects will be discussed separately below. The sympathetic innervation of the intestine has been extensively reviewed (FURNESS and COSTA 1974a, b) and only a summarised version of those reviews together with the relevant work published since then will be dealt with in this section.

1. Sympathetic Pathways: Noradrenergic Effects

The action of the sympathetic pathways on the movements of nonsphincteric parts of the intestine is inhibitory while the sympathetic innervation of the sphincters is excitatory. The sympathetic pathway to the intestine involves preganglionic cholinergic neurones with cell bodies located in the thoracolumbar spinal cord. Their axons pass in the ventral root, to join the ganglia of the sympathetic chain, via white rami. They pass through these ganglia without synapsing to enter the thoracic and lumbar splanchnic nerves. These preganglionic cholinergic neurones establish excitatory synapses through nicotinic receptors with postganglionic sympathetic neurones located in the prevertebral ganglia, i.e. coeliac, superior and inferior mesenteric ganglia, and pelvic ganglia. Recently, LUNBERG et al. (1978) demonstrated by retrograde transport of horseradish peroxidase that some neurones located in the cervical sympathetic ganglia project to the upper small intestine via the vagus nerve (see also Chap. 2).

The postganglionic sympathetic neurones to the intestine are noradrenergic (FURNESS and COSTA 1974a; Chap. 11). Most of them run in the superior mesenteric and inferior mesenteric (colonic) nerves to supply the small and large intestine. Some noradrenergic fibres also reach the distal rectum and internal anal sphincter via the pudendal nerves and the postganglionic nerve trunks arising from the pelvic ganglia. The nerve fibres form nerve trunks which run parallel to the blood vessels to reach the intestine. These are called paravascular nerves while those noradrenergic axons which innervate the mesenteric and intestinal blood vessels are called perivascular. Most of the noradrenergic axons, which can be easily visualized with fluorescence histochemical techniques for catecholamines, branch within the intestinal wall to end around the nerve cell bodies of the myenteric and submucous plexuses. A small proportion supply the circular musculature in nonsphincteric regions and they also innervate the blood vessels in the submucosa and supply the mucosa (FURNESS and COSTA 1974a). There is a substantial noradrenergic supply to sphincter smooth muscle.

From the work published up to 1972, it was apparent that a low frequency of activity the noradrenergic nerves act mainly on the enteric ganglia where they inhibit the release of acetylcholine from the excitatory cholinergic neurones by acting through α-adrenoceptors (FURNESS and COSTA 1974a). Soon afterwards it was shown that noradrenaline depressed the cholinergic excitatory postsynaptic potentials (e.p.s.p.) in the myenteric neurones (NISHI and NORTH 1973b) and that α-adrenoceptor agonists inhibited the firing of myenteric neurones (SATO et al. 1973). JULÉ and GONELLA (1972) and GARDETTE and GONELLA (1974) found that in the rabbit colon stimulation of the noradrenergic nerves at low frequencies caused the disappearance of the cholinergic e.j.p. elicited by pelvic nerve stimulation. HIRST and MCKIRDY (1974) found that mesenteric nerve stimulation abolished the cholinergic e.p.s.p. that are recorded from neurones in response to distension of the intestine or electrical stimulation. Since the e.p.s.p. were abolished in an all-or-nothing fashion and no changes in the electrical properties of the myenteric neurones were detected, they concluded that the inhibitory effect of the noradrenergic axons was mediated presynaptically. This conclusion is consistent with the recent ultrastructural studies of noradrenergic axons in the enteric ganglia in which no synapses were found between noradrenergic axons and myenteric nerve cell bodies (LLEWELLYN-SMITH et al. 1980). Recent evidence by GILLESPIE and KHOYI (1977) confirms the findings of BEANI et al. (1969), HULTÉN (1969) and ROSTAD (1973) that the sympathetic nerve can inhibit the cholinergic excitatory pelvic nerve responses at frequencies of stimulation which have little effect on the spontaneous muscle activity.

The only direct evidence available suggests that the noradrenergic nerves do not have a significant effect on the enteric inhibitory neurones since stimulation of noradrenergic nerves does not affect the amplitude of the i.j.p. elicited by transmural stimulation of the rabbit colon (JULÉ and GONELLA 1972). At high frequencies of activity noradrenergic nerves can have a direct inhibitory effect on the muscles of the circular layer (GONELLA and LECCHINI 1971; JULÉ and GONELLA 1972; HIRST and MCKIRDY 1974; for earlier references see FURNESS and COSTA 1974a). All these results confirm the view that the major action of the sympathetic nerves which affect motility in the intestine is to isolate the intestinal smooth muscle from its neural motor control.

There is no convincing evidence that noradrenergic nerves can be excitatory to nonsphincteric parts of the intestine. The reversal by ergotamine of the contractions of the colon and rectum elicited by stimulation of the colonic and the hypogastric nerves in dogs (WELLS et al. 1942) deserves reinvestigation. The excitatory effects of the noradrenergic nerves to the sphincters are discussed in Sects. E.IV and E.V.

2. Non-Noradrenergic Effects of Stimulating Sympathetic Pathways

The presence of excitatory fibres in sympathetic pathways to the nonsphincteric parts of the intestine has been recognised by many authors (see CAMPBELL 1970). In most cases excitation of the intestinal muscle following electrical stimulation of these nerves could only be unmasked by using drugs such as reserpine and guanethidine which abolish transmission from noradrenergic nerves (VARAGIC 1956; DAY and RAND 1961; GILLESPIE and MACKENNA 1961; BENTLEY 1962; BOYD et al. 1962; VAN HARN 1963; AKUBUE 1966; FUKUDA 1966; NG 1966; KEWENTER et al. 1970; GONELLA and LECCHINI 1971; SZOLCSANYI and BARTHO 1978), although with low strengths and durations of pulses some workers could demonstrate excitatory responses mixed with the inhibitory noradrenergic responses (FINKLEMAN 1930; KEN KURE et al. 1931; NEWMAN and THIENES 1933; MUNRO 1953; SEMBA 1956; SEMBA and HIRAOKA 1957; KEWENTER et al. 1970). Such excitatory responses could be due to the orthodromic stimulation of preganglionic vagal or pelvic fibres which reach the intestine via the mesenteric nerves or to the antidromic stimulation either of spinal sensory neurones or of intrinsic cholinergic neurones projecting from the enteric plexuses to the prevertebral ganglia (see Sect. K.I.1.a).

a) Antidromic Stimulation of Sensory Fibres

Stimulation of thoracic and lumbar splanchnic nerves elicits intestinal contractions which are not abolished but are indeed potentiated by blockade of nicotinic transmission through sympathetic ganglia (KEN KURE et al. 1931; SEMBA 1956; SEMBA and HIRAOKA 1957; FUKUDA 1966). This suggests that these fibres do not synapse in the prevertebral sympathetic ganglia. Since similar excitatory responses can be elicited by stimulation of the dorsal but not of the ventral roots of the spinal cord (SEMBA and HIAOKA 1957; OHYA 1969), it is most likely that they result from the antidromic stimulation of spinal sensory neurones which supply the intestine.

The excitatory responses obtained by stimulating the superior mesenteric nerves after vagotomy (DAY and RAND 1961; VAN HARN 1963; GONELLA and LECCHINI 1971) are probably due to the activation of sensory fibres, although it cannot be excluded that they are due to stimulation of postganglionic parasympathetic fibres with their cells of origin in the prevertebral ganglia. In most species and preparations tested, these excitatory responses were abolished by atropine (SEMBA 1956; SEMBA and HIRAOKA 1957; DAY and RAND 1961; BOYD et al. 1962; FUKUDA 1966; OHYA 1969). There is no evidence for spinal cholinergic sensory neurones; the likely explanation for a cholinergic involvement in these antidromic sensory responses has been provided by the work of BARTHO and SZOLCSANYI (1978) and SZOLCSANYI and BARTHO (1978). These authors found that capsaicin contracted the guinea-pig small intestine by stimulating the nerve endings of extrinsic nerves which released an unidentified substance which in turn excited intrinsic cholinergic neurones. Because capsaicin itself blocked such responses and also blocked the cholinergic responses of mesenteric nerve stimulation, they concluded that stimulation of mesenteric nerves activates noncholinergic sensory fibres that are sensitive to capsaicin; these fibres in turn excite, via a collateral branch, intrinsic cholinergic neurones. This interpretation is supported by the observations that nicotinic receptor antagonists neither antagonise such excitatory responses (DAY and RAND 1961; AKUBUE 1966; SZOLCSANYI and BARTHO 1978) nor do they antagonise the excitatory responses recorded extracellularly in some neurones following stimulation of mesenteric nerves (OKHAWA and PROSSER 1972b; TAKAYANAGI et al. 1977). With intracellular recording from myenteric neurones, excitatory transmission in response to stimulation of the mesenteric nerves has been described by HIRST and MCKIRDY (1974). However, no antinicotinic drugs were tested on these responses. It is likely that the cholinergic responses to mesenteric nerve stimulation, which are more easily demonstrated in early development (DAY and RAND 1961; BURN 1968; GERSHON and THOMPSON 1973; GULATI and PANCHAL 1978) when the noradrenergic effects are absent, are also due to antidromic stimulation of extrinsic sensory nerves.

b) Admixed Parasympathetic Fibres

Not all these cholinergic responses appear to be due to the antidromic stimulation of sensory nerves. For instance, the excitatory cholinergic response to lumbar colonic nerve stimulation appears to be due to the activation of sacral parasympathetic fibres which reach the colon via the colonic nerves (GILLESPIE and MACKENNA 1961). A number of authors also found that cholinergic responses to mesenteric nerve stimulation were abolished or reduced by antinicotinic drugs (GILLESPIE and MACKENNA 1961; BENTLEY 1962; PATON and VANE 1963; AKUBUE 1966; NG 1966), suggesting that they may be due to true parasympathetic preganglionic cholinergic fibres running in the mesenteric nerves, as had been suggested by TELFORD and STOPFORD (1934).

There is evidence that stimulation of the colonic nerves can activate enteric inhibitory nerves (COSTA et al. 1975). It is possible that such non-noradrenergic pathways are part of the sacral parasympathetic inhibitory pathways to the intestine which reach the intestine with the colonic nerves. Other excitatory responses which are neither cholinergic nor adrenergic have been described in both small and large intestine (MUNRO 1953; GRAY et al. 1955; BENTLEY 1962; FÜLGRAFF et al. 1964; RAND and RIDEHALGH 1965; HULTÉN 1969; GOLDENBERG and BURNS 1971; ROSTAD 1973; FASTH et al. 1978, 1980) but not investigated further.

III. Pelvic Nerves

The pelvic nerves supply most of the large intestine, including the internal anal sphincter. Studies in different species have shown that there are cholinergic, noradrenergic and unidentified neurones in the pelvic nerve pathways. Most authors agree that the prevailing response of the large intestine to electrical stimulation of the pelvic nerves is excitatory. Contractions of the large intestine including proximal, distal colon and rectum following stimulation of pelvic nerves have been recorded in dogs (BAYLISS and STARLING 1900; CARLSON 1930; WELLS et al. 1942; GRAY et al. 1955; GOLDENBERG and BURNS 1971), cats (COURTADE and GUYON 1897; LANGLEY and ANDERSON 1895; FÜLGRAFF et al. 1964; HULTÉN 1969; ROSTAD 1973; GARDETTE and GONELLA 1974; FASTH et al. 1980), rabbits (LANGLEY and ANDERSON 1895; BAYLISS and STARLING 1900; GARRY and GILLESPIE 1955; LEE 1960; GILLESPIE 1962; BOYD et al. 1962; FURNESS 1969a; JULÉ and GONELLA 1972; GARDETTE and GONELLA 1974), guinea-pigs (LEE 1960; RAND and RIDEHALGH 1965; DEL TACCA et al. 1968; BIANCHI et al. 1968; BEANI et al. 1969; FURNESS 1969a), monkeys and pigs (WELLS et al. 1942).

The excitatory effects of pelvic nerve stimulation are more prominent in the distal colon and rectum but the proximal colon also responds with some delay and the distal ileum may even respond. There is, therefore, overlapping of the fields of innervation of the vagus and the pelvic nerves where they supply the proximal to middle portion of the colon. The pelvic nerves are very effective in contracting both circular and longitudinal muscle layers. GILLESPIE (1962) and FURNESS (1969a) found that in guinea-pigs and rabbits e.j.p. could be recorded in almost all the longitudinal smooth muscle cells while in the circular muscle e.j.p. could be recorded only in 20% of cells. The e.j.p. summate and trigger action potentials which result in the contraction of the muscle (GILLESPIE 1962; FURNESS 1969a; JULÉ and GONELLA 1972).

In most cases where antimuscarinic drugs have been tested the excitatory responses are either abolished (WELLS et al. 1942; GARRY and GILLESPIE 1955; LEE 1960; RAND and RIDEHALGH 1965; BIANCHI et al. 1968; FURNESS 1969a; JULÉ and GONELLA 1972; GARDETTE and GONELLA 1974) or significantly reduced or delayed (LANGLEY and ANDERSON 1895; GRAY et al. 1955; FÜLGRAFF et al. 1964; DEL TACCA et al. 1968; HULTÉN 1969; GOLDENBERG and BURNS 1971; ROSTAD 1973; FASTH et al. 1980). The remaining excitatory responses were interpreted by many authors as due to noncholinergic excitatory neurones. Such atropine-resistant responses are particularly prominent in dogs. SEMBA and MIZONISHI (1978) have traced the origin of these pathways to various areas of the grey matter of the spinal cord. YA-

MAMOTO et al. (1978) found nerve cell bodies in the cord projecting to the cat rectum. An adequate pharmacological evaluation of these responses has not yet been performed. Although several workers did not detect any inhibitory effect to pelvic stimulation, it is interesting that many others described a transient initial relaxation of the large intestine (LANGLEY and ANDERSON 1895; COURTADE and GUYON 1897; BAYLISS and STARLING 1900; RAND and RIDEHALGH 1965; FÜLGRAFF et al. 1964). Following administration of atropine or hyoscine, several authors recorded either relaxation or transient hyperpolarisation (i.j.p.) in the large intestine (JULÉ and GONELLA 1972; GARDETTE and GONELLA 1974; FASTH et al. 1980). From the characteristic time course of the i.j.p. and the insensitivity of transmission of adrenergic neurone-blocking drugs it is deduced that these inhibitory pathways involve the enteric inhibitory neurones (see Sect. B.IV.2).

In the few experiments in which antinicotinic drugs were used, both inhibitory and excitatory responses to pelvic nerve stimulation were abolished. With stimulation at the posterior pelvic plexus, which includes small ganglia (BIANCHI et al. 1968), antinicotinic drugs only reduced the response, suggesting that some of the postganglionic neurones in the sacral parasympathetic excitatory pathway lie outside the intestine (LANGLEY and ANDERSON 1895). Electrophysiological studies by DE GROAT and KRIER (1976) confirmed that these extramural pelvic ganglia contain cell bodies of the postganglionic neurones to the intestine. All these results indicate that, in the sacral parasympathetic pathways there are excitatory and inhibitory pathways, both involving cholinergic preganglionic neurones, which synapse either in the small ganglia of the posterior pelvic plexus or within the enteric plexuses with cholinergic excitatory or enteric inhibitory postganglionic neurones. While the sacral excitatory pathways are involved in defecation (Sect. K.II) the sacral inhibitory pathways are likely to be related to the function of the large intestine as a reservoir which requires, like the accommodation of the stomach to food intake, a set of inhibitory nerves involved in a reflex relaxation (FASTH et al. 1980).

IV. Extrinsic Nerves to the Ileocaecal Sphincter

There is general agreement that there is a zone of elevated pressure at the ileocaecal junction separating the small and large intestine (ALVAREZ 1940; Kelley et al. 1965). Stimulation of the vagus nerve in the cat elicits a contraction of the sphincter muscle which is abolished by atropine, indicating that excitatory cholinergic nerves are involved (PAHLIN and KEWENTER 1976b; RUBIN et al. 1979). HINRICHSEN and IVY (1931) and JARRETT and GAZET (1966) described a relaxation of the sphincter that often preceded the contraction, particularly at low frequencies of stimulation, in dogs and cats; this result suggests that in the vagus there are inhibitory pathways to the sphincter similar to those to the small intestine and proximal colon. The inhibitory pathways were more easily unmasked by HINRICHSEN and IVY (1931) when the sphincter was contracted reflexly. Stimulation of splanchnic, mesenteric and colonic nerves elicits contraction of the sphincter and adrenergic neurone-blocking drugs or drugs wich antagonise α-adrenoreceptors abolish these contractions, indicating that the noradrenergic nerves are excitatory to the sphincter (ELLIOTT 1904; DALE 1906; HINRICHSEN and IVY 1931; SMETS 1936; JARRETT and GAZET 1966; PAHLIN and KEWENTER 1976a).The action of the extrinsic noradrenergic nerves appears to be a direct one on the sphincter muscle through excitatory α-adrenoceptors, unlike the nonsphincteric parts of the intestine where these nerves act primarily on the enteric ganglia (see FURNESS and COSTA 1974a; and Sect. E.II.1).

V. Extrinsic Nerves to the Internal Anal Sphincter

The internal anal sphincter receives inputs via the colonic nerves, the hypogastric nerves, the pelvic nerves and the pudendal nerves. LANGLEY and ANDERSON described contractions of the internal anal sphincter in the cat following stimulation of either the hypogastric or the colonic nerves (LANGLEY and ANDERSON 1895) and found that stimulation of the second, third and fourth lumbar spinal nerves elicited a contraction of the sphincter. The excitatory action of the nerves of sympathetic origin on the internal anal sphincter has since been confirmed in this and several other species including humans (see GARRY 1934; FURNESS and COSTA 1974a; COSTA and FURNESS 1974; GARRETT et al. 1974; RAYNER 1979). There are some exceptions. In the rabbit, stimulation of colonic or hypogastric nerves elicits mainly a relaxation of the sphincter (LANGLEY and ANDERSON 1895; FRANKL-HOCHWART and FRÖHLICH 1900) and in humans, stimulation of the hypogastric nerves elicits relaxation of the sphincter (SHEPHERD and WRIGHT 1968). The sympathetic excitatory responses appear to be mediated by noradrenergic nerves which act directly on the sphincteric musculature through excitatory α-adrenoreceptors (COSTA and FURNESS 1974; FURNESS and COSTA 1974a; GARRETT et al. 1974; RAYNER 1979). Relaxation of the sphincter to sympathetic nerve stimulation was obtained in some cases after blockade of α-receptors and was mediated by inhibitory β-receptors (LEARMONTH and MARKOWITZ 1934; GARRETT et al. 1974). Excitatory noradrenergic nerves reaching the sphincter via the pudendal nerves have been described in the guinea-pig by COSTA and FURNESS (1974) but their presence in other species has not been investigated.

From the few studies performed on the types of extrinsic nerves which supply the internal anal sphincter, it appears to be supplied, as is the rest of the large intestine, by excitatory cholinergic pathways and inhibitory pathways which involve the enteric inhibitory nerves, both pathways running with the sacral parasympathetic nerves. Unlike the colon and rectum, the internal anal sphincter receives some excitatory noradrenergic fibres from the pelvic and sacral nerves. These pathways together with connections from the rectum are involved in the maintenance of continence and in the process of defecation (see Sect. J.I, K.II).

LANGLEY and ANDERSON (1895) and LANGLEY (1911) reported that stimulation of the roots of the sacral nerves caused the sphincter of the rabbit to relax and rarely to contract while in the dog FRANKL-HOCHWART and FRÖLICH (1900) recorded only constriction of the sphincter. In the cat, LANGLEY and ANDERSON (1895), FRANKL-HOCHWART and FRÖLICH (1900) and GARRETT et al. (1974) recorded a mixture of contraction and relaxation. In the guinea-pig, COSTA and FURNESS (1974) found that stimulation of pelvic nerves elicited primarily a strong contraction of the sphincter; RAYNER (1979) observed in monkeys either a small contraction or no contraction during stimulation and a rebound contraction at the end of the stimulation. It appears that at least three different pathways can be activated by stimulation of these nerves. In the guinea-pig, hyoscine abolishes the contraction of the sphincter to pelvic nerve stimulation, indicating that cholinergic excitatory nerves are involved in the response. After administration of hyoscine the response is reversed into a non-noradrenergic relaxation, suggesting that the enteric inhibitory neurones are involved. Stimulation at high frequency elicited a noncholinergic contraction which was abolished by antiadrenergic drugs, indicating that a few excitatory noradrenergic fibres reach the sphincter via the pelvic nerves (COSTA and FURNESS 1974). In the cat, hyoscine and blockade of α- and β-adrenoceptors reversed the contraction into a relaxation, indicating that these three pathways are also present in this species (GARRETT et al. 1974). In monkeys, the presence of a rebound contraction at the end of the stimulation strongly suggest that an inhibitory pathway is also activated in the sacral parasympathetic outflow (RAYNER 1979). The finding that hexamethonium did not affect the inhibitory responses of the cat sphincter when the sacral nerves were stimulated (GARRETT et al. 1974) strongly suggests that many of the cell bodies of these unidentified inhibitory neurones are located outside the intestinal wall.

F. Conclusions on Neuromuscular and Ganglionic Transmission in the Intestine

Intestinal smooth muscle is continuously undergoing regular fluctuations of excitability (slow waves) which synchronise the contractions of large groups of smooth muscle cells. In the small intestine the slow waves propagate anally in most cases, while in the large intestine they are more or less stationary. The nerves exert their inhibitory and excitatory actions on these periodic fluctuations of smooth muscle excitability, their effectiveness depending on the phase of the slow wave cycle. However, action potentials and the resulting contractions initiated spontaneously or by nerves do not propagate for more than about 5 mm. Thus, any propagated contractions result from the propagation of nerve activity along the intestine. Although there are no specialised junctions between nerves and muscle in the intestine, transmission from intrinsic excitatory and inhibitory neurones is very effective and even single nerve pulses elicit either transient depolarisations (excitatory junction potentials e.j.p.) or hyperpolarisations (inhibitory junction potentials, i.j.p.). Nerve-mediated contractions result from action potentials triggered by summated excitatory junction potentials. It appears that, with few exceptions, only intrinsic neurones in the myenteric plexus act directly on the nonsphincteric muscle while the extrinsic nerves act indirectly by inhibiting or exciting enteric neurones. In contrast, noradrenergic nerves have a direct excitatory action on sphincteric muscle.

The two main types of enteric neurones which act on the intestinal muscle are the *enteric excitatory cholinergic neurones* and the *enteric inhibitory neurones*. The enteric excitatory cholinergic neurones receive inputs from other cholinergic interneurones and also receive cholinergic inputs from preganglionic nerves via the vagus and the pelvic nerves. Some of the enteric cholinergic neurones project centripetally to the prevertebral ganglia where they excite postganglionic sympathetic noradrenergic neurones. The cholinergic neurones are inhibited by the extrinsic noradrenergic nerves which act either directly on them or indirectly by presynaptic inhibition of excitatory interneurones. It will be shown in Sect. J that these cholinergic neurones form the final neurone in the enteric ascending excitatory reflex. The enteric inhibitory neurones receive excitatory inputs from cholinergic interneurones and from extrinsic preganglionic cholinergic neurones via the vagus and the pelvic nerves. These enteric inhibitory neurones form the efferent neurone of the descending inhibitory reflex (see Sect. J).

There is direct evidence that, apart from the two types of efferent neurones and cholinergic interneurones, there are also *intrinsic sensory neurones* and a number of other possible inhibitory and excitatory interneurones which may utilise other transmitter substances. The intestine is also supplied by sets of extrinsic sensory nerves which detect the presence of different nutrients in the lumen, mechanical distension, contractions of the muscle and abrasion of the mucosa during the passage of contents.

G. Patterns of Motility of the Intestine

The progression of food from the stomach to the rectum is the result of the mechanical activity of the two external muscle layers of the intestine. A description of the pattern of this mechanical activity is necessary to investigate the role that nerves play. The main difficulty that investigators in this field face is that the accurate description of the patterns of behaviour of the intestine in its natural environment requires the use of recording devices that often interfere with the physiological pattern, and the records of these studies are often difficult to interpret. The following description of the patterns of intestinal motility if the result of observations made by several authors and obtained by a variety of methods, including electrical and mechanical recordings of intestinal muscle activity, radiographic and cineradiographic studies, direct observations of intestinal motility through abdominal windows and the use of various markers for intestinal transit.

Differences in the pattern of motility are found in different species with different feeding habits and diet, in different states of feeding and in different parts of the intestine. The luminal contents undergo dramatic changes in volume, in physical state and in chemical composition as they progress through the digestive tube. The amount of digesta that enters the intestine depends on the rate of gastric emptying which is under a complex neurohormonal control (see Chap. 9).

I. Patterns of Motility of the Small Intestine

Since in the omnivorous and carnivorous species such as human, dog, cat, and rat the motility in the fasted state differs significantly from that in the fed state, the two patterns of motility will be described separately, while the pattern of motility in herbivores is not significantly affected by feeding.

1. Pattern in Fasted Human, Dog, Cat, and Rat: The Migrating Myoelectric Motor Complex

The presence of cyclical changes in the pattern of intestinal motility in fasted dogs and humans was noticed by some early workers (LEGROS and ONIMUS 1869; CASTLETON 1934; DOUGLAS and MANN 1939a; KYFOREVA 1948; FOULK et al. 1954;

Reinke et al. 1967) as periods of intestinal quiescence followed by periods of irregular and then regular changes in intraluminal pressure.

In more recent years this cyclical pattern has been studied by recording the electrical activity of the intestinal muscle in animals and humans with electrodes permanently implanted in the intestinal wall and by recording the corresponding changes in intraluminal pressure resulting from circular muscle contractions (Szurszewski 1969; Grivel and Ruckebusch 1972; Code and Schlegel 1974; Code and Marlett 1975; Ruckebusch and Fioramonti 1975; Vantrappen et al. 1977). The term "interdigestive" has been introduced to indicate that this cyclical pattern in these species only occurs in the fasted state. The following sequential phases in the cycle have been described: phase I in which the intestine is almost completely quiescent with no action potentials superimposed on the slow waves of the muscle; phase II with irregular contractions of the circular muscle with increasing proportions of slow waves showing action potentials (irregular spiking activity); phase III in which each slow wave has superimposed action potentials (regular spiking activity) resulting in regular contractions of the circular muscle; and phase IV, a very brief phase of prompt disappearance of action potentials with return to the quiescence of phase I (Szurszewski 1969; Carlson et al. 1972; Grivel and Ruckebusch 1972; Code and Marlett 1975). The term "myoelectric complex" is sometimes used, more precisely, to refer to the active part of the cycle (phases II and III).

a) Propagation of the Myoelectric Complex

Szurszewski (1969) was the first to demonstrate that the interdigestive myoelectric complex in dogs propagates slowly from the duodenum to the ileocaecal valve and that the repeated appearance and propagation of these complexes resulted at any one point of the small intestine in the cyclical pattern of activity. The migration of the myoelectric complex has been confirmed in fasted dogs (Grivel and Ruckebusch 1972; Carlson et al. 1972; Code and Marlett 1975), humans (Fleckenstein et al. 1978; Stoddart et al. 1978; Vantrappen et al. 1977, 1978) and rats (Ruckebusch and Fioramonti 1975; Ruckebusch and Ferré 1974; Pascaud et al. 1978; Scott and Summers 1976). In cats, there is a more irregular cycle of motor activity represented by bursts of action potentials which elicit a tonic contraction which migrates anally (Weisbrodt 1974).

b) Origin of the Migrating Complex

Code and Marlett (1975) found that the gastric antrum in dogs displayed the same cyclical recurrent sequence of action potential activity as the duodenum, and that activity at the two sites was nearly simultaneous. Aizawa et al. (1978) and Itoh et al. (1978a) observed an almost synchronous interdigestive cyclical contraction of the lower oesophageal sphincter and the rest of the stomach and this was confirmed by Diamant et al. (1979a, b). Thomas and Kelly (1979) and Thomas et al. (1979) concluded that the migratory myoelectric complex in dogs appears first in the proximal stomach and then migrates caudally to the small intestine. Also in humans, the interdigestive myoelectric migrating complex appears in the fundic region of the stomach well correlated with lower oesophageal sphincter and antral activity (Lux et al. 1979b).

c) Frequency and Duration of the Cycle

Most authors have reported that in dogs the interdigestive motor pattern recurs at fairly constant intervals of 80–170 min at reasonably constant times of the daily cycle (CARLSON et al. 1972; GRIVEL and RUCKEBUSCH 1972; CODE and MARLETT 1975; SUMMERS et al. 1976; ORMSBEE and MIR 1978; BUENO et al. 1975). For humans, most authors have found a cycle of 110–120 min (FLECKENSTEIN et al. 1978; VANTRAPPEN et al. 1978; ARCHER et al. 1979; READ 1979). In the rat, the frequency is much higher; the pattern recurs every 10–20 min (RUCKEBUSCH and FERRÉ 1974; RUCKEBUSCH and FIORAMONTI 1975; PASCAUD et al. 1978; SCOTT and SUMMERS 1976). The cycle is not interrupted during sleep (CODE and MARLETT 1975; COCCAGNA et al. 1977) and it has no apparent relation with any of the sleep phases (COCCAGNA et al. 1977, VANTRAPPEN et al. 1977; FLECKENSTEIN et al. 1978; FINCH and CATCHPOLE 1979).

In humans, the quiescent phase lasts for about 75% of the cycle while the irregular phase II and the regular phase III comprise, respectively, 13% and 11% of the cycle (STODDART et al. 1978). Phase III, with regular spiking activity, lasts 4.5–11 min in humans (STODDART et al. 1978, VANTRAPPEN et al. 1977, 1978; ARCHER et al. 1979). In the dog the quiescent phase corresponds to about 49% of the cycle, and phases II and III, 42% and 9% respectively (CODE and MARLETT 1975); the regular spiking phase III lasts 5–8 min and depends on the distance along the intestine, being longer in the upper intestine (SZURSZEWSKI 1969; CARLSON et al. 1972; BUENO et al. 1975; CODE and MARLETT 1975).

d) Speed of Propagation of the Migrating Complex

The speed at which the myoelectric complex propagates along the small intestine in humans varies from 12 cm/min in the duodenum to 6–8 cm/min in the lower ileum (FLECKENSTEIN et al. 1978; VANTRAPPEN et al. 1977, 1978). The length of the segment of intestine that is undergoing the regular spiking activity during the passage of the myoelectric complex has been calculated to be about 40 cm in the duodenum and 25–35 cm in the jejunum (VANTRAPPEN et al. 1977, 1978; FLECKENSTEIN et al. 1978; HELLEMANS et al. 1978).

The speed of caudal propagation of the myoelectric complex in the dog also decreases with the distance from the pylorus, being 5.3–11.7 cm/min in the upper duodenum and 1–2 cm/min in the lower ileum (SZURSZEWSKI 1969; GRIVEL and RUCKEBUSCH 1972; CARLSON et al. 1972; CODE and MARLETT 1975). The resulting length of the segment of small intestine which is undergoing the regular phase of the cycle during the passage of the migrating complex is 32 cm in the duodenum according to SZURSZEWSKI (1969) and 42–62 cm according to CODE and MARLETT (1975) and 5–10 cm in the lower ileum.

When one complex reaches the ileum another is already developing in the proximal small intestine (SZURSZEWSKI 1969). A variable proportion of migrating myoelectric complexes are not propagated for the whole length of the small intestine (SZURSZEWSKI 1969; CARLSON et al. 1972; GRIVEL and RUCKEBUSCH 1972; CODE and MARLETT 1975). In rats, the myoelectric complex migrates at about 2–3 cm/min (RUCKEBUSCH and FERRÉ 1974) while in cats the more irregular bursts of activity migrate at 5–12 cm/min (WEISBRODT 1974).

e) Relation Between Migrating Complex and Propulsion

Several authors have investigated canine and human propulsive ability of the different motor patterns which occur in the different phases of the motor cycle. During phase I the intestine is quiescent and little or no movements of the intestinal contents occurs (BUENO et al. 1975; SUMMERS et al. 1976; SUMMERS and DUSDIEKER 1979). Phase II is characterised by the irregular appearance of contractions caused by the occurrence of bursts of action potentials on a few of the slow waves. At the beginning of phase II the intermittent contractions are localised (SARR and KELLY 1979; SUMMERS and DUSDIEKER 1979) and produce a movement of the luminal contents to and fro, similar to the movements of rhythmic segmentation described by CANNON (1911) and CODE and SCHLEGEL (1974).

As spike activity increases, clusters of 8–15 successive contractions propagate for short lengths of intestine and appear every 3–6 min (SUMMERS and DUSDIEKER 1979). These short propagated contractions propel the contents for short distances as has been demonstrated by cinefluorography (MCLAREN et al. 1950; CODE and SCHLEGEL 1974). The contractions gradually become more rhythmic and powerful and propagate for longer distances as they become progressively more propulsive. The speed of these propagated contractions is of the order of 2–5 cm/s. In the human jejunum these contractions appeared regularly at intervals of about 60–80 s (minute rhythm) and propagated for up to 100 cm (FLECKENSTEIN et al. 1978; RUPPIN et al. 1979). In some cases shorter bursts of action potentials migrate for a short section of the small intestine and were interpreted as short peristaltic rushes (MCLAREN et al. 1950; FLECKENSTEIN et al. 1978).

Phase III of the cycle consists of a distinct band of 50 or more consecutive contractions at the frequency of the slow waves which propagate uniformly for the whole length of the active band (about 40 cm in the upper intestine) at the speed of the propagation of the slow wave (from 17 cm/s in the duodenum to 0.5 cm/s in the lower ileum (CODE et al. 1968). These rhythmic propagated contractions are very effective in clearing the active segment of its contents with the maximal speed of propulsion of the intestinal contents occurring just in front of the advancing band of regular contractions (CODE and SCHLEGEL 1974; SUMMERS et al. 1976; SUMMERS and DUSDIEKER 1979).

In the rat, the maximal propulsion also occurs ahead of the advancing segment of regular activity, i.e. during the late phases of the irregular propagated contractions (RUCKEBUSCH and FIORAMONTI 1975; SCOTT and SUMMERS 1976; GUSTAVSSON 1978).

f) Role of the Migrating Complex

One of the remarkable features of the migrating myoelectric complex is that it represents an intense motor activity at a time when the luminal contents of the small intestine consist mainly of secretory products. It is significant in this respect that a secretory component of the interdigestive complex has been demonstrated by VANTRAPPEN et al. (1979) and LUX et al. (1979a). The occurrence of the activity front of the migrating myoelectric complex in the duodenum is preceded by an increase in bile acid, gastric acid and pepsin output and followed by a peak of HCO_3 and amylase secretion. The motor component in the fasted state would act by sweeping the secretions that occur between meals towards the large intestine. The contractions that occur in the stomach and intestine during the interdigestive state, and correspond to the hunger contractions described in humans, would flush the excess mucous and nonmucous secretions as was originally proposed by REINKE et al. (1967).

2. Effect of Feeding in Human, Dog, Cat, and Rat

a) Disruption of the Interdigestive Pattern by Feeding

In human, dog, cat, and rat, food intake disrupts the interdigestive motor pattern and, irrespective of the phase of the cycle, a uniformly irregular motor activity is initiated throughout the small intestine (McCOY and BAKER 1968; WEISBRODT 1974; BUENO et al. 1975; CODE and MARLETT 1975; CARLSON et al. 1977; HELLEMANS et al. 1978; STODDART et al. 1978; SUMMERS and DUSDIEKER 1979; THOMAS and KELLY 1979). An increase in the motor activity of the intestine after a meal has been noted by many authors (CASH 1886; HERTZ 1913; BARCROFT and ROBINSON 1929; CASTLETON 1934; PUESTOW 1937; DOUGLAS and MANN 1939a, b, 1940; GREGORY 1950; CODE et al. 1957; BASS et al. 1961; CONNELL 1961; BÁRÁNY and JACOBSON 1964; RAMORINO and COLAGRANDE 1964; MISIEWICZ et al. 1966; PRESHAW and KNAUFF 1966; DAHLGREN and SELKING 1972; TANSY and KENDALL 1973; SNAPE et al. 1979). Because of the short delay between ingestion of a meal and the increase of motor activity even in the ileocaecal region, the term "gastroileocolic reflex" was introduced (HERTZ 1913; CONNELL 1961; TANSY and KENDALL 1973). There seem to be two phases in the initiation of the postprandial pattern: an early increase of motility which occurs almost simultaneously throughout the small intestine down to the ileocaecal region and a delayed, maintained phase of increased motor activity which appears to be related to the arrival of food in the intestine (CODE et al. 1957; McCOY and BAKER 1968; CODE and MARLETT 1975; BALFOUR and HARDCASTLE 1975; CARLSON et al. 1977).

b) Duration of the Postprandial Pattern

In dogs, the postprandial pattern of motility lasts for 6–19 h after a meal of mixed food, the time depending on both the composition and the volume of the meal (BUENO et al. 1975; BUENO and RUCKEBUSCH 1976; CARLSON et al. 1977; RUCKEBUSCH and BUENO 1977a; DE WEVER et al. 1978). CODE and MARLETT (1975) found that after ingestion of 400 ml milk the fed pattern lasted for only 4 h before the reappearance of the interdigestive pattern. EECKHOUT et al. (1978a) and DE WEVER et al. (1978) investigated the effects of mixed foods of different caloric value and found that 30 kcal/kg initiated a postprandial pattern that lasted 325 min while 60 and 90 kcal/kg initiated fed patterns of 561 and 800 min, respectively. When equicaloric meals were supplied, medium triglyceride chains induced the longest periods of the postprandial pattern followed in order of efficacy by peanut oil, sucrose, and milk proteins. Similar results were reported by SCHANG et al. (1978). It appears therefore that the caloric value of food, more than the volume of ingested material, determines the length of the postprandial disruption of the cyclic motor pattern and that some nutrients are more effective than others.

c) Electrical Activity, Contraction, and Propulsion in the Fed State

During the postprandial period the incidence of slow waves associated with action potentials, and thus with contractions of the circular muscle, is higher on average than that occurring during phase II of irregular spiking activity of the interdigestive motor cycle, but is lower on average than the maximal incidence (100%) that occurs during phase III of regular spiking activity. Apart from an initial period of high incidence of contractions in the dog duodenum, about 25% of slow waves show action potentials during the fed pattern (CODE and MARLETT 1975; CARLSON et al. 1977), resulting in about 4–5 contractions/min.

It is interesting that this frequency corresponds well to the frequency of slow waves in the antrum which, after feeding, all have superimposed action potentials and are seen as propagated contractions propelling the food towards the pylorus (KELLY 1977). From the records of CARLSON et al. (1977) it is possible to recognise the synchrony between antral contraction and duodenal contractions after feeding. The presence of an antral rhythm of contractions in the duodenum superimposed on the faster slow wave duodenum rhythm has been described (ALVAREZ and MAHONEY 1922; MESCHAM and QUIGLEY 1938; DUTHIE et al. 1972). In humans, milk induces a pattern of motility in the duodenum in which 30%–80% of slow waves (11–12 cycles/min) show action potentials (STODDART et al. 1978), and the antral rhythm (3 cycles/min) is superimposed on the duodenal slow wave rhythm (COUTURIER et al. 1970; DUTHIE et al. 1972; WATERFALL et al. 1973). This synchrony disappears in the distal duodenum where the incidence of contractions becomes more irregular (MESCHAM and QUIGLEY 1938; LEWIS et al. 1979). In the dog, these postprandial waves of contractions usually propagate over the whole length of the duodenum at the speed of the slow waves, propelling the luminal contents into the jejunum (DOUGLAS 1941; VAN LIERE et al. 1945; ARMSTRONG et al. 1956; CONNELL 1961; QUIGLEY and LOUCKES 1962). In humans, the duodenum is slowly filled and nonpropagated irregular contractions appear, followed by caudally propagated contractions (CANNON 1907; FOULK et al. 1954; LAGERLÖF et al. 1974).

The patterns of motility that occur during the postprandial phase are highly variable and led most authors in the early twentieth century to attempt to identify simplified patterns. The most common classification was introduced by CANNON (1902, 1911) who identified two main patterns of motility – rhythmic segmentation and peristalsis. Despite this simplified classification, CANNON was aware of its inadequacy and described several variations of patterns of motility with more irregularities, incomplete segmentations and shorter and longer propagations of contractions. DOUGLAS and MANN (1939a) observed that, in the dog, rhythmic segmentation as described by CANNON is a rare event while a completely irregular pattern of circular muscle contractions with some contractions propagated for short distances is more common.

A description of the patterns of motility observed in conscious animals through an abdominal window demonstrates the complexity of motility in fed animals.

> The predominant motility seen through the abdominal window was the short rapid progressive peristalsis characteristic of the upper small bowel with its scanty liquid contents. This peristalsis conveyed the liquid contents rapidly and continuously in small spurts in the distal direction. The liquid contents coming from above accumulated in the lower small bowel which gradually became somewhat distended, like a roundish sausage. Here the per-

istalsis was intermittent, infrequent and tubular in type. It began as circular contractions. Each successive contraction started somewhat more distally and proceeded somewhat further than the preceding one. Finally, segments several centimetres long contracted forcefully around the solid contents and slowly propelled them distal. The contractions were repeated a few times until the visible bowel loops were empty [PARKKULAINEN 1962).

Most authors agree that the rate of propulsion is faster in the upper small intestine, where the contents are more liquid, and slower in the distal part of the small intestine where the contents become more viscous (INGELFINGER and ABBOT 1940; VAN LIERE et al. 1945; REYNELL and SPRAY 1956; CONNELL 1961; DERBLOM et al. 1966; SUMMERS et al. 1970; CANTOR and REYNOLDS 1957).

In the fed dog, the contents are quickly transported caudally so that a marker injected in the jejunum would pass through a loop of the lower ileum within 30–120 s (SUMMERS et al. 1976). This rate of transport is about three times greater than the average rate in the interdigestive state (BUENO et al. 1975; BUENO and RUCKEBUSCH 1978). The electrical and mechanical activity recorded during the digestive period in humans confirms that the small intestine undergoes scattered nonpropagated segmental contractions mixed with isolated or small groups of 3–6 contractions which propagate rapidly and propel the chyme caudally for up to 30 cm (SUMMERS and DUSDIEKER 1979).

3. Pattern of Motility in Herbivores

A cyclic motor pattern which migrates from the stomach to the ileocaecal valve is also present in sheep, rabbits (GRIVEL and RUCKEBUSCH 1972; RUCKEBUSCH and BUENO 1977a, b) calves (RUCKEBUSCH and BUENO 1973), horses (RUCKEBUSCH 1973) and pigs (BUENO and RUCKEBUSCH 1978). The major difference between herbivorous, carnivorous and omnivorous species is that in herbivores the cyclic motor pattern persists during normal feeding which lasts for long periods. This finding has led RUCKEBUSCH and his collaborators to query the use of the term "interdigestive housekeeper" to describe the migrating motor complexes. In both the rabbit and the pig, feeding is accompanied by an increase of spiking activity throughout the small intestine without disruption of the migrating complex (RUCKEBUSCH et al. 1971; BUENO and RUCKEBUSCH 1978).

The duration of the cycle in the sheep and rabbit (77–88 min and 70–120 min, respectively) is similar to that of the dog. In these species as in the dog, a new complex starts when the other arrives at the ileocaecal junction. This implies that the speed of propagation along the sheep small intestine, which is longer, is several times higher than that of the dog. The phases of the cycle are similar to those described in nonherbivores (Sect. G.I.1). In sheep jejunum, phase I of inactivity represents 33% of the cycle; phase II of irregular spiking activity, 60%; and phase III of regular spiking activity, 7% (RUCKEBUSCH and BUENO 1975). Closer to the pylorus, where the migrating motor complex appears to originate (RUCKEBUSCH and BUENO 1977b), the period of quiescence is very short (BUENO and RUCKEBUSCH 1978). During phase II of irregular activity isolated or regular sequences of propagated contractions of the circular muscle occur and travel for up to 150 cm at a speed of about 5–12 cm/s. These propagated contractions which propel the contents as discrete boluses occur, at regular intervals of about 1 min (minute rhythm) in the upper jejunum and 0.5 cycles/min in the jejunum, towards the end of phase II just before the regular phase (Ruckebusch and Bueno 1977 b). From the records of Ruckebusch and Bueno (1975) and Bueno et al. (1975) it is clear that most propulsive movements occur during this late phase II in synchrony with propagated bursts of action potentials. In contrast the phase of regular activity is characterised by a sustained increase in intramural pressure with shallow superimposed contractions at the same frequency as slow waves. The results is that the whole length of intestine undergoing this spiking activity appears tonically constricted with complete occlusion of the lumen and thus with no flow of digesta in it. This long constricted segment of intestine slowly travels anally and represents the front of the migrating myoelectric complex.

In the rabbit, a similar pattern of activity occurs although fewer studies have been made in this species (Grivel and Ruckebusch 1972; Burns et al. 1979). It is interesting that description of the rapidly propagated contractions observed in the normal unanaesthetised rabbit (Rödén 1937; Parkkulainen 1962; Burns et al. 1979) correspond very well to the description of the peristaltic rush observed in nonphysiological conditions by early authors (Alvarez 1940). It is ironic that very little is known about the pattern of motility of the guinea-pig small intestine which has often been taken as a model in investigations of the neuronal mechanisms of peristalsis in vitro (Sect. J I.3d). Bozler (1939) and Thouvenot and Harichaux (1963) have recorded groups of slow waves with superimposed action potentials in the guinea-pig. These groups recur cyclically every 22 min. They may represent the cyclic periods of regular activity of the migrating complex.

4. Conclusion and Summary

Despite the great variability of the patterns of motility in different species with different feedings habits, some common simple patterns of motility can be identified in the small intestine:

1) A quiescent state in which the slow waves do not have superimposed action potentials and therefore no mechanical activity occurs; this represents phase I of the interdigestive migratory motor complex and is rarely observed during the fed state.
2) An irregular pattern of nonpropagated contractions of the circular muscle wich results in movements to and fro of the intestinal contents with some net aboral transport of chyme. This pattern with its electrical correlate of variable incidence of action potentials on slow waves is found in both phase II of the interdigestive motor complex and during the postprandial state. (This pattern of activity appears to correspond to the rhythmic segmentation of Cannon.)
3) Irregular patterns of contractions which propagate for short distances at the speed of the slow waves and propel the contents. This pattern is also observed during phase II of irregular spiking activity of the migrating complex and during the fed state. The short propagated contractions have been described as short peristaltic rushes.
4) A regular pattern of propagated contractions at the speed of the slow waves. Two types can be distinguished: (a) During phase III of the migrating motor complex each slow wave is accompanied by action potentials and each resulting con-

traction is propagated in the caudal direction for the whole length of the active segment (about 30–40 cm in dogs) resulting in the complete emptying of the segment; (b) In the late part of phase II of the migrating complex, bursts of action potentials occur about every 80 s (minute rhythm) and give rise to contractions which are propagated for up to 100 cm. These propagated contractions are very effective in propelling the chyme anally. Similar rapidly propagated contractions which correspond to the peristaltic rushes described in the older literature occur during the fed state.

5) The slowly migrating front of activity of the interdigestive motor complex (phase III) which in dogs and humans can be the site of the rapidly propagated regular contractions already described or which in sheep can be composed of fused contractions of the whole length of the active segment (20–30 cm) which is contracted like a narrow tube. The contracted area slowly progresses caudally. Slowly travelling waves of contraction observed in the fed state correspond to the typical slow peristalsis described by BAYLISS and STARLING (1899) and CANNON (1911) and appear to be associated with propulsion of more solid intestinal contents. The relation between peristalsis and the migrating complex will be discussed in Sect. J.I.7.

II. Patterns of Motility of the Large Intestine

It is not possible to give a simple description of the patterns of motility of the large intestines of different species, first because there is not enough information in all cases and second because the interspecies morphological and functional differences in the large intestine are much greater than the differences in the small intestine. Excellent reviews on motility of the large intestine have been published (CANNON 1911; LARSON and BARGEN 1933; GARRY 1934; PUESTOW 1937; ALVAREZ 1940; TRUELOVE 1966; CONNELL 1968; MISIEWICZ 1975; DAVENPORT 1977). CHRISTENSEN (1978) provided a review of the problem. The absorption of water and electrolytes which begins in the small intestine is completed in the caecum and proximal colon. Thus, the semiliquid contents entering the large intestine in carnivores and omnivores are transformed into soft faecal matter which is further hardened and stored for relatively long periods, being intermittently transported towards the distal part of the large intestine to be evacuated. In herbivores the faecal matter remains soft and is transported and excreted in a more or less continuous fashion.

Thus, the large intestine must possess both storage and efficient propulsive capabilities. It is not surprising therefore that herbivores with higher proportions of undigested residues have a sacculated caecum with a large storing capacity while carnivores and omnivores have a rather straight and simple large intestine (GARRY 1934; CHRISTENSEN 1978). Despite the differences, there are some common features in the colons of different species. In all species, a pacemaker area can be identified from which spontaneous contractions originate and travel both towards the caecum (antiperistalsis) and towards the anus (peristalsis). In herbivores, such as guinea-pig, rabbit, and horse, this pacemaker area is located in a flexure of the large intestine which clearly separates the proximal part from the distal part (ELLIOTT and BARCLAY-SMITH 1904; CANNON 1902; HUKUHARA and NEYA 1968; SELLERS et al. 1979), while in the rat, cat, dog, and human it is located in the middle portion of the transverse colon (ELLIOTT and BARCLAY-SMITH 1904; BOEHM 1911;

Case 1914; Barcroft and Steggerda 1932; Hukuhara and Neya 1968; Christensen 1978). In the herbivores, this area clearly separates the thick homogeneous faecal material which fills the whole caecum and proximal colon from the hardened faeces of the distal colon which are formed into ovoid or ball-like pellets. In omnivores and carnivores, the distinction between soft and hard faecal matter is less well defined and cylindrical irregular faecal masses are usually formed. The large intestine can be quiescent for long periods while at other times it undergoes nonpropulsive contractions or some form of propulsive contractions. Elliott and Barclay-Smith (1904) described three basic types of movements in the colon of various species: *antiperistaltic* waves which drive the semifluid contents of the proximal colon towards the caecum; *peristaltic* waves which occur when the contents become harder and which drive the faecal matter to the distal colon; and powerful peristaltic waves which evacuate the distal colon and rectum during *defecation* (see review by Larson and Bargen 1933).

1. Antiperistalsis

In the rabbit caecum, after long intervals of immobility, antiperistalsis occurs in the proximal colon where haustrations partially separate the faecal matter. The haustral indentations are due to contractions of circular muscle which slowly travel towards the caecum at about 10 cm/min (Meltzer and Auer 1907; Rödén 1937). In the horse, antiperistaltic waves travel at about 4 cm/s und occur at the rate of about 4 cycles/min (Sellers et al. 1979). In the cat, the antiperistaltic waves occur at a rate of 6–7 cycles/min (Cannon 1902; Alvarez 1940), a frequency which corresponds to that of the slow waves (Christensen and Hauser 1971; Daniel 1975).

In humans, Case (1914) and Hurst (1919) described successive antiperistaltic waves which originate in the transverse colon near the hepatic flexure and travel towards the caecum. The point of origin corresponded to the ring of contraction in the transverse colon described by Boehm (1911) who stated that the contents proximal to this ring are usually undivided segmented masses are found between the ring and sigmoid colon. Antiperistaltic waves have been described in dogs (Barcroft and Steggerda 1932) and rats (Elliott and Barclay-Smith 1904; Hukuhara and Neya 1968).

A problem that has puzzled many investigators is the origin, in some species including humans, of the sacculations which appear to separate the faecal content, even in periods of quiescence (haustra). There is little doubt that there are two different types of haustration, one with a clear anatomical basis and the other entirely functional (Hawkins and Hardy 1950). The first type occurs, as a result of the state of relative contraction of the taeniae, where anatomically identifiable folds of the intestinal wall are formed, usually at points where blood vessels run between two taeniae and there is corresponding infolding of the circular muscle and mucosa. These fixed indentations have been clearly identified in rabbits (Rödén 1937), humans (Hawkins and Hardy 1950; Williams 1967) and dogs (Puestow 1937) and are also found after death (see Connell 1968).

The second type of haustration is due to deep indentations produced by rings of contraction of the circular muscle around the whole circumference (Williams

1967). The formation of these haustra was demonstrated by the same author who stimulated electrically an excised segment of human normal sigmoid colon distended with a barium enema and examined the formation of haustra radiologically. Functional haustral constrictions can either appear almost static or they slowly migrate, anally in the distal colon and in either direction in the proximal colon (RITCHIE 1968). Propulsive and nonpropulsive contractions in the large intestine thus have a similar appearance.

2. Peristalsis

In the human colon, periods of quiescence occupy 38% of the total time, while, for 60% of the time, some mechanical activity occurs, only 10% of which can be regarded as propulsive (ADLER et al. 1941; CONNELL 1968; RITCHIE 1968). The pattern of motility changes from nonpropulsive to propulsive in a continuous way and it is very difficult to decide when the transition occurs. RITCHIE (1968, 1971) who attempted a classification of such movements in humans, described the progression of single haustra as well as multihaustral progression. Several haustra may even fuse into a larger progressing segment containing faecal matter and finally the transport of a long segment of colonic content which was originally described by HOLTZKNECHT (1909) as a mass movement or peristalsis may occur. Most authors agree that in humans, the colonic content progresses distally by a series of infrequent large movements. According to the description by HOLTZKNECHT (1909) a long segment of the quiescent colon with deep rounded interhaustral folds suddenly loses its haustra and is converted in a featureless tube. A constriction develops at the proximal end and a wave sweeps the contents rapidly forward into the next portion of the bowel. Then the haustral pattern re-forms and the intestine returns to quiescence. Several authors subsequently reported very similar behaviour (CASE 1914; JORDAN 1914; HURST 1919; LARSSON and BARGEN 1933; BARCLAY 1936). These propulsive contractions were reported to start anywhere from the caecum to the pelvic colon (JORDAN 1914; HURST 1919). BARCLAY (1936) believed that the most common sites were near each end of the transverse colon. Such propulsive movements were reported to occur from almost imperceptible speeds up to 8 cm/s, and the distance travelled could be the whole extent of the proximal and distal colon or more commonly from the ascending colon to the transverse or from the transverse colon to the descending colon (CASE 1914; JORDAN 1914; HURST 1919; BARCLAY 1936; WILLIAMS 1967; HARDCASTLE and MANN 1968; RITCHIE 1968; TORSOLI et al. 1971). HOLDSTOCK et al. (1970) calculated the distance travelled by a marker capsule as 5–22 cm for each movement. Such mass movements, which have all the characteristics of peristalsis of solid or semisolid contents, occur at intervals of about 1 h but because they do not necessarily initiate defecation, there is no proper record of their incidence over long periods. In fact in humans, most peristaltic movements push the faecal contents only as far as the sigmoid colon, leaving the rectum empty. Despite the lack of a zone of high pressure in the rectosigmoid region (HILL et al. 1960), there is probably some functional mechanism for the slowing of the contents at this junction (CONNELL 1968; CHOWDHURY et al. 1976).

WILLIAMS (1967) gave convincing evidence that the mass movements (or mass peristalsis) are the result of propagated anular contractions of the circular muscle with concomitant contractions of the longitudinal muscle. This process of transport of faecal contents appears identical with the process of evacuation of enemas, except that the acute distension of the large intestine by the enema appears to trigger the propulsive wave of contraction which is often referred to as a "stripping wave" (WILLIAMS 1967). Detailed descriptions of the propulsive waves induced by barium enemas have been published (WILLIAMS 1967; HARDCASTLE and MANN 1968; TORSOLI et al. 1971). Also during the propulsive contractions elicited by enemas both the circular and longitudinal muscle layers contract (WILLIAMS 1967) pushing forward the contents while, in front of the advancing column of contents, the colonic wall loses all the circular indentations, even those which may have appeared to be stable structures (WILLIAMS 1967). Whether or not in the spontaneous mass movements or the stripping waves there is an active relaxation of the muscle in front of the advancing contraction has not been established, although both WILLIAMS (1967) and TORSOLI et al. (1971) maintained that this occurs. From the detailed study of WILLIAMS (1967) with enemas it appears that propulsive contractions can be initiated anywhere in the human large intestine, indicating that the mechanism for starting the waves is present in all areas. Once started these propulsive waves travel variable distances, usually from one part of the colon to the next, but they rarely reach the main body of the rectum. Like the spontaneous progression of faecal contents, the speed of the propulsive waves induced by enemas varied from barely perceptible to 8 cm/s.

In dogs, slowly travelling segmental contractions have been recorded by TEMPLETON and LAWSON (1931) and by PUESTOW (1937). In rabbits, vigorous peristaltic waves which push the contents forward have been described by RÖDÉN (1937), while in cats, deep circular muscle contractions appear following accumulation of contents in the proximal colon and slowly push the contents forward, separating them into scybala (CANNON 1902). In most species, peristalsis originates from all parts of the large intestine, but mainly from the proximal colon and from the upper part of the distal colon (ELLIOTT and BARCLAY-SMITH 1904). Antiperistaltic waves appear to give way to peristaltic waves when the contents become harder, as observed in rats by ELLIOTT and BARCLAY-SMITH (1904). These and other authors claim that these propulsive waves of contractions differ from the very strong contractions which appear during defecation.

3. Defecation

The significant differences in the manner in which different species deal with the gastrointestinal excretory products is exemplified by the formation of faeces in humans and in guinea-pigs. In humans, the frequency of bowel action is 3–11 times per week (CONNELL and SMITH 1974; MARTELLI et al. 1978) while the guinea-pig produces an average of over 100 faecal pellets per day in an almost continuous fashion (M. COSTA and J. B. FURNESS unpublished work 1974). Since the process of defecation involves not only the intestinal musculature but also the striated musculature of the external anal sphincter and of the pelvic floor and also involves specific postures of the body as well as being, in humans, particularly highly controlled by social and behavioural habits, it is not surprising that a significant amount of literature is devoted to this function (see GARRY 1934; ALVAREZ 1940; TRUELOVE 1966; SCHUSTER 1968; MENDELOFF 1968; IHRE 1974; DAVENPORT 1977).

Table 2. Increased colonic motility by feeding

Species	Type of response	Reference
Human	Mass peristalsis	MacEwen (1904)
	Mass peristalsis and defecation	Holzknecht (1909)
	Defecation	Cannon (1911)
	Peristalsis	Hertz (1913)
	Peristalsis	Hertz and Newton (1913)
	Peristalsis	Hurst (1919)
	Propulsive contractions	Adler et al. (1941)
	Increased pressure waves	Davidson et al. (1956)
	Segmentation	Ramorino and Colagande (1964)
	Increased pressure waves	Connell et al. (1965)
	Increased motility index	Deller and Wangel (1965)
	Increased pressure waves	Misiewicz et al. (1966)
	Propulsive waves	Ritchie (1968)
	Propulsion and segmentation	Holdstock and Misiewicz (1970)
	Propulsion and segmentation	Holdstock et al. (1970)
	Increased pressure waves	Dahlgren and Selking (1972)
	Increased spike electrical activity	Wienbeck and Janssen (1974)
	Increased pressure wave and electrical activity	Snape et al. (1978, 1979)
	Increased pressure wave and electrical activity	Sullivan et al. (1978)
Monkey	Increased pressure waves	Sillin et al. (1978)
Horse	Increased pressure waves	Sellers et al. (1979)
Rabbit	Increased spike electrical activity	Ruckebusch et al. (1971)
Dog	Peristalsis	Zondek (1920)
	Increase in pressure and defecation	Welch and Plant (1926)
	Increased pressure waves	Hinrichsen and Ivy (1931)
	Propulsive contractions	Barcroft and Steggerda (1932)
	Propulsive contractions	Larsen and Bargen (1933)
	Increased pressure waves	Galapeaux and Templeton (1937)
	Increased electrical activity	Daniel et al. (1960)
	Defecation	Reinke et al. (1967)
Cat	Defecation	Cannon (1902)

The process is well described in humans by Schuster (1975). Following a voluntary increase of abdominal pressure the external anal sphincter, which is composed of striated musculature under voluntary and reflex control, is closed. Segmenting motor activity of the colon is inhibited, producing disappearance of the haustral sacculations. When the faecal bolus reaches the upper rectum the pelvic muscles relax which results in descent of the pelvic floor and straightening of the previously angulated rectum. After building up pressure the internal anal sphincter relaxes and subsequently also the external anal sphincter. The faecal bolus is expelled and when this has been completed the pelvic floor rises to its normal position and again obliterates the lumen, when both the internal and external anal sphincters contract.

The propulsive contractions which initiate the process of defecation do not differ fundamentally from the other colonic peristaltic contractions or from those initiated by enemas which evoke defecation (WHITE et al. 1940; HALLS 1965; TRUELOVE 1966; WILLIAMS 1967; CONNELL 1968). The intrinsic and extrinsic nerve control of defecation is discussed in Sect. J.I, K.II.

4. Response of the Large Intestine to Feeding

As was mentioned in Sect. G.I.2, feeding initiates an irregular pattern of motility in the small intestine with disruption of the interdigestive migrating motor complex. The colon usually contains faecal matter, even in the interdigestive periods, and no cyclic motor patterns have been reported in this part of the intestine. The effect of feeding is less dramatic than in the small intestine but has been well described by a number of workers (see TANSY and KENDALL 1973; Table 2). Feeding initiates increases of motility of variable intensity in the large intestine, from increased segmental contractions in the proximal colon, to the appearance of propulsive contractions in the proximal and distal colon, to defecation (Table 2). Feeding increases the peristaltic activity in the distal ileum and ileal contents pass into the colon (HERTZ 1913; SHORT 1919; DOUGLAS and MANN 1939b; CODE et al. 1957; KELLEY and DE WEESE 1969) with a resulting increase in colonic motility (CANNON 1911). It is therefore possible that the increased motor activity of the large intestine is due to the entry of material from the ileum. In order for food to initiate propulsive activity the colon must be full (WELCH and PLANT 1926). Also physical activity facilitates the motor response to food (HOLDSTOCK et al. 1970). The possible neuronal and hormonal mechanisms for these responses are discussed in Sect. H.IV.

H. Nervous Control of Intestinal Motility

I. Role of Nerves

Since the end of nineteenth century, it has been known that the intestine, isolated from the central nervous system, is capable of performing coordinated movements that are controlled by the nerve plexuses embedded within the intestinal wall, i.e. by the enteric nervous system (MALL 1896; BAYLISS and STARLING 1899; LANGLEY and MAGNUS 1905, 1906; CANNON 1911, 1912; THOMAS and KUNTZ 1926b; GARRY 1934; DOUGLAS and MANN 1939a; ALVAREZ 1940; CONNELL et al. 1963; ROSS et al. 1963; JOHNSON 1972; GUSTAVSSON 1978). Most of the studies on the role of the intrinsic nerves in determining motility patterns have been performed in isolated segments of the intestine (KOSTERLITZ and LEES 1964; KOSTERLITZ 1968; FRIGO et al. 1972; COSTA and FURNESS 1976; HIRST 1979). Most of the studies on the role of the extrinsic nerves are based on the often misleading effects elicited by electrical stimulation (See Sect. E) or rely on analysis of changes in motility following surgical interruption of extrinsic nerves (YOUMANS 1952; GRAY et al. 1955; CONNELL 1961; KOSTERLITZ 1968; FURNESS and COSTA 1974a). The studies show what nerves are capable of doing but sometimes give little direct information on their involvement in physiological processes. In other words, we know the repertoire of intestinal motor patterns of activity and the acting ability of each nerve type, but we do not know how they perform the play. Investigations on conscious animals could clarify the physiological role of intestinal nerves but these studies are limited by the difficulty of probing both nerve and motor functions in vivo.

The use of specific drugs which antagonise cholinergic transmission indicates that intestinal motor functions are heavily dependent on cholinergic nerves (YOUMANS 1952). Thus, in dogs, atropine has been shown to inhibit small intestinal motility (CANNON 1912; QUIGLEY et al. 1934; PUESTOW 1937; YOUMANS et al. 1943; REINKE et al. 1967; HALADAY et al. 1958; RUWART et al. 1979), to abolish all action potentials superimposed on slow waves (WINGATE et al. 1976) and to inhibit colonic motility (GRAY et al. 1955). Similar results were obtained in human small intestine (QUIGLEY and SALOMON 1930; PUESTOW 1937; HOLT et al. 1947; BROWN et al. 1948; CHAPMAN et al. 1948; DODDS et al. 1948; POSEY et al. 1948; CODE et al. 1957; REINKE et al. 1967) and large intestine (CHOWDHURY et al. 1976). The action potentials of intestinal muscle in conscious cats are abolished by atropine (WEISBRODT and CHRISTENSEN 1972). Atropine slows the progression of a barium meal from the stomach to the caecum in humans (CHAPMAN et al. 1953; 1955 a, b) and significantly reduces the transit of intestinal contents in dogs and guinea-pigs (RUWART et al. 1979). Furthermore, the movements of the small intestine in all the phases of the interdigestive motor complex is abolished by atropine in both humans and dogs (CHEY et al. 1978; ORMSBEE and MIR 1978; GONELLA 1978; YOU et al. 1980). Hexamethonium also abolishes all motor activity in the dog duodenum (REINKE et al. 1967) and nicotine reduces motility in the small intestine (CANNON 1912; BOZLER 1948) and abolishes all action potentials in the cat jejunum (ABBADESSA et al. 1969). Hexamethonium also blocks all the phases of the interdigestive motor complex in the dog small intestine (ORMSBEE and MIR 1978; GONELLA 1978). These actions of nicotinic blocking agents indicate that cholinergic interneurones are involved in generating the normal patterns of motility. Since the pattern of motility in the small intestine is highly dependent on the cyclic migration of the migrating myoelectric motor complex (see Sect. G.I) the role of extrinsic and intrinsic nerves in the control of this cyclic pattern of motility will be considered first.

II. Neuronal and Non-Neuronal Mechanisms of the Migrating Myoelectric Motor Complex

The finding that atropine abolishes both the irregular and the regular active phases of the migrating myoelectric complex (see Sect. H.I) indicates that the patterns of motility present in the intestine during the cycle must be dependent on the activation of cholinergic neurones which release acetylcholine onto the muscle (see Sect. B.IV.1). The question, therefore, is not whether the migratory complex requires nerve activity, but what is the mechanism that initiates the complex, what makes it propagate caudally, what causes the activity to be localised to a relatively short segment of intestine at one time, and what mechanisms are involved in the disruption of the cyclic activity and initiation and maintenance of the postprandial activity.

1. Mechanisms of Initiation

Removal of extrinsic neuronal influences by cutting the vagus does not prevent the appearance of the migratory complex in the small intestine (MARIK and CODE 1975;

WEISBRODT et al. 1975; RUCKEBUSCH and BUENO 1977 a; ORMSBEE and MIR 1978). Moreover, the complex still occurs in the sheep and dog after the removal of the sympathetic outflow by cutting the splanchnic nerve (RUCKEBUSCH and BUENO 1975; GONELLA 1978). However, DIAMANT et al. (1979 a) reported that cold blockade of the vagus nerve prevents the initiation of the migrating complex in the lower oesophageal sphincter and the stomach without affecting the appearance of the complex in the small intestine. Similar disruption of the fasting antral motor activity following vagotomy had been reported by KHAN and BEDI (1972) and by WALKER et al. (1974). This dissociation between the gastric and intestinal migrating motor complex suggests that the two processes in the stomach and in the small intestine are independent (CATCHPOLE and DUTHIE 1978; FINCH and CATCHPOLE 1979) but that under normal circumstances they appear in a temporally coordinated sequence. There is evidence that a humoral factor is involved in the initiation of the migrating complex in the stomach (ITOH et al. 1978 c; THOMAS and KELLY 1979). These authors have shown that an isolated and denervated fundic pouch still displays the typical pattern of cyclical motor activity in synchrony with the rest of the stomach even when it is autotransplanted to an ectopic position. These results suggest that the initiation and spread of the migrating complex in the stomach is under both vagal and hormonal control.

The peptide motilin has been shown to initiate migrating complexes indistinguishable from those occurring spontaneously in both dogs and humans (ITOH et al. 1975; WINGATE et al. 1976; AIZAWA et al. 1978; LUX et al. 1978, 1979 b; ORMSBEE and MIR 1978) and was found to increase in plasma in both species at the time of the appearance of a new spontaneous migrating complex (AIZAWA et al. 1978; LUX et al. 1978; ITOH et al. 1978 b, 1979; LEE et al. 1978; VANTRAPPEN et al. 1978; PEETERS et al. 1979; THOMAS et al. 1979; CHUL et al. 1980; YOU et al. 1980). The plasma concentration of motilin capable of initiating a migrating complex is well within the physiological range (AIZAWA et al. 1978). Both the motilin-induced and the naturally occurring migrating complexes in the stomach are antagonised by atropine and hexamethonium (ORMSBEE and MIR 1978) and motilin has been shown to act by stimulating cholinergic neurones in the gastrointestinal tract (WINGATE et al. 1976; MORGAN et al. 1979). These results taken together suggest that the initiation of a motor complex in the stomach is due to a peak of circulating motilin which acts on a neuronal chain which involves cholinergic neurones. However, the physiological stimulus for the release of motilin from the duodenal mucosa is not known. It has been shown that in humans, acidification of the duodenal mucosa is a good stimulus for motilin release (MITZNEGG et al. 1976; FOX et al. 1979). This led LEWIS et al. (1979) to postulate that a bolus of Cl reaching the duodenal mucosa releases motilin which in turn initiated the intestinal migrating motor complex. However, in dogs, alkalinisation of the duodenal mucosa (but not acidifaction) released motilin (FOX et al. 1979). The final demonstration of the humoral factors responsible for the initiation of the migrating motor complex in different species remains to be done.

2. Mechanisms of Migration

An intact vagus nerve is not required for the caudal propagation of the motor complex, either in dogs or sheep (MARIK and CODE 1975; WEISBRODT et al. 1975; BUENO and RUCKEBUSCH 1976; RUCKEBUSCH and BUENO 1975, 1977 a). However, the reported effects of vagotomy on the frequency, speed, propagation and regularity of the phases of the cycle vary from decrease in frequency in the motor complex (MARIK and CODE 1975) to no change (WEISBRODT et al. 1975) to an increase (BUENO and RUCKEBUSCH 1976; RUCKEBUSCH and BUENO 1977a) and from no change in speed of propagation (WEISBRODT et al. 1975) to reduced speed of propagation (RUCKEBUSCH and BUENO 1977a). MARIK and CODE (1975) reported an increase of the quiescent phase at the expense of phases II and III; RUCKEBUSCH and BUENO (1975, 1977a) reported a decrease in phase II while WEISBRODT et al. (1975) found no changes in the phases of the cycle. The sympathetic outflow does not appear to be essential for the migration of the complex (RUCKEBUSCH and BUENO 1975; GONELLA 1978). Finally, even after both vagotomy and splanchnicectomy, regular migrating complexes occur in the sheep (RUCKEBUSCH and BUENO 1975). It is unlikely that the mechanism of migration involves circulating hormones for they could not act on localised areas of the intestine and in a sequential fashion. POITRAS et al. (1979) found that the ileal propagation of the migrating complex is independent of blood levels of motilin.

The blocking effect of atropine and hexamethonium on the migrating complex in the small intestine (ORMSBEE and MIR 1978; ORMSBEE et al. 1979) indicates that the irregular contractions of phase II and the regular contractions of phase III involve enteric cholinergic interneurones and enteric cholinergic motor neurones. The nerve pathways that may be involved in the mechanisms of progression of the complex from one segment of intestine to the next caudal segment may be located in the enteric plexuses of the intestinal wall, the extrinsic nerves that reach the intestinal segment through the mesenteric nerves or both.

The pattern of the migrating complex in a loop of intestine of the sheep or the dog which has been extrinsically denervated is not different from normal, although the speed of migration is slower (RUCKEBUSCH and BUENO 1975; GONELLA 1978; BUENO et al. 1979). This suggests that in the intestine there is a neuronal mechanism independent of extrinsic nerves which is responsible for the slow propagation of the myoelectric motor complex in the caudal direction. On the other hand, a loop which has been isolated from the rest of the intestine but that retains its extrinsic innervation (Thiry–Vella loop) still shows migrating complexes with a good degree of synchrony with the rest of the intestine (CARLSON et al. 1972; GRIVEL and RUCKEBUSCH 1972; BUENO et al. 1979). However, with time the cyclic motor complexes in isolated loops become dissociated from the rest of the intestine (ORMSBEE et al. 1979; PEARSE and WINGATE 1979). This suggests that the continuity of the intestine is necessary for a normal migration of the complex. With extrinsic denervation of the isolated loop, the motor complex is completely disrupted and an intermittent pattern of bursts of activity at a frequency of about one every 22 min appears (BUENO et al. 1979). Their records indicate that even the short bursts of spiking activity are slowly propagated caudally within the isolated and denervated loop. This would indicate that although isolated denervated loops retain some intrinsic mechanism for initiation and propagation of a motor complex, the continuity with the rest of the intestine and the extrinsic nerve connections are essential for correct timing and coordination of the migrating complex.

The function of the contents, at least in the herbivores, in the maintenance of the propagation of the complex is not established. The migrating complex appears to "jump" an anastomosis of the intestine with a delay that led RUCKEBUSCH and collaborators to postulate that there is a reorganisation of a new migrating complex below the anastomosis produced by extrinsic excitatory reflexes arising from the oral side and acting on the anal side of the anastomosis and by intrinsic reflexes elicited by the arrival of the intraluminal contents. In support of the interpretation that the contents play an important role in the migration of the complex, BUENO and RUCKEBUSCH (1978) have shown that an increase or decrease of the induced intestinal flow of ingesta results in a corresponding increase or decrease in the phase of irregular contraction (phase II), where most of propulsive contractions occur. Furthermore, following obstruction of a jejunal loop in the sheep, the irregular activity of the cycle in the region above the obstruction becomes almost continuous while below the obstruction, where the intestine is empty, the irregular phase is replaced by the quiescent phase. The phase of regular spiking activity (phase III) persists and propagates normally across the obstruction (RUCKEBUSCH and BUINO 1975). The dissociation between the irregular and the regular phases of the motor cycle suggests that they are probably dependent on different neuronal mechanisms, the irregular phase being more dependent on the stimulation of the intestine by its contents, while the regular phase is less dependent on contents.

These results, taken together, suggest that the intestine is able to initiate motor acticity which propagates in the caudal direction. The propagation depends on both the continuity with the rest of the intestine and on an intact mesenteric nerve supply which transforms this basic activity into a regular, recurrent organised pattern of motor activity which migrates caudally. The migrating myoelectric complex is also influenced by the intestinal contents. The evidence for intrinsic neuronal mechanisms which spontaneously migrate anally is discussed in Sect. H.V.3.b.

III. Mechanisms Involved in the Response of the Small Intestine to Food

Following the ingestion of food, the motility of the small intestine increases and the interdigestive cyclic motor pattern is disrupted (Sect. G.I.2). It is difficult to identify the numerous factors involved in the initiation, maintenance and termination of the postprandial intestinal pattern of motility. It is proposed here that since these factors come into action in a sequential manner, depending on the site of action of the nutrients along the digestive tract, three phases of motor responses to food can be described, i.e. *cephalic*, *gastric* and *intestinal* phases by analogy with similar phases of gastric secretion. GREGORY (1950) had reached the conclusion that the prompt onset of the effect of feeding on intestinal motility probably involved reflexes initiated by the act of feeding (cephalic phase) or by the presence of the meal in the stomach (gastric phase) and that the fact that these changes continued for some hours after the meal suggested that reflex or humoral influences of intestinal origin were also involved (delayed or intestinal phase).

1. Cephalic Phase

A short transient decrease, followed by an increase in motor activity was observed in the upper small intestine of rabbits and dogs by McCOY and BAKER (1968) and by RUCKEBUSCH et al. (1971), even before the ingestion of food, thus confirming the earlier report by CASH (1886). GREGORY (1950) observed that sham feeding increased the motility of an isolated jejunal loop and that this increase died away within a few minutes of discontinuing sham feeding. Since this phenomenon was abolished by vagotomy, efferent excitatory pathways in the vagus (most likely cholinergic, see Sect. E.I) are responsible for the cephalic phase of intestinal motility. The increase in intestinal movements during this phase is transient and fails to interrupt the interdigestive cyclic migratory motor complexes (RUCKEBUSCH and BUENO 1977a).

2. Gastric Phase

The arrival of food in the stomach appears to be a prerequisite for the disruption of the interdigestive motor complex in the small intestine, since parenteral feeding does not initiate the postprandial pattern of motility (WEISBRODT et al 1975; BUENO and RUCKEBUSCH 1976). Distension of the stomach by a balloon or by ingestion of isotonic saline, liquid paraffin or an agar-agar meal disrupts the interdigestive pattern of motility but is less effective than a nutrient meal (GREGORY 1950; CODE and MARLETT 1975; RUCKEBUSCH and FIORAMONTI 1975; EECKHOUT et al. 1978a; see Sect. G.I.2). The increase in jejunal contractions caused by gastric distension is probably mediated in part by the vagus nerve (GREGORY 1950). The response of the stomach to feeding is also suppressed by vagotomy or cold blockade of this nerve (DIAMANT et al. 1979a) and vagotomy impairs bud does not prevent the appearance of the irregular motor activity typical of the postprandial period (MARIK and CODE 1975; STODDART and DUTHIE 1973; WEISBRODT et al. 1975; BUENO and RUCKEBUSCH 1976; RUCKEBUSCH and BUENO 1977a; REVERDIN et al. 1979). The early motor response which occurs throughout the small intestine, including the ileocaecal region, when the food is still in the stomach also appears to be nerve mediated (GREGORY 1950; DAHLGREN and SELKING 1972; BALFOUR and HARDCASTLE 1975) being often referred to as the gastroileal or gastrocolic reflex. This indicates that the arrival of food in the stomach triggers vagal reflexes which initiate the postprandial motor activity in the stomach and small intestine.

3. Intestinal Phase

The arrival of nutrients in the upper small intestine is certainly a powerful factor in eliciting the full postprandial prolonged motor pattern in the small intestine. Perfusion of the duodenum with amino acids, glucose or arachidonic acid initiates the typical long-lasting postprandial pattern (CODE and MARLETT 1975; RUCKEBUSCH and FIORAMONTI 1975; EECKHOUT et al. 1978a, 1979; DE WEVER et al. 1979). The delayed and prolonged motor responses of the small intestine are not affected by vagotomy (TANSY and KENDALL 1973) and it is therefore likely that this response depends on either humoral factors or on local intestinal reflexes initiated by the arrival of digesta, as was suggested by GRINDLAY and MANN (1941) and by GREGORY (1950). When an isolated loop of intestine is perfused with nutrients including glucose or with acid or hypertonic solutions, the interdigestive migrating motor complex was disrupted in the loop but not in the rest of the small intestine (PEARSE and WINGATE 1979; EECKHOUT et al. 1979). On the other hand, the isolated segment was more active when the digesta was in the portion of intestine from which the segment had been removed (PUESTOW 1937). Therefore, the digesta appear to act mainly locally but there seems also to be a nerve-mediated interaction between adjacent portions of the small intestine that is elicited by nutrients, possibly through extrinsic reflexes. Since local application of acid or hypertonic solution to the mucosa of an isolated and extrinsically denervated segment still evokes an increase in motility in the loop (GREGORY 1950), it is most likely that local reflexes in the enteric nervous system are also involved in the response to nutrients. This conclusion is reinforced by the results of DOUGLAS and MANN (1939b, 1940) who found that a loop of extrinsically denervated intestine completely isolated from the rest of the intestine did not show a motor response to feeding but that a similar loop of intestine which had been reanastomosed to the intact intestine showed a motor response which must have been initiated by the arrival of digesta in the loop.

Glucose and peanut oil disrupt the interdigestive pattern only when applied to the upper small intestine but noth the ileum (RUCKEBUSCH and FIORAMONTI 1975; DE WEVER et al. 1979). CODE and MARLETT (1975) and RUCKEBUSCH and FIORAMONTI (1975) suggest that in a mixture of nutrients, each food component acts individually by coming in contact with specific sites in the intestinal mucosa and therefore the net effect of eating on motility represents a summation of their different effects. According to RUCKEBUSCH and FIORAMONTI (1975), the absorption of substances by the epithelium is an adequate stimulus for the initiation of the fed pattern in the corresponding section of intestine. Differences in the rate of caudal flow of intestinal contents depend on the food composition (LAGERLÖF et al. 1974; JOHANSSON and EKELUND 1975) and have been attributed by these authors to differences in gastric emptying, in the degree of dilution or concentration of digesta along the intestine and in the rate of propulsion along the intestine. The view that chemical stimuli in the lumen, more than the distending volumes are responsible for the control of progression was put forward originally by CANNON (1911). The prevalence of chemical over mechanical effects of food in disrupting the interdigestive motor pattern may explain the persistence of the cyclic motor pattern in herbivores whose food chemistry differs from other groups. It is interesting, in this respect, that an increased ingestion of food in herbivores leads first to an increase of the irregular activity of phase II of the motor complex (the phase with the highest incidence of propulsive contractions) and then, if the amount of food is excessive, to a complete disruption of the cyclic pattern with associated diarrhoea (RUCKEBUSCH and BUENO 1975; BUENO and RUCKEBUSCH 1978).

The case for the involvement of humoral factors in initiating and maintaining the fed pattern has been made for gastrin, cholecystokinin and insulin. However, their physiological role has been seriously questioned. While gastrin has been shown to disrupt the interdigestive pattern in the stomach and duodenum (MARIK and CODE 1975; EECKHOUT et al. 1978b; HELLEMANS et al. 1978), it has little effect on the ileum (WINGATE et al. 1978) unlike the effect of food. Furthermore, in patients with Zollinger–Ellison syndrome or pernicious anaemia associated with high levels of serum gastrin, the interdigestive pattern was still present (WINGATE et al. 1978). Also, while serum gastrin has been reported to increase after feeding, the serum level decreases well before the reappearance of the interdigestive pattern (WINGATE et al. 1977). The effect of cholecystokinin is similar to that of gastrin

(MUKHOPADHYAY et al. 1977) in that it produces disruption of the interdigestive pattern only in the upper intestine (WINGATE et al. 1978). Exogenous insulin disrupts the interdigestive pattern in dogs (BUENO and RUCKEBUSCH 1976) but little or no increase in the plasma concentration of insulin is associated with food intake (REVERDIN et al. 1979; EECKHOUT et al. 1978b).

IV. Mechanisms Involved in the Response of the Large Intestine to Food

As shown in Sect. G.II.4 feeding elicits a variety of motor responses in all parts of the large intestine (see Table 2). There has been little systematic investigation on the psychical component of the cephalic phase of the colonic response to food. In the rabbit, RUCKEBUSCH et al. (1971) found a consistent motor response of the small intestine to food presentation but no motor response in the caecum. HERTZ and NEWTON (1913) reported that in one patient, the mere sight of food elicited a prompt motor response of the colon, but more recently RAMORINO and COLAGRANDE (1964) could not detect such responses in their patients. There is, however, little doubt that certain emotional states are associated with increased motor activity of the colon (ALVAREZ 1940; DELLER and WANGEL 1965). WELCH and PLANT (1926) originally proposed that the increase in motor activity of the colon was a response to feeding and did not require gastric stimulation; they suggested that the term "feeding reflex" should be used for this phenomenon. However, for the full motor response of the colon to occur the food must be ingested and reach the stomach and duodenum. In most cases this elicits an almost immediate increase in motor activity of the colon (GALAPEAUX and TEMPLETON 1937; WELCH and PLANT 1926; REINKE et al. 1967; HOLDSTOCK et al. 1970; RUCKEBUSCH et al. 1971; DAHLGREN and SELKING 1972; WIENBECK and JANSSEN 1974; SULLIVAN et al. 1978; SNAPE et al. 1979). The motor activity reaches its maximum within 39 min and continues for up to 1–3 h (GALAPEAUX and TEMPLETON 1937; HOLDSTOCK et al. 1970; WIENBECK and JANSSEN 1974; SULLIVAN et al. 1978; SELLERS et al. 1979; SNAPE et al. 1979). It is difficult to establish whether these responses were initiated in the stomach as well as in the duodenum. Motor responses of the colon are still elicited by ingested food following gastrectomy (CONNELL and LOGAN 1967; HOLDSTOCK and MISIEWICZ 1970) and distension of the stomach is a poor stimulus compared with distension of the duodenum (IVY and MCILVAIN 1923; HINRICHSEN and IVY 1931; TANSY et al. 1972). Placing food directly into the duodenum is a very good stimulus for the colonic motor response (HINRICHSEN and IVY 1931). Furthermore, perfusing the duodenum with magnesium sulphate, amino acids or sodium oleate initiates a prompt motor response in the colon (HARVEY and READ 1973; MESHKINPOUR et al. 1974).

It appears therefore that food can elicit motor responses in the colon by acting on the duodenum both mechanically and chemically, the intensity and the duration of the motor responses being related to the amount and chemical nature of the food (SULLIVAN et al. 1978; SNAPE et al. 1979). It is not known, however, whether there is a reflex effect on the colon or whether the motor responses are due to the propagation of waves from the ileum which occur immediately after feeding (WIENBECK and JANSSEN 1974). The prompt initial colonic response to feeding is likely to be mediated by nerves since anticholinergic drugs abolish the response (SULLIVAN et al. 1978; SNAPE et al. 1979).

In the rabbit, vagotomy abolishes the colonic motor response to food (RUCKEBUSCH et al. 1971). SEMBA (1954) has shown that distension of the stomach of the dog with a hypertonic salt solution elicited an increase in colonic motility which was not abolished by extirpation of the coeliac, superior and inferior mesenteric ganglia but was abolished by vagotomy, by cutting the pelvic nerves or by blocking transmission through pelvic ganglia with nicotine or cocaine. This pathway thus appears to involve a vagal afferent, the medulla, the spinal cord and the pelvic nerves. However, in the same species TANSY and KENDALL (1973) reached a different conclusion, having found that the colonic response to gastric distension was not abolished by transection of the ileum, by pelvic nerve transection or by transection of the spinal cord at the lumbosacral junction but was abolished by vagotomy and by spinal transection at the thoracic level. These authors distinguished three reflex components of their responses: the first a cholinergic propulsive reflex; the second a cholinergic segmenting reflex; and the third a noradrenergic segmenting reflex which they attributed to the activity of the muscularis mucosa (TANSY and KENDALL 1973). The early observation by WELCH and PLANT (1926) that the effectiveness of feeding in determining the motor response of the colon depends on the state of the colon itself was confirmed by the findings of SEMBA (1954). When there is an intracolonic pressure of less than 6 cm H_2O, distension of the stomach does not elicit colonic responses while with pressures of 7–18 cm H_2O in the colon, distension of the stomach elicits a motor response. This suggest that distension of the colon excites the intrinsic cholinergic neurones (see Sect. J.I) facilitating the effects of extrinsic excitatory reflexes activated by gastric distension. Physical exercise also facilitates the motor responses of the colon to feeding (HOLDSTOCK et al. 1970). The lack of excitatory influences such as feeding and somatic motor activity may partly explain why the motility of the large intestine is reduced during sleep (DAVIDSON et al. 1956; HOLDSTOCK et al. 1970; WIENBECK et al. 1978).

Because the increased motor activity of the colon following feeding is prolonged, some investigators suggested that humoral factors are involved (HOLDSTOCK and MISIEWICZ 1970; SNAPE et al. 1979). This concept was supported by the observations of CONNELL and MCKELVEY (1970) and SELLERS et al. (1979) that even after vagotomy there is still a motor response of the colon to feeding. It is quite conceivable, however, that the nutrients progressing along the gastrointestinal tract would be able to affect the colonic motility directly through intramural reflexes. Amongst the humoral candidates are gastrin and cholecystokinin (HARVEY 1975) which increase colonic motility when infused. However, despite the demonstration by CONNELL and LOGAN (1967) that gastrin elicits the colonic motor responses by acting on cholinergic nerves, subsequent experiments found only a weak effect on the colon (see HARVEY 1975). Furthermore, SNAPE et al. (1979) detected a late increase in plasma gastrin following feeding but its level remained high while colonic motility returned to the baseline. Therefore, if gastrin is involved in this response, it is not the sole factor responsible for the delayed response.

Cholecystokinin (CCK) also elicits a motor response and is released in the circulation by feeding with a delay similar to the delayed colonic motor response (SNAPE et al. 1979). Furthermore, those substances which are known to release

CCK from the duodenum (HARVEY 1975) are very effective in eliciting the colonic motor response. However, the final identification of the neuronal and humoral mechanisms involved in the response of the large intestine to feeding must await more extensive investigations. From the results summarised it is possible to postulate different phases of the feeding response for large intestine, i.e. a *cephalic phase* which is not very prominent or absent; a *gastroduodenal phase;* a minor *intestinal phase;* and even a *colonic phase.*

V. Role of Intrinsic Nerves in Patterns of Motility of the Small Intestine in the Fasted State

Periods of quiescence, of irregular nonpropagated contractions and of irregular and regular short propagated contractions occur in fasted animals (Sects. G, H). In this section a simplified hypothesis based on that of CODE et al. (1968) is presented in which an attempt is made to identify those factors sufficient to explain the diverse patterns of motility.

1. Quiescence of the Small Intestine

During these periods the slow waves are still present but do not have superimposed action potentials (Sect. G.I.1). Since some smooth muscle cells are spontaneously active (Sect. B.III.2.b) and since there is ample evidence in the literature that smooth muscle in vitro can contract rhythmically in the absence of any neural activity, the important question arises as to whether in vivo mechanical quiescence of the intestine results from an active inhibitory influence or whether resting muscle in vivo does not spontaneously reach the threshold for contraction during the slow wave cycle. It is difficult to investigate the presence of inhibitory influences that act on the muscle in vivo. However, the dramatic abolition of all mechanical activity when anticholinergic drugs are injected (GONELLA 1978; ORMSBEE and MIR 1978; ORMSBEE et al. 1979) strongly suggests that the intestinal smooth muscle only contracts when there is sufficient acetylcholine being released from the excitatory enteric cholinergic neurones to initiate action potentials on the slow wave crest. The hypothesis that there is a continuous inhibitory tone on the intestinal muscle by intrinsic enteric inhibitory neurones (WOOD 1972, 1975) is difficult to evaluate at present since no specific antagonist of transmission from these neurones is available (see Sect. B.IV.2) and since the experiments on which this hypothesis was based were performed in vitro. The finding that antinicotinic drugs also induce a period of quiescence (ORMSBEE et al. 1979) appears to conflict with the accepted view, based on earlier work by CANNON and many others that the rhythmic nonpropagated contractions which occur in the intestine are entirely myogenic since they remained unchanged after administration of antinicotinic drugs. It is interesting that several authors, including some of those who reached the conclusion that rhythmic segmentations are of myogenic origin, observed that with higher doses of nicotine or other antinicotinic drugs all mechanical activity of the circular muscle ceased (THOMAS and KUNTZ 1926a; ABBADESSA et al. 1969; GONELLA 1971). It is quite likely, therefore, that spontaneously active pacemaker smooth muscle cells in the intestine are involved in providing a "resting tone" while independent phasic contractions, which occur at the crest of the slow waves, only occur when cholinergic nerves are active.

Thus, the period of quiescence during phase I of the migrating motor complex is postulated to be due to silence of the cholinergic neurones. The state of quiescence of the enteric cholinergic neurones could be due either to the lack of continuous local reflex activity in the empty intestine or to an active inhibition exerted on the neuronal pathways.

Periods of silence of intestinal muscle have been reported even in loops of intestine in vitro following the passage of peristaltic waves (see Nakayama 1962), and in vivo the quiescent phase still occurs in a loop which has ben extrinsically denervated (Ruckebusch and Bueno 1975; Gonella 1978). In sheep, obstruction to the passage of intestinal contents leads ot an increase of propagated contractions above the obstruction during phase II. Below the obstruction no contractions occurred during phase II of the uninterrupted migrating motor complex. The recent report by Ormsbee et al. (1979) that in dogs, guanethidine abolished the quiescent phase which was replaced by a prolonged phase II of irregular contractions suggests that noradrenergic nerves act to maintain the intestine quiescent. This interpretation is supported by the finding that phentolamine also abolishes the quiescent phase in dog stomach (Papasova et al. 1979). However, guanethidine also blocks transmission from nonadrenergic inhibitory nerves in the submucosa and from excitatory nerves, which release an unknown transmitter, in the proximal colon of the guinea-pig (Chap. 11). Phentolamine is also an effective antagonist of 5-HT receptors on intestinal muscle (Chap. 11). Since a quiescent phase is still present in an extrinsically denervated loop of intestine (see Sect. H.II.2) the extrinsic sympathetic nerves are not the only mechanism responsible for the quiescence of the intestine in phase I.

2. Nonpropagated Contractions

Since contractions superimposed on slow waves only occur as a result of cholinergic nerve activity, it is possible to postulate how a stationary ring of rhythmic or tonic contraction can occur. If there is a low level of activity in cholinergic neurones in a short segment of myenteric plexus which controls a ring of circular muscle, the resulting activity will be a ring of pulsating contraction of the same width as the area of cholinergic activity. Action potentials will only occur when the crests of the travelling slow waves pass through the excited area. The frequency of pulsation will be the frequency of the slow wave. Since the slow waves, but not the action potentials, are propagated (Sect. B.III.2.a), the active area will remain stationary. If the neuronal cholinergic activity should increase within the same area, the smooth muscle membrane will be above threshold even between slow waves and thus stationary rings of tonic contraction as described in the literature will occur. Such nonpropagated rhythmic or tonic stationary rings of contraction are observed mainly during the initial part of phase II of the migrating complex. It could be postulated that they represent either transient escape of localised groups of cholinergic enteric neurones from the inhibitory influences acting during the quiescent phase I or excitation of small groups of these neurones by chemical or mechanical stimuli. Such localised initiation of activity of cholinergic neurones is not propagated and occurs sporadically in different parts of the intestine. The appearance of the rhythmic segmentation described by CANNON (1911) would arise, according to this hypothesis, from stationary localised areas of cholinergic neuronal activity. If the localised neuronal activity is only transient, lasting less than a slow wave cycle, then the resulting contraction will be a single phasic nonpropagated contractions; these also occur sporadically.

3. Propagated Contractions

There are two types of propagated contractions which (a) fast propagated contractions occur after feeding, and towards the end of phase II and during phase III in the fasted state; and (b) slow progression of the front of phase III of the migrating complex.

a) Fast Propagated Contractions

If the localised areas of cholinergic neuronal activity merge into a larger area, contractions associated with slow waves can propagate. When the neuronal activity is mild, action potentials occur only as shown, on the crest of the slow waves. The travelling crest of the slow wave, which is only a few centimetres long in the upper intestine (Sect. B.III.1.c; CODE et al. 1968), will have superimposed action potentials as it travels through the whole segment excited by cholinergic nerves. Thus, within this segment a 2–4 cm segment of circular muscle will show action potentials and contractions will propagate from oral to anal at the speed and frequency of the slow waves. This is observed during phase III of the migrating complex, the length of intestine undergoing such activity being in the range of 30–40 cm (Sect. G.I.1).

This situation, in which acetylcholine is continuously released over a long segment of intestine, is consistent with BORTOFF's account of a myogenic peristalsis (BORTOFF 1969; 1972, 1976). It can be predicted, in fact, that following nerve paralysis with tetrodotoxin, the application of muscarinic or other muscle stimulants will reveal propagated waves of contractions riding the slow waves just as BORTOFF and SACCO (1974) found in isolated cat intestine. Indeed, if in vivo any hypothetical excitatory substance were to act on long segments of intestine, one would observe similar rapidly propagated "myogenic" contractions. However, this does not appear to happen in vivo, and therefore "myogenic" peristalsis in the small intestine does not occur, except against a background of excitation mediated through cholinergic nerves. During the late part of phase II, intermittent contractions which are rapidly propagated for several centimetres are recorded and are likely to be triggered by transient increases of cholinergic neuronal activity over long segments of intestine. BORTOFF and SACCO (1974) and CODE et al. (1968) correctly predicted that such rapidly propagated contractions riding slow waves would propagate to a maximum of the length of the frequency plateau of the slow wave (Sect. G.I.1) where the reduced excitability of the muscle, owing to the reduced amplitude of the slow waves in the area of transition between the plateaus, would prevent them travelling further (BORTOFF 1969). When cholinergic nerve activity is high over a long segment then the circular muscle will show action potentials irrespective of the point in the cycle of the travelling slow waves. The result will be a tonic contraction of the active segment. If the neuronal activity is not maximal, propagated contractions superimposed on the tonic constriction will still ride the slow waves. This situation is observed in vivo during phase III of the migrating motor complex in herbivorous animals. Thus, several different patterns of motility observed in vivo can be explained by considering only the interaction between cholinergic nerve activity and the slow waves.

b) Slow Propagated Contractions

There is considerable evidence that slowly moving areas of heightened muscle excitability, which appear to be equivalent to the migrating complex seen in fasted animals, can occur in segments of intestine which are isolated from the remainder of the intestine in vivo or are maintained in vitro. Therefore, it can be postulated that within the enteric plexuses there is an inbuilt mechanism of unidirectional migration of "spontaneous" neuronal activity.

A number of experiments suggest the existence of such a neuronal mechanism. As early as 1899 Bayliss and Starling reported that electrical stimulation of the small intestine in the dog evoked a local contraction which in some cases moved slowly down the intestine. Similarly, contractions elicited by a bolus in the intestine also travelled slowly in the anal direction and passed over the bolus when its progression was prevented. In dog isolated small intestine, waves of contractions appeared either spontaneously or after reflex activation and slowly travelled anally (Nakayama 1962). The records of Streeten and Vaughan-Williams (1951) show that in isolated loops of dog small intestine in vivo, the rapidly propagated contractions which travel at the speed of slow waves undergo periodic fluctuations every 1–2 min and that the periodic increase in the amplitude of the rhythmic contraction appeared to travel slowly in the anal direction, i.e. a migrating complex travelled anally.

In isolated segments of cat small intestine in vivo Klinge (1951), Perkins (1971), Weisbrodt and Christensen (1972) and Weisbrodt (1974) recorded rings of constriction of 4–16 mm long which spontaneously travelled anally at about 6 cm/min. Similar spontaneous rings of contractions were observed by Cowie and Lashmet (1928) and by Gonella and Vienot (1972) in the rabbit small intestine in vitro or in isolated loops in vivo.

Such spontaneous slowly migrating contractions have also been recorded in the large intestines of different species. Raiford and Mulinos (1934b) found that in dog colon, circular muscle contractions, once initiated, slowly migrated down the intestine. Hukuhara and Miyake (1959) also found that spontaneous contractions in extrinsically denervated dog colon travelled anally although they also observed some which travelled orally. Recently Kingma et al. (1978) and Kocylowski et al. (1979) described in the isolated and perfused dog colon, prolonged large contractions associated with high electrical activity which, after administration of neostigmine, slowly propagated anally. These slowly migrating spontaneous waves recurred every 0.2–1 min. Similar propagaged contractions were reported in the same species by El-Sharkawy (1978) in the public discussion of his paper.

In the cat colon in vitro a slowly migrating spike activity was described by Christensen et al. (1974) and attributed to a neuronal mechanism. In the proximal colon of the rat, long rings of contractions were also described by HUKUHARA and NEYA (1968) to migrate slowly anally. In the guinea-pig distal colon in vitro, COSTA and FURNESS (1976) found that contractions of the circular muscle can slowly migrate anally. First, spontaneous bursts of contractions recurring every 4-min migrated along the segment in the anal direction at speeds of 0.6–6 cm/min. These bursts were abolished by hyoscine and never occurred after administration of tetrodotoxin. Second, contractions elicited by the ascending excitatory reflex, or as a rebound contraction following the descending inhibitory reflex (see Sect. J.I.2), often initiated a wave of contractions which slowly migrated anally. Also when a artificial pellet, propelled by the contraction of the circular muscle above it, was allowed to leave the segment before reaching the anal end, the contraction continued to travel although it decreased in amplitude and speed. Slowly travelling constrictions were also observed visually to travel anally during the spontaneous emptying of the colon, often passing over the scybala.

All these results taken together are highly suggestive of a common basic intrinsic mechanism of initiation and migration of neuronal activity within the intestine. It is impossible at this stage to ascertain whether this neural activity can arise spontaneously or whether neurones must be activated to initiate and maintain the propagated contractions. This intrinsic migrating neural activity is almost certainly involved in the maintenance of the migrating myoelectric complex in extrinsically denervated segments of small intestine (Sect. H.II.2) and in generating the slowly advancing active area of the migrating complex in the normal intestine.

J. Role of Intrinsic Nerves in the Pattern of Intestinal Motility in the Freely Fed State

It is apparent that the contents of the stomach and intestine are mainly responsible for the disruption of the cyclic pattern of motor activity (see Sects. G.I.2, H.III). In the large intestine, of course, luminal contents are normally present. In this section the ways in which the contents elicit intrinsic intestinal reflexes which modify intestinal motility are discussed.

I. Intrinsic Intestinal (Enteric) Reflexes

1. Historical Perspective

The ability of the intestine to transport its contents aborally, even when it is disconnected from the central nervous system, has been fully recognised since the second half of the nineteenth century (see CANNON 1911). Many authors have observed that a solid bolus in a segment of intestine isolated from the central nervous system is transported in the aboral direction (e.g. LEGROS and ONIMUS 1869; CASH 1886; LÜDERITZ 1889; MALL 1896; BAYLISS and STARLING 1899; LANGLEY and MAGNUS 1905; QUIGLEY et al. 1934; HUKUHARA et al. 1936; AUER and KRUEGER 1947; BOZLER 1949b; NAKAYAMA 1962; CREMA et al. 1970; COSTA and FURNESS 1976). Similarly, the classic experiments of TRENDELENBURG (1917) showed that a loop of small intestine filled with liquid in vitro empties its contents by way of a propagated contraction of the circular muscle which sweeps the segment from oral to anal. Since then most of the studies on the mechanisms of the propulsive mechanisms of the intestine have been performed by using modified versions of the Trendelenburg method (HOLMAN 1975; KOSTERLITZ and WATT 1975; see Chap. 8). There is a general consensus that the caudal propulsion of contents in isolated segments of intestine is due to the coordinated contractions of the external musculature which is in turn due to the intrinsic nerve reflexes. At the end of the nineteenth century it was realised that polarised reflexes are involved in the peristaltic reflexes and that propulsion can be explained in terms of successive activation of these reflexes. Chemical stimulation by NaCl applied to the rabbit intestinal mucosa first disclosed a polarity of responses (NOTHNAGEL 1882) and it was MALL in 1896 who first articulated the possible role of the intrinsic reflexes in the propulsion of intestinal contents: "A foreign body within the intestine causes contraction above the bolus and the contractions necessarily move the body downwards. A new portion of mucous membrane is now irritated, which causes renewed contractions but farther down. It therefore seems that Nothnagel's NaCl experiment brings a normal reflex circle."

The classic work of BAYLISS and STARLING (1899, 1900, 1901) better characterised these reflexes involved in peristalsis. Their work has been often simplistically summarised on the basis of their major conclusion that "excitation at any point of the gut excites contraction above, inhibition below. This is the law of the intestine" (BAYLISS and STARLING 1899). This dual polarised reflex response was named the "myenteric reflex" by CANNON. Several authors argued against the validity of this law, mainly because of their inability to detect the relaxation of the circular muscle in front of the advancing content (ALVAREZ and ZIMMERMAN 1927; BAUR 1928; HENDERSON 1928; ALVAREZ 1940; BOZLER 1949b). The concept that the progression of digesta is due to the sequential activation of the intrinsic reflexes along the intestine was thus already well developed when CANNON (1911) pointed out that the mechanism of initiation of the reflexes should be distinguished from the mechanism of their progressive involvement as material passes along the intestine.

2. Properties of the Enteric Reflexes

In both small and large intestine, a variety of stimuli elicit intrinsic reflexes which have a distinct polarity. Mechanical stimulation, such as stroking the mucosa or distending the intestinal wall, elicits nerve-mediated responses which result in contraction of the circular muscle above and relaxation below the point of stimulation. Similar responses can be elicited by chemical stimuli acting on the mucosa.

The responses are abolished by tetrodotoxin, indicating that they are nerve mediated (FUKUDA 1968b; CREMA et al. 1970; FRIGO and LECCHINI 1970; COSTA and FURNESS 1976, FURNESS and COSTA 1976). The two nerve reflexes, one directed orally the other anally, are referred to here as the *enteric ascending excitatory reflex* and the *enteric descending inhibitory reflex* (see COSTA and FURNESS 1976). Both ascending and descending reflexes are antagonised by antinicotinic drugs (BAYLISS and STARLING 1899; HUKUHARA et al. 1958; NAKAYAMA 1962; CREMA et al. 1970; COSTA and FURNESS 1976; FURNESS and COSTA 1976), and this has been taken as evidence that cholinergic excitatory interneurones are involved in these reflexes. Since atropine and hyoscine abolish or antagonise the ascending excitatory reflex (RAIFORD and MULINOS 1934a, b; Hukuhara and MIYAKE 1959; KOSTERLITZ and LEES 1964; CREMA et al. 1970; COSTA and FURNESS 1976; FURNESS and COSTA 1976), the final neurone in this ascending pathway is probably cholinergic and acts on the circular muscle via muscarinic receptors. The final inhibitory neurone in the descending reflex pathway is the enteric inhibitory neurone described in Sect. B.IV.2. None of the antagonists to known transmitter substances selectively antagonise the descending inhibitory reflex (FUKUDA 1968b; CREMA et al. 1970; BURNSTOCK 1972; FURNESS and COSTA 1973, 1977; HIRST et al. 1975; COSTA and FURNESS 1976).

The existence of sensory neurones in both pathways is postulated although the nature of their transmitter substances is still unknown. Cocaine or xylocaine applied to the intestinal mucosa have been shown to abolish the reflexes elicited by chemicals applied to the mucosa and by stroking the mucosa (ALVAREZ 1940; GREGORY 1950; HUKUHARA et al. 1961; ABBADESSA et al. 1969) and the integrity of the mucosa was found to be essential for eliciting these reflexes (HUKUHARA et al. 1958), suggesting that the nerve terminals of the sensory neurones of these reflexes are located in the mucosa. The reflex elicited by distension of the intestine with liquids or solids is not affected by cocaine applied to the mucosa (HUKUHARA et al. 1961) or by removal of the mucosa and submucosa (GINZLER 1959; DIAMANT et al. 1961; COSTA and FURNESS 1976). Thus, the nerve terminals of the sensory neurones which are activated by distension of the intestine are located in the external musculature or in the myenteric plexus itself.

The final neurones of the two reflexes, the cholinergic excitatory neurones in the ascending reflex and the enteric inhibitory neurones in the descending reflex, are located in the myenteric ganglia and the nerve pathways of both reflexes run within the myenteric plexus and then supply the circular muscle (COSTA and FURNESS 1976). The fact that the drugs which antagonise the enteric reflexes also antagonise the propagation of the coordinated muscle contractions (peristalsis, Sect. J.I.4) has been taken as good evidence that these reflexes are involved in the propulsion of a bolus.

3. Enteric Reflexes Initiated by Mechanical Stimuli

a) Stroking the Mucosa

In the dog small intestine, stroking the mucosa elicits an excitation on the oral side and an inhibition on the anal side (HUKUHARA et al. 1958; DANIEL et al. 1960; FUKUDA 1968b). BOZLER (1949b), while observing the oral excitation, failed to record any changes on the anal side. In the rabbit, these responses could be elicited by stroking the mucosa in vivo under local anaesthesia (HUKUHARA et al. 1961) but not in vitro (BOZLER 1949a). In all cases the magnitude of the response was related to the strength of the stimulus. In the dog and rabbit intestine, in which the slow waves trigger rhythmic contractions of the circular muscle, the stimulus first increased the amplitude of the rhythmic contractions on the oral side. With stronger stimulation these contractions began to merge to produce a tonic stationary ring of contraction from which weak rings of contraction propagated for short distances within the excited area in both oral and anal directions. On the anal side of the stroking the amplitude of the rhythmic contractions decreased and sometimes ceased completely. In the guinea-pig jejunum, which is usually in a state of mechanical quiescence, stroking elicited either single or sequential contractions on the oral side. Stronger stimuli elicited a series of contractions that appeared to propagate orally (Hukuhara et al. 1961).

Stroking the mucosa of an exteriorised loop of colon in unanaesthetised dogs elicited a contraction of the circular muscle on the oral side (Raiford and Mulinos 1934a). A contraction of the longitudinal muscle below the stimulus was also noted by these authors, but they do not mention inhibitory reflexes and attribute the relaxation of the circular muscle which was observed to a mechanical effect of reflex contraction of the longitudinal muscle. In anaesthetised dogs, stroking of the mucosa of the proximal colon elicited excitation above and inhibition below the stimulation (Hukuhara and Miyake 1959). Hukuhara and his collaborators named these reflexes initiated from the mucosa as "the mucosal reflex".

b) Localised Distension of the Intestine by a Solid Bolus or Balloon

Localised distension of the *small intestine* by a solid bolus or by a balloon also elicits an ascending excitatory and a descending inhibitory reflex in dogs and rabbits (Cash 1866; Lüderitz 1889; Bayliss and Starling 1899, 1900, 1901; Bozler 1939, 1949a; Hukuhara et al. 1958; Daniel et al. 1960; Nakayama 1962). The typical response oral to a distension in the dog can be described as follows: if rhythmic contractions were absent, they are initiated immediately above by the localised distension; if rhythmic contractions were present, their amplitude is augmented, the effect being marked between 1 and 2 cm but detected for up to 6 cm. The increase in amplitude of the rhythmic contractions is due to the increase in number and frequency of action potentials superimposed on every slow wave. There is a slight increase in the frequency of the slow waves in the excited area and from within this area contractions propagate anally, riding the slow waves and may pass over the bolus and fade away. If the distension is more marked, then action potentials occur between as well as on top of the slow waves, resulting in a stationary ring of tonic constriction. About 1–2 cm above the distension, the slow waves often increase in frequency and appear to propagate both orally and anally. It is most likely that the appearance of this new pacemaker area in the excited area is induced by the acetylcholine released from enteric cholinergic nerves which are the final neurones of this reflex pathway. Hukuhara et al. (1958) have indeed shown that local application of acetylcholine initiates a tonic or pulsating stationary ring of constriction from which waves propagate for short distances in both directions (the effects of exogenous acetylcholine are further described in Sect. B.III.1.b). According to the observations of Nakayama (1962), "the contraction waves (initiated by acetylcholine) originating close to the region oral to the bolus strongly propagate to the oral side and weakly to the anal side over the bolus."

On the anal side of the localised distension, the existing contraction of the circular muscle in the dog small intestine ceases almost instantly; the effect is obvious at 10 cm although it can be recorded up to 90 cm anally (Bayliss and Starling 1899; Nakayama 1962). Correspondingly, the action potentials disappear leaving only the slow waves (Nakayama 1962; Fukuda 1968b).

In the rabbit, BOZLER (1949 a), who could not elicit reflexes by stroking the mucosa in vitro, could initiate the ascending excitatory reflex by stretching the gut with the fingers, and the introduction of a solid bolus initiated action potentials above the bolus at a frequency of 7–10 Hz. With a balloon in the rabbit small intestine BOZLER (1939), BAYLISS and STARLING (1900, 1901) and LÜDERITZ (1889) elicited both excitation above and inhibition below the point of stimulation. The inhibition was more transient and less marked than in the dog. In the cat small intestine, these reflexes appear to be less prominent than in the dog (BAYLISS and STARLING 1899), ALTHOUGH BOZLER (1939) found that a solid bolus elicited strong rhythmic contractions on one or both sides. In the guinea-pig, distension by a small balloon elicits responses above and below the stimulation; these have been investigated electrophysiologically by HIRST and collaborators (see HIRST (1979). On the oral side, rare excitatory inputs were recorded in myenteric neurones while, on the anal side, most of the myenteric neurones were activated synaptically, and i.j.p. were recorded from the circular muscle, followed after 2–3 s by e.j.p.

After cocaine was applied to the mucosa or the mucosa was removed, distension of the dog small intestine with haemostats or hooks elicited relaxation of the circular muscle on both sides (NAKAYAMA 1962; HUKUHARA et al. 1958; 1960). These results prompted these authors to postulate the existence of a "muscular reflex". However, these mechanical stimuli are quite different from the physiological stimuli which the intestine receives, and so the responses may be abnormal. Pinching of the intestinal wall from the serosal side, utilised by many investigators, also cannot be regarded as a physiological stimulus to initiate and study reflex activity in the intestine.

Distension of the *colon* by a balloon or a solid bolus also elicited polarised responses, namely an ascending excitatory response and a descending inhibitory response (BAYLISS and STARLING 1900; ELLIOTT and BARCLAY-SMITH 1904; LYMAN 1913; AUER and KRUEGER 1947; CREMA et al. 1970; FRIGO and LECCHINI 1970). In the guinea-pig distal colon, the first response of the colonic musculature to distension by a balloon is a phasic contraction of the longitudinal muscle, during which a contraction of the circular muscle above the bolus appears; on the anal side a relaxation of the circular muscle is recorded (FRIGO and LECCHINI 1970; CREMA et al. 1970). The same authors found similar reflexes in the isolated cat colon.

The literature on the intrinsic intestinal reflexes acting on the *ileocaecal sphincter* is scanty. It was shown in Sect. B.IV that this portion of the circular musculature is innervated by intrinsic excitatory cholinergic nerves and by the enteric inhibitory nerves. The sphincter is also supplied by extrinsic excitatory cholinergic pathways and by extrinsic excitatory noradrenergic pathways (Sect. E.IV). It is most likely that the relaxation of the sphincter following distension of the ileum (LYMAN 1913; HUKUHARA and MIYAKE 1959; KELLEY et al. 1966; COHEN et al. 1968) is due to the activation of the intrinsic descending inhibitory reflex. This reflex is also likely to be activated during the emptying of the ileal contents into the colon by peristaltic waves, allowing the contents to pass through the sphincter which is normally tonically closed.

Contraction of the ileocaecal sphincter is elicited by distension of the colon (see PEARCY and VAN LIERE 1926; ALVAREZ 1940; HUKUHARA and MIYAKE 1959; PAHLIN and KEWENTER 1975) and appears to be mediated by an extrinsic excitatory reflex with noradrenergic nerves as the efferent limb of the reflex (Pahlin and KEWENTER 1975). Activation of this reflex slows the passage of the ileal contents through the sphincter. It is not known whether intrinsic ascending excitatory reflexes from the colon also act on the sphincter.

There is also clear evidence from the literature that the *internal anal sphincter*, which is the thickened continuation of the circular muscle of the rectum, is under the influence of the descending inhibitory reflex. The first evidence in humans was provided by GOWERS (1877) who showed that distension or irritation of the mucosa of the human rectum produces a relaxation of the internal anal sphincter, even after lesions of the sacral roots. GOWERS was also the first to suspect that the relaxation of the sphincter elicited in this manner is equivalent to the relaxation of the circular muscle which occurs during peristalsis in the rest of the intestine. Injection of air into the rectum resulted in rhythmic relaxations, followed by contractions of the sphincter, and he interpreted this sequence as the arrival of peristaltic waves, with relaxation preceding the propulsive contraction. He concluded: "the effect therefore, of the presence in the intestine of a mass of faeces or other contents would be to cause ... complete relaxation of the next portion of the intestinal wall into which the contents of the into which the contents of the intestine could pass, and ... a strong contraction behind, sustained, and moving on the stimulating body." Several authors since then have confirmed the existence of this intrinsic rectoanal reflex in humans (GARRY 1934; DENNY-BROWN and ROBERTSON 1935; SCHUSTER et al. 1963; KERREMANS 1968; SCHUSTER 1968; USTACH et al. 1970; IHRE 1974; FRENCKNER and IHRE 1976; MEUNIER and MOLLARD 1977; DICKINSON 1978; MARTELLI et al. 1978), in monkeys (RAYNER 971), in cats (Garry 1933; Garrett and HOWARD 1972; GARRETT et al. 1974) and in guinea-pigs (COSTA and FURNESS 1974). This reflex, similar to the descending inhibitory reflex of other parts of the intestine, is not antagonised by antiadrenergic drugs but is antagonised by hexamethonium (GARRETT et al. 1974). The test for the rectoanal inhibitory reflex has been used clinically to detect aganglionic segments in the lower rectum (SCHUSTER 1968; MARZIO et al. 1979; MEUNIER et al. 1979) since the reflex can be elicited in infants from the sixth day after birth (HOWARD and NIXON 1968). The internal anal sphincter is also innervated by extrinsic nerves (Sect. E.V) and is relaxed by intrinsic and extrinsic reflexes during the process of defecation (Sect. K.II.2).

c) Localised Distension of the Intestine by a Hook

COSTA and FURNESS (1976) introduced a method for studying intrinsic reflexes which allows the application of localised and graded distensions of the intestine with a hook and the recording of the mechanical responses of the circular muscle on both oral and anal sides.

In the guinea-pig *small intestine* the localised distension elicited single or multiple transient contractions orally for up to 6 cm while on the anal side in the vast majority of cases no response was recorded; in a few cases there was a small transient contraction and in some cases a small transient relaxation occurred. The presence of the two intrinsic reflex pathways, the *ascending excitatory reflex* and the *descending inhibitory reflex* were thus demonstrated (FURNESS and COSTA 1976).

Similar nerve-mediated reflexes can be elicited in isolated segments of guinea-pig *distal colon* (COSTA and FURNESS 1976). Thus, a localised distension elicits graded contractions of the circular muscle on the oral side and graded relaxation on the anal side. The ascending contraction lasted up to 20 s and could be recorded up to 7 cm orally. The descending relaxation could be recorded for up to 10 cm and at a closer distance was well maintained for up to 3–5 min. A contraction usually occurred following the removal of the stretch stimulus.

While in the small intestine, antimuscarinic drugs abolish the ascending excitatory reflex, in the colon they antagonise the ascending reflex only partially. This finding is consistent with the results of CREMA et al (1970) and suggests that non-cholinergic excitatory neurones may be involved in the reflexes in the guinea-pig colon (see COSTA and FURNESS 1976; and Chap. 11). When the ascending and the descending reflexes in the colon are activated simultaneously, the inhibitory reflex appears to prevail, at least under conditions which give near maximal response (COSTA and FURNESS 1976).

In contrast to the guinea-pig small intestine, contractions elicited by distension in the guinea-pig colon sometimes appear to migrate anally (COSTA and FURNESS 1976). Progression of these slowly migrating waves of contraction appears to be nerve mediated since pinching the wall or stimulating the muscle directly after administration of tetrodotoxin elicits a contraction which does not propagate.

d) Distension of the Intestine by Liquid

Of all mechanical stimuli, liquid distension is most appropriate to mimic physiological stimuli in the *small intestine*. Several authors, beginning with TRENDELENBURG (1917), demonstrated the reflex contraction of the circular muscle in response to liquid distension of isolated segments of small intestine (see KOSTERLITZ and LEES 1964; KOSTERLITZ 1968). In the guinea-pig, the typical sequence is as follows. When the pressure in the isolated segment is increased by lifting the liquid reservoir the segment slowly shortens. The membrane of the longitudinal muscle cells depolarises without action potentials being recorded (TSUCHIYA 1972). The segment, previously quiescent, begins to show shallow rhythmic contractions and slow oscillations of the muscle membrane potential are recorded (preparatory phase) (HENDERSON 1928; HUKUHARA and FUKUDA 1968; FRIGO et al. 1972). When the pressure reaches 1.5–2 cm H_2O, a strong contraction of the circular muscle occurs at the oral end of the segment with a simultaneous contraction of the longitudinal muscle. The contraction of the circular muscle propagates anally, expelling the fluid (emptying phase) and this event can be repeated for many hours. The electrophysiological recordings show that every contraction is due to the appearance of a slow wave of depolarisation with superimposed action potentials (HUKUHARA and FUKUDA 1968; TSUCHIYA 1972). The number and frequency of the action potentials increases with pressure. If the segment is held rigidly at constant length and the emptying prevented, then the cycle of intermittent propagated contractions is interrupted and action potentials are recorded continuously at 6–7 Hz. A similar effect was observed when acetylcholine was added to the bath (HUKUHARA and FUKUDA 1968). This suggests that the intermittency of the reflex is linked to the periodic emptying of the loop with subsequent cycles of nerve stimulation by distension.

It has been argued that the initial shortening of the longitudinal muscle is a passive phenomenon due to the filling of the loop (HUKUHARA and FUKUDA 1965). However, most authors agree that there is also a nerve-mediated reflex excitation of the longitudinal muscle during the preparatory phase (KOSTERLITZ and WATTS 1964; KOSTERLITZ 1968; TSUCHIYA 1972). Indeed, KOSTERLITZ and collaborators suggested that a noncholinergic excitatory neurone may be involved in the response of the longitudinal muscle (see KOSTERLITZ 1967). The apparent relaxation of the longitudinal muscle which occurs at the onset of the strong circular muscle contraction has been interpreted as a reciprocal inhibitory reflex to the longitudinal muscle (KOTTEGODA 1969, 1970). However, this relaxation almost certainly results from the passive elongation of the longitudinal muscle, caused by the contraction of the circular muscle (see Sect. B.III.3). Because the reflex contraction of the circular muscle propagates anally as soon as it is initiated, it is difficult to analyse the mechanisms of initiation and progression separately. Radial distension appears to be an adequate stimulus for this reflex whereas increases in tension or stretching in the longitudinal direction are not (KOSTERLITZ 1968).

Similar reflexes elicited by liquid distension have been observed in isolated preparations of rabbit, cat and rat small intestine, although in these species less thorough investigation was performed (TRENDELENBURG 1917; ALVAREZ 1940; BÜLBRING and LIN 1958; BÜLBRING et al. 1958; FRIGO et al. 1972; TONINI et al. 1974). ALVAREZ (1940) noted that injection of fluid into the anal end of the rabbit ileum resulted in a contraction at some distance above the distending column. In isolated and extrinsically denervated loops of dog small intestine in vivo perfusion of the lumen with isotonic saline elicited waves that propagated anally; if the waves were weak, they emptied the loops in small spurts (GREGORY 1950) and, if strong, each wave swept the whole segment and emptied it (HUKUHARA et al. 1958; NAKAYAMA 1962). These waves of contraction occurred at regular intervals of 12.5 cycles/min (GREGORY 1950) which is a similar frequency to that of the slow waves. Also in the *colon*, liquid distension elicits a reflex contraction of the circular muscle which propagates and empties the segment (HUKUHARA et al. 1961; LEE 1960; MACKENNA and MCKIRDY 1972; PESCATORI et al. 1979).

Very little work has been done on the reflex responses elicited by longitudinal distension of a small intestine. In the guinea-pig, FLEISCH and WYSS (1923) and FLEISCH (1928) described ascending and descending cholinergic contractions of the longitudinal muscle following stretch of the longitudinal muscle itself. However, because the longitudinal muscle is much thinner than the circular muscle, it is difficult to distinguish direct nerve-mediated effects on this muscle from those induced indirectly by mechanical interference by the more powerful circular muscle.

e) Intestinal Adaptation

It had already been shown that localised distension produces opposite responses above and below the point of stimulation but it was not clear where the boundary between the ascending excitation and the descending inhibition occurs. In order to investigate this problem FURNESS and COSTA (1977) have recorded in the guinea-pig colon the degree of distension produced by a given tension and found that the distensibility was reduced by tetrodotoxin but not by anticholinergic or antiadrenergic drugs. HENDERSON (1928) had earlier reported that the guinea-pig intestine was less distensible after administration of nicotine. This indicates that distension of the colon elicits an inhibitory reflex which acts on the distended area, allowing its further distension. These results have been confirmed more recently by DAVISON and

PEARSON (1979), who found however a small excitatory cholinergic component. This local inhibitory reflex is apparently involved in a process of local accommodation in the colon. Indications that such accommodation may also occur in human large intestine during distension can be found in the work of WHITE et al. (1940) and LIPKIN et al. (1962). Although no specific experiments have been performed to determine whether the inhibitory reflex also produces an adaptation of the small intestine to its contents, some early evidence may suggest that such adaptation occurs. BAYLISS and STARLING (1901) observed that the circular muscle surrounding a solid bolus was relaxed but was contracted above it. More experiments are required to establish the effects of the intrinsic reflexes at the point of distension in the small intestine.

4. Role of the Enteric Reflexes Initiated by Mechanical Stimuli in Propulsion

We have shown that, as originally pointed out by MALL (1896) and by CANNON (1911), the contractions or relaxations elicited by enteric reflexes do not progress by themselves when initiated by a localised fixed stimulus. However, both solid boluses and liquids are transported anally in isolated segments of small and large intestine. The drugs that antagonise the enteric reflexes also abolish propulsion in intestinal segments. Thus, in the small intestine tetrodotoxin, atropine or hyoscine and antinicotinic drugs abolish the propulsion of both solids and liquids in isolated segments (BAYLISS and STARLING 1899; CANNON 1911; TRENDELENBURG 1917; QUIGLEY et al. 1934; HUKUHARA and FUKUDA 1968; KOSTERLITZ 1968; KOTTEGODA 1969; TSUCHIYA 1972; FONTAINE et al. 1973; TONINI et al. 1974; WEEMS and SEYGAL 1979) and the same drugs abolish the propulsion of solids and liquids in isolated preparations of large intestine (LEE 1960; HUKUHARA et al. 1961; CREMA et al. 1970; McKENNA and McKIRDY 1972; COSTA and FURNESS 1976; NAKAYAMA et al. 1979; PESCATORI et al. 1979). These results strongly suggest that the enteric reflexes are essential for the propulsion of intestinal contents.

There is consensus among authors that when a solid bolus is introduced into the small intestine of the dog, rabbit or cat, the reflex contraction elicited by the bolus initiates the transport of the bolus anally and that therefore the contraction also travels in this direction. BOZLER (1949b) noted that the speed of propulsion of a bolus in the dog intestine "varies widely and increases within limits with the size of the bolus," confirming similar ovservations by CASH (1886). From the description of the events in this species made by CASH (1886), BAYLISS and STARLING (1899), QUIGLEY et al. (1934), BOZLER (1939, 1949a, b), DANIEL et al. (1960), NAKAYAMA (1962) and BASS (1968) the following description emerges. Weak mechanical stimuli by the bolus elicit a weak ascending excitatory reflex which results in a mean increase in amplitude of the rhythmic contractions above the bolus but does not necessarily propel the bolus. In such a case the excited area above the bolus only extends for a short distance, corresponding to the width of a ring of rhythmic contraction (about 0.5 cm in the cat, 0.5 cm in the rabbit, 4–8 cm in the dog upper intestine). However, the excited area above the advancing bolus usually extends for greater distances: 4–5 cm in the cat, 5–10 cm in the rabbit and 8 cm or more in the dog (BAYLISS and STARLING 1899; MELTZER and AUER 1907; CANNON 1912; PARKKULAINEN 1962). These pulsating waves, which appear to sweep anally over the bolus and fade away, are not capable of pushing the solid bolus at their own speed because it offers some resistance. In such circumstances, progression of the bolus is slow and every contraction produces only a small advancement. According to Bozler (1939), each contraction pushes the bolus as little as 1 mm while Nakayama (1962) calculated that each contraction pushes the bolus 1.3 cm.

If the progression of the bolus is prevented, the rhythmic contractions above the bolus do not reach the anal side. A bigger bolus elicits a stronger reflex excitation, in which the contractions in the oral area summate and fuse to produce a tonic ring of constriction on which rhythmic superimposed contractions can be observed. The frequency of slow waves with the accompanying contractions increase and a new local slow wave pacemaker, induced by acetylcholine released from the nerves, may appear from which slow waves propagate in both oral and anal directions. Despite the superimposed waves of contraction spreading both orally and anally, the effective force that pushes the bolus is now the tonic contraction. If the stimulus is even stronger, then action potentials in the excited area occur as a continuous discharge producing a strong tonic contraction which can push the bolus at 0.4 cm/s (Nakayama 1962). It is quite clear that in the upper intestine, progression of the bolus is much slower than the fast sweeping contractions that ride the slow waves in the excited areas. In the lower small intestine, slow waves propagate at similar speeds to the bolus (0.5–2 cm/s for slow waves; 1 cm/s for peristalsis, Bayliss and Starling 1899). It could be argued that the propagation speed of the slow waves is more suitable to reinforce the neurogenic excitatory reflex effectively and slowly transport solids in the lower small intestine than it is in the upper intestine.

The wide range of speeds of transport of solid boluses (1–24 cm/min) is thus likely to depend on the effectiveness of the bolus in eliciting excitation above it, through the ascending excitatory reflex and on the resistance to progression of the bolus itself. In addition, there are influences mediated by extrinsic nerves (Sect. K). The resistance to the progression of a solid bolus results from friction between the bolus and the walls of the intestine and from the state of contraction of the segment in front of the advancing bolus.

Several authors have recorded a reduction or cessation of action potentials with resulting mechanical relaxation of the circular muscle ahead of the advancing bolus in the dog small intestine (Bayliss and Starling 1899; Mescham and Quigley 1938; Daniel et al. 1960; Nakayama 1962; Bass 1968). Failure by some authors to record a descending inhibition (see for example Bozler 1949b) may depend on several factors: the stimulus may have been insufficient to elicit an effective descending inhibitory reflex, or the activity of the circular muscle below the bolus may have already been low and no further reduction could be detected or the effect could be too short-lived to be easily detected in some parts of the small intestine. In dogs, the recorded wave of inhibition was present only 1–2 cm ahead of the bolus (Bayliss and Starling 1899; Daniel et al. 1960). However, the intestine was shown in Nakayama's records (1962) to be inhibited up to 14 cm ahead of the advancing bolus. In the rabbit small intestine, Bayliss and Starling (1900, 1901) and Bozler (1939) described the rapid propulsion of a bolus of cotton wool and petroleum jelly or a distended baloon. The contraction that pushed the bolus was composed of partially fused rhythmic contractions and the bolus was preceded by an area of inhibition. BAYLISS and STARLING (1899) were aware of the difficulty of recording the transient descending inhibitory reflex.

With *liquid distension*, the speed of propulsion of the column of liquid by the advancing ring of circular muscle constriction is much higher than with a solid bolus. In the guinea-pig, speeds of 1–7 cm/s have been recorded (BOZLER 1939; HUKUHARA and FUKUDA 1965; HUKUHARA et al. 1969), in the rabbit 1.5–17 cm/s (ALVAREZ and HOSOI 1930; GONELLA 1971) and in the dog several centimeters per second (QUIGLEY et al. 1934; GREGORY 1950; NAKAYAMA 1962). Since these speeds are within the range of the speed of propagation of the slow waves, it is interesting to establish whether the contractions travel anally in synchrony with slow waves.

The observation that with increasing pressure the speed of propulsion also increases (HUKUHARA et al. 1969) suggests that the propagated contractions do not ride slow waves which would usually propagate at a fixed speed. In the rabbit, the work of most authors indicates that the propagated contractions during peristalsis travel slower than the slow waves (BAYLISS and STARLING 1900, 1901; BÜLBRING et al. 1958; GONELLA 1971; FRIGO et al. 1972). In dogs, liquid transport has been studied only in isolated loops of upper small intestine in vivo. NAKAYAMA (1962) observed that each of the rhythmic waves which occurred at the slow wave frequency and travelled anally at the speed of the slow waves, emptied the loop. Other authors, such as GREGORY (1950), observed that each contraction only travelled for a short distance and that the emptying of the loop was effected by several contractions, each of which started a little bit further down the segment than the preceding one. From the records of STREETEN and VAUGHAN-WILLIAMS (1951) it appears that periods in which small amplitude propagated contractions failed to propel the fluid effectively alternated every 2 min with periods in which each of a series of large amplitude contractions propelled fluid out of the anal end of the loop. From these results it would appear that in some circumstances the speed of the advancing column of liquid approaches the speed of propagation of the slow waves. In this case the speed of progression of the enteric reflex coincides with the speed of propagation of the slow waves.

The question now arises as to whether the aboral propagation of the reflex contraction of the circular muscle is dependent on the continuous presence of the column of liquid or whether, once initiated, the neuronal mechanism will propagate by itself. ALVAREZ (1940) has shown that if the column of liquid set in motion by the peristaltic wave is caused to leave the intestine through an opening in the wall, the wave does not progress beyond that point. Similar results were obtained by HENDERSON (1928) who used an isolated segment of guinea-pig ileum in which passage of fluid between the two halves of the segment could be prevented. When pressure was allowed to build up only in the oral segment the peristaltic wave travelled only in the oral half.

These experiments indicate that the propagation of the contraction in the small intestine elicited by liquid distension is the result of the sequential activation of the ascending excitatory reflex and not of an intrinsic propagation of neuronal activity. The interpretation of HENDERSON (1928) was that "the impulse to peristalsis does not travel far in the plexus but only the immediately adjacent muscle fibres respond. Their contraction displaced fluid to a lower level where the internal pressure is almost sufficient to produce activity and hence the wave is reiterated at successive lower levels." However, it is not clear why the reflex contraction should appear first at the oral end of the segment or loop, despite the fact that the whole length is equally distended by the liquid, or in other words, why the rest of the loop should fail to contract simultaneously when distended. Two possible mechanisms can be proposed to explain why the muscle should be more excitable at the oral end: (a) distension of the loop elicits the excitatory reflex which increases the excitability of the muscle of the whole segment, but the first contraction appears at the oral end because the slow waves are initiated orally and propagate anally; (b) the simultaneous activation of the polarised ascending excitatory and descending inhibitory reflexes result in the unequal activation of the muscle within the segment because the oral end is not affected by the inhibitory reflex. The second interpretation was first suggested by HUKUHARA and FUKUDA (1965). According to this view, every point of the distended segment or loop except its ends is under the influence of both reflexes which therefore act against each other. The oral part of the segment wich lacks the inhibitory influence from above is under an unopposed influence of the ascending excitatory reflex. Passage of the liquid from the oral end will cause the successive activation of the two reflexes as already described, and the liquid will be propelled anally. According to this hypothesis, the descending inhibitory reflex plays a key role in unidirectional transport in the small intestine.

Several experiments suggest that there are neuronal mechanisms for a descending excitation (see Sect. H.V.3.b). Some authors (BAYLISS and STARLING 1899; NAKAYAMA 1962) reported that when an advancing bolus is stopped or is allowed to leave the lumen through an incision, the contraction which was pushing the bolus often appears to continue to travel slowly for a short distance before fading away. Similarly, fast descending contractions originating from the constriction oral to a bolus appear to travel over the bolus and fade away at some distance below (BAYLISS and STARLING 1899, 1900, 1901; BOZLER 1949 a; DANIEL et al. 1960). This is consistent with the observation that, in a small proportion of cases, radial stretching of the guinea-pig small intestine elicits a small contraction on the anal side (FURNESS and COSTA 1976). HIRST et al. (1975) recorded e.j.p. in the circular muscle, anal to a distension immediately after the expected i.j.p. It is quite possible that in vivo these descending excitatory mechanisms may play a significant role.

The propulsive wave of contraction in most cases travels slower than the slow waves and in isolated segments the *migrating neuronal mechanisms* (see Sect. H.V.3.b) appear to play little or no role in the progression of the contents in the small intestine. Progression is maintained by continued activation of the reflexes as the contents move anally.

For the *large intestine* there is a remarkable similarity between the caudal transport of solid boluses or balloons in isolated preparations and the transport of faecal matter in vivo. Introduction of a solid bolus in the isolated large intestine elicits an ascending reflex contraction which causes the propulsion of the bolus towards the anal end in the dog, rabbit, cat, and guinea-pig (BAYLISS and STARLING 1901; LANGLEY and MAGNUS 1905; AUER and KRUEGER 1947; BOZLER 1949 b; CREMA et al. 1970; FRIGO and LECCHINI 1970; COSTA and FURNESS 1976; NAKAYAMA et al. 1979). In those species in which scybala are present in the distal colon, the ability of the intrinsic mechanisms to coordinate the emptying of the whole colon in vitro can be clearly demonstrated (CURRIE and HENDERSON 1926; AUER and KRUEGER 1947). Since the rate at which scybala are extruded at the anal end is constant, simple tests of the effects of antagonist drugs on such a complex process can be done (COSTA and FURNESS unpublished work). In those experiments in which the mechanical activity of the circular muscle war recorded (LANGLEY and MAGNUS 1905; FRIGO and LECCHINI 1970; CREMA et al. 1970; COSTA and FURNESS 1976) it was clear that the circular muscle just above the advancing bolus was strongly contracted while the area in front of the bolus was relaxed. While there is no doubt about the essential role of the ascending excitatory reflex in the progression of the bolus, the role of the descending inhibitory reflex is more difficult to determine since it is not possible specifically to block transmission from the final inhibitory neurone (see Chap. 11).

The speed at which the bolus is transported in vitro is similar to the speed of the propagated peristaltic contraction observed in vivo. In vitro, the bolus is transported at about 3–10 cm/min (FRIGO and LECCHINI 1970; COSTA and FURNESS 1976). In some cases, the balloon was transported without any apparent distension of the wall, suggesting that stroking the mucosa is sufficient to initiate propulsive reflexes.

Little work has been done on the role of the consistency of luminal contents on the speed of the reflex propulsion. Some writers, however, used the Trendelenburg technique of liquid distension to elicit peristalsis in the isolated colon of the guinea-pig and rabbit (LEE 1960; HUKUHARA et al. 1961; MACKENNA and MCKIRDY 1972; PESCATORI et al. 1979). In all instances, the speed of propagation was found to be slower than that of the small intestine, but was faster than the speed of propulsion of a solid bolus. The situation is thus similar to the small intestine, in which liquids are propelled at a faster speed than solids. When the colon empties in vitro, the first activities which develop in the organ bath and can be associated with propulsion are rings of contraction which appear at irregular intervals between adjacent pellets (COSTA and FURNESS 1976). Succeeding contractions become stronger and travel towards the anal end. When contractions occur above the most anal pellets they are separated from the more proximal ones and are expelled either singly or in groups of two or three. Contractions which begin on the oral side of larger groups of pellets do not propel the whole group. Sometimes these contractions just fade out. On other occasions, they push the most oral pellet forward, so that it partly overlaps the next pellet. The contractions often move in an anal direction and pass over some of the pellets without pushing them along. Rings of contraction which pass over the more oral pellets often become effective towards the anal end and are able to expel the last 1–3 pellets.

All these results taken together strongly suggest that propulsion in the large intestine that is disconnected from the central nervous system occurs as a result of the sequential activation of the enteric ascending excitatory reflex whose effect is facilitated by the activation of the descending inhibitory reflex. Superimposed on this mechanism a *spontaneous migrating neuronal mechanism* appears to scan the colon, reinforcing, by its slow anal propagation, the ascending excitatory reflex to become effective in propelling the contents (COSTA and FURNESS 1976).

5. Reflexes Evoked by Chemical Stimuli and Their Role in Propulsion

Although the influence of the chemical content of the intestine on motility was recognised by CANNON (1911) little systematic work has been done on the possible intrinsic nerve pathways which may be activated by the large variety of chemicals that contact the mucosa during digestion.

The effect of localised application of chemicals to the mucosa in a loop of small intestine was studied by HUKUHARA et al. (1958) and NAKAYAMA and NANBA (1961) in vivo in an isolated and extrinsically denervated loop. They found that when care was taken to avoid mechanical stimulation, application of HCl, copper sulphate solution, crystals of NaCl or mustard paste, elicited polarised reflex responses in the dog jejunum, similar to those obtained by stroking the mucosa, namely an excitation on the oral side and an inhibition on the anal side. The effects on luminal pressure of injecting chemicals through a cannula placed in the dog duodenum have been studied (THOMAS and BALDWIN 1971; BALDWIN and THOMAS 1975). Following injections of HCl, potassium oleate or sodium deoxycholate, a marked reduction of the rhythmic contractions below and an increase above the injection site was observed. Similar responses were observed in extrinsically denervated intestines. The descending inhibition was more constantly present than was the ascending contraction. The excitation and the inhibition appeared to be slowly propagated caudally. Cold block between the stimulus and the recording prevented the responses.

In similar experiments STEWARD and BASS (1976a, b) administered oleic and ricinoleic acid intraduodenally and recorded the mechanical activity from several estraluminal strain gauge transducers placed along the small and the large intestine. The *cis* forms of both fatty acids, but not their *trans* isomers, produced an excitation of the jejunal area and the responses appeared to travel caudally at 2 cm/s. After the initial brief stimulation lasting 2 min the contractile activity of the jejunum was inhibited for 45 min (oleic acid) and 4–6 h (ricinoleic acid). No responses were detected in the ileum or colon. Similar effects were obtained after oral ingestion of the fatty acids. The responses to these substances were abolished by atropine and reduced by hexamethonium. The authors suggested that the effect is mediated by the release of cholecystokinin from the mucosal endocrine cells by the fatty acids; cholecystokinin then acts directly or indirectly on the enteric excitatory cholinergic neurones. ALVAREZ (1940) refers to the work of BORCHARDT who, using Thiry-Vella loops in dogs found that introduction of dilute HCl at the oral end of the loop produced a spasm which rapidly travelled caudally. After application of cocaine to the mucosa the response was inhibited. GREGORY (1950) introduced diluted HCl, NaCl solutions (2% and 10%), bile salts, soap, pancreatic digest of proteins and neutralised duodenal content and found that all these substances induced increases in contractions of the loop with a reduction in the amount of liquid flowing through it. Since the responses were abolished by cocaine and were present in an extrinsically denervated loop, they most likely involved intrinsic reflexes. NAKAYAMA (1962) confirmed GREGORY's results by using HCl (0.025 *M*). The tone of the isolated loop of dog small intestine increased, the lumen narrowed, sporadic waves occurred all over the segment and transport of fluid was reduced.

These results have been confirmed in the small intestine of unanaesthetised dogs by SUMMERS (1978) who found that infusion of dilute HCl into the duodenum increased the number of action potentials in the neighbouring muscle, with marked reduction of fluid propulsion. HUKUHARA et al. (1961) found that a very weak HCl solution (0.0125 *M*) reduced the threshold pressure for liquid transport in an isolated loop of anaesthetised dog small intestine, the effect being abolished by application of xylocaine to the mucosa. LEWIS et al. (1979) found that acid application onto the duodenal mucosa of unanaesthetised fasted dogs initiated a new migrating myoelectric complex. BENNETT and ELEY (1976) found that increased concentration of H^+ increased peristalsis in the isolated guinea-pig ileum.

In a perfused loop of cat duodenum or jejunum ABBADESSA et al. (1969) found that infusion of increasing concentrations of NaCl or of HCl increased the number of action potentials in the muscle with resulting increased amplitude of muscle contractions. Infused bile also elicited similar responses. Infusion of glucose was without effect with 5% solution while with 10% solution the number of action potentials increased and with a 15% solution the action potentials disappeared. All these responses were blocked by a local anaesthetic and by nicotine. The results with glucose strongly suggest that glucose stimulates not only extrinsic sensory nerves during absorption (Sect. D.II) but also intrinsic sensory neurones. Pretreatment with phloridzine, a specific antagonist of active transport of glucose, abolished these responses and passively transported sugars (xylose and saccharose) failed to produce them. In humans, administration of sodium myristate reduces the number of propulsive waves in the duodenum without changing the stationary rhythmic contractions (BORGSTROM and ARBORELIUS 1975).

Although the information about chemical stimulation of enteric reflexes is still scanty, it shows that chemical stimulation of the mucosa can increase or decrease intestinal contractions by acting locally and that most of the substances tested appear to increase the nonpropulsive movements of the small intestine and delay propulsion of the contents. These findings are consistent with the fact that, following arrival of digesta, irregular contractions in the small intestine occur which interrupt the interdigestive migrating complexes (Sects. G.I.2, H.III).

An enormous field of investigation is open to establish the presence, properties, pathways, nerve types and role of chemically induced intrinsic intestinal reflexes. Chemical substances may not only act on sensory nerve endings during the process of absorption but may also stimulate specific endocrine cells to release amines, such as 5-HT, and polypeptides, such as substance P, cholecystokinin, gastrin, motilin and secretin which may act locally on enteric nerves to affect motility.

Not surprisingly, less work than in the small intestine has been done on the chemical stimulation in the large intestine. HUKUHARA and MIYAKE (1959) have shown that in anaesthetised dogs, dilute HCl, mustard paste and NaCl crystals elicited reflex responses similar to those evoked by mechanical stimulation of the colon and HUKUHARA et al. (1961) found that HCl (0.01 *M*) increases the rate of movement of fluid through an isolated segment of dog colon. Local application of chemical substances to stimulate colonic motility has been used rather commonly in studies of humans. Substances such as bisacodyl, oxyphenisatin and derivatives of senna have been used with barium enemas to increase the occurrence of progressive waves of contractions which empty long segments of colon; these have been called stripping waves and can be regarded as a form of induced peristalsis similar to that occurring under normal circumstances (WILLIAMS 1967; HARDCASTLE and MAN 1968, 1970; RITCHIE 1972). Since this response is prevented by local anaesthetics, these substances are likely to activate sensory neurones. It is not known, however, whether the effect is mediated by intrinsic or by extrinsic neural mechanisms. HARDCASTLE and MANN (1970) investigated the effect of acid phosphate buffer, concentrated NaCl and glycerin suppositories on colon peristalsis in healthy patients. The effect of acid was poor; increasing the osmolarity with NaCl had no effect, but glycerine induced peristalsis. Infusion of the secondary bile acid, sodium deoxycholate, which is found in the colon of normal rabbits, stimulated colonic motility in this species. This effect was attributed to its surfactant properties (FALCONER et al. 1978). There is no doubt that the reflexes evoked by chemical stimulation from the contents of the large intestine require far more investigation.

6. Antiperistalsis

The nature of the antiperistaltic waves observed in the proximal colon of different species, even when isolated from the central nervous system (Sect. G.II.1) has not been established. The relative roles of myogenic and neurogenic mechanisms have not been investigated and for a long time the interpretation of ELLIOTT and BARCLAY-SMITH (1904) and CANNON (1911) that the mechanism is myogenic, prevailed. However, in rats and guinea-pigs, HUKUHARA and NEYA (1968) found that antinicotinic drugs antagonised the antiperistaltic waves but not the pulsating ring, suggesting that the antiperistaltic wave propagates by a neuronal mechanism which involves cholinergic interneurones.

In vitro mild distension of a segment of rat colon which included parts proximal and distal to the pulsating ring elicited both shallow antiperistaltic waves and peristaltic waves originating from the pulsating ring. With increased distension, the antiperistalsis was supplanted by a strong peristaltic wave of contractions which started at the oral end and swept the whole segment, travelling anally. In a similar way, when the distension in the proximal colon is increased by the arrival of material from the ileum, antiperistalsis gives way to peristalsis (see Sect. G.II).

7. Role of the Enteric Reflexes in Propulsion and Their Relation to the Migrating Neuronal Mechanisms

The weight of evidence presented indicates that intrinsic reflexes can be evoked by mechanical stimulation in all parts of the small and the large intestine. These enteric reflexes can be divided according to their polarity into *ascending excitatory* and *descending inhibitory reflexes*. Both reflexes involve sensory neurones sensitive to stretch located in the mucosa and in the external musculature and cholinergic interneurones. In the ascending reflex there is a final excitatory cholinergic neurone. In the descending inhibitory reflex the final enteric inhibitory neurones release an unidentified transmitter substance. These reflex pathways run up and down within the myenteric plexus for variable lengths depending on the species and for distances usually equivalent to several intestinal diameters.

Although it appears to be present in all species, in some parts of the intestine the descending inhibitory reflex is not easily detected. The descending inhibitory reflex is more prominent in the large intestine where is ensures unopposed progress of the contents while in the small intestine this reflex ensures the unidirectional anal propulsion of the liquid contents, but its mechanical effects are not always recorded.

Propulsion occurs as a result of the sequential activation of the reflexes in a cascade fasion along the intestine. The speed of propulsion of the contents depends on the rate of progression of the sequential activation of the reflexes and the resistance offered by the contents. The speed of propulsion varies with the consistency of the contents, liquids being propelled faster than solids in any given segment. Different parts of the intestine appear to have inbuilt differences in the speed at which the progression of the refluxes can occur, being faster in the small intestine than in the large intestine. In most circumstances, the speed of propulsion in the small intestine appears to be slower than the speed of propagation of the slow waves. Advancement of the contents therefore occurs by a sequence of rhythmic contractions which occur in an area of excitation on the oral side of the bolus, each contraction corresponding to a crest of a slow wave and pushing the contents a little bit. If the stimulus is stronger, there is a travelling tonic contraction in which the muscle is continuously activated throughout the slow wave cycle by the overriding ascending excitatory reflex. Only in some cases, in dogs and perhaps in humans, the speed of progression of the column of liquid may sometimes coincide with the slow wave speed, causing a wave of contraction which appears to ride a slow wave for some distance (CODE et al. 1968).

Since peristalsis is due to a sequential activation of ascending excitatory and descending inhibitory reflexes, it must be possible to increase the speed and force of peristalsis by potentiating nerve transmission within the pathways. Drugs which potentiate cholinergic transmission do in fact increase the speed of peristalsis in low doses (MACKENNA and MCKIRDY 1972; M. COSTA and J. B. FURNESS, unpublished work 1974) while in excessive doses they produce a paralysis of peristalsis owing to a tonic contraction of the muscle layers (COSTA and FURNESS 1976). Another consequence of the involvement of the reflexes in propulsion is that the rate at which peristalsis can occur depends on several factors. These factors include the time required for a segment to empty which is a function of the force and speed of the contraction, the time required for the empty segment to be distended again, and the intensity of stimulation by stretching.

There are obvious similarities between the propulsion of liquids in isolated segments of small intestine in vitro and in the small intestine in vivo. The drugs which abolish the propulsion in vitro have also been shown to abolish the fast propagated contractions observed in vivo in the small intestine (MELTZER and AUER 1907; CANNON 1911; ALVAREZ 1940; BURNS et al. 1979). Similarly, the propulsion of semisolid or solid boluses in segments of colon in vitro is similar to that of the faecal masses which occurs in vivo (see Sect. G.II).

In vivo, and even in vitro, the intrinsic propulsive reflexes are not evoked continuously; otherwise, the ingested substances would be transported anally before proper digestion and absorption could take place. One of the likely mechanisms by which the progress of digesta in the small intestine is slowed down is the activation of local reflexes by specific chemicals produced during digestion. In the large intestine, the interaction of inhibitory and excitatory reflexes initiated from different sites by the solid contents may slow down the propulsion.

There are neuronal mechanisms that periodically modify the excitability of the enteric reflexes. One modifying factor is the area of increased neuronal activity which slowly migrates along the intestine. In the small intestine, this mechanism is evident during the interdigestive periods in carnivores and omnnivores while it is always present in herbivores (Sect. G.I). In the large intestine, a similar mechanism is present although it has been less well investigated. The migrating complexes rely on neuronal mechanisms built into the enteric plexuses; they are under the influence of extrinsic nerves which in the small intestine time and coordinate the initiation and propagation of these migrating motor complexes. The elegant experiments of CODE and SCHLEGEL (1974) show that the excitability of the ascending excitatory reflex in the small intestine changes with the phases of the interdigestive migrating myoelectric complex. In conscious dogs, introduction of a bolus of X-ray-opaque liquid into the lumen does not elicit reflex propulsion during the quiescent phase (phase I). When injected during phase II, the liquid initiates fast propagated contractions which run for short distances and if injected just before phase III, a bolus of liquid is rapidly transported anally. In the colon, the migrating neuronal activity also facilitates the enteric reflexes by reinforcing the ascending excitatory reflex which therefore becomes more effective in propulsion.

The intrinsic reflexes therefore represent a basic neuronal mechanism on which modulatory influences of intrinsic and extrinsic origin act. It is most likely that the different types of nerve present in the enteric plexuses, which are not known to be directly involved in the reflex pathways (Chap. 11), and some of the inhibitory mechanisms detected electrophysiologically in the enteric plexus (Chap. 6) may be involved in modulating the intrinsic reflexes. The identification of intrinsic modulatory mechanisms represents a great challenge which, when met, will considerably enhance our understanding of the role of the enteric nerves in the control of intestinal motility.

K. Extrinsic Intestinal (Intestinointestinal) Reflexes

Intestinal motility is affected by extrinsic reflexes which can be initiated from the digestive tract or from extradigestive organs. In this section only the reflexes initiated from the digestive tract are considered. The term intestinointestinal reflex is taken to encompass all reflexes which originate in and act on the digestive tract although more precise terms, e.g. gastrocolic reflex, may be used. These reflexes can be subdivided into inhibitory and excitatory reflexes. The afferent and the efferent limbs of the intestinointestinal reflexes run in the extrinsic nerves as was described in Sects. D.II and E.

I. Inhibitory Intestinointestinal Reflexes

1. Inhibitory Sympathetic Reflexes

The presence of reflexes initiated from the intestine which inhibit intestinal motility via the sympathetic inhibitory pathways is well established (HERMAN and MORIN 1934; YOUMANS 1952, 1968). FURNESS and COSTA (1974a, b) in reviewing the properties and role of these reflexes, distinguished two reflexes pathways, one going through the spinal cord (the long, or central reflex) and the other through prevertebral sympathetic ganglia (the short, or peripheral reflex). More recent investigations have further clarified the pathways, the characteristics and the suggested roles for these reflexes.

a) Nerve Pathways

The pathways of the long intestinointestinal reflexes through the spinal cord have been well established (see FURNESS and COSTA 1974a, b). These pathways have sensory neurones with cell bodies in the spinal ganglia, and their peripheral sensory endings located in the external musculature of the intestine. Their peripheral processes run in the mesenteric and splanchnic nerves and their central processes synapse within the spinal cord through at least one interneurone which in turn excites the preganglionic cholinergic neurones of the efferent part of the pathway. This neurone synapses with postganglionic noradrenergic neurones in the prevertebral ganglia and the axons of the noradrenergic neurones terminate mainly around enteric neurones (see Sect. E.II.1; Chap. 2).

The short extrinsic intestinointestinal nerve pathways were initially investigated by KUNTZ and collaborators (see KUNTZ 1953) and more recently by SZURSZEWSKI and collaborators (see SZURSZEWSKI 1977a). The afferent limb of this reflex is formed by cholinergic neurones with their cell bodies in the enteric ganglia. These project outside the intestine in the mesenteric nerves and synapse with postganglionic sympathetic neurones in the prevertebral ganglia. Since many of the neurones in the prevertebral ganglia receive excitatory inputs from both sympathetic preganglionic cholinergic and intestinofugal cholinergic axons, these two reflex pathways share their final noradrenergic neurone (KREULEN and SZURSZWESKI 1979). As was pointed out by SZURSZEWSKI (1977a) in the discussion of his paper, it would be very difficult to investigate the role of the short extrinsic reflex after removal of the inputs from the long reflex pathway because all inputs contribute to the normal excitability of the final neurone. The sensory receptors of the short reflexes are very sensitive to both distension of the intestine and contraction of the intestinal muscle (CROWCROFT et al. 1971; SZURSZEWSKI and WEEMS 1976).

b) Effects of Sympathetic Reflexes on Intestinal Motility

The noradrenergic nerves to the intestine act mainly by reducing the excitability of the enteric cholinergic neurones (Sect. E.II.1) which are the common final neurones of excitatory nerve pathways involved in the control of intestinal motility. The activation of the sympathetic pathways therefore results in inhibition of all types of intestinal movement. Thus, sympathetic stimulation inhibits the enteric reflexes involved in peristalsis (KOSTERLITZ and LEES 1964; FURNESS and COSTA 1974a, b) and also inhibits the irregular nonpropulsive contraction of phase II of the interdigestive motor complex as well as the regular contractions of phase III which occur in the stomach and small intestine (YOUMANS 1952; CODE and MARLETT 1975; RUCKEBUSCH and BUENO 1975). Colonic antiperistalsis can also be inhibited by stimulation of the sympathetic pathways (ELLIOTT and BARCLAY-SMITH 1904). Thus, the main effect of the sympathetic nerves is to inhibit transport of the intestinal contents and this is achieved not only by inhibiting the nerve activity responsible for propulsion in the nonsphincteric parts of the intestine but also by closure of the sphincters which receive a direct excitatory noradrenergic innervation (see FURNESS and COSTA 1974a, b).

YOUMANS (1952, 1968) maintained that the long intestinal reflexes are diffused and not segmental. While FRIEDMANN (1975) agrees that a generalised inhibition of the stomach and intestine can be produced by strong distension of the intestine, he reported that moderate distension in the middle of the small intestine inhibits motility on the oral side while on the anal side excitation of motility was recorded. More investigations are required to substantiate the existence of an oral polarity of the extrinsic sympathetic reflex.

c) Physiological Activation of Sympathetic Reflexes

The inhibitory sympathetic nerves are activated in a number of circumstances which require the digestive processes to be delayed, such as during heavy physical exercise when blood is diverted to other vascular beds (FURNESS and COSTA 1974a). In intestinal obstruction, excessive distension of the intestine oral to the obstacle initiates a generalised reflex inhibition of gastrointestinal motility. Peritoneal irritation which occurs in handling of viscera during abdominal surgery also activates these reflexes which are partially responsible for the ensuing adynamic ileus (FURNESS and COSTA 1974a, b). Further confirmation that the reflex activation of the noradrenergic fibres to the intestine is responsible for the postoperative adynamic ileus comes from the work of ÖHRN and collaborators (ÖHRN and RENTZHOG 1976; ÖHRN et al. 1976) and of DUBOIS (1977) who found that chemical sympathectomy with 6-hydroxydopamine, and pretreatment with bretylium or phenoxybenzamine prevented the retardation of small bowel propulsion in rats after laparotomy.

The question remains as to whether the sympathetic inhibitory reflexes are activated by the normal stimuli associated with the passage of contents along the intestine. Several workers using radiographic methods observed that when one part of the intestine was active the others were quiescent (BARCLAY 1936; PUESTOW 1937). This may suggest that distension of the intestine by food or contractions involved in its propulsion may be capable of evoking a sympathetic reflex inhibition of other parts of the intestine. Propagated contractions in the small intestine were shown to initiate intestinofugal nerve activity in the mesenteric nerves (Sect. D.II) and SZURSZEWSKI and WEEMS (1976) demonstrated that both distension of the colon and its contraction activate the intestinofugal cholinergic neurones, which are the afferent limb of the short sympathetic reflexes. These authors also recorded excitatory inputs to neurones of the sympathetic ganglia which originated from empty segments of intestine. These experiments suggest that the spontaneous activity of the intestine is sufficient to activate the postganglionic nordadrenergic neurones. It follows that there is probably continuous sympathetic activity to the intestine. Some experimental evidence supports this hypothesis, although many experimental procedures themselves lead to a reflex activation of the sympathetic inhibitory reflexes to the intestine. Continuous sympathetic activity to the cat colon has been inferred by HULTÉN (1969) who could elicit a reflex contraction of the rectum to distension only when the noradrenergic transmission had been blocked with guanethidine. DE GROAT and KRIER (1979) demonstrated continuous activity in sympathetic pathways in the colonic nerves which was not dependent on supraspinal inputs or afferent inputs to the spinal cord. It is interesting that JULÉ and SZURSZEWSKI (1979) found a small proportion of neurones in isolated cat inferior mesenteric ganglia which showed spontaneous action potentials that were not produced by synaptic inputs.

Recently ORMSBEE et al. (1979) have shown that administration of guanethidine to conscious dogs abolished the quiescent phase of the interdigestive migrating myoelectric complex, which was replaced by an irregular motor activity. Similarly, injection of phentolamine in conscious dogs elicited uninterrupted irregular contractions of the stomach that replaced the quiescent phase of the interdigestive cycle (PAPASOVA et al. 1979). If guanethidine and phentolamine only affected noradrenergic neurones and α-adrenoceptors, respectively, then these experiments suggest that noradrenergic nerves are activated during the interdigestive motor cycle to produce a state of quiescence (phase I) by inhibition of more orally located enteric cholinergic neurones. Since splanchnicectomy does not significantly modify the phases of the motor complex (RUCKEBUSCH and BUENO 1975; GONELLA 1978; see Sect. H.II), the long sympathetic reflexes through the spinal cord are unlikely to be involved, leaving the short sympathetic reflexes through prevertebral ganglia as likely pathways responsible for the inhibition of motility which follows the passage of phase III of the migrating complex. It is possible that the strong contractions which occur in the slowly migrating area of phase III of the cycle themselves initiate the reflex inhibition of the gastrointestinal tract behind the advancing front of activity. More investigation is needed to clarify the subtle roles of sympathetic reflexes in the modulation of the pattern of motility associated with the normal passage of food in the intestine.

2. Inhibitory Parasympathetic Reflexes

The efferent pathways running in the vagus and pelvic nerves which are involved in the parasympathetic inhibitory reflexes have enteric inhibitory neurones (see Sect. B.IV.2) as the common final neurones which act on the muscles. These neurones do not release noradrenaline; their transmitter is unknown (Chap. 11).

In the stomach, the mechanism of adaptation to the arrival of food involves both an intrinsic and an extrinsic inhibitory reflex. The extrinsic inhibitory reflex is mediated by vagal afferent fibres and the vagal inhibitory pathway which relaxes the stomach through the final enteric neurone following distension (JANSSON 1969; OHGA et al. 1970; ABRAHAMSON 1973; MARTINSON 1975). No specific studies on the possible existence of similar vagovagal inhibitory reflexes to the small intestine have been performed. However, ABRAHAMSON et al. (1979) have shown that nociceptive stimulation of the duodenum in anaesthetised cats elicits a vagally mediated relaxation of the stomach which involves the enteric inhibitory neurones. Furthermore, excessive distension of the duodenum produces nausea and an associated transient relaxation of the duodenum itself which appears to be mediated by the vagus nerve (GREGORY 1946, 1947; SLAUGHTER and GRANT 1974; see Sect. K.II.1).

The large intestine, which functions in many species as a reservoir of a faecal matter, also appears to have intrinsic (Sect. J.I.3.e) and extrinsic inhibitory reflexes which allow it to adapt to its contents. Distension of the distal colon and rectum, when it does not initiate defecation (see Sect. K.II.2), activates an extrinsic reflex with afferent fibres most likely running in the pelvic nerves and with the inhibitory sacral parasympathetic pathways as the efferent pathways which also run in the pelvic nerves (MCFADDEN et al. 1935; SCHUSTER 1968; DAVENPORT 1977; FASTH et al. 1980).

II. Excitatory Intestinointestinal Reflexes

It has been shown in a previous section (Sect. H.III) that the passage of food along the gastrointestinal tract elicits reflexes which increase the motility of the digestive tract at considerable distances ahead of the advancing ingested food. Thus, distension of the stomach and duodenum or the arrival of food increase the motility in the distal ileum with emptying of the ileal contents into the colon. Arrival of food in the small intestine faciliates propulsive movements in the colon, including defecation. Some of these responses described in Sects. G.I.2 and G.II.4 are regarded as part of the gastroileocolic reflex. FRIEDMAN (1975) has shown that moderate distension of the small intestine in unanaesthetised dogs elicits in most cases an increase in propulsive and nonpropulsive contractions in parts of the intestine more than 20 cm anally. He also confirmed that distension of the ileum induces defecation. SEMBA and MISHIMA (1958) have shown in dogs that when hypertonic salt solution was dripped onto the stomach there was a reflex increase in the movements of the distal colon. The afferent fibres of this reflex are mainly in the vagus nerve while the efferent pathway is in the pelvic nerves. These authors found that mechanical or chemical stimulation of the small intestine also increased colonic motility via the pelvic nerves. GARDETTE and GONELLA (1974) found that distension of the rabbit ileum elicits contractions of the colon. These reflexes are still present after interruption of the continuity of the intestine but are abolished by section of the extrinsic nerves (GREGORY 1950). Although an adequate pharmacological analysis has not been performed in all cases, the types of response strongly suggest that both the vagal excitatory pathways and the sacral parasympathetic excitatory pathways represent the efferent limb of these reflexes.

1. Vomiting

Vomiting is a complex pattern of motor functions which involves gastrointestinal movements and movements of the somatic musculature. Vomiting can be induced by central or by peripheral stimulation and involves a "vomiting centre" located in the medulla in close relation to the nucleus of the tractus solitarius (KURU and SUGIHARA 1955). Distension of the stomach and duodenum or application of emetic substances to the duodenal mucosa can initiate nausea and vomiting (ALVAREZ 1940; SLAUGHTER and GRANT 1974). During nausea a transient relaxation of the small intestine mediated by the vagus nerve occurs (GREGORY 1946, 1947; MONGES et al. 1974; SLAUGHTER and GRANT 1974). If vomiting follows the transient relaxation during which the intestinal slow waves disappear, a strong contraction of the circular muscle starts at the lower ileum and progresses orally throughout the whole small intestine at a speed of 2–10 cm/s (ALVAREZ 1925; MATHUR et al. 1948; SMITH and BRIZZEE 1961; WEISBRODT and CHRISTENSEN 1972; MONGES et al. 1974; SLAUGHTER and GRANT 1974). Vagotomy abolishes the initial relaxation which occurs during nausea and also abolishes the antiperistaltic wave which procedes retching and vomiting (GREGORY 1947; STEWART et al. 1977). Atropine abolishes the antiperistaltic wave (STEWART et al. 1977).

These results suggest that during nausea and vomiting there is a transient activation of the *vagal parasympathetic inhibitory pathway* which involves a final enteric inhibitory neurone. A sequential anal to oral activation of the *vagal parasym-*

pathetic excitatory pathway to the small intestine which involves a final enteric cholinergic neurone precedes vomiting. This mechanism illustrates the ability of a programmed pattern of extrinsic neuronal influences to override the primarily intrinsic polarised mechanisms of anal transport of contents.

2. Defecation

The process of defecation involves autonomic and somatic events (SCHUSTER 1975; DAVENPORT 1977). The stimulus for the initiation of defecation is distension of the rectum by faeces with a parallel increase in pressure or irritation of the rectal mucosa. The increase in pressure can be achieved voluntarily by increasing abdominal pressure. Distension of the rectum evokes both intrinsic and extrinsic reflexes which will culminate in defecation.

a) Intrinsic Reflexes

Distension of the rectum elicits the ascending excitatory reflex which causes contraction of the circular muscle on the oral side, and the descending inhibitory reflex, which relaxes the circular muscle and the internal anal sphincter itself (recto-anal reflex; Sects. J.I.3, 4). These reflexes are observed in isolated preparations and after lesion of the sacral spinal cord or sacral nerves in different species including humans, (DENNY-BROWN and ROBERTSON 1935; SCHUSTER et al. 1963; NAKAYAMA et al. 1979) and are responsible for what has been termed by DENNY-BROWN and ROBERTSON (1935) *autonomic defecation*. The sequential activation of this reflex by a bolus or a balloon results in its propulsion and expulsion. The intrinsic reflexes by themselves do not appear to be powerful enough to expel solid contents in humans (DENNY-BROWN and ROBERTSON 1935), the contractions of the rectum being weaker than normal in patients whith severed connections to the spinal cord.

b) Extrinsic Reflexes

Distension of the rectum evokes extrinsic reflexes via the sacral region of the spinal cord. During slow distension of the rectum, the extrinsic inhibitory parasympathetic reflex (Sect. K.I.2) is activated, allowing the rectum to be filled and distended before the extrinsic excitatory reflex is activated which results in a powerful propulsive contraction and defecation. The process of defecation with the sacral spinal cord intact has been termed *reflex defecation* by DENNY-BROWN and ROBERTSON (1935). DENNY-BROWN and ROBERTSON (1935) described the events as follows: "It would appear that spinal connections mildly checked the local autonomic activity up to a point beyond which a reinforcing activity appeared." In an extrinsically denervated rectum, the contraction elicited by distension fades away immediately after cessation of the stimulus whereas in the normal innervated rectum, once the contraction has started it continues, in spite of an early cessation of the stimulus (DENNY-BROWN and ROBERTSON 1935). These authors also noticed that for the extrinsic spinal reflex to be effective a certain degree of rectal distension was needed, suggesting that an effective force of contraction of the rectum is achieved by summation of the intrinsic and extrinsic excitatory reflexes.

The facilitatory role of these extrinsic excitatory reflexes on the autonomic defecation is also demonstrated by the sluggishness and lack of power of the rectum, following denervation or lesion of the spinal cord (GARRY 1934; DENNY-BROWN and ROBERTSON 1935; TRUELOVE 1966; NAKAYAMA et al. 1979). KOCK et al. (1972) demonstrated that the contraction of the whole left portion of the human colon in response to rectal stimulation is mediated by spinal excitatory reflexes. MEUNIER and MOLLARD (1977) also demonstrated the reflex excitation of the rectum to distension. The afferent pathway for this excitatory reflex probably runs in the pelvic nerves to the sacral segment of the spinal cord and the efferent pathway is the sacral parasympathetic excitatory pathway which activates a final cholinergic neurone (Sect. E.III). Electrical stimulation of the pelvic nerves can also trigger the evacuation of the colon and rectum (ELLIOTT and BARCLAY SMITH 1904; GARRY 1934; HULTÉN 1969; NAKAYAMA et al. 1979). DE GROAT and KRIER (1978) have shown that all cholinergic contractions of the rectum are associated with efferent activity of the extrinsic sacral nerves.

The question arises as to whether extrinsic reflexes are also acting on the internal sphincter which is merely the continuation of the circular muscle coat. In Sect. E.V it was shown that the internal anal sphincter receives extrinsic cholinergic nerves and extrinsic inhibitory pathways involving enteric inhibitory neurones. It is most likely that these pathways are reflexly activated by distension of the rectum as are similar pathways to the rectum. There is no direct evidence for the involvement of the extrinsic excitatory cholinergic pathways to the sphincter during defecation. However, since the propagation of the propulsive contraction which reaches the anal sphincter could not be due to the activation of the intrinsic enteric ascending excitatory reflex, it is probably due to the activation of the extrinsic reflex. There is some evidence from the work of SCHUSTER et al. (1963) that the extrinsic inhibitory pathway to the sphincter is activated by rectal distension and its effect summates with that of the intrinsic descending inhibitory reflex; in some children with aganglionosis of the anal canal, including the sphincter, distension of the rectum elicited relaxation of the sphincter. It should be noted that some of the cell bodies of the final neurones of the extrinsic inhibitory pathway to the sphincter are located outside the intestine, scattered in the pelvic plexuses (Sect. E.III), and it is possible that this pathway was intact in these patients. There is no evidence for a necessary involvement of sympathetic pathways in defecation, although they may contribute to continence by maintaining a tonic closure of the sphincter.

L. General Summary and Conclusions

The progress of the contents through the digestive tract is dependent on the ordered control of the movements of intestinal muscle, particularly of the circulary arranged muscle. Waves of depolarisation, known as slow waves, sweep down segments of small intestine, but if nerves in the intestine are inactive, these depolarisations do not elicit action potentials and so their passage has no mechanical effect. On the other hand, when the level of excitability in a section of intestine is raised by the action of cholinergic nerves, the slow waves in this section become effective and an area in which contractile waves sweep from oral to anal is created. If the action of the nerves is sufficiently intense, the whole area can become tonically constricted and the contractions elicited by individual slow waves can no longer be recognised.

In the large intestine, the slow waves propagate little and thus favour the formation of stationary rings of constriction of the circular muscle. The enteric cholinergic nerves represent the final common pathways for excitatory reflexes of intrinsic origin and for extrinsic excitatory pathways to the nonsphincteric parts of the intestine.

The cholinergic neurones are activated by short reflexes (the enteric reflexes) that arise locally and run in the wall of the intestine, by long intestinointestinal reflexes that are initiated in the intestine and follow extrinsic pathways to reach their sites of effect, by extrinsic nerves activated from outside the intestine, and by intrinsic circuits that generate an area of excitation which migrates slowly from oral to anal along the small intestine. The enteric inhibitory nerves also supply the smooth muscle; their roles are to aid in accommodating the intestine to increased volume of the luminal contents and to facilitate the passage of the contents from oral to anal.

I. Enteric Reflexes

Distension of a small area of intestine, the local application of some chemicals, and gentle irritation of the mucosa all elicit a response which can be divided into two components: the enteric ascending excitatory reflex and the enteric descending inhibitory reflex. Because these reflexes cause the circular muscle to contract on the oral side and relax on the anal side of the point of initiation, they move the contents in an anal direction. Together, the ascending excitatory and descending inhibitory reflexes constitute the peristaltic reflex which was first studied in detail by BAYLISS and STARLING (1899, 1900, 1901).

The pathways of the ascending excitatory and descending inhibitory reflexes both contain sensory neurones, an undetermined number of interneurones and final efferent neurones to the muscle. The chemical natures of the transmitters of the sensory neurones and of some of the presumed interneurones are unknown. In each reflex there are excitatory cholinergic interneurones which act through nicotinic receptors. The final neurones in the excitatory reflexes release acetylcholine which depolarises the muscle sufficiently for the slow waves to generate action potentials. The final neurones of the descending reflex, the enteric inhibitory neurones, release a transmitter that is yet to be identified, which hyperpolarises the smooth muscle.

These enteric reflexes project to all areas of intestine including the ileocaecal sphincter, except that only the descending inhibitory reflex acts on the internal anal sphincter. The inhibitory reflex appears less prominent and less important for propulsion of the liquid contents of the small intestine than it does for propulsion of more solid material in the large intestine. The relaxation of gastrointestinal sphincters in advance of the moving contents is a consequence of the extension of the descending inhibitory reflexes into the sphincters.

II. Slow Migration of Excitation

In interdigestive periods in carnivores and omnivores, the small intestine undergoes cyclic changes in activity as motor complexes migrate slowly from oral to anal. At each point, four successive phases occur: phase I, during which the intestine is quiescent; phase II, when there are irregular contractions of the circular muscle, some of which propel the contents; phase III, when each slow wave is associated with a contraction which sweeps from oral to anal through the area of excitation; and phase IV, a brief period of irregular contractions before the intestine again becomes quiescent. The area of excitatory activity which travels slowly down the intestine is known as the migrating myoelectric complex. Underlying the migrating motor complex is a moving area of activation of enteric cholinergic neurones. The enteric nervous system has an inherent ability to generate moving areas of excitatory activity which can be observed in the decentralised small and large intestine, or even in vitro. The appearance and propagation of the migrating complex in the small intestine is timed and coordinated by extrinsic nerves.

III. Nonpropulsive Movements

In a number of circumstances in both small and large intestine, contractions of the circular muscle occur which do not result in any significant propulsion of the contents. In the small intestine, nonpropulsive contractions occur when nutrient substances activate local reflexes which result in slowing down of propulsion. Nonpropulsive contractions also occur during phase II of the interdigestive cycle. In both cases these contractions occur where there is a stationary area of excitation of enteric cholinergic nerves; as slow waves move through these stationary and restricted regions, local rhythmic nonpropagated pulses of contractions occur.

IV. Propulsive Movements

The term peristalsis is commonly used to describe all propulsive movements. However, the sequential activation of the ascending and descending enteric reflexes (the peristaltic reflex) although important, is not the only mechanism for propulsion. The waves of contractions which sweep, riding the slow waves, through the segment which is undergoing phase III are also propulsive. Both the slowly migrating motor complex in the small intestine and the slow spontaneous migration of nerve excitation in the large intestine also contribute to propulsion by facilitating the enteric reflexes. Thus, the unidirectionality of propulsion is ensured by the anally migrating slow waves, by the polarity of the enteric reflexes and by the anal migration of areas of neuronal excitability within the enteric plexuses. Reverse peristalsis, that is propulsion in an oral direction, occurs in the proximal large intestine although the underlying mechanism is not yet established.

V. Extrinsic Nerves

Extrinsic nerve pathways run to and from the intestine with the vagus, with the splanchnic and mesenteric nerves, and with the pelvic nerves. The vagus supplies the intestine down to the mid-colon where it overlaps with the pelvic nerves which also supply the remainder of the large intestine. In the vagus and pelvic nerves there are both excitatory and inhibitory pathways. The excitatory pathways activate enteric cholinergic neurones and the inhibitory pathways activate enteric inhibitory neurones. The splanchnic and mesenteric nerves form the major route for the sympathetic pathways to the intestine. The noradrenergic postganglionic sympathetic nerves inhibit intestinal motility by inhibiting the excitability of the enteric cholinergic neurones. They also constrict the sphincter muscles.

In these extrinsic nerve pathways there are also intestinofugal nerves which are activated by mechanical and chemical stimuli in the intestine. Some of these pathways form the afferent limbs of long intestinointestinal reflexes through the central nervous system and some form the afferent limbs of short peripheral, intestinointestinal reflexes with their reflex centres in the prevertebral sympathetic ganglia.

Extrinsic excitatory and inhibitory reflexes are evoked by the normal passage of food through the gastrointestinal tract. The excitatory pathways of the vagus are mainly involved in the cephalic and gastric phases of heightened intestinal activity which occur immediately before and during a meal. The role of the inhibitory pathways in the vagus which run to the small intestine has not been determined. However, both excitatory and inhibitory pathways are activated during nausea. Vagal excitatory pathways are involved in the reverse peristalsis which sweeps the small intestine from the ileocaecal region to the duodenum preceding vomiting.

The excitatory pathways in the pelvic nerves are involved in the initiation of the transient propulsive peristalsis of the large intestine including defecation. The inhibitory pathways are mainly involved, together with the intrinsic inhibitory reflexes, in adaptation of the large intestine to store bulky faeces for long periods. The inhibitory pathways in the pelvic nerves also aid the relaxation of the internal anal sphincter during defecation.

VI. Intestinointestinal Sympathetic Reflexes

The sympathetic inhibitory intestinointestinal reflexes are activated by mechanical activity, either distension or contraction of the intestine. There are two types of reflex pathways mediated by noradrenergic neurones: a long reflex through the central nervous system which involves the well-known sympathetic efferent pathways to the intestine which consists of a preganglionic cholinergic and a postganglionic noradrenergic neurone; and a shorter reflex which involves only the postganglionic noradrenergic neurone located in the prevertebral ganglia and which receives its afferent input from cholinergic neurones located within the intestinal wall.

When the reflexes are only mildly activated by the passage of the digesta, they show some polarity so that mechanical activation in one part of the intestine leads to the reflex inhibition of more oral parts of the intestine to delay transport. If these reflexes are activated more intensely, which in normal circumstances seldom occurs, a diffuse inhibition of the intestine whith contraction of the sphincters occurs. Generalised discharge of sympathetic pathways is more often a consequence of events occurring outside the intestine (e.g. heavy exercise) or abnormal circumstances (e.g. peritoneal irritation).

Acknowledgements. Work from the authors' laboratory was supported by the Australian Research Grants Committee and the National Health and Medical Research Council of Australia. We thank Sally Fiebig and Laima Visockis for the typing of the chapter and Judy Gotch for the patient translation of the original draft into a typescript. We are indebted to Venetta Esson and Pat Vilimas for their skill and dedication in the experimental work and in the preparation of the chapter. The continuous collaboration and fruitful discussions with Mr. Rony Franco and Mr. Alan Wilson is gratefully acknowledged. We should like to thank Dr. I. J. Llewellyn-Smith for careful reading the first draft and for the substantial improvement which resulted from suggestions made.

Note Added in Proof, see p. 461

References

Abbadessa S, Digregorio I, Gravante G (1969) Analyse de quelques caractéristiques des réflexes intrinséques intestinaux. Arch Int Physiol Biochim 77:787–796
Abrahamsson H (1973) Studies on the inhibitory nervous control of gastric motility. Acta Physiol Scand [Suppl] 390
Abrahamsson H, Glise H, Glise K (1979) Reflex suppression of gastric motility during laparotomy and gastroduodenal nociceptive stimul. Scand J Gastroenterol 14:101–106
Adler HF, Atkinson AJ, Ivy ACA (1941) A study of the motility of the human colon. An explanat. of dysynergia of the colon, or of the "unstable colon". Am J Dig Dis 8:197–202
Aizawa I, Hiwatashi K, Takahashi I, Itoh Z (1978) Control of motor activity in the lower oesophageal sphincter by motilin. In: Duthie HL (ed) Gastrointestinal motility in health and man. MTP Press, Lancaster, pp 101–109
Akubue PI (1966) A periarterial nerve-circular muscle preparation from the caecum of the guinea-pig. J Pharm Pharmacol 18:390–395
Alvarez WC (1925) Reverse peristalsis in the bowel, a precursor of vomiting. JAMA 85:1051–1054
Alvarez WC (1940) An introduction to gastro-enterology, 3rd edn. Hoeber, New York London
Alvarez WC, Bennett MF (1931) Inquiries into the structure and function of the myenteric plexus. I. Differences in the reaction of the muscle and nerves of the bowel to constant and interupted currents. Am J Physiol 99:179–198
Alvarez WC, Hosoi K (1930) Conduction in different parts of the small intestine. Am J Physiol 94:448–458
Alvarez WC, Mahoney LJ (1922) (a) Action currents in stomach and intestine. Am J Physiol 58:476–493
Alvarez WC, Zimmermann A (1927) The absence of inhibition ahead of peristaltic rushes. Am J Physiol 83:52–59
Ambache N (1951) Unmasking, after cholinergic paralysis by botulinium toxin, of a reversed action of nicotine on the mammalian intestine, revealing the probable presence of local inhibitory ganglion cells in the enteric plexuses. Br J Pharmacol 6:51–67
Ambache N (1955) The use of limitations of atropine for pharmacological studies on autonomic effectors. Pharmacol Rev 7:467–494
Ambache N, Freeman MA (1968) Atropine-resistant longitudinal muscle spasms due to excitation of non-cholinergic neurones in the Auerbach's plexus. J Physiol 199:705–727
Ambache N, Verney J, Zar M (1970) Evidence for the release of two atropine-resistant spasmogens from Auerbach's plexus. J Physiol (Lond) 207:761–782
Anuras S, Christensen J, Cooke AR (1977) A comparison of intrinsic nerve supplies of two muscular layers of duodenum. Am J Physiol 223:E28–31
Anuras S, Faulk DL, Christensen J (1979) Effect of some autonomic drugs on duodenal smooth muscle. Am J Physiol 236:E33–38
Archer L, Benson MJ, Green WJ, Hardy RJ, Thompson DG, Wingate DL (1979) Radiotelemetric measurement of normal human small bowel motor activity during prolonged fasting. J Physiol (Lond) 296:53P
Armstrong HIO, Milton GW, Smith AWM (1956) Electropotential changes of the small intestine. J Physiol (Lond) 131:147–153
Auer J, Krueger H (1947) Experimental study of antiperistaltic and peristaltic motor and inhibitory phenomena. Am J Physiol 148:350–357
Baldwin MV, Thomas JE (1975) The intestinal intrinsic mucosal reflex; a possible mechanism of propulsive motility. In: Friedman MHF (ed) Functions of the stomach and intestine. University Park Press, Baltimore, pp 75–91
Balfour TW, Hardcastle JD (1975) The myoelectrical activity of the canine ileo-caecal region. The response of feeding and gastrointestinal hormones. In: Vantrappen G (ed) Proceedings of the Fifth International Symposium on Gastrointestinal Motility. Typoff, Herentals

References to Note Added in Proof, see p. 462

Balfour TW, Hardcastle JD (1978) The identification of an electrically silent zone at the ileocaecocolic junction. In: Duthie HL (ed) Gastrointestinal motility in health and disease. MTP Press, Lancaster, pp 407–408
Bárány F, Jacobson B (1964) Endoradiosonde study of propulsion and pressure activity induced by test meals, prostigmine and diphenoxylate in the small intestine. Gut 5:90–95
Barclay AE (1936) The digestive tract; a radiological study of its anatomy, physiology and pathology. Cambridge University Press, Cambridge, pp 395
Barcroft J, Robinson CJ (1929) A study of some factors influencing intestinal movement. J Physiol (Lond) 67:211–220
Barcroft J, Steggerda FR (1932) Observations of the proximal portion of the exteriorized colon. J Physiol (Lond) 76:460–471
Barlett V, Stewart RR, Nakatsu K (1979) Evidence for two adenine derivative receptors in rat ileum which are not involved in the nonadrenergic, noncholinergic response. Can J Physiol Pharmacol 57:1130–1137
Bartho L, Szolcsanyi J (1978) The site of action of capsaicin on the guinea-pig isolated ileum. Naunyn-Schmiedeberg Arch Pharmacol 305:75–81
Bass P (1968) In vivo electrical activity of the small bowel. In: Code CF (ed) Alimentary canal American Physiological Society, Washington, DC (Handbook of physiology, vol IV, sect 6, pp 2051–2074
Bass P, Wiley JN (1965) Electric and extraluminal contractile-force activity of the duodenum of the dog. Am J Dig Dis 10:183–200
Bass P, Code CF, Lambert EH (1961) Motor and electric activity of the duodenum. Am J Physiol 201:287–291
Baur M (1928) Die Peristaltik des isolierten Meerschweinchendünndarms im Filmversuch. Arch Exp Pathol Pharmakol 133:69–83
Bayliss WM, Starling EH (1899) The movements and innervation of the small intestine. J Physiol (Lond) 24:99–143
Bayliss WM, Starling EH (1900) The movements and innervation of the large intestine. J Physiol (Lond) 26:107–118
Bayliss WM, Starling EH (1901) The movements and innervation of the small intestine. J Physiol (Lond) 26:125–138
Beani L, Bianchi C, Crema A (1969) The effect of catecholamines and sympathetic stimulation on the release of acetylcholine from the guinea-pig colon. Br J Pharmacol 36:1–17
Beani L, Bianchi C, Crema A (1971) Vagal non-adrenergic inhibition of guinea-pig stomach. J Physiol (Lond) 217:259–279
Benelli G, Santini V (1974) Analisi dell'azione di alcune sostanze sulle terminazioni sensitive viscerali. Boll Chim Farm 113:291–298
Bennett A, Eley KG (1976) Intestinal pH and propulsion: an explanation of diarrhoea in lactase deficiency and laxation by lactulose. J Pharm Pharmacol 28:192–195
Bennett A, Fleshler B (1969) A hyoscine-resistant excitatory nerve pathway in guinea-pig colon. J Physiol (Lond) 203:62–63P
Bennett A, Stockley HL (1975) The intrinsic innervation of the human alimentary tract and its relation to function. Gut 16:443–453
Bennett MR (1972) Autonomic Neuromuscular Transmission. Cambridge University Press, Cambridge
Bennett MR (1966a) A model of the membrane of the smooth muscle cells of the guinea-pig taenia coli during transmission from inhibitory and excitatory nerves. Nature 211:1149–1152
Bennett MR (1966b) Rebound excitation of the smooth muscle cells of the guinea-pig taenia coli after stimulation of intramural inhibitory nerves. J Physiol (Lond) 185:124–131
Bennett MR, Burnstock G (1968) Electrophysiology of the innervation of intestinal smooth muscle. In: Code CF (ed) Alimentary canal. American Physiological Society, Washington, DC (Handbook of physiology, vol IV, sect 6, pp 1709–1732)
Bennett MR, Burnstock G, Holman ME (1966) Transmission from intramural inhibitory nerves to the smooth muscle of the guinea-pig taenia coli. J Physiol (Lond) 182:541–558
Bentley GA (1962) Studies on sympathetic mechanisms in isolated intestinal and vas deferens preparations. Br J Pharmacol Chemother 19:85–98

Bessou P, Perl ER (1966) A movement receptor of the small intestine. J Physiol (Lond) 18:404–426
Bianchi C, Beani L, Frigo GM, Crema A (1968) Further evidence for the presence of non-adrenergic inhibitory structures in the guinea-pig colon. Eur J Pharmacol 4:51–61
Biber B, Fara J (1973) Intestinal motility increased by tetrodotoxin, lidocaine, and procaine. Experientia 29:551–552
Blair EL, Harper AA, Kidd C, Scratcherd T (1959) Post activation potentiation of gastric and intestinal contractions in response to stimulation of the vagus nerves. J Physiol (Lond) 148:437–449
Boehm S (1979 a) Die spatische Obstipation und ihre Beziehungen zur Antiperistaltik. Dtsch Arch Klin Med 102:431–450
Bolton TB (1979 a) Cholinergic mechanisms in smooth muscle. Br Med Bull 35:275–283
Bolton TB (1979 b) Mechanisms of action of transmitter and other substances on smooth muscle. Physiol Rev 59:606–718
Borgstrom S, Arborelius M Jr (1975) Influence of a fatty acid on duodenal motility. Scand J Gastroenterol 10:599–601
Bortoff A (1961) Slow potential variations of small intestine. Am J Physiol 201:203–208
Bortoff A (1969) Medical intelligence: current concepts – intestinal motility. N Engl J Med 280:1335–1337
Bortoff A (1972) Digestion-motility. Ann Rev Physiol Toxicol 34:261–290
Bortoff A (1976) Myogenic control of intestinal motility. Physiol Rev 56:418–434
Bortoff A, Sacco J (1974) Myogenic control of intestinal peristalsis. In: Daniel EE (ed) Proceedings of the 4th International Symposium on Gastrointestinal Motility. Mitchell, Vancouver, pp 53–60
Bortoff A, Sachs F (1970) Electronic spread of slow waves in circular muscle of small intestine. Am J Physiol 218:576–581
Boyd G, Gillespie JS, Mackenna BR (1962) Origin of the cholinergic response of the rabbit intestine to stimulation of its extrinsic sympathetic nerves after exposure to sympathetic blocking agents. Br J Pharmacol 19:258–270
Bozler E (1939) Electrophysiological studies on the motility of the gastrointestinal tract. Am J Physiol 127:301–307
Bozler E (1948) Conduction, automaticity, and tonus of visceral muscle. Experientia 4:213–218
Bozler E (1949 a) Myenteric reflex. Am J Physiol 157:329–337
Bozler E (1949 b) Reflex peristalsis of the intestine. Am J Physiol 157:338–342
Brading A, Bülbring E, Tomita T (1969) The effect of sodium and calcium on the action potential of the smooth muscle of the guinea-pig taenia coli. J Physiol (Lond) 200:637–654
Brown GL, Gray JAB (1948) Some effects of nicotine-like substances and their relation to sensory nerve endings. J Physiol (Lond) 107:306–317
Brown HS, Posey EL, Gambill EE (1948) Studies on the effect of tetra ethylammonium chloride on gastric motor and secretory function in patients with duodenal ulcer. Gastroenterology 10:837–847
Bucknell A (1965) Effects of direct and indirect stimulation on isolated colon. J Physiol (Lond) 177:58–59P
Bueno L, Ruckebusch M (1976) Insulin and jejunal electrical activity in dogs and sheep. Am J Physiol 230:1538– 1544
Bueno L, Ruckebusch Y (1978) Migrating myoelectrical complexes: disruption, enhancement and disorganization. In: Duthie HL (ed) Gastrointestinal motility in health and disease. MTP Press, Lancaster, pp 83–90
Bueno L, Fioramonti J, Ruckebusch Y (1975) Rate of flow of digesta and electrical activity of the small intestine in dogs and sheep. J Physiol (Lond) 249:69–85
Bueno L, Praddaude F, Ruckebusch Y (1979) Propagation of electrical spiking activity along the small intestine: intrinsic versus extrinsic neural influences. J Physiol (Lond) 292:15–26
Bülbring E, Crema A (1959) The action of 5-Hydroxytryptamine, 5-hydroxytryptophan and reserpine on intestinal peristalsis in anaesthetized guinea-pigs. J Physiol (Lond) 146:29–53

Bülbring E, Lin RCY (1958) The effect of intraluminal application of 5-hydroxytryptamine and 5-hydroxytryptophan on peristalsis; the local production of 5-HT and its release in relation to intraluminal pressure and propulsive activity. J Physiol (Lond) 140:381–407
Bülbring E, Tomita T (1967) Properties of the inhibitory potential of smooth muscle as observed in the response to field stimulation of the guinea-pig taenia coli. J Physiol (Lond) 189:299–315
Bülbring E, Tomita T (1969) Suppression of spontaneous spike generation by catecholamines in the smooth muscle of the guinea-pig taenia-coli. Proc Soc Lond (Biol) 172:103–119
Bülbring E, Tomita T (1970) Effects of Ca removal on the smooth muscle of the guinea-pig taenia coli. J Physiol (Lond) 210:217–232
Bülbring E, Lin RCY, Schofield G (1958) An investigation of the peristaltic reflex in relation to anatomical observations. Q J Exp Physiol 43:26–37
Bunch WL (1898) On the origin, course and cell connections of the viscero-motor nerves of the small intestine. J Physiol (Lond) 22:357–378
Bunker CE, Johnson LP, Nelson TS (1967) Chronic in situ studies of the electrical activity of the small intestine. Arch Surg 95:259–268
Burleigh DE, Damello A, Parks AG (1979) Reponses of isolated human internal anal sphincter to drugs and electrical field stimulation. Gastroenterology 77:484–490
Burn JH (1968) The development of the adrenergic fibre. Br J Pharmacol Chemother 32:575–582
Burns TW, Sinar DR, Gilmore CJ (1979) A comparison of the migrating action potential complex of cholera and the peristaltic rush of normal small intestine. Gastroenterology 76:1109
Burnstock G (1970) Structure of smooth muscle and its innervation. In: Bulbring E, Brading A, Jones A, Tomita T (eds) Smooth muscle. Arnold, London, pp 1–69
Burnstock G (1972) Purinergic nerves. Pharmacol Rev 24:509–581
Burnstock G, Campbell G, Bennett M, Holman ME (1964) Innervation of the guinea-pig taenia coli: Are there intrinsic inhibitory nerves which are distinct from sympathetic nerves? Int J Neuropharmacol 3:163–166
Burnstock G, Campbell G, Rand MJ (1966) The inhibitory innervation of the taenia of the guinea-pig caecum. J Physiol (Lond) 182:504–526
Burnstock G, Satchell DG, Smyths A (1972) A comparison of the excitatory and inhibitory effects of non-adrenergic, non-cholinergic nerve stimulation and exogenously applied ATP on a variety of smooth muscle preparations from different vertebrate species. Br J Pharmacol 46:234–242
Bywater RAR, Taylor GS (1979) Atropine-resistant junction potentials in the guinea-pig small intestine. Proc Aust Physiol Pharmacol Soc 10:233P
Campbell G (1966a) Nerve-mediated excitation of the taenia of the guinea-pic caecum. J Physiol (Lond) 185:148-159
Campbell G (1966b) The inhibitory nerve fibres in the vagal supply to the guinea-pig stomach. J Physiol (Lond) 185:600–612
Campbell G (1970) Autonomic nervous supply to effector tissues. In: Bülbring E, Brading A, Jones A, Tomita T (eds) Smooth muscle. Arnold, London, pp 451–495
Campbell G, Burnstock G (1968) Comparative physiology of gastrointestinal motility. In: Code CF (ed) Alimentary canal American Physiological Society, Washington, DC (Handbook of physiology, vol IV, sect 6, pp 2213–2266)
Cannon WB (1902) Movements of the intestine studied by the means of roentgen rays. Am J Physiol 6:251–277
Cannon WB (1907) The acid control of the pylorus. Am J Physiol 20:283–322
Cannon WB (1911) The mechanical factors of digestion. In: Hill L, Bulloch W (eds) International medical monographs. Longmans & Green, New York
Cannon WB (1912) Peristalsis, segmentation and the myenteric reflex. Am J. Physiol 30:114–128
Cantor MO, Reynolds RP (1957) Gastrointestinal obstruction. Williams & Wilkins, Baltimore
Caprilli R, Onori L (1972) Origin, transmission and ionic dependence of colonic electrical slow waves. Scand J Gastroenterol 7:65–74

Carlson AJ (1930) The extrinsic nervous control of the large bowel. JAMA 94:78–79
Carlson GM, Bedi BS, Code CF (1972) Mechanism of propagation of intestinal interdigestive myoelectric complex. Am J Physiol 222:1027–1030
Carlson GM, Mathias JR, Bertiger G (1977) Nervous control of the response of the gut to meals. In: Brooks FP, Evers PW (eds) Nerves and the gut. Slack, pp 261–271
Case JT (1914) X-ray observations on colonic peristalsis and antiperistalsis with special reference to the function of the ileocolic valve. Med Rec 85:415–426
Cash JT (1886) Contribution to the intestinal rest and movement. Proc R Soc Lond 41:212–231
Castleton KB (1934) An experimental study of the movements of the small intestine. Am J Physiol 107:641–646
Catchpole BN, Duthie HL (1978) Postoperative gastrointestinal complexes. In: Duthie HL (ed) Gastrointestinal motility in health and disease. MTP Press, Lancaster, pp 33–41
Chapman WP, Stanbury JB, Jones CM (1948) The effect of tetraethylammonium on the small bowel of man. J Clin Invest 27:34–38
Chapman WP, Wyman SM, Moro LO, Gillis MA, Jones CM (1953) Barium studies of the comparative action of banthine, tincture of belladonna and placebos on the motility of the gastrointestinal tract in man. Gastroenterology 23:234–243
Chapman WP, Wyman SM, Gagnon JO, Benson JA, Jones CM, Sexton C (1955a) Comparative effects of pamine, bathine, and placebos on gastrointestinal motility. I. Radiographic studies in eight adults. Subjects tested when fasting and after three weeks administration of agents. Gastroenterology 28:500–509
Chapman WP, Wyman SM, Gagnon JO, Jones CM, Sexton C (1955b) Comparative effects of pamine, banthine, and placebos on gastrointestinal motility. II. Radiographic studies in eight adult subjects tested when fasting and following the administration of a standard meal. Gastroenterology 28:510–518
Chey WY, Lee KY, Tai HH (1978) Endogenous plasma motilin concentration and interdigestive myoelectric activity in the canine duodenum. In: Bloom SR (ed) Gut hormones. Churchill Livingstone, London Edinburgh, pp 355–358
Chowdhury AR, Dinoso VP, Lorber SH (1976) Characterisation of a hyperactive segment at a rectosigmoid junction. Gastroenterology 71:584–588
Christensen J (1978) Colonic motility. In: Duthie HL (ed) Gastrointestinal motility in health and disease. MTP Press, Lancaster, pp 267–377
Christensen J, Hauser RL (1971) Longitudinal axial coupling of slow waves in proximal cat colon. Am J Physiol 221:246–250
Christensen J, Schedl HP, Clifton JA (1966) The small intestinal basic electrical rhythm (slow wave) frequency gradient in normal men and in patients with a variety of diseases. Gastroenterology 50:309–315
Christensen J, Caprilli R, Lund CF (1969) Electric slow waves in circular muscle of cat colon. Am J Physiol 217:771–776
Christensen J, Anuras S, Hauser RL (1974) Migrating spike burst and electrical slow waves in the cat colon: effect of sectioning. Gastroenterology 66:240–247
Chul HY, Chey WY, Lee KL (1980) Studies on plasma motilin concentration and interdigestive motility of the duodenum in humans. Gastroenterology 79:62–66
Clarke GD, Davison JS (1978) Mucosal receptors in the gastric antrum and small intestine of the rat with afferent fibres in the cervical vagus. J Physiol (Lond) 284:55–67
Coccagna G, Moschen R, Vela A, Cirignotta F, Gallassi R, Lugaresi E (1977) Studio poligrafico di alcune funzioni vegetative durante il sonno nell uomo. Riv Neurol 47:491–506
Code CF, Marlett JA (1975) The interdigestive myoelectric complex of the stomach and small bowel of dogs. J Physiol (Lond) 246:289–309
Code CF, Schlegel J (1974) The gastrointestinal interdigestive housekeeper: motor correlates of the interdigestive myoelectric complex of the dog. In: Daniel EE (ed) Proceedings of the 4th International Symposium on Gastrointestinal Motility. Mitchell, Vancouver, pp 631–634
Code CF, Rogers AG, Schlegel J, Hightower NC, Bargen JA (1957) Motility patterns in the terminal ileum; studies on two patients with ulcerative colitis and ilea stomas. Gastroenterology 32:651– 665

Code CF, Szurszewski JH, Kelley KA, Smith IB (1968) A concept of control of gastrointestinal motility. In: Code CF (ed) Alimentary canal. American Physiological Society, Washington, DC (Handbook of physiology, vol IV, sect 6, pp 2881–2896)
Cohen S, Harris LD, HSU FY (1968) Manometric characteristics of the human ileocecal junctional zone. Gastroenterology 54:72–75
Conklin JL, Christensen J (1975) Local specialization at ileocecal junction of the cat and opossum. Am J Physiol 228:1075–1081
Connell AM (1961) The motility of the small intestine. Postgrad Med J 37:703–716
Connell AM (1968) Motor action of the large bowel. In: Code CF (ed) Alimentary canal. American Physiological Society, Washington, DC (Handbook of physiology, vol IV, sect 6, pp 2075–2091)
Connell AM, Logan CJH (1967) The role of gastrin in gastroileocolic responses. Am J Dig Dis 12:227–284
Connell AM, McKelvey STD (1970) The influence of vagotomy on the colon. Proc R Soc Med 63:7
Connell AM, Smith CL (1974) The effect of dietary fibre on transit time. In: Daniel EE (ed) Proceedings of the 4th International Symposium on Gastrointestinal Motility. Mitchell, Vancouver, pp 365–368
Connell AM, Frankel H, Guttmann L (1963) The motility of the pelvic colon following complete lesions of the spinal cord. Paraplegia 1:98–115
Connell AM, Avery Jones F, Rowlands EN (1965) The motility of the pelvic colon IV – abdominal pain associated with colonic hypermotility after meals. Gut 6:105–112
Connor C, Prosser CL (1974) Comparison of ionic effects on longitudinal and circular muscle of cat jejunum. Am J Physiol 226:1212–1218
Connor JA, Kreulen D, Prosser CL, Weigel R (1977) Interaction between longitudinal and circular muscle in intestine in cat. J Physiol (Lond) 273:665–689
Costa M, Furness JB (1972) Slow contraction of the guinea-pig proximal colon in response to the stimulation of an identified type of nerve. Br J Pharmacol 45:151P–152P
Costa M, Furness JB (1974) The innervation of the internal anal sphincter of the guinea-pig. In: Daniel EE (ed) Proceedings of the 4th International Symposium on Gastrointestinal Motility. Mitchell, Vancouver, pp 681–689
Costa M, Furness JB (1976) The peristaltic reflex: an analysis of the nerve pathways and their pharmacology. Naunyn-Schmiedeberg Arch Pharmacol 294:47–60
Costa M, Furness JB (1979) On the possibility that an indoleamine is a neurotransmitter in the gastrointestinal tract. Biochem Pharmacol 28::565–571
Costa M, Furness JB, Dawson K (1975) Potentiation by cocaine of relaxations of the guinea-pig colon caused by noradrenaline and by stimulation of adrenergic nerves. Aust J Exp Biol Med Sci 53:223–232
Costa M, Furness JB, Llewellyn-Smith IJ, Davies B, Oliver J (1980a) An immunohistochemical study of the projections of somatostatin containing neurons in the guinea-pig intestine. Neuroscience 5:841–852
Costa M, Furness JB, Llewellyn-Smith IJ, Cuello C (1980b) Projections of substance P neurons within the guinea-pig small intestine. Neuroscience 6:411–424
Courtade D, Guyon JF (1897) Influence motrice du grand sympathique et du nerf érecteur sacre sur les gros intestin. Arch Physiol 9:880–890
Couturier D, Roze C, Couturier-Turbin MH, Debray C (1969) Electromyography of the colon situ. An experimental study in man and in the rabbit. Gastroenterology 56:317–322
Couturier D, Roze C, Debray C (1970) Activité motrice du duodénum chez l'homme: corrélation avec la contraction antrale et avec l'activité électrique. J Physiol (Paris) 62:387–405
Cowie DM, Lashmet FH (1928) Studies on the function of the intestinal musculature. II. Longitudinal muscle of the rabbit. Am J Physiol 88:369–389
Creed KE (1979) Functional diversity of smooth muscle. Br Med Bull 35:243–247
Crema A (1970) On the polarity of the peristaltic reflex in the colon. In: Bülbring E, Brading AF, Jones AW, Tomita T (eds) Smooth muscle. Arnold, London, pp 542–548

Crema T, Tacca MD, Frigo GM, Lecchini S (1968) Presence of a nonadrenergic inhibitory system in the human colon. J Br Soc Gastroenterol 9:633–637

Crema A, Frigo GM, Lecchini S (1970) A pharmacological analysis of the peristaltic reflex in the isolated colon of the guinea-pig or cat. Br J Pharmacol 39:334–345

Crowcroft PJ, Holman ME, Szurszewski JH (1971) Excitatory input from the distal colon to the inferior mesenteric ganglion in the guinea-pig. J Physiol (Lond) 219:443–461

Currie GC, Henderson VE (1926) A study of the movements of the large intestine in the guinea-pig. Am J Physiol 78:287–298

Dahlgren S, Selking O (1972) Motility of the human digestive tract under resting conditions and after ingestion of food. A study with endoradiosondes. Ups J Med Sci 77:167–174

Dale HH (1906) On some physiological actions of ergot. J Physiol (Lond) 34:163–206

Daniel EE (1968) The electrical activity of the alimentary tract. Am J Dig Dis 13:297–319

Daniel EE (1975) Electrophysiology of the colon. Gut 16:298–306

Daniel EE (1977) Nerves and motor activity of the gut: In: Brooks FP, Evers PW (eds) Nerves and the gut. Slack, pp 154–199

Daniel EE, Chapman KM (1963) Electrical activity of the gastrointestinal tract as an indication of mechanical activity. Am J Dig Dis 8:54–103

Daniel EE, Sarna S (1978) The generation and conduction of activity in smooth muscle. Ann Rev Pharmacol Toxicol 18:145–166

Daniel EE, Honour AJ, Bagoch A (1960) Electrical activity of the longitudinal muscle of dog small intestine studied in vitro using microelectrodes. Am J Physiol 198:113–118

Davenport HW (1977) Physiology of the digestive tract, 4th edn Year Book Medical Publishers, Chicago

Davidson M, Sleisenger MH, Almy TP, Levine SZ (1956) Studies of distal colonic motility in children. Pediatrics 17:807–819

Davison JS, Pearson GT (1979) The role of intrinsic, non-adrenergic, non-cholinergic, inhibitory nerves in the regulation of distensibility of the guinea-pig colon. Pfluegers Arch 381:73–77

Day MD, Rand MJ (1961) Effect of guanethidine in revealing cholinergic sympathetic fibres. Br J Pharmacol 17:245–260

Day M, Vane JP (1963) An analysis of the direct and indirect actions of drugs on the isolated guinea-pig ileum. Br J Pharmacol 20:150–170

Day MD, Warren PR (1968) A pharmacological analysis of the responses to transmural stimulation in isolated intestinal preparations. Br J Pharmacol Chemother 32:227–240

De Groat WC, Krier J (1976) An electrophysiological study of the sacral parasympathetic pathway to the colon of the cat. J Physiol (Lond) 260:425–445

De Groat WC, Krier J (1978) The sacral parasympathetic reflex pathway regulating colonic motility and defecation in the cat. J Physiol (Lond) 276:481–500

De Groat WC, Krier J (1979) The central control of the lumbar sympathetic pathway to the large intestine of the cat. J Physiol (Lond) 289:449–468

Deller DJ, Wangel AG (1965) Intestinal motility in man. I. A study combining the use of intraluminal pressure recording and cineradiography. Gastroenterology 48:45–57

Del Tacca M, Lecchini S, Frigo GM, Crema A, Benzi G (1968) Antagonism of atropine towards endogenous and exogenous acetylcholine before and after sympathetic system blockade in the isolated distal guinea-pig colon. Eur J Pharmacol 4:188–197

Denny-Brown D, Robertson EG (1935) An investigation of the nervous control of defaecation. Brain 58:256–310

Derblom H, Johansson H, Nylander G (1966) A simple method of recording certain gastrointestinal motility functions in the rat. Acta Chir Scand 132:154–165

De Wever I, Eeckhout C, Vantrappen G, Hellemans J (1978) Disruptive effect of test meals on interdigestive motor complex in dogs. Am J Physiol 235:E661–665

De Weveer I, Eeckhout C, Vantrappen G, Hellmans J (1979) How does oil disrupt the interdigestive myoelectric complex? Gastroenterology 76:1120

Diamant NE, Bortoff A (1969) Nature of the intestinal slow wave frequency gradient. Am J Physiol 216:301–307

Diamant NE, Hall K, Mui H, El-Sharkawy TY (1979a) Vagal control of the feeding motor pattern in the lower esophageal sphincter, stomach and small intestine in dog. In: Christensen J (ed) Gastrointestinal motility. 7th Int Symp Gastrointest Motility. Raven, New York, pp 365–370
Diamant NE, Mui H, El-Sharkawy TY, Hall K (1979b) The vagus controls the lower esophageal sphincter and gastric components of the migrating motor complex in the dog. Gastroenterology 76:1122
Diament ML, Kosterlitz HW, McKenzie J (1961) Role of the mucous membrane in the peristaltic reflex in the isolated ileum of the guinea-pig. Nature 190:1205–1206
Dickinson VA (1978) Maintenance of anal continence: a review of pelvic floor physiology. Gut 19:1163–1174
Dodds DC, Ould CL, Dailey ME (1948) The effect of tetraethylammonium chloride on gastric motility in man. Gastroenterology 10:1007–1009
Douglas DM (1941) The activity of the duodenum. J Physiol (Lond) 107:472–478
Douglas DM, Mann FC (1939a) An experimental study of the rhythmic contractions in the small intestine of the dog. Am J Dig Dis 6:318–322
Douglas DM, Mann FC (1939b) The activity of the lower part of the ileum of the dog in relation to the ingestion of food. Am J Dig Dis 6:434–439
Douglas DM, Mann FC (1940) The gastro-ileac reflex further experimental studies. Am J Dig Dis 7:53–57
Dubois A (1977) Études physiopathologiques du l'ileus postoperatoire. Acta Chir Belg 76:141–166
Duthie HL, Brown BH, Robertson-Dunn B, Kwong NK, Whittaker GE, Waterfall W (1972) Electrical activity in the gastroduodenal area – slow waves in the proximal duodenum. A comparison of man and dog. Am J Dig Dis 17:344–351
Eeckhout C, De Wever I, Hellemans J, Vantrappen G (1978a) The effect of different test meals on their interdigestive myoelectrical complex (MMC) in dogs. In: Duthie HL (ed) Gastrointestinal motility in health and disease. MTP Press, Lancaster, pp 43–45
Eeckhout C, De Wever I, Peeters T, Hellemans J, Vantrappen G (1978b) Role of gastrin and insulin in the postprandial disruption of migrating complex in dogs. Am J Physiol 235:E666–669
Eeckhout C, De Wever I, Vantrappen G (1979) Effect of glucose perfusion on the migrating complex of a Thiry-Vella loop. In: Christensen J (ed) Gastrointestinal motility. 7th International Symposium on Gastrointestinal motility. Raven, New York, pp 289–293
Elliott TR (1904) On the innervation of the ileo-colic sphincter. J Physiol (Lond) 31:157–168
Elliott TR, Barclay-Smith E (1904) Antiperistalsis and other muscular activities of the colon. J Physiol (Lond) 31:272–304
El Quazzani T, Mei N (1979) Vagal thermoreceptors in the gastrointestinal area. Their role in the regulation of the digestive motility. Exp Brain Res 34:419–434
El-Sharkawy TY (1978) Electrophysiological control of motility in canine colon. In: Duthie HL (ed) Gastrointestinal motility in health and disease. MTP Press, Lancaster, pp 387–369
Falconer JD, Smith AN, Eastwood MA (1978) Effects of bile salts and prostaglandins on the colonic motility in the rabbit. In: Duthie HL (ed) Gastrointestinal motility in health and disease. MTP Press, Lancaster, pp 607–615
Fasth S, Hultén L, Johnson JB, Nordgren S, Zeitlin IJ (1978) Mobilization of colonic kallikrein following pelvic nerve stimulation in the atropinized cat. J Physiol (Lond) 285:471–478
Fasth S, Hultén L, Nordgren S (1980) Evidence for a dual pelvic nerve influence on large bowel motility in the cat. J Physiol (Lond) 298:159–169
Finch P, Catchpole B (1979) The relationship of sleep stage to the migrating gastrointestinal complex of man. In: Christensen J (ed) Gastrointestinal motility. 7th International Symposium on Gastrointestinal Motility. Raven, New York
Finkleman B (1930) On the nature of inhibition in the intestine. J Physiol (Lond) 70:145–157
Fleckenstein P, Krough F, Øigaard A (1978) The interdigestive myoelectrical complex and other migrating electrical phenomena in the human small intestine. In: Duthie HL (ed) Gastrointestinal motility in health and disease. MTP Press, Lancaster, pp 19–27

Fleisch A (1928) Der Verkürzungsreflex des Darmes. Pfluegers Arch 220:512–523
Fleisch A, Wyss WH (1923) Zur Kenntnis der visceralen Tiefensensibilität. Pfluegers Arch 200:290–312
Fontaine J, Van Neuten JM, Janssen PAJ (1973) Analysis of the peristaltic reflex in vitro: effects of some antagonists. Arch Int Pharmacodyn Ther 203:396–398
Foulk WT, Code CF, Morlock CG, Bargen JA (1954) A study of the motility patterns and the basic rhythm in the duodenum and upper part of the jejunum of human beings. Gastroenterology 26:601–611
Fox JET, Daniel EE, Collins SM, Lewis TD, Track NS (1979) Motilin release, differences between man and dog and relationship to migrating motor complexes. Gastroenterology 76:1134
Franco R, Costa M, Furness JB (1979 a) Evidence for the release of endogenous substance P from intestinal nerves. Naunyn-Schmiedeberg Arch Pharmacol 306:185–201
Franco R, Costa M, Furness JB (1979 b) Evidence that axons containing substance P in the guinea-pig ileum are of intrinsic origin. Naunyn-Schmiedeberg Arch Pharmacol 307:57–63
Franco R, Costa M, Furness JB (1979 c) The presence of a cholinergic excitatory input to substance P neurons in the intestine. Proc Aust Physiol Pharmacol Soc 10:255P
Frankl-Hochwart L von, Frölich A (1900) Über Tonus und Innervation der Sphinkteren des Anus. Arch Ges Physiol 81:420–482
Frenckner B, Ihre T (1976) Influence of autonomic nerves on the internal anal sphincter in man. Gut 17:306–312
Friedman MHF (1975) The entero-enteric reflexes. In: Friedman MHF (ed) Functions of the stomach and intestine. University Park Press, Baltimore, pp 57–73
Frigo GM, Lecchini S (1970) An improved method for studying the peristaltic reflex in the isolated colon. Br J Pharmacol 39:346–456
Frigo GM, Torsoli A, Lecchini S, Falaschi CF, Crema A (1972) Recent advances in the pharmacology of peristalsis. Arch Int Pharmacol Ther [Suppl] 196:9–23
Fülgraff G, Schmidt L (1963) Die Wirkung elektrischer Reizung sympathischer und parasympathischer Nerven auf das proximale und distale Colon und ihre pharmacologische Beeinflußbarkeit. Naunyn-Schmiedebergs Arch Exp Phatol Pharmakol 245:106–107
Fülgraff G, Schmidt L, Azokwu P (1964) Über die atropinresistente neuro-muskulare Übertragung am Pelvicus-Colon-Präparat der Katze. Arch Int Pharmacodyn Ther 149:537–551
Fujita T, Kobayashi S (1974) The cells and hormones of the GEP endocrine system – The current of studies. In: Fujita T (ed) Gastro-entero-pancreatic endocrine system. A cell biological approach, 1st edn. Williams & Wilkins, Baltimore, pp 1–16
Fukuda H (1966) Mechanism underlying the augmentation of the intestinal motility produced by the stimulation of the splanchnic nerve. J Physiol Soc Jpn. 28:45–52
Fukuda H (1968 a) On the relationship of the inhibitory neurone concerned with the intestinal intrinsic reflex with vagal inhibition. J Physiol Soc Jpn 30:702–709
Fukuda H (1968 b) On the inhibitory efferent neurone concerned with the intestinal intrinsic reflexes. J Physiol Soc Jpn 30:697–701
Furness JB (1969 a) An electrophysiological study of the smooth muscle of the colon. J Physiol (Lond) 205:549–562
Furness JB (1969 b) The presence of inhibitory nerves in the colon after sympathetic denervation. Eur J Pharmacol 6:349–352
Furness JB (1970) An examination of nerve-mediated, hyoscine-resistant excitation of the guinea-pig colon. J Physiol (Lond) 207:803–821
Furness JB (1971) Secondary excitation of intestinal smooth muscle. Br J Pharmacol 41:213–226
Furness JB, Costa M (1973) The nervous release and the action of substances which affect intestinal muscle through neither adrenoreceptors nor cholinoreceptors. Philos Trans Soc Lond [Biol] 265:123–133
Furness JB, Costa M (1974 a) The adrenergic innervation of the gastrointestinal tract. Ergeb Physiol 69:1–51
Furness JB, Costa M (1974 b) Adynamic ileus, its pathogenesis and treatment. Med Biol 52:82–89

Furness JB, Costa M (1976) Ascending and descending enteric reflexes in the isolated small intestine of the guinea-pig. Proc Physiol Pharmacol Soc 7:172P
Furness JB, Costa M (1977) The participation of enteric inhibitory nerves in accommodation of the intestine to distension. Clin Exp Pharmacol Physiol 4:37–41
Furness JB, Costa M (1979) Projections of intestinal neurons showing immunoreactivity for vasoactive intestinal polypeptide are consistent with these neurons being the enteric inhibitory neurons. Neurosci Lett 15:199–204
Furness JB, Costa M (1980) Types of nerve in the enteric nervous system. Neuroscience 5:1–20
Furness JB, Costa M, Franco R, Llewellyn-Smith IJ (1980) Neuronal peptides in the intestine: distribution and possible functions. Adv Biochem Psychopharmacol 21:601–617
Gabella G (1979a) Smooth muscle cell junctions and structural aspects of contraction. Br Med Bull 35:213–218
Gabella G (1979b) Innervation of the gastrointestinal tract. Int Rev Cytol 59:129–193
Galapeaux EA, Templeton RD (1937) The influence of filling the stomach on colon motility in the dog. Am J Physiol 119:312–313
Garcia-Rodrígues M, Menéndez-Cepero F, Sainz-Guevara F, Reyes Diaz JM (1971) Biophasic response of the transmurally stimulation rat duodenum. Can J Physiol Pharmacol 49:370–372
Gardette B, Gonella J (1974) Étude électromyographique in vivo de la commande nerveuse orthosympathique du côlon chez le chat. J Physiol (Paris) 68:671–692
Garrett JR, Howard JR (1972) Effects of rectal distension on the internal anal sphincter of cats. J Physiol (Lond) 222:85P
Garrett JR, Howard ER, Jones W (1974) The internal anal sphincter in the cat: A study of nervous mechanisms affecting tone and reflex ability. J Physiol (Lond) 243:153–166
Garry RC (1933) The nervous control of the caudal region of the large bowel in the cat. J Physiol (Lond) 77:422–431
Garry RC (1934) The movement of the large intestine. Physiol Rev 14:103–132
Garry RC, Gillespie JS (1955) The responses of the musculature of the colon of the rabbit to stimualtion, in vitro, of the parasympathetic and the sympathetic outflows. J Physiol (Lond) 128:557–576
Gernandt B, Zotterman Y (1946) Intestinal pain; an electrophysiological investigation on mesenteric nerves. Acta Physiol Scand 12:56–72
Gershon MD, Thompson EB (1973) The maturation of neuromuscular function in a multiple innervated structure: Development of the longitudinal smooth muscle of the foetal mammalian gut and its cholinergic excitatory, adrenergic inhibitory and non-cholinergic inhibitory innervation. J Physiol (Lond) 234:257–277
Gillespie JS (1962) Spontaneous mechanical and electrical activity of stretched and unstretched intestinal smooth muscle cells and their response to sympathetic-nerve stimulation. J Physiol (Lond) 162:54–75
Gillespie JS (1968) Electrical activity in the colon. In: Code CF (ed) alimentary canal. American Physiological Society, Washington, DC (Handbook of physiology, vol IV, sect 6, pp 2093–2120)
Gillespie JS, Khoyi MA (1977) The site and receptors responsible for the inhibition by sympathetic nerves of intestinal smooth muscle and its parasympathetic motor nerves. J Physiol (Lond) 267:767–789
Gillespie JS, Mackenna BR (1961) The inhibitory action of the sympathetic nerves on the smooth muscle of the rabbit gut, its reversal by reserpine and restoration by catecholamines. J Physiol (Lond) 156:17–34
Ginzler KH (1959) Are mucosal nerve fibres essential for the peristaltic reflex? Nature 184:1235–1236
Golden R (1945) Radiologic examination of the small intestine. Lippincott, Philadelphia
Goldenberg MM, Burns RH (1971) Atropine-resistant spasm of the dog colon induced by intermittent pelvic nerve stimulation. Life Sci 10:591–600
Golenhofen K (1976) Spontaneous activity and functional classification of mammalian smooth muscle. In: Bülbring E, Shuba MF (eds) Physiology of smooth muscle. Raven, New York, pp 91–97

Gonella J (1964) Étude de l'activité électrique des fibres musculaires longitudinales du duodénum in vivo. Action de la stimulation des nerfs vagues. C R Soc Biol (Paris) 158:2409–2413
Gonella J (1970) Étude de l'activité électrique de la couche musculaire longitudinale du duodénum de lapin. J Physiol (Paris) 62:447–476
Gonella J (1971) Étude électromyographique des contractions segmentaires et péristaltiques du duodénum de lapin. Pfluegers Arch 322:217–234
Gonella J (1978) La motricité digestive et sa regulation nerveuse. J Physiol (Paris) 74:131–140
Gonella J, Lecchini S (1971) Inhibition de l'activité électrique de la couche circulaire du duodénum de lapin, in vitro, par stimulation des fibres sympathiques périarterielles du mesentère. C R Acad Sci [D] (Paris) 273:214–217
Gonella J, Vienot J (1972) Action des ganglioplégiques sur la propagation du péristaltisme duodénal. J Physiol (Paris) 64:623–630
Gowers WR (1877) The autonomic action of the sphincter ani. Proc R Soc Lond 26:77–84
Grafe P, Wood JD, Mayer CJ (1979) Fast excitatory postsynaptic potentials in AH (Type 2) neurons of guinea-pig myenteric plexus. Brain Res 163:349–352
Gray WC, Hendershot C, Whitrock RM, Seevers MH (1955) Influence of parasympathetic nerves and their relation to the action of atropine in the ileum and colon of the dog. Am J Physiol 181:679–687
Gregory JE (1950) Some factors influencing the passage of fluid through intestinal loops in dogs. J Physiol (Lond) 111:119–137
Gregory JE, Bentley GA (1968) The peristaltic reflex in the isolated guinea-pig ileum during drug-induced spasm of the longitudinal muscle. Aust J Exp Biol Med Sci 46:1–16
Gregory RA (1946) Changes in intestinal tone and motility associated with nausea and vomiting. J Physiol (Lond) 105:58–65
Gregory RA (1947) The nervous pathways of intestinal reflexes associated with nausea and vomiting. J Physiol (Lond) 106:95–103
Grindlay JH, Mann FC (1941) effect of liquid and solid meals on intestinal activity. Am J Dig Dis 8:324–327
Grivel ML, Ruckebusch Y (1972) The propagation of segmental contractions along the small intestine. J Physiol (Lond) 227:611–625
Gulati OD, Panchal DI (1978) Some observations on the development of adrenergic innervation in rabbit intestine. Br J Pharmacol 64:247–251
Gustavsson S (1978) Propulsion and mixing of small bowel contents. Acta Univ Ups 296
Haladay DA, Volk H, Mandell J (1958) Electrical activity of the small intestine with special reference to the origin of rythmicity. Am J Physiol 195:505–515
Halls J (1965) Bowel content shift during normal defection. Proc R Soc Med 58:859–860
Hardcastle J, Hardcastle PT, Sanford PA (1978) Effect of actively transported hexoses on afferent nerve discharges from rat small intestine. J Physiol (Lond) 285:71–84
Hardcastle JD, Mann CV (1968) Study of large bowel peristalsis. Gut 9:512–520
Hardcastle JD, Mann CV (1970) Physical factors in the stimulation of colonic peristalsis. Gut 11:41–46
Harper AA, Kidd C, Scratcherd T (1959) Vago-vagal reflex effects on gastric and pancreatic secretion and gastrointestinal motility. J Physiol (Lond) 148:417–436
Harvey RF (1975) Hormonal control of gastrointestinal motility. Dig Dis 20:523–539
Harvey RF, Read AE (1973) Effect of oral magnesium sulphate on colonic motility in patients with the irritable bowel syndrome. Gut 14:983–987
Hawkins CF, Hardy TL (1950) On the nature of haustration of the colon. J Fac Radiol (Lond) 2:95–98
Hellemans J, Vantrappen G, Janssens J, Peeters T (1978) Effect of feeding and of gastrin on the interdigestive myoelectrical complex in man. In: Duthie HL (ed) Gastrointestinal motility in health and disease. MPT Press, Lancaster, pp 29–30
Henderson VE (1928) The mechanism of intestinal peristalsis. Am J Physiol 86:82–98
Herman H, Morin G (1934) Mise en évidence d'un réflexe inhibiteur intestino-intestinal. C R Soc Biol (Paris) 115:529–531
Hertz AF (1913) The ileo-caecal sphincter. J Physiol (Lond) 47:54–56

Hertz AF, Newton A (1913) The normal movements of the colon in man. J Physiol (Lond) 47:57–65
Hiatt RB, Goodman I, Sandler M, Cheskin H (1976) The effects of coherin on the basic electrical rhythm of the dog ileum in vivo. Am J Dig Dis 22:108–112
Hill JR, Kelley ML, Schlegel JF (1960) Pressure profile of the rectum and anus of healthy persons. Dis Colon Rectum 3:203–209
Hinrichsen J, Ivy AC (1931) Studies on the ileo-cecal sphincter of the dog. Am J Physiol 96:494–507
Hirst GDS (1979) Mechanisms of peristalsis. Br Med Bull 35:263–268
Hirst GDS, McKirdy HC (1974) Presynaptic inhibition at mammalian peripheral synapse. Nature 250:430–431
Hirst GDS, Holman ME, Spence I (1974) Two types of neurones of the myenteric plexus of duodenum in the guinea-pig. J Physiol (Lond) 236:303–326
Hirst GDS, Holman ME, McKirdy HC (1975) Two dsescending nerve pathways activated by distension of guinea-pig small intestine. J Physiol (Lond) 244:133–127
Holaday DA, Volk H, Mandell J (1958) Electrical activity of the small intestine with special reference to the origin of rhythmicity. Am J Physiol 195:505–515
Holdstock DJ, Misiewicz JJ (1970) Factors controlling colonic motility: colonic pressures and transit after meals in patients with total gastrectomy, pernicious anaemia of duodenal ulcer. J Br Soc Gastroenterol 11:100–110
Holdstock DJ, Misiewicz JJ, Smith T, Rowlands EN (1970) Propulsion (mass movements) in the human colon and its relationship to meals and somatic activity. Gut 11:91–99
Holman ME (1968) An introduction to electro-physiology of visceral smooth muscle. In: Code CF (ed) Alimentary canal. American Physiological Society, Washington, DC (Handbook of physiology, vol IV, sect 6, pp 1665–1708)
Holman ME (1975) Excitation of nerves. In: Daniel EE, Paton DM (eds) Methods in pharmacology, vol 3, chap 14. Plenum, New York, pp 299–311
Holman ME, Hughes J (1965) Inhibition of intestinal smooth muscle. Aust J Exp Biol Med Sci 43:277–290
Holt JF, Lyons RH, Neligh RB, Moe GK, Hodges FJ (1947) X-ray signs of altered alimentary function following autonomic blockade with tetraethylammonium. Radiology 49:603–610
Holzknecht G (1909) Die normale Peristaltic des Kolon. MMW 56:2401–2403
Howard ER, Nixon HH (1968) Internal anal sphincter. Arch Dis Child 43:569–578
Hukuhara T, Fukuda H (1965) The motility of the isolated guinea-pig small intestine. Jpn J Physiol 15:125–139
Hukuhara T, Fukuda H (1968) The electrical activity of guinea-pig small intestine with special reference to the slow wave. Jpn J Physiol 18:71–86
Hukuhara T, Miyake T (1959) The intrinsic reflexes in the colon. Jpn J Physiol 9:49–55
Hukuhara T, Neya T (1968) The movements of the colon of rats and guinea-pig. Jpn J Physiol 18:551–562
Hukuhara T, Kinose S, Masuda K (1936) Beiträge zur Physiologie der Bewegung des Duodenums. Arch Ges Physiol 238:124–134
Hukuhara T, Yamagami M, Nakayama S (1958) On the intestinal intrinsic reflexes. Jpn J Physiol 8:9–20
Hukuhara T, Nakayama S, Nanba R (1960) Locality of receptors concerned with the intestino-intestinal extrinsic and intestinal muscular intrinsic reflexes. Jpn J Physiol 10:414–419
Hukuhara T, Nakayama S, Namba R (1961) The role of the intrinsic mucosal reflex in the fluid transport through the denervated colonic loop. Jpn J Physiol 11:71–79
Hukuhara T, Neya T, Tsuchiya K (1969) The effect of the intrinsic mucosal reflex upon the propagation of intestinal contractions. Jpn J Physiol 19:824–833
Hultén L (1969) Extrinsic nervous control of colonic motility and blood flow. Acta Physiol Scand [Suppl] 335
Hunt JN, Knox MT (1968) Regulation of gastric emptying. In: Code CF (ed) Alimentary canal. American Physiological Society, Washington, DC (Handbook of physiology, vol IV, sect 6, pp 1911–1936

Hurst AF (1919) Constipation and allied disorders, 2nd edn., Froude, London
Iantsev AV (1979) Interrelationship of afferent impulses and the activity of smooth muscles of the small intestine. Fiziol Zh SSSR 65:741–746
Iggo A (1957) Gastro-intestinal tension receptors with unmyelinated afferent fibres in the vagus of the cat. Q J Exp Physiol 42:130–142
Iggo A (1966) Physiology of visceral afferent systems. Acta Neuroveg (Wien) 28:121–134
Ihre T (1974) Studies on anal function in continent and incontinent patients. Scand J Gastroenterol 9:1–80
Ingelfinger FJ, Abbott WO (1940) Incubation studies in the human small intestine. XX. The diagnostic significance of motor disturbances. Am J Dig Dis 7:468
Irving JT, McSwiney BA, Suffolk SF (1937) Afferent fibres from the stomach and small intestines. J Physiol (Lond) 89:407–420
Itina LV, Sergeev VA (1978) Autonomic nerve firing and small intestinal motoricity following introduction of a sucrose solution into the lumen. Fiziol Zh SSSR 64:1027–1034
Ito Y, Kuriyama H (1973) Membrane properties and inhibitory innervation of the circular muscle cells of guinea-pig caecum. J Physiol (Lond) 231:455–470
Itoh Z, Aizawa I, Takeuchi S, Couch EF (1975) Hunger contractions and motilin. In: Vantrappen G (ed) Proceedings of the 5th International Symposium on Gastrointestinal Motility. Typoff, Herentals, pp 48–55
Itoh Z, Honda R, Aizawa I, Takeuchi S, Hiwatashi K, Copuch EF (1978a) Interdigestive motor activity of the lower esophageal sphincter in the conscious dog. Am J Dig Dis 23:239–247
Itoh Z, Takayanagi R, Takeuchi S, Isshiki S (1978b) Interdigestive motor activity of Heidenhain pouches in relation to main stomach in conscious dogs. Am J Physiol 234:E333–338
Itoh Z, Honda R, Takeuchi S, Aizawa I (1979) Regular and irregular control of interdigestive contractions by motilin and intraduodenal pH. In: Christensen J (ed) 7th International Symposium on Gastrointestinal Motility. Raven, New York, pp 279–286
Ivy AC, McIlvain GB (1923) The excitation of gastric secretion by application of substances to the duodenal and jejunal mucosa. Am J Physiol 67:124–140
Jansson G (1969) Extrinsic nervous control of gastric motility. Acta Physiol Scand [Suppl] 326
Jarrett RJ, Gazet JC (1966) Studies in vivo of the ileocaeco-colic sphincter in the cat and dog. Gut 7:271–275
Johansson C Ekelund K (1975) Jejunal fluid flows and intestinal transit times in relation to the gastric emptying of different meals. In: 5th International Symposium on Gastrointestinal Motility (Leuven)
Johnson L (1972) Propulsive motility and intraluminal pressure variations in isolated homologously perfused small intestine. Acta Chir Scand 138:834–843
Jordan AC (1914) The peristalsis of the large intestine. Arch Roentg Ray 18:328–339
Julé Y (1974) Étude in vitro de l'activité électromyographique du côlon proximal et distal du lapin. J Physiol (Paris) 68:305–329
Julé Y (1975) Modification de l'activité électrique du côlon proximal de lapin in vivo, par stimulation des nerfs vagues et splanchniques. J Physiol (Paris) 70:5–26
Julé Y (1980) Nerve-mediated descending inhibition in the proximal colon of the rabbit. J Physiol (Lond) 309:487–498
Julé Y, Gonella J (1972) Modifications de l'activité électrique du côlon terminal de lapin par stimulation des fibres nerveuses pelviennes et sympathiques. J Physiol (Paris) 64:599–621
Julé Y, Szurszewski JH (1979) Occurrence of spontaneous oscillatory neurons in the cat inferior mesenteric ganglia: relationship to ileus? Gastroenterology 76:1163
Kajirsuka T (1979) Phasic and tonic contraction of rabbit intestinal muscle. Jpn J Physiol 29:159–177
Kelly KA (1977) Neural control of gastric electric and motor activity. In: Brooks FP, Evers PW (eds) Nerves and the gut. Slack, pp 223–231
Kelley ML Jr, De Weese JA (1969) Effects of eating and intraluminal filling on the ileocolonic junctional zone pressures. Am J Physiol 216:1491–1495

Kelley ML Jr, Gordon EA, De Weese JA (1965) Pressure studies of the ileocolonic junctional zone of dogs. Am J Physiol 209:333–339
Kelley ML, Gordon EA, De Weese JA (1966) Pressure responses of canine ileocolonic junctional zone to intestinal distension. Am J Physiol 211:614–618
Ken Kure K, Ichiko K-I, Ishikawa K (1931) On the spinal parasympathetic physiological significance of the spinal parasympathetic system in relation to the digestive tract. J Exp Physiol 21:1–19
Kerremans R (1968) Electrical activity and motility of the internal anal sphincter. An in vivo electrophysiological study in man. Acta Gastroenterol Belg 31:465–482
Kewender J, Pahlin PE, Storm B (1970) The effect of periarterial nerve stimulation on the jejunal and ileal motility in cat. Acta Physiol Scand 80:353–359
Khan IH, Bedi BS (1972) Effect of vagotomy and pyloroplasty on the interdigestive myoelectrical complex of the stomach. Gut 13:841–842
King CE, Glass LC, Townsend SE (1947) The circular components of the muscularis mucosae of the small intestine of the dog. Am J Physiol 148:667–674
Kingma YJ, Durdle N, Bowes KL, Kocylovski M, Szmidt J (1979) Size of electrical oscillating regions in the canine colon. In: Christensen J (ed) Gastrointestinal motility. 7th International Symposium on Gastrointestinal Motility. Raven, New York, pp 425–431
Klee P (1912) Der Einfluß der Vagusreizung auf den Ablauf der Verdauungsbewegungen. Pfluegers Arch 145:557–594
Klinge FW (1951) Behaviour of isolated intestinal segments without one or both plexuses. Am J Physiol 164:284–293
Kobayashi M, Nagai T, Prosser CL (1966) Electrical interaction between muscle layers of cat intestine. Am J Physiol 211:1281–1291
Kock NG, Kewenter J, Sundin T (1972) Studies on the defecation reflex in man. Scand J Gastroenterol 7:689–693
Kocylowski M, Bowes KL, Kingma YJ (1979) Electrical and mechanical activity in the ex vivo perfused total canine colon. Gastroenterology 77:1021–1026
Kosterlitz HW (1967) Intrinsic intestinal reflexes. Am J Dig Dis 12:245
Kosterlitz HW (1968) Intrinsic and extrinsic nervous control of motility of the stomach and the intestines. In: Code CF (ed) Alimentary canal. American Physiological Society, Washington, DC (Handbook of physiology, vol IV, sect 6, pp 2147–2172)
Kosterlitz HW, Lees GM (1964) Pharmacological analysis of intrinsic intestinal reflexes. Pharmacol Rev 16:301–339
Kosterlitz HW, Lydon RJ (1968) Spontaneous electrical activity and nerve mediated inhibition in the innervated longitudinal muscle strip of the guinea-pig ileum. J Physiol (Lond) 200:126–128
Kosterlitz HW, Lydon RJ (1971) Impulse transmission in the myenteric plexus – longitudinal muscle preparations of the guinea-pig ileum. Br J Pharmacol 43:74–85
Kosterlitz HW, Watt AJ (1975) The peristaltic reflex. In: Daniel EE, Paton DM (eds) Methods in pharmacology, vol 3, chap 21. Plenum, New York, pp 391–401
Kosterlitz HW, Pirie VW, Robinson JA (1956) The mechanism of the peristaltic reflex in the isolated guinea-pig ileum. J Physiol (Lond) 133:681– 694
Kottegoda SR (1969) An analysis of possible nervous mechanisms involved in the peristaltic reflex. J Physiol (Lond) 200:687–712
Kottegoda SR (1970) Peristalsis of the small intestine. In: Bülbring E, Brading AF, Jones AW, Tomita T (eds) Smooth muscle. Arnold, London, pp 525–541
Krantis A, Costa M, Furness JB, Orbach J (1980) Gamma-aminobutyric acid stimulates intrinsic inhibitory and excitatory nerves in the guinea-pig intestine. Eur J Pharmacol 67:461–468
Krishnan BT (1932) Studies of the function of the intestinal musculature. 1. The normal movements of the small intestine and the relations between the action of the longitudinal and circular muscle fibres in those movements. Q J Exp Physiol 22:37–63
Kuntz A (1953) The autonomic nervous system, 3rd edn. Lea & Febiger, Philadelphia
Kuriyama H, Tomita T (1970) Action potential in smooth muscle of guinea-pig taenia coli and ureter studied by double sucrose-gap method. J Gen Physiol 55:147–162

Kuriyama H, Osa T, Toida N (1967) Nervous factors influencing the membrane activity of intestinal smooth muscle. J Physiol (Lond) 191:257–270

Kuru M, Sugihara (1955) Contributions to the knowledge of Bulbar Autonomic Centres. II. Relationship of the vagal nuclei to the gastro-jejunal motility. Jpn J Physiol 5:21–36

Kyforeva O (1948) Duodenal function in man. Motor and secretory function of the duodenum and their relation in healthy subjects. Klin Med (Mosk) 26:45

Lagerlöf HO, Johansson C, Ekelund K (1974) Studies of gastrointestinal interactions. VI. Intestinal flow, mean transit time and mixing after composite meals in man. Scand J Gastroenterol 9:261–270

Lang RJ (1979) Temperature and inhibitory junctional transmission in guinea-pig ileum. Br J Pharmacol 66:355–357

Langley JN (1898) On inhibitory fibers in the vagus for the end of the esophagus and the stomach. J Physiol (Lond) 23:407–414

Langley JN (1911) The effect of various poisons upon the response to nervous stimuli chiefly in relation to the bladder. J Physiol (Lond) 43:125–181

Langley JN (1921) The autonomic nervous system. Cambridge

Langley JN, Anderson HK (1895) On the innervation of the pelvic and adjoining viscera. J Physiol (Lond) 18:67–105

Langley JN, Dickinson WL (1889) On the local paralysis of the peripheral ganglia and on the connexion of different classes of nerve fibres with them. Proc R Soc 46:423–431

Langley JN, Magnus R (1905/1906) Some observations of the movements of the intestine before and after degenerative section of the mesenteric nerves. J Physiol (Lond) 33:34–51

Larson LM, Bargen JA (1933) Physiology of the colon. Arch Surg 27:1–50

Lee CY (1960) The effect of stimulation of extrinsic nerves on peristalsis and on the release of 5-hydroxytryptamine in the large intestine of the guinea-pig and of the rabbit. J Physiol (Lond) 152:405–418

Lee KY, Chey WY, Tai HH, Yajima H (1978) Radioimmunoassay of motilin, validation and studies on the relationship between plasma motilin and interdigestive myoelectric activity of the duodenum in dog. Am J Dig Dis 23:788–795

Leek BF (1972) Abdominal visceral receptors. In: Neil E (ed) Enteroceptors. Springer, Berlin Heidelberg New York (Handbook of sensory physiology, vol III/1, pp 113–160)

Leek BF (1977) Abdominal and pelvic visceral receptors. Br Med Bull 33:163–168

Legros et Onimus (1869) Recherchers experimentales sur les movements de l'intestin. J Anat Physiol (Paris) 6:37–66

Lewis TD, Collins SM, Fox J-AE, Daniel ED (1979) Initiation of duodenal acid-induced motor complexes. Gastroenterology 77:1217–1224

Lipkin M, Almy TP, Bell BM (1962) Pressure volume characteristics of the human colon. J Clin Invest 41:1831–1839

Liu J, Prosser CL, Job DD (1969) Ionic dependence of slow waves and spikes in intestinal muscle. Am J Physiol 217:1542–1547

Llewellyn-Smith IJ, Wilson AJ, Furness JB, Costa M, Rush RA Ultrastructural identification of noradrenergic axons and their distribution within the enteric plexuses of the guinea-pig small intestine. Neurocytol 10:331–352

Lord MG, Hutton M, Wingate DL (1979) Patterns of slow wave and spike activity in the colon of the conscious dog. In: Christensen J (ed) Gastrointestinal motility. 7th International Symposium on Gastrointestinal Motility. Raven, New York

Lüderitz C (1889) Experimentelle Untersuchungen über die Entstehung der Darmperistaltik. Virchows Arch [Pathol Anat] 118:19–36

Lundberg JM, Dahlstrom A, Larsson I, Patterson G, Alderman H, Kewenter J (1978) Efferent innervation of the small intestine by adrenergic neurons from the cervical sympathetic and stellate ganglia, studied by retrograde transport of peroxidase. Acta Physiol Scand 104:33–42

Lux GU, Strunz U, Domschke J, Femppel J, Rosch W, Domschke W (1978) Motilin and interdigestive myoelectric and motor activity of small intestine in man: lack of causal relationship. Scand J Gastroenterol [Suppl 49] 13:118

Lux G, Femppel J, Lederer P, Rösch W, Domschke W (1979 a) Increased duodenal alkali load associated with the interdigestive myoelectric complex. Acta Hepatogastroenterol 26:166–169
Lux G, Lederer P, Femppel J, Rosch W, Domschke W (1979 b) Spontaneous and 13-NLE-motilin-induced interdigestive motor activity of esophagus, stomach and small intestine in man. In: Christensen J (ed) Gastrointestinal motility. 7 th International Symposium on Gastrointestinal Motility. Raven, New York, pp 269–277
Lyman H (1913) The receptive relaxation of the colon. Am J Physiol 32:61–64
MacEwen W (1904) The function on the caecum and appendix. Lancet 82:995–1000
Mackenna BR, McKirdy HC (1972) Peristalsis in the rabbit distal colon. J Physiol (Lond) 220:33–54
Maddison S, Horrell RI (1979) Hypothalamic unit responses to alimentary perfusions in the anesthetised rat. Brain Res Bull 4:259–266
Mall F (1896) A study of the intestinal contraction. Johns Hopkins Hosp Rep 1:37–75
Marik F, Code CF (1975) Control of the interdigestive myoelectric activity in dogs by the vagus nerve and pentagastrin. Gastroenterology 69:387–395
Martelli H, Devroede G, Arhan P, Duguoy C (1978) Mechanisms of idiopathic constipation: outlet obstruction. Gastroenterology 75:623–631
Martinson J (1965) Studies on the efferent vagal control of the stomach. Acta Physiol Scand [Suppl 255] 65
Martinson J (1975) Nervous control of gastroduodenal motility and emptying. Scand J Gastroenterol [Suppl] 10:31–44
Martinson J, Muren A (1963) Excitatory and inhibitory effects of vagus stimulation on gastric motility in the cat. Acta Physiol Scand 57:309–316
Marzio L, Lanfranchi GA, Trento L, Campieri M, Brignola C (1979) Rectoanal inhibitory reflex pattern in normal subjects and patients with idiopatic constipation. Ital J Gastroenterol 11:142
Mashima H, Yoshida T, Handa M (1966) Contraction and relaxation of the guinea-pig's taenia coli in relation to spike discharges. Jpn J Physiol 16:304–315
Mathur PD, Grindlay JH, Mann FC (1948) Observations on duodenal motility in dogs with special reference to activity during vomiting. Gastroenterology 10:866–879
Matsuo Y, Seki A (1978) The coordination of gastrointestinal hormones and the autonomic nerves. Am J Gastroenterol 69:21–50
Mayer CJ, Wood JD (1975) Properties of mechanosensitive neurons within Auerbach's plexus of the small intestine of the cat. Pfluegers Arch 357:35–49
McCoy EJ, Baker RD (1968) Effects of feeding on electrical activity of dog's small intestine. Am J Physiol 214:1291–1295
McCoy EJ, Baker RD (1969) Intestinal slow waves: Decrease in propagation velocity along the upper small intestine. Am J Dig Dis 14:9–13
McFadden GDS, Loughride JS, Milroy TH (1935) The nerve control of the distal colon. Q J Exp Physiol 25:315–327
McKirdy HC (1972) Functional relationship of longitudinal andd circular layers of the muscularis externa of the rabbit large intestine. J Physiol (Lond) 227:839–853
McLaren JW, Ardran GM, Sutcliffe J (1950) Radiographic studies of duodenum and jejunum in man. J Fac Radiol Lond 2:148–164
Mei N (1970) Mécanorécepteurs vagaux digestifs chez le chat. Exp Brain Res 11:502–514
Mei N (1978) Vagal glucoreceptors in the small intestine of the cat. J Physiol (Lond) 282:485–506
Meltzer SJ, Auer J (1907) Peristaltic rush. Am J Physiol 20:259–281
Mendel C, Pousse A, Schang JC, Dauchel J, Grenier JF (1978) Longitudinal contractions in the jejunum of fasting dogs. In: Duthie HL (ed) Gastrointestinal motility in health and disease. MTP Press, Lancaster, pp 61–69
Mendeloff AI (1968) Defecation. In: Code CF (ed) American Physiological Society, Washington, DC (Handbook of physiology, vol IV, sect 6, pp 2140–2146)
Mescham I, Quigley JP (1938) Spontaneous motility of the pyloric sphincter and adjacent regions of the gut in the unanaesthetized dog. Am J Physiol 121:350–357

Meshkinpour H, Dinoso VP, Lorber SH (1974) Effect of intraduodenal administration of essential amino acids and sodium oleate on motor activity of the sigmoid colon. Gastroenterology 66:373–377
Meunier PO, Marechall JM, De Beaujeu MJ (1979) Rectoanal pressures and rectal sensitivity studies in chronic childhood constipation. Gastroenterology 77:330–336
Meunier P, Mollard P (1977) Control of the internal anal sphincter (monometric study with human subjects). Pfluegers Arch 370:233–239
Mir SS, Mason R, Ormsbee HS III (1977) An inhibitory innervation at the gastroduodenal junction in anaesthetized dogs. Gastroenterology 73:432–434
Mir SS, Mason GR, Ormsbee HS (1978) Vagal influence on duodenal motor activity. Am J Surg 135:97–101
Misiewicz JJ (1975) Colonic motility. Gut 16:311–314
Misiewicz JJ, Waller SL, Eisner M (1966) Motor responses of human gastrointestinal tract to 5-hydroxytryptamine in vivo and in vitro. Gut 7:208–216
Mitchell GAG (1953) Anatomy of the autonomic nervous system. Livingstone, Edinburgh
Mitznegg P, Bloom SR, Domschke W, Domschke S, Wunsch E, Demling L (1976) Release of motilin after duodenal acidification. Lancet 1:888–889
Monges H, Salducci J, Naudy B (1974) Electrical activity of the gastrointestinal tract in dog during vomiting. In: Daniel EE (ed) Proceedings of the 4th International Symposium on Gastrointestinal Motility. Mitchell, Vancouver, pp 479–488
Monges H, Salducci J, Nauudy B, Ranieri F, Gonella J, Bouvier M (1979) The electrical activity of the internal anal sphincter: a comparative study in man and in cat. In: Christensen J (ed) Gastrointestinal motility. 7th International Symposium on Gastrointestinal Motility. Raven, New York
Morgan KG, Go VLW, Szurszewski JH (1979) Motilin increases the influence of excitatory myenteric plexus neurons on gastric smooth muscle in vitro. In: Christensen J (ed) Gastrointestinal motility. 7th International Symposium on Gastrointestinal Motility. Raven, New York
Mukhopadhyay AK, Thor PJ, Copeland EM, Johnson LR, Weisbrodt NW (1977) Effect of cholecystokinin on myoelectric activity of small bowel of dog. Am J Physiol 232:E44–47
Mulinos MG (1927) Gastrointestinal motor reponse to vagus stimulation after nicotine. Proc Soc Exp Biol Med 25:49–53
Munro AF (1953) Effect of autonomic drugs on the responses of isolated preparations from the guinea-pig intestine to electrical stimulation. J Physiol (Lond) 120:41–52
Nakayama S (1962) Movements of the small intestine in transport of intraluminal contents. Jpn J Physiol 12:522–533
Nakayama S (1965) Effects of stimulation of the vagus nerve on the movements of the small intestine. Jpn J Physiol 15:243–252
Nakayama S, Nanba R (1961) Electrophysiological studies on the intestinal intrinsic reflex. Jpn J Physiol 11:499–505
Nakayama S, Neya T, Yamasato T, Takaki M, Mizutani M (1979) Activity of the spinal defecation centre in the guinea-pig. Ital J Gastroenterol 11:168–173
Newman M, Thienes CH (1933) On the sympathetic innervation of guinea-pig intestine. Am J Physiol 104:113–116
Ng KKF (1966) The effect of some anticholinesterases on the response of the taenia to sympathetic nerve stimulation. J Physiol (Lond) 182:233–243
Ninomiya I, Irishawa H, Woolley G (1974) Intestinal mechanoreceptor reflex effects on sympathetic nerve activity to intestine and kidney. Am J Physiol 227:584–591
Nishi S, North RA (1973a) Intracellular recording from the myenteric plexus of the guinea-pig ileum. J Physiol (Lond) 231:471–491
Nishi S, North RA (1973b) Presynaptic action of noradrenaline in the myenteric plexus. J Physiol (Lond) 231:29–30P
North RA, Williams JT (1977) Extracellular recording from the guinea-pig myenteric plexus and the action of morphine. Eur J Pharmacol 45:23–33
Nothnagel H (1882) Zur chemischen Reizung der glatten Muskeln; zugleich als Beitrag zur Physiologie des Darms. Arch Pathol Anat Physiol Klin Med 88

Öhrn P-G, Rentzhog L (1976) Effect of adrenergic blockade on gastrointestinal propulsion after laparotomy. Acta Chir Scand [Suppl] 461:53–64
Öhrn P-G, Rentzhog L, Winkstrom S (1976) Effects of sympathoadrenal activity and pharmacological treatment on gastrointestinal propulsion in the early postoperative period. Acta Chir Scand [Suppl] 461:65–76
Ohga A, Nakazato Y, Saito K (1970) Considerations of the efferent nervous mechanism of the vago-vagal reflex relaxation of the stomach in the dog. Jpn J Pharmacol 20:116–130
Ohkawa H, Prosser CL (1972a) Electrical activity in myenteric and submucous plexus of cat intestine. Am J Physiol 222:1412–1419
Ohkawa H, Prosser CL (1972b) Functions of neurons in enteric plexuses of cat intestine. Am J Physiol 222:1420–1426
Ohya S (1969) On the responses of colon motility to stimulation of dog's lumbar cord. Jpn J Smooth Muscle Res 5:100–107
Ormsbee HS III, Mir SS (1978) The role of the cholinergic nervous system in the gastrointestinal response to motilin in vivo. In: Duthie HL (ed) Gastrointestinal motility in health and disease. MTP Press, Lancaster, pp 113–122
Ormsbee HS, Eisenstat TE, Mason GR (1976) Vagal stimulation of canine duodenal motor activity. Surg Forum 27:392–394
Ormsbee HS, Telford GL, Mason GR (1979) Required neural involvement in control of canine migrating motor complex. Am J Physiol 237:E451–E456
Pahlin PE, Kewenter J (1975) Reflexogenic contraction of the ileocecal sphincter in the cat following small or large intestinal distension. Acta Physiol Scand 95:126–132
Pahlin PE, Kewenter J (1976a) Sympathetic nervous control of cat ileocecal sphincter. Am J Physiol 231:296–305
Pahlin PE, Kewenter J (1976b) The vagal control of the ileo-cecal sphincter in the cat. Acta Physiol Scand 96:433–442
Paintal AS (1957) Responses from mucosal mechanoreceptors in the small intestine of the cat. J Physiol (Lond) 139:353–368
Paintal AS (1963) Vagal afferent fibres. Ergeb Physiol 52:74–156
Paintal AS (1973) Vagal sensory receptors and their reflex effects. Physiol Rev 53:159–227
Papasova M, Velkova V, Atanassova E (1979) Character of the gastric and duodenal electrical activity after blocking of the adrenoreactive structures. Acta Physiol Pharmacol Bulg 5:3–10
Parkkulainen KV (1962) Simple low small bowel obstruction. Acta Chir Scand [Suppl] 290:1–97
Pascaud XB, Genton MJH, Bass P (1978) Gastroduodenal contractile activity in fed and fasted unrestrained rats. In: Duthie HL (ed) Gastrointestinal motility in health and disease. MTP Press, Lancaster, pp 637–645
Paton WDM (1955) The response of the guinea-pig ileum to electrical stimulation by coaxial electrodes. J Physiol (Lond) 127:40P–41P
Paton WDM, Vane JR (1963) An analysis of the responses of the isolated stomach to electrical stimulation and to drugs. J Physiol (Lond) 165:10–46
Paton WDM, Zar A (1968) The origin of acetylcholine released from guinea-pig intestine and longitudinal muscle strips. J Physiol (Lond) 194:13–33
Pearce EA, Wingate DL (1979) The role of the myenteric plexuses? Gastroenterology 76:1215
Pearcy JF, Van Liere EJ (1926) Studies on the visceral nervous system. XVII. Reflexes from the colon. Reflexes to the stomach. Am J Physiol 78:64–73
Peeters TL, Vantrappen G, Janssens J (1979) Fluctuations of motilin and gastrin levels in relation to the interdigestive motility complex in man. In: Christensen J (ed) Gastrointestinal motility. 7th International Symposium on Gastrointestinal Motility. Raven, New York, p 287
Perkins WE (1971) A method for studying the electrical and mechanical activity of isolated intestine. J Appl Physiol 30:768–771
Persson CGA (1976) Inhibitory innervation of cat sphincter of oddi. Br J Pharmacol 58:479–482

Pescatori M, Grassetti F, Ronzoni G, Mancinelli R, Bertuzzi A, Salinari S (1979) Peristalsis in distal colon of the rabbit: an analysis of mechanical events. Am J Physiol 236:464–472
Pick J (1970) The autonomic nervous system; morphological, comparative, clinical and surgical aspects. Lippincott, Philadelphia
Poitras P, Steinbach J, Van Deventer G, Walsh JH, Code CF (1979) Effects of somatostatin on interdigestive myoelectric complexes and motilin blood levels. Gastroenterology 76:1218
Pomeroy AR, Raper C (1971) Maximal responses in guinea-pig isolated ileum preparations: influence of longitudinal and circular muscle. J Pharm Pharmacol 23:796–798
Posey EL, Brown HS, Bargen JA (1948) The response of human intestinal motility to tetraethylammonium chloride. Gastroenterology 11:83–89
Preshaw AM, Knauff RS (1966) The effect of sham feeding on the secretion and motility of canine duodenal pouches. Gastroenterology 51:193–199
Prosser CL (1974) Diversity of electrical activity in gastrointestinal muscle. In: Daniel EE (ed) Proceedings of the 4th International Symposium on Gastrointestinal Motility. Mitchell, Vancouver, pp 21–37
Prosser CL, Bortoff A (1968) Electrical activity of intestinal muscle under in vitro conditions. In: Code CF (ed) Alimentary canal. American Physiological Society, Washington, DC (Handbook of physiology, vol IV, sect 6, pp 2025–2074)
Provenzale L, Pisano M (1971) Methods for recording electrical activity of the human colon in vivo. Am J Dig Dis 16:712–722
Puestow CB (1937) Intestinal motility in the dog and man. Ill Med Dent Monogr 2:1–69
Quigley JP, Highstone WH, Ivy AC (1934) A study of the propulsive activity of the Thiry-Vella loop of intestine. Am J Physiol 108:151–158
Quigley JP, Louckes MS (1962) Gastric emptying. Am J Dig Dis 7:672–676
Quigley JP, Solomon EI (1930) Action of insulin on the motility of the gastrointestinal tract. V.a. Action on the human duodenum. b. Action on the colon of dogs. Am J Physiol 91:488–495
Raiford T, Mulinos MG (1934a) Intestinal activity in the exteriorized colon of the dog. Am J Physiol 110:123–128
Raiford T, Mulinos MG (1934b) The myenteric reflex as exhibited by the exteriozed colon of the dog. Am J Physiol 110:129–136
Ramorino ML, Colagrande C (1964) Intestinal motility preliminary studies with telemetering capsule and synchronized fluorocinematrography. Am J Dig Dis 9:64–71
Rand MJ, Ridehalgh A (1965) Actions of hemicholinium and triethylcholine on responses of guinea-pig colon to stimulation of autonomic nerves. J Pharm Pharmacol 17:144–156
Ranieri F, Crousillat J, Mei N (1975) Étude électrophysiologique et histologique des fibres afferentes splanchniques. Arch Ital Biol 113:354–373
Rayner V (1971) Observations on the functional internal anal sphincter of the vervet monkey. J Physiol (Lond) 213:27–28P
Rayner V (1979) Characterisation of the internal anal sphincter and the rectum of the vervet monkey. J Physiol (Lond) 286:383–399
Read NW (1979) The migrating motor complex and spontaneous fluctuations of transmural potential difference in the human small intestine. In: Christensen J (ed) Gastrointestinal motility. 7th International Symposium on Gastrointestinal motility. Raven, New York, pp 299–306
Reinke DA, Rosenbaum AH, Bennett DR (1967) Patterns of dog gastrointestinal contractile activity monitored in vivo with extraluminal force transducers. Am J Dig Dis 12:113–141
Rentz E (1938) Über den Einfluß der Vagusreizung auf die Bewegungen des Dickdarms. Arch Exp Pathol Pharmakol 191:172–182
Reverdin N, Hutton M, Ling A, Wingate DL, Ritchie HD, Christofides N, Adrian TE, Bloom SR (1979) The motor responses to food: vagus-dependent or independent? Gastroenterology 76:1225
Reynell PC, Spray GH (1956) The simultaneous measurement of absorption and transit in the gastrointestinal tract of the rat. J Physiol (Lond) 131:452–462
Rikimaru A (1971a) Contractile properties of organ-cultured intestinal smooth muscle. Tohoku J Exp Med 103:317–327

Rikimaru A (1971 b) Relaxing response to transmural stimulation of isolated taenia coli of the chimpanzee and the pig. Tohoku J Exp Med 103:115–116
Rikimaru A, Suzuki T (1971) Neural mechanism of the relaxing responses of guinea-pig taenia coli. Tohoku J Exp Med 103:303–315
Ritchie JA (1968) Colonic motor activity and bowel function. Part 1. Normal movement of contents. Gut 9:442–456
Ritchie JA (1971) Movement of segmental constrictions in the human colon. Gut 12:350–355
Ritchie J (1972) Mass peristalsis in the human colon after contact with oxyphenisatin. Gut 13:211–219
Rödén SH (1937) An experimental study on intestinal movements; particularly with regard to ileus conditions in cases of trauma and peritonitis. Acta Chir Scand [Suppl 51] 80:1–146
Ross B, Watson BW, Kay AW (1963) Studies on the effect of vagotomy on small intestinal motility using the radio-telemetering capsule. Gut 4:77–83
Rostad H (1973) Central and peripheral nervous control of colonic motility in the cat. University of Oslo
Rubin MR, Snape WJ Jr, Cohen S (1979) The vagal and cholinergic control of the ileocecal sphincter. Gastroenterology 76:1230
Ruckebusch Y (1973) Les particularites motrices de jejuno-ileum chez le cheval. Cah Med Vet 42:128–143
Ruckebusch Y, Bueno L (1973) The effect of the small intestine in the calf. Br J Nutr 30:491–499
Ruckebusch Y, Bueno L (1975) Electrical activity of ovine jejunum and changes due to disturbances. Am J Dig Dis 20:1027–1034
Ruckebusch Y, Bueno L (1977 a) Migrating myoelectrical complex of the small intestine. Gastroenterology 73:1309–1314
Ruckebusch Y, Bueno L (1977 b) Origin of migrating myoelectric complex in sheep. Am J Physiol 233:E483–487
Ruckebusch M, Ferré JP (1974) Origine alimentaire des variation nycthemerales de l'activité électrique de l'intestin grêle chez le rat. C R Soc Biol (Paris) 167:2005–2009
Ruckebusch M, Fioramonti J (1975) Electrical spiking activity and propulsion in small intestine in fed and fasted rats. Gastroenterology 68:1500–1508
Ruckebusch Y, Grivel ML, Fargeas MJ (1971) Activité électrique de l'intestin et prise de nourriture conditionnelle chez le lapin. Physiol Behav 6:359–365
Ruppin H, Soergel KH, Dodds JW, Wood CM, Domschke W (1979) Effects of the interdigestive motor complex (IMC) and 13-Norleucine motilin (NLEM) on fasting intestinal flow rate and velocity in man. Gastroenterology 76:1231
Ruwart MJ, Klepper MS, Rush BD (1979) Evidence for non-cholinergic mediation of small intestinal transit in the rat. J Pharmacol Exp Ther 209:462–465
Sancholuz AG, Croley TE II, Christensen J, Macagno EO, Glover JR (1975) Phase lock of electrical slow waves and spike bursts in cat duodenum. Am J Physiol 229:608–612
Sanford PA (1976) Effect of actively transported hexoses on afferent nerve discharge from rat small intestine in vivo. J Physiol (Lond) 254:75–76P
Sarna SK, Bardakjian BL, Waterfall WE, Lind JF (1980) Human colonic electrical control activity (ECA). Gastroenterology 78:1526–1536
Sarna SK, Daniel EE, Kingma YJ (1971) Stimulation of slow wave electrical activity of small intestine. Am J Physiol 221:166–175
Sarr MG, Kelly KA (1979) Jejunal transit of liquids and solids during jejunal interdigestive and digestive motor activity. In: Christensen J (ed) Gastrointestinal motility. 7 th International Symposium on Gastrointestinal Motility. Raven, New York, p 309
Sato T, Takayanagi I, Takagi K (1973) Pharmacological properties of electrical activities obtained from neurons in Auerbach's plexus. Jpn J Pharmacol 23:665–671
Schang JC, Dauchel J, Sava P, Angel F, Bauchet P, Lambert A, Grenier JF (1978) Specific effects of different food components on intestinal motility. Electromyographic study in dogs. Eur Surg Res 10:425–432
Schaumann O, Jochum K, Schmidt H (1953) Analgetika und Darmmotorik. III. Zum Mechanismus der Peristaltik. Arch Expt Pathol Pharmakol 219:302–309

Schofield GC (1968) Anatomy of muscular and neural tissues in the alimentary canal. In: Code CF (ed) Alimentary canal. American Physiological Society, Washington, DC (Handbook of physiology, vol IV, sect 6, pp 1579–1628)
Schuster MW (1968) Motor action of rectum and anal sphincters in continence and defecation. In: Code CF (ed) Alimentary canal. American Physiological Society, Washington, DC (Handbook of physiology, vol IV, sect 6, pp 2121–2139)
Schuster MM (1975) The riddle of the sphincters. Gastroenterology 69:249–262
Schuster MM, Hendrix TR, Mendeloff AI (1963) The internal anal sphincter response: manometric studies on its normal physiology, neural pathways and alteration in bowel disorder. J Clin Invest 42:196–207
Scott LD, Summers RW (1976) Correlation of contractions and transit in rat small intestine. Am J Physiol 230:132–137
Sellers AF, Lowe JE, Brondum J (1979) Motor events in equine large colon. Am J Physiol 237:E457–464
Semba T (1954) Studies on the gastro-colic reflexes. Hiroshima J Med Sci 2:329–333
Semba T (1956) Motor effect of the distal colon caused by stimulating the dorsal roots of the dog's lumbar nerves. Jpn J Physiol 6:321–326
Semba T, Hiraoka T (1957) Motor responses of the stomach and small intestine caused by stimulation of the peripheral end of the splanchnic nerve. Jpn J Physiol 7:64–71
Semba T, Mishima H (1958) Studies on the motor reflexes of the distal colon. J Hiroshima Med Assoc 11:11–20
Semba T, Mizonishi T (1978) Atropine-resistant excitation of motility of the dog stomach and colon induced by stimulation of the extrinsic nerves and their centers. Jpn J Physiol 28:239–248
Sharma KN (1967) Receptor mechanisms in the alimentary tract: their excitation and functions. In: Code CF (ed) Alimentary canal. American Physiological Society, Washington, DC (Handbook of physiology, vol I, sect 6, pp 225– 237
Sharma KN, Nasset ES (1962) Electrical activity in mesenteric nerves after perfusion of gut lumen. Am J Physiol 202:725–730
Shearin NL, Bowes KL, Kingma YJ (1978) In vitro electrical activity in canine colon. Gut 20:780–786
Shepherd JJ, Wright POG (1968) The response of the internal anal sphincter in man to stimulation of the presacral nerve. Am J Dig Dis 13:421–427
Short AR (1919) Observations on the ileo-cecal valve. Br Med J 2:164–165
Sillin LF, Condon RF, Shulte WJ, Woods JH, Bass PO, Go VWL (1978) The relationship between gastric inhibitory peptide and right colon electro-mechanical activity after feeding. In: Duthie HL (ed) Gastrointestinal motility in health and disease. MTP Press, Lancaster, pp 361–362
Slaughter RL, Grant EE (1974) Small intestine motor response of the dog during vomiting with demonstration of a reverse peristaltic contraction. Gastroenterology 66:779
Small RC (1971) Transmission from cholinergic neurones to circular smooth muscle obtained from the rabbit caecum. Br J Pharmacol 42:656–657P
Small RC (1972) Transmission from intramural inhibitory neurones to circular smooth muscle of the rabbit caecum and the effects of catecholamines. Br J Pharmacol 45:149P
Small RC, Weston AH (1979) Intramural inhibition in rabbit and guinea-pig intestine. In: Baer HP, Drummond GI (eds) Physiological and regulatory functions of adenosine and adenine nucleotides. Raven, New York, pp 45–61
Smets W (1936) L'activité réflexe de la valvule iléo-caecale et du segment terminal de grêle. C R Soc Biol (Paris) 123:106–107
Smith CC, Brizzee KR (1961) Cineradiographic analysis of vomiting in the cat. Gastroenterology 40:654–664
Snape WJ Jr, Carlson GM, Cohen S (1976) Colonic myoelectric activity in the irritable bowel syndrome. Gastroenterology 70:326–330
Snape WJ Jr, Matarazzo SA, Cohen S (1978) Effects of eating and gastrointestinal hormones on human colonic myoelectric and motor activity. Gastroenterology 75:373–378
Snape WJ Jr, Wright SH, Battle WM, Cohen S (1979) The gastro-colic response: evidence for a neural mechanism. Gastroenterology 77:1235–1240

Stavney SL, Kato T, Griffith CA, Nyhus LM, Harkins HN (1963) A physiologic study of motility changes following selective gastric vagotomy. J Surg Res 3:390–394
Stewart JJ, Bass P (1976a) Effect of intravenous C-terminal octapeptide of cholecystokinin and intraduodenal ricinoleic acid on contractile activity of the dog intestine. Proc Soc Exp Biol Med 152:213–217
Stewart JJ, Bass P (1976b) Effects of ricinoleic and oleic acids on the digestive contractile activity of the canine small and large bowel. Gastroenterology 70:371–376
Stewart JJ, Weisbrodt NW, Burks TF (1977) Evidence for centrally mediated drug effects on the myoelectric activity of the feline small intestine. In: Brooks FP, Evers PW (eds) Nerves and the gut. Slack, pp 272–284
Stockley HL, Bennett A (1974) The intrinsic innervation of human sigmoid colon muscle. In: Daniel EE (ed) Proceedings of the 4th International Symposium on Gastrointestinal Motility. Mitchell, Vancouver, pp 165–176
Stockley HL, Bennett A (1977) Relaxations mediated by adrenergic and non-adrenergic nerves in human isolated taenia coli. J Pharm Pharmacol 29:533–537
Stoddart CJ, Duthie HL (1973) The changes in gastroduodenal myoelectrical activity after varying degrees of vagal denervation. Gut 14:824
Stoddard CJ, Smallwood RH, Duthie HL (1978) Migrating myoelectrical complexes in man. In: Duthie HL (ed) Gastrointestinal motility in health and disease. MTP Press, Lancaster, pp 9–15
Stoddart CJ, Duthie HL, Smallwood RH (1979) Colonic myoelectrical activity in man's comparison of recording techniques and methods of analysis. Gut 20:476–483
Streeten DHP, Vaughan-Williams EM (1951) The influence of intraluminal pressure upon the transport of fluid through cannulated thiry-vella loops in dogs. J Physiol (Lond) 112:1–21
Sullivan MA, Cohen S, Snape WJ Jr (1978) Colonic myoelectrical activity in irritable bowel syndrome. Effects of eating and anticholinergics. N Engl J Med 298:878–883
Summers RW (1978) Hydrogen ions inhibit jejunal flow. In: Duthie HL (ed) Gastrointestinal motility in health and disease. MTP Press, Lancaster, pp 625–633
Summers RW, Dusdieker NS (1979) Computer-generated display of longitudinal spike burst spread in the small intestine. In: Christensen J (ed) Gastrointestinal motility. 7th International Symposium on Gastrointestinal Motility. Raven, New York, pp 339–344
Summers RW, Kent TH, Osborne JW (1970) Effects of drugs, ileal obstruction, and irradiation on rat gastrointestinal propulsion. Gastroenterology 59:731–739
Summers RW, Helm J, Christensen J (1976) Intestinal propulsion in the dog. Its relation to food intake and the migrating myoelectric complex. Gastroenterology 70:753–758
Suzuki T, Inomata H (1964) The inhibitory post-synaptic potential in intestinal smooth muscle investigated with intracellular microelectrode. Tohoku J Exp Med 82:48–51
Szolcsányi J, Barthó L (1978) New type of nerve-mediated cholinergic contractions of the guinea-pig small intestine and its selective blockade by capsacin. Nauny-Schmiedeberg Arch Pharmacol 305:83–90
Szurszewski JH (1969) A migrating complex of the canine small intestine. Am J Physiol 217:1757
Szurszewski JH (1977a) Towards a new view of prevertebral ganglion. In: Brooks FP, Evers PW (eds) Nerves and the gut. Slack, pp 244–258
Szurszewski JH (1977b) Modulation of smooth muscle by nervous activity: a review and a hypothesis. Fed Proc 36: 2456–2461
Szurszewski JH, Weems WA (1976) A study of peripheral input to and its control by postganglionic neurones of the inferior mesenteric ganglion. J Physiol (Lond) 256:541–556
Takayanagi I, Sato T, Takagi K (1977) Effects of sympathetic stimulation on electrical activity of Auerbach's plexus and intestinal smooth muscle tone. J Pharm Pharmacol 29:376–377
Tansy MF, Kendall FM (1973) Experimental and clinical aspects of gastrocolic reflexes. Am J Dig Dis 18:521–531
Tansy MF, Kendall FM (1977) Systemic effects of visceral afferent fiber stimulation. In: Brooks FP, Evers PW (eds) Nerves and the gut. Stack, pp 334–349

Tansy MF, Kendall FM, Murphy JJ (1972) The reflex nature of the gastrocolic propulsive response in the dog. Surg Gynecol Obstet 135:404–410
Tansy MF, Martin JS, Landin WE, Kendall FM (1979) Evidence of reflexive beta adrenergic motor stimulation in the canine stomach and small intestine. Surg Gynecol Obstet 148:905–912
Taylor GS, Daniel EE, Tomita T (1975a) Origin and mechanism of intestinal slow waves. In: 5th International Symposium on Gastrointestinal Motility.
Taylor I, Duthie HL, Smallwood R, Linkens D (1975b) Large bowel myoelectrical activity in man. Gut 16:808–814
Telford ED, Stopford JSB (1934) The autonomic nerve supply of the distal colon. An anatomical and clinical study. Br Med J 1:572–574
Telford GL, Mir SS, Mason GR, Ormsbee HS III (1979) Neural control of the canine pylorus. Am J Surg 137:92–98
Templeton RD, Lawson H (1931) Studies in the motor activity of the large intestine. Am J Physiol 96:667–676
Thomas JE (1957) Mechanics and regulation of gastric emptying. I. Mech Physiol Rev 37:453–474
Thomas JE, Baldwin MV (1971) The intestinal mucosal reflex in the unanesthetizes dog. Am J Dig Dis 16:642–647
Thomas JE, Kuntz A (1926a) A study of the vagoenteric mechanism by means of nicotine. Am J Physiol 76:598–605
Thomas JE, Kuntz A (1926b) A study of gastrointestinal motility in relation to the enteric nervous system. Am J Physiol 76:606–626
Thomas PA, Kelly KA (1979) Hormonal control of interdigestive motor cycles of canine proximal stomach. Am J Physiol 236:192–197
Thomas PA, Kelly A, Go VWL (1979) Hormonal regulation of gastrointestinal interdigestive motor cycles. In: Christensen J (ed) Gastrointestinal motility. 7th International Sympsium on Gastrointestinal Motility. Raven, New York, pp 267–268
Thouvenot J, Harichaux P (1963) Activité électrique spontanée de l'intestine. Étude chez le cobaye anesthésié. J Physiol (Paris) 55:344–345
Tomita T (1972) Conductance change during the inhibitory potential in the guinea-pig taenia coli. J Physiol (Lond) 255:693–703
Tonini M, Lecchini S, Frigo G, Crema A (1974) Action of tetrodotoxin on spontaneous electrical activity of some smooth muscle preparations. Eur J Pharmacol 29:236–240
Torsoli A, Ramorino ML, Ammaturo MV, Capurso L, Paoluzzi P, Anzini F (1971) Mass movements and intracolonic pressures. Am J Dig Dis 16:693–696
Trendelenburg P (1917) Physiologische und pharmakologische Versuche über die Dünndarmperistaltik. Naunyn-Schmiedeberg Arch Exp Pathol Pharmakol 81:55–129
Truelove SC (1966) Movements of the large intestine. Physiol Rev 46:457–512
Tsuchiya K (1972) Electrical and mechanical activities of the longitudinal muscle in the peristaltic wave elicited by the intraluminal pressure raising. Rend Gastroenterol 4:115–125
Ustach TJ, Tobon F, Hambrecht T, Bass DD, Schuster MM (1970) Electrophysiological aspects of human sphincter function. J Clin Invest 49:41–48
Van Harn GL (1963) Responses of muscles of cat small intestine to autonomic nerve stimulation. Am J Physiol 204:352–358
Van Liere EJ, Stickney JC, Northup DW (1945) The rate of progress of inert material through the small intestine. Gastroenterology 5:37–42
Van Merwyk AJ, Duthie HL (1979) Characteristics of human colonic smooth muscle in vitro. In: Christensen J (ed) Gastrointestinal motility. 7th International Symposium on Gastrointestinal Motility. Raven, New York, pp 473–478
Vantrappen G, Janssen SJ, Hellemans J, Ghoos Y (1977) The interdigestive motor complex of normal subjects and patients with bacterial overgrowth of the small intestine. J Clin Invest 59:1158–1166
Vantrappen G, Janssen SJ, Hellemans J, Christofides N, Bloom S (1978) Studies on the interdigestive (migrating) motor complex in man. In: Duthie HL (ed) Gastrointestinal motility in health and disease. MTP Press, Lancaster, pp 3–8

Vantrappen G, Peeters TL, Janssen SJ (1979) The secretory component of the interdigestive complex. In: Christensen J (ed) Gastrointestinal motility. 7th International Symposium on Gastrointestinal Motility. Raven, New York

Varagic V (1956) The effect of tolazoline and other substances on the response of the isolated colon of the rabbit to nerve stimulation. Arch Int Pharmacodyn Ther 106:141–150

Vermillon DL, Gillespie JP, Cooke AR, Wood JD (1979) Does 5-hydroxytryptamine influence "purinergic" inhibitory neurons in the intestine? Am J Physiol 236:198–202

Walker GD, Stewart JJ, Bass P (1974) The effect of parietal cell and truncal vagotomy on gastric and duodenal contractile activity of the unanaesthetized dog. Ann Surg 179:853–858

Wankling WJ, Brown BH, Collins CD, Duthie HL (1968) Basal electrical activity in the anal canal in man. Gut 9:457–460

Waterfall WE, Duthie HL, Brown BH (1973) The electrical and motor actions of gastrointestinal hormones on the duodenum in man. Gut 14:689–696

Weems WA, Seygal GE (1979) Intestinal propulsion: studies employing a method for its quantitative evaluation. In: Christensen J (ed) Gastrointestinal motility. 7th International Symposium on Gastrointestinal Motility. Raven, New York, pp 331–338

Weisbrodt NW (1974) Electrical and contractile activities of the small intestine of the cat. Am J Dig Dis 19:93–99

Weisbrodt NW, Christensen J (1972) Electrical activity in the cat duodenum in fasting and vomiting. Gastroenterology 63:1004–1010

Weisbrodt NW, Copeland EM, Moore EP, Kearley RW, Johnson LR (1975) Effect of vagotomy on electrical activity of the small intestine of the dog. Am J Physiol 228:650–654

Welch PB, Plant OH (1926) A graphic study of the muscular activity of the colon with special reference to its response to feeding. Am J Med Sci 172:261–268

Wells JA, Mercer TH, Gray JS, Ivy AC (1942) The motor innervation of the colon. Am J Physiol 138:83–93

Weston AH (1973a) Nerve mediated inhibition of mechanical activity in rabbit duodenum and the effects of desensitization to adenosine and several of its derivatives. Br J Pharmacol 48:302–308

Weston AH (1973b) The effect of desensitization to adenosine triphosphate on the peristaltic reflex in guinea-pig ileum. Br J Pharmacol 47:606–608

White JC, Verlot MG, Ehrentheil O (1940) Neurogenic disturbances of the colon and their investigation by the colon metrogram. Ann Surg 112:1042–1058

Wienbeck M (1972) The electrical activity of the cat in vivo. I. The normal electrical activity and its relation to contractile activity. Res Exp Med 158:268–279

Wienbeck M, Altaparmacov I (1979) Is the internal anal sphincter controlled by a myoelectrical mechanism? In: Christensen J (ed) Gastrointestinal motility. 7th International Symposium on Gastrointestinal Motility. Raven, New York, pp 487–493

Wienbeck M, Janssen H (1974) Electrical control mechanisms at the ileo-colic junction. In: Daniel EE (ed) Proceedings of the 4th International Symposium on Gastrointestinal Motility. Mitchell, Vancouver, pp 97–107

Wienbeck M, Janssen H, Kreuzpainter G (1978) Nycthemeral variation of ileocolic myoelectrical activity in the cat. In: Duthie HL (ed) Gastrointestinal motility in health and disease. MTP Press, Lancaster, pp 399–404

Williams I (1967) Mass movement (mass peristalsis) and diverticular disease of the colon. Br J Radiol 40:2–14

Wingate DL, Ruppin H, Green WER et al. (1976) Motilin-induced electrical activity in the canine gastrointestinal tract. Scand J Gastroenterol 11:111–118

Wingate DL, Thompson HH, Pearce EA, Dand A (1977) Quantitative analysis of the effects of oral feeding on canine intestinal myoelectrical activity. Gastroenterology 72:1151

Wingate DL, Thompson HH, Pearce EA, Dand A (1978) The effects of exogenous cholecystokinin and pentagastrin on myoelectrical activity in the small intestine of the conscious fasted dog. In: Duthe HL (ed) Gastrointestinal motility in health and disease. MTP Press, Lancaster, pp 47–58

Wood JD (1970) Electrical activity from single neurons in Auerbach's plexus. Am J Physiol 219:159–169

Wood JD (1972) Excitation of intestinal muscle by atropine, tetrodotoxin, and xylocaine. Am J Physiol 222:118–125
Wood JD (1973) Electrical discharge of single enteric neurons of guinea-pig small intestine. Am J Physiol 225:1107–1113
Wood JD (1975) Neurophysiology of Auerbach's plexus and control of intestinal motility. Physiol Rev 55:307–324
Wood JD, Mayer CJ (1978 a) Intracellular study of electrical activity of Auerbach's plexus in guinea-pig small intestine. Pfluegers Arch 374:265–275
Wood JD, Perkins WE (1970) Mechanical interaction between longitudinal and circular axes of the small intestine. Am J Physiol 281:762–768
Yamamoto T, Satomi H, Ise H, Takatamo H, Takahashi K (1978) Sacral spinal innervations of the rectal and vesical smooth muscles and the sphincteric striated muscules as demonstrated by the horseradish peroxidase method. Neurosci Lett 7:41–47
Yokoyama S (1966) Aktionspotentiale der Ganglienzelle des Auerbachschen Plexus im Kaninchendünndarm. Pfluegers Arch 288:95–102
Yokoyama S, Ozaki T (1978) Functions of Auerbach's plexus. Jpn J Smooth Muscle Res 14:173–187
Yokoyama S, Ozaki T, Kajitsuka T (1977) Excitation conduction in Auerbach's plexus of rabbit small intestine. Am J Physiol 232:E100–108
You CH, Chey WY, Lee KY (1980) Studies on plasma motilin concentration and interdigestive motility of the duodenum in humans. Gastroenterology 79:62–66
Youmans WB (1952) Neural regulation of gastric and intestinal motility. Am J Med 13:209–226
Youmans WB (1968) Innervation of the gastrointestinal tract. In: Codde CF (ed) Alimentary canal. American Physiological Society, Washington, DC (Handbook of physiology, vol IV, sect 6, pp 1655–1664)
Youmans WB, Karstens AI, Aumann KW (1943) Relation of the extrinsic nerves of the intestine to the inhibitory action of atropine and scopalamine on intestinal motility. J Pharmacol Exp Ther 777:266–273
Zondek B (1920) Über Dickdarmperistaltik. Beobachtungen am experimentellen Bauchfenster. Arch Verdauungskr 27:18–23

CHAPTER 11

Identification of Gastrointestinal Neurotransmitters

J. B. FURNESS and M. COSTA

A. Criteria for Transmitter Identification and Their Application to the Intestine

There have been a number of attempts to define criteria which, if satisfied, would indicate that a substance is a neurotransmitter (e.g. PATON 1958; CURTIS 1961; MCLENNAN 1963; ECCLES 1964; WERMAN 1966, 1972; PHILLIS 1970; ORREGO 1979). However, the underlying criteria that need to be met are that release is detectable and that there is identity of action. In experimental terms this means the following:

That a transient increase in the concentration of the substance can be detected at the neuroeffector junction when the nerve is activated and that when the substance is applied at the junction, in the same concentration and with the same time course, the postjunctional events associated with activation of the nerve are reproduced.

Other criteria which have been brought forward are subsidiary to these central criteria of *release* and *identitiy of action* and are, in some cases, expectations based on experience with the longer established neurotransmitters, acetylcholine and noradrenaline. These subsidiary criteria include the following conditions:

1) There should be an inactivating enzyme (or enzymes) for the transmitter
2) The transmitter should be stored in the neurone
3) There should be a synthesising system for the transmitter
4) Appropriate precursors should be present
5) Pharmacological agents which modify the action of the transmitter should modify the effect of stimulating the nerve in an identical or predictable way

None of these subsidiary conditions is a necessary criterion for transmitter identification. Inactivation need not be enzymatic: diffusion from the point of action may be sufficient (WERMAN 1966; ORREGO 1979). The transmitter may not even be present, except transiently, or be present in amounts not greater than in cells not utilising it as a neurotransmitter. It is possible, as WERMAN (1966) points out, that synthesis could keep pace with release, obviating the need for storage. If prostaglandins turn out to be neurotransmitters, their formation on demand is conceivable. In the case of certain amino acids, neither the third nor fourth criteria need apply, their presence in neurones could depend on uptake from interstitial fluid. The final condition should apply, but the experimental difficulties in its application are often formidable, especially in the intestine.

In practice, the criteria used in transmitter identification are empirical. Certain clues, which may accumulate more or less randomly, lead to the proposal that a certain substance is a transmitter. The known properties of the substance then suggest tests of the likelihood of its being the transmitter. For example, if the substance is known to be synthesised in a particular way in another organ or tissue, the appropriate synthesising enzymes might be sought. If, on the other hand, the substance is common to all or most cells it might not be appropriate to examine its synthesis. Thus, detailed consideration of criteria will be left until the individual transmitter candidates are discussed later in this chapter.

We now come back to a consideration of the central criteria of release and identity of action. In the intestine, the problem of demonstrating release (and indeed of demonstrating storage) can be a very difficult one. Primary reasons for this difficulty are the large number of pharmacologically active substances present in the intestine (Table 1) and our frequent inability to stimulate one nerve type at a time. The nerves of the intestine, and most of their cell bodies of origin, are embedded in its wall[1] (see Chap. 2). The nerve types are closely intermingled, so that each enteric ganglion contains nerve cell bodies of several types and each connective between ganglia contains many different types of axon. Moreover, a number of nerve types is found in the extrinsic connections (e.g., vagus, mesenteric and pelvic nerves; see Chap. 10). The selective electrical stimulation of one nerve type is unlikely ever to be achieved, even though responses which appear to be dominated by one type can be obtained (e.g. FURNESS 1975). Some selectivity of stimulation is achieved when reflexes are activated, but the amount of material released can be considerably less than the amount released by drugs or electrical stimulation. Furthermore, the substances are released from interneurones, from final neurones, and in the case of intrinsic reflexes, from sensory neurones.

If an active substance is collected during nerve stimulation, it must then be decided where it really came from. Of course, this decision is easier if the substance can be identified and its storage sites are known. However, in the intestine things are seldom so simple. Many of the substances now proposed to be transmitters are found in several locations. For example, 5-hydroxytryptamine (5-HT) is found in much greater quantities in enterochromaffin cells in the mucosa and submucosa than in nerves. Furthermore, peptides which have been proposed as intestinal transmitters are also found in endocrine-like cells of the mucosa (Table 1). Moreover, other cell types, for example muscle cells, may release biologically active substances as a consequence of activity induced in them by nerve stimulation.

There are a number of problems in studying the ability of a proposed transmitter to mimic a nerve-mediated response and of drugs to modify the effects of both the endogenous and exogenous agents in parallel. These problems are particularly apparent when the response measured is simply a change in tension in the muscle, but there are also difficulties in experiments where electrical events in nerve or muscle are monitored. Muscle tension can be modified directly by an exogenous substance acting on the muscle cells, or indirectly in a number of ways: by stimulating intrinsic neurones; by facilitating or inhibiting release of transmitter from the nerves; or by releasing other active substances from nerves, muscle, endocrine cells, or perhaps even from hitherto little suspected sources such as glial cells or the interstitial cells of Cajal. The case of 5-hydroxytryptamine might be considered as

1 Those neurones whose cell bodies of origin are embedded in the wall of the intestine are known as *enteric* neurones, whereas neurones whose cell bodies lie outside the intestine but whose processes end in the intestine are known as *extrinsic* neurones of the intestine

Table 1. Endogenous substances which can influence gastrointestinal motility[a]

Substance	Site(s) of storage
Amines	
Acetylcholine	Nerves
Dopamine	Gastric mucosal cells
Histamine	Mast cells, gastric enterochromaffin-like cells
5-Hydroxytryptamine	Enterochromaffin cells, mast cells (in some species) submucosal cells, nerves
Noradrenaline	Nerves
Peptides	
Angiotensin-like	Nerves
Bombesin-like	Mucosal cells, nerves
Cholecystokinin (CCK)	Mucosal cells (primarily duodenal)
CCK subfragments	Nerves
Coherin	Mucosal cells
Enkephalins (Leu- and Met-)	Nerves, mucosal cells
Enterogastrone	Mucosal cells
Gastric inhibitory peptide	Mucosal cells
Gastrin	Antral mucosal cells
Motilin	Mucosal cells
Neurotensin	Mucosal cells, nerves (?)
Pancreatic polypeptide	Nerves
Secretin	Duodenal mucosal cells
Somatostatin	Mucosal cells, nerves
Substance P	Mucosal cells, nerves
Vasoactive intestinal polypeptide (VIP)	Nerves
Amino Acids	
γ-Aminobutyric acid	Nerves (?)
Nucleotides	
Adenosine triphosphate and related nucleotides	Nerves, mast cells, mucosal cells, muscle and other cell types
Lipids	
Darmstoff	Unknown
Prostaglandins	Formed in muscle and other cells
Electrolytes	
Inorganic phosphate	Released from active cells
Potassium	

[a] This a list of known substances which may be released when intestinal nerves are stimulated or could be included in extracts from the gut. There are significant regional and species differences in distribution which are not dealt with in the table

an example (see Table 5 in Sect. E.I). This substance has a direct stimulating action on intestinal smooth muscle, although its effectiveness varies between species and between areas of the intestine (GADDUM and PICARELLI 1957; BROWNLEE and JOHNSON 1963; DANIEL 1968; DRANKONTIDES and GERSHON 1968; COSTA and FURNESS 1979a). It also stimulates both cholinergic nerves (GADDUM and PICARELLI 1957) and enteric inhibitory nerves (BÜLBRING and GERSHON 1967). Furthermore, it reduces the output of acetylcholine from cholinergic nerves making connections in the myenteric plexus (NORTH et al. 1980). From most of the vast number of experiments in which this drug has been used in the intestine, only cautious statements can be made about its site of action. And experiments with exogenously applied agents only indicate what the agent is capable of doing; they do not decide for us how the substance might act if released from axons within the organ. It is conceivable that sites of action of transmitters released from nerve endings might be so sequestered in the tissue that they are not readily available to exogenous compounds. Alternatively, exogenous compounds may have such strong actions at spurious sites that actions at neuroeffector junctions are overridden.

There is an assumption made in the following sections that transmitter types to not vary significantly from one area of intestine to another or from one species of mammal to another. For example, it is assumed that the sympathetic postganglionic nerves to the stomach and the colon, whether in the monkey or in the mouse, all release noradrenaline, even though this has not been demonstrated for each individual case. Likewise, it is assumed that a group of functionally identified nerves, such as the enteric inhibitory nerves would, even if the actual substance is in dispute, release the same transmitter in all mammals. Therefore it has been deemed reasonable to draw together data from different areas of intestine and different species in trying to assemble the evidence for each of the proposed transmitter types. This assumption may not be precisely correct in the case of peptides in that slight variations in amino acid sequence between species, as is the case for peptide hormones, are likely to occur. However, it is assumed that the biologically active portions are virtually identical. It should be kept in mind that there is considerable variation in dietary habits and digestive physiology between species and that there might exist considerable differences in the prominence of different nerve types and in the details of their projections within the gut wall.

B. Evidence that a Number of Different Neurotransmitters are Released in the Gastrointestinal Tract

Two substances, acetylcholine and noradrenaline, have been established as neurotransmitters in peripheral nerves (see Sect. C and D). Altough, with insight from very recent discoveries, early reports can be reinterpreted to show evidence for other gastrointestinal transmitter substances, it was not until the early 1960s that unequivocal data began to accumulate. Such data first came from studies of transmission from inhibitory nerves to gastrointestinal smooth muscle (BURNSTOCK et al. 1964, 1966; HOLMAN and HUGHES 1965; MARTINSON 1965; for reviews see CAMPBELL and BURNSTOCK 1968; CAMPBELL 1970; BURNSTOCK 1972; FURNESS and COSTA 1973; Chap. 10).

Although this early work showed quite clearly that these enteric inhibitory nerves release neither acetylcholine nor noradrenaline, the actual transmitter is still not known (see Sect. H). It was shown that transmission from the nerves was very little influenced by drugs which antagonise the release or actions of noradrenaline or acetylcholine. Furthermore, acetylcholine contracts rather than relaxes intestinal muscle. In some areas, notably on sphincter muscle, noradrenaline also has the opposite effect to the inhibitory nerves. Moreover, transmission from these enteric inhibitory nerves is normal after noradrenaline-containing nerves have degenerated following surgical or chemical procedures to destroy them.

Table 2. Examples of the sensitivity of intestinal muscle and nerves to different agents[a]

Substance	Area of test	Threshold (M)	Reference
Acetylcholine	Rat stomach strip	6×10^{-10}	VANE (1957)
	Rabbit small intestine	3×10^{-10}	OZAKI (1979)
Adrenaline	Rabbit stomach muscle	10^{-8}	FURCHGOTT (1959)
	Human stomach muscle	2.5×10^{-8}	GRAHAM (1949)
Angiotensin	Rat colon	10^{-10}	REGOLI and VANE (1966)
Bradykinin	Cat jejunum	10^{-10}	FERRIERA and VANE (1967)
Histamine	Guinea-pig ileum	10^{-8}	GADDUM (1953a)
5-Hydroxytryptamine	Rat stomach strip	6×10^{-11}	VANE (1957)
Neurotensin	Guinea-pig myenteric neurones	10^{-10}	WILLIAMS et al. (1979)
Noradrenaline	Rabbit stomach	10^{-8}	FURCHGOTT (1959)
	Rat stomach strip	5×10^{-9}	VANE (1964)
Substance P	Guinea-pig ileum	$<10^{-10}$	FRANCO et al. (1979a)
Somatostatin	Guinea-pig ileum	2×10^{-10}	FURNESS and COSTA (1979a)
Vasoactive intestinal polypeptide	Various	$<5\times10^{-9}$	Table 6

[a] These are some of the maximum sensitivities reported – there are in fact considerable differences in the potency of many of these compounds in different species and areas of gut

Evidence for the release of excitatory substances other than acetylcholine from intestinal nerves is convincing, but nevertheless it provides a dilemma: these substances are clearly released from enteric nerves and act on the smooth muscle or on neurones under experimental conditions, but are they released in sufficient concentrations to reach and have actions on these apparent effectors under physiological conditions? We do not know. In most cases, the release of these substances has been assumed because residual responses to nerve stimulation have been detected after the actions of cholinergic nerves have presumably been blocked by excessive concentrations of appropriate antagonists. Resistance to antagonists of acetylcholine action does not demonstrate, by itself, that transmission is not cholinergic (see AMBACHE 1955). However, in cases discussed later in this section, the effects of stimulating the nerves were modified by drugs which do not have comparable effects on cholinergic transmission. It is thus on two pharmacological bases – resistance to antagonists of acetylcholine and unique effects of other drugs – that release from noncholinergic excitatory nerves in the intestine is characterised.

It should be emphasised that to stimulate nerves electrically, even with a single shock, in an isolated tissue maintained without vascular perfusion is far removed from physiological conditions. Such experiments can indeed reveal valuable information, but they can also exhibit phenomena which might never occur without the intervention of the experimenter. It happens that the muscle cells of the gastrointestinal tract, particularly the longitudinal muscle, as well as the nerves, are extremely sensitive to a number of substances (Table 2). It would not be surprising if a synchronous volley in a large number of axons could release sufficient transmitter to diffuse to a nonphysiological target and have an action there, particularly in vitro when normal pathways for inactivation, for example by being swept up in the circulation or by circulating enzymes, might not operate. AMBACHE and his colleagues, who produced some of the first clear evidence for the release of excitatory substances other than acetylcholine from intestinal nerves (AMBACHE and FREEMAN 1968; AMBACHE et al. 1970), realised this and in fact proposed that the substances they showed to affect the longitudinal muscle were released within the myenteric plexus and probably have their physiological sites of action within the plexus.

AMBACHE et al. (1970) concluded that two unidentified spasmogens were released from nerves in the ileum during transmural stimulation. This they based on the observation that responses obtained with stimulation of 5 and 50 Hz could be distinguished pharmacologically. The response at 50 Hz was antagonised by histamine after H_1 receptors had been blocked by mepyramine whereas the response at 5 Hz was unaffected by histamine and was enhanced by diphenhydramine. Later experiments show that responses at low frequencies and a significant part of those at 50 Hz are due to the release of substance P, or a similar compound (see Sect. J.I).

Nerve-mediated contractions which are not blocked by hyoscine or atropine have been reported in the cat, dog, guinea-pig and human colon (FÜLGRAFF et al. 1964; BENNETT and FLESCHLER 1969; GOLDENBERG and BURNS 1968, 1971; COSTA and FURNESS 1972, 1976; FURNESS and COSTA 1973; ROSTAD 1973; STOCKLEY and BENNETT 1974; FASTH et al. 1980). Noncholinergic responses to transmural stimulation in the proximal colon (COSTA and FURNESS 1972; FURNESS and COSTA 1973) and noncholinergic contractions which are reflexly evoked in the distal colon of the guinea-pig (COSTA and FURNESS 1976) are blocked by methysergide and by exposure of the intestine to 5-HT. Because 5-HT causes a rapid and profound tachyphylaxis of its own response, these observations suggest that the substance released might be 5-HT, or at least act through the same receptors as 5-HT (COSTA and FURNESS 1979 b). However, other interpretations of these results should also be considered (see Sect. E). The contractions recorded in the large intestines of other species are not affected by 5-HT or its antagonists and seem to be elicited by a different compound (FÜLGRAFF et al. 1964; GOLDENBERG and BURNS 1971; STOCKLEY and BENNETT 1974).

A further problem that arises in considering atropine-resistant excitatory responses is the propensity of intestinal muscle to contract after the action of an inhibitory agent is ended. This phenomenon, usually referred to as rebound excitation, has been investigated by a number of authors (BENNETT 1966; CAMPBELL 1966; FURNESS 1971) and in some cases may explain atropine-resistant contractions which have been reported (see CAMPBELL and BURNSTOCK 1968).

Both intracellular and extracellular recordings from enteric neurones indicate that there are excitatory and inhibitory inputs to these neurones involving as yet unidentified neurotransmitters. Slow excitatory postsynaptic potentials (e.p.s.p.) associated with bursts of action potentials are recorded from some neurones when connectives in the myenteric plexus are stimulated (DINGLEDINE and GOLDSTEIN 1976; KATAYAMA and NORTH 1978; WOOD and MAYER 1979 a, b). The excitation is not seen in calcium-free solutions and is therefore likely to be due to the release of a neurotransmitter. The slow e.p.s.p. are associated with a transient increase in membrane resistance, probably due to a decrease in potassium conductance. They are not affected by hexamethonium, atropine, curare or physostigmine. The suggestion that the excitatory transmitter is 5-HT is discussed in Sect. E.

Inhibitory postsynaptic potentials have been recorded from neurones in the submucous plexus and some experimental evidence suggests that the transmitter could be dopamine or a pharmacologically similar substance (HIRST and MCKIRDY 1975; HIRST and SILINSKY 1975; and see Sect. F). Similar slow i.p.s.p. have been recorded from myenteric neurones, but their pharmacological properties have yet to be explored (WOOD and MAYER 1978b).

Also in the intestine are enteric vasodilator nerves which release neither acetylcholine nor noradrenaline (HULTEN et al. 1969; BIBER 1973; FASTH et al. 1977b, 1980). These nerves are activated as part of a reflex vasodilation which occurs when the muscosa is irritated and can be also triggered by stimulation of pelvic nerves. There is evidence that these nerves release VIP (FAHRENKRUG et al. 1978b; Sect. I.IV). There is also evidence that 5-HT and plasma kinins might contribute to the vasodilator reflex (BIBER et al. 1974; FASTH et al. 1977a).

C. Identification of Acetylcholine as an Intestinal Transmitter

The assumption that acetylcholine is a transmitter released by excitatory nerves to intestinal smooth muscle, by nerve fibres in the vagus and pelvic nerves which form excitatory connections with neurones in the enteric plexuses, and by excitatory interneurones in the plexuses has been accepted without question in almost all modern investigations of intestinal nerves. This acceptance is based on a vast pharmacological literature which shows the parallel actions of drugs on the effects of nerve stimulation and of acetylcholine, on the demonstration of acetylcholine and its synthesising enzyme, choline acetyltransferase (EC 2.3.1.6), in the intestine and on the demonstration of acetylcholine release in response to nerve stimulation.

Pharmacological evidence for an action of cholinergic nerves on the muscle includes the many observations that stimulation of extrinsic or intrinsic nerves causes contractions which, like those caused by acetylcholine, are blocked by antagonists acting on muscarinic receptors and are enhanced by anticholinesterases (LANGLEY 1898; HARRISON and MCSWINEY 1936; AMBACHE 1955; PATON and VANE 1963). Evidence for cholinergic axons impinging on other neurones in the enteric plexuses rests almost entirely on pharmacological studies which show blocking effects of drugs acting at nicotinic receptors (KOSTERLITZ and LEES 1964). At a cellular level, fast e.p.s.p. recorded in enteric neurones with intracellular microelectrodes are blocked by nicotinic antagonists such as hexamethonium and tubocurarine (NISHI and NORTH 1973a; HIRST et al. 1974). However, the concentrations used in these experiments were quite high ($1–4 \times 10^{-4}$ g/ml) and at these concentrations it is uncertain whether the drugs would act specifically. In *Aplysia* neurones fast excitatory responses, carried by a sodium current increase, in response to the close iontophoretic application of acetylcholine, dopamine or histamine are all blocked by tubocurarine (10^{-4} *M*; CARPENTER et al. 1977). The authors suggested that tubocurarine acts on the sodium conductance channels and that responses evoked through receptors linked to these channels will be blocked by tubocurarine independent of the agonist–receptor interaction. It must therefore be considered that although all fast e.p.s.p. in the myenteric plexus are blocked by tubocurarine they might not all be caused by acetylcholine.

Table 3. Gastrointestinal concentrations of acetylcholine

Species	Tissue	Acetyl-choline (μg/g)	Reference
Rat	Duodenum	3.4	TUCEK and KOUDELKOVA (1966)
	Ileum	6.4	TUCEK and KOUDELKOVA (1966)
Guinea-pig	Duodenum	7.9	TUCEK and KOUDELKOVA (1966)
	Ileum	6–10	FELDBERG and LIN (1950)
	Ileum	7.1	TUCEK and KOUDELKOVA (1966)
	Ileum, longitudinal muscle and myenteric plexus	27.8	PATON and ZAR (1968)
	Ileum, longitudinal muscle and myenteric plexus	14.3	HUTCHINSON et al. (1976)
	Ileum, plexus-free longitudinal muscle	0.4	PATON and ZAR (1968)
Rabbit	Distal colon	8.8	BEANI et al. (1969)
	Stomach	0.9, 1.1	CHANG and GADDUM (1933)
	Small intestine	2.3, 3.2	CHANG and GADDUM (1933)
	Small intestine	2.3–2.6	FELDBERG and LIN (1950)
Dog	Stomach	0.6	CHANG and GADDUM (1933)
	Small intestine	1.4–3.0	FELDBERG and LIN (1950)
	Small intestine, muscle layers	1.5	CHANG and GADDUM (1933)
	Small intestine, mucosa	1.4	CHANG and GADDUM (1933)
	Large intestine	1.6	CHANG and GADDUM (1933)
Cat	Oesophagus	0.3	CHANG and GADDUM (1933)
	Small intestine	1.2–3.0	FELDBERG and LIN (1950)
Monkey	Jejunum	5.2	FELDBERG and LIN (1950)
	Ileum	11.0	FELDBERG and LIN (1950)
Horse	Small intestine	1.5, 1.7	CHANG and GADDUM (1933)

Choline acetyltransferase is present in the intestine (DIKSHIT 1938; FELDBERG and LIN 1950; CHUJYO 1953; TUCEK and KONDELKOVA 1966; BEANI et al. 1969; FILOGAMO and MARCHISIO 1970; SZERB 1975, 1976). DIKSHIT's (1938) original studies, and the more recent work of FILOGAMO and MARCHISIO (1970) and SZERB (1975), indicate that the myenteric plexus is a major site of synthesis of acetylcholine. The rate of synthesis in preparations of longitudinal muscle with attached myenteric plexus is 10–15 times that in longitudinal muscle alone (DIKSHIT 1938; SZERB 1975). Furthermore, when the myenteric plexus and hence nerve endings in the external muscle are destroyed selectively by a freezing technique, choline acetyltransferase can no longer be detected in the external muscle layers (FILOGAMO and MARCHISIO 1970). In the developing foetus, the enzyme is absent before, but present after, the development of the plexus (DIKSHIT, 1938). That the synthesis occurs in nerves of the plexus is indicated by experiments which show that acetylcholine ^{3}H synthesised from choline ^{3}H is released by nerve stimulation (SZERB, 1976). However, it is likely that choline acetyltransferase in the intestine is not confined to nerves. FELDBERG and LIN (1950) found that the distribution of enzyme activity does not correspond to the distribution of nerves in a predictable way. Particularly high rates of synthesis were found in the mucosa. The cells responsible for this synthesis have yet to be identified. More recent evidence suggests that choline acetyltransferase may also be associated with smooth muscle cells in the intestine (MAJCEN and BRZIN 1979).

Release of acetylcholine from gastrointestinal nerves was first demonstrated by DALE and FELDBERG (1934) who detected its presence in gastric perfusates following stimulation of the vagus nerve and since that time a number of studies have shown the release of acetylcholine in response to stimulation of intestinal nerves (BACQ and GOFFART 1939; GREEFF and HOLTZ 1956; PATON 1957; HARRY 1962; PATON and ZAR 1968; BEANI et al. 1969; PATON et al. 1971; SZERB 1975, 1976). The release is blocked or substantially reduced by tetrodotoxin, by botulinus toxin and by removing calcium ions from the medium bathing the tissue.

Some of the many measurements of acetylcholine content of tissue from the gastrointestinal tract are listed in Table 3. Acetylcholine is present in higher concentrations than any of the other substances which are possible gastrointestinal transmitters.

D. Noradrenaline as an Intestinal Transmitter

The similar effects that stimulating sympathetic pathways to the intestine and adrenaline have on gastrointestinal movements were noted in the early twentieth century and it was generally assumed that adrenaline or a similar substance was the transmitter from postganglionic sympathetic nerves (ELLIOTT 1904, 1905; STARLING 1906; CANNON and ROSENBLUETH 1937; DALE 1938). Actual release of an adrenaline-like substance was demonstrated by FINKLEMAN (1930) who found that stimulation of the mesenteric nerves supplying an isolated segment of rabbit duodenum liberated a substance which relaxed a second, unstimulated, piece of intestine. The inhibition caused by nerve stimulation was mimicked by adrenaline and relaxations, whether caused by stimulation or by adrenaline, were antagonised by ephedrine or by desensitisation of the muscle to adrenaline. ASTRÖM (1949) made similar comparisons of the effects of stimulating mesenteric nerves and of the actions of sympathomimetic amines and, on the basis of this work and the observation that the mesenteric nerves contained 12–17 times more noradrenaline than adrenaline, conluded that the transmitter was noradrenaline. Assays of intestinal perfusates following stimulation of mesenteric nerves were made soon after this (MANN and WEST 1951; MIRKIN and BONNYCASTLE 1954). Both noradrenaline and adrenaline were detected in the perfusate, with noradrenaline accounting for some 75%–95% of total catecholamines.

Significant concentrations of noradrenaline have been detected in assays of various areas of the gastrointestinal tract (Table 4) and histochemical studies show that the noradrenaline is contained in axons which, in almost all areas of the gastrointestinal tract, arise from cell bodies in extrinsic (prevertebral) ganglia (see FURNESS and COSTA 1974; and Chap. 2). Thus when nerves reaching the intestine are cut and allowed to degenerate the stores of noradrenaline normally detectable in the intestine disappear (FURNESS and COSTA 1971 a, 1974; JUORIO and GABELLA

Table 4. Gastrointestinal concentrations of noradrenaline and adrenaline[a]

Species	Tissue	Nor-adren-aline (μg/g)	Adren-aline (μg/g)	Reference
Rat	Distal oesophagus	0.251		TAUBIN et al. (1972)
	Nonglandular (proximal) stomach	0.261		TAUBIN et al. (1972)
	Glandular stomach	0.226		TAUBIN et al. (1972)
	Stomach (whole)	0.360		BRODIE et al. (1964)
	Stomach (antrum)	0.42		KLINGMAN et al. (1964)
	Duodenum	0.21		KARKI and PAASONEN (1959)
	Duodenum	0.22		GÖRÖG and SZPORNY (1961)
	Duodenum	0.296		FARRANT et al. (1964)
	Duodenum	0.61		KLINGMAN et al. (1964)
	Duodenum	0.326		TAUBIN et al. (1972)
	Jejunum	0.47		KLINGMAN et al. (1964)
	Jejunum	0.241		TAUBIN et al. (1972)
	Ileum	0.35		BRODIE et al. (1964)
	Ileum	0.37		COLLINS and WEST (1968)
	Ileum	0.248		TAUBIN et al. (1972)
	Proximal ileum	0.37		KLINGMAN et al. (1964)
	Distal ileum	0.28		KLINGMAN et al. (1964)
	Caecum	0.40		KLINGMAN et al. (1964)
	Caecum	0.98	<0.1	STRANDBERG et al. (1966)
	Colon	0.340		TAUBIN et al. (1972)
Guinea-pig	Longitudinal muscle and myenteric plexus			JUORIO and GABELLA (1974)
	Stomach	0.85		
	Duodenum	0.75		
	Ileum	0.56		
	Proximal colon	1.33		
	Distal colon	0.91		
	Taenia coli	0.82		
	Circular muscle, submucosa and mucosa			
	Stomach	0.35		
	Duodenum	0.52		
	Ileum	0.53		
	Proximal colon	0.56		
	Distal colon	0.72		
	Ileum longitudinal muscle and myenteric plexus	0.5–0.9	0	GOVIER et al. (1969)
Rabbit	Gastric mucosa	0.19		SHORE (1959)
	Ileum	0.30		SHORE (1959)
	Ileum	0.41		COLLINS and WEST (1968)
Dog	Stomach			
	Muscle, lesser curvature	0.053	0.029	KAZAROVA and ESAYAN (1964)
	Mucosa, lesser curvature	0.213	0.037	
	Muscle, greater curvature	0.097	0.020	
	Mucosa, greater curvature	0.281	0.021	
	Jejunum	0.003		SCHUMANN (1959)
	Jejunum, muscle layers	0		SCHUMANN (1959)
	Jejunum, mucosa	0.003		SCHUMANN (1959)
	Colon	0.02		SCHUMANN (1959)

Table 4 (continued)

Species	Tissue	Nor-adren-aline (μg/g)	Adren-aline (μg/g)	Reference
Cattle	Duodenum	0.1		BERTLER et al. (1959)
	Jejunum	0.04		SCHUMANN (1959)
	Jejunum, muscle layers	0.06		SCHUMANN (1959)
	Jejunum, mucosa	0.05		SCHUMANN (1959)
Sheep	Jejunum	0.01		SCHUMANN (1959)
	Jejunum, muscle layers	0.003		
	Jejunum, mucosa	0.007		
	Colon	0.004		
Goat	Third stomach	0.1		BERTLER et al. (1959)
	Duodenum	0.2, 0.4		
	Jejunum	0.2		
	Colon	0.2		

[a] This table is an extension of that published by HOLZBAUER and SHARMAN (1972)

1974). The noradrenergic axons in the intestine give their most prolific supply to the myenteric and submucous ganglia, to arterioles in nonsphincteric regions, and to the circular smooth muscle of the sphincters. There are sparser supplies to non-sphincteric muscle and to the mucosa.

The synthesising enzymes for noradrenaline, tyrosine hydroxylase (EC 1.14.16.2) and dopamine β-hydroxylase (EC 1.14.17.1) are both present in the intestine (FURNESS et al. 1979). Tyrosine hydroxylase activity was measured by a radioenzymatic method in the different layers of the small intestine and found to be concentrated in the myenteric plexus and submucosa, those areas where noradrenergic nerves are most prevalent. Dopamine β-hydroxylase was localised immunohistochemically in nerves having the same distribution as those giving a histochemical reaction for noradrenaline. After extrinsic denervation, tyrosine hydroxylase activity fell to undetectable levels and neither dopamine β-hydroxylase nor noradrenaline could be detected histochemically. It should perhaps be pointed out that, although intrinsic noradrenergic neurones are absent from most parts of the gut, there are exceptions, notably the proximal colon of the guinea-pig where noradrenaline-containing cell bodies are found (FURNESS and COSTA 1974; see Chap. 2).

Drugs which interfere with transmission from noradrenergic nerves have predictable actions on the responses to stimulating sympathetic pathways to the intestine. Responses are abolished by pretreating animals with reserpine (which depletes intestinal stores of noradrenaline: COSTA and FURNESS 1971; JUORIO and GABELLA 1974), and the inhibitory actions of nerve stimulation can be restored if noradrenaline or its precursors (dopa or dopamine) are supplied for a period before stimulation is again attempted (GILLESPIE and MACKENNA 1961). Both α- and β-receptors are often involved in the mediation of responses to nerve stimulation or to sympathomimetic amines in the gastrointestinal tract (FURNESS and BURNSTOCK

Table 5. Actions of exogenous 5-Hydroxytryptamine in the gastrointestinal tract

Action	References
Direct stimulation of smooth muscle	GADDUM and PICARELLI (1957) VANE (1957)
Stimulation of cholinergic nerves supplying muscle	BROWNLEE and JOHNSON (1963)
Stimulation of noncholinergic excitatory nerves	COSTA and FURNESS (1979a)
Stimulation of enteric inhibitory nerves	BULBRING and GERSHON (1967) COSTA and FURNESS (1979a)
Depolarisation of unidentified neurones	HIRST and SILINSKY (1975) WOOD and MAYER (1979b)
Hyperpolarisation of unidentified neurones	JOHNSON et al. (1980)
Antagonism of release of acetylcholine at synapses in the myenteric plexus	NORTH et al. (1980)
Enhancement of peristaltic reflex by mucosal application	BULBRING and CREMA (1958)
Inhibition of peristaltic reflex by serosal application	KOSTERLITZ and ROBINSON (1957)
Vasodilation and vasoconstriction	BIBER et al. (1973b)
Inhibition of gastric secretion	HÅKANSON et al. (1967)
Enhanced water and electrolyte secretion in the small intestine	KISLOFF and MOORE (1976) DONOWITZ et al. (1977)

1975). If a combination of antagonists acting on these receptors, or drugs such as guanethidine or bretylium, which block noradrenaline release, are used then the relaxations caused by the stimulation of sympathetic nerves are substantially reduced or abolished (BOURA and GREEN 1959; BOYD et al. 1962; BOWMAN and HALL 1970; WESTON 1973b).

Studies with intracellular microelectrodes show that both exogenous noradrenaline and stimulation of sympathetic nerves cause a presynaptic inhibition of excitatory transmission within the myenteric plexus (NISHI and NORTH 1973b; HIRST and MCKIRDY 1974a). These observations are consistent with previous work which showed that both sympathetic nerves and noradrenaline act on enteric nerves to decrease the output of acetylcholine and hence inhibit intestinal movements. In both cases, antagonists acting at α-receptors reduce the extent of inhibition (FURNESS and COSTA 1974; FURNESS and BURNSTOCK 1975).

E. Evidence For and Against 5-Hydroxytrypamine

I. Pharmacology and Sites of Action of 5-HT

A substantial problem in considering 5-HT as a potential neurotransmitter in the gastrointestinal tract is its multiplicity of actions (Table 5). Hence there is considerable uncertainty attached to experiments in which attempts are made to mimic its proposed action as a neurotransmitter and thus satisfy one of the fundamental criteria for transmitter identification (Sect. A). This leads to the unusual situation that many publications which have dealt with the possibility that it is a transmitter have ignored discussion of the physiological role it might play.

GADDUM and PICARELLI (1957) discovered that the contraction of the longitudinal muscle of the guinea-pig ileum caused by 5-HT could be divided into two components, one blocked by morphine and the other by dibenamine. They proposed that two receptor types are present: the M (morphine-sensitive) and D (dibenamine-sensitive) receptors. It is now realised that the two components result from actions of 5-HT at two different sites, on the cholinergic neurones and on the muscle. Morphine blocks the contraction caused by stimulation of cholinergic nerves by antagonising the release of acetylcholine (see Sect. J.V), not by acting on receptors for 5-HT. That is, there are no M receptors where 5-HT and morphine compete. Furthermore, dibenamine is not a drug which acts specifically at receptors for 5-HT and has been replaced as a blocking agent by other drugs. So the designation of M and D receptors is no longer useful. Most authors now refer to muscle and nerve receptors for 5-HT, as is done in this chapter, with the explicit warning that there are receptors on a number of nerve types as well as on nerve endings (Table 5) and it is not known whether these different groups of receptors have comparable pharmacological properties, almost no systematic work having been done in this area. In molluscs, six distinct receptor mechanisms for the action of 5-HT on neurones, with different susceptibilities to antagonist drugs, have been discovered (GERSCHENFELD and PAUPARDIN-TRITSCH 1974). It is conceivable that similar complexity will be revealed when responses of enteric neurones are examined in detail. The unsatisfactory situation at the moment is that no drugs are known to block the actions of 5-HT on enteric neurones selectively. On the other hand, there appears to be but one receptor type on intestinal muscle, on which methysergide is a sufficiently discriminating antagonist to be experimentally useful (VANE 1959; GERSHON 1977; COSTA and FURNESS 1979b). It bears emphasising that this drug has a spectrum of effects, which become more apparent as its concentration is raised and, at concentrations which are effective at neural receptors for 5-HT, it is comparably effective in antagonising the stimulant actions of carbachol, dimethylphenylpiperazinium and histamine on the intestine (M. COSTA and J. B. FURNESS unpublished work 1979). It has been suggested that attempts to use this drug in analysis of 5-HT actions on intestinal nerves might profitably be abandoned (NORTH et al. 1980).

It might be thought that the well-known tachyphylaxis caused by 5-HT could be exploited to examine whether a nerve-mediated response could be due to the release of 5-HT. This is based on the following assumption: if after a period during which 5-HT is in contact with the tissue, 5-HT no longer produces an effect and a nerve-mediated action mimicked by 5-HT is also blocked, then the neurotransmitter and 5-HT act through the same receptors. An obvious problem here is the large number of compounds which are also agonists at receptors responsive to 5-HT (BARLOW and KAHN 1959; VANE 1959; INNES and KOHLI 1969) and thus could be prospective transmitter candidates. Not so obvious is the possibility that 5-HT may antagonise the release of unknown transmitters in the intestine as it does for acetylcholine at neuroneuronal synapses (NORTH et al. 1980). Furthermore, antagonists acting at receptors for 5-HT, perhaps particularly if they are partial agonists, could mediate their effects by acting prejunctionally. These points are well illustrated by considering the hyoscine-resistant reflex contractions of circular muscle which are elicited by distension of the guinea-pig colon (COSTA and FUR-

NESS 1976). These contractions are blocked by tetrodotoxin but not by hyoscine, even in concentrations up to 10^{-4} g/ml. They are therefore due to the release of an excitatory substance other than acetylcholine which contracts the circular muscle. The contractions are blocked both by methysergide and by exposing the tissue to 5-HT and so it was originally suggested that 5-HT or a compound acting through the same receptors as 5-HT was responsible (COSTA and FURNESS 1976). However, it was later discovered that the circular muscle of the colon is almost completely insensitive to 5-HT, virtually excluding this substance as the transmitter (COSTA and FURNESS 1979 a). It is still not understood how 5-HT blocks these non-cholinergic excitatory responses.

II. The Presence of 5-HT Neurones in the Intestine

The amine-handling properties of certain neurones of the enteric nervous system are consistent with the idea that they synthesise 5-HT. These neurones actively take up indoleamines and, with less avidity, catecholamines; they contain the synthesising enzymes aromatic(-)-amino acid decarboxylase (AADC, EC 4.1.1.26) and tryptophan hydroxylase (TPH,EC 1.14.16.4), and the degradative enzyme monoamine oxidase (MAO, EC 1.4.3.4). It is only an assumption that it is one class of neurones which share these properties in common. The distributions of neurones which take up aromatic amines and those which contain AADC seem to be the same (FURNESS and COSTA 1978), but there is not direct evidence that all AADC-containing neurones also contain TPH or that all TPH neurones contain AADC. Histochemical studies indicate that MAO is contained in many more neurones than can be demonstrated to take up and synthesise amines (compare FURNESS and COSTA 1971 b, 1978).

The normal intestine contains noradrenergic axons which are, in most areas, the processes of nerve cells in sympathetic, prevertebral, ganglia. However, it is relatively simple to eliminate these noradrenergic axons (whose presence could compromise studies of the enteric amine-handling neurones) by surgical denervation, by drugs such as 6-hydroxydopamine, by examining the foetal intestine before the ingrowth of noradrenergic axons, or by using isolated intestine maintained in tissue culture long enough for extrinsic nerves to degenerate.

Uptake of 5-HT into intrinsic neurones of the intestine was shown by ROBINSON and GERSHON (1971) who used a histochemical technique to examine the intestine from guinea-pigs which had been injected with 5-HT and a MAO inhibitor after depletion of noradrenaline. High affinity uptake of 5-HT by preparations of longitudinal muscle and myenteric plexus has been reported (GERSHON and ALTMAN 1971; GERSHON et al. 1976). The uptake was not competitively inhibited by noradrenaline, and noradrenaline and 5-HT uptake were differently affected by drugs, chlorimipramine being the most effective against 5-HT and desmethylimipramine showing selectivity for inhibition of the noradrenaline uptake system. Axons which can take up 5-HT are found in the foetal intestine before axons which concentrate noradrenaline become apparent (ROTHMAN et al. 1976). Further studies, using sections of intestine grown in tissue culture, confirmed that the elements which take up 5-HT are intrinsic to the intestine and, by electron microscope autoradiography, these elements were shown to be neuronal (DREYFUS et al. 1977 b; JONAKAIT et al. 1979). The close analogue of 5-HT, 6-hydroxytryptamine, has been shown histochemically to be taken up and retained by intrinsic nerve cell bodies and axons, although these cells and fibres also take up dopamine (COSTA et al. 1976; FURNESS and COSTA 1978).

Certain analogues of 5-HT, 5,6-dihydroxytryptamine (5,6-DHT) and 5,7-DHT, are taken up by neurones which take up 5-HT in the central nervous system and cause these neurones to degenerate (BJÖRKLUND et al. 1974). Electron microscope observations indicate that these drugs cause similar degeneration of intrinsic axons in the enteric nervous system (GERSHON and ROSS 1973; FEHER 1977; GERSHON and SHERMAN 1978).

Two enzymes required for 5-HT synthesis, AADC and TPH, have been demonstrated in intestinal nerves. AADC has been detected in histochemical experiments in which the products of the decarboxylation of 5-hydroxytryptophan or dopa, that is 5-HT or dopamine, were detected in enteric neurones (ROBINSON and GERSHON 1971; COSTA et al. 1976; FURNESS and COSTA 1978). Amine fluorescence was observed in both cell bodies and axons after the injection of the precursor in vivo, or after incubation of segments of intestine with the precursor in vitro. The inhibitor of AADC, benseraside (RO 4-4602), prevented the conversion of the precursor. In order to detect the product, it was necessary to inhibit MAO, implying that this enzyme is also contained in the neurones. The distributions of enteric neurones which contain AADC and MAO and which take up indoleamines and catecholamines in the guinea-pig small intestine seem the same (FURNESS and COSTA 1978). In the submucous plexus, about 10% of nerve cell bodies showed amine-handling properties, compared with only about 0.5% in the myenteric plexus. The axons of these neurones ramify extensively amongst the nerve cell bodies of both the myenteric and submucous plexuses. They also contribute to the fine nerve trunks in the deep muscular plexus, implying that they might act on the circular muscle. A few axons followed submucous arterioles and some were found around the bases of the crypts in the mucosa of the small intestine.

Immunohistochemical studies using antisera raised in rabbits against purified rat brain TPH indicate that this enzyme is present in the cell bodies of intrinsic neurones of the submucous and myenteric plexus of guinea-pigs, mice and rats (GERSHON et al. 1977). Further evidence for the presence of TPH, and AADC in the same neurones, is that a diffuse yellow fluorescence reaction was found throughout the myenteric plexus, possibly in fine axons, in segments of intestine removed after the injection of l-tryptophan and an inhibitor of MAO (DREYFUS et al. 1977a). If the animals were also injected with *p*-chlorophenylalanine, an inhibitor of TPH, the diffuse reaction was suppressed and fluorescence in nerve cell bodies could be seen. The authors suggest that this observation can be attributed to a greater sensitivity of TPH in axons compared with that in cell bodies to inactivation by *p*-chlorophenylalanine. DREYFUS et al. (1977a) have produced biochemical evidence for the presence of TPH. They incubated strips of longitudinal muscle and attached myenteric plexus from the guinea-pig ileum with l-tryptophan ^{3}H. After incubation for 0.5–3 h the tissue was homogenised and radiolabelled material in the extract was separated by thin layer chromatography or high voltage electrophoresis. Radiolabelled material comigrated with tryptophan, 5-HT and 5-hydroxyindolacetic acid. The production of a 5-HT-like compound was antagonised by *p*-chlorophenylalanine. There was also evidence for the production of 5-HT ^{3}H from tryptophan ^{3}H by cultured explants of mouse intestine which contained the myenteric plexus. Other investigators have attempted to make determinations of TPH activity in intestinal nerves by radio-enzymatic assays, but have not been able to detect any TPH (LEES et al. 1979; H.G. BAUMGARTEN and W.M. LOVENBERG, unpublished work 1979). Both these groups measured TPH in the central nervous system in parallel experiments and obtained values similar to those in the literature.

A protein which selectively binds 5-HT has recently been isolated from the intestine (JONAKAIT et al. 1977). Binding of 5-HT to this protein was not inhibited by antagonists of 5-HT uptake into neurones or by receptor-blocking drugs, but it was inhibited by reserpine, which suggests that the protein might be involved in 5-HT storage within nerves. After incubation of segments of intestine with 5-HT ^{3}H, the labelled amine and this binding protein were released together by nerve stimulation (JONAKAIT et al. 1979). In both cases release was prevented by removing calcium.

For a number of years, one of the difficulties faced in the examination of proposed 5-HT neurones in the intestine has been that they are not revealed by histochemical reactions which rely on the formation of a fluorescent product by condensation with aldehydes (see COSTA and FURNESS 1979a). However, neurones containing a 5-HT-like substance, possibly 5-HT itself, have now been revealed by immunohistochemical techniques (COSTA et al. 1980e; and see note added in proof p. 440). These studies indicate that a 5-HT-like antigen is found in nerve cell bodies and axons in the myenteric plexus but, surprisingly, no positive cell bodies were found in the submucosa, where there are cells capable of taking up and retaining indoleamines. The 5-HT-like substance in these neurones was depleted by reserpine or by 5,7-DHT. The absence of an aldehyde-induced fluorescence of 5-HT in enteric neurones, confirmed by many observations (ROBINSON and GERSHON 1971; COSTA and FURNESS 1971; AHLMAN et al. 1973; AHLMAN and ENERBACK 1974; DUBOIS and JACOBOWITZ 1974; FURNESS and COSTA 1974, 1978; DIAB et al. 1976), remains something of a puzzle: it is not clear whether 5-HT is present in such low concentrations or is stored in such a way that it cannot be revealed histochemically.

In assessing measurements of 5-HT concentrations in extracts of intestine one must consider the very high concentrations of 5-HT in the intestinal mucosa which make it necessary to separate the mucosa from the layers containing the nerves. In the dog small intestine, the nonmucosal layers contain 47 ± 5 ng/g whereas the mucosa contains $5{,}400 \pm 700$ ng/g (ABE et al. 1973). In preparations of longitudinal muscle, with myenteric plexus attached, from the guinea-pig ileum, concentrations of 5-HT from 80 to 110 ng/g wet weight are found (ROBINSON and GERSHON 1971; JUORIO and GABELLA 1974). However, 6,350–8,900 ng/g are found in the remaining layers (GABELLA and JUORIO 1973; JUORIO and GABELLA 1974). Therefore, if there was 1%–2% contamination from these other layers in the myenteric plexus preparations, the observed concentrations could be explained. Contamination could arise because of inadvergent inclusion of small fragments of mucosa or because 5-HT was released during dissection and absorbed by the outer layers of intestine. If the 5-HT in the longitudinal muscle–myenteric plexus was already present in this layer it could have been in one or more of a variety of cell types, for example glial cells, the interstitial cells of Cajal, smooth muscle cells or vascular endothelial cells.

In fact, DIAB et al. (1976) have reported small fluorescent cells in longitudinal muscle–myenteric plexus preparations from the guinea-pig ileum after treatment with formaldehyde vapour. These cells lay outside the ganglia and had excitation and emission maxima (400 and 520 nm) consistent with the fluorophore formed by 5-HT. DIAB et al. (1976) postulated that the cells were mast cells or fibroblasts. Although these cells would have been expected to appear in preparations examined in other studies (e.g. AHLMAN and ENERBACK 1974; FURNESS and COSTA 1978), the observations have yet to be confirmed.

There are a number of parallels between the properties of amine-handling neurones in the intestine and neurones with similar properties in the retina (for review see FLOREN 1979). Like the intestinal neurones, those in the retina do not normally contain any histochemically detectable 5-HT, but they take up and retain 5-HT and related indoleamines, and with less efficiency also take up catecholamines. Like those in the intestine, the neurones degenerate after pretreatment with 5,6- or 5,7-DHT. TPH activity in retinal neurones is extremely low or absent. FLOREN (1979) concludes that the retinal amine-handling neurones do not utilise 5-HT as a transmitter.

III. Release of Endogenous 5-HT, or a Similar Compound, from Intestinal Nerves

Pharmacological studies of responses of intestinal muscle or nerve cells to nerve stimulation have, in a number of cases, suggested that 5-HT or a 5-HT-like compound might be released (BÜLBRING and GERSHON 1967; BIANCHI et al. 1970; SINGH and SINGH 1970; COSTA and FURNESS 1972, 1976; FURNESS and COSTA 1973; DINGELDINE and GOLDSTEIN 1976; RATTAN and GOYAL 1978; WOOD and MAYER 1978a, 1979b).

1. 5-HT and Vagal Inhibitory Pathways

The hypothesis has been advanced that there are two groups of preganglionic neurones, one which releases 5-HT and another which releases acetylcholine, in the vagal pathway which causes gastric relaxation (BÜLBRING and GERSHON 1967). It was found that acetylcholine, 5-HT and vagus nerve stimulation all activated enteric inhibitory nerves to the guinea-pig stomach. The effects of acetylcholine and its analogues were selectively blocked by pentolinium or hexamethonium while the effects of 5-HT were selectively blocked by desensitising the receptors with 5-HT itself. Pentolinium or hexamethonium by themselves, or exposure to 5-HT by itself, only partly blocked the inhibitory effect of vagal stimulation. Complete block was only achieved when antagonists of acetylcholine and exposure to 5-HT were combined. It was also found that transmural stimulation of the isolated stomach of the mouse, after the mucosa had been largely destroyed by asphyxiation, caused the release of a 5-HT-like substance and that this release was prevented by tetrodotoxin (BÜLBRING and GERSHON 1967). BEANI et al. (1971) repeated these experiments using broad (15 mm) strips of stomach supplied by the vagus. The mucosa was removed from these strips. They found that relaxations in response to vagal stimulation were only partly blocked by hexamethonium but, unlike BÜL-

BRING and GERSHON, they found no reduction of the hexamethonium-insensitive component when receptors for 5-HT in the stomach were desensitised.

More recently, RATTAN and GOYAL (1978) have provided evidence for 5-HT as a transmitter in preganglionic vagal axons impinging on enteric inhibitory nerves supplying the lower oesophageal sphincter in the possum. They found that stimulation of the vagus relaxed the sphincter in the combined presence of hexamethonium and atropine. 5-Methoxydimethyltryptamine, which antagonised the action of 5-HT, abolished this relaxation of the sphincter caused by vagal stimulation. However, attempts to detect 5-HT axons in the vagus have not been successful. DREYFUS et al. (1977b) were able to detect noradrenaline fluorescence of axons proximal to a ligature of the vagus in the guinea-pig. However, after the fluorescence of the noradrenergic axons had been abolished by 6-hydroxydopamine or α-methylparatyrosine, no 5-HT-containing axons were revealed, even when monoamine oxidase was inhibited and the animals were injected with tryptophan. Uptake experiments showed that the vagus nerves did not concentrate 5-HT. Furthermore, no tryptophan hydroxylase could be found in the nerves with immunohistochemical methods.

An explanation for the apparently contradictory results of BÜLBRING and GERSHON (1967) and BEANI et al. (1971) and for the apparent inconsistency between the pharmacological results (which imply that 5-HT fibres are in the vagus) and the neurochemical results (which show that there are probably no vagal 5-HT fibres) is suggested by the observation that stimulation of noradrenergic axons running with the vagus releases 5-HT from enterochromaffin cells (see Sect. E.III.5). Perhaps, in BÜLBRING and GERSHON's experiments, 5-HT was released from the enterochromaffin cells and diffused to the inhibitory neurones. In the experiments of BEANI et al., this would not have been possible as the mucosa containing the enterochromaffin cells was removed. The mucosa was present in the experiments of RATTAN and GOYAL (1978).

2. Pharmacological Evidence for Release of Endogenous 5-HT from Intramural Nerves

Contractions elicited by transmural stimulation of circular muscle strips taken from the most distal 1.5 cm of the canine oesophagus were blocked by bromolysergic acid and by exposure to 5-HT (SINGH and SINGH 1970), suggesting that the transmitter released from these nerves acts on the same receptors as 5-HT. The contractions were not modified by either atropine or eserine. These experiments do not appear to have been repeated by other investigators, although a review of the literature confirms that transmission to oesophageal circular muscle is atropine resistant in other species (CHRISTENSEN 1975).

Nicotine might release 5-HT or a similar substance from intestinal nerves (BIANCHI et al. 1970). Nicotine caused relaxations of the guinea-pig distal colon which were blocked by tetrodotoxin. These relaxations were also blocked by desensitising receptors for 5-HT and by metoclopramide. 5-HT had no direct effect on the muscle. It did, however, stimulate inhibitory nerves to the muscle and this effect was blocked by 5-HT desensitisation of its own receptors as well as by metoclopramide. A simple explanation of these results would be that nicotine causes the release of 5-HT which then stimulates enteric inhibitory nerves which relax the muscle. The source of the 5-HT-like substance is still not known.

Transmural stimulation of the proximal colon of the guinea-pig at frequencies of 5–50 Hz causes the release of an excitatory substance which might be 5-HT. Such stimulation in the presence of hyoscine resulted in a relaxation of the longitudinal muscle which was followed, after the end of stimulation, by a rapid and then by a slow contraction (COSTA and FURNESS 1972). The slow contraction is nerve mediated in that it is blocked by tetrodotoxin and was considered to be non-cholinergic because it was unaffected by hyoscine in concentrations up to 10^{-5} g/ml. In the presence of tetrodotoxin, 5-HT causes similar contractions, indicating that it acts directly on the muscle. Methysergide and phentolamine were both effective antagonists of the slow contraction and of the action of 5-HT on the muscle (FURNESS and COSTA 1973). After the muscle was desensitised to the contractile action of 5-HT by sustained exposure, the slow contraction in response to transmural stimulation was abolished, although the excitability of the muscle was not impaired. These experiments suggest that the substance which causes the slow contraction might act through the same receptors as 5-HT. However, 5-HT could also interfere with transmission by a prejunctional inhibition of release (see Sect. E.I).

When recordings of electrical activity in enteric neurones are made with suction electrodes, direct activation of the neurones can be distinguished from synaptic activation by differences in latency, in ability to follow a train of stimuli, and by the susceptibility of synaptic transmission to removal of calcium (DINGLEDINE and GOLDSTEIN 1976). Brief trains of stimuli (40 Hz for 0.5 s) caused a repetitive synaptic activation of neurones lasting 10–30 s. This barrage of activity which followed stimulation was abolished by exposing the tissue to 10^{-5} *M* 5-HT. The repetitive activity was not significantly affected by hexamethonium (10^{-4} *M*), although hexamethonium and curare, another antagonist acting at nicotinic receptors, are effective antagonists of excitatory synaptic transmission in the intestine when the stimuli are single pulses (see Sect. C). The repetitive synaptic activation was examined in preparations devoid of mucosa which therefore did not contain mucosal enterochromaffin cells which could be sources of 5-HT.

More recently, intracellular microelectrodes have been used to examine responses which seem to be essentially the same as those observed by DINGLEDINE and GOLDSTEIN (1976) (WOOD and MAYER 1978a, 1979b; GRAFE et al. 1979). These authors reported that slow e.p.s.p. which can be elicited in enteric neurones by repetetive stimulation of fibre tracts were mimicked by 5-HT and blocked by methysergide (3×10^{-6}–3×10^{-5} *M*) or by exposure to high concentrations of 5-HT (1–2.5×10^{-6} *M*). These e.p.s.p. are associated with an increase in membrane resistance caused by a decrease in potassium conductance. The mimicking action of 5-HT is obviously central to the proposition made by these authors that 5-HT is the transmitter. However 5-HT does not always mimic. In Fig. 2b of the paper by WOOD and MAYER (1978a), the depolarisation elicited by 5-HT was associated with a *decreased* membrane resistance; increased resistance was only apparent as the depolarisation waned. In another study, it was found that 5-HT depolarised 37% and hyperpolarised 33% of the cells in which slow e.p.s.p. were recorded (JOHNSON et al. 1980). The antagonism of the slow e.p.s.p. by 5-HT occurred in all cells, independent of the effect of 5-HT on the neurones.

On the basis of mimicry, substance P and somatostatin are at least as good candidates as 5-HT; both these compounds are present in intestinal nerves and cause depolarisations associated with increased membrane resistance (KATAYAMA and NORTH 1978, 1980). Methysergide needs to be tested further, to see if, at the high concentrations used, it is a selective antagonist of 5-HT receptors on the neurones. This has been done in part; at a concentration of 2×10^{-5} *M* it does not affect the action of substance P (GRAFE et al. 1979). However, methysergide (3×10^{-5} *M*) causes a significant reduction in the input resistance of myenteric neurones (WOOD and MAYER 1979 b). It has not been determined whether the action of 5-HT (or for that matter methysergide) in blocking the slow e.p.s.p. is due to presynaptic or postsynaptic action. 5-HT does cause a presynaptic inhibition of the release of acetylcholine within the myenteric plexus (NORTH et al. 1980) and a similar presynaptic inhibition of release of the transmitter mediating the slow e.p.s.p. would explain the block of this potential, independent of the effects of applied 5-HT on the nerve cell body. R. A. NORTH (personal communication 1980) has found that substance P, which is known to cause a relatively specific desensitisation of its own receptors (see Sect. J.II) also reduces the amplitude of the slow e.p.s.p.

3. Possible Involvement of Nerves Releasing 5-HT in Peristalsis

The pharmacology of the events underlying the peristaltic reflex has been examined by COSTA and FURNESS (1976) using a system in which the contraction which occurs on the oral side of a point of distension and the relaxation occurring on the anal side are recorded separately. The reflex which results in a contraction on the oral side has been called the ascending excitatory reflex. In the guinea-pig colon, it was found that the ascending excitatory reflex was partly blocked by hyoscine and was also partly blocked by methysergide or by exposing the preparation to 5-HT. Complete block was obtained when hyoscine and either methysergide or 5-HT desensitisation were combined. The reflex was also entirely blocked by tetrodotoxin. These results indicate that when the ascending excitatory reflex is activated a stimulant of the smooth muscle, other than acetylcholine, is released and suggest that it could be 5-HT. However, when exogenous 5-HT was applied to the circular muscle in concentrations up to 10^{-4} g/ml it failed to have any direct effect on the muscle in more than 90% of experiments (COSTA and FURNESS 1979 a). When a direct response was observed it was a weak contraction. In the ileum, the ascending excitatory reflex was completely blocked by hyoscine (FURNESS and COSTA 1976; and unpublished work 1976). However, methysergide and exposure to 5-HT also caused a partial antagonism. This would suggest that, in this pathway, the final transmitter which acts on the smooth muscle is acetylcholine, and that another transmitter whose action or release is sensitive to 5-HT acts at a neuroneuronal synapse. The action of methysergide is difficult to explain, in that it is not an antagonist of the action of 5-HT on intestinal neurones (see Sect. E.I). The effect of drugs on the excitatory reflex could be partly explained by assuming that an unidentified transmitter, not necessarily 5-HT, is released at neuroneuronal synapses in the pathway. In the isolated colon, the unknown transmitter is released in sufficient quantitiy to diffuse to the circular muscle which also contains receptors for the transmitter. In the ileum, either the substance is not released in sufficient quantity, or there are not receptors for it in the circular muscle. Antagonism of reflex contractions of the circular muscle by 5-HT has also been reported in investigations using other methods of recording peristalsis (KOSTERLITZ and ROBINSON 1957; BÜLBRING and CREMA 1958; LEMBECK 1958).

If nerves releasing 5-HT were essential for peristalsis, it would be expected that the peristaltic reflex would be inhibited when 5-HT stores were depleted. When reserpine was used to reduce 5-HT to 2% or less of control levels, the reflex was unimpaired (BÜLBRING and CREMA 1959). Similarly, when tryptophan-free diets were used to reduce 5-HT to undetectable levels the peristaltic reflex remained intact (BOULLIN 1964). In both these studies the assays were of mucosal 5-HT and it is not known if 5-HT concentrations in the muscle layers were similarly reduced, although reserpine was found in another study to deplete intestinal nerves of a 5-HT-like antigen (COSTA et al. 1980e).

4. Evidence for 5-HT Release from Nerves Impinging on Intestinal Vasodilator Nerves

Blood flow through the decentralised cat jejunum, in vivo, is increased when the mucosa is irritated (BIBER et al. 1971). This increase was blocked by injecting tetrodotoxin into the arterial supply to the jejunum, indicating that the dilation was nerve mediated. The increase in blood flow was also blocked by the 5-HT receptor antagonist, bromolysergic acid, suggesting that the transmitter might be 5-HT. The responses were not affected by atropine, by blockade of α- or β-receptors for noradrenaline or by blocking nicotinic receptors. Similar vasodilator responses in the cat small intestine in response to electrical stimulation of intramural nerves have been reported (BIBER et al. 1973a). The vasodilator responses to both mucosal irritation and transmural stimulation were blocked by a continuous intra-arterial infusion of 5-HT which was shown to block responses to 5-HT (BIBER et al. 1974). It was found that the close intra-arterial injection of a bolus of 5-HT caused an increase in intestinal blood flow. This increase was blocked and a decrease was observed in the presence of tetrodotoxin, indicating that 5-HT stimulates intramural vasodilator nerves as well as having a direct constrictor action (BIBER et al. 1973b). Transmural stimulation was associated with a release of 5-HT which could be detected in the venous outflow from the intestine. However, it was not determined whether the release was affected by tetrodotoxin. The results so far have failed to identify the cell type from which a 5-HT-like substance which stimulates intrinsic vasodilator nerves is released. The release might be from enterochromaffin cells which would be expected to release 5-HT when irritated or depolarised by electrical stimulation. On the other hand release might be from nerves, in which case the transmitter seems to act through the same receptors as 5-HT.

The vessels of the proximal colon of the cat dilated when the mucosa was irritated or when 5-HT was injected into the superior mesenteric artery (FASTH et al. 1977b). Both effects were blocked by dihydroergotamine. These results suggest that a similar reflex to that in the small intestine also exists for the cat colon. There is now evidence to suggest that the transmitter released from the final neurones in these vasodilator pathways is vasoactive intestinal polypeptide (see Sect. J.IV).

5. Nerve-Mediated Release of 5-HT from Enterochromaffin Cells

In both cats and guinea-pigs, stimulation of the vagus decreases the 5-HT content of enterochromaffin cells in the small intestine (TANSY et al. 1971; AHLMAN et al. 1976). The release appears to be due to the stimulation of sympathetic, noradrenergic nerves which have been shown in fluorescence histochemical studies to run with the vagus (MURYOBAYASHI et al. 1968; BAUMGARTEN and LANGE 1969; LIEDBERG et al. 1973). The response was prevented by destroying noradrenergic nerves by the injection of 6-hydroxydopamine or by injecting antibodies to nerve growth factor and was also prevented by prior removal of the superior cervical ganglia. These experiments illustrate that 5-HT can be released through nerve activity, even though the source of 5-HT is not the nerves themselves.

6. Collection of Assayable 5-HT in Response to Nerve Stimulation

A substance which contracts the rat fundus and whose action on the fundus is blocked by methysergide is released from the mouse stomach in response to stimulation of intramural nerves (BÜLBRING and GERSHON 1967). The collections were made from stomachs in which the mucosa had sloughed off and had been removed following 3 h asphyxiation. Histological examination showed only small remnants of mucosa. These experiments leave several doubts. First, the 5-HT-like substance might have come from enterochromaffin or mast cells. This cannot be excluded, but it seems unlikely because the release was blocked by tetrodotoxin and even in the remnants of mucosa, no argentaffin (enterochromaffin) cells could be detected. Second, 5-HT, liberated from the asphyxiated mucosa, might have been taken up into nerves and subsequently released by nerve stimulation. This is possible because intestinal nerves have been shown to take up 5-HT ^{3}H and release it in response to electrical stimulation, the release being blocked by tetrodotoxin (SCHULZ and CARTWRIGHT 1974; GERSHON and JONAKAIT 1979; JONAKAIT et al. 1979). The active substance which was released was assayed on rat fundic strip. It is known that many substances can act through the same receptors as 5-HT (see Sect. E.I); in order to demonstrate whether or not the substance was 5-HT, chemical identification is required.

Release of endogenous 5-HT into the solution bathing the serosal surface of everted segments of guinea-pig small intestine has been detected (GERSHON and TAMIR 1979). The authors examined diffusion of 5-HT through the tissue and concluded that the source could not have been the mucosa. Release in response to electrical stimulation was calcium dependent.

IV. Conclusions

Within the gastrointestinal tract there is a class of intrinsic nerves which can take up and retain indoleamines, which can synthesise indoleamines utilising the enzymes TPH and AADC, which contain a binding protein for 5-HT, and which can inactivate indoleamines through MAO. 5-HT in these neurones is not revealed by histochemical methods relying on an aldehyde-induced conversion of 5-HT to a fluorescent product. However, recent immunohistochemical studies have revealed a 5-HT-like substance in intestinal nerves.

Pharmacological experiments show that some intestinal neurones release a substance or substances which have excitatory effects on intestinal neurones and/or smooth muscle. Transmission from these nerves can be blocked by a variety of drugs which block receptors for 5-HT; transmission can also be blocked by 5-HT itself. There is doubt as to whether the blocking effects are pre- or postjunctional. If the effects are prejunctional, the observations do not provide any clues to the nature of the transmitters. If the effects are postjunctional, then the experiments provide good evidence for the release of a substance which acts through the same receptors as 5-HT and which might be 5-HT or a related indoleamine.

F. Evidence For and Against Dopamine

Slow i.p.s.p. were recorded from 30% to 40% of neurones in the submucous plexus of the guinea-pig ileum when interganglionic fibre tracts were stimulated (HIRST and MCKIRDY 1975). A similar frequency of observation was made after extrinsic denervation, indicating that the inputs were not noradrenergic. The i.p.s.p. were not affected by tubocurarine, atropine or bicuculline but were blocked by guanethidine (10^{-5} g/ml), bromolysergic acid ($1–10 \times 10^{-7}$ g/ml) and methysergide (10^{-5} g/ml; HIRST and MCKIRDY 1975; HIRST and SILINSKY 1975). In these cell 5-HT caused a depolarisation, but dopamine and noradrenaline produced hyperpolarisations. The hyperpolarisations caused by the i.p.s.p. and those caused by dopamine or noradrenaline were all associated with an increased membrane conductance and had similar reversal potentials. Both were blocked by methysergide. Dopamine did not cause any membrane potential changes in neurones with no demonstrable inhibitory input. HIRST and SILINSKY (1975) have made the tentative suggestion that the inhibitory transmitter is a catecholamine, possibly dopamine.

It has been suggested that dopamine might be the transmitter released from enteric inhibitory nerves supplying the stomach (VALENZUELA 1976; BURNSTOCK et al. 1979). This was based on results which showed that gastric volume increased in dogs in response to intravenous dopamine (VALENZUELA 1976). The relaxation of gastric tone caused by dopamine and the reflex relaxation caused by gastric distension or feeding were both antagonised by metoclopramide or pimozide. Because these reflexes are mediated by transmission from enteric inhibitory nerves (see Chap. 10), it was suggested that the enteric inhibitory nerves release dopamine.

In both the situations summarised above, the most compelling reason for excluding dopamine as a likely transmitter candidate is its apparent absence, or presence in extremely low concentrations, in intestinal nerves, and the absence also of its synthesising enzyme, tyrosine hydroxylase. Dopamine, like noradrenaline, can be readily detected with fluorescence histochemical techniques. However, when segments of the gastrointestinal tract are extrinsically denervated by cutting sympathetic pathways, all histochemically detectable catecholamines in nerves disappear (FURNESS and COSTA 1974; see Sect. E). Microspectrofluorometric studies indicated that noradrenaline, but no detectable dopamine, is contained in the nerves (DIAB et al. 1976). Furthermore, at least in the guinea-pig ileum, tyrosine hydroxylase activity can no longer be detected after the degeneration of extrinsic nerves. Very low concentrations of dopamine are found by radioenzymatic assay after extrinsic denervation, but this dopamine is contained in a nonvesicular, reserpine-insensitive store (HOWE et al. 1979; FURNESS et al. 1980 b). It is presumed that this small residual store represents dopamine absorbed from the circulation or formed by enteric AADC (see COSTA et al. 1976) from circulating dopa.

In the stomach, the actions of dopamine have been further examined by VAN NUETEN and JANSSEN (1978) who showed that the effect of dopamine was substantially reduced by tetrodotoxin and that doperamide was much less effective as an antagonist of relaxations mediated by enteric nerves than it was of dopamine: 50% inhibition of the response to dopamine was caused by 0.3 μg/ml doperamide whereas about 1.2 μg/ml was required for 50% inhibition of the response to sympathetic stimulation and about 5 μg/ml was required to halve the vagal inhibitory response. Furthermore, although the enteric inhibitory nerves are present throughout the gastrointestinal tract, dopamine does not always cause relaxation. In fact, in many areas it has little or no effect and in some regions it causes contraction (RATTAN and GOYAL 1976; LANFRANCHI et al. 1978; MIACHON et al. 1978).

In conclusion, dopamine is probably not the transmitter released at inhibitory synapses in the submucous plexus, although it is probable that dopamine and the transmitter act through the same receptors. Dopamine is almost certainly not the transmitter released from the enteric inhibitory nerves, either in the stomach or elsewhere in the gastrointestinal tract; the possible identity of this transmitter is discussed in Sect. H.

G. Evidence that γ-Aminobutyric Acid is a Neurotransmitter in the Intestine

It has recently been reported that a small proportion of myenteric neurones take up and retain γ-aminobutyric acid (GABA), and that the synthesising enzyme for GABA, glutamic acid decarboxylase (GAD, EC 4.1.1.15) and low concentrations of GABA are found in the intestine (JESSEN et al. 1979). The uptake is restricted to about 5% of the neurones and is inhibited by *cis*-1,3-aminocyclohexane carboxylic acid but not by β-alanine. In the central nervous system, specific uptake of GABA seems to be confined to neurones which are likely to utilise it as a neurotransmitter (ROBERTS et al. 1976). The activity of GAD in the intestine is quite low, compared for example with the cerebellum, although this is reasonable considering the small proportion of intestinal neurones which take up and therefore might also synthesise GABA. It is still necessary to determine that the GAD activity is nerve related. An interesting sidelight on these experiments is the observation that glutamic acid is also converted to homocarnosine (γ-aminobutyryl-l-histidine) in the intestine, homocarnosine being a proposed central neurotransmitter (NG et al. 1977).

GABA seems to have only two substantial effects on intestinal motility. In some experiments it has been found to stimulate the smooth muscle indirectly by exciting cholinergic nerves and in areas of intestine where relaxations can be recorded easily, GABA stimulates enteric inhibitory nerves (HOBBIGER 1958; INOUYE et al. 1960; LEWIS et al. 1972; KRANTIS et al. 1980). GABA seems to have no direct action on intestinal muscle (KRANTIS et al. 1980). It thus seems feasible that GABA could be a transmitter released at neuroneuronal synapses in the myenteric plexus.

H. Identification of the Transmitter Released from Enteric Inhibitory Nerves

The enteric inhibitory nerves are found throughout the digestive tracts of mammals (BURNSTOCK 1972, 1975, 1979; FURNESS and COSTA 1973; Chap. 10). They supply the circular smooth muscle of all regions and in many places also supply the longitudinal muscle. They also innervate the anal accessory muscles (rectococcygeus and anococcygeus) which are continuations of the longitudinal muscle layer of the rectum. Smooth muscle in the lung, which is embryologically derived from the foregut, is supplied by similar nerves (COBURN and TOMITA 1973; COLEMAN and LEVY 1974; RICHARDSON and BELAND 1976). In the gastrointestinal tract, the cell bodies of enteric inhibitory neurones are contained in the myenteric plexus; these neurones are involved in local accommodation to distension and in descending inhibitory reflexes (see Chap. 10). In the stomach and rectum, the neurones are innervated by axons of the vagus and pelvic nerves respectively. Stimulation of enteric inhibitory nerves causes relaxation, accompanied by hyperpolarisation of the muscle. For a substance to be considered the transmitter of these nerves it should hyperpolarise and relax the muscle in all areas of intestine where the nerves are found, it should be released when the nerves are activated reflexly or by electrical stimulation, and, if it can be detected histochemically in neurones, these should have cell bodies in the myenteric plexus and axons should supply the circular smooth muscle in both sphincteric and nonsphincteric regions. Adenosine 5′-triphosphate (ATP) and vasoactive intestinal peptide (VIP) have both been suggested to be the transmitter; the evidence for each is discussed in Sect. H. Dopamine, which has also been proposed, is almost certainly not the enteric inhibitory transmitter (see Sect. F).

I. Identity of Action: ATP

One of the principal criteria for transmitter identification is that there should be an identity of action; that is, the proposed transmitter when applied to the tissue, should reproduce the effects of nerve stimulation (Sect. A). In the intestine, as in most organs, it is not possible to apply an endogenous substance so that it has the same spatial and temporal distribution within the tissue as does the transmitter released from the nerves. However, some degree of parallelism should be expected: the endogenous transmitter and the exogenous agent should cause conductance changes and membrane potential in the same direction and they should also alter tension in the same sense. If a drug is found which alters the response to the exogenous compound, then a change in the response to nerve stimulation in the same direction should be found; that is, there should continue to be mimicry. In a majority of regions of the gastrointestinal tract which have been examined, ATP and the inhibitory transmitter cause similar changes in muscle tension (BURNSTOCK 1972; see Table 6) and in membrane potential and conductance (TOMITA and WATANABE 1973; DEN HERTOG and JAGER 1975). However, there are a sufficient number of regions where mimicry is not observed or is inconsistent (MACKAY and MCKIRDY 1972; COSTA and FURNESS 1974; OHGA and TANEIKE 1977; BURKS and GRUBB 1978; DANIEL 1979; NORTHWAY and BURKS 1980; RATTAN and GOYAL

Table 6. Comparison of potencies of ATP and VIP on gastrointestinal smooth muscle[a]

Organ	Threshold (M or as indicated)	Reference
Adenosine triphosphate		
Stomach		
Guinea-pig	10^{-6}	BURNSTOCK et al. (1972)
Guinea-pig	2×10^{-5}	BURNSTOCK et al. (1970)
Rabbit	2×10^{-5}	BURNSTOCK et al. (1970)
Rat	4×10^{-6}	BURNSTOCK et al. (1970)
Pig	5×10^{-6} (contraction)	OHGA and TANEIKE (1977)
Small intestine		
Rabbit	4×10^{-5}	DRURY and SZENT-GYÖRGYI (1929)
Rabbit	4×10^{-6}	GILLESPIE (1934)
Duodenum		
Rabbit	10^{-7}	WESTON (1973b)
Rat	4×10^{-6}	BURNSTOCK et al. (1970)
Ileum		
Guinea-pig	2×10^{-6}	BURNSTOCK et al. (1970)
Rabbit	2×10^{-6}	BURNSTOCK et al. (1970)
Rat	10^{-5}	BURNSTOCK et al. (1972)
Mouse	10^{-5}	BURNSTOCK et al. (1970)
Taenia coli		
Guinea-pig	10^{-7}	AXELSSON and HOLMBERG (1969)
Colon		
Guinea-pig	10^{-6}	BURNSTOCK et al. (1972)
Rabbit	2×10^{-5}	BURNSTOCK et al. (1970)
Rabbit (longitudinal muscle)	2×10^{-4} (contraction)	MACKAY and MCKIRDY (1972)
Rabbit (circular muscle)	2×10^{-4}	MACKAY and MCKIRDY (1972)
Rat	4×10^{-5}	BURNSTOCK et al. (1970)
Mouse	10^{-5}	BURNSTOCK et al. (1970)
Human	4×10^{-4}	BURNSTOCK et al. (1972)
Internal anal sphincter		
Guinea-pig	2×10^{-5} (contraction)	COSTA and FURNESS (1974)
Human	10^{-6}	BURLEIGH et al. (1979)
Vasoactive intestinal polypeptide		
Lower oesophageal sphincter		
Cat	0.25 μg/kg (intravenous) (7×10^{-8} mol/kg)	BEHAR et al. (1979)
Cat	3×10^{-9}	UDDMAN et al. (1978)
Opossum	0.25 μg/kg (intravenous) (7×10^{-8} mol/kg)	RATTAN et al. (1977)
Human	1.5×10^{-10} (plasma concentration)	DOMSCHKE et al. (1978)
Stomach		
Rat	$< 2\times10^{-7}$ (based on result with impure extract)	PIPER et al. (1970)
Dog	5×10^{-9}	MORGAN et al. (1978)

Table 6. (continued)

Organ	Threshold (M or as indicated)	Reference
Jejunum		
Dog	10^{-12}	KACHELHOFFER et al. (1976)
Taenia coli		
Guinea-pig	3×10^{-9}	COCKS and BURNSTOCK (1979)
Internal anal sphincter		
Human	6×10^{-10}	BURLEIGH et al. (1979)
Gall bladder		
Guinea-pig	3×10^{-10}	RYAN and RYAVE (1978)

[a] Unless otherwise indicated, the responses obtained were relaxations. The concentrations or doses quoted as thresholds are the lowest concentrations or doses stated to be effective in the references cited. Some references do not make it clear if lesser amounts were tested

1980) or where pharmacological manipulations have not affected responses to ATP and enteric inhibitory nerve stimulation in parallel (RIKIMARU et al. 1971; SUZUKI et al. 1971; SAITO 1972; WESTON 1973b; SPEDDING et al. 1975; AMBACHE et al. 1977; OHGA and TANEIKE 1977; HUNT et al. 1978; STOCKLEY 1978; BAER and FREW 1979; BARTLETT et al. 1979; RATTAN and GOYAL 1980) to cast doubt on the hypothesis that ATP is the transmitter.

1. Comparison of Effects of ATP and of Inhibitory Nerve Stimulation

In most areas of intestine, ATP in concentrations from about 10^{-4}–10^{-7} M causes relaxations (Table 6). However, in considering the possible role of ATP as a neurotransmitter, it is valuable to examine those instances which are discussed in this section where ATP does not mimic the effect of stimulating the enteric inhibitory nerves.

The rabbit distal colon and rectum are supplied by enteric inhibitory nerves and when these nerves are stimulated inhibitory junction potentials (i.j.p.) can be recorded in both muscle layers (FURNESS 1969). The rectum can be arranged as a flat preparation in which tension is recorded independently from the two muscle layers, which have about the same thickness (MACKENNA and MCKIRDY 1970; MCGROARTY and MCKIRDY 1971). Under these conditions, ATP (2–6×10^{-4} M) relaxed the circular muscle but contracted the longitudinal muscle; the response was not blocked by tetrodotoxin (MACKAY and MCKIRDY 1972). It thus seems that ATP, in contrast to the enteric inhibitory nerves, has a direct excitatory action on the longitudinal muscle.

The internal anal sphincter in all species seems to be supplied by enteric inhibitory nerves (e.g. FURNESS and COSTA 1973; COSTA and FURNESS 1974; Chap. 10). In the guinea-pig, relaxation of the muscle is recorded mechanically and i.j.p. are recorded electrically when the nerves are stimulated (COSTA and FURNESS 1974). The sphincter muscle can be set up as a strip, to which the relatively thin longitudinal muscle adheres. Responses to enteric inhibitory nerve stimulation can be recorded in such strips, but when ATP (2×10^{-5} M) was applied the muscle either contracted or gave no response. With 2×10^{-4} M ATP the responses were still weak, in some cases slight contraction and in others slight relaxation was observed. In another sphincter, that of the lower oesophagus, which also receives an enteric inhibitory innervation, similar results were obtained (DANIEL 1979). The spincter in the opossum was only affected by high concentrations of ATP (2×10^{-4}–2×10^{-3} M). In about 50% of trials only a contraction occurred and in the other experiments there was a brief relaxation followed by contraction. RATTAN and GOYAL (1980) found that the opossum lower oesophageal sphincter gave a contraction followed by relaxation; tetrodotoxin partly antagonised the relaxation and potentiated the contraction caused by ATP.

Many investigations have shown that the mammalian stomach receives an innervation by enteric inhibitory nerves (see Chap. 10). Stimulation of these nerves supplying strips of porcine stomach elicits relaxations accompanied by i.j.p. in the smooth muscle cells (OHGA and TANEIKE 1977). The nerve-mediated responses were blocked by tetrodotoxin, but were not affected by α- or β-adrenoceptor blockade or by guanethidine. However, ATP (5–20×10^{-6} M) caused only a tonic contraction. At higher concentrations of ATP (2.5–20×10^{-5} M) biphasic responses, slight relaxations followed by sustained contractions, were observed in 35% of experiments. Electrical recording showed that ATP (5–10×10^{-5} M) caused depolarisation and initiated action potentials in the smooth muscle. The excitatory effects of ATP were not affected by tetrodotoxin. No hyperpolarising responses to ATP were observed. To explain these results, while still retaining the hypothesis that ATP is the inhibitory transmitter, it could be postulated that exogenously applied ATP has access to one set of receptors, through which depolarisation and excitation is mediated, whereas ATP released from the nerves diffuses only, or predominantly, to receptors mediating hyperpolarisation and relaxation. There is no adequate way to test this postulate. However, it should be pointed out that the axons supplying intestinal smooth muscle do not form discrete, close junctions but lie, usually in small bundles, at distances of 0.1 μm or more from the smooth muscle cells and it is generally considered that the transmitter must diffuse a considerable distance to have its effect (BENNETT and ROGERS 1967; BURNSTOCK 1970; GABELLA 1979; Chap. 2). There is no evidence that other drugs have restricted access to receptors within the muscle layers.

Intra-arterial injections of ATP caused the circular muscle of the dog small intestine to contract (BURKS and GRUBB 1978; NORTHWAY and BURKS 1980). This contraction was substantially reduced by tetrodotoxin or atropine, but the remaining response was still a contraction. Brief relaxations were observed when the tone of the muscle was raised with substance P (NORTHWAY and BURKS 1980). It has been suggested that responses to ATP which do not mimic responses to stimulation of nerves presumed to utilise ATP as a neurotransmitter arise because the ATP is dephosphorylated in the tissue and the degradation products have different actions than ATP (BROWN et al. 1979). However, α- and β-methylene ATP, which are resistant to dephosphorylation, also contracted the dog small intestine, as did the degradation products ADP, AMP and adenosine. It has also been suggested that contractions caused by exogenous ATP in other areas might result from the release of prostaglandins which mask its inhibitory effects (BURNSTOCK 1979). However, the contractions of the dog small intestine were not affected by an inhibitor of prostaglandin synthesis, indomethacin.

The results of electrophysiological studies are not entirely consistent with ATP being the inhibitory transmitter (TOMITA and WATANABE 1973). When the nerves are stimulated, profound hyperpolarising potential changes (i.j.p.) are observed. However, concentrations of ATP (10^{-6}–10^{-5} M) which caused muscle relaxation blocked spontaneous action potential discharge without causing membrane hyperpolarisation or increased conductance; concentrations above 10^{-4} M were required to cause electrical changes and 10^{-3} M was required to match the amplitude of the i.j.p. When the muscle was depolarised and partly contracted by carbachol, the threshold concentration for relaxation was 10^{-8} M and for hyperpolarisation was between 10^{-7} and 10^{-6} M (JAGER and SCHEVERS 1980).

2. Effects of Drugs Which Antagonise Responses to ATP

In this section the effects of a number of drugs which reduce the responses to ATP in the intestine are discussed. Some of these drugs reduce responses to a variety of unrelated agonists. Nevertheless, such drugs are of interest if, although they antagonise responses to ATP and a variety of other agonists, they differentially affect responses to ATP and to stimulation of enteric inhibitory nerves. No drugs which are fully specific as antagonists of ATP and other substances acting through the same receptors are currently available.

The compounds which have been used as antagonists of the relaxing actions of ATP and other purine nucleotides in the intestine include imidazole and related compounds (BUEDING et al. 1967; RIKIMARU et al. 1971; SATCHELL et al. 1973), 2-2′-pyridylisatogen (PIT; SPEDDING et al. 1975; STOCKLEY 1978), ATP itself (BURNSTOCK et al. 1970; WESTON 1973b), phentolamine (RIKIMARU et al. 1971; SATCHELL et al. 1973), quinidine (BURNSTOCK et al. 1970) and theophylline (ALLY and NAKATSU 1976; HOOPER et al. 1978; BAER and FREW 1979; SMALL and WESTON 1979a, b). Of these drugs, imidazole and imidazole-like compounds (e.g. antazoline, talazoline), yohimbine, phentolamine and quinidine seem least specific in their actions, for example having similar effects on the actions of ATP and catecholamines (SPEDDING et al. 1975).

Both imidazole (3.5×10^{-3} g/ml) and phentolamine (5×10^{-6} g/ml) are reported to block relaxations of intestinal muscle caused by ATP without significantly affecting transmission from enteric inhibitory nerves (RIKIMARU et al. 1971; SUZUKI et al. 1971). Results obtained with intracellular microelectrodes lead to essentially similar conclusions (SAITO 1972). In contrast, SATCHELL et al. (1973) showed that relaxations caused by ATP and by stimulation of enteric inhibitory nerves were both substantially reduced by imidazole or phentolamine. SATCHELL et al. (1973) used low frequencies of stimulation (0.1–5 or 0.1–1.5 Hz) and consider this might be the reason for the difference between their results and those of RIKIMARU et al. (1971) who used 5 and 10 Hz. However, SAITO (1972) examined inhibitory junction potentials in response to single stimuli 5 s or more apart and found these to be unaltered when responses to ATP were abolished. Likewise, AMBACHE et al. (1977) found that phentolamine substantially reduced relaxations caused by ATP, but did not affect relaxations caused by single stimuli applied to enteric inhibitory nerves. None of these authors constructed concentration–response or frequency–response curves; until this is done for a range of agonist and antagonist concentrations and a range of frequencies of stimulation the inconsistency in the observations cannot be adequately resolved. Nevertheless, it has been shown that under some circumstances transmission from enteric inhibitory nerves is unaffected while responses to ATP are substantially reduced or abolished.

The report by OKWUASABA et al. (1977), that theophylline selectively antagonised relaxations elicited by ATP and enteric inhibitory nerves in the fundus of the guinea-pig stomach, provided evidence in support of the role of ATP as transmitter, but later studies failed to confirm this observation. In the original report, theophylline (5×10^{-5} *M*) was said to shift the concentration response curve for ATP in guinea-pig stomach strips 5.8-fold to the right, whereas relaxations caused by noradrenaline, isoprenaline and papaverine were slightly augmented. At the same concentration, theophylline was claimed to cause substantial reductions, to less than 10% of control, in relaxations mediated by enteric inhibitory nerves stimulated at frequencies of 1–20 Hz. When these experiments were repeated by two of the original authors (COOK and HAMILTON personal communication in SMALL and WESTON 1979a), specific antagonism could not be demonstrated. Other investigators found that theophylline (10^{-4} *M*) did not antagonise the actions of ATP, noradrenaline or enteric inhibitory nerves in the taenia coli or fundic strip from the guinea-pig or in the rabbit duodenum (HOOPER et al. 1978; BAER and FREW 1979; SMALL and WESTON 1979a, b). In view of these later observations, it seems prudent to ignore the original report of OKWUASABA et al. (1977).

SPEDDING et al. (1975) found that PIT (5×10^{-5} *M*) shifted the concentration-response curve for ATP ten-fold to the right but had no effect on responses to stimulation of enteric inhibitory nerves with single pulses or at frequencies of 0.2–10 Hz. When responses to stimulation of enteric inhibitory nerves at 2 Hz were matched in amplitude with ATP (2×10^{-5} *M*), the responses to ATP were completely blocked by PIT whereas relaxations caused by stimulation of the enteric inhibitory nerves were not affected. STOCKLEY (1978) reported similar results obtained on strips of human taenia coli. She found that 2-2′-pyridylisatogen tosylate (4.5×10^{-5} g/ml) reduced the responses to a test concentration of ATP (4×10^{-4} *M*) to about 20% of control whereas responses to the stimilation of enteric inhibitory nerves at frequencies of 1–16 Hz were not affected. HUNT et al. (1978) have found that the relaxations of rat gastric muscle caused by ATP were antagonised by PIT, whereas relaxations caused by stimulation of enteric inhibitory nerves at 0.5–20 Hz were unaffected. Likewise, in the lower oesophageal sphincter, responses to stimulating enteric inhibitory nerves were unaffected by PIT whereas effects of ATP were substantially reduced (RATTAN and GOYAL 1980). The results of SPEDDING et al. (1975), HUNT et al. (1978), STOCKLEY (1978) and RATTAN and GOYAL (1980) could be interpreted in several ways: that ATP is not the transmitter, or that the antagonist did not reach the receptors activated by the transmitter, or that PIT enhanced the release of transmitter, thus overcoming its antagonism of postjunctional receptors. In the absence of experimental evidence of receptor inaccessibility, or of enhanced release caused by PIT, the observations cast considerable doubt on the hypothesis that ATP is the enteric inhibitory transmitter.

Attempts have been made to exploit the desensitisation of its own action caused by ATP as a tool to investigate its possible role as an intestinal neurotransmitter

(BURNSTOCK et al. 1970; WESTON 1973b; OHGA and TANEIKE 1977; OKWUASABA et al. 1977; BAER and FREW 1979; BARTLETT et al. 1979). WESTON (1973b) showed that exposure of longitudinal muscle strips from the rabbit duodenum to ATP (10^{-4} *M*) substantially reduced the amplitudes of relaxations caused by subsequent applications of ATP. He constructed concentration–response curves for ATP in the range $0.16–4 \times 10^{-6}$ *M*. After exposure to ATP the concentration–response relation was shifted to the right and the maximum depressed. Because of the substantial reduction of the maximum response, to about 5% of control, and the associated change in slope, it is not possible to give a really significant measure of the shift to the right. However, the published figure suggests a shift in the EC_{50} (which compares the equivalent concentrations at which 50% of the agonist's effect is achieved) of at least 40-fold. This marked desensitisation seems specific for agonists acting at receptors for ATP; responses to noradrenaline, phenylephrine, isoprenaline or stimulation of perivascular (noradrenergic) nerves were not affected, whereas responses to ATP, ADP, AMP and adenosine were all antagonised. However, there was no significant reduction in responses to stimulation of enteric inhibitory nerves at frequencies of 1–32 Hz.

In an earlier paper, BURNSTOCK et al. (1970) described tachyphylaxis to ATP in the rabbit ileum and stated that tachyphylaxis was associated with a blockade or reduction of responses to stimulation of enteric inhibitory nerves; however, no concentration–response or frequency–response relationships were published and in the absence of quantitative data it is difficult to reconcile or even compare these results with those of WESTON (1973b). Also in contrast to WESTON's data are the results of OKWUASABA et al. (1977) who worked on fundic strips from the guinea-pig (this paper has been discussed earlier in this section where it was concluded that some of its results are in error). The authors reported that a desensitising concentration of ATP of 5×10^{-5} *M* moved the concentration–response curve for ATP 4.2-fold to the right without affecting the relaxations caused by noradrenaline or papaverine. This concentration was claimed to cause a depression, to less than 25% of control at frequencies of 1–20 Hz, of the relaxations caused by stimuli to the enteric nerves. Relaxations in response to stimulation of noradrenergic nerves were not affected. However, when these experiments of OKWUASABA et al. (1977) were repeated, it was found that desensitisation with ATP (5×10^{-5} *M*) caused no change in the frequency–response curve for stimulation of enteric inhibitory nerves (BAER and FREW 1979). OHGA and TANEIKE (1977) found that responses to ATP were abolished by exposing pig gastric muscle to ATP ($5–10 \times 10^{-5}$ *M*), but that responses to stimulation of enteric inhibitory nerves were not reduced. In the rat ileum, desensitisation with ATP (3×10^{-6} *M*) reduced the maximum response to ATP to less than 50%, but did not alter the responses to stimulation of enteric inhibitory nerves at frequencies of 0.5–32 Hz (BARTLETT et al. 1979). RATTAN and GOYAL (1980) reduced the response of the lower oesophageal sphincter to a supramaximal dose of ATP to less than 20% by desensitisation, but no reduction in the effectiveness of stimulating enteric inhibitory nerves was observed.

It can be concluded from these investigations that desensitisation with ATP does substantially reduce responses to ATP without affecting responses to the stimulation of enteric inhibitory nerves (WESTON 1973b; OHGA and TANEIKE 1977; BAER and FREW 1979; BARTLETT et al. 1979; RATTAN and GOYAL 1980). Three explanations seem possible. First, the transmitter released from the enteric inhibitory nerves might not be ATP. Second, exogenous ATP might not reach the same receptors that are reached by ATP released from the nerves. Third, ATP might enhance its own release from the enteric inhibitory nerves and thus overcome a postjunctional block. The second explanation implies that all mimicking experiments with ATP are of doubtful validity, as the exogenous ATP does not reach the appropriate receptors. It might be pointed out that acetylcholine seems to be able to penetrate adequately, in that desensitisation with acetylcholine blocks the peristaltic reflex (SCHAUMANN et al. 1953; WESTON 1973a). Moreover, desensitisation with substance P blocks the action of substance P released from nerves in the guinea-pig ileum (FRANCO et al. 1979a, b) and exposure to somatostatin, proposed as a transmitter in descending pathways in the intestine (see Chap. 10), antagonises the descending inhibitory component of the peristalic reflex (FURNESS and COSTA 1979a). Furthermore, ATP released when the enteric inhibitory nerves are stimulated reaches the bathing medium (RUTHERFORD and BURNSTOCK 1978; BURNSTOCK et al. 1978b; see Sect. H.II), so it can be presumed that diffusion in the opposite direction, from the bathing medium to the points of release, is similarly effective. There is no information with which to evaluate the third possibility, that the release of ATP is enhanced by ATP. However, in other systems, transmitters are thought to work by negative feedback to antagonise their own release (WESTFALL 1977) and BURNSTOCK (see BURNSTOCK et al. 1979) has implied that ATP and other purines act on nerve endings to reduce the release of ATP. Further investigation of the nature and specificity of the desensitisation of the intestine to ATP is required. Nevertheless, the results of desensitisation experiments suggest that ATP is unlikely to be the enteric inhibitory transmitter.

II. Release of ATP from Enteric Inhibitory Nerves

ATP is released when some noradrenergic (SU et al. 1971; SU 1975; WESTFALL et al. 1978; BURNSTOCK et al. 1979), cholinergic (SILINSKY and HUBBARD 1973; SILINSKY 1975) or sensory nerve endings (HOLTON 1959) are stimulated. However, there is evidence that not all noradrenergic and cholinergic nerves release sufficient ATP for it to be detected (BURNSTOCK et al. 1970; BLASCHKE and UVNAS 1979). ATP is also released along with catecholamines from adrenal chromaffin cells (DOUGLAS and POISNER 1966; DOUGLAS 1968) and from skeletal and smooth muscle cells when these are excited (ABOOD et al. 1962; LUCHELLI-FORTIS et al. 1979). It is therefore difficult to verify the special nature of any release of ATP from enteric inhibitory nerves as a transmitter rather than as an accessory to transmitter release.

Release of ATP or its metabolites (ADP, AMP, adenosine and inosine) associated with the stimulation of enteric inhibitory nerves has been detected in a number of studies (BURNSTOCK et al. 1970; SATCHELL and BURNSTOCK 1971; SU et al. 1971; KUCHII et al. 1973a, b; RUTHERFORD and BURNSTOCK 1978; BURNSTOCK et al. 1978a, b). BURNSTOCK et al. (1970) and SATCHELL and BURNSTOCK (1971) found that adenosine and inosine could be detected in perfusates of toad stomach after prolonged stimulation of vagal pathways impinging on enteric inhibitory nerves. In contrast, stimulation of cholinergic pathways to the stomach did not increase the release of nucleosides. BURNSTOCK et al. (1970) also found that adenosine and inosine were released into the perfusate of the guinea-pig stomach when the vagus was stimulated and that release could be detected from the isolated myenteric plexus of the chicken gizzard when this was stimulated directly. Unfortunately, the authors did not report whether release from the isolated plexus was inhibited by tetrodotoxin or lowered Ca^{2+}, which would be expected for release from nerves.

SU et al. (1971) showed that most of the adenosine ^3H taken up by the guinea-pig taenia coli was stored as ATP ^3H and when the preparation was stimulated with transmural electrodes, radioactive material was liberated in association with the activation of enteric inhibitory nerves. Further studies of the release of labelled compounds were made by RUTHERFORD and BURNSTOCK (1978) who found that 33% of the release of label by transmural stimulation after loading with 10^{-7} *M* adenosine ^3H was blocked by tetrodotoxin. The remaining label was presumed to be released from non-neuronal sources. The responses to enteric inhibitory nerves were reduced or abolished by tretrodotoxin. As 60%–70% of the label appears to be stored as ATP and more than 20% of the remainder is ADP, AMP or adenosine (SU et al. 1971; KUCHII et al. 1973 b), which also relax the taenia coli, it is puzzling that relaxation could be blocked with only one-third reduction in the release of label. KUCHII et al. (1973 a) also showed a lack of correlation between the release of label and relaxations obtained in the taenia coli when enteric inhibitory nerves were stimulated. They found comparable release of label by nicotine or by transmural stimulation in fresh tissue and in tissue stored in the cold for 8 days although the relaxations in response to these stimuli were stubstantially reduced or abolished after 8 days. The resting tension in stored preparations was comparable to that in fresh tissue and the stored tissue was still capable of being relaxed by phenylephrine (FUKUDA and SHIBATA 1972; KUCHII et al. 1973 a). After incubating the taenia coli with 10^{-7} *M* tritiated adenosine, KUCHII et al. (1973 a, b) found that release of label was associated with relaxations caused by nitroglycerine, noradrenaline, papaverine, nicotine and transmural stimulation. For the same degree of relaxation to noradrenaline, papaverine and transmural stimulation, the release was approximately the same. In contrast, RUTHERFORD and BURNSTOCK (1978) report that noradrenaline caused only 15% of the release of ATP obtained with transmural stimulation.

In the experiments of BURNSTOCK et al. (1978 b) ATP release from the taenia coli was measured using the luciferin–luciferase firefly reaction, without preincubating the tissue with adenosine. Transmural stimulation in the presence of atropine and guanethidine caused a release of unmetabolised ATP of 4–5 times the prestimulation release into the solution bathing the taenia coli. The increased release and the relaxation associated with nerve stimulation were blocked by tetrodotoxin. Noradrenaline did not increase ATP release.

From these experiments, particularly those of BURNSTOCK et al. (1970, 1978 b), it seems that stimulation of enteric inhibitory nerves is associated with the release of ATP or related nucleotides and nucleosides. At least some of this release is from nerves, but the proportion which comes from the enteric inhibitory nerves cannot be determined because several of the many other types of nerve in the intestine (see FURNESS and COSTA 1980; and Sects. A,B) would be stimulated at the same time as the enteric inhibitory nerves. No calculation can be made of the interstitial concentration of ATP which is achieved when the nerves are stimulated. A tetrodotoxin-sensitive release of ATP or its metabolites could be detected during periods of stimulation of 15–30 s at 5–30 Hz (RUTHERFORD and BURNSTOCK 1978; BURNSTOCK et al. 1978 b) which indicates that ATP could readily diffuse from the sites of release to the surrounding medium. BURNSTOCK et al. (1978 b) have shown that some of the ATP reaches the bathing medium without being metabolised and therefore ATP released elsewhere in the tissue, or added to the bathing medium in sufficient concentration, should reach the sites of nerve release and hence the receptors activated by transmitter released from the nerves. However, as KUCHII et al. (1973 a) and RUTHERFORD and BURNSTOCK (1978) have shown, significant release of ATP can occur without causing the tissue to relax. As discussed in detail in Sect. H.I, exogenous ATP, which from the release experiments would be expected to diffuse to the receptors activated by the enteric inhibitory nerves, does not always mimic the effects of enteric inhibitory nerves. Furthermore, although ATP causes a substantial desensitisation of accessable receptors for ATP in the muscle it does not antagonise transmission from the inhibitory nerves. Taken together, the observations summarised and considered in this paragraph suggest that ATP is not the transmitter released by the enteric inhibitory nerves. It is possible that it is released from the nerves in conjunction with the true transmitter.

III. Other Evidence Presented for a Transmitter Role for ATP

The review of the literature presented in the preceding sections indicates that the two primary criteria to identify ATP as the enteric inhibitory transmitter (identity of action and correlation between release and action), have not been adequately satisfied. Evidence for the synthesis and degradation of ATP in the intestine does not assist in deciding whether it is a neurotransmitter. More recently it has been suggested that quinacrine binds to enteric inhibitory nerves and may serve as a histochemical marker for these nerves (BURNSTOCK et al. 1978 a; BURNSTOCK 1979; OLSON and ALUND 1979). BURNSTOCK (1979) has proposed that the binding of quinacrine occurs because the nerves contain high levels of ATP. This proposal really does not add evidence for or against the hypothesis of ATP transmission. If quinacrine does indeed bind specifically to enteric inhibitory nerves, this will be of great benefit in studying the morphology and arrangement of these nerves. But even if the binding is specific and is due to a reaction with ATP, this does not indicate that ATP is the transmitter. It is possible that quinacrine binds to many acidic compounds in tissue as well as to storage granules rich in ATP. In fact, ALUND and OLSON (1979) have shown that quinacrine binds to renal juxtaglomerular cells, pancreatic islet cells, adrenal chromaffin cells and a number of other cell types.

IV. Identity of Action: VIP

In most work, the 28 amino acid sequence of VIP (Table 7) has been used to study VIP actions on intestinal muscle. However, there is evidence that not all the VIP-like immunoreactivity in intestinal nerves is due to this form (see Sect. H.VI); the experiments described may have to be repeated with the other forms of VIP, which are probably fragments of the 28 amino acid peptide, when these are identified and become available.

1. Comparison of Effects of VIP and Inhibitory Nerve Stimulation

VIP has been shown to relax muscle of the stomach, small and large intestines and gastrointestinal sphincters in concentrations of about 10^{-9} *M* or less (Table 6; FURNESS and COSTA 1981). As with ATP, mimicry by itself merely shows a consistency with the hypothesis that VIP is the transmitter, and it is sensible to examine instances where it does not mimic (JAFFER et al. 1974; KACHELHOFFER et al. 1976; COHEN and LANDRY 1980). JAFFER et al. (1974) and COHEN and LANDRY (1980) found that VIP (10^{-9}–10^{-7} *M*) contracted the longitudinal muscle of the guinea-pig small intestine. The contraction was blocked by tetrodotoxin and reduced in amplitude by atropine, which indicates that VIP can stimulate cholinergic nerves in the intestine. Although it was not stated whether any relaxation was seen after administration of tetrodotoxin, the guinea-pig ileum has little or no tone and relaxations are not readily recorded. COHEN and LANDRY (1980) also reported contraction of rabbit small intestinal longitudinal muscle in response to VIP, but did not report the actions of tetrodotoxin or atropine on these responses. KACHELHOFFER et al. (1976) showed that VIP caused biphasic responses, relaxation followed by contraction, in isolated segments of dog jejunum which were perfused by the extracorporeal circulation of blood from an anaesthetised dog. The nature of their experiments limited the examination of the secondary contractions; the authors considered they were probably rebound contractions following the initial relaxations.

COCKS and BURNSTOCK (1979) have pointed out that the relaxations caused by VIP in vitro are very slow in onset, reaching their peak only 4–5 min after injection into an organ bath. This finding might be expected of such a large molecule, and does not indicate that VIP is not the transmitter. Injection of VIP into the perfusate of gastrointestinal preparations can result in substantial relaxation within 20 s (KACHELHOFFER et al. 1976). As it is unlikely that any substance can be applied to the tissue with the same efficiency as transmitter released from the nerves it is probably futile to look for exact duplication of the time course of responses.

2. Antagonism of Responses to VIP

There are no drugs which are known to act specifically at receptors for VIP, so comparisons of sensitivities of responses to VIP and inhibitory nerve stimulation have not been made. However, GOYAL et al. (1979) have performed the ingenious experiment of attempting to antagonise responses to VIP with the infusion of antisera. They found that responses to VIP were reduced by about 50% when the perfusing solution contained antiserum to VIP. Responses to stimulating enteric inhibitory nerves via vagal pathways at 2 Hz were reduced by 40% and responses elicited by transmural stimulation were similarly antagonised. Normal serum had no significant effects. Relaxations caused by isoprenaline were not reduced by infusion of the antiserum.

V. Release of VIP from Enteric Inhibitory Nerves

FAHRENKRUG et al. (1978 b) have shown that VIP-like immunoreactive material appears in the venous effluent from the cat stomach when vagal pathways impinging on enteric inhibitory nerves are stimulated electrically. These authors also found that when a receptive relaxation of the stomach was elicited by distending the oesophagus, a VIP-like substance appeared in the venous effluent. The receptive relaxation is due to a reflex whose final neurone is the enteric inhibitory neurone (see Chap. 10; FURNESS and COSTA 1981). EKLUND et al. (1979) showed that the receptive relaxation of the stomach could be mimicked by the intra-arterial infusion of VIP. FAHRENKRUG et al. (1978 a) showed that release of VIP in response to stimulation of the porcine vagus was completely blocked by hexamethonium and that the released material coeluted with highly purified porcine VIP on gel chromatography. Release of VIP into the portal blood in response to vagal stimulation has been shown in the calf by BLOOM and EDWARDS (1980). These experiments clearly show that VIP-like material is released as a consequence of vagal activity, but they do not definitely show that nerves are the source of this release; immunohistochemical studies indicate that VIP is in endocrine cells as well as nerves in the intestine (e.g. BRYANT et al. 1976; LARSSON et al. 1976).

On the other hand, FOX et al. (1979) could not detect any increase in VIP outflow from strips of lower oesophageal sphincter during stimulation of enteric inhibitory nerves. Increase in outflow only occurred when the stimuli were sufficient to excite the muscle directly. This outflow occurred in the presence of tetrodotoxin. These experiments indicate that if VIP is the transmitter, it is not released in sufficient quantity to diffuse to the bathing medium and be detected. The source of the VIP released in the presence of tetrodotoxin is uncertain. Immunohistochemical observations indicate that VIP in the lower oesophageal sphincter is in nerves (ALUMETS et al. 1978 b; UDDMAN et al. 1978). It is feasible that the stimuli may have been strong enough to release VIP by direct depolarisation of the axons. It would be valuable to repeat these experiments using perfused preparations.

VI. Other Evidence that VIP Could be the Enteric Inhibitory Transmitter

Enteric inhibitory nerves supply the sphincters and circular smooth muscle of the gastrointestinal tract and also supply the anal accessory muscles; their cell bodies of origin are in the myenteric plexus (Chap. 10). The nerves are involved in descending inhibitory reflexes in the intestine and would therefore be expected to project in an anal direction. Immunohistochemical observations indicate that VIP-containing neurones have this distribution. Nerve cell bodies are found in the myenteric plexus (LARSSON et al. 1976; SCHULTZBERG et al. 1978, 1980; FURNESS and COSTA 1979 b; COSTA et al. 1980 b; FURNESS et al. 1981). Axons are found in the circular smooth muscle of nonsphincteric (LARSSON et al. 1976; FURNESS and COSTA 1979 b; COSTA et al. 1980 b; SCHULTZBERG et al. 1980) and sphincteric regions (ALUMETS et al. 1978 b; UDDMAN et al. 1978; EDIN et al. 1979).

FURNESS and COSTA (1979 b) made lesions of the wall of the guinea-pig small intestine to study the projections of VIP neurones. They found that nerve cell bodies in the myenteric plexus gave rise to axons which ran for short distances in an anal direction to supply the circular smooth muscle.

The report of JESSEN et al. (1980) contains on observation which seems inconsistent with VIP being the transmitter. They reported that no VIP-containing cell bodies are found in the myenteric plexus beneath the taenia of the guinea-pig caecum, although earlier studies indicate the presence of the cells of origin of enteric inhibitory neurones in the myenteric plexus in other regions (see earlier in this section). Recent work from the authors' laboratory is at variance with the observations of JESSEN et al. (1980). We have found nerve cell bodies with VIP immunoreactivity in the myenteric plexus beneath the taenia; lesion experiments indicated that the VIP axons in the taenia arise from cell bodies in the myenteric plexus (FURNESS et al. 1981).

LARSSON et al. (1976) have shown that immunoreactive VIP-like material extracted from intestinal nerves of the cat elutes with highly purified porcine VIP on column chromatography and DIMALINE and DOCKRAY (1978) have used gel filtration and ion exchange chromatography to show that the immunoreactive form of VIP in nerves of the human colon corresponds to the full 28 amino acid sequence of VIP. However, the proportion of VIP-like immunoreactivity attributable to authentic VIP varies from species to species. In porcine intestinal nerves it is about 80%, whereas it is about 50% in the dog and 30% in the rat (DIMALINE and DOCKRAY 1979). The variants have not been sequenced and it is therefore not known whether they have similar pharmacological properties to VIP, although the radioimmunoassay results indicate they have structural resemblances to VIP.

VII. Conclusions

The published evidence places serious doubt on the contention that ATP or a related nucleotide is the transmitter released from enteric inhibitory nerves. The doubts arise from observations that ATP does not always mimic responses to nerve stimulation, that in some cases responses to ATP can be antagonised, while responses to nerve stimulation are unaffected, and that release of ATP is not necessarily associated with relaxation of intestinal muscle.

Evidence for VIP being the enteric inhibitory transmitter, while consistent with this hypothesis, comes from a limited number of studies. It is still necessary to characterise the different forms of VIP immunoreactive material and to demonstrate unequivocally the release of VIP from the nerves. Progress in this area will be facilitated if a specific antagonist of VIP is discovered.

J. Peptides as Intestinal Transmitters

I. The General Problem

1. Identification of Neural Peptides

A number of different peptides are contained in intestinal nerves (Table 7, and elsewhere in this section). The peptides have been localised in the nerves by immunohistochemical methods and it is this localisation that has prompted their consideration as possible neurotransmitters. In most cases, the true identities of the peptides have not been established and the convention has evolved to refer to the immunoreactivity as substance P-like, enkephalin-like, etc. Comparison of the distribution of nerves showing immunoreactivity for the different peptides indicates that they are in different sets of nerves (FURNESS and COSTA 1980; SCHULTZBERG

Table 7. Amino acid sequences of peptides suggested to be present in intestinal nerves

1) Substance P	H-Arg-Pro-Lys-Pro-Gln-Gln-Phe-Phe-Gly-Leu-Met-NH_2
2) Somatostatin	H-Ala-Gly-Cys-Lys-Asn-Phe-Phe-Trp-Lys-Thr-Phe-Thr-Ser-Cys-OH
3) Vasoactive intestinal polypeptide (VIP)	H-His-Ser-Asp-Ala-Val-Phe-Thr-Asp-Asn-Tyr-Thr-Arg-Leu-Arg-Lys-Gln-Met-Ala-Val-Lys-Lys-Tyr-Leu-Asn-Ser-Ile-Leu-Asn-NH_2
4) Met-enkephalin	H-Tyr-Gly-Gly-Phe-Met-OH
5) Leu-enkephalin	H-Tyr-Gly-Gly-Phe-Leu-OH
6) Gastrin/CCK 4	H-Trp-Met-Asp-Phe-NH_2
7) CCK 8	H-Asp-Tyr(SO_3)-Met-Gly-Trp-Met-Asp-Phe-NH_2
8) Nonantral gastric bombesin-like peptide	H-Ala-Pro-Val-Ser-Val-Gly-Gly-Gly-Thr-Val-Leu-Ala-Lys-Met-Tyr-Pro-Arg-Gly-Asn-His-Trp-Ala-Val-Gly-His-Leu-Met-NH_2
9) Bombesin (amphibian)	H-Pyr-Gln-Arg-Leu-Gly-Asn-Gln-Trp-Ala-Val-Gly-His-Leu-Met-NH_2
10) Neurotensin	P-Glu-Leu-Tyr-Glu-Asn-Lys-Pro-Arg-Arg-Pro-Tyr-Ile-Leu-OH
11) Pancreatic polypeptide (bovine)	H-Ala-Pro-Leu-Glu-Pro-Gln-Tyr-Pro-Gly-Asp-Asp-Ala-Thr-Pro-Glu-Gln-Met-Ala-Gln-Tyr-Ala-Ala-Glu-Leu-Arg-Arg-Tyr-Ile-Asn-Met-Leu-Thr-Arg-Pro-Arg-Tyr-NH_2
12) Angiotensin II	H-Asp-Arg-Val-Tyr-Ile-His-Pro-Phe-OH
13) Angiotensin I	H-Asp-Arg-Val-Tyr-Ile-His-Pro-Phe-His-Leu-OH

1) CHANG et al. (1971); 2) BRAZEAU et al. (1973); 3) MUTT and SAID (1974); 4), 5) HUGHES et al. (1975); 6), 7) MUTT and JORPES (1971); 8) McDONALD et al. (1979); 9) ANASTASI et al. (1971); 10) CARRAWAY and LEEMAN (1975); 11) FLOYD et al. (1977); 12), 13) SCHWARZ et al. (1957)
In a number of cases it is uncertain whether the actual peptide present in the nerves has a sequence identical to that given. The sequence of pancreatic polypeptide varies slightly amongst species. The following variations are known: Ovine, 2 = Ser; human, 6 = Val, 10 = Asn, 11 = Asn, 23 = Asp; Porcine, 6 = Val

et al. 1980; Table 8), with the exception that cholecystokinin-like and somatostatin-like activity is in some cases found in the same nerve cell bodies in the submucosa (SCHULTZBERG et al. 1980).

Uncertainty about the identity of the peptides arises because the antisera recognise only some of the amino acid groups of the peptide against which they were raised. Therefore the antisera can cross-react with other known or unknown peptides which have some antigenic determinant sites in common with the immunogen, but differ in other respects. The degree of cross-reactivity is often assessed using radioimmunoassay (RIA). However, extrapolation from the results of RIA to immunohistochemistry may not be appropriate. In RIA, low concentrations of antisera are used to detect low concentrations of peptides, whereas local concentrations of peptides in tissue used for immunohistochemistry may be very high and high concentrations of antisera are used. For example, an antiserum used at 1:20,000 in RIA to detect picomolar amounts of peptide in a homogenate, could be used at 1:100 in immunohistochemical experiments and be detecting peptides contained in storage vesicles at micromolar or greater concentrations. An antibody

Table 8. Localisation of neuronal peptides in guinea-pig ileum[a]

Peptide	Presence in Varicose Axons					Polarity of axons in myenteric plexus		Proportion of all neurones (%)	
	Myenteric plexus	Submucous plexus	Circular muscle[b]	Mucosa	Arterioles	Directed orally	Directed anally	Myenteric plexus	Submucous plexus
Substance P	+	+	+	+	+	+	+	3.6 (2.8)	11.3 (5.1)
Somatostatin	+	+	−	+	Rare	−	+	4.7 (3.8)	17.4 (19.7)
VIP	+	+	+	+	Rare	−	+	2.5 (7.8)	45 (27.3)
Enkephalins	+	+	+	−	−	+	+	24.5 (15.2)	0 (0)
Bombesin	+	+	+	−	−	−	+	0.5	0
Cholecystokinin	+	+	−	?	−	−	+	+	+

[a] Data from COSTA et al. (1980b, c, d); FURNESS and COSTA (1980); SCHULTZBERG et al. (1980). The proportions of neurones found by the present authors are given, with the data of SCHULTZBERG et al. (1980) in parentheses. This table indicates only presence or absence of axons: when distributions are examined in detail further differences between the peptide axons are discerned (e.g. FURNESS et al. 1980a; SCHULTZBERG et al. 1980)
[b] Includes the deep muscular plexus
+ = present; − = absent

which is present as a minor component of the antiserum as determined by RIA could conceivably bind to an antigen, present in high local concentration in the tissue, even though it is different from the peptide against which the antiserum was raised. Therefore tests of specificity should be made directly in the histochemical experiments.

To identify a peptide with certainty, it needs to be extracted and its amino acid sequence determined. However, almost certain identity with a peptide of known structure can be demonstrated using chromatographic separation, particularly high pressure liquid chromatography (HPLC), combined with RIA. The difficulties that arise in isolating peptides are two-fold. First, most of the peptides found in nerves in the gastrointestinal tract are also found in mucosal endocrine or paracrine cells (BLOOM 1978; Table 1). It is therefore necessary to separate the layers of the intestine before extractions are undertaken. The second problem is that there may be multiple forms of some peptides. A good example is cholecystokinin (CCK), which seems to exist in at least four forms (CCK 39, CCK 33, CCK 8, and CCK 4) and probably also exists as part of a larger precursor molecule (DOCKRAY 1979; LARSSON and REHFELD 1979; REHFELD and LARSSON 1979). VIP also appears to exist in multiple forms (DIMALINE and DOCKRAY 1978, 1979). If more than one form can be extracted from the nerves, we need to ask which form or forms are released and which are biologically active. It is even possible that a peptide is released in one form from nerves and converted extraneuronally to the biologically active form. A further problem in determining the natural forms of peptides is the changes which may occur during processing. An example of this is the ratios of β-lipotropin to β-endorphin found with different extraction procedures applied to the anterior pituitary (LIOTTA et al. 1978). Extraction of fresh tissue with 0.2 *M* HCl at 90 °C yielded 98% β-lipotropin whereas extraction with 1 *M* acetic acid yielded 52% and with 0.2 *M* acetic acid, 29%. Freezing and thawing the tissue before extraction with HCl reduced the proportion of β-lipotropin to 20%.

Many active peptides, whether hormones or potential neurotransmitters, show interspecific differences, which could arise from simple gene mutations, but which do not affect biological activity. For example, vasopressins of ovine and bovine origins differ in one amino acid (DU VIGNEAUD et al. 1953). In the intestine there are similar instances, such as pancreatic polypeptide (Table 7) and gastrin (BARRINGTON and DOCKRAY 1976). Although the term gastrin, for example, can be used to cover all gastrins when discussing its physiological role, in some cases it may be necessary to specify the species of origin. These differences mean that biologically equivalent peptides (isopeptides) may behave differently on chromatographs. It is therefore necessary to look beyond small chemical differences and to investigate the biological effects of extracts and to compare these effects with those of synthetic peptides, which by immunohistochemical techniques appear to be in intestinal nerves, although their sequences may have been determined in other organs or species.

In summary, the initial immunohistochemical evidence suggesting the presence of a peptide in nerves should be followed by experiments to extract the peptide and to characterise it chemically, ideally by determining its amino acid sequence, but at least by examining its properties by chromatography. The peptide should also be characterised in terms of its biological activity.

2. Roles of Neural Peptides

The demonstration of a particular peptide within a nerve leads one to further questions: what is its role, is it a neurotransmitter in the normally accepted sense of a substance which is released by a propagated action potential and has an acute effect on a neighbouring cell? In the case of neural peptides in the intestine, evidence for neurotransmitter roles is scant. Some, such as substance P, enkephalin and VIP, will probably be shown to be transmitters. On the other hand, peptides are sometimes contained in the same neurones as other biologically active substances. For example, there are central neurones which contain both 5-HT and substance P-like immunoreactivity (CHAN-PALAY et al. 1978; HÖKFELT et al. 1978), there is evidence for somatostatin in some sympathetic neurones which also contain noradrenaline (HÖKFELT et al. 1977) and evidence for VIP in some, but not all, postganglionic autonomic cholinergic neurones (LUNDBERG et al. 1979). In the intestine, certain neurones with cell bodies in the submucosa contain both somatostatin-like and CCK-like immunoreactivity (SCHULTZBERG et al. 1980). Very little is known of the significance of the presence of two biologically active substances within the same neurone. They could be released together as cotransmitters, both having postjunctional actions. Another hypothesis is that one substance is released as a modulator, that modifies the release or action of the first substance. It could also be postulated that some neuropeptides do not have acute effects on excitability but influence the growth or metabolism of cells with which they interact.

The understanding of the properties and roles of the peptide neurones of the gastrointestinal tract is still very rudimentary. Histochemical studies show that axons which are immunoreactive for a particular peptide may ramify extensively in the intestine. For example, substance P axons arising from cell bodies in the myenteric plexus project orally and anally, to the circular muscle, and to submucous ganglia; axons from the submucous plexus project to the mucosa; and axons of extrinsic origin supply submucous ganglia and intestinal arterioles (COSTA et al. 1980c). The observations indicate that substance P neurones may be involved in the control of different aspects of intestinal function: motility, blood flow, absorption, and visceral sensation are suggested by their distribution. Thus we can postulate separate classes of substance P neurones involved in independent pathways, just as different sets of noradrenergic neurones influence intestinal motility and blood flow (Sect. D) and, in a wider context, the same transmitter may be involved in quite different roles in central and peripheral nerves. It is a measure of our ignorance that few suggestions for the roles of peptides in intestinal nerves can be made and those that are made are based on very little data.

The fundamental criteria for transmitter identification discussed in Sect. A have not been established satisfactorily for any of the neural peptides of the gastrointestinal tract. In testing for release, it should be remembered that most, if not all, of the peptides discussed are contained in mucosal cells as well as in nerves.

3. Peptide Synthesis

Enkephalin is the only peptide whose synthesis in intestinal nerves has been shown (Sect. J.V). However, peptides are not taken up into nerves as far as has been determined, and it is likely that the general model of peptide synthesis and transport in neurones advanced by GAINER et al. (1977) for central neurones will also apply to intestinal neurones. The peptides are thought to be synthesised as part of larger precursor molecules in the cell bodies and transported by an active process along the axon to the sites of release. Axonal transport of peptides in intestinal nerves has been demonstrated by their accumulation in damaged axons (FURNESS et al. 1980a).

II. Substance P

The peptide which is now defined as substance P and whose structure is given in Table 7 was isolated from bovine hypothalamus and its amino acid composition determined by CHANG and LEEMAN (1970). It was subsequently sequenced by CHANG et al. (1971). Previously substance P had been defined in terms of its pharmacological and apparent chemical properties in extracts of tissue. These extracts were originally made from intestine and brain of the horse (VON EULER and GADDUM 1931) and later from other tissues (e.g. PERNOW 1953). Activity was expressed in units, which became known as Euler units, defined on the basis of the ability of extracts to contract the isolated rabbit jejunum (VON EULER 1942). It was shown that the peptide identified in extracts of bovine hypothalamus had the same biological and chemical characteristics as the substance P of different tissue extracts (CHANG and LEEMAN 1970; TREGEAR et al. 1971). It has been shown that 1 mg pure substance P has about 2.6×10^6 Euler units of activity. STUDER et al. (1973) isolated substance P from one of the original sources, horse intestine, and found it to have the same amino acid sequence as bovine hypothalamic substance P. It is not known whether the substance P isolated by STUDER et al. was of primarily endocrine cell origin or whether it came from nerves. FRANCO et al. (1979a, b) have shown that substance P isolated from the muscle layers of guinea-pig small intestine, which contain immunoreactive nerves but not endocrine cells, is identical to synthetic substance P in its pharmacological properties and sensitivity to peptidases. The biological activity of substance P depends primarily on its COOH terminal pentapeptide amide. It can therefore be concluded from the work discussed here that immunohistochemical procedures localise authentic substance P or an isopeptide in intestinal nerves. There may be variations in one or more amino acids near the NH_2 terminus in different species. There is also a possibility that some of the nerves which have been described as containing substance P-like immunoreactivity contain a bombesin-like peptide which does not share the biological actions of substance P, but does have the same COOH terminal dipeptide amide (see Table 7; Sect. J.VII).

Substance P-like immunoreactivity is present in intestinal nerves of the many species that have been examined, baboon, dog, guinea-pig, human, mouse, pig and rat (PEARSE and POLAK 1975; NILSSON et al. 1975; SUNDLER et al. 1977d; SCHULTZBERG et al. 1978, 1980; FRANCO et al. 1979a; COSTA et al. 1980a, c). In the case of the guinea-pig ileum examined by COSTA et al. (1980a, c) absorption tests have been made with bombesin and none of the immunoreactivity seems to be due to this peptide (J. B. FURNESS unpublished work 1980). In the intestine of all species, substance P axons supply the myenteric and submucous ganglia, the circular smooth muscle and the mucosa. The majority of the axons appear to be of intrinsic origin: this is certainly so for the guinea-pig small intestine (COSTA et al. 1980c), and is probably so for other species (SCHULTZBERG et al. 1980). Axons supplying arterioles and some of those to submucous ganglia in the guinea-pig ileum arise from extrinsic sources. Nerve cell bodies are found in both the myenteric and submucous plexuses (COSTA et al. 1980a; SCHULTZBERG et al. 1980). Substance P axons form a particularly dense network around nerve cell bodies of the myenteric plexus. As yet, detailed studies of the distribution and projections of the nerves have only been made in guinea-pig small intestine. It is possible that in other areas or species differences may be found.

The release of substance P from intestinal nerves has been demonstrated by FRANCO et al. (1979a, b). These authors exploited the desensitisation of substance P receptors which occurs when intestinal muscle is exposed to this peptide (GADDUM 1953b; LEMBECK and ZETLER 1971). It was shown that desensitisation was specific for substance P; the concentration–response curves for carbachol, dimethylphenylpiperazinium (DMPP), 5-HT and bradykinin were not affected by a concentration of substance P (7.5×10^{-8} *M*) which moved the concentration–response curve for substance P 300-fold to the right. Responses to structurally related peptides, eledoisin and physalemin, which are presumed to act through the same receptors as substance P, are reduced by exposure of the intestine to substance P (LEMBECK and FISCHER 1967). When the receptors for substance P were desensitised, transmission from noncholinergic excitatory nerves in the guinea-pig ileum was blocked, whereas transmission from cholinergic nerves and from enteric inhibitory nerves was not affected (FRANCO et al. 1979a). Transmission from the noncholinergic excitatory nerves was also blocked by tetrodotoxin. UVNÄS-WALLENSTEIN (1978) detected substance P by RIA in the feline gastric lumen after stimulation of the vagus. However, substance P is found in gastrointestinal mucosal cells as well as in nerves (NILSSON et al. 1975; PEARSE and POLAK 1975; HEITZ et al. 1976) and as vagal stimulation is known to degranulate enterochromaffin cells (see Sect. E.III.5), the source of the substance P detected by UVNÄS-WALLENSTEIN is uncertain.

It is possible that substance P released from intestinal nerves contributes to slow e.p.s.p. in myenteric neurones. These e.p.s.p. are elicited by repetitive stimuli applied to adjacent nerve strands and are associated with transient decreases in membrane conductance (Sect. E.III.2). Substance P mimics the e.p.s.p. as do 5-HT (in some of the neurones) and somatostatin. Substance P, which desensitises the neurones to its own action, reduces the amplitude of the e.p.s.p. (R. A. NORTH personal communication 1980).

The available evidence thus shows that substance P is contained in intestinal neurones and is released from the neurones by a mechanism which depends on the action potential. However, the role of substance P is uncertain. It is excitatory to both neurones and smooth muscle (LEMBECK and ZETLER 1971; KATAYAMA et al. 1979). Substance P neurones project both orally and anally within the myenteric plexus (FRANCO et al. 1979b; COSTA et al. 1980c) and end around myenteric neurones and in the circular muscle coat. Substance P axons also supply submucous neurones, submucous arterioles and the mucosa (see COSTA et al. 1980a, c). Substance P neurones in the myenteric plexus receive cholinergic inputs (FRANCO et al. 1979c). It seems probable that different groups of substance P neurones may have different roles.

III. Somatostatin

Somatostatin-like immunoreactivity has been found in intrinsic intestinal neurones in guinea-pig, mouse and rat (HÖKFELT et al. 1975; COSTA et al. 1977; SCHULTZBERG et al. 1978, 1980). The amino acid sequence Thr-Phe-Thr-Ser (somatostatin 10–13) also occurs in glucagon and secretin and similar series (Thr-Phe-Ile-Ser and Val-Phe-Thr-Asp) occur in gastric inhibitory polypeptide (GIP) and VIP, respectively. Thus, it is possible that not all the immunoreactivity reported in these studies is due to authentic somatostatin. On the other hand, the distribution of somatostatin-containing and VIP-containing cell bodies and axons in the intestine seem quite distinct (FURNESS et al. 1980a; SCHULTZBERG et al. 1980; Table 8) and although glucagon, secretin and GIP have been detected by immunohistochemical methods in gastrointestinal mucosal cells, these peptides have not been reported in intestinal nerves. The antisera used by COSTA et al. (1977) have had their antigenic determinant sites defined. For one antiserum these were Asn(5), Phe(7), Trp(8), Lys(9), and Phe(11) and for the other they were Asn(5), Phe(6), Phe(7), Trp(8), and Phe(11). With the qualifications discussed above, about extrapolation from RIA results to immunohistochemical findings, it is still not certain that the immunoreactivity was due to somatostatin, although the histochemical reaction no longer occurred if the antisera were preincubated with synthetic somatostatin. More recently, extracts from the muscularis externa and submucosa of the guinea-pig small intestine (which contain immunoreactive nerves, but not other immunoreactive cells) have been shown to contain a peptide which has the same retention time on HPLC as somatostatin and which is detected as somatostatin in RIA (FURNESS et al. 1980c). It thus seems likely that the nerves located histochemically in the guinea-pig small intestine do contain somatostatin and do not contain the related intestinal peptides already discussed.

Release of somatostatin from intestinal nerves has not been reported. In spite of this, a speculative proposal concerning the possible role of somatostatin neurones in the myenteric plexus has been made (FURNESS et al. 1980a). Examination of the somatostatin-containing nerves after lesioning pathways in the myenteric plexus shows that the nerve cell bodies give rise to axons which project in an anal direction to end around other neurones (COSTA et al. 1980d). This implies that somatostatin-containing neurones are interneurones in descending pathways. Examination of the responses of the intestine in vitro show that somatostatin has two actions: to inhibit the output of acetylcholine and to stimulate enteric inhibitory nerves (GUILLEMIN 1976; COHEN et al. 1978, 1979; FURNESS and COSTA 1979a).

Intracellular microelectrode studies show that somatostatin depolarises some neurones and hyperpolarises others, although the majority of neurones are unaffected (KATAYAMA and NORTH 1980). The actions of somatostatin in vitro show tachyphylaxis. When the intestine is distended, a relaxation occurs on the anal side, which is the end result of a reflex which activates the enteric inhibitory nerves (HIRST and MCKIRDY 1974b; COSTA and FURNESS 1976). It was speculated that somatostatin might be released from interneurones in this pathway. Consistent with this possibility is the observation that exposure of the intestine to relatively high concentrations of somatostatin blocks the descending inhibitory reflex in vitro.

There is incomplete information on the effects of somatostatin on motility in vivo (Table 9). It appears that somatostatin interacts with other gut hormones which influence motility and that the effects of somatostatin depend on digestive state. In the human stomach, BLOOM et al. (1975) have reported a slowing of gastric emptying, whereas JOHANSSON et al. (1978) found that gastric emptying which had been suppressed by a glucose load was enhanced. The increase in contractile activity caused by gastric distension was diminished by infusion of somatostatin (STADAAS et al. 1978). In monkeys, basal emptying was unaffected but there was inhibition of the enhanced emptying induced by a water load (DUBOIS et al. 1979). GUSTAVSSON et al. (1979) found no effects of somatostatin on gastric emptying in fasted rats. In the human small intestine, suppression of motility and slowing of transit time has been found (EFENDIC and MATTSSON 1978; JOHANSSON et al. 1978). In the fasted dog, THOR et al. (1978) found that somatostatin (0.6–5 μg kg^{-1} h^{-1}) increased the frequency of occurrence of the interdigestive complex, whereas ORMSBEE et al. (1978) reported that infusion at 5 μg kg^{-1} h^{-1} delayed or abolished the complex. Spike activity in fed dogs was suppressed and the activity was converted to that typical of the fasted state (THOR et al. 1978). Observations by BYBEE et al. (1979) indicate that somatostatin has no direct effect on the lower oesophageal sphincter in baboon, but that increases in pressure induced by intragastric alkali or glycine were suppressed.

In addition to the neurones and their projections in the myenteric plexus, there are many immunoreactive cell bodies in the submucosa and varicose axons ramify around submucous ganglia and also run in the mucosa. Some of the submucous neurones also show CCK-like immunoreactivity (SCHULTZBERG et al. 1980). It is therefore probable that a number of roles are filled by different groups of intestinal neurones showing somatostatin-like immunoreactivity.

IV. Vasoactive Intestinal Polypeptide

VIP-like immunoreactivity has been localised in cell bodies and axons of neurones throughout the gastrointestinal tract in the cat, guinea-pig, human, mouse, pig and rat (BRYANT et al. 1976; LARSSON et al. 1976; FUXE et al. 1977; LARSSON 1977; SUNDLER et al. 1977b; ALUMETS et al. 1978b; UDDMAN et al. 1978; SCHULTZBERG et al. 1978, 1980; EDIN et al. 1979; COSTA et al. 1980b; FURNESS and COSTA 1979b, 1981; FURNESS et al. 1981). Axons are numerous in the circular smooth muscle, both of the sphincters and of nonsphincteric regions, and varicose axons also supply both the myenteric and submucous plexuses. The innervation of the mucosa is dense in the small and large intestines but axons are less numerous in the gastric mucosa. VIP axons are associated with small blood vessels, mostly in the mucosa, throughout the gastrointestinal tract. The distribution of fibres is not detectably affected by vagotomy or by cutting nerves running to the intestine through the mesentery. Therefore, the majority of axons arise from intrinsic nerve cell bodies which have been detected in both the myenteric and submucous plexuses. VIP-containing cell bodies are numerous in the submucous plexus of the small intestine and it is likely that they are the source of the dense mucosal innervation.

The immunoreactive material contained in the intestinal nerves seems to consist of different proportions of the 28 amino acid peptide along with other im-

Table 9. Effects of somatostatin on gastrointestinal motility

Area and species	Effect	References
Lower oesophageal sphincter		
Baboon	No direct action	BYBEE et al. (1979)
Stomach		
Human	Gastric emptying slowed	BLOOM et al. (1975)
	Contractile activity inhibited	STADAAS et al. (1978)
	Enhancement of gastric emptying which had been suppressed by glucose	JOHANSSON et al. (1978)
Monkey	No effect on basal rate of emptying; physiologically enhanced emptying slowed	DUBOIS et al. (1979)
Rat	Emptying unaffected	GUSTAVSSON et al. (1979)
Cat, dog, rabbit, rat	Contractile activity inhibited	TANSY et al. (1979)
Small intestine		
Human	Suppressed motor activity	EFENDIC and MATTSON (1978)
	Slowed propulsion	JOHANSSON et al. (1978)
Cat, rabbit, rat	Contractile activity diminished	TANSY et al. (1979)
Dog	External muscle inhibited, muscularis mucosae contracted	TANSY et al. (1979)
	Increased frequency of interdigestive complex, suppression of fed activity	THOR et al. (1978)
	Suppression of initiation of interdigestive complex	POITRAS et al. (1979)
	Reduced motor activity	ORMSBEE et al. (1978)
Guinea-pig	Suppression of acetylcholine release from excitatory nerves	GUILLEMIN (1976) COHEN et al. (1978) FURNESS and COSTA (1979a)
	Stimulation of enteric inhibitory nerves	FURNESS and COSTA (1979a)
	Some neurones depolarised, others hyperpolarised in myenteric plexus	KATAYAMA and NORTH (1980)
Rabbit	Stimulation of enteric inhibitory nerves	COHEN et al. (1979)
Rat	Propulsion unaffected in vivo	GUSTAVSSON et al. (1979)
Large intestine		
Cat, rabbit, rat	Inhibition of activity	TANSY et al. (1979)
Dog	Inconsistent minor effects	TANSY et al. (1979)
Guinea-pig	Stimulation of enteric inhibitory nerves, antagonism of peristaltic reflex	FURNESS and COSTA (1979a)
Gall bladder		
Human	Inhibition of activity	CREUTZFELD et al. (1975)
Cat	Reduction of CCK-induced contraction	ALBINUS et al. (1977)

munoreactive forms (see Sect. H.VI). The nature of these other forms has not been determined; some may be precursor molecules, while others could be naturally occurring fragments of VIP or might be cleavage products produced during extraction and separation of the peptides.

The evidence that VIP could be the transmitter released from enteric inhibitory nerves has been reviewed (Sect. H.IV–H.VII). However, there are also VIP neurones which project to intestinal blood vessels, to enteric ganglia and to the mucosa. The VIP neurones supplying the blood vessels might be the enteric vasodilator nerves which are reflexly activated when the mucosa is irritated (BIBER 1973) and which, in the colon, can be excited by stimulation of the pelvic nerves (HULTEN et al. 1969). Vasodilation is not affected by drugs which block transmission from cholinergic or noradrenergic nerves. FAHRENKRUG et al. (1978 b) found that when these nerves were excited either reflexly or by electrical stimulation of the pelvic nerves, dilation was associated with the appearance of VIP in the venous effluent from the cat colon. In this species VIP is contained in colonic nerves, but not in endocrine-like cells (LARSSON et al. 1976). A number of investigations have shown VIP to cause intestinal vasodilation (SAID and MUTT 1970 a, b; THULIN and OLSSEN 1973; KACHELHOFFER et al. 1974; EKLUND et al. 1979). VIP stimulates the secretion of water and electrolytes by the intestinal mucosa (BARBEZAT and GROSSMAN 1971; SCHWARTZ et al. 1974; KREJS et al. 1978; EKLUND et al. 1979) and it is thus possible that the role of the numerous VIP-containing axons in the mucosa could be to enhance secretion, although there is no direct evidence to support this suggestion. There is no evidence of specific release of VIP from axons in the enteric plexuses, although if VIP is released it might be expected to be excitatory, since recordings with intracellular microelectrodes show that VIP excites nerve cells in the myenteric plexus of the guinea-pig ileum (WILLIAMS and NORTH 1979).

As discussed in the preceding paragraphs and in Sect. H, FAHRENKRUG et al. (1978 a, b) have shown the release of VIP into gastric and colonic veins. BITAR et al. (1979) reported that oxytocin caused a substantial rise in VIP levels in the portal vein of the anaesthetised dog. This release was substantially reduced by tetrodotoxin, leading the authors to conclude that it was of neural origin. VIP has also been detected in the portal blood of pigs and calves in response to stimulation of the vagus, but the source of the VIP has not been determined (SCHAFFALITZKY DE MUCKADELL et al. 1977; BLOOM and EDWARDS 1980).

V. Enkephalins

1. Presence, Synthesis, and Release of Enkephalins

Immunohistochemical evidence shows enkephalins to be present in intestinal nerves of cat, guinea-pig, human, monkey, mouse, opossum, pig, rat, and rabbit (ELDE et al. 1976; POLAK et al. 1977; ALUMETS et al. 1978 a; LINNOILA et al. 1978; SCHULTZBERG et al. 1978; 1980; FURNESS and COSTA 1980; FURNESS et al. 1980 a; LARSSON et al. 1979; UDDMAN et al. 1980). In most species examined, and in different parts of the intestine, enkephalin nerves are confined to the myenteric plexus and circular muscle, although a few fibres have been reported in the muscularis mucosae (ELDE et al. 1976; ALUMETS et al. 1978 a; FURNESS et al. 1980 b). Axons are found in submucous ganglia in the cat (LARSSON et al. 1979). LINNOILA et al. (1978) have reported nerves in the submucosa of the guinea-pig small intestine, but examination of their published micrographs indicate that they mistook nerves in the deep muscular plexus as being in the submucosa. There are no nerves in the mucosa, although enkephalin-containing mucosal cells occur in monkeys, mice, pigs and rats but not in cats, guinea-pigs or humans (ALUMETS et al. 1978a). Enkephalins are contained in a high proportion, about 25%, of nerve cell bodies in the myenteric plexus in the guinea-pig ileum, but none are found in submucous ganglia (FURNESS and COSTA 1980). Met- and Leu-enkephalin have been separated and measured in extracts from the guinea-pig and rabbit ileum (HUGHES et al. 1977). The assays showed that most of the enkephalin is in extracts of longitudinal muscle and myenteric plexus (about 560 pmol/g in guinea-pigs and 380 pmol/g in rabbits) whereas the remaining layers, containing circular muscle, submucosa and mucosa and designated "circular muscle" in the original paper, contained about 60 and 45 pmol/g in guinea-pigs, and rabbits, respectively. There was about 1.5–5 times more Met- than Leu-enkephalin. This preponderance of Met-enkephalin is consistent with the observations of LARSSON et al. (1979) who reported that the two enkephalins are in separate axons, the Met-enkephalin axons outnumbering the Leu-enkephalin axons by 4–5-fold.

SOSA et al. (1977) demonstrated the synthesis of enkephalins in the isolated small intestine of the guinea-pig by examining the incorporation of tyrosine ^{3}H. There was a delay of 1–2 h before label was detected in Met- and Leu-enkephalin, suggesting that tyrosine was first incorporated into a precursor molecule from which the enkephalins were later formed. Incorporation of label was substantially reduced by protein synthesis inhibitors.

The release of enkephalins at rest and in response to stimulation of enteric nerves of the guinea-pig ileum has been detected by SCHULZ et al. (1977) who identified the released enkephalins by chromatography and RIA. However, HUGHES et al. (1978a) were not able to repeat these observations. Indirect evidence for enkephalin release has been obtained by WATERFIELD and KOSTERLITZ (1977) who found that the output of acetylcholine during stimulation was enhanced by opioid antagonists. Further indirect evidence is the observation that a depressant compound whose action was blocked by naloxone was released during the fatigue of the peristaltic reflex in the guinea-pig ileum (VAN NUETEN et al. 1976). The fatigue was also reversed by naloxone. Brief bursts of high frequency stimulation cause a sustained inhibition of cholinergic twitch contractions in the guinea-pig ileum; the inhibition is blocked by naloxone (PUIG et al. 1977; HUGHES et al. 1978a). Because similar inhibition caused by enkephalin was blocked by naloxone and carboxypeptidase but the effects of nerve stimulation were not antagonised by carboxypeptidase, HUGHES et al. (1978a) postulated that the sustained inhibition was not due to enkephalins released during stimulation persisting in the bath, but to the sustained firing of the enkephalin neurones. HUGHES et al. (1978b) found that transmural stimulation of the guinea-pig ileum in the presence of an inhibitor of protein synthesis, cyclohexamide, significantly reduced the content of enkephalin. This reduction was prevented by tetrodotoxin, implying that it was due to an action-potential-dependent release of enkephalins from the nerves. Thus, except for the failure of HUGHES et al. (1978a) to confirm the results of SCHULZ et al. (1977), there is convincing evidence for the release of enkephalins when intestinal nerves are activated.

2. Actions of Enkephalins on the Gastrointestinal Tract

Enkephalins act through the same classes of opiate receptors, which may be of several kinds (GILBERT and MARTIN 1976; LORD et al. 1977; WÜSTER et al. 1979), as morphine and related alkaloids. However, the relative potencies for the actions of morphine and the enkephalins vary between species and between organs. Nevertheless, any evaluation of the possible roles of enkephalins in the gastrointestinal tract should take into account experiments with morphine, its related drugs, and with enkephalins and similar peptides, although it must be acknowledged that lack of effect of morphine in a particular species or organ might occur because it is a poor agonist on enkephalin receptors in that species or organ.

In humans, the medical use of opium, and later of morphine, for the control of diarrhoea or dysentery preceded its use for analgesia and investigations of the mechanism by which morphine exerts its constipating action date back to nineteenth century. These studies have been reviewed several times (e.g. KREUGER 1937; VAUGHAN WILLIAMS 1954; KOSTERLITZ and LEES 1964; DANIEL 1968; WEINSTOCK 1971; AMBINDER and SCHUSTER 1979). A number of factors complicate analysis of the actions of opiate agonists. First, propulsion is modified both by central actions of morphine and by its direct effect on the intestine. In fact, morphine injected into the cerebrospinal fluid is considerably more potent than is the same dose given peripherally and its primary action is certainly central (MARGOLIN 1954, 1963; BORISON 1959; PAROLARO et al. 1977; SCHULZ et al. 1979). However, the peripheral link in the constipating action of centrally administered morphine appears not to involve opiate receptors (SCHULZ et al. 1979), and so in contemplating the possible roles of enkephalin neurones in the intestine the effects of morphine in vivo should not necessarily be taken into account. In the work discussed here only peripheral effects are considered. Second, although morphine seems to have a constipating effect in all mammals, the direct effects of morphine and enkephalins on gastrointestinal motility in different species and regions vary (Table 10). Third, the ileum of the guinea-pig, which is the organ most extensively examined (e.g. KOSTERLITZ and LEES 1964; NORTH and TONINI 1977; NORTH et al. 1979) is not typical of mammalian intestine. In fact, the small intestine of the guinea-pig is the only one of a number of regions studied in a variety of species which consistently is not excited by opiate agonists (Table 9). For no easily discernible reason, the results of DANIEL et al. (1959), who found that morphine had no effect, or in high concentration inhibited the activity of the muscle of the small intestine of a number of species, stand in contrast to those of other workers.

The mechanism by which morphine contracts intestinal muscle has been examined in the dog intestine by BURKS and his colleagues (see BURKS 1973). Their results indicate that morphine has an indirect effect mediated by the release of 5-HT (from an unknown source) which in turn stimulates cholinergic nerves which act upon the muscle. The involvement of cholinergic nerves is deduced from the ob-

Table 10. Peripheral effects of opiate agonists on intestinal motility

Area and function	Effect[a]	References
A) Effects on muscle tension		
Lower oesophageal sphincter		
Cat	0	UDDMAN et al. (1980)
Opossum	+	MCCALLUM et al. (1980)
Stomach		
Cat	+(Morphine) −(Enkephalin-amide)	EDIN et al. (1980)
Dog	+	KONTUREK et al. (1978)
Pyloric sphincter		
Cat	+	EDIN et al. (1980)
Small intestine		
Cat	+	GRUBER et al. (1935)
Dog	+	BURKS and LONG (1967) BURKS and GRUBB (1974)
Guinea-pig	−	GRUBER et al. (1935) SCHUMANN (1955)
Rabbit	+ (− in 12%)	GRUBER et al. (1935)
Rat	+	GRUBER et al. (1935) MATTILA (1962) BURKS (1976) GILLON and POLLOCK (1980)
Terminal ileum		
Dog, guinea-pig, human, rabbit, rat	0 (− at high concentration)	DANIEL et al. (1959)
Bile duct		
Ox, cat	+	CREMA et al. (1965)
Colon and rectum		
Rat	+	HUIDOBRO-TORA and WAY (1976) NIJKAMP and VAN REE (1978)
B) Effects on the peristaltic reflex		
Small intestine		
Dog	Inhibit	VAUGHAN WILLIAMS and STREETEN (1950)
Guinea-pig	Inhibit	TRENDELENBERG (1917) SCHAUMANN (1955)
Rabbit	No effect except at very high concentration	VAUGHAN WILLIAMS and STREETEN (1950) SCHAUMANN (1955)
C) Effects on acetylcholine output		
Small intestine		
Guinea-pig	Antagonise	PATON (1957) SCHAUMANN (1957)
Rabbit	No effect (morphine) Antagonism (enkephalin)[b]	GREENBERG et al. (1970) OKA (1980)
Rat	Antagonise[b]	MATTILA (1962) OKA (1980)
Mouse	Antagonise[b]	OKA (1980)
Colon		
Rat	Antagonise[b]	SHAW (1979) GILLON and POLLOCK (1980)

Table 10 (continued)

Area and function	Effect	References
D) Effect on transmission from enteric inhibitory nerves		
Taenia coli		
Guinea-pig	Antagonise	SHIMO and ISHII (1978)
E) Effect on nonadrenaline release		
Lower oesophageal sphincter		
Cat	Antagonise	UDDMAN et al. (1980)

[a] + = contracts or enhances contractile activity; − = relaxes or diminishes contractions induced by other agents; 0 = no significant or consistent effect
[b] Indirect evidence

servations that atropine and tetrodotoxin cause similar substantial reductions in the responses to morphine. That the nerves are stimulated by 5-HT is deduced from the observation that 5-HT appears in the perfusate of morphine-stimulated intestine and that responses to morphine are diminished by prior treatment with reserpine or with the 5-HT antagonists, cinanserin and cyproheptidine. BURKS (1976) has found evidence for a similar indirect mechanism of morphine action in the rat small intestine. However, the contractile action of morphine on the circular muscle of the bile duct, lower oesophagus and on the rat colon probably does not involve cholinergic nerves, because it is resistant to atropine and tetrodotoxin (CREMA et al. 1965; MCCALLUM et al. 1980; GILLON and POLLOCK 1980). In the guinea-pig ileum, opiate agonists inhibit peristalsis and reduce the spontaneous and stimulated release of acetylcholine (Table 9: KOSTERLITZ and LEES 1964). The relaxation of the muscle sometimes seen in this preparation in probably due to the inhibition of acetylcholine release. In the rabbit, morphine does not reduce the output of acetylcholine or inhibit peristalsis in vitro. However, enkephalin does inhibit transmission from cholinergic nerves (OKA 1980), suggesting that morphine is not an effective agonist on enkephalin receptors in the rabbit intestine.

The arrangement of axons which run in the fine nerve bundles supplying the circular muscle (FURNESS et al. 1980a) suggests that the antagonism of acetylcholine release might be through a prejunctional axoaxonal interaction at the cholinergic nerve endings in the circular muscle. Pharmacological observations also suggest that opiate agonists cause prejunctional inhibition of release from the final cholinergic neurones which run to the muscle (NORTH et al. 1979). However, enkephalin axons also form a sparse network within the myenteric plexus, where enkephalin acts to hyperpolarise nerve cell bodies (NORTH et al. 1979). It is therefore conceivable that a component of enkephalin action is to lower the excitability of myenteric neurones. It remains to be determined whether endogenous enkephalin could cause excitation of intestinal muscle through the mechanism described by BURKS (1973) for the dog intestine. In the dog, as in other species, the majority of 5-HT is present in the mucosa (see Sect. E.II). Because enkephalin-containing axons are absent from the mucosa and because of the long diffusion path for mucosal 5-HT to reach the myenteric plexus and stimulate cholinergic nerves, it is almost certain that enteric enkephalin nerves do not influence motility via the release of mucosal 5-HT.

3. Possible Role of Enkephalin as an Excitatory Transmitter to the Pyloric Sphincter

Recent evidence suggests that enkephalin could be an excitatory transmitter released from nerves supplying the circular muscle of the pyloric sphincter in the cat (EDIN et al. 1980). It was found that atropine-resistant contractions of the sphincter in response to stimulation of the vagus were blocked by naloxone. Met and Leu-enkephalin and the enkephalin analogue, (+)-Ala^2, Met^5-enkephalinamide also caused a contraction which was blocked by naloxone. Immunohistochemical studies showed that enkephalin-like immunoreactivity was in nerve cell bodies in the myenteric plexus at the level of the sphincter and in axons in the sphincter.

VI. Cholecystokinin, COOH-Terminal Fragments of Cholecystokinin and Gastrin

Cholecystokinin occurs in two forms, one with 39 amino acids (CCK 39) and the other with 33 amino acids (CCK 33), which is equivalent to CCK 39 without its NH_2-terminal hexapeptide, whereas gastrin occurs as a 17 amino acid peptide, G 17, and a number of larger forms, including big gastrin, G 34. Gastrin and cholecystokinin have the same COOH terminal pentapeptide amide (CCK/G 5) which is also found in the peptides cerulein and phyllocerulein, isolated from amphibian skin (see BERTACCINI 1976). The ceruleins and CCK have a sulphated tyrosyl which occurs at the same position relative to the COOH terminal, residue 2 in CCK 8, and whose presence in sulphated form is essential for these peptides to show CCK-like activity. The minimum sequence for gastrin-like activity is the common COOH terminal tetrapeptide amide, CCK/G 4.

Two main approaches have been made to distinguish gastrin, CCK and their COOH terminal fragments in the intestine, the use of region-specific antisera and the use of chromatographic separation combined with RIA (REHFELD 1978a, b; LARSSON and REHFELD 1979). These studies indicate that CCK-like immunoreactivity may be due to five or more components, namely large molecules which are presumed precursors, CCK 39 and 33, a peptide of similar size to CCK 12, CCK 8 and CCK/G 4. By chromatographic separation combined with RIA using region-specific antisera, UVNÄS-WALLENSTEIN et al. (1977) found high levels (50–270 pmol/g) of immunoreactivity attributed to G 17 in the abdominal vagus, but did not find CCK-like immunoreactivity. This is a very puzzling observation in that almost no activity (less than 2 pmol/g) was found in the cervical or thoracic parts of the vagus and no or very few gastrin-containing axons were found with immunohistochemical methods in the stomach or duodenum. Numerous immunoreactive axons were in the colon and there were some in the lower small intestine. The antibody (4562) used by UVNÄS-WALLENSTEIN et al. (1977) for their immunohistochemical studies has been reported by LARSSON and REHFELD (1979) to react with both gastrin and CCK in tissue, and so it is uncertain what peptide the nerves contained. LARSSON and REHFELD (1979) used different antisera to distinguish gastrin and CCK-like immunoreactivity and described nerves with CCK-like activity with a similar distribution to that described by UVNÄS-WALLENSTEIN et al. (1977). Nerves were found in the colon, rectum and ileum of the guinea-pig, but not in the duodenum or stomach. They were most numerous in the colon. Axons were found around nerve cell bodies in both the myenteric and submucous plexuses, in the muscularis mucosae and in the mucosa. Positive nerve cell bodies were found in the submucosa. A similar distribution was described by SCHULTZBERG et al. (1980), who in addition found some positive nerve cell bodies in the myenteric plexus and a few axons in the duodenum. SCHULTZBERG et al. (1980) also reported that some nerve cell bodies in submucous ganglia contained both somatostatin-like and CCK-like immunoreactivity.

Extracts from the guinea-pig colon, which contains immunoreactive nerves but no immunoreactive endocrine-like cells have been examined by chromatography and RIA (LARSSON and REHFELD 1979). It was found that the predominant immunoreactive species coeluted with CCK/G 4 and that lesser amounts corresponding to CCK 8, CCK 12 (presumed) and higher molecular weight forms were present. They also reported the presence of true gastrins but did not publish any details of this finding. The CCK/G 4 is probably normally present and is not a fragment formed during the extraction procedure because synthetic CCK 8 could be recovered in similar experiments without loss (REHFELD 1978 b). However, the ratios of material successfully extracted do depend on the extraction procedure; REHFELD (1978 b) found that in acetic acid extracts of brain, 20% of activity eluted with CCK 33, 10% with CCK 12 and 70% with CCK 8, whereas elution of boiling water extracts revealed 90% CCK 8 and 10% CCK 4. The best extraction seems to be boiling water followed by acetic acid, which is the method used by LARSSON and REHFELD (1979). Some doubt must be attached to the identification of CCK/G 4 in these assays. The assay was performed with CCK 8 as a standard and the antiserum recognised CCK/G 4 with a molar potency of $^1/_{30}$ that of CCK 8. Therefore, all values for CCK/G 4 represent the actual value obtained multiplied by 30. In addition, the CCK/G 4 activity eluted as a broad peak which could have included a number of fragments. It must be considered that the substance detected was not in fact CCK/G 4. In a separate study using different antisera, CCK 8, but little or no CCK/G 4-like activity was detected in extracts from the nerve-containing layers of the guinea-pig ileum (G. J. DOCKRAY personal communication 1981).

Until it is determined under what conditions and in which forms these peptides are released from the nerves, it seems presumptuous to discuss their possible roles as neurotransmitters, particularly in view of the different pharmacological actions of the related peptides (see BERTACCINI 1976).

VII. Bombesin-Like Peptide

Bombesin is a 14 amino acid peptide which is one of a family of peptides with similar amino acid sequences that have been isolated from amphibian skin (ERSPAMER and MELCHIORRI 1973; BERTACCINI 1976). In studies of antisera raised against synthetic amphibian skin bombesin, bombesin-like immunoreactivity has been found in mucosal endocrine cells throughout the gut (POLAK et al. 1976). A peptide which, in common with bombesin (BERTACCINI et al. 1974), has gastrin-releasing activity has been isolated recently from porcine nonantral stomach and its amino acid sequence has been determined (MCDONALD et al. 1978, 1979). The authors have called this peptide "gastrin-releasing peptide." It has 9 amino acids of its COOH terminal decapeptide in common with amphibian skin bombesin (Table 7). POLAK et al. (1978) and BLOOM et al. (1979) initially reported the presence of bombesin-like immunoreactivity in intestinal nerves and, soon after, DOCKRAY et al. (1979) described the distribution of bombesin-like immunoreactivity in nerves of the rat gastrointestinal tract. In the stomach, axons were found in the myenteric and submucous plexuses and in the mucosa. There were few nerves in the muscle. In small and large intestine, most axons were in the myenteric plexus, with a sparse supply to the submucous plexus. There were few axons in the muscle or mucosa. No nerve cell bodies were reported. Acidified boiling water extracts eluted on Sephadex G 50 resin revealed two molecular forms with the bombesin-like immunoreactivity. The smaller form had a similar elution profile to amphibian bombesin. This small form accounted for about 90% of the activity recognised by RIA in extracts of intestinal muscle including the myenteric plexus. There is thus good evidence for the presence of a peptide with some similarity to bombesin in intestinal nerves, but there is no evidence for its release as a neurotransmitter.

VIII. Neurotensin

Neurotensin has been isolated from extracts of bovine hypothalamus and small intestine and its sequence determined (CARRAWAY and LEEMAN 1975; CARRAWAY et al. 1978). In both cases an identical 13 amino acid sequence was obtained (Table 7). Experiments with extracts made from rat intestine suggest that an identical peptide occurs in this species (CARRAWAY and LEEMAN 1976). The neurotensin-like material which was identified by RIA behaved as bovine neurotensin on chromatography columns and had similar susceptibility to peptidases. There is one report of neurotensin-like immunoreactivity in nerves in the intestine (SCHULTZBERG et al. 1980). Axons were found in the myenteric plexus throughout the gastrointestinal tract, in the circular muscle of the stomach and caecum and in the gastric mucosa of guinea-pig and rats. No nerve cell bodies showing a positive reaction were found.

The observations of SCHULTZBERG et al. (1980) require confirmation. In other immunohistochemical studies of neurotensin localisation in the gastrointestinal tract it was only found in mucosal endocrine cells (ORCI et al. 1976; SUNDLER et al. 1977a, c; FRIGERIO et al. 1977; HELMSTAEDTER et al. 1977; BUCHAN et al. 1978). HELMSTAEDTER et al. (1977) and SUNDLER et al. (1977c) stated explicitly that they were unable to detect immunoreactive nerves in the intestine. Measurements by RIA in humans and in rats indicate that most of the extractable neurotensin-like activity is in the mucosa of the jejunum and ileum and that relatively little activity is in the oesophagus, stomach, duodenum or colon (CARRAWAY and LEEMAN 1976; BUCHAN et al. 1978).

IX. Pancreatic Polypeptide

Pancreatic polypeptide was originally isolated from the pancreas and its sequence determined (KIMMEL et al. 1975). In the pancreas it is contained in a subpopulation of islet cells (LARSSON et al. 1974, 1975). LOREN et al. (1979) have now reported immunoreactivity resembling that of pancreatic peptide in axons and nerve cell bodies of the stomach and intestines of cats, mice and rats. Axons were most numerous in the myenteric plexus and were also in the smooth muscle, around submucosal blood vessels and in the villi of the small intestine. Immunoreactive cell bodies were seen in both ganglionated plexuses. The precise identity of the peptide located in this work is yet to be determined.

X. Angiotensin

Angiotensin-like immunoreactivity was briefly reported in intestinal neurones by GANTEN et al. (1979). We have confirmed that angiotensin-like activity can be detected (M. COSTA and J. B. FURNESS unpublished work 1979). However, angiotensin has not been extracted from the external, nerve-containing layers, so the identity of the tissue antigen remains unresolved (but see FURNESS et al. 1982).

XI. Other Peptides

There are a number of additional peptides that affect gastrointestinal function and that are apparently present in gut tissues, but have yet to be definitely located in nerves. Amongst these is thyrotropin-releasing hormone (TRH). TRH-like immunoreactivity has been found in extracts from rat stomach, intestine and pancreas (MORLEY et al. 1977; LEPPÄLUOTO et al. 1978). However, the immunoreactive material is present in low concentrations in gut compared with central nervous system and the intestinal concentrations are also low compared with those of peptides which have been clearly demonstrated in intestinal nerves. There is some doubt whether the material is authentic TRH (YOUNGBLOOD et al. 1979) and it has yet to be demonstrated in gut nerves. There is also RIA evidence for low concentrations of motilin in the muscle layers of the gastrointestinal tract (CHEY et al. 1980; CHEY and LEE 1980; FOX et al. 1980). CHEY and collaborators have also reported the immunohistochemical localisation of motilin in enteric nerves, but attempts to confirm this in our laboratory have so far been unsuccessful. Earlier immunohistochemical studies reported motilin in gastrointestinal endocrine cells, but not in nerves (PEARSE et al. 1974; POLAK et al. 1975; DEMLING and DOMSCHKE 1976; TOBE et al. 1976; HELMSTAEDTER et al. 1979), although the authors seemed primarily concerned with the endocrine location. Extracts of rat and human intestine contain a peptide similar to hydra-head-activating peptide, but the cell type which is the source of this peptide is unknown (SCHALLER et al. 1977; BÖDENMÜLLER et al. 1980).

K. Summary

Summarising comments on 16 biologically active substances whose presence within intestinal nerves has been demonstrated or proposed are given below.

I. Acetylcholine

Acetylcholine fulfils, beyond reasonable doubt, the primary and secondary criteria which allow its acceptance as a neurotransmitter. Its presence, synthesis and action-potential-dependent release have been demonstrated. Acetylcholine mimics the effect of the endogenous transmitter and its action is similarly affected by drugs.

II. Noradrenaline

Noradrenaline is also, beyond significant doubt, an intestinal neurotransmitter. This amine and its sythesising enzymes are present, it is released when the nerves are activated and it mimics the effect of nerve stimulation.

III. 5-Hydroxytryptamine

5-Hydroxytryptamine or a similar substance is likely to be an intestinal transmitter. A reservation in fully accepting it as a neurotransmitter is the failure to identify a role for 5-HT nerves. On the other hand, nerves containing synthesising and degradative enzymes for 5-HT and which take up and retain 5-HT are present. There is evidence that both 5-HT and its binding protein are released when intestinal nerves are stimulated.

IV. Dopamine

This amine is unlikely to be an intestinal transmitter. Its presence, or the presence of its synthesising enzyme, tyrosine hydroxylase, cannot be demonstrated in intrinsic intestinal nerves which have been postulated to utilise dopamine as a neurotransmitter.

V. γ-Aminobutyric Acid

The evidence for GABA neurones is so far circumstantial, based on observations that some intestinal neurones accumulate radiolabelled GABA and that low levels of the synthesising enzyme, glutamic acid decarboxylase, is present in extracts from the intestine.

VI. Adenosine 5′-Triphosphate

ATP was originally proposed as the transmitter utilised by the enteric inhibitory nerves which are involved in descending reflexes causing relaxation of gastrointestinal muscle. However, more recent observations on its release and ability to mimic nerve stimulation seem inconsistent with its being the transmitter.

VII. Substance P

This peptide is contained in intestinal nerves and released by action potentials. It seems likely to be an excitatory neurotransmitter to neurones and smooth muscle, but its physiological roles at these sites is unknown. Other populations of substance P neurones with different projections are likely to fulfil different roles.

VIII. Somatostatin

Although this peptide is contained in specific populations of intestinal nerves, its release from these nerves has yet to be demonstrated. There is speculation that it is a transmitter in interneurones of a descending reflex pathway.

IX. Vasoactive Intestinal Polypeptide

VIP is contained in intestinal nerves and is released when the nerves are stimulated either electrically or reflexly. It is possibly the transmitter released by the enteric inhibitory nerves. Certain intrinsic intestinal vasodilator nerves may also utilise VIP as a transmitter.

X. Enkephalins

Enkephalins have been shown to be contained in, synthesised and released by enteric nerves. However, the way or ways in which their effects on the intestine are manifested in vivo are not clear.

XI. Gastrin

The evidence for true gastrin in intestinal nerves is still controversial.

XII. Cholecystokinin COOH Terminal Peptides

COOH terminal peptides of CCK, either one or a combination of CCK 8 and CCK 4 are present in intestinal nerves. Their release has not been shown.

XIII. Bombesin-Like Peptide

The form of this peptide found in intestinal nerves has not been elucidated, nor has its release been demonstrated.

XIV. Neurotensin

It remains to be confirmed that this peptide is indeed present in intestinal nerves.

XV. Pancreatic Polypeptide

Immunohistochemical evidence for the presence of pancreatic polypeptide in gut nerves has been obtained, but the molecular form in the nerves is not known, and its release has not been detected.

XVI. Angiotensin

The presence of angiotensin in enteric nerves requires confirmation.

XVII. Conclusions

It is emphasised that this summary represents an assessment of the current literature: we have now entered a new and exciting phase of research on nerves of the gastrointestinal system and it is likely that many of the present conclusions will be rendered obsolete by observations made in the near future.

For only two of the active compounds in intestinal nerves, actylcholine and noradrenaline, does the experimental data justify concluding that they are neurotransmitters. It seems likely that 5-HT is a transmitter, but the role of nerves utilising this amine are not known. There is considerable evidence that ATP is not the transmitter of the enteric inhibitory nerves, where it has been proposed to be. It is quite unlikely that dopamine is a transmitter. For the other substances there are too few data. The peptides can perhaps be divided into three groups: those whose roles as chemical messengers in the intestine seem likely, that is, substance P, VIP and enkephalins; those peptides whose presence seems beyond reasonable doubt, but whose possible roles as neurotransmitters are little explored, that is, COOH terminal fragments of CCK, somatostatin and bombesin-like peptide; and those whose presence and/or molecular form is not yet sufficiently well established (gastrin, neurotensin, pancreatic polypeptide and angiotensin). All the peptides which are confirmed or suspected to be present in intestinal nerves are biologically active substances. However, there is no certainty that these act as neurotransmitters in the generally accepted sense of substances which are released when an action potential invades a nerve terminal and which have an acute effect on the excitability of an adjacent cell. For the moment, these peptides should be looked upon as biologically active substances which are present in nerves, but have unknown roles.

Note Added in Proof

More publications continue to appear in this field. Of those to appear since the submission of the manuscript, the following are of particular relevance.

MACKENZIE and BURNSTOCK (1980) have compared the relaxing effect of ATP, VIP and stimulation of enteric inhibitory nerves in the presence of the proteolytic enzyme α-chymotrypsin and a component of bee venom, apamin. Chymotrypsin significantly reduced the effect of VIP of the muscle, without altering the actions of ATP and nerve stimulation. On the other hand, apamin significantly reduced or abolished relaxations caused by ATP or nerve stimulation while only slightly reducing those of VIP. These results suggest that VIP is not the transmitter released by enteric inhibitory nerves (see Sect. H). A possible interpretation of other observations in the light of these new findings is that VIP is released from the nerves along with another substance which relaxes the muscle.

LAZARUS et al. (1980) used radioimmunoassay to show physalemin-like activity in various parts of the gastrointestinal tract of guinea-pig, mouse, pig, rabbit, and rat. With immunohistochemical methods, physalemin-like activity was found in axons of the myenteric and submucous plexuses and in axons of the circular muscle of the lower oesophageal sphincter.

MORITA K, NORTH RA, KATAYAMA Y (1980): Both the slow e.p.s.p. in nerve cells of the myenteric plexus in guinea-pigs and the slow depolarisation caused by substance P were reversibly depressed by chymotrypsin, an enzyme that degrades substance P, although responses to other agonists were not affected. This provides further evidence that substance P could be an excitatory transmitter in the myenteric plexus.

KRANTIS A, KERR DIB (1981): GABA was found to be concentrated by a small proportion of nerve cell bodies in the myenteric plexus and to be also taken up by nerve fibres.

COSTA M, FURNESS JB, CUELLO AC, VERHOFSTAD AAJ, STEINBUSCH HWM, ELDE RP (1982): Three different antibody preparations raised against 5-HT revealed a population of reactive cell bodies in the myenteric plexus and of nerve fibres in the myenteric and submucous plexuses in guinea-pig, mice, rabbit and rat. The properties of the neurons were studied in guinea-pig. The immunoreactivity was depleted by reserpine but not by guanethidine, 6-hydroxydopamine or 5,7-dihydroxytryptamine. Immunoreactivity was restored with 5-HT or 5-hydroxytryptophan. A comparison of the distribution of nerves with 5-HT-like immunoreactivity and of enteric amine-handling neurons revealed by L-dopa loading indicated that these are separate populations. Thus there are two classes of aromatic amine neuron in the guinea-pig small intestine: the enteric 5-HT neurons and enteric, non 5-HT, amine handling neurons.

FURNESS JB, COSTA M (1982): Microsurgical lesions were used to interrupt pathways and the tissue was examined histochemically to determine the projections of enteric 5-HT neurons. 5-HT nerve cell bodies in the myenteric plexus send processes in an anal direction. Some of these processes pass through myenteric ganglia, giving rise to small varicose branchlets which divide within the ganglia. These processes average about 15 mm in length. Other processes of 5-HT cell bodies in the myenteric ganglia enter the submucosa to supply varicosities to the ganglia.

HUTCHISON JB, DIMALINE R, DOCKRAY GJ (1981): This study indicates that VIP in nerves of the guinea-pig intestine is chemically somewhat different from the standard porcine VIP. The major form of cholecystokinin in the nerves seems to be CCK 8. Bombesin-like activity is due to a small form similar to amphibian skin bombesin and an unidentified larger form.

GERSHON MD (1981): This review deals in particular with the evidence that 5-HT is a neurotransmitter in the enteric nervous system.

FURNESS JB, COSTA M, MURPHY R, BEARDSLEY AM, OLIVER JR, LLEWELLYN-SMITH IJ, ESKAY RL, SHULKES AA, MOODY TW, MEYER DK (1982): Relative concentrations of fourteen potential neurotransmitters have been compared in the guinea-pig small intestine. It was concluded that angiotensin and neurotensin are unlikely to be neurotransmitters in this species.

TONINI M et al. (1981): Desensitization of receptors for ATP in the longitudinal muscle of the rabbit distal colon caused a 20-fold shift to be right of the concentration response curve for ATP, but did not alter the frequency response relationship obtained when the enteric inhibitory nerves were stimulated at 0.5 to 5 Hz.

CREMA A et al. (1982): Desensitization of receptors for ATP in the circular muscle of the rabbit distal colon caused an eight fold shift to the right of the dose response curve for ATP, but did not affect the relaxations caused by either the reflex or the electrical stimulation of enteric inhibitory nerves.

Acknowledgements. Work from the authors' laboratory was supported by the Australian Research Grants Committee and the National Health and Medical Research Council of Australia. It was brought to fruition in large measure through the dedicated assistance of VENETTA ESSON and PAT VILLIMAS and the collaboration of Dr. IDA LLEWELLYN-SMITH, Dr. RONY FRANCO and Dr. ALAN WILSON. This chapter was written during the tenure of a Fulbright Fellowship awarded to JBF.

References

Abe H, Appert H, Carballo J, Howard JM (1973) Nonmucosal serotonin in motility of the small bowel when subjected to acute anoxic trauma. Arch Surg 106:183–187

Abood LG, Koketsu K, Miyamoto S (1962) Outflux of various phosphates during membrane depolarisation of excitable tissues. Am J Physiol 202:469–474

Ahlman H, Enerback L (1974) A cytofluorometric study of the myenteric plexus in the guinea-pig. Cell Tissue Res 153:419–434

Ahlman H, Enerback L, Kewenter J, Storm B (1973) Effects of extrinsic denervation on the fluorescence of monoamines in the small intestine of the cat. Acta Physiol Scand 89:429–435

Ahlman H, Lundberg J, Dahlstrom A, Kewenter J (1976) A possible vagal adrenergic release of serotonin from enterochromaffin cells in the cat. Acta Physiol Scand 98:366–375

Albinus M, Blair EL, Case RM et al. (1977) Comparison of the effect of somatostatin on gastrointestinal function in the conscious and anaesthetized cat and on the isolated cat pancreas. J Physiol (Lond) 269:77–91

Ally AI, Nakatsu K (1976) Adenosine inhibition of isolated rabbit ileum and antagonism by theophylline. J Pharmacol Exp Ther 19:208–215

Alumets J, Håkanson R, Sundler F, Chang K-J (1978a) Leuenkephalin-like material in nerves and enterochromaffin cells in the gut. Histochemistry 56:187–196

Alumets J, Håkanson R, Sundler F, Uddman R (1978b) VIP innervation of sphincters. Scand J Gastroenterol [Suppl 49] 13:6

Alund M, Olson L (1979) Quincrine affinity of endocrine cell systems containing dense core vesicles as visualized by fluorescence microscopy. Cell Tissue Res 204:171–186

Ambache N (1955) The use and limitations of atropine for pharmacological studies on autonomic effectors. Pharmacol Rev 7:467–494

Ambache N, Freeman MA (1968) Atropine-resistant longitudinal muscle spasms due to excitation of non-cholinergic neurons in Auerbach's plexus. J Physiol (Lond) 199:705–727

Ambache N, Verney J, Zar MA (1970) Evidence for the release of two atropine-resistant spasmogens from Auerbach's plexus. J Physiol (Lond) 207:761–782

Ambache N, Daly S, Killick SW, Woodley JP (1977) Differentiation of neurogenic inhibition from ATP-responses in guinea-pig taenia caeci. Br J Pharmacol 61:113–114P

Ambinder RF, Schuster MM (1979) Endorphins: new gut peptides with a familiar face. Gastroenterology 77:1132–1140

Anastasi A, Erspamer V, Bucci M (1971) Isolation and structure of bombesin and alytesin, two analogous active peptides from the skin of the European amphibians bombina and alytes. Experientia 27:166–167

Aström A (1949) Anti-sympathetic action of sympathomimetic amines. Acta Physiol Scand 18:295–307

Axelsson J, Holmberg B (1969) The effects of extracellularly applied ATP and related compounds on electrical and mechanical activity of the smooth muscle taenia coli from the guinea-pig. Acta Physiol Scand 75:149–156

Bacq ZM, Goffart M (1939) L'acetylcholine libre du sang veneux du tube digestif chez le chien. Arch Int Physiol 49:179–188

Baer HP, Frew R (1979) Relaxation of guinea-pig fundic strip by adenosine, adenosine triphosphate and electrical stimulation: lack of antagonism by theophylline or ATP treatment. Br J Pharmacol 67:293–300

Barbezat GO, Grossman MI (1971) Intestinal secretion: stimulation by peptides. Science 174:422–423

Barlow RB, Khan I (1959) Actions of some analogues of tryptamine on the isolated rat uterus and on isolated rat fundus strip preparations. Br J Pharmacol 14:99–107

Barrington EJW, Dockray GJ (1976) Gastrointestinal hormones. J Endocrinol 69:299–325

Bartlett V, Stewart RR, Nakatsu K (1979) Evidence for two adenine derivative receptors in rat ileum which are not involved in the nonadrenergic, noncholinergic response. Can J Physiol Pharmacol 57:1130–1137

Baumgarten HG, Lange W (1969) Adrenergic innervation of the oesophagus in the cat (felis domestica) and rhesus monkey (macacus rhesus). Z Zellforsch 95:529–545

Beani L, Bianchi C, Crema A (1969) The effect of catecholamines and sympathetic stimulation on the release of acetylcholine from the guinea-pig colon. Br J Pharmacol 36:1–17
Beani L, Bianchi C, Crema A (1971) Vagal non-adrenergic inhibition of guinea-pig stomach. J Physiol (Lond) 217:259–279
Behar J, Field S, Marin C (1979) Effect of glucagon, secretion, and vasoactive intestinal polypeptide on the feline lower esophageal sphincter: mechanisms of action. Gastroenterology 77:1001–1007
Bennett A, Fleshler B (1969) A hyoscine-resistant excitatory nerve pathway in guinea-pig colon. J Physiol (Lond) 203:62–63 P
Bennett MR (1966) Rebound excitation of the smooth muscle cells of the guinea-pig taenia after stimulation of intramural inhibitory nerves. J Physiol (Lond) 185:124–131
Bennett MR, Rogers DC (1967) A study of the innervation of the taenia coli. J Cell Biol 33:573–596
Bertaccini G (1976) Active polypeptides of non-mammalian origin. Pharmacol Rev 28:127–177
Bertaccini G, Erspamer V, Melchiorri P, Sopranzi N (1974) Gastrin release by bombesin in the dog. Br J Pharmacol 52:219–225
Bertler A, Falck B, Hillarp NA, Rosengren E, Torp A (1959) Dopamine and chromaffin cells. Acta Physiol Scand 47:251–258
Bianchi C, Beani L, Crema A (1970) Effects of metoclopramide on isolated guinea-pig colon. 2. Interference with ganglionic stimulant drugs. Eur J Pharmacol 12:332–341
Biber B (1973) Vasodilator mechanisms in the small intestine. Acta Physiol Scand [Suppl] 401:1–31
Biber B, Lundgren O, Svanvik J (1971) Studies on the intestinal vasodilation observed after mechanical stimulation of the mucosa of the gut. Acta Physiol Scand 82:177–190
Biber B, Fara J, Lundgren O (1973 a) Intestinal vasodilation in response to transmural electrical field stimulation. Acta Physiol Scand 87:277–282
Biber B, Fara J, Lundgren O (1973 b) Intestinal vascular response to 5-hydroxytryptamine. Acta Physiol Scand 87:526–534
Biber B, Fara J, Lundgren O (1974) A pharmacological study of intestinal vasodilatory mechanisms in the cat. Acta Physiol Scand 90:673–683
Bitar KN, Zfass AM, Saffouri B, Said SI, Makhlouf GM (1979) Release of VIP from nerves in the gut. Gastroenterology 76:1101
Björklund A, Baumgarten HG, Nobin A (1974) Chemical lesioning of central monoamine axons by means of 5,6-dihydroxytryptamine and 5,7-dihydroxytryptamine. Adv Biochem Psychopharmacol 10:13–33
Blaschke E, Uvnas B (1979) Effect of splenic nerve stimulation on the contents of noradrenaline, ATP and sulphomucopolysaccharides in noradrenergic vesicle fractions from the cat spleen. Acta Physiol Scand 105:496–507
Bloom SR (ed) (1978) Gut hormones. Churchill Livingstone Edinburgh London
Bloom SR, Edwards AV (1980) Effects of autonomic stimulation on the release of vasoactive intestinal peptide from the gastrointestinal tract in the calf. J Physiol (Lond) 299:437–452
Bloom SR, Ralphs DN, Besser GM, Hall R, Coy DH, Kastin AJ, Schally AV (1975) Effect of somatostatin on motilin levels and gastric emptying. Gut 16:834
Bloom SR, Ghatei MA, Wharton JW, Polak JM, Brown MR (1979) Distribution of bombesin in human alimentary tract. Gastroenterology 76:1103
Bodenmüller H, Schaller HC, Darai G (1980) Human hypothalamus and intestine contain a hydra neuropeptide. Neurosci Lett 16:71–74
Borison HL (1959) Effect of oblation of medullary chemoreceptor trigger zone on vomiting responses to cerebral intraventricular injection of adrenaline, amorphine and pilocarpine in the cat. J Physiol (Lond) 147:172–177
Boullin DJ (1964) Observations on the significance of 5-hydroxytryptamine in relation to the peristaltic reflex of the rat. Br J Pharmacol 23:14–33
Boura ALA, Green AF (1959) The actions of bretylium: adrenergic neurone blocking and other effects. Br J Pharmacol 14:536–548
Bowman WC, Hall MT (1970) Inhibition of rabbit intestine mediated by α- and β-adrenoceptors. Br J Pharmacol 38:399–415

Boyd G, Gillespie JS, MacKenna BR (1962) Origin of the cholinergic response of the rabbit intestine to stimulation of its extrinsic sympathetic nerves after exposure to sympathetic blocking agents. Br J Pharmacol 19:258–270
Brazeau P, Vale W, Burgus R, Ling N, Butcher M, Rivier J, Guillemin R (1973) Hypothalamic polypeptide that inhibits the secretion of immunoreactive pituitary growth hormone. Science 179:77–79
Brodie BB, Bogdanski DF, Bonomi L (1964) Formation, storage and metabolism of serotonin and catecholamines in lower vertebrates. In: Richter D (ed) Comparative neurochemistry. Macmillan, New York, pp 367–377
Brown C, Burnstock G, Cocks T (1979) Effects of adenosine 5′-triphosphate (ATP) and β-γ-methylene ATP on the rat urinary bladder. Br J Pharmacol 65:97–102
Brownlee G, Johnson ES (1963) The site of 5-hydroxytryptamine receptor on the intramural nervous plexus of the guinea-pig isolated ileum. Br J Pharmacol 21:306–322
Bryant MG, Polak JM, Modlin I, Bloom SR, Albuquerque RH, Pearse AGE (1976) Possible dual role for vasoactive intestinal peptide as gastrointestinal hormone and neurotransmitter substance. Lancet 1:991–993
Buchan AMJ, Polak JM, Sullivan S, Bloom SR, Brown M, Pearse AGE (1978) Neurotensin in the gut. In: Bloom SR (ed) Gut hormones. Churchill Livingstone, Edinburgh London, pp 544–549
Bueding E, Bülbring E, Gercken G, Hawkins JT, Kuriyama H (1967) The effect of adrenaline on the adenosine triphosphate and creatinine phosphate content of intestinal smooth muscle. J Physiol (Lond) 193:187–212
Bülbring E, Crema A (1958) Observations concerning the action of 5-hydroxytryptamine on the peristaltic reflex. Br J Pharmacol 13:444–457
Bülbring E, Crema A (1959) The action of 5-hydroxytryptamine, 5-hydroxytryptophan and reserpine on intestinal peristalsis in anaesthetized guinea-pigs. J Physiol (Lond) 146:29–53
Bülbring E, Gershon MD (1967) 5-Hydroxytryptamine participation in the vagal inhibitory innervation of the stomach. J Physiol (Lond) 192:823–846
Burks TF (1973) Mediation by 5-hydroxytryptamine of morphine stimulant actions in dog intestine. J Pharmacol Exp Ther 185:530–539
Burks TF (1976) Acute effects of morphine on rat intestinal motility. Eur J Pharmacol 40:279–283
Burks TF, Grubb MN (1974) Sites of acute morphine tolerance in the intestine. J Pharmacol Exp Ther 191:518–526
Burks TF, Grubb MN (1978) Stimulatory actions of adenosine triphosphate in dog intestine. In: Duthie HL (ed) Gastrointestinal motility in health and disease. University Park Press, Baltimore, pp 151–159
Burks TF, Long JP (1967) Responses of isolated dog small intestine to analgesic agents. J Pharmacol Exp Ther 158:264–271
Burleigh DE, D'Mello A, Parks AG (1979) Responses of isolated human internal anal sphincter to drugs and electrical field stimulation. Gastroenterology 77:484–490
Burnstock G (1970) Structure of smooth muscle and its innervation. In: Bülbring E, Brading A, Jones A, Tomita T (eds) Smooth muscle. Arnold, London, pp 1–69
Burnstock G (1972) Purinergic nerves. Pharmacol Rev 24:509–581
Burnstock G (1975) Purinergic transmission. In: Iversen LL, Iversen S, Snyder S (eds) Handbook of Psychopharmacology, vol 5. Plenum, New York, pp 131–194
Burnstock G (1979) Past and current evidence for the purinergic nerve hypothesis. In: Baer HP, Drummond GI (eds) Physiological and regulatory functions of adenosine and adenine nucleotides. Raven, New York, pp 3–32
Burnstock G, Campbell G, Bennett M, Holman ME (1964) Innervation of the guinea-pig taenia coli: are there intrinsic inhibitory nerves which are distinct from sympathetic nerves? Int J Neuropharmacol 3:163–166
Burnstock G, Campbell G, Rand MJ (1966) The inhibitory innervation of the taenia of the guinea-pig caecum. J Physiol (Lond) 182:504–526
Burnstock G, Campbell G, Satchell D, Smythe A (1970) Evidence that adenosine triphosphate or a related nucleotide is the transmitter substance released by non-adrenergic nerves in the gut. Br J Pharmacol 40:668–688

Burnstock G, Satchell DG, Smythe A (1972) A comparison of the excitatory and inhibitory effects of non-adrenergic, non-cholinergic nerve stimulation and exogenously applied ATP on a variety of smooth muscle preparations from different vertebrate species. Br J Pharmacol 46:234–242

Burnstock G, Cocks T, Crowe R (1978 a) Evidence for purinergic innervation of the anococcygeus muscle. Br J Pharmacol 64:13–20

Burnstock G, Cocks T, Kasakov L, Wong HK (1978 b) Direct evidence for ATP release from non-adrenergic, non-cholinergic ("purinergic") nerves in the guinea-pig taenia coli and bladder. Eur J Pharmacol 49:145–149

Burnstock G, Hökfelt T, Gershon MD, Iversen LL, Kosterlitz HW, Szurszewski JH (1979) Non-adrenergic, non-cholinergic autonomic neurotransmission mechanisms. Neurosci Res Program Bull 17:379–519

Bybee DE, Brown FL, Georges LP, Castell DO, McGuigan JE (1979) Somatostatin effects on lower esophageal sphincter function. Am J Physiol 237:E 77–81

Campbell G (1966) Nerve-mediated excitation of the taenia of the guinea-pig caecum. J Physiol (Lond) 185:148–159

Campbell G (1970) Autonomic nervous supply to effector tissues. In: Bülbring E, Brading A, Jones A, Tomita T (eds) Smooth muscle. Arnold, London, pp 451–495

Campbell G, Burnstock G (1968) Comparative physiology of gastrointestinal motility. In: Code CF (ed) Alimentary canal. American Physiological Society, Washington, DC (Handbook of physiology, vol IV, sect 6, pp 2213–2266

Cannon WB, Rosenblueth A (1937) Autonomic neuroeffector systems. MacMillan, New York

Carpenter DO, Swann JW, Yarowsky PJ (1977) Effect of curare on responses to different putative neurotransmitters in aplysia neurons. J Neurobiol 8:119–132

Carraway R, Leeman SE (1975) The amino acid sequence of a hypothalamic peptide, neurotensin. J Biol Chem 250:1907–1911

Carraway R, Leeman SE (1976) Characterization of radioimmunoassayable neurotensin in the rat: its differential distribution in the central nervous system, small intestine and stomach. J Biol Chem 251:7045–7052

Carraway R, Kitabgi P, Leeman SE (1978) The amino acid sequence of radioimmunoassayable neurotensin from bovine intestine. J Biol Chem 253:7996–7998

Chang HC, Gaddum JH (1933) Choline esters in tissue extracts. J Physiol (Lond) 79:255–285

Chang MM, Leeman SE (1970) Isolation of a sialogogic peptide from bovine hypothalamic tissue and its characterization as substance P. J Biol Chem 245:4784–4790

Chang MM, Leeman SE, Niall HD (1971) Amino-acid sequence of substance P. Nature New Biol 232:86–87

Chan-Palay V, Jonsson G, Palay SL (1978) Serotonin and substance P coexist in neurons of the rat's central nervous system. Proc Natl Acad Sci USA 75:1582–1586

Chey WY, Lee KY (1980) Motilin. Clin Gastroenterol 9:645–656

Chey WY, Escoffery R, Roth F, Chang TM, Yajima HL (1980) Motilin-like immunoreactivity (MLI) in the gut and neurones of peripheral and central nervous system. Gastroenterology 78:1150

Christensen J (1975) Pharmacology of the esophageal motor function. Annu Rev Pharmacol Toxicol 15:243–258

Chujyo N (1953) Site of acetylcholine production in the wall of the intestine. Am J Physiol 174:196–198

Coburn RF, Tomita T (1973) Evidence for noradrenergic inhibitory nerves in the guinea-pig trachealist muscle. Am J Physiol 224:1072–1080

Cocks T, Burnstock G (1979) Effects of neuronal polypeptides on intestinal smooth muscle: a comparison with non-adrenergic, non-cholinergic nerve stimulation and ATP. Eur J Pharmacol 54:251–259

Cohen ML, Landry AS (1980) Vasoactive intestinal polypeptide: increased tone, enhancement of acetylcholine release, and stimulation of adenylate cyclase in intestinal smooth muscle. Life Sci 26:811–822

Cohen ML, Rosing E, Wiley KS, Slater IH (1978) Somatostatin inhibits adrenergic and cholinergic neurotransmission in smooth muscle. Life Sci 23:1659–1664

Cohen ML, Wiley KS, Yaden E, Slater IH (1979) In vitro actions of somatostatin d-Val[1], d-Trp[8]-somatostatin and glucagon in rabbit jejunum and guinea-pig ileum. J Pharmacol Exp Ther 211:423–429

Coleman RA, Levy GP (1974) A noradrenergic inhibitory nervous pathway in guinea-pig trachea. Br J Pharmacol 52:167–174

Collins GGS, West GW (1968) Some pharmacological actions of diethyldithiocarbamate on rabbit and rat ileum. Br J Pharmacol 32:402–409

Costa M, Furness JB (1971) Storage, uptake and synthesis of catecholamines in the intrinsic adrenergic neurones in the proximal colon of the guinea-pig. Z Zellforsch 120:364–385

Costa M, Furness JB (1972) Slow contraction of the guinea-pig proximal colon in response to the stimulation of an unidentified type of nerve. Br J Pharmacol 45:151 P–152 P

Costa M, Furness JB (1974) The innervation of the internal and sphincter of the guinea-pig. In: Daniel EE (ed) Proceedings of the 4th International Symposium on Gastrointestinal Motility. Mitchell, Vancouver, pp 681–689

Costa M, Furness JB (1976) The peristaltic reflex: an analysis of nerve pathways and their pharmacology. Naunyn-Schmiedeberg Arch Pharmacol 294:47–60

Costa M, Furness JB (1979 a) On the possibility that an indoleamine is a neurotransmitter in the gastrointestinal tract. Biochem Pharmacol 28:565–571

Costa M, Furness JB (1979 b) The sites of action of 5-hydroxytryptamine in nervemuscle preparations from the guinea-pig small intestine and colon. Br J Pharmacol 65:237–248

Costa M, Furness JB, McLean JR (1976) The presence of aromatic l-amino acid decarboxylase in certain intestinal nerve cells. Histochemistry 48:120–143

Costa M, Patel Y, Furness JB, Arimura A (1977) Evidence that some intrinsic neurons of the intestine contain somatostatin. Neurosci Lett 6:215–222

Costa M, Cuello C, Furness JB, Franco R (1980 a) Distribution of enteric neurons showing immunoreactivity for substance P in the guinea-pig ileum. Neuroscience 5:323–331

Costa M, Furness JB, Buffa R, Said I (1980 b) Distribution of enteric neurons showing immunoreactivity for vasoactive intestinal polypeptide (VIP) in the guinea-pig intestine. Neuroscience 5:587–596

Costa M, Furness JB, Llewellyn-Smith IJ, Cuello C (1980 c) Projections of substance P neurons within the guinea-pig small intestine. Neuroscience 6:411–424

Costa M, Furness JB, Llewellyn-Smith IJ, Davies B, Oliver J (1980 d) An immunohistochemical study of the projections of somatostatin containing neurons in the guinea-pig intestine. Neuroscience 5:841–852

Costa M, Furness JB, Verhofstad A, Steinbusch HWM, Elde R (1980 e) Presence and distribution of 5-hydroxytryptamine-like immunoreactivity in enteric neurons in the guinea-pig small intestine. Proc Aust Physiol Pharmacol Soc 11:18 P

Costa M, Furness JB, Cuello AC, Verhofstad AAJ, Steinbusch HWM, Elde RP (1982) Neurons with 5-hydroxytryptamine-like immunoreactivity in the enteric nervous system: their visualization and reactions to drug treatment. Neuroscience 7:351–363

Crema A, Benzi G, Frigo GM, Berte F (1965) The responses of the terminal bile duct to morphine and morphine like drugs. J Pharmacol Exp Ther 149:373–378

Crema A, D'Angelo L, Frigo GM, Lecchini S, Onori L, Tonini M (1982) Effects of desensitization to adenosine 5′-triphosphate and adenosine on non-adrenergic inhibitory responses in the circular muscle of rabbit colon. Br J Pharmacol (in press)

Creutzfeld W, Lankisch PG, Fölsch UR (1975) Hemmung der Secretin- und Cholecystokinin-Pancreozyminideuzierten Saft- und Enzymsekretion des Pankreas und der Gallenblasenkontraktion beim Menschen durch Somatostatin. Dtsch Med Wochensch 100:1135–1138

Curtis DR (1961) The identification of mammalian inhibitory transmitters. In: Florey E (ed) Neurons and Inhibition. Oxford, Pergamon, pp 342–349

Dale HH (1938) Natural chemical stimulators. Edinburgh Med J 45:461–480

Dale HH, Feldberg W (1934) The chemical transmitter of vagus effects to the stomach. J Physiol (Lond) 81:320–334

Daniel EE (1968) Pharmacology of the gastrointestinal tract. In: Code CF (ed) Alimentary canal. American Physiological Society, Washington, DC. Handbook of physiology, vol 4, sect 6, pp 2267–2324

Daniel EE (1979) Distribution of nonadrenergic inhibitory nerves in the intestine, their structural identification, and the role of prostaglandins in their function. In: Baer HP, Drummond GI (eds) Physiological and regulatory functions of adenosine and adenine nucleotides. Raven, New York, pp 61–68

Daniel EE, Sutherland WH, Bogoch A (1959) Effects of morphine and other drugs on motility of the terminal ileum. Gastroenterology 36:510–523

Demling L, Domschke W (eds) (1976) Motilin, origins, chemistry and actions. Scand J Gastroenterol [Suppl 39] 11:1–120

Den Hertog A, Jager LP (1975) Ion fluxes during the inhibitory junction potential in the guinea-pig taenia coli. J Physiol (Lond) 250:681–691

Diab IM, Dinerstein RJ, Watanabe M, Roth LJ (1976) [^{3}H]-Morphine localization in myenteric plexus. Science 193:689–691

Dikshit BB (1938) Acetylcholine formation by tissues. Q J Exp Physiol 28:243–251

Dimaline R, Dockray GJ (1978) Multiple immunoreactive forms of vasoactive intestinal polypeptide in human colonic mucosa. Gastroenterology 75:387–392

Dimaline R, Dockray GJ (1979) Molecular variants of vasoactive intestinal polypeptide in dog, rat and hog. Life Sci 25:1893–1900

Dingledine R, Goldstein A (1976) Effect of synaptic transmission blockade on morphine action in the guinea-pig myenteric plexus. J Pharmacol Exp Ther 196:97–106

Dockray GJ (1979) Evolutionary relationships of the gut hormones. Fed Proc 38:2295–2301

Dockray GJ, Vaillant C, Walsh JH (1979) The neuronal origin of bombesin-like immunoreactivity in the rat gastrointestinal tract. Neuroscience 4:1561–1568

Domschke W, Lux G, Domschke S, Strunz V, Bloom SR, Wunsch E (1978) Effects of vasoactive intestinal peptide on resting and pentagastrin-stimulated lower esophageal sphincter pressure. Gastroenterology 75:9–12

Donowitz M, Charney AN, Heffernan JM (1977) Effect of serotonin treatment on intestinal transport in the rabbit. Am J Physiol 232:E 85–E 94

Douglas WW (1968) Stimulus-secretion coupling: the concept and clues from chromaffin and other cells. Br J Pharmacol 34:451–474

Douglas WW, Poisner AM (1966) Evidence that the secreting adrenal chromaffin cell releases catecholamines directly from ATP-rich granules. J Physiol (Lond) 183:236–248

Drakontides AB, Gershon MD (1968) 5-Hydroxytryptamine receptors in the mouse duodenum. Br J Pharmacol 33:480–492

Dreyfus CF, Bornstein MB, Gershon MD (1977 a) Synthesis of serotonin by neurons of the myenteric plexus in situ and in organotypic tissue culture. Brain Res 128:125–139

Dreyfus CF, Sherman DL, Gershon MD (1977 b) Uptake of serotonin by intrinsic neurons of the myenteric plexus grown in organotypic tissue culture. Brain Res 128:109–123

Drury AN, Szent-Györgyi A (1929) The physiological activity of adenine compounds with especial reference to their action upon the mammalian heart. J Physiol (Lond) 68:213–237

Dubois A, Jacobowitz DM (1974) Failure to demonstrate serotonergic neurons in the myenteric plexus of the rat. Cell Tissue Res 150:493–496

Dubois A, Shea-Donohue PT, Myers L, Czerwinsky C, Castell DO (1979) Effect of somatostatin on gastric emptying, gastric output and serum gastrin. Gastroenterology 76:1121

Du Vigneaud V, Lawler HC, Popenoe EA (1953) Enzymatic cleavage of glycinamide from vasopressin and a proposed structure for this pressor-antidiuretic hormone of the posterior pituitary. J Am Chem Soc 75:4880–4881

Eccles JC (1964) The physiology of synapses. Springer, Berlin Heidelberg New York

Edin R, Lundberg JM, Ahlman H, Dahlström A, Fahrenkrug J, Hökfelt T, Kewenter I (1979) On the VIP-ergic innervation of the feline pylorus. Acta Physiol Scand 107:185–188

Edin R, Lundberg J, Terenius L, Dahlström A, Hökfelt T, Kewenter J, Ahlman H (1980) Evidence for vagal enkephalinergic neural control of the feline pylorus and stomach. Gastroenterology 78:492–497

Efendic S, Mattsson O (1978) Effect of somatostatin on intestinal motility. Acta Radiol [Diagn] (Stockh) 19:348–352

Eklund S, Jodal M, Lundgren O, Sjöqvist A (1979) Effects of vasoactive intestinal polypeptide on blood flow, motility and fluid transport in the gastrointestinal tract of the cat. Acta Physiol Scand 105:461–468
Elde R, Hökfelt T, Johansson O, Terenius L (1976) Immunohistochemical studies using antibodies to leucine enkephalin: initial observations on the nervous system of the rat. Neuroscience 1:349–351
Elliott TR (1904) On the action of adrenalin. J Physiol (Lond) 31:XX–XXI
Elliott TR (1905) The action of adrenaline. J Physiol (Lond) 32:401–467
Erspamer V, Melchiorri P (1973) Active polypeptides of the amphibian skin and their synthetic analogues. Pure Appl Chem 35:463–494
Euler US von (1942) Herstellung und Eigenschaften von Substanz P. Acta Physiol Scand 4:373–375
Euler US von, Gaddum JH (1931) An unidentified depressor substance in certain tissue extracts. J Physiol (Lond) 72:74–87
Fahrenkrug J, Galbo H, Holst JJ, Schaffalitzky de Muckadell OB (1978a) Influence of the autonomic nervous system on the release of vasoactive intestinal polypeptide from the porcine gastrointestinal tract. J Physiol (Lond) 280:405–422
Fahrenkrug J, Haglund U, Jodal M, Lundgren O, Olbe L, Schaffalitzky de Muckadell OB (1978b) Nervous release of vasoactive intestinal polypeptide in the gastrointestinal tract of cats: possible physiological implications. J Physiol (Lond) 284:291–305
Farrant J, Harvey JA, Pennefather JN (1964) The influence of phenoxybenzamine on the storage of noradrenaline in rat and cat tissues. Br J Pharmacol 22:104–112
Fasth S, Hulten L, Johnson BJ, Zeitlin IJ (1977a) Changes in the kinin system in cat colon during the atropine-resistant response to pelvic nerve stimulation. J Physiol (Lond) 265:56P
Fasth S, Hulten L, Lundgren O, Nordgren S (1977b) Vascular responses to mechanical stimulation of the mucosa of the cat colon. Acta Physiol Scand 101:98–104
Fasth S, Hulten L, Nordgren S (1980) Evidence for a dual pelvic nerve influence on large bowel motility in the cat. J Physiol (Lond) 298:159–170
Feher E (1974) Effect of monoamine oxidase inhibitor on the nerve elements of the isolated cat ileum. Acta Morphol Acad Sci Hung 22:249–263
Feher E (1977) Effect of 5,6-dihydroxytryptamine on the structure of nerve fibres in the chronically isolated cat ileum. Acta Anat 98:83–90
Feldberg W, Lin RCY (1950) Synthesis of acetylcholine in the wall of the gastrointestinal tract. J Physiol (Lond) 111:96–118
Ferreira SH, Vane JR (1967) The detection and estimation of bradykinin in the circulation. Br J Pharmacol 29:367–377
Filogamo G, Marchisio PC (1970) Choline acetyltransferase activity of rabbit ileum wall. The effects of extrinsic and intrinsic denervation and of combined experimental hypertrophy. Arch Int Physiol Biochem 78:141–152
Finkleman B (1930) On the nature of inhibition in the intestine. J Physiol (Lond) 70:145–157
Floren I (1979) Arguments against 5-hydroxytryptamine as neurotransmitter in the rabbit retina. J Neural Transm 46:1–15
Floyd JC, Fajans SS, Pek S, Chance RE (1977) A newly recognized pancreatic polypeptide; plasma levels in health and disease. Rec Prog Horm Res 33:519–570
Fox J, Said SI, Daniel EE (1979) Is vasoactive intestinal polypeptide (VIP) an inhibitory neurotransmitter in the lower esophageal sphincter (LES) in the north american opossum? Gastroenterology 76:1134
Fox JET, Track NS, Daniel EE (1980) Motilin: its presence and function in muscle layers of the gastrointestinal tract. In: Christensen J (ed) Gastrointestinal motility. Raven, New York, pp 59–65
Franco R, Costa M, Furness JB (1979a) Evidence for the release of endogenous substance P from intestinal nerves. Naunyn-Schmiedeberg Arch Pharmacol 306:185–201
Franco R, Costa M, Furness JB (1979b) Evidence that axons containing substance P in the guinea-pig ileum are of intrinsic origin. Naunyn-Schmiedeberg Arch Pharmacol 307:57–63
Franco R, Costa M, Furness JB (1979c) The presence of a cholinergic excitatory input to substance P neurons in the intestine. Proc Aust Physiol Pharmacol Soc 10:255P

Frigerio B, Ravazzola M, Ito S, Buffa R, Capella C, Solcia E, Orci L (1977) Histochemical and ultrastructural identification of neurotensin cells in the dog ileum. Histochemistry 54:123–132
Fukuda H, Shibata S (1972) The effect of cold storage on the inhibitory action of isoprenaline, phenylephrine and nicotine on the mechanical and membranal activities of guienea-pig taenia caecum. Br J Pharmacol 46:438–448
Fülgraff G, Schmidt L, Azokwu P (1964) Über die atropinresistente neuromuskulare Übertragung am Pelvicus-Colon-Präparat der Katze. Arch Int Pharmacodyn Ther 149: 537–551
Furchgott RF (1959) The receptors for epinephrine and norepinephrine (adrenergic receptors). Pharmacol Rev 11:429–441
Furness JB (1969) An electrophysiological study of the innervation of the smooth muscle of the colon. J Physiol (Lond) 205:549–562
Furness JB (1971) Secondary excitation of intestinal smooth muscle. Br J Pharmacol 41:213–226
Furness JB (1975) Nerve-muscle preparations of large intestine. In: Daniel EE, Paton DM (eds) Methods in pharmacology, vol 3. Plenum, New York, pp 385–390
Furness JB, Burnstock G (1975) Role of circulating catecholamines in the gastrointestinal tract. In: Handbook of physiology, vol 6, sect 7. American Physiological Society, Washington, DC, pp 515–536
Furness JB, Costa M (1971 a) Morphology and distribution of intrinsic adrenergic neurones in the proximal colon of the guinea-pig. Z Zellforsch 120:346–363
Furness JB, Costa M (1971 b) Monoamine oxidase histochemistry of enteric neurones of the guinea-pig. Histochemie 28:324–336
Furness JB, Costa M (1973) The nervous release and the action of substances which affect intestinal muscle through neither adrenoreceptors nor cholinoreceptors. Philos Trans R Soc Lond [Biol] 265:123–133
Furness JB, Costa M (1974) The adrenergic innervation of the gastrointestinal tract. Ergeb Physiol 69:1–51
Furness JB, Costa M (1976) Ascending and descending enteric reflexes in the isolated small intestine of the guinea-pig. Proc Aust Physiol Pharmacol Soc 7:172P
Furness JB, Costa M (1978) Distribution of intrinsic nerve cell bodies and axons which take up aromatic amines and their precursors in the small intestine of the guinea-pig. Cell Tissue Res 188:527–543
Furness JB, Costa M (1979 a) Actions of somatostatin on excitatory and inhibitory nerves in the intestine. Eur J Pharmacol 56:69–74
Furness JB, Costa M (1979 b) Projections of intestinal neurons showing immunoreactivity for vasoactive intestinal polypeptide are consistent with these neurons being the enteric inhibitory neurons. Neurosci Lett 15:199–204
Furness JB, Costa M (1980) Types of nerves in the enteric nervous system. Neuroscience 5:1–20
Furness JB, Costa M (1981) The enteric inhibitory nerves and VIP. In: Said SI (ed) Vasoactive intestinal peptide. Raven, New York (Advances in peptide hormone research vol 1)
Furness JB, Costa M (1982) Neurons with 5-hydroxytryptamine-like immunoreactivity in the enteric nervous system: their projections in the guinea-pig small intestine. Neuroscience 7:341–349
Furness JB, Costa M, Freeman CC (1979) Absence of tyrosine hydroxylase activity and dopamine β-hydroxylase immunoreactivity in intrinsic nerves of the guinea-pig ileum. Neuroscience 4:305–310
Furness JB, Costa M, Franco R, Llewellyn-Smith IJ (1980 a) Neuronal peptides in the intestine: distribution and possible functions. Adv Biochem Psychopharmacol 22:601–617
Furness JB, Costa M, Howe PRC (1980 b) Intrinsic amine-handling neurons in the intestine. In: Eränkö O (ed) Histochemistry and cell biology of autonomic neurones, sif cells and paraneurons. Raven, New York, pp 367–373
Furness JB, Costa M, Murphy R, Beardsley AM, Oliver JR, Llewellyn-Smith IJ, Eskay RL, Shulkes AA, Moody TW, Myer DK (1982) Detection and characterisation of neurotransmitters, particularly peptides, in the gastrointestinal tract. Scand J Gastroenterol, in press

Furness JB, Eskay RL, Brownstein MJ, Costa M (1980c) Characterization of somatostatin-like immunoreactivity in intestinal nerves by high pressure liquid chromatography and radioimmunoassay. Neuropeptides 1:97–104
Furness JB, Costa M, Walsh JH (1981) Evidence for and significance of the projection of VIP neurons from the myenteric plexus to the taenia coli in the guinea-pig. Gastroenterology 80:1557–1561
Gabella G (1979) Innervation of the gastrointestinal tract. Int Rev Cytol 59:129–193
Gabella G, Juorio AV (1973) Monoamines in the guinea-pig intestine. Br J Pharmacol 47:635P
Gaddum JH (1953a) The technique of superfusion. Br J Pharmacol 8:321–326
Gaddum JH (1953b) Tryptamine receptors. J Physiol (Lond) 119:363–368
Gaddum JH, Picarelli ZP (1957) Two kinds of tryptamine receptor. Br J Pharmacol 12:323–328
Gainer H, Sarne Y, Brownstein MJ (1977) Biosynthesis and axonal transport of rat neurohypophysial proteins and peptides. J Cell Biol 73:366–381
Ganten D, Fuxe K, Phillips MI, Mann JFE, Ganten V (1978) The brain isorenin-angiotensin system: biochemistry, localization, and possible role in drinking and blood pressure regulation. In: Ganong WF, Martini L (eds) Frontier in Neuroendocrinology 5, 61–99. Raven, New York
Gerschenfeld HM, Paupardin-Tritsch D (1974) Ionic mechanisms and receptor properties underlying the responses of molluscan neurones to 5-hydroxytryptamine. J Physiol (Lond) 243:427–456
Gershon MD (1977) Biochemistry and physiology of serotonergic transmission. In: Handbook of physiology, vol 1, sect 1. American Physiological Society, Washington, DC, pp 573–624
Gershon MD (1981) The enteric nervous system. Ann Rev Neurosci 4:227–272
Gershon MD, Altman RF (1971) An analysis of the uptake of 5-hydroxytryptamine by the myenteric plexus of the small intestine of the guinea-pig. J Pharmacol Exp Ther 179:29–41
Gershon MD, Jonakait GM (1979) Uptake and release of 5-hydroxytryptamine by enteric 5-hydroxytryptaminergic neurones: effects of fluoxetine (Lilly 110140) and chlorimipramine. Br J Pharmacol 66:7–9
Gershon MD, Ross LL (1973) Selective destruction of axons in the myenteric plexus which take up 5-hydroxytryptamine. Anat Rec 175:328
Gershon MD, Sherman D (1978) Effects of neurotoxins on peripheral serotonergic neurons. Anat Rec 190:401
Gershon MD, Tamir H (1979) Serotonin (5-HT) release from stimulated peripheral neurons. Proc Int Soc Neurochem p 349
Gershon MD, Robinson RG, Ross LL (1976) Serotonin accumulation in the guinea-pig myenteric plexus: ion dependence, structure activity relationship and the effect of drugs. J Pharmacol Exp Ther 198:548–561
Gershon MD, Dreyfus CF, Pickel VM, Joh TH, Reis DJ (1977) Serotonergic neurons in the peripheral nervous system: identification in gut by histochemical localization of tryptophan hydroxylase. Proc Natl Acad Sci USA 74:3086–3089
Gilbert PE, Martin WR (1976) The effects of morphine- and nalorphine-like drugs in the nondependent, morphine-dependent and cyclazocine-dependent chronic spinal dog. J Pharmacol Exp Ther 198:66–82
Gillespie JH (1934) The biological significance of the linkages in adenosine triphosphoric acid. J Physiol (Lond) 80:345–359
Gillespie JS, MacKenna BR (1961) The inhibitory action of the sympathetic nerves on the smooth muscle of the rabbit gut, its reversal by reserpine and restoration by catecholamines and by dopa. J Physiol (Lond) 156:17–34
Gillon MGC, Pollock D (1980) Acute effects of morphine and opioid peptides on the motility and responses of rat colon to electrical stimulation. Br J Pharmacol 68:381–392
Görög P, Szporny L (1961) Effect of Vincamin on the noradrenaline content of rat tissue. Biochem Pharmacol 8:259–262
Goldenberg MM, Burns RH (1971) Atropine-resistant spasm of the dog colon induced by intermittent pelvic nerve stimulation. Life Sci 10:591–600

Goldenberg MM, Burns RH (1968) Effect of atropine on parasympathetic responses of the gastrointestinal tract of the dog. Arch Int Pharmacodyn Ther 174:342
Govier WC, Sugrue MF, Shore PA (1969) On the inability to produce supersensitivity to catecholamines in intestinal smooth muscle. J Pharmacol Exp Ther 165:71–77
Goyal RK, Said SI, Rattan S (1979) Influence of VIP antiserum on lower esophageal sphincter relaxation: possible evidence for VIP as the inhibitory neurotransmitter. Gastroenterology 76:1142
Grafe P, Mayer LJ, Wood JD (1979) Evidence that substance P does not mediate slow synaptic excitation within the myenteric plexus. Nature 279:720–721
Graham JDP (1949) The effect of drugs on the motility of isolated strips of human stomach muscle. J Pharm Pharmacol 1:95–102
Greeff K, Holtz P (1956) Über die Natur des Übertragerstoffes vagal-motorischer Magenerregungen. Arch Exp Pathol Pharmakol 227:559–565
Greenberg R, Kosterlitz HW, Waterfield AA (1970) The effects of hexamethonium, morphine and adrenaline on the output of acetylcholine from the myenteric plexuslongitudinal muscle preparation of the ileum. Br J Pharmacol 40:553–554 P
Gruber CM, Brundage JT, De Note A, Heiligman R (1935) A comparison of the actions of dilaudid hydrochloride and morphine sulfate upon segments of excised intestine and uterus. J Pharmacol Exp Ther 55:430–434
Guillemin R (1976) Somatostatin inhibits the release of acetylcholine induced electrically in the myenteric plexus. Endocrinology 99:1653
Gustavsson S, Johansson H, Jung B, Lundqvist G (1979) Gastrointestinal motility and absorption of calcium during infusion of somatostatin in the rat. Digestion 19:170–174
Håkanson R, Lilja B, Owman C, Thunell S (1967) Changes in gastric secretion induced by certain amines, amine precursors and related enzyme inhibitors. Eur J Pharmacol 1:425–433
Harrison JS, McSwiney BA (1936) The chemical transmitter of motor impulses to the stomach. J Physiol (Lond) 87:79–86
Harry J (1962) Effect of cooling, local anaesthetic compounds and botulinum toxin on the responses of and the acetylcholine output from the electrically transmurally stimulated isolated guinea-pig ileum. Br J Pharmacol 19:42–55
Heitz P, Polak JM, Timson CM, Pearse AGE (1976) Enterochromaffin cells as the endocrine source of gastrointestinal substance P. Histochemistry 49:343–347
Helmstaedter V, Taugner C, Feurle GE, Forssmann WG (1977) Localization of neurotensin-immunoreactive cells in the small intestine of man and various mammals. Histochemistry 53:35–41
Helmstaedter V, Kreppein W, Domschke W, Mitznegg P, Yanaihara N, Wünsch E, Forssmann WG (1979) Immunohistochemical localization of motilin in endocrine non-enterochromaffin cells of the small intestine of humans and monkey. Gastroenterology 76:897–902
Hirst GDS, McKirdy HC (1974a) Presynaptic inhibition at mammalian peripheral synapse? Nature 250:430–431
Hirst GDS, McKirdy HC (1974b) A nervous mechanism for descending inhibition in guinea-pig small intestine. J Physiol (Lond) 238:129–144
Hirst GDS, McKirdy HC (1975) Synaptic potentials recorded from neurones of the submucous plexus of guinea-pig small intestine. J Physiol (Lond) 249:369–385
Hirst GDS, Silinsky EM (1975) Some effects of 5-hydroxytryptamine, dopamine and noradrenaline on neurones in the submucous plexus of guinea-pig small intestine. J Physiol (Lond) 251:817–832
Hirst GDS, Holman ME, Spence I (1974) Two types of neurones in the myenteric plexus of duodenum in the guinea-pig. J Physiol (Lond) 236:303–326
Hobbiger F (1958) Effects of γ-aminobutyric acid on the isolated mammalian ileum. J Physiol (Lond) 142:147–164
Hökfelt T, Johansson O, Efendig S, Luft A, Arimura A (1975) Are there somatostatin containing nerves in the rat gut? Immunohistochemical evidence for a new type of peripheral nerve. Experientia 31:852–854
Hökfelt T, Elfvin LG, Elde R, Schultzberg M, Goldstein M, Luft R (1977) Occurrence of somatostatin-like immunoreactivity in some peripheral sympathetic noradrenergic neurons. Proc Natl Acad Sci USA 74:3587–3591

Hökfelt T, Ljungdahl A, Steinbusch H et al. (1978) Immunohistochemical evidence of substance P-like immunoreactivity in some 5-hydroxytryptamine-containing neurons in the rat central nervous system. Neuroscience 3:517–538

Holman ME, Hughes JR (1965) Inhibition of intestinal smooth muscle. Aust J Exp Biol Med Sci 43:277–290

Holton P (1959) The liberation of adenosine triphosphate on antidramatic stimulation of sensory nerves. J Physiol (Lond) 145:494–504

Holzbauer M, Sharman DF (1972) The distribution of catecholamines in vertebrates. In: Blaschko H, Muscholl E (eds) Catecholamines. Springer, Berlin Heidelberg New York, pp 110–185

Hooper M, Spedding M, Sweetman AJ, Weetman DF (1978) The effects of 2-2′-pyridylisatogen on responses to adenosine 5′-triphosphate in isolated smooth muscle preparations. In: Baer HP, Drummond G (eds) Physiological and regulatory functions of adenosine and adenine nucleotides. Raven, New York

Howe PRC, Provis JC, Furness JB, Costa M (1979) Are there intrinsic stores of catecholamines in the intestine? Proc Aust Physiol Pharmacol Soc 10:305P

Hughes J, Smith TW, Kosterlitz HW, Fothergill LA, Morgan BA, Morris HR (1975) Identification of two related pentapeptides from the brain with potent opiate agonist activity. Nature 258:577–579

Hughes J, Kosterlitz HW, Smith TW (1977) The distribution of methionine-enkephalin and leucine-enkephalin in the brain and peripheral tissues. Br J Pharmacol 61:639–647

Hughes J, Kosterlitz HW, McKnight AT, Sosa RP, Lord JAH, Waterfield AA (1978a) Pharmacological and biochemical aspects of the enkephalins. In: Hughes J (ed) Centrally acting peptides. University Park Press, Baltimore, pp 179–193

Hughes J, Kosterlitz HW, Sosa P (1978b) Enkephalin release from the myenteric plexus of the guinea-pig small intestine in the presence of cycloheximide. Br J Pharmacol 63:397P

Huidobro-Toro JP, Way EL (1976) Studies on opiate effects in the terminal colon of the rat. Red Proc 35:265

Hulten L, Jodal M, Lundgren O (1969) The effects of graded electrical stimulation or reflex activation of the sympathetic and the parasympathetic nerve supply on the regional flow in cat colon. Bibl Anat 10:312–315

Hunt WB, Parsons DG, Wahid A, Wilkinson J (1978) Influence of 2-2′-pyridylisation tosylate on responses produced by ATP and by neural stimulation on the rat gastric corps. Br J Pharmacol 63:378–379P

Hutchison JB, Dimaline R, Dockray GJ (1981) Neuropeptides in the gut: quantification and characterization of cholecystokinin octapeptide-, bombesin- and vasoactive intestinal polypeptide-like immunoreactivities in the myenteric plexus of the guinea-pig small intestine. Peptides 2:23–30

Hutchinson M, Kosterlitz HW, Gilbert JC (1976) Effects of physostigmine and electrical stimulation on the acetylcholine content of the guinea-pig ileum. Eur J Pharmacol 39:221–235

Innes IR, Kohli JD (1969) Exitatory actions of sympathomimetic amines on 5-hydroxytryptamine receptors of gut. Br J Pharmacol 35:383–393

Inouye A, Fukuya M, Tsuchiya K, Tsujioka T (1960) Studies on the effects of γ-aminobutyric acid on the isolated guinea-pig ileum. Jpn J Physiol 10:167–182

Jaffer SS, Farrar JT, Yau WM, Makhlouf AM (1974) Mode of action and interplay of vasoactive intestinal peptide (VIP), secretion and octa-peptide of cholecystokinin (OCTA-CCK) on duodenal and ileal muscle in vitro. Gastroenterology 66:A-62/716

Jager LP, Schevers JAM (1980) A comparison of effects evoked in guinea-pig taenia caecum by purine nucleotides and by "purinergic" nerve stimulation. J Physiol (Lond) 299:75–83

Jessen KR, Mirsky R, Dennison ME, Burnstock G (1979) GABA may be a neurotransmitter in the vertebrate peripheral nervous system. Nature 281:71–74

Jessen KR, Polak JM, Van Noorden S, Bloom SR, Burnstock G (1980) Peptide-containing neurones connect the two ganglionated plexuses of the enteric nervous system. Nature 283:391–393

Johansson C, Efendic S, Wisen O, Uvnäs-Wallenstein K, Luft R (1978) Effects of short-time somatostatin infusion on the gastric and intestinal propulsion in humans. Gastroenterology 13:481–483

Johnson SM, Katayama Y, North RA (1980) Multiple actions of 5-hydroxytryptamine on myenteric neurones of the guinea-pig ileum. J Physiol (Lond) 304:459–470
Jonakait GM, Tamir H, Rapport M, Gershon MD (1977) Detection of a soluble serotonin-binding protein in the mammalian myenteric plexus and other peripheral sites of serotonin storage. J Neurochem 28:277–284
Jonakait GM, Tamir H, Gintzler AR, Gershon MD (1979) Release of ^{3}H-serotonin and its binding protein from enteric neurons. Brain Res 174:55–69
Juorio AV, Gabella G (1974) Noradrenaline in the guinea-pig alimentary canal: regional distribution and sensitivity to denervation and reserpine: J Neurochem 22:851–858
Kachelhoffer J, Eloy MR, Pousse A, Hohmatter D, Grenier JF (1974) Mesenteric vasomotor effects of vasoactive intestinal polypeptide: study on perfused isolated canine jejunal loops. Pfluegers Arch Ges Physiol 352:37–46
Kachelhoffer J, Mendel C, Dauchel J, Hohmatter D, Grenier JF (1976) The effect of VIP on intestinal motility. Study on ex vivo perfused isolated canine jejunal loops. Am J Dig Dis 21:957–962
Karki NT, Paasonen MK (1959) Selective depletion of noradrenaline and 5-hydroxytryptamine from rat brain and intestine by rauwolfia alkaloids. J. Neurochem 3:352–357
Katayama Y, North RA (1978) Does substance P mediate slow synaptic excitation within the myenteric plexus? Nature 274:387–388
Katayama Y, North RA (1980) The action of somatostatin on neurones of the myenteric plexus of the guinea-pig ileum. J Physiol (Lond)
Katayama Y, North RA, Williams JT (1979) The action of substance P on neurones of the myenteric plexus of the guinea-pig small intestine. Proc R Soc Lond [Biol] 206:191–208
Kazarova EK, Esayan NA (1966) Noradrenaline and adrenaline of dog stomach wall and gastric juice (in Armenian). Biol Zh Arm 19:58–64
Kimmel JR, Hayden LJ, Pollock HG (1975) Isolation and characterization of a new pancreatic polypeptide hormone. J Biol Chem 250:9369–9376
Kisloff B, Moore EW (1976) Effect of serotonin on water and electrolyte transport in the in vivo rabbit small intestine. Gastroenterology 71:1033–1038
Klingman GI, Kardaman S, Haber J (1964) Amine levels, monoamine oxidase and dopa-decarboxylase activities in the gastrointestinal tract of the rat. Life Sci 3:1355–1360
Konturek SJ, Pawlik W, Tasler J et al. (1978) Effects of enkephalin on the gastrointestinal tract. In: Bloom SR (ed) Gut hormones. Churchill Livingstone, Edinburgh London, pp 507–512
Kosterlitz HW, Lees GM (1964) Pharmacological analysis of intrinsic intestinal reflexes. Pharmacol Rev 16:301–339
Kosterlitz HW, Robinson JA (1957) Inhibition of the peristaltic reflex of the isolated guinea-pig ileum. J Physiol (Lond) 136:249–262
Krantis A, Kerr DIB (1981) Autoradiographic localization of [^{3}H] gamma-aminobutyric acid in the myenteric plexus of the guinea-pig ileum. Neurosci Lett 23:263–268
Krantis A, Costa M, Furness JB, Orbach J (1980) Gamma-aminobutyric acid stimulates intrinsic inhibitory and excitatory nerves in the guinea-pig intestine. Eur J Pharmacol 67:461–468
Krejs GJ, Brakley RM, Read NW, Fordtran JS (1978) Intestinal secretion induced by vasoactive intestinal polypeptide. J Clin Invest 61:1337–1345
Kreuger H (1937) The action of morphine on the digestive tract. Physiol Rev 17:619–645
Kuchii M, Miyahara JT, Shibata S (1973 a) [^{3}H]-adenine nucleotide and [^{3}H]-noradrenaline release evoked by electrical field stimulation, perivascular nerve stimulation and nicotine from the taenia of the guinea-pig caceum. Br J Pharmacol 49:258–267
Kuchii M, Miyahara JT, Shibata S (1973 b) [^{3}H]-adenosine nucleotide and [^{3}H]-noradrenaline uptake by cold stored guinea-pig taenia caecum; mechanical effects and release of [^{3}H]-adenosine nucleotide by noradrenaline, papaverine and nitroglycerine. Br J Pharmacol 49:642–650
Lanfranchi GI, Marzio L, Cortini C, Osset ME (1978) Motor effect of dopamine on human signoid colon: Evidence for specific receptors. Am J Dig Dis 23:257–263
Langley JN (1898) The inhibitory fibres in the vagus for the end of the oesophagus and the stomach. J Physiol (Lond) 23:407–414
Larsson L-I (1977) Ultrastructural localization of a new neuronal peptide (VIP). Histochemistry 54:173–176

Larsson LI, Rehfeld JF (1979) Localization and molecular heterogeneity of cholecystokinin in the central and peripheral nervous system. Brain Res 165:201–218
Larsson LI, Sundler F, Håkanson R, Pollock HG, Kimmel JR (1974) Localization of APP, a postulated new hormone, to a pancreatic endocrine cell type. Histochemistry 42:377–382
Larsson LI, Sundler F, Håkanson R (1975) Immunohistochemical localization of human pancreatic polypeptide to a population of islet cells. Cell Tissue Res 156:167–171
Larsson L-I, Fahrenkrug J, Schaffalitzky de Muckadell O, Sundler F, Håkanson R, Rehfeld JF (1976) Localization of vasoactive intestinal polypeptide (VIP) to central and peripheral neurons. Proc Natl Acad Sci USA 73:3197–3200
Larsson LI, Childers S, Snyder SH (1979) Met and leu enkephalin immunoreactivity in separate neurons. Nature 282:407–410
Lazarus LH, Linnoila LI, Hernandez O, Di Augustine RP (1980) A neuropeptide in mammalian tissues with physalaemin-like immunoreactivity. Nature 287:555–558
Lees GJ, Furness JB, Costa M (1979) Apparent absence of tryptophan hydroxylase from the myenteric plexus of the guinea-pig ileum. Proc Aust Physiol Pharmacol Soc 10:304 P
Lembeck F (1958) Die Beeinflussung der Darmmotilität durch Hydroxytryptamin. Pfluegers Arch Ges Physiol 265:567–574
Lembeck F, Fischer G (1967) Crossed tachyphylaxis of peptides. Naunyn-Schmiedeberg Arch Pharmacol 258:452–456
Lembeck F, Zetler G (1971) Pharmacology of naturally occurring polypeptides and lipid-soluble acids. In: Walker JM (ed) International encyclopedia of pharmacology and therapeutics, vol 1, sec 72. Pergamon, Oxford, pp 29–71
Leppäluoto J, Koivusalo F, Kraama R (1978) Thyrotropin-releasing factor: distribution in neural and gastrointestinal tissues. Acta Physiol Scand 104:175–179
Lewis CP, McMartin C, Rosenthal SR, Yates C (1972) Isolation and identification of pharmacologically active amino acids in skin and their structure-activity relationship on the guinea-pig ileum. Br J Pharmacol 45:104–117
Liedberg G, Nielsen KC, Owman C, Sjoberg NO (1973) Adrenergic contribution to the abdominal vagus nerves in the cat. Scand J Gastroenterol 8:177–180
Linnoila RI, Di Augustine RP, Miller RJ, Chang KJ, Cuatrecasas P (1978) An immunohistochemical and radioimmunological study of the distribution of (met^5) and (leu^5)-enkephalin in the gastrointestinal tract. Neuroscience 3:1187–1196
Liotta AS, Suda S, Krieger DT (1978) β-Lipotropin is the major opioid-like peptide of human pituitary and rat pars distalis: lack of significant β-endorphin. Proc Natl Acad Sci USA 75:2950–2954
Lord JAH, Waterfield AA, Hughes J, Kosterlitz HW (1977) Endogenous opioid peptides: multiple agonists and receptors. Nature 267:495–499
Loren I, Alumets J, Håkanson R, Sundler F (1979) Immunoreactive pancreatic polypeptide (PP) occurs in the central and peripheral nervous system: preliminary immunocytochemical observations. Cell Tissue Res 200:179–186
Luchelli-Fortis MA, Fredholm BB, Langer SZ (1979) Release of radioactive purines from cat nictitating membrane labelled with ^{3}H-adenine. Eur J Pharmacol 58:389–397
Lundberg JM, Hökfelt T, Schultzberg M, Uvnäs-Wallenstein K, Köhler C, Said SI (1979) Occurrence of vasoactive intestinal polypeptide (VIP)-like immunoreactivity in certain cholinergic neurons of the cat: evidence from combined immunohistochemistry and acetylcholinesterase staining. Neuroscience 4:1539–1560
MacKay D, McKirdy HC (1972) Effect of vasopressin and of adenosine triphosphate on the flat preparation of rabbit rectum. Br J Pharmacol 44:366–367 P
MacKenna BR, McKirdy HC (1970) A simple method for investigating the functional relationship of the longitudinal and circular layers of the muscularis externa of the rabbit bowel, using a "flat" preparation. J Physiol (Lond) 211:18 P
MacKenzie I, Burnstock G (1980) Evidence against vasoactive intestinal polypeptide being the non-adrenergic, non-cholinergic inhibitory transmitter released from nerves supplying the smooth muscle of the guinea-pig taenia coli. Eur J Pharmacol 67:255–264
Majcen Z, Brzin M (1979) Cholinesterases and choline acetyltransferase in the longitudinal muscle of the guinea-pig ileum. Histochemistry 63:295–302
Mann M, West GB (1951) The nature of uterine and intestinal sympathin. Br J Pharmacol 6:79–82

Margolin S (1954) Decreased gastrointestinal propulsive activity after intracranial morphine. Fed Proc 13:383
Margolin S (1963) Centrally mediated inhibition of gastrointestinal propulsive motility by morphine over a non-neural pathway. Proc Soc Exp Biol Med 112:311–315
Martinson J (1965) Vagal relaxation of the stomach. Experimental re-investigation of the concept of the transmission mechanism. Acta Physiol Scand 64:453–462
Mattila M (1962) The effects of morphine and nalorphine on the small intestine of normal and morphine-tolerant rat and guinea-pig. Acta Pharmacol Toxicol 19:47–52
McCallum RW, Dodds J, Osborne HP, Biancani P (1980) IN vitro effect of enkephalin and other opiates on opossum lower esophageal sphincter. In: Christensen J (ed) Proceedings of the 7th International Congress on Gastrointestinal Motility. Raven, New York, pp 9–10
McDonald TJ, Nilsson G, Vagne M, Ghatei M, Bloom SR, Mutt V (1978) A gastrin releasing peptide from the porcine non-antral gastric tissue. Gut 19:767–774
McDonald TJ, Jörnvall H, Nilsson G, Vagne M, Ghatei M, Bloom SR, Mutt V (1979) Characterization of gastrin releasing peptide from porcine nonantral gastric tissue. Biochem Biophys Res Commun 90:227–233
McGroarty FC, McKirdy HC (1971) Preliminary observations on the functional relationship of the longitudinal and circular layers of the muscularis externa of the rabbit rectum using a flat preparation. J Physiol (Lond) 214:19–20P
McLennan H (1963) Synaptic transmission. Saunders, Philadelphia
Miachon S, Peyrin L, Cier JF, Legheand J (1978) Action de la dopa et de la dopamine sur la motricite du duodenum isole de rat, in vitro. CR Soc Biol (Paris) 172:110–116
Mirkin BL, Bonnycastle DD (1954) A pharmacological and chemical study of humoral mediators in the sympathetic nervous system. Am J Physiol 178:529–534
Morgan KG, Schmalz PF, Szurszewski JH (1978) The inhibitary effects of vasoactive intestinal polypeptide on the mechanical and electrical activity of canine antral smooth muscle. J Physiol (Lond) 282:437–450
Morita K, North RA, Katayama Y (1980) Evidence that substance P is a neurotransmitter in the myenteric plexus. Nature 287:151–152
Morley JE, Garvin TJ, Pekary AE, Hershman JM (1977) Thyrotropinreleasing hormone in the gastrointestinal tract. Biochem Biophys Res Commun 79:314–318
Muryobayashi T, Mori J, Fujiwara M, Shimamoto K (1968) Fluorescence histochemical demonstration of adrenergic nerve fibers in the vagus nerve of cats and dogs. Jpn J Pharmacol 18:285–293
Mutt V, Jorpes JE (1971) Hormonal polypeptides of the upper intestine. Biochem J 125:57–58P
Mutt V, Said SI (1974) Structure of the porcine vasoactive intestinal octacosapeptide. Eur J Biochem 42:581–589
Ng RH, Marshall FD, Henn FA, Sellstrom A (1977) Metabolism of carnosine and homocarnosine in subcellular fractions of neuronal and glial cell enriched fractions of rabbit brain. J Neurochem 28:449–452
Nijkamp FP, Van Ree JM (1978) Effects of endorphins on different parts of the gastrointestinal tract in vitro. In: Van Ree JM, Terenius L (eds) Characteristics and functions of opioids. Elsevier, Amsterdam, pp 179–180
Nilsson G, Larsson LI, Håkanson R, Brodin E, Pernow B, Sundler F (1975) Localization of substance P-like immunoreactivity in mouse gut. Histochemistry 43:97–99
Nishi S, North RA (1973a) Intracellular recording from the myenteric plexus of the guinea-pig ileum. J Physiol (Lond) 231:471–491
Nishi S, North RA (1973b) Presynaptic action of noradrenaline in the myenteric plexus. J Physiol (Lond) 231:29–30P
North RA, Tonini M (1977) The mechanism of action of narcotic analgesics in the guinea-pig ileum. Br J Pharmacol 61:541–549
North RA, Katayama Y, Williams JT (1979) On the mechanism and site of action of enkephalin on single myenteric neurons. Brain Res 165:67–77
North RA, Henderson G, Katayama Y, Johnson SM (1980) Presynaptic inhibition of acetylcholine release by 5-hydroxytryptamine in the enteric nervous system. Neuroscience 5:581–586

Northway MN, Burks TF (1980) Stimulation of cholinergic nerves in dog intestine by adenine nucleotides. Eur J Pharmacol 65:11–19
Ohga A, Taneike T (1977) Dissimilarity between the responses to adenosine triphosphate or its related compounds and non-adrenergic inhibitory nerve stimulation in the longitudinal smooth muscle of pig stomach. Br J Pharmacol 60:221–231
Oka T (1980) Enkephalin receptor in the rabbit ileum. Br J Pharmacol 68:193–195
Okwuasaba FK, Hamilton JT, Cook MA (1977) Relaxations of guinea-pig fundig strip by adenosine, adenine nucleotides and electrical stimulation: antagonism by theophylline and desensitization to adenosine and its derivatives. Eur J Pharmacol 46:181–198
Olson L, Alund M (1979) Quinacrine-binding nerves: presence in the mouse anococcygeus muscle, disappearance after muscle transection. Med Biol 57:182–186
Orci L, Baetens O, Ruffener C, Brown M, Vale W, Guillemin R (1976) Evidence for immunoreactive neurotensin in dog intestinal mucosa. Life Sci 19:559–562
Ormsbee HS, Koehler SL, Telford GL (1978) Somatostatin inhibits motilin-induced interdigestive contractile activity in the dog. Am J Dig Dis 23:781–788
Orrego F (1979) Criteria for the identification of central neurotransmitters, and their application to studies with some nerve tissue preparations in vitro. Neuroscience 4:1037–1057
Ozaki T (1979) Effects of stimulation of auerbach's plexus on both longitudinal and circular muscles. Jpn J Physiol 29:195–209
Parolaro D, Sala M, Gori E (1977) Effect of intracerebroventricular administration of morphine upon intestinal motility in rats and its antagonism with naloxone. Eur J Pharmacol 46:329–338
Paton WDM (1957) The action of morphine and related substances on contraction and on acetylcholine output of coaxially stimulated guinea-pig ileum. Br J Pharmacol 11:119–127
Paton WDM (1958) Central and synaptic transmission in the nervous system. Annu Rev Physiol 20:431–470
Paton WDM, Vane JR (1963) An analysis of the responses of the isolated stomach to electrical stimulation and to drugs. J Physiol (Lond) 165:10–46
Paton WDM, Vizi ES, Zar MA (1971) Mechanism of acetylcholine release from parasympathetic nerves. J Physiol (Lond) 215:819–848
Paton WDM, Zar MA (1968) The origin of acetylcholine released from guinea-pig intestine and longitudinal muscle strips. J Physiol (Lond) 194:13–33
Pearse AGE, Polak JM (1975) Immunocytochemical localization of substance P in mammalian intestine. Histochemistry 41:373–375
Pearse AGE, Polak JM, Bloom SR, Adams C, Dryburgh JR, Brown JC (1974) Enterochromaffin cells of the mammalian small intestine as the source of motilin. Virchows Arch [Cell Pathol] 16:111–120
Pernow B (1953) Studies on substance P. Purification, occurrence and biological actions. Acta Physiol Scand [Suppl 105] 29:1–90
Phillis JW (1970) The pharmacology of synapses. Pergamon, New York
Piper PJ, Said SI, Vane JR (1970) Effects on smooth muscle preparations of unidentified vasoactive peptides from intestine and lung. Nature 225:1144–1146
Poitras P, Steinbach J, Vandenventer G, Walsh JH, Code CF (1979) Effect of somatostatin on interdigestive myoelectric complexes and motilin blood levels. Gastroenterology 76:1218
Polak JM, Pearse AGE, Heath CM (1975) Complete identification of endocrine cells in the gastrointestinal tract using semi-thin sections to identify motilin cells in human and animal intestine. Gut 16:225–229
Polak JM, Bloom SR, Hobbs S, Solcia E, Pearse AGE (1976) Distribution of a bombesin-like peptide in human gastrointestinal tract. Lancet 1:1109–1110
Polak JM, Sullivan SN, Bloom SR, Facer P, Pearse AGE (1977) Enkephalin-like immunoreactivity in the human gastrointestinal tract. Lancet 1:972–974
Polak JM, Buchan AMJ, Czykowska W, Solcia E, Bloom SR, Pearse AGE (1978) Bombesin in the gut. In: Bloom SR (ed) Gut hormones. Churchill Livingstone, Edinburgh London, pp 541–543
Puig MM, Gascon P, Graviso GL, Musachio JM (1977) Endogenous opiate receptor ligand: Electrically induced release in the guinea-pig ileum. Science 195:419–420

Rattan S, Goyal RK (1976) Effect of dopamine on the esophageal smooth muscle in vivo. Gastroenterology 70:377–381
Rattan S, Goyal RK (1978) Evidence of 5-HT participation in vagal inhibitory pathway to opossum LES. Am J Physiol 234:E273–276
Rattan S, Goyal RK (1980) Evidence against purinergic inhibitory nerves in the vagal pathway to the opossum lower esophageal sphincter. Gastroenterology 78:898–904
Rattan S, Said SI, Goyal RK (1977) Effect of vasoactive intestinal polypeptide (VIP) on the lower esophageal sphincter pressure (LESP). Proc Soc Exp Biol Med 155:40–43
Regoli D, Vane JR (1966) The continuous estimation of angiotensin formed in the circulation of the dog. J Physiol (Lond) 183:513–531
Rehfeld JF (1978a) Immunochemical studies on cholecystokinin. I. Development of sequence-specific radioimmunoassays for porcine triacontatriapeptide cholecystokinin. J Biol Chem 253:4016–4021
Rehfeld JF (1978b) Immunocytochemical studies on cholecystokinin. II. Distribution and molecular heterogeneity in the central nervous system and small intestine of man and hog. J Biol Chem 253:4022–4030
Rehfeld JF, Larsson LI (1979) The predominating molecular form of gastrin and cholecystokinin in the gut is a small peptide corresponding to their COOH-terminal tetrapeptide amide. Acta Physiol Scand 105:117–119
Richardson J, Beland J (1976) Noradrenergic inhibitory nervous system in human airways. J Appl Physiol 41:764–771
Rikimaru A, Fukushi Y, Suzuki T (1971) Effects of imidazole and phentolamine on the relaxant responses of guinea-pig taenia coli to transmural stimulation and to adenosine triphosphate. Tohoku J Exp Med 105:199–200
Roberts E, Chase TN, Tower DB (eds) (1976) GABA in nervous system function. Raven, New York
Robinson RG, Gershon MD (1971) Synthesis and uptake of 5-hydroxytryptamine by the myenteric plexus of the guinea-pig ileum: a histochemical study. J Pharmacol Exp Ther 178:311–324
Rostad H (1973) Colonic motility in the cat. II. Extrinsic nervous control. Acta Physiol Scand 89:91–103
Rothman TP, Ross LL, Gershon MD (1976) Separately developing axonal uptake of 5-hydroxytryptamine and norepinephrine in the fetal ileum of the rabbit. Brain Res 115:437–456
Rutherford A, Burnstock G (1978) Neuronal and non-neuronal components in the overflow of labelled adenyl compounds from guinea-pig taenia coli. Eur J Pharmacol 48:195–202
Ryan JP, Ryave S (1978) Effect of vasoactive intestinal polypeptide on gall bladder smooth muscle in vitro. Am J Physiol 234:E44–46
Said SI, Mutt V (1970a) Polypeptide with broad biological activity: isolation from small intestine. Science 169:1217–1220
Said SI, Mutt V (1970b) Potent peripheral and splanchnic vasodilator peptide from normal gut. Nature 225:863–864
Saito K (1972) Effects of extracellularly applied ATP and its related nucleotides on the membrane potential of the guinea-pig taenia coli. Jpn J Smooth Muscle Res 8:32–39
Satchell DG, Burnstock G (1971) Quantitative studies of the release of purine compounds following stimulation of non-adrenergic inhibitory neurones in the stomach. Biochem Pharmacol 20:1694–1697
Satchell DG, Burnstock G, Dann P (1973) Antagonism of the effects of purinergic nerve stimulation and exogenously applied ATP on the guinea-pig taenia coli by 2-substituted imidazolines and related compounds. Eur J Pharmacol 23:264–269
Schaffalitzky de Muckadell DB, Fahrenkrug J, Holst JJ (1977) Release of vasoactive intestinal polypeptide (VIP) by electrical stimulation of the vagal nerves. Gastroenterology 72:373–375
Schaller HC, Flick K, Darai G (1977) A neurohormone from hydra is present in brain and intestine of rat embryos. J Neurochem 29:393–394
Schaumann O, Jochum K, Schmidt H (1953) Analgetica und Darmmotorik. III. Zum Mechanismus der Peristaltik. Arch Exp Pathol Pharmakol 81:55–129

Schaumann W (1955) The paralysing action of morphine on the guinea-pig ileum. Br J Pharmacol 10:456–461
Schaumann W (1957) Inhibition by morphine of the release of acetylcholine from the intestine of the guinea-pig. Br J Pharmacol 12:115–118
Schultzberg M, Dreyfus CF, Gershon MD et al. (1978) VIP-enkephalin-substance P-, and somatostatin-like immunoreactivity in neurons intrinsic to the intestine. Immunohistochemical evidence from organotypic tissue cultures. Brain Res 155:239–248
Schultzberg M, Hökfelt T, Nilsson G et al. (1980) Distribution of peptide and catecholamine neurons in the gastrointestinal tract of rat and guinea-pig: immunohistochemical studies with antisera to substance P, VIP, enkephalins, somatostatin, gastrin, neurotensin and dopamine β-hydroxylase. Neuroscience 5:689–744
Schulz R, Cartwright C (1974) Effect of morphine on serotonin release from myenteric plexus of the guinea-pig. J Pharmacol Exp Ther 190:420–430
Schulz R, Wüster M, Simantov R, Snyder S, Herz A (1977) Electrically stimulated release of opiate-like material from the myenteric plexus of the guinea-pig ileum. Eur J Pharmacol 41:347–348
Schulz R, Wüster M, Herz A (1979) Centrally and peripherally mediated inhibition of intestinal motility by opioids. Naunyn-Schmiedeberg Arch Pharmacol 308:255–260
Schumann HF (1959) Über den Hydroxytyramingehalt der Organe. Naunyn-Schmiedeberg Arch Exp Pathol Pharmakol 236:474–482
Schwartz CJ, Kimberg DV, Scheerin HE, Field M, Said SI (1974) Vasoactive intestinal peptide stimulation of adenylate cyclase and active electrolyte secretion in intestinal mucosa. J Clin Invest 54:536–544
Schwarz H, Bumpus FM, Page IH (1957) Synthesis of a biologically active octapeptide similar to natural isoleucine angiotonin octapeptide. J Am Chem Soc 79:5697–5703
Shaw JS (1979) Characterization of opiate receptors in the isolated rat rectum. Br J Pharmacol 67:428–429 P
Shimo Y, Ishii T (1978) Effects of morphine on non-adrenergic inhibitory responses of the guinea-pig taenia-coli. J Pharm Pharmacol 30:596–597
Shore PA (1959) A simple technique involving solvent extraction for the estimation of norepinephrine and epinephrine in tissues. Pharmacol Rev 11:276–277
Silinsky EM (1975) On the association between transmitter secretion and the release of adenine nucleotides from mammalian motor nerve terminals. J Physiol (Lond) 247:145–162
Silinsky EM, Hubbard JI (1973) Release of ATP from rat motor nerve terminals. Nature 243:404–405
Singh I, Singh A (1970) Transmission hydroxytryptaminergique des nerfs intrapariétaux au muscle lisse de l'oesophage du chien. J Physiol (Paris) 62:421–429
Small RC, Weston AH (1979 a) Theophylline is not an antagonist of intramural inhibitory nerve activity in rabbit and guinea-pig intestine. Br J Pharmacol 66:84 P
Small RC, Weston AH (1979 b) Theophylline antagonizes some effects of purines in the intestine but not those of intramural inhibitory nerve stimulation. Br J Pharmacol 67:301–308
Sosa RP, McKnight AT, Hughes J, Kosterlitz HW (1977) Incorporation of labelled amino acids into the enkephalins. FEBS Lett 84:195–198
Spedding M, Sweetman AJ, Weetman DF (1975) Antagonism of adenosine 5′-triphosphate induced relaxation by 2-2′-pyridylisatogen in the taenia of guinea-pig caecum. Br J Pharmacol 53:575–583
Stadas JO, Schrumpf E, Hanssen KF (1978) Somatostatin inhibits gastric motility in response to distension. Scand J Gastroenterol 13:145–148
Starling EH (1906) Recent advances in the physiology of digestion. Constable, London
Stockley HL (1978) 2-2′pyridylisatogen antagonizes adenosine 5′-triphosphate but not nerve mediated relaxations in human isolated taenia coli. In: Duthie HL (ed) Gastrointestinal motility in health and disease. University Park Press, Baltimore, pp 145–150
Stockley HL, Bennett A (1974) The intrinsic innervation of human sigmoid colon muscle. In: Daniel EE (ed) Proceedings of the 4th International Symposium on Gastintestinal Motility. Mitchell, Vancouver, pp 165–176
Strandberg K, Sedvall G, Midtvedt T, Gustafsson B (1966) Effect of some biologically active amines on the caecum wall of germ free rats. Proc Soc Exp Biol 121:699–702

Studer RO, Trzeciak A, Lergier W (1973) Isolierung und Aminosäuresequenz von Substanz P aus Pferdedarm. Helv Chim Acta 56:860–866
Su C (1975) Neurogenic release of purine compounds in blood vessels. J Pharmacol Exp Ther 195:159–166
Su C, Bevan JA, Burnstock G (1971) [^{3}H]-adenosine triphosphate: release during stimulation of enteric nerves. Science 173:336–339
Sundler F, Alumets A, Håkanson R, Carraway R, Leeman SE (1977a) Ultrastructure of gut neurotensin cells. Histochemistry 53:25–34
Sundler F, Alumets J, Håkanson R, Ingermansson S, Fahrenkrug J, Schaffalitzky de Muckadell O (1977b) VIP innervation of the gallbladder. Gastroenterology 72:1375–1377
Sundler F, Håkanson R, Hammer RA, Alumets A, Carraway R, Leeman S, Zimmerman EA (1977c) Immunohistochemical localization of neurotensin in endocrine cells of the gut. Cell Tissue Res 178:313–321
Sundler F, Håkanson R, Larsson LI, Brodin E, Nilsson G (1977d) Substance P in the gut: an immunochemical and immunohistochemical study of its distribution and development. In: Euler US von, Pernow B (eds) Substance P. Raven, New York, pp 59–65
Suzuki T, Fukushi Y, Rikimaru A (1971) Relaxant effect of adenosine triphosphate and its related nucleotides on the guinea-pig taenia coli. Jpn J Smooth Muscle Res 7:207–212
Szerb JC (1975) Endogenous acetylcholine release and labelled acetylcholine formation from [^{3}H] choline in the myenteric plexus of the guinea-pig ileum. Can J Physiol Pharmacol 53:566–574
Szerb JC (1976) Storage and release of labeled acetylcholine in the myenteric plexus of the guinea-pig ileum. Can J Physiol Pharmacol 54:12–22
Tansy MF, Rothman G, Bartlett J, Farber P, Hohenleitner FJ (1971) Vagal adrenergic degranulation of enterochromaffin cell system in guinea-pig duodenum. J Pharm Sci 60:81–84
Tansy MF, Martin JS, Landin WE, Kendall FM (1979) Species difference in GI motor response to somatostatin. J Pharm Sci 68:1107–1113
Taubin HL, Djahanguiri D, Landsberg L, Lerner E (1972) Noradrenaline concentration and turnover in different regions of the gastrointestinal tract of the rat. Gut 13:790–795
Thor P, Krol R, Konturek SJ, Coy DH, Schally AV (1978) Effect of somatostatin on myoelectrical activity of small bowel. Am J Physiol 235:E 249–254
Thulin L, Olsson P (1973) Effects of intestinal peptide mixture G 2 and vasoactive intestinal peptide VIP on splanchnic circulation in the dog. Acta Chir Scand 139:691–697
Tobe T, Chen ST, Yajima H, Kai Y, Kawatani H (1976) Localization of motilin in human and dog gastrointestinal tract by immunofluorescence. In: Fujita T (ed) Endocrine gut and pancreas. Elsevier, Amsterdam, pp 113–118
Tomita T, Watanabe H (1973) A comparison of the effects of adenosine triphosphate with noradrenaline and with the inhibitory potential of the guinea-pig taenia coli. J Physiol (Lond) 231:167–177
Tonini M, Onori L, Frigo GM, Lecchini S, D'Angelo L, Crema A (1981) Non-adrenergic inhibition of the longitudinal muscle of rabbit distal colon may not be mediated by purinergic nerves. J Pharm Pharmacol 33:536–537
Tregear GW, Niall HD, Potts JT, Leeman SE, Chang MM (1971) Synthesis of substance P. Nature New Biol 232:87–89
Trendelenberg P (1917) Physiologische und pharmakologische Versuche über die Dünndarmperistaltik. Naunyn-Schmiedeberg Arch Exp Pathol Pharmakol 81:55–129
Tucek S, Kondelkova Z (1966) Choline acetyltransferase and acetylcholine in brain, heart atria, intestine, uterus and diaphragm of rats and guinea-pigs. Arch Int Physiol Biochem 74:123–134
Uddman R, Alumets J, Edvinsson L, Håkanson R, Sundler F (1978) Peptidergic (VIP) innervation of the esophagus. Gastroenterology 75:5–8
Uddman R, Alumets J, Håkanson R, Sundler F, Walles B (1980) Peptidergic (enkephalin) innervation of the mammalian esophagus. Gastroenterology 78:732–737
Uvnäs-Wallenstein K (1978) Release of substance P-like immunoreactivity into the antral lumen of cats. Acta Physiol Scand 104:464–468
Uvnäs-Wallenstein K, Rehfeld JF, Larsson LI, Uvnäs B (1977) Heptadecapeptide gastrin in the vagal nerve. Proc Natl Acad Sci USA 74:8707–8710

Valenzuela JE (1976) Dopamine as a possible neurotransmitter in gastric relaxation. Gastroenterology 71:1019–1022
Vane JR (1957) A sensitive method for the assay of 5-hydroxytryptamine. Br J Pharmacol 12:344–349
Vane JR (1959) The relative activities of some tryptamine analogues on the isolated rat stomach strip preparation. Br J Pharmacol 14:87–98
Vane JR (1964) The use of isolated organs for detecting active substances in the circulating blood. Br J Pharmacol 23:360–373
Van Nueten JM, Janssen PA (1978) Is dopamine an endogenous inhibitor of gastric emptying? In: Duthie HL (ed) Gastrointestinal motility in health and disease. University Park Press, Baltimore, pp 173–181
Van Nueten JM, Janssen PAJ, Fontaine J (1976) Unexpected reversal effects of naloxone on the guinea-pig ileum. Life Sci 18:803–810
Vaughan Williams EM (1954) Mode of action of drugs upon intestinal motility. Pharmacol Rev 6:159–190
Vaughan Williams EM, Streeten DHP (1950) The action of morphine, pethidine, and amidone upon the intestinal motility of conscious dogs. Br J Pharmacol 5:584–603
Waterfield AA, Kosterlitz HW (1977) Stereospecific increase by narcotic antagonists of evoked acetylcholine output in guinea-pig ileum. Life Sci 16:1787–1792
Weinstock M (1971) Sites of action of narcotic analgesic drugs – peripheral tissues. In: Clouet DH (ed) Narcotic drugs – biochemical pharmacology. Plenum, New York, pp 394–407
Werman R (1966) A review-criteria for identification of a central nervous system transmitter. Comp Biochem Physiol 18:745–766
Werman R (1972) CNS cellular level: membranes. Annu Rev Physiol 34:337–374
Westfall TC (1977) Local regulation of adrenergic neurotransmission. Physiol Rev 57:659–728
Westfall DP, Stitzel RE, Rowe JN (1978) The postjunctional effects and neural release of purine compounds in the guinea-pig vas deferens. Eur J Pharmacol 50:27–38
Weston AH (1973a) The effect of desensitization to adenosine triphosphate on the peristaltic reflex in guinea-pig ileum. Br J Pharmacol 47:606–608
Weston AH (1973b) Nerve-mediated inhibition of mechanical activity in rabbit duodenum and the effects of desensitization to adenosine and several of its derivatives. Br J Pharmacol 48:302–308
Williams JT, North RA (1979) Vasoactive intestinal polypeptide excites neurones of the myenteric plexus. Brain Res 175:174–177
Williams JT, Katayama Y, North RA (1979) The actions of neurotensin on single myenteric neurons. Eur J Pharmacol 59:181–186
Wood JD, Mayer CJ (1978a) Slow synaptic excitation mediated by serotonin in Auerbach's plexus. Nature 276:836–837
Wood JD, Mayer CJ (1978b) Intracellular study of electrical activity of Auerbach's plexus in guinea-pig small intestine. Pfluegers Arch Eur J Physiol 374:265–275
Wood JD, Mayer CJ (1979a) Intracellular study of tonic-type enteric neurons in guinea-pig small intestine. J Neurophysiol 43:569–581
Wood JD, Mayer CJ (1979b) Serotonergic activation of tonic-type enteric neurons in guinea-pig small bowel. J Neurophysiol 42:582–593
Wüster M, Schulz R, Herz A (1979) Specificity of opioids towards the μ, δ- ε-opiate receptors. Neurosci Lett 15:193–198
Youngblood WW, Humm J, Kizer JS (1979) TRH-like immunoreactivity in rat pancreas and eye, bovine and sheep pineals, and human placenta: non-identity with authentic pyroglu-his-pro-NH_2(TRH). Brain Res 163:101–110

Note Added in Proof to Chapter 10

Since this chapter was submitted a number of publications have appeared in this field. Some of the most relevant are mentioned below under the appropriate heading.

E. DELBRO and LISANDER (1980) demonstrated that stimulation of the greater splanchnic nerves in the cat elicited an atropine-resistant contraction of the stomach and reached the conclusion that this excitatory effect was due to the antidromic stimulation of sensory fibres which had a collateral ending on enteric cholinergic neurones.

In his thesis, DELBRO (1981) describes the presence of similar fibres in the vagus nerve and provides some evidence that the branching of sensory nerves in the gastrointestinal wall may be involved in local axo-axonal reflexes to nociceptive stimuli.

G. An excellent comprehensive review on the natural history of the migratory complex has been published by WINGATE (1981).

H. Two excellent papers (AEBERHARD et al. 1980; ITOH et al. 1981) fully confirm the conclusions reached in this chapter that the migrating motor complex is basically a neuronal mechanism which is intrinsic to the intestinal wall. Both groups of workers showed that in isolated and extrinsically denervated loop of dog small intestine a motor complex could appear at the proximal end of the loop and migrate anally in the loop. Both groups concluded that there must be an integrative network within the enteric plexuses able to initiate a motor activity which slowly migrates anally.

Another elegant paper (SARNE et al. 1981) fully confirmed these findings by recording migrating complexes in the dog small intestine after complete extrinsic denervation. By perfusing individual loops with antimuscarinic and antinicotinic drugs or with tetrodotoxin they were able to block the propagation of the migrating complex concluding, like other authors quoted in this chapter, that the mechanism of migration of the complex is inbuilt in the enteric nervous system and that it involves cholinergic nerves.

Neurotensin has been implicated in the disruption of interdigestive myoelectric migrating complex by feeding on the basis that, in man, ingestion of fat releases neurotensin in the circulation (ROSELL and RÖKAEUS, 1979) and that neurotensin infused in man during the interdigestive period inhibits motor complexes (THOR et al. 1980).

J. An important paper by TONINI et al. (1981) provides evidence that noncholinergic excitatory nerves might be involved in the peristalsis of the guinea-pig small intestine. They showed that even after blockade of cholinergic transmission to the muscle, propulsion does still occur in isolated segments of small intestine.

K. Three papers and parts of Glise thesis (1980) confirm that abdominal trauma leads to inhibition of gastric motility via sympathetic and vagal reflexes (GLISE and ABRAHAMSSON 1980a; 1980b; GLISE et al. 1980).

References to Note Added in Proof

Aeberhard PF, Magnenat LD, Zimmerman WA (1980) Nervous control of migratory myoelectric complex of the small bowel. Am J Physiol G 238:102–108

Delbro D (1981) Gastric excitatory motor responses conveyed by collaterals of thin efferent fibres in the vagal and splanchnic nerves. Thesis, Goteborg

Delbro D, Lisander B (1980) Non-ganglionic cholinergic excitatory pathways in the sympathetic supply to the feline stomach. Acta Physiol Scand 110:137–144

Glise H (1980) Reflex inhibition of gastric motility in response to abdominal trauma. Thesis, Goteborg

Glise H, Abrahamsson H (1980a) Reflex vagal inhibition of gastric motility by intestinal nociceptive stimulation in the cat. scand J Gastroenterol 15:769–774

Glise H, Abrahamsson H (1980b) Spino-vagal noradrenergic inhibition of gastric motility elicited by abdominal nociceptive stimulation in the cat. Scand J Gastroenterol 15:665–672

Glise H, Lindahl B-O, Abrahamsson H (1980) Reflux adrenergic inhibition of gastric motility by nociceptive intestinal stimulation and peritoneal irritation in the cat. Scand J Gastroenterol 15:673–681

Itoh Z, Aizawa I, Takeuchi S (1981) Neural regulation of interdigestive motor activity in canine jejunum. Am J Physiol G 240:324–330

Rosell S, Rökaeus Å (1979) The effect of ingestion of amino acids, glucose and fat on circulating neurotensin-like immunoreactivity (NTLI) in man. Acta Physiol Scand 107:263–267

Sarna S, Stoddard C, Belbeck L, McWade D (1981) Intrinsic nervous control of migrating myoelectric complexes. Am J Physiol G 241:16–23

Thor K, Rökaeus Å, Kager L, Folkers K, Rosell S (1980) (Glu^4)-Neurotensin inhibits the interdigestive migrating motor complex in man. Regulatory Peptides Suppl 1 Abs 114

Tonini M, Frigo G, Leuchini S, D'Angelo L, Creme A (1981) Hyoscine-resistant peristalsis in guinea-pig ileum. Eur J Pharmacol 71:375–381

Wingate DL (1981) Backwards and forwards with the migrating complex. Digest Dis Sci 26:641–666

Subject Index

Handbook of Experimental Pharmacology

Continuation of "Handbuch der experimentellen Pharmakologie"

Springer-Verlag
Berlin
Heidelberg
NewYork

Volume 17
Ions alcalino-terreux

Part 1
Systèmes isolés

Part 2
Organismes entiers

Volume 18
Histamine and Anti-Histaminics

Part 1
Histamine. Its Chemistry, Metabolism and Physiological and Pharmacological Actions

Part 2
Histamine II and Anti-Histaminics

Volume 19
5-Hydrocytryptamine and Related Indolealkylamines

Volume 20: Part 1
Pharmacology of Fluorides I

Part 2
Pharmacology of Fluorides II

Volume 21
Beryllium

Volume 22: Part 1
Die Gestagene I

Part 2
Die Gestagene II

Volume 23
Neurohypophysial Hormones and Similar Polypeptides

Volume 24
Diuretica

Volume 25
Bradykinin, Kallidin and Kallikrein

Volume 26
Vergleichende Pharmakologie von Überträgersubstanzen in tiersystematischer Darstellung

Volume 27
Anticoagulantien

Volume 28: Part 1
Concepts in Biochemical Pharmacology I

Part 3
Concepts in Biochemical Pharmacology III

Volume 29
Oral wirksame Antidiabetika

Volume 30
Modern Inhalation Anesthetics

Volume 32: Part 2
Insulin II

Volume 34
Secretin, Cholecystokinin, Pancreozymin and Gastrin

Volume 35: Part 1
Androgene I

Part 2
Androgens II and Antiandrogens/Androgene II und Antiandrogene

Volume 36
Uranium - Plutonium - Transplutonic Elements

Volume 37
Angiotensin

Volume 38: Part 1
Antineoplastic and Immunosuppressive Agents I

Handbook of Experimental Pharmacology

Continuation of "Handbuch der experimentellen Pharmakologie"

Part 2
Antineoplastic and Immunosuppressive Agents II

Volume 39
Antihypertensive Agents

Volume 40
Organic Nitrates

Volume 41
Hypolipidemic Agents

Volume 42
Neuromuscular Junction

Volume 43
Anabolic-Androgenic Steroids

Volume 44
Heme and Hemoproteins

Volume 45: Part 1
Drug Addiction I

Part 2
Drug Addiction II

Volume 46
Fibrinolytics and Antifibrinolytics

Volume 47
Kinetics of Drug Action

Volume 48
Arthropod Venoms

Volume 49
Ergot Alkaloids and Related Compounds

Volume 50: Part 1
Inflammation

Part 2
Anti-Inflammatory Drugs

Volume 51
Uric Acid

Volume 52
Snake Venoms

Volume 53
Pharmacology of Ganglionic Transmission

Volume 54: Part 1
Adrenergic Activators and Inhibitors I

Part 2
Adrenergic Activators and Inhibitors II

Volume 55: Part 1
Antipsychotics and Antidepressants I

Part 2
Antpsychotics and Antidepressants II

Part 3
Alcohol and Psychotomimetics, Psychotropic Effects of Central Acting Drugs

Volume 56, Part 1 + 2
Cardiac Glycosides

Volume 57
Tissue Growth Factors

Volume 58
Cyclic Nucleotides
Part 1: **Biochemistry**
Part 2: **Physiology and Pharmacology**

Volume 59
Mediators and Drugs in Gastrointestinal Motility
Part 1: **Morphological Basis and Neurophysiological Control**
Part 2: **Endogenis and Exogenis Agents**

Volume 60
Pyretics and Antipyretics

Springer-Verlag
Berlin
Heidelberg
New York

MIX
Papier aus verantwortungsvollen Quellen
Paper from responsible sources
FSC® C105338

If you have any concerns about our products,
you can contact us on
ProductSafety@springernature.com

In case Publisher is established outside the EU,
the EU authorized representative is:
Springer Nature Customer Service Center GmbH
Europaplatz 3, 69115 Heidelberg, Germany

Printed by Libri Plureos GmbH
in Hamburg, Germany